Paleobiology of the Polycystine Radiolaria

Books in the **Topics in Paleobiology** series will feature key fossil groups, key events, and analytical methods, with emphasis on paleobiology, large-scale macroevolutionary studies, and the latest phylogenetic debates.

The books will provide a summary of the current state of knowledge and a trusted route into the primary literature, and will act as pointers for future directions for research. As well as volumes on individual groups, the Series will also deal with topics that have a cross-cutting relevance, such as the evolution of significant ecosystems, particular key times and events in the history of life, climate change, and the application of new techniques such as molecular paleontology.

The books are written by leading international experts and will be pitched at a level suitable for advanced undergraduates, postgraduates, and researchers in both the paleontological and biological sciences.

The Series Editor is Mike Benton, Professor of Vertebrate Palaeontology in the School of Earth Sciences, University of Bristol.

The Series is a joint venture with the Palaeontological Association.

Paleobiology of the Polycystine Radiolaria

David Lazarus
Museum für Naturkunde
Invalidenstrasse 43
Berlin, 10115, Germany

Noritoshi Suzuki
Tohoku University
Aoba-Ku, Sendai City
980-8578, Japan

Yoshiyuki Ishitani
University of Tsukuba
Tsukubashi, Ibaraki, 305-8577, Japan

Kozo Takahashi
Kyushu University
744 Motooka Nishi-ku, Fukuoka
819-0395, Japan

WILEY Blackwell

This edition first published 2021

Registered Office(s)
John Wiley & Sons, Inc., 111 River Street, Hoboken, NJ 07030, USA
John Wiley & Sons Ltd, The Atrium, Southern Gate, Chichester, West Sussex, PO19 8SQ, UK

Editorial Office
9600 Garsington Road, Oxford, OX4 2DQ, UK

For details of our global editorial offices, customer services, and more information about Wiley products visit us at www.wiley.com.

Wiley also publishes its books in a variety of electronic formats and by print-on-demand. Some content that appears in standard print versions of this book may not be available in other formats.

Library of Congress Cataloging-in-Publication Data

Names: Lazarus, David B., 1954- author. | Suzuki, Noritoshi, author. | Ishitani, Yoshiyuki, 1982- author. | Takahashi, Kozo, author.
Title: Paleobiology of the polycystine radiolaria / David Lazarus, Noritoshi Suzuki, Yoshiyuki Ishitani, Kozo Takahashi.
Description: Hoboken, NJ : Wiley-Blackwell, 2021. | Series: Topics in paleobiology | Includes index.
Identifiers: LCCN 2020020074 (print) | LCCN 2020020075 (ebook) | ISBN 9780470671443 (paperback) | ISBN 9781119697527 (adobe pdf) | ISBN 9781119697466 (epub)
Subjects: LCSH: Radiolaria, Fossil. | Radiolaria.
Classification: LCC QE773 .L39 2021 (print) | LCC QE773 (ebook) | DDC 561/.995--dc23
LC record available at https://lccn.loc.gov/2020020074
LC ebook record available at https://lccn.loc.gov/2020020075

Cover Design: Wiley
Cover Images: World map © NASA, Callimitra Haeckel Transparent from Report on the scientific results of the voyage of H.M.S. Challenger during the years 1873-76 under the command of Captain George S. Nares … and the late Captain Frank Tourle Thomson, R.N.

Set in 9/11.5pt TrumpMediaevalLTStd by SPi Global, Chennai, India

Contents

Preface

This book summarizes current understanding of the marine planktonic protist group Polycystina, or informally, the polycystine radiolaria, both in living and fossil form. Although the organisms are small and the number of researchers working on them at any moment in time is also not large, their study spans a great range of specialties, from molecular biology to marine geology. For this reason, a team of specialists has worked together to create this volume. Our approach to the book also takes into consideration both the intellectual challenges of understanding this broad, highly interdisciplinary subject, as well as the changing nature of communications in the modern world. In contrast to many other specialty volumes, and to some extent to the few other older existing books on radiolarians (summarized below), we have chosen to emphasize concepts and basic principles, rather than to compile a large number of facts. One of the most important things to understand about a new subject are the underlying concepts that guide workers in the field. These are often so widely accepted that they are usually assumed, and rarely articulated, yet are essential to understanding the activities, data, and interpretations that make up the visible content of the field; nor are these concepts always the same as those used in other areas of intellectual activity, creating major barriers to understanding between disciplines. By contrast, it is becoming increasingly easy for individuals to themselves assemble enormous amounts of raw information – piles of data and interpretations of it – by electronic means. This includes searching the internet, using local and online databases of scientific publications, and searching archives of published literature, both local and online. We have, therefore, deliberately emphasized providing the relatively hard-to-find explanations for conceptual issues, and give only a subset of citations to the factual data behind them, to serve as a starting point for further self-guided study by the reader.

What are polycystine radiolaria, and why study them? The term radiolaria is itself in need of definition. Historically, it was created to refer to several superficially similar groups of marine zooplanktonic protists who share a similar structure of a central nucleus and cytoplasm surrounded by a halo of radiating, partially stiffened, filamentous pseudopods; and, with few exceptions, possessing a mineralised skeleton or shell. Three major biologic groups come under this definition: the Polycystina, the Acantharia, and the Phaeodaria. Recent molecular work has established that these three groups are not all that closely related to each other; further, only the Polycystina's massive siliceous shells are commonly found in fossil form. Thousands of fossil species have been described, from rocks dating back to the base of the Phanerozoic – e.g. the entire time period in which fossils are at all common in the geologic record. In the last few years it is also becoming apparent that polycystine radiolarians are an important component of the living plankton. These are the basic justifications for the focus on Polycystina in this book, as it is their biology and geologic record that currently is of greatest interest to science. Indeed, given the dominance of interest in this one group, we, as is common to many other workers, will usually simply use "radiolaria" to refer to the polycystines, only using the latter term when a differentiation to acantharians or phaeodarians is needed. There are in addition two other reasons for a book on the paleobiology of radiolarians. The lesser reason is simply that it has been nearly 20 years since the previous standard reference book on radiolarians (De Wever et al., 2001) appeared, and a new work summarizing research developments in the last two decades is urgently needed. The more important reason though is that radiolarian research is undergoing a major change in orientation. New technologies, particularly molecular biology, automatic data collection using AI computer systems, and global database networks are opening vast new opportunities for biologically oriented research using both living and fossil radiolarians. While classic themes such as biostratigraphy and paleoceanography are not in danger of dissapearing, the focus nonetheless in the next decades will increasingly shift towards (paleo)biologic studies, including biologically based higher systematics, micro and macroevolution, and integrated studies of diversity and environmental change. This book we hope will provide a starting point to help train the next generation of radiolarian workers in these subjects.

Radiolarians are exclusively marine and are found in all ocean waters, from polar regions to the tropics, and at all water depths, from the well-mixed surface layer to abyssal waters. There are approximately 600 distinct described living species and several thousand fossil species of polycystines. Radiolarians in general, and polycystines in particular, were until recently not thought to be a major component of the living plankton and were not so intensively studied as groups, e.g. diatoms or coccolithophores, with perceived greater abundance and/or impact on marine ecosystems and geochemical cycles. As fossils, radiolarians are however fairly common, and often occur in sediments where other types of fossils are absent. This has made them very valuable for certain types of geologic research, particularly estimating the geologic age of the sediments containing them, and as guides to past oceanic water conditions. As our current understanding of the biology, and even taxonomy of the living fauna is still very incomplete, evolutionary studies based on living polycystines are still rare. However, the common occurrence of numerous specimens for many species, and in a wide variety of oceanic environments, also provides an excellent opportunity to study biologic evolution in the fossil record. Thus, the structure of the book is approximately three-fold: after a historical introduction, three chapters on living radiolaria; two chapters on general information (a large chapter on taxonomy with a summary of the fossil record, and a shorter chapter on preservation and methods); and a concluding three chapters on research applications: paleoceanography, stratigraphy, and evolution. Given the title of this series of volumes ("Topics in Paleobiology...") we have presented evolution in more detail than our coverage of subjects in the other chapters. We also have, in contrast to earlier books on radiolaria, clearly separated general observations of radiolarian change over time (Chapter 5) from the use of these data in understanding evolutionary processes (Chapter 9). Of particular interest are two chapters which are the first such to appear in book form. Our book begins with a historical review of this field of research. This is of interest for anyone studying the history of science, but also gives a context for the rest of the book's content. Much of what we know, and do not know about radiolarians, depend on the historical constraints of how this field of knowledge developed. The chapter on the newly developing subject of radiolarian genetics and molecular systematics is also a first to appear in book form.

Certain assumptions are made about the readers of the book. It is expected that the reader has completed a bachelor's degree or equivalent in one of the natural sciences, ideally either biology or the earth sciences, and, as stated above, is willing to find additional factual data themselves, using our text as a starting point. Since not all readers will have the same degree of interest in all aspects of this subject, and many will primarily need to consult specific parts for reference, we have tried to write the text so that each chapter can, to some degree, stand alone, although we provide cross references to other relevant parts of the volume whenever possible. This does mean also a certain degree of overlap and/or duplication, but in most cases only at the most superficial level, with detailed exposition, hopefully, being confined to a single, most logical place in the text. We also specifically wish to make radiolarians accessible to the widest possible range of specialists, and thus have deliberately included a fair amount of required background on many subjects in biology and the earth sciences, as specialists in one rarely have much background in the other.

Since we stress the need to use information sources beyond the content of this book it is important to note some more specific sources here, beyond those general resources known to everyone in the internet age. There are several such available to beginning students of radiolaria, including both books and online databases with taxonomic, geographic, and geologic age distributional information.

There are three other reasonably modern book-length treatments of radiolarians available in the western literature. Each of these has its own special strengths which complement the content of this volume.

Radiolaria, by Anderson (1983). This book is the first book published in the twentieth century in the West specifically devoted to radiolarians. The focus is strongly on radiolarian biology and the fossil record is covered somewhat sketchily, primarily in the context of evolutionary studies. Dr. Anderson's speciality of radiolarian cellular biology is excellently summarized and in most respects this volume is still the best, most detailed reference for this aspect of radiolarian biology available.

Radiolarians in the Fossil Record by De Wever et al. (2001). As the title suggests, this book was seen as a complement to the earlier work by Anderson, focusing on the geologic aspects of radiolarians rather than the biologic ones. In addition to chapters on the use of radiolarians in geologic research, this volume presents, for the first time, a complete family-level classification of radiolarians, both fossil and living. In all, 131 families are defined, and short but adequate summaries of the key morphologic characters are given for all of these. This taxonomic section, at 208 pages, makes up nearly half of the De Wever volume (ignoring references and indices). It is thus far more detailed than any that we could attempt in our own work, and still leave adequate room for coverage of other topics. It is partially for this reason that the taxonomic section of our own book makes no attempt to be comprehensive but instead concentrates on principles and specific examples (mostly for the Cenozoic geologic time interval); as for more detailed treatment of other families the reader can make use of the De Wever volume. We do, however, provide a critique of this, and other attempts to synthesize radiolarian taxonomy, to enable the reader to better evaluate the often conflicting details of taxonomy present in the literature.

Lastly, the monographic-length special issue of Afanasieva et al. (2005), provides a recent summary of radiolarian studies from the viewpoint of mostly Russian workers. This work is available both in the original Russian, and in English translation, which makes it much more accessible than earlier monographic publications from the Soviet Union (e.g. Petrushevskaya, 1971, 1981, 1986) which were published only in Russian. As explained in our chapter on the history of radiolarian research, distinctly different "schools" of radiolarian studies developed in the Cold War years in the West and the Soviet Bloc, and many differences in both content and basic approach to the subject still exist. In our book we refer to this body of research when appropriate, but do not attempt to fully synthesize or harmonize the sometimes still strongly disparate ideas and information.

In addition to these more general reference works, and a variety of more specialized resources cited within the book as appropriate, e.g. for biostratigraphy, there are other resources which give more comprehensively the specifics of radiolarian taxonomy and distribution at the level of species and genera, albeit without the interpretive aspects found in the above books. These resources are essential since as yet, there are no comprehensive reference catalogs for radiolarian species and generic taxonomy. These resources include:

The NSB (Neptune) database (www.nsb-mfn-berlin.de) which gives basic taxonomic information plus name status (e.g. valid, synonym) for most of the several thousand species mostly of Cenozoic age, plus (as of 2019) over 135,000 species occurrence records derived from the deep-sea drilling project cores detailing their distribution in space and time;

The Paleobiology Database (https://paleobiodb.org) – the main general paleontology database, it has substantial radiolarian occurrence data (as of January 2020 ca. 48,000 occurrence records, of which however only about 27,000 are identified to species level), mostly for the Mesozoic;

Several papers in a special issue of *Geodiversitas* (O'Dogherty et al. 2009) which collectively comprehensively cover the genus taxonomy of Mesozoic radiolarians, plus a similar compendium for the Paleozoic (Caridroit et al., 2017);

The radiolarian taxonomic database based on Paleotax (Suzuki, 2016) and the RadWorld taxonomic database (Caulet et al.), neither of which as yet are available to the general public;

The multi-organism taxonomic database at http://bio-world.org (note – the content of this database is quite extensive – about 20,000 radiolarian names, but decidedly mixed quality; also, the database's name and web address changes frequently;

And lastly, the radiolarian community website www.radiolaria.org. This site provides general information about radiolarians to the world of internet users, news and announcements for the community of radiolarian researchers, and an online catalog of several hundred (primarily) Cenozoic to Recent species, with description, images, and synonymies.

In addition to the above, two other projects are in progress but have not yet appeared in public form. Caulet et al. (2017) are preparing a catalog of Cenozoic genera, and Cenozoic workers are beginning to prepare a more comprehensive online catalog of both species and genera using the Mikrotax system (http://www.mikrotax.org/index.php).

Lastly we note the authorship of this multi-author effort. Suzuki wrote initial drafts of Chapter 2 and the Paleozoic and Mesozoic parts of Chapters 5 and 8; Takahashi contributed to Chapters 3 and 6, and Ishitani wrote the draft of Chapter 4. Lazarus revised and substantially extended these initial drafts and created all other content, and is thus primarily responsible for the interpretations and conclusions of this book.

Acknowledgements

We wish to thank those who helped us bring this book into existence: Mike Benton for the invitation to contribute this volume; Norbert Meier for careful work in preparing plates; Luis O'Dogherty for providing his latest list of radiolarian families; Daniel Bernoulli, Timo Mappes, Ted Moore, Pieter van der Paverd, Atsushi Matsuoka, Peter Miller, Annika Sanfilippo and Marina Afanasieva for photographs and images; and Fabrice Not, Johan Renaudie, Gabrielle Faria, Gayane Asatryan, and Volkan Özen for reviewing specific chapters. All errors of commission and omission remain those of the authors.

1 History

Introduction

The history of radiolarian research begins in the mid-nineteenth century, with research activity continuing with widely varying intensity and focus up to the present day. Unlike large primary divisions of science which have their own internal dynamics, radiolarian studies have always been carried out by relatively small numbers of individuals. Their work has been very strongly shaped by the development of the larger scientific subjects within which radiolarian research has been embedded. It is thus not possible to understand the historical development of the field without also understanding something of the history of these larger fields within which radiolarian studies have been carried out. These include biology, both in general, and specifically protistan biology and marine biology (particularly plankton); and earth sciences: paleontology, stratigraphy, and paleoenvironmental studies – e.g. paleoclimatology and paleoceanography. All of these disciplines have significantly affected the type and intensity of radiolarian research. This chapter thus covers not only radiolarian research but, to the degree appropriate, developments in these other disciplines as they affected the development of radiolarian research.

Bibliographic summaries of radiolarian research provide an overview of activity in the field. Because different bibliographies,

Paleobiology of the Polycystine Radiolaria, First Edition. David Lazarus, Noritoshi Suzuki, Yoshiyuki Ishitani and Kozo Takahashi.

each with different degrees of completeness and detail, are available, it is best to look at data from multiple sources (Figures 1.1, 1.2). Species descriptions (Figure 1.1a) provide one measure of research activity as taxonomic work is a foundation for other types of research. After a modest degree of publication in initial period of study from ca. 1840–1870, there is a huge spike (due to Haeckel's Challenger monograph, described in "E. Haeckel and his disciples") in the late 1880s, followed by a period of moderate activity up to ca. 1910 or so, after which taxonomic research on radiolarians dropped to very low values until the late 1950s, other than a couple of "spikes" due to some monographic papers in the late 1920s and mid-1940s. Taxonomic work on radiolarians then, from ca. 1960 or so, increased gradually to levels higher than seen in earlier times, followed by a decline in annual species numbers from around 1980 on. Interestingly, a rather parallel pattern of initial interest up to the beginning of the twentieth century, then a period of decline until ca. the 1950s, is also seen in the history of diatom research (Stoermer and Smol, 2010).

If, instead of using numbers of species, one plots numbers of papers (Figures 1.1b, 1.2), a rather different picture emerges. Although there are still two peaks in publication activity, one in the late nineteenth century, and the other in the 1980s, the relative sizes of the peaks are now dramatically different, with many more papers being published in the twentieth, vs. nineteenth century. More papers with time reflects a general pattern in the history of science, where scientific activity and numbers of publications increase rapidly in the twentieth century – indeed, for academic publishing as a whole, the trend has been more or less one of continuous geometric increase with doubling times of 15–24 years since the early nineteenth century, measured either by publications (Larsen and von Ins, 2010) or numbers of journals (Jinha, 2010). Several aspects of the radiolarian publication history however differ from the general pattern of scientific publication.

First, the ratio between numbers of new species and numbers of papers has changed dramatically over time, with nineteenth-century work being dominated by many new species descriptions in a small number of papers, while twentieth-century studies published, per paper, many fewer species, even if due to the larger total number of papers the total taxonomic output is similar. There is thus either a fundamental change in the style of taxonomic publishing, and/or a shift in publication to non-taxonomic research. Both are probably significant factors. It is possible even today for a student to encounter a nearly unknown radiolarian assemblage which needs to be sorted into species categories without the help of prior literature. This effort, vastly eased by modern imaging technology, can now be done for hundreds of species in a few months' work. However, without the need to reference prior literature, early workers could publish simple descriptions of large numbers of species relatively easily. There is of course also a shift toward applied work with radiolarians over time, so that, unlike in early years, more recent literature is dominated by papers that contain no new taxonomy.

Second, there is a substantial decline in outputs in both taxonomic description and general numbers of publications in the last ca. 20 years. Although this is in part due to incomplete recording in the last ca. 10 years, the similarity between different data sources, and in particular in the relatively comprehensive Paleotax data, suggests that the decline in output is real. This is not a general feature of science publication, which continues to grow rapidly (Larsen and von Ins, 2010; Jinha, 2010). Lastly, there are noticable differences in activity between subdisciplines. Subdisciplines in radiolarian research are, due to many differences in material and the research context that are linked to geologic age, conveniently expressed by the geologic time interval that individual studies examined (living; Cenozoic; Mesozoic; Paleozoic: Figure 1.2). Research on Cenozoic/living radiolarians dominates nineteenth-century studies, and increases relatively rapidly after WWII to a plateau by around 1970, before gradually declining in the 1990s to the present day. Mesozoic and Paleozoic studies by contrast begin to increase in the 1970s and 1980s, respectively, before declining in the beginning years of the twenty-first century. These differences create different time intervals of dominance, with Cenozoic/living being the dominant area of research up to ca. 1980, and Mesozoic research being the dominant activity after this time. The causes of these patterns will be described in our historical review.

These patterns suggest convenient time intervals in which to frame a review of radiolarian reseach: early studies up to around

1910, the low activity interval between 1910 and ca. 1950, the growth period of Cenozoic/living studies up to ca. 1970, and the "modern" phase of research beginning in the 1970s and continuing to the present day. The small (third) peak in publication seen in all data beginning in the late 2000s may represent a new phase of research, though this is more open to interpretation, and will be discussed together with "outlook – future developments" at the end of this chapter.

Within the field of radiolarian studies, the number of species descriptions by author (see Table 5.5) provides at least a rough guide to the importance of individual workers, particularly in earlier times when descriptive work dominated research activity.

Scientific Context

The first of the broader issues that influenced early studies of radiolarians are the basic framework ideas on the development of science itself, as early radiolarian research grew up during this period of general scientific development. Modern science, with its distinguishing characteristics of emphasis on non-metaphysical material mechanisms as the cause of the patterns and structure of the natural world, formal hypotheses, and testability, is generally agreed to have originally developed primarily in Europe over several centuries, but maturing in the eighteenth and nineteenth centuries (Wikipedia contributors, 21.6.2018). By the mid-1800s natural sciences were firmly established in western countries and divided into the major divisions we still use today, with numerous professional positions in universities, supported by professional societies and other coordinating organizations. Even the name was modern, with the term "scientist" (coined by W. Whewell in 1833) gradually replacing the earlier, less distinctive term "natural philosopher" (https://en.wikipedia.org/wiki/William_Whewell; accessed 3.9.2017). In other parts of the world descriptive and sometimes quite extensive studies of nature analogous to the descriptive studies of biodiversity in the western world were also developing, e.g. in Japan (Marcon, 2015). These activities were linked, as in western countries, to other interests (medicine, agriculture, hobby gardens, and others) but they were not systematically linked to developing formal paradigms of scientific hypotheses and testing as in the West. Nor, lacking widespread access to microscopes and other suitable technology, was the systematic study of microorganisms possible.

The development of science in Europe was gradual and not uniform between disciplines, with physical science developing more rapidly than biologic. At the beginning of the nineteenth century however, the transition in biology from a descriptive style to one where description was informed, and thus influenced by scientific theories of process, was already underway. Initial steps in this transformation included a re-examination of the "scala natura" description of the natural biologic world that had been inherited from antiquity (traceable back to Plato), and had been overlain with religious-metaphysical mechanisms (Mayr, 1982; https://en.wikipedia.org/wiki/Great_chain_of_being, accessed 21.6.2018). Observations of the diversity of life had been until then pressed into a

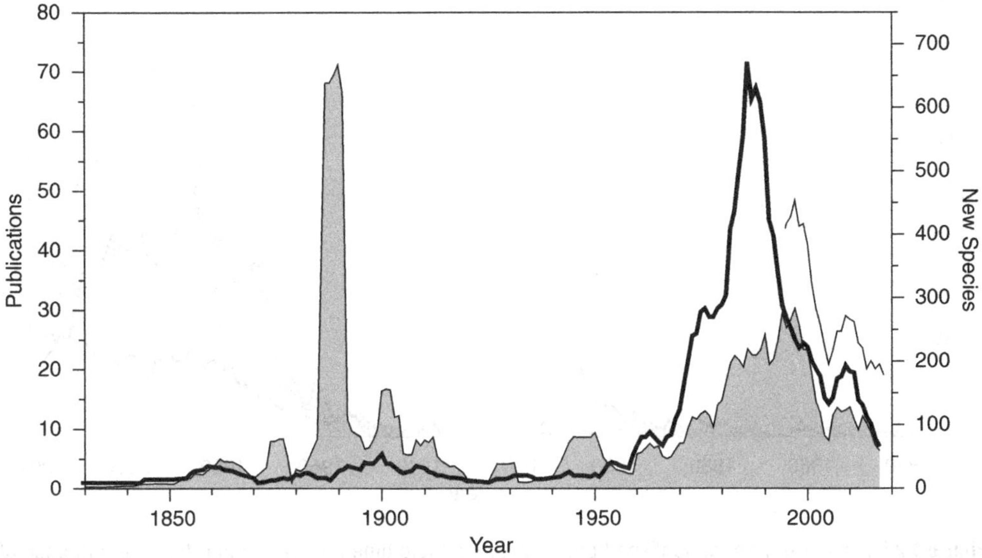

Figure 1.1 Publishing activity over time for radiolarian research. a) shaded – number of new species per year (3-point smoothing); b) bold line – number of publications per year (Suzuki database with 3-point smoothing); c) thin line – number of publications per year with "radiolaria" in the title in Google Scholar, 1988–2018 (searched August 2019). The Suzuki database primarily records taxonomic literature and thus underestimates applied radiolarian research where radiolarians are used but no new taxonomy is published. However, a simple Google Scholar search for radiolarian papers, not specifically taxonomy, gives a very similar trend, albeit with higher total numbers of papers.

description of a continuous chain of forms, from the simplest to the most complex, perfect forms (man, and above this the metaphysical beings of angels and god). The simplest forms at the base merged into and continuously arose from the inorganic world below via spontaneous generation – an idea which, although long under attack, was still present well into the early nineteenth century (https://en.wikipedia.org/wiki/Spontaneous_generation, accessed 22.6.18). More objective study showed numerous flaws in this description, weakening the metaphysical/religious explanations of mechanism tied to this description. Cuvier for example demonstrated in 1812 that there was not a continuous series of forms among metazoans (Mayr, 1982). A series of tests, culminating in Pasteur's 1859 experiment, finally disproved the idea of spontaneous generation. New scientific theories, based on these more objective descriptions of nature, and using materialistic mechanisms, were proposed in to replace the scala natura. Several theories of evolution were also developed, such as those of Lacépède and Lamarck, although these lacked an adequate mechanistic basis (Mayr, 1982). These earlier ideas were replaced later in the century by Darwin's evolutionary theory (Darwin, 1859).

The idea that all life was made of cells was published by Schleiden (1839) and Schwann (1839), and that all cells arise from other cells (and thus not, at the base of the chain, from the inorganic world) was proposed by Raspail in 1825 but only popularized by Virchow in 1858 (Enerson, 2018).

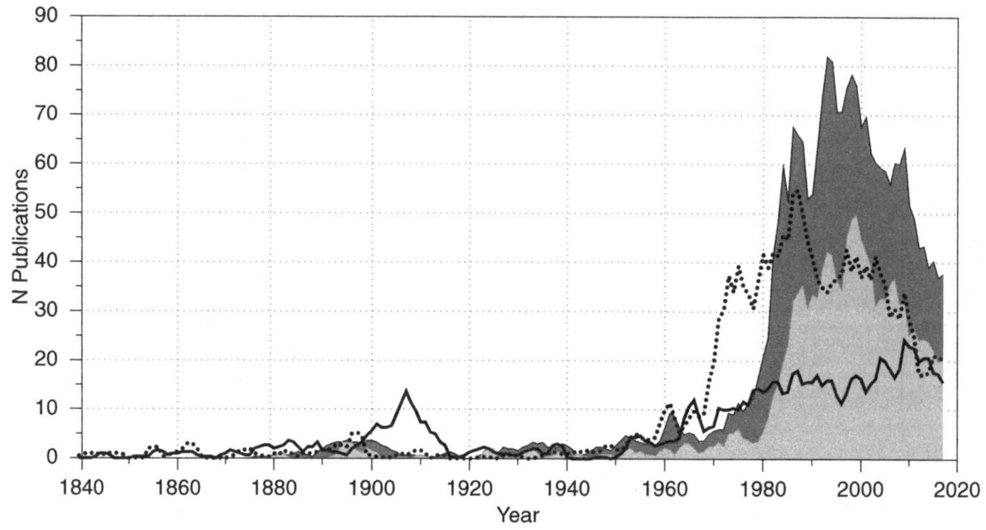

Figure 1.2 Numbers of papers published per year by geologic time interval, from reference database of radiolarian literature distributed at the Cadiz Interrad meeting (2012). Light grey: Paleozoic; Dark grey: Mesozoic; Dotted line: Cenozoic; Solid line: Living. Data smoothed (3-point moving average) for clarity. Cadiz literature database includes many applied papers and papers without the word "radiolaria" in the title but does not include as many titles per year for pure taxonomy as the Suzuki database.

Early studies in Europe using primitive microscopes had documented the existence of microscopic life forms, (e.g. J. Hill 1652, van Leeuwenhoek 1674: Corliss, 1975), and O.F. Müller (1770s and 1780s), but little was known about their biology or diversity even in the early nineteenth century (Corliss, 1978). Improved microscopes in the mid–late nineteenth century, however, were enabling new insights on microscopic life, in particular from various innovations in lens and illumination design in the 1830s to 1870s (Croft, 2006).

Increasing wealth and technologic power from the industrial revolution meant increased resources were available to more systematically explore remote environments, including the open oceans. Scientific studies of these had begun already in the 1700s, but the number of expeditions increased substantially in the later 1800s, most notably with the Challenger expedition of 1872–1876 (Lalli and Parsons, 1993; Figure 1.3). The Challenger Expedition grew out of the first systematic studies of marine organisms by E. Forbes and the discovery in 1860 of abundant life growing on a deep-sea telegraph cable. This disproved Forbes' 1843 Azoic Theory of no life in the deep oceans and stimulated new research on the vast, scientifically unknown biologic realm of the open oceans. Plankton studies would become a regular part of the research program on such expeditions, though "plankton" as a distinct concept was not coined until 1887 by V. Henson, who subsequently, in 1889, would lead the German "Plankton Expedition" – the first specifically devoted to study of planktonic organisms. (The study of the radiolarians from this expedition by Brandt were however mostly on the poorly fossilizing Collodaria.) This expedition's financing provides an example of the growing commercial interests in ocean environments, as a significant contribution to the costs was donated by the German

Figure 1.3 The H.M.S. Challenger. Primarily known for its seminal ocean research expedition (1872–1876) which circumnavigated the globe, covered 127,580 km, and explored all major ocean basins from the tropics to polar regions. A small shipboard staff of six scientists led by C.W. Thomson took physical oceanographic measurements, and collected water, plankton, and deep sea samples during the cruise which were distributed to numerous other scientists when the expedition ended, including E. Haeckel, for radiolarian studies. Source: Wikimedia Commons.

commercial industry's Fisheries Association (Deustcher Fischereiverein) (https://de.wikipedia.org/wiki/Deutscher_Fischerei-Verband, accessed 6.9.2017). In a parallel development, the late 1800s saw the establishment of several marine biology laboratories, e.g. the Naples laboratory in 1872, and the Marine Biologic Laboratory in the USA in 1873 (moved to Woods Hole in 1888; Lalli and Parsons, 1993). The Naples Zoological Station was particularly important to the development of marine biologic research. Founded by A. Dohrn, a zoologist and later evolutionist who had studied in part under the famous radiolarian researcher E. Haeckel in Jena, the Naples Station was an "open" facility, which could be used by visiting scientists from any country, who paid (an innovation itself) "bench fees" for using the facility (Cortese, 2014).

Geology in the first half of the nineteenth century was still primarily a practical, descriptive science, and considerable energy was devoted to stratigraphic categorization, mapping, and similar activities, though major debates on tectonics and other geologic themes were present (Gohau, 1990). Theoretical work was supported by the work of Lyell and others who had largely established the required basic principles such as uniformitarianism (the geologic record is to be interpreted as resulting from processes known in recent times). Historical geology however was still hampered, despite the spread of Lyellian uniformitarianism, by religious requirements for a very young age for the earth. The basics of stratigraphy were well established, even if the number and names of stratigraphic divisions were still in a state of flux. Biostratigraphy was also an established tool, having been founded by W. Smith (e.g. "Strata Identified by Organized Fossils," published 1816–1819). A. d'Orbigny in 1840 introduced the use of microfossils (specifically foraminifera) for biostratigraphic purposes. Micropaleontology as a distinct field had however not developed – indeed, d'Orbigny himself at first saw foraminifera simply as very small cephalopods. This was corrected by Dujardin's 1835 observation that micro-organisms are unicellular (Corliss, 1978; Vénec-Peyré and Bartolini, 2013). Toward the end of the nineteenth century, geology became more closely allied to expanding industrial and national interests, with the founding of numerous geological surveys and extensive field geologic studies to map and identify natural resources. The United States Geological Survey, for example, was founded in 1879 (Rabbitt, 1975). These activities, along with the discovery and increasing use of oil deposits at the end of the century, would also have a major impact on radiolarian research, but not until well into the twentieth century.

Figure 1.4 Christian Gottfried Ehrenberg (1795–1876). Source: Museum für Naturkunde historical document archives.

Early Studies (First Half of the Nineteenth Century)

C.G. Ehrenberg and J. Müller

Radiolarians during the first half of the nineteenth century were known only from relatively inconsequential, minor notes in early biologic surveys, e.g. Meyen (1834), and had not yet been studied in any significant way in fossil form. Radiolarian study (and micropaleontology in general) really began with the work of C.G. Ehrenberg, who systematically described numerous fossil forms (but very little living material), and J. Müller, who initiated study of living radiolarians in 1855 (Lazarus, 2014).

Christian Gottfried Ehrenberg (1795–1876, Figure 1.4) grew up in Delitzsch, Germany, and developed an interest in natural history while still a Gymnasium student. This led him to break off his initial study of theology and take his degree (1818) in medicine at the university in Berlin, as at this time the study of biology was still closely connected to the field of medicine (Wetzel, 1966; Mayr, 1982). Ehrenberg at first studied a variety of plant and animal species, and as a biologist took part in two expeditions: one to the Near East (1820–1826, under the leadership of the

Prussian general J.H. v. Minutoli), and a second to Russia (1829). The former expedition was in many ways disastrous, as it was underfunded and poorly planned, and the majority of the participants died of illness before it was completed (Siesser, 1981). For Ehrenberg, who lost his friend Hemprich to illness during the expedition and only narrowly survived himself, it was a double disaster. His painfully collected scientific materials (ca. 80,000 specimens that had been mailed to Berlin) had been mishandled. Many of the specimens had been sold to other scientists or collectors, and most of the labels on those still remaining had been lost. Ehrenberg soon turned to the study of microorganisms. Some he collected himself in Berlin or in Germany; many more were sent to him by other naturalists as his fame as "the" microorganism specialist grew with the years. Ehrenberg's exceptional knowledge of microorganisms was known as well beyond the realm of scientists and naturalists; he was not infrequently invited to meetings of high society (on one occasion the King of Prussia) to set up his microscope and show off his microorganisms to those assembled (Jahn and Landsberg, 2000).

Ehrenberg's fame was deserved. Although protists had been described sporadically for many decades, even the more detailed studies, e.g. by O.F. Müller (1730–1784, posthumous publication 1786), tended to mix protist and metazoans together. Initial definitions of them as a group, while recognizing their distinctiveness, also implied their primitive, simple nature, as the word "protozoa," coined by Goldfuss in 1817–1818 clearly suggests. Goldfuss placed Protozoa at the base of the scala natura, emerging from the "Urschleim" [primeval slime] (Entzeroth and Hausmann, 1993). Ehrenberg was the first to systematically study a broad range of living and fossil microorganisms and to document them as widespread, diverse, complex, highly developed forms of life. He thus can be considered the father of protozoology and, together with A. d'Orbigny, the cofather of micropaleontology (Corliss, 1978; Siesser, 1981). He described thousands of new species, including many of the type species for common genera in important groups of microorganisms, such as the diatoms (Jahn, 1998; Lazarus, 1998). Ehrenberg did not have the ability to understand the geologic aspect of the fossils he studied as most of the geologic age assignments at that time were still extremely imprecise (e.g. Tertiary vs. Cretaceous) and even these were frequently wrong (Zittel, 1876, p. 76). His importance is that, other than the short, one-page description of a single species (the collodarian colony *Sphaerozoum fuscum* by Meyen, 1834), he was the first to systematically study and describe radiolarians.

Ehrenberg's doctoral thesis on fungi had demonstrated that they developed by normal biologic means of reproduction, and not, as the still commonly believed medieval (or older) theory of scala natura (also known as "the great chain of being") suggested, by development from lower forms. Finding evidence to refute the scala natura theory remained important to Ehrenberg's later scientific development, as in his studies of microorganisms he stressed their complex nature and constancy in characters, both characteristics that he felt refuted the predictions of scala natura, which claimed that lower forms of life were ever simpler, less developed, and had a tendency to transform into other life forms along the "scale"; and at the base (where microorganisms were assigned), a seamless transition occurred to the inorganic realm (Mayr, 1982). The scala natura was however, regardless of popular belief, already largely disproven to many other contemporary biologists, e.g. from Cuvier's (1812–1817) studies of animal systematics (Mayr, 1982). Ehrenberg went so far in his effort to establish the complexity of microscopic life as to claim that the organelles of living protist cells were in fact miniature organs, fully homologous with those of macroscopic, multicellular life forms. In this he rejected the discoveries of other biologists on the unicellular nature of microorganisms (e.g. Dujardin, 1835). His biases on the constancy of species (based in part on comparisons using broadly defined genera, e.g. Ehrenberg, 1839, p. 154 on *Haliomma*, which he found living in ocean water and in "Cretaceous" rocks from Greece) and gaps between forms similarly led him later to reject Darwin's evolution theory. These contrarian views resulted in much of his work being, if not rejected in its entirety, nonetheless being neglected by both contemporary and later generations of biologists (Wetzel, 1966; Siesser, 1981). T.H. Huxley, for example, described Ehrenberg's research (and before the publication of Darwin's *Origin*) as "wonderful monuments of intense and unremitting labour, but as least as wonderful illustrations of what zoological and physiological reasoning should *not* be" (Huxley, 1851, p. 436 – with several following pages rebutting Ehrenberg's ideas). Ehrenberg also persisted in using the same microscope throughout his career (Figure 1.5; pers. comm. T. Mappes, 6.8.19) even though it was technically increasingly obsolete, and did not yield the same quality images as newer models (Siesser, 1981).

Despite these weaknesses, Ehrenberg's strength lay in precise observation and documentation of the morphology of the organisms he was studying, including remarkably objective, nearly photographic drawings, and careful preservation and labeling of his source materials (Figures 1.6, 1.7), in which work he was aided by his longtime assistant, his daughter Clara. This documentation of his descriptive studies has permitted most of his taxonomic legacy to survive and be used in modern scientific work. It is possible that his care in documentation was rooted in his earlier, disastrous experience with the materials from his Near East expedition.

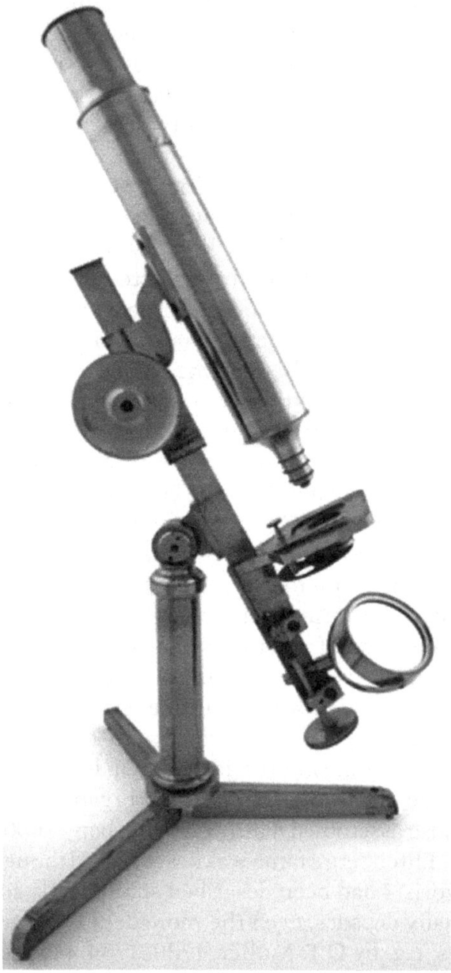

Figure 1.5 Microscope similar to that used by Ehrenberg. Made by the firm of Schieck in Berlin, they were considered to be of very good quality when first introduced. Note the long working distance to the specimen stage and round area for holding preparations. Ehrenberg's micas fit this type of instrument well. Source: Photo and instrument courtesy of Timo Mappes.

Ehrenberg defined several of the higher level categories of radiolarians (e.g. Polycystina, Spumellaria, Nassellaria) that are still used today. In all, Ehrenberg

Figure 1.6 Ehrenberg Collection sample of Eocene radiolarian chalk, Barbados. One of several samples from Barbados in the Ehrenberg Collection. Determining the geologic context of historical material is challenging and requires historical research, as this sample illustrates. The material was given to Ehrenberg by Sir Robert Schomburgk (Ehrenberg, 1846, 1847), who like many naturalists explored the Americas in the early 1800s. Although Schomburgk was originally German he spent most of his career in English-speaking countries and he may well have personally collected the material even though the label in the box: "Locality uncertain" is in English, as is the scrap of paper it was written on (the back, shown in the inset, apparently refers to a stock offering in English Pounds). Schomburgk was at the time stationed on British government business in Barbados (Robert Hermann Schomburgk, Wikipedia, accessed 17.8.19) and may simply have failed to note precisely where the material was collected. No details are given in the Ehrenberg publications as to the extent of the material, other than stating that "several fist-sized" pieces were included. The presence of multiple individual geologic objects, sometimes with clearly different geologic characteristics in a single "sample," is common in such historic collections. Anna Jobst, the archivist who databased the collection, by matching different published sources and collection records suggested that the sample(s) can be assigned to the peak of Mount Hillaby (Ehrenberg Collection database, accessed 17.8.19). This would place them several km away from the better known material from modern studies of Barbados microfossils of the Bath Cliffs (Saunders et al., 1985), though still part of the Eocene nappes of pelagic sediments that are present in the island's geology (Donovan, 2005).

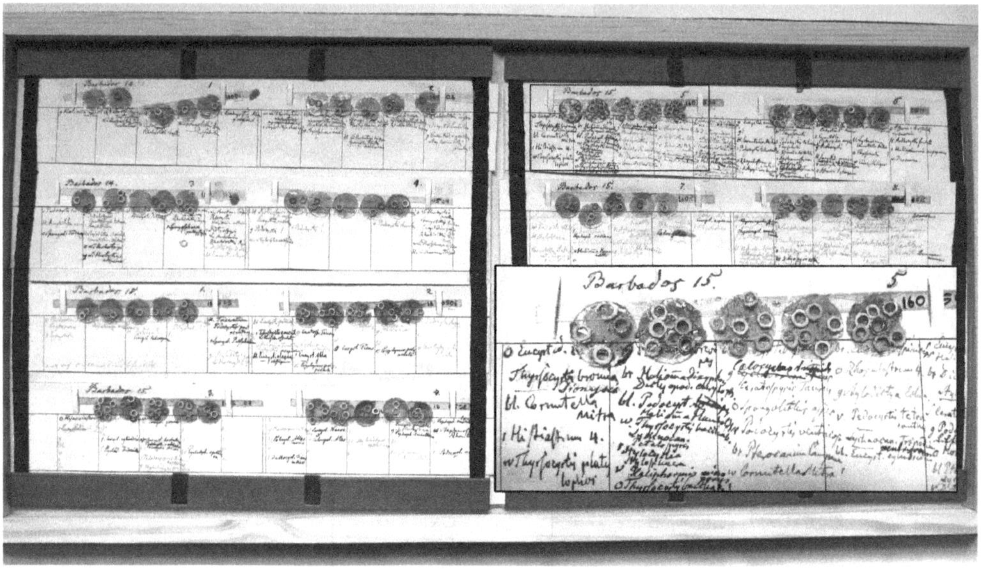

Figure 1.7 Ehrenberg mica microscopic preparations (here cabinet 15, tray 5). Ehrenberg made his preparations on small round discs of mica, on which he melted a thin layer of Canada Balsam then sprinkled the source material over. He generally made at least 5, and often more than 30 such discs from a single sample (though note that "sample" is somewhat ambiguous – see caption to Figure 1.6). For storage, groups of 5 discs were glued, using more balsam, onto a thin mica strip and inserted into slots cut into a broad paper mounting strip. Specimens of interest were marked by small colored paper rings on the micas and the taxonomic name noted with an abbreviation to the ring color on the strip of paper underneath, with blocked areas corresponding to the sequence of micas. Strips were bundled into cardboard folders, and in turn held in larger cardboard boxes resembling books. With the passing of years, the micas began to loosen from the mica strips, and break during opening or closing the folders. They were thus transferred to the tray system used today. Given the very limited space Ehrenberg's taxonomic names were often abbreviated and are frequently hard to read, even in close-up (see inset). As Ehrenberg worked prior to the development of the concept of type specimens, there is no indication in the collection which mica or ring holds the specimens used in his original species description publications. This makes finding the ca. 2,000 type specimens for his published species (all taxonomic groups included) challenging, as there are ca. 40,000 micas in the collection. Given the rough surface of the micas and the optical flaws of the mica backing, imaging the specimens once found is also sometimes difficult. After Lazarus (1998) and Lazarus and Jahn (1998).

described over 500 radiolarian species and more than 70 genera (Suzuki, 2009), and of these at least 159 species and most of his genera are known to be accepted and in current use (Lazarus et al., 2015). The taxonomic status of many more of his species are under revision, based on a foundational re-examination of Ehrenberg's original collections (individual chapters in Tanimura and Aita, 2009; see also Chapter 5: Taxonomy and Fossil Record).

Johannes Müller (1801–1858, Figure 1.8) was a very influential biologist (see, e.g. Otis, 2007) who carried out the first extensive

Figure 1.8 Johannes Müller. Source: Image archives of the Humboldt University, Berlin.

studies of living radiolarians, though the very first usable description is that of Huxley (1851), who briefly described (three text pages and a plate) collosphaerid colonies and the large solitary species *Thalassicolla nucleata* Huxley. Although Müller's interest in the diversity of nature was lifelong, for much of his early career he concentrated on medicine, where he substantially advanced the field of physiology (summarized in his multivolume *Handbuch der Physiologie des Menschen*, published between 1833–1840). Beginning in the 1840s and until his premature death in 1858, he studied many groups of invertebrates, ending with five published studies of living radiolarians from coastal seas in the Mediterranean (1855–1858). The last of these in particular introduced the word "radiolaria" for the group and gave detailed descriptions and illustrations of numerous living species (with dozens of new species and genera) of acantharians, spumellarians, nassellarians, and collodarians. His text discussed details of both cytoplasm and skeleton, and together with his illustrations provided a starting point for future biologic studies, particularly those of his student Ernst Haeckel.

Second Half of the Nineteenth Century to ca. 1920

E. Haeckel and his disciples

Ernst Haeckel (1834–1919, Figure 1.9) was without question the single most important individual ever to study radiolarians: measured both by his radiolarian studies, and by his general importance to human

Figure 1.9 Ernst Haeckel, ca. 1860. Source: Wikimedia Commons.

history, as Haeckel was also very active in other areas of biology, science popularization, and in politics (Gasman, 2004; Richards, 2008). Building on the foundations of shell morphologic descriptions by Ehrenberg and cell biology by his doctoral advisor J. Müller, Haeckel named (although often failed to adequately describe) literally thousands of radiolarian species, both fossil and living. He also popularized radiolarians (and other groups of organisms), and explicitly linked radiolarian taxonomy to Darwinian evolutionary theory. The only areas of radiolarian research he did not profoundly influence were the study of pre-Cenozoic forms, or the use of radiolarians in geologic research, e.g. for biostratigraphy.

Haeckel came from an upper middle-class family and initially studied medicine. He soon shifted to studying zoology under, among others, Karl Gegenbauer and J. Müller. During an extended trip to Italy, Haeckel began in 1859 to study radiolarians from the coastal waters off Messina, as Müller had done earlier. These studies were extended into his Habilitation work (formal qualification for a professorship in classical European academic systems), which was published (Haeckel, 1862) as his first major work on radiolarians. He was appointed to the first professorship in Germany dedicated to the new discipline of zoology in 1865 (Entzeroth and Hausmann, 1993). Haeckel carried out extensive studies of many types of organisms including vertebrates, sponges, and coelenterates over the next two decades and became very active in promoting, via talks and publications, Darwin's evolutionary theory. Based on his 1862 publication he was given the radiolarians from the recent Challenger Expedition to analyze. His report, published in 1887, was one of the largest published in the Challenger series, and remains to this day the most extensive single work ever published on radiolarians, with 2,750 pages and 140 plates in 3 large volumes. Haeckel concentrated more and more on political and popular science activities from around 1900 until his death in 1919.

Haeckel is primarily known both to scientists and to historians not for his radiolarian studies, massive though they were, but for his other scientific work, his popularization of biology including biodiversity and evolution, and his promotion of a political version of evolution. Haeckel was an important figure in biology in the second half of the nineteenth century, producing major syntheses of general metazoan morphology, developmental biology, and evolution, e.g. his *Generelle Morphologie* (Haeckel, 1866). He is famous in biology for, among other ideas, his theory that "ontogeny recapitulates phylogeny," as well as for coining many biologic terms, such as "ecology." Haeckel attempted to insert evolutionary biology into politics and sociology, promoting racial theories and eugenics. Haeckel also founded a new school of philosophy – Monism – many members of which also promoted his political views, and not a few became active members of the Nazi Party. Haeckel was however also cited by communist activists. Because of his major role not only in nineteenth-century biology but his complex, problematic contribution to political ideas in Germany, he is a controversial figure and a regular subject for historical study. See, for example, Gasman (2004) and Richards (2008) for decidedly different interpretations of his person and historical role. Haeckel's popular works e.g. *Kunstformen der Natur* (Haeckel, 1904) introduced radiolarians to the general public, played a role in linking the biology of form to the then-current Art Deco movement (Kockerbeck, 1986), and to this day remain popular and continue to inspire artists and students of biologic form.

Haeckel provided the first comprehensive description of all aspects of radiolarian diversity – species descriptions, higher level taxonomic concepts, morphologic characteristics. His studies of living radiolarians were devoted approximately equally to all major groups of marine rhizarians, all of which at that time were called "radiolaria": Spumellaria and Nassellaria (Polycystina), Acantharia and Phaeodaria; the latter two groups today are no longer considered radiolarians in the modern sense of the word. Haeckel described in all over 3,000 new species, several hundred genera (counting subgenera, over 1,000), and dozens of families alone for the Polycystina (precise numbers are not available as current databases partially still mix information on Polycystina, Acantharia, and Phaeodaria together. See also Chapter 5: Taxonomy and Fossil Record). Haeckel's first major work on radiolarians in 1862 covered, in great detail, every aspect of radiolarian biology in over 200 text pages and 38 plates, including cellular morphology, behavior, biogeography, reproduction, and taxonomy – a quantum leap in our understanding of the group. One reason for this is that Haeckel, unlike Ehrenberg, who used the same early nineteenth-century model throughout his life, used then state-of-the-art microscopes (his first, custom-made for him in 1853 from the Berlin manufacturer Schieck; Otis, 2007) which allowed him to see much more cellular detail than earlier workers. After becoming professor, Haeckel could amplify his interest in radiolarians through his students and followers – notably Hertwig (1876, 1879); Bütschlii (1882); Dreyer (1889); Jorgensen (1905); Popofsky (1908); and Mast (1910) – protozoologists who continued his biologically oriented examination of the group. Many of these studies were, similar to Haeckel's (1887) work, connected to the exploratory marine research expeditions which characterized the latter nineteenth and early twentieth centuries. Haeckel's (1887) monograph however remains unique in the scope and detail of its results, although many of his described taxa are problematic, as discussed in Chapter 5 (Taxonomy and Fossil Record).

Roughly in parallel to the growth of these early biologic studies, other studies describing only fossil forms continued to be produced in Europe, often more similar in style to Ehrenberg than Haeckel, e.g. Stöhr (1880), Bütschli (1882), Dreyer (1889), and Tan (1927) on Cenozoic materials. An important extension to study of pre-Cenozoic materials also occurred, e.g. Pantanelli (1880), Hinde (1890; Fig. 1.10), Rüst (1892), and Squinabol (1904) on Mesozoic and Paleozoic materials. These latter studies,

Figure 1.10 George J. Hinde (1839–1918). Son of a factory owner in England, Hinde began his career as a farmer in England, then worked for several years as a sheep rancher in Argentina. After further years of wandering and doing geologic research in North America, Hinde returned to England in 1874 and was elected a Fellow of the Geological Society of London. From then on, he concentrated on studies of various invertebrate, mostly Paleozoic microfossil groups, including sponges, conodonts, and radiolaria (Woodward, 1918). He described in all 231 new radiolarian species (Suzuki Paleotax database) Source: illustration provided by M. Afanasieva.

although generally hindered by the difficulty in extracting and obtaining high-quality images of radiolarians from cherts and other lithified rock, nonetheless demonstrated the great geologic range of radiolarians and documented that the forms of the individual shells were at least in part clearly different from those in the Cenozoic. It should be noted though that these early studies primarily documented the faunas, but did not significantly develop them as applied tools for geologic research.

Another significant development was the gradual spread of radiolarian studies outside of Western Europe. Brief mentions of radiolarians began to appear from authors based in, e.g. North America (Dana, 1863 – cited in Mivart, 1878; Baker and Bowman 1917; Aberdeen, 1940) and more detailed studies eventually also began to appear, e.g. a biologic study by the Russia-based Cienkowski (1871), and geologic papers by the Hungarian worker Hojnos (1916, and subsequent papers). A first mention of radiolaria in Japan comes from geologic papers by S. Yamane (1924) and more detailed radiolarian reports from Mesozoic cherts in Japan by S. Yehara appeared in 1926 and 1927. J. Khudyaev (1931) published the first Russian [soviet] paper, on fossils. The spread of radiolarian research around the globe was due in part to the growth in the use of modern scientific methods in many non-western countries, e.g. in Japan as part of the Meiji Restoration and subsequent industrialization. Western Europe, however, remained overwhelmingly the center of radiolarian research.

Legacy of Early Studies

The first phase of radiolarian research in the nineteenth century laid the foundations for later work, and in particular provided a broad, detailed set of studies of living radiolarian morphology, the extent of which has not since been duplicated. The taxonomic legacy is decidedly more mixed. As noted in Lazarus (2014) much of the taxonomic legacy of early workers is difficult to use. This is particularly true for German studies, but also true for most other early studies. There are a variety of reasons for this, including artificial, overly elaborate higher level taxonomy; inadequate species descriptions and illustrations; failure to designate or deposit type specimens or other reference material; and loss of many German collections in World War II (see this section and also Chapter 5: Taxonomy and Fossil Record). These problems affect Cenozoic and living material much more than older geologic materials, simply due to the much greater number of Cenozoic-living taxa named in early research. The confused nature of radiolarian taxonomy made it much more difficult to use radiolarian species (or other taxa) in applied research. Generic names were so poorly delineated or documented according to the Rules of Nomenclature that generic name use was for most species almost arbitrary, with several equally unclear choices available for any given species. The artificial nature of taxa definitions extended ranges artificially over great intervals of geologic time, hindering the development of stratigraphy (see also "The origins of radiolarian biostratigraphy: 1940s to 1950s"). Furthermore, the failure to deposit type materials, or their subsequent loss, meant that many of these issues could only be resolved with great difficulty. As discussed in Chapter 5, despite a great deal of recent progress, there is still much further work to do, and taxonomic problems with early studies remain a barrier to modern research. Lastly, nineteenth-century studies demonstrated the great geologic range of radiolarians and documented that the forms of the individual shells were at least in part clearly different from each other on the scale of geologic periods, even if the taxonomy did not yet adequately reflect these differences.

Early Twentieth Century (ca. 1920–1940)

In the early decades of the twentieth century there was a gradual shift in the development of scientific research on living and fossil protists. The growth of scientific medicine led to a more focused interest on microorganisms (bacterial and eukaryotic) responsible for disease, with relatively fewer resources being devoted to general field-based surveys of protist diversity (Corliss, 1978). Presaged by the shift in the latter part of the nineteenth century from "natural history" toward physiology, biology in general was in the early twentieth century becoming more experimental, quantitative, and laboratory-based, with a greater focus on cellular and subcellular research (Mayr, 1982; https://en.wikipedia.org/wiki/History_of_biology, accessed 6.7.2018). In the ocean sciences, marine laboratories were gradually shifting from an early focus on biodiversity and organismal research to a broader range of more quantitative studies of physical and chemical oceanography (a trend that, particularly after WWII, in part was driven by increased military interest in the oceans: Ocean Studies Board NRC, 2000), or to more focused studies of fisheries science (Lalli and Parsons, 1993). In the USA for example, the Scripps Institution for Biological Research, originally founded in 1903 by marine biologists as the Marine Biological Association of San Diego, was renamed the Scripps Institution of Oceanography in 1925 to reflect the changing focus of its research (https://scripps.ucsd.edu/about/timeline; https://en.wikipedia.org/wiki/Scripps_Institution_of_Oceanography, both accessed 6.7.2018); while on the US East Coast, the older (established 1888) Marine Biological Laboratory on Cape Cod was supplemented, and eventually overshadowed by the new Woods Hole Oceanographic Institution, established in 1930 explicitly, in the words of the National Academy Of Sciences, to "prosecute oceanography in all its branches" (http://www.whoi.edu/main/history-legacy, accessed 6.7.2018). The growth of ecology in the early twentieth century assisted the development of biological oceanography as a field distinct from marine biology. Biological oceanography shifted plankton studies more toward understanding general patterns of plankton such as abundance and productivity, dominated by phytoplankton, and their relationship to physical oceanography; descriptive studies of individual taxa by contrast declined in importance. These changes were in part due to the need by fisheries scientists to understand the food chains of fish that began with patterns of plankton productivity (Lalli and Parsons, 1993). This shift towards medical and applied economic research affected the biologic study of many groups of organisms. Diatom research, despite having a connection to plankton productivity and ecology, became increasingly marginalized, particularly for freshwater diatoms. Even in the late 1950s there were no university-based researchers in the USA specializing in freshwater diatoms (Stoermer and Smol, 2010). Experimental, laboratory-based research on open-ocean, unculturable zooplankton such as radiolarians, which in addition were of no known significance for applied biological oceanographic research (e.g. fisheries) was even less attractive. Radiolarians thus, by not being optimal material to study the priority questions of early twentieth-century biology, became relatively neglected by biologists. Biologic studies of radiolarians in the early twentieth century were published, but had relatively little impact on the main trend of development in the field. In addition to workers such as Popofsky (see "E. Haeckel and his disciples"), others such as Hertwig (1932) and Bernstein (1932) continued the tradition of plankton studies established by Haeckel.

Sugano (1937) described many species of radiolarians in the plankton near Japan, also using the artificial nomenclature and concepts of Haeckel. Enriques (1919 and subsequent papers) began a more modern, biologically based taxonomic revision of living radiolarians which would have been influential, but unfortunately died early in a traffic accident and his work was largely unknown (Cortese, 2014).

While biological research on radiolarians gradually declined, changes in geology and paleontology were taking place, equally gradually, that would eventually lead to a resurgence in radiolarian research, though primarily on fossil forms. This took place as part of the gradual growth of the discipline of micropaleontology.

The study of microfossils is nearly as old as the study of microorganisms, as they were extensively described by Ehrenberg, d'Orbigny, and others already in the early nineteenth century. Micropaleontology however only began to develop as an independent discipline when microfossils were understood not just as biologic remains, as had been largely done by e.g. Ehrenberg, but as material that could be used to study geologic questions. The main focus, at least initially, was as a means to correlate between geologic sections (biostratigraphy), an idea originally introduced by William Smith in 1816 (Hay and Southam, 1978). d'Orbigny was certainly the first to use microfossils (foraminifera) this way, although microfossils were simply one of many fossil groups used for his stratigraphic work. In the latter part of the nineteenth century, although still partially based on concepts of catastrophism, biostratigraphy developed rapidly. During this time period it was first used primarily as a way to distinguish the major geologic periods, but increasingly to assist basin level correlations by using refinements in concepts, such as stages (d'Orbigny) or zones (Oppel), developments that were fully in line with Smith's original use in local to regional correlations (Hancock, 1977; Hay and Southam, 1978; Balini et al., 2017). Microfossils however only began to be used extensively for biostratigraphic work at the end of the nineteenth and beginning of the twentieth century, when improved technology allowed numerous wells to be drilled, resulting in the recovery of rock samples much too small to contain useful amounts of larger fossil organisms (Martin, 2013). European workers such as Grzybowski discovered their utility in oil wells in the 1890s (Czarniecki, 1993; Martin, 2013), and somewhat later, in the early 1920s, micropaleontologists (mostly young women, particularly E. Applin, A. Ellisor, and H. Kniker) demonstrated this in the USA (Gries, 2017). Joseph Cushman's 1928 book on foraminifera established and spread this new type of microfossil use worldwide (Cushman, 1928; Huber et al., 2013). By the 1930s the dramatic increase in demand for oil supported the rapid expansion of the new applied science field of micropaleontology, with micropaleontology laboratories being established by many oil companies, universities, and government agencies worldwide (Russell, 1970; Yakovleva-O'Neill and Kozlova, 2013; Martin, 2013), courses in micropaleontology being added to university curricula (Martin, 2013), and even popular articles on micropaleontology being published (Figure 1.11: *Popular Mechanics*, 1932). Initial studies on foraminifera spread quickly by the 1930s to other fossil groups such as dinoflagellates, which were studied by European workers e.g. Deflandre, O. and W. Wetzel (Dietz et al., 1999). While these developments established the importance of, and general financial support for, micropaleontology, studies of several other fossil groups did not initially benefit. Radiolarians were not initially targets of study, and diatoms were only marginally involved. As was also true for radiolarians, initial studies of fossil diatoms in the nineteenth century primarily documented their diversity and morphology. Biostratigraphically oriented

Figure 1.11 An early popular science article portrayal of micropaleontologists. The tendency of industrial micropaleontologists to refer to their microfossils as "bugs" persists to this day Source: *Popular Mechanics*, 1932. Reproduced with permission of *Popular Mechanics*.

studies, in connection to oil exploration, developed gradually in the 1920s, particularly in papers on Cenozoic California sections by G.D. Hanna (Burckle, 1978). An even slower start is seen to the study of fossil coccoliths, with limited morphologic studies in the nineteenth century, and biostratigraphic work appearing, in this case, only well into the twentieth century, with the landmark study of Bramlette and Riedel (1954) (Olney, 2002: www.ucl.ac.uk/GeolSci/micropal/calcnanno.html, accessed 8.2017; Agnini et al., 2017).

One major change in the scientific context of radiolarian research that occurred between ca. 1930 and 1950 was the shift from Europe to America as the main center of both academic and applied scientific research. As a consequence, after World War II English gradually became the global standard for publication in the sciences. Scientists whose native language was English were increasingly likely not to learn other scientific languages (which could fairly be argued was for these workers at least culturally a trend towards provincialism, not globalization), and for non-native speakers of English, the only scientific foreign language learned would be English. In practice however, this meant that, unlike the situation in the late nineteenth century and earliest twentieth century, scientists could not publish in one of several other languages (particularly French or German) and expect their work to be fully known. In Western Europe, most workers soon adapted to this new communication standard. In countries such as Japan or in the Soviet Bloc, the large number of radiolarian co-workers led to several large bodies of "own language" literature, particularly Russian and Japanese. This development certainly made it easier for workers in these countries to publish and read research papers in their field, and thus helped radiolarian research to develop in these countries. However, these papers were not widely read outside of their home countries. These new barriers to communication increasingly affected the flow of ideas, and even the awareness of taxonomic publications between countries.

The Early New Period (ca. 1940–1970)

The Second World War period was a disaster for radiolarian research in that many of the most important collections of radiolarian materials, all in Germany, were destroyed by bombing or fires (Lazarus, 2014). This includes those of Rüst and Popofsky. Further, both the surviving material from Haeckel and Ehrenberg soon became inaccessible to almost all radiolarian researchers as they were located in East German cites – behind the Iron Curtain for western workers, and not easily accessible even for many workers inside the Soviet system. However, the massive technological changes that WWII triggered soon led in the post-war years to major changes in science and radiolarian research. In the years during and following WWII many of these developments occurred in parallel, so that it is easier to treat them thematically. Themes include biostratigraphy, the deep sea fossil record, taxonomy, biogeography and ecology, and cell biology.

The Origins of Radiolarian Biostratigraphy: 1940s to 1950s

The development of radiolarians for stratigraphic studies occurred gradually. It was clear at the beginning of the twentieth century from the many individual studies of faunas from widely different age intervals e.g. Stöhr, Rüst, Yehara, Hinde, Khabakov, and others (see "E. Haeckel and his disciples") that radiolarian species compositions were not constant over time, and thus they had stratigraphic potential. Individual authors, though noting these differences (e.g. a summary for the entire Phanerozoic by Lipman, 1959) did not however develop this into practical biostratigraphic use. The stratigraphic utility of radiolarians was obscured by the inappropriate use of Haeckelian generic taxonomy (and the artificial nature of this taxonomy) which, together with the poor preservation, technical limits to specimen extraction (thin sections were often used on older material), and poor quality imagery of the material, sedimentary reworking, geologic age misidentifications, and other issues, led many to think that at least at the generic level, little change had occurred in radiolarians over the entire span of the Phanerozoic (Moore et al., 1952; Campbell, 1954; Sanfilippo and Riedel, 1985; Sanfilippo et al., 1985). Aberdeen (1940) for example used Haeckelian genera defined from living material to assign all species found in the Devonian material she studied. Similarly, in a dispute over the age of carbonate nodules embedded in Japanese "pre-Cambrian" metamorphic rocks, a revised age estimate of Mesozoic by use of a Lyellian percent similarity to modern taxa (Fujimoto, 1938) was debunked by the assertion that most radiolarian genera and even species had already appeared by the Paleozoic (Ichikawa, 1946). Even at the level of gross comparisons of taxonomic composition between major geologic intervals, there were also doubts as to the validity of taxonomic differences in lists created by different authors, as published illustrations were often inadequate. More careful, high stratigraphic resolution studies of radiolarians in series of samples from more continuous stratigraphic successions in single studies, e.g. Clark and Campbell (1942), had by the mid-twentieth century reported substantial differences even over short periods of geologic time, which led them to suggest that more comprehensive studies might yield a very useful stratigraphic tool. This would only be realized by the availability of new material, obtained due to new political and economic developments in marine science.

The scientific and technical resources invested by the militaries of all nations during World War II rapidly increased the technical capabilities available for marine research. Ships became more capable of longer voyages and supporting heavier marine sampling equipment; radar, sonar, and satellites (the latter initially only for the military, but by the 1960s openly available) improved navigation; and new or improved technologies (e.g. seismic profiling of the sea bottom) were developed which would assist in detailed sampling of ocean sediments (Lalli and Parsons, 1993; Ocean Studies Board, 2000). The military presence in ocean research remained strong after the war ended, and continued throughout the decades of the Cold War (Ocean Studies Board, 2000; Hamblin, 2005): "The current format of oceanography, which involves an interdisciplinary grouping of marine physicists, biologists, engineers, chemists, and geologists, was largely an invention of the [US] Navy [in World War II] to meet its specific needs" (McNutt, p. 52 in Ocean Studies Board, 2000). Parallel to this, commercial interests increased in offshore and open ocean regions, in particular from the oil industry. The resultant dramatic increases in resources available transformed ocean science from small-scale, individual research with episodic expeditions into a systematic, global, "big-science" activity. This large enterprise would provide the materials needed to develop the first major phase of radiolarian biostratigraphic research.

After WWII W.R. Riedel (Figure 1.12) in 1950s studied deep-sea sediments recovered with the aid of the new piston core technology (Kullenberg, 1947), as large numbers of these cores were being collected by newly empowered oceanographic institutions. He realized that detailed species-level stratigraphy was possible; also that one reason for the belief of "no change" in radiolarians over time was due to reworking in surface sediments in some ocean basins, particularly in the tropical Pacific (e.g. Riedel (1951) where he compared the largely not reworked tropical Indian Ocean record to the more reworked record in the tropical Pacific). He also corrected key older mis-interpreted publications by redescribing the "Pliocene" fauna of Tan (1927) from Rotti as being actually two different assemblages, originally from different samples – one Cretaceous, the other Pliocene-Pleistocene (Riedel, 1953). By 1957, Riedel (working with U.S. Navy funding) was able to use a combination of land section samples and deep-sea piston cores to establish the first outline of a Cenozoic biostratigraphy (Riedel, 1957); by 1959, related species could in some cases be assigned to evolutionary lineages, many extending over large parts of the Cenozoic (Riedel, 1959). In parallel, Lipman (1948 and many subsequent papers, Figure 1.13) in Soviet Russia, and Nakaseko (1954 and numerous subsequent papers; Figure 1.14) worked on the rich, if complex record of radiolarians in land sections. Nakaseko established a regional Cenozoic stratigraphy for Japan. He particularly focused on robust taxa (e.g. the thick-shelled species of the spherical spumellarian *Thecosphaera*) which were common even in poorly preserved, highly diagenetically altered Neogene rocks studied in the – successful – search for oil deposits.

Deep-Sea Drilling

By the late 1950s the oil industry had developed technologies capable of drilling far offshore, using drill ships and other platforms that did not require anchoring to the seabed (Bascom, 1961). As scientists became aware of these capabilities they were employed to examine major scientific questions. Project Mohole, for example, tried to drill through the thin crust of the ocean floor to determine the nature of the mantle beneath (Bascom, 1961). This project failed, primarily due to management problems, but also as drilling a crust-penetrating hole through ocean basalts was effectively beyond the technology of the time (Winterer, 2000). Other scientists however saw that, employed as designed to penetrate extended

(a) (b)

Figure 1.12 a) William R. (Bill) Riedel (1927–) and b) Annika Sanfilippo. Riedel began working with radiolarians in the early 1950s in his native Australia as assistant paleontologist at the South Australia Museum, and continued his work with early piston cores taken by Swedish and U.S. expeditions, in particular those collected by Scripps Institution of Oceanography where he became curator of the core collection in 1955. Riedel co-authored the proposal to start the Deep Sea Drilling Program and remained deeply involved with deep-sea drilling throughout his career. Riedel also helped found the Interrad radiolarian society and the first fully online paleontologic journal *Paleontologica Electronica*. He carried out research not only on radiolarians but also fossil fish teeth (ichthyoliths). Riedel largely rekindled interest in radiolarians as important microfossils in applied research after WWII. In collaboration with his longtime co-worker at Scripps, Annika Sanfilippo, he described numerous lineages of Cenozoic radiolarian species, which provided the biostratigraphic events for the now-standard low-latitude Cenozoic radiolarian zonation. Riedel retired to Australia in 2000, while Sanfilippo is still active and associated with Scripps. Source: Courtesy of Annika Sanfilippo.

sediment sections in deep water, such tools could recover what was increasingly being realized was an unprecedented archive of high-resolution climate and ocean change held in deep-sea sediments. Emiliani's 1963 pilot deep-sea drilling project "LOCO" (Winterer, 2000) provided an example of the technical possibilities. Seismic studies had shown that deep-sea sediments were in some cases more than 1 km thick, while studies of the uppermost layers from piston cores had made clear their rich content of fossils and geochemical indicators could be used to understand the history of the earth's climate with unparalled detail. These piston core studies included those of Riedel and particularly research by Emiliani, who obtained paleoclimate data from the geochemistry of foraminiferal carbonate (Emiliani, 1955). It was clear that it would also be possible to use the fossils to date the sediments directly overlying the

Figure 1.13 R. Kh. Lipman. Source: courtesy of Marina Afanasieva.

ocean crust at different locations, and thus provide a definitive test of the newly developing but extremely controversial paradigm in the earth sciences of plate tectonics and continental drift (see also "Mid New Period 1970s–2000s").

The Deep Sea Drilling Project (DSDP) began recovering deep-sea sediments in 1968, after a few years of planning and preparation. In order to reliably recover sediments at depth, and in all ocean basins, a large, dedicated drilling ship was commissioned – the Glomar Challenger (Figure 1.15). This was a major investment and moved studies of marine sediments into the realm of large, well-funded science activities. From the outset, DSDP studies relied extensively on planktonic marine microfossils to provide geologic age estimates for the recovered sediments, and more material accumulated, as probes into past climate conditions via their distinctive, climate-sensitive compositions of species in the sediments. DSDP had been initiated and was managed by a small group of oceanographic institutions, and as Riedel at Scripps was actively involved (e.g. Riedel, 1963), radiolarian research was further linked to the rapidly developing study of deep-sea sediments using interdisciplinary, big-science methods. Within a few years at least 15–20 radiolarian papers were being published annually based on study of the new materials from the drilling ship by a new generation of marine micropaleontologists at several institutions, an output level that would continue until the 1990s. The DSDP, and its successor programs ODP and IODP, significantly affected radiolarian research in other ways. Because the large majority of recovered sediments were Cenozoic in age, study of this time interval became relatively more important. Secondly, as both the (then) Soviet Union and Japan were participating members in the international DSDP program, at least some level of contact was maintained between these – based on language, training, and political barriers – very different groups of workers. Research done on DSDP, and piston core sediments, also furthered the highly interdisciplinary approach to science that originated during WWII. Biostratigraphy in particular was normally integrated for individual drilled sections, with multiple microfossil groups. Radiolarian workers thus shared methods and data intensively with specialists working on other groups, and with many other types of geologic specialists working, on or off the drilling ship, on the same sections – sedimentologists, geochemists, geophysicists, and others.

Biostratigraphy from land sections also developed between ca. 1950 and 1970. Workers in several countries began to develop initial zonations based on what were still dominantly taxonomic studies

Figure 1.14 Kojiro Nakaseko (1925–2011). After studies in Kyoto, Nakaseko took a position at the University of Osaka (his hometown) and studied Neogene geology in Japan and, beginning in the 1950s, increasingly studied radiolarians. He later introduced the HF method for preparing radiolarians from indurated rocks into Japan, founded radiolarian meetings in Japan, contributed to the replacement of geosyncline theory by modern plate tectonic theory via radiolarian biostratigraphy and paleobiogeography, and was a successful geologic consultant to the oil and engineering professions Source: Takemura and Hori, 2015. Reproduced with permission of Elsevier.

describing assemblages of species, e.g. Clark and Campbell (1942; Campbell and Clark, 1944a, b) in the USA, Lipman (1960a, b)in the Soviet Union, and Nakaseko (1959, 1960) in Japan. Although the median number of papers published per year in Japan in the 1950s–1970s was still only about 1 (Yao et al., 2001), Nakaseko's work was particularly important in that he trained a large number of students who would contribute to the expansion of radiolarian research in Japan in the subsequent decades. Most importantly however, many workers were active in biostratigraphic research in the Soviet Union, and indeed, other than the substantial, more biologically oriented output from French workers, Soviet biostratigraphic publications dominated the radiolarian literature from the 1950s until the mid-1960s, when North American research became more active.

Figure 1.15 The Glomar Challenger (1968–1983). Named after the original nineteenth-century Challenger expedition ship, prefixed by the portmanteau word Glomar, for global marine, the Glomar Challenger was used in the Deep Sea Drilling Project – the first phase of what became a series of deep-sea drilling programs. Built by the Levingston Shipbuilding Company for the DSDP in Texas, it was operated by the Scripps Institute of Oceanography for the U.S. National Science Foundation. It eventually recovered nearly 100 km of deep-sea sediment cores from 624 sites, covering over 375,000 nm. Material recovered by the program provided the abundant material for studies by dozens of marine micropaleontologists, including many early radiolarian workers such as Riedel, Sanfilippo, Nigrini, and Moore. The Glomar Challenger was replaced with the larger, more capable Joides Resolution with the onset of the Ocean Drilling Program, the successor to DSDP. Source: Peterson et al. 1970. Photo from Peterson et al. (1970).

Taxonomy

Although monographs on radiolarian taxonomy were published in the 1950s (Deflandre, 1953; Campbell, 1954, and similar volumes in the Soviet Union), these were still largely summaries of the highly artificial genus and higher level taxonomy inherited from Haeckel, and contributed little to advancement of the field. Campbell's (1954) volume in fact was a step backward – it introduced so many problematic changes that it was soon heavily criticized by Riedel and others (Deflandre, 1960), and has in fact virtually never been cited since. Most activity instead was in the description of new species and genera, as new material from deep-sea sediments, and ever more studies of land sections around the world, yielded enormous diversities of undescribed forms. As it became increasingly apparent that much of the diversity of form in the Mesozoic and Paleozoic could not easily be fit into Haeckelian concepts, a new taxonomy began to be created for older materials, both at genus and family level (e.g. Deflandre, 1953; Foreman, 1968; Pessagno, 1969; Dumitrica, 1970: Figure 1.16). Haeckelian taxonomy was retained for Cenozoic forms but substantially modified as more material data on natural variation, biology, and stratigraphic change accumulated. Riedel (1967a, b) revised many of the Haeckelian families and introduced new ones. At lower taxonomic levels the study of lineages in longer sections from DSDP led in the 1960s and 1970s to the revision of many artificial Haeckelian generic concepts and rejection of the idea of extremely long generic ranges in radiolarians (Sanfilippo et al., 1985). Unfortunately, the great diversity of material, limited manpower, and pre-digital technical resources (no databases, digital literature, etc.) hindered the development of new catalogs to replace those of the Haeckelian era (e.g. Campbell, 1954). Lack of comprehensive taxonomic reference sources continues to dog the field even today (see Chapter 5: Taxonomy and Fossil Record). Crucial groundwork however was being laid for future taxonomic revision via the collection of much of the scattered literature into private libraries, which in turn were to be the foundation that many subsequent students of radiolarians used to learn the subject. Most important were the efforts of Riedel and his coworker Sanfilippo, who, often in collaboration with Nigrini in the 1950s to 1980s collated much of the older radiolarian literature, funded translation of much of the Russian language literature into English, and made this library available in the form of microfiche, which until the advent of graphical personal computers in the 1980s was the most sophisticated technology available. A copy of the Riedel Microfiche was a hallmark of most of the major radiolarian research groups. Unfortunately, the microfiche technology was not able to reproduce grayscale photographs, so only the text and illustrations done with line drawings could be effectively disseminated. A first volume including line drawings of the type species for each genus in literature up to 1900 was published (Foreman and Riedel, 1972), but due to lack of manpower not continued.

Figure 1.16 Paulian Dumitrica (1937–) Dumitrica grew up and obtained a Masters' degree in 1960 in Romania, where he began his studies of radiolarians. He completed his doctoral studies in 1974 (on silicoflagellates) and has been based at the University of Lausanne since 1993 (Cortese, 2014). Dumitricia has worked for decades to provide a consistent framework for the higher-level systematics of all Phanerozoic radiolarians, which culminated in the extensive (>200 page-long, plus index) chapter on radiolarian taxonomy in De Wever et al. (2001). He is also, with 461 new species in the Suzuki database, among the 10 most prolific authors of descriptive radiolarian taxonomy in the history of the field. Source: Courtesy of Paulian Dumitrica.

Biology

In the years following WWII and into the mid-1970s, research on the cellular biology of living polycystines was largely done by a single team in France – Hollande (Figure 1.17), J. Cachon, and M. Cachon-Enjumet (the latter also published as just Cachon or Enjumet) (e.g. Hollande, 1953; Hollande and Enjumet, 1960). Using a combination of detailed drawings and observations of whole, living cells and carefully stained thin sections, they discovered a wealth of new morphologic details of radiolarian cells, including details of the nucleus, capsular wall, axopods, and other organelles. Substantial new observations and amplifications of earlier reports on asexual reproduction were done, and several studies of symbionts and parasites as well. Although some observations of shells and shell growth were reported, these were fairly limited in scope, other than as characters for classical species and higher level taxonomy. In general the cellular biology was not integrated into the structure and possible functions of the shell, which, being the only part of the organism available to paleontologists, meant that most of the biologic details observed could not be directly used in paleontologic research. Indirect use, via linking of cellular morphology to cell ecology and function in the environment were also very limited, as it was, then as now, not possible to maintain most radiolarians for long in culture, or to observe individuals "in the wild" due to size and logistic limitations. Lastly, as virtually all the studies were published French, which was, as noted above, in the post-war years rapidly losing importance as a major scientific language, many scientists, particularly younger ones, did not read the studies in any detail. The difficulties of linking biologic observations in a meaningful way to the fossil record continues to be a challenge for radiolarian research even today.

Studies of radiolarian geographic distributions have been published since the earliest studies of Ehrenberg, who went to

Figure 1.17 André Hollande (left, talking to colleague J.B. Jadin; ca. 1998). Biologist who, working with colleagues Monique Enjumet-Cachon (also with last names used singly) and Jean Cachon, carried out extensive studies of radiolarian cellular biology in the 1950s to 1970s. He began his career in Algiers but in later years worked in Paris. Source: Mignot, 1998. Reproduced with permission of Elsevier.

considerable lengths in his many papers on microorganisms to compile tables of individual species distributions in different types of samples, as well as those from different geographic regions. Studies of modern species distributions date back at least to Haeckel, Cleve (1899 and subsequent papers), and Popofsky, who also recorded distributions in both sediment and plankton samples. Nonetheless, the very patchy nature of sampling meant that biogeographic patterns were not very discernable. With the availability after WWII of a rapidly expanding archive of both surface sediments samples from coring, and water plankton samples from net tows, it became possible to produce a much clearer picture of radiolarian biogeography. The work was carried out in parallel to numerous other studies of components in ocean plankton: sediment geographic distributions, e.g. phytoplankton in water samples, bulk lithologies, and other fossil groups such as foraminifera in sediments. Workers such as Petrushevskaya (1967, 1971a; Figure 1.18) described many new species from extensive sampling of ocean regions, using both plankton and ocean surface sediments. Hays (1965) and Nigrini (1967) published the first regional distribution maps for species (Southern and Indian Oceans, respectively). Numerous other studies followed in the 1970s, including statistical methods to define coherent assemblages that could be correlated to surface water masses (Nigrini, 1970; Sachs, 1973a,b) which were later used to develop paleoceanographic studies, as well as conceptual interpretations of these interrelated patterns (Casey, 1971).

Figure 1.18 Maria G. Petrushevska. Productive researcher working in Soviet Russia who in the 1960s to 1980s studied Recent and Cenozoic radiolarian taxonomy, biogeography, and distribution in the water column. In addition to over 90 new species descriptions she revised numerous species and genera and proposed a variety of higher-level taxonomic systems. She also published two books on radiolaria (Petrushevskaya, 1981, 1986) and other monographic studies. Source: courtesy of Marina Afanasieva.

Mid New Period (1970–2000)

The ca. 30-year period between 1970 and 2000 yielded more radiolarian papers than all other time intervals before or since – more than two-thirds of the known literature, as well as peak rates of publication, reaching rates of more than 200 papers/year, particularly between 1980 and 2000 (Figure 1.1). This period can be divided into a shorter interval between 1970 and 1980 in which the field was dominated by Cenozoic studies, and the subsequent 20-year-long interval in which Mesozoic research predominated, with, particularly since the late 1980s, a substantial number of studies on Paleozoic radiolarians.

The field of marine micropaleontology, with a focus on the (mostly Cenozoic) microfossil content of deep-sea sediments, became a distinct discipline in the early 1970s, with the landmark volume *The Micropaleontology of Oceans* (Funnell and Riedel, 1971), and the founding of the field-specific journal *Marine Micropaleontology* in 1976. Over 50 radiolarian papers, mostly on Cenozoic or Recent/living materials appeared in this journal in the initial 10-year period of publication. The first deep-sea drilling sections with radiolarian-rich sections were recovered by the Deep Sea Drilling Program in the late 1960s (e.g. Riedel and Sanfilippo, 1970). These reports marked the beginning of three decades of intense study of radiolarians from Cenozoic deep sea sediments. Numerous researchers were also continuing to work on the materials initially studied by Lipman in the 1950s, i.e. radiolarian bearing sedimentary rocks in the Soviet Union, both from Paleogene siliceous deposits in the continental interior, and radiolarians of all ages from accreted deep-sea sediments along the eastern Pacific margin. Nearly 300 papers on radiolarians, including also studies of living materials, were published by Soviet authors from 1970 until the collapse of the USSR beginning in 1989. Particularly noteworthy were researchers such as Petrushevskaya (e.g. 1967, 1971, 1981, 1986) who used plankton materials, surface sediments, and Cenozoic assemblages to extend knowledge of radiolarian morphology and taxonomy.

Radiolarians played an important role in the development of the new field of paleoceanography in the 1970s and 1980s. This field had developed gradually from the integrated, interdisciplinary research efforts of marine science teams initially assembled for military purposes during and after WWII (see "The origins of radiolarian biostratigraphy: 1940s to 1950s"). Piston coring of the oceans had become so extensive that literally thousands of locations in the oceans had been sampled, so that, despite problems with sediment disturbance, variable preservation of microfossils, and other issues, it became possible by the 1960s to create detailed geographic distribution maps of individual species of microfossils in surface sediments, which confirmed, but provided much more complete data than similar biogeographic studies of microorganisms in the water column. Early workers on these radiolarian faunas such as Hays (1965; Figure 1.19) and Nigrini (1967) teamed up with researchers working on other microfossil groups in the CLIMAP project (Imbrie and Imbrie, 1979). They used Imbrie's newly developed method of combining factor analysis with multiple regression (Imbrie and Kipp, 1971) to convert quantitative counts of relative species abundances in samples into ocean surface water temperature estimates, then applied the regression functions to glacial aged microfossil assemblages to reconstruct ocean water temperatures globally during the last glaciation. Similar quantitative methods with microfossils, when combined with geochemical data, were used to demonstrate the validity of the Milankovitch theory that the ice ages were regulated by orbital variation in the earth's absorption of solar radiation (Hays et al., 1976a). The motivation for this early work on climate change was mostly academic. In the early 1970s there was some concern among these research groups that the earth might, in a few thousand years, descend into a new ice age. There was awareness also of rising CO_2 concentrations and global warming, but only as a possibility; the first expert assessment of the climate consequences of continued CO_2 release, the Charney report, was only created (from a workshop held at the Woods Hole Oceanographic Institution) in 1979 (McNutt, 2019). In the early 1970s many other environmental issues such as ozone layer destruction attracted more attention. That changing atmospheric CO_2 concentrations would soon lead to severe global warming only became the dominant environmental concern among scientists in the 1980s.

Figure 1.19 James D. Hays. Although his radiolarian training was done partially with Riedel, Hays' career was centered at the Lamont Observatory of Columbia University. Early work on Antarctic radiolarian biogeography and stratigraphy was soon extended into other ocean regions though mostly with a focus on late Neogene polar assemblages. Hays worked with John Imbrie to pioneer the use of quantitative analysis of microfossil assemblages to reconstruct ocean temperatures over glacial-interglacial cycles, and together with Nick Shackleton confirmed the Milankovitch hypothesis of orbital control over the timing of ice ages (Hays et al., 1976a). Orbital variations in sediment archives would become a key method to develop high-resolution paleoceanographic studies in Cenozoic and older time intervals in the 1990s and beyond. He also maintained an interest in radiolarian evolution, working with evolutionary biologists at Columbia, and paleobiologists such as Niles Eldredge at the American Museum of Natural History, to train students such as Kellogg and Lazarus in evolutionary studies. Source: Lamont-Doherty Earth Observatory. Reproduced with permission of Lamont-Doherty Earth Observatory

Another aspect of CLIMAP and a series of follow-up projects (SPECMAP, CENOP, etc.) was to focus the training of Quaternary radiolarian workers on quantitative applications to earth science questions. Taxonomic work was relegated to designated team specialists (often C. Nigrini and J.P. Caulet, Figure 1.20) with the goal of standardizing a small but sufficient number of species concepts to allow quantitative analysis of common species in Quaternary radiolarian assemblages. Given the vast amount of work needed to untangle the legacy of artificial taxonomy from late nineteenth-century workers this was practical, but also limited both the usable taxonomic foundation for future applied research, and the number of taxonomically oriented workers who would in the future work to improve the situation. This problem of allocating human resources between foundational taxonomy and the application to other hypothesis testing is a common one in biodiversity-based sciences. For example, an analogous situation developed in invertebrate paleontology much later in the 1990s, as growth in fundable, high-profile research in quantitative paleobiology to some extent came at the expense of primary taxonomic work. The consequences however can be very long lasting – there is still, four decades later, no accepted catalog or even numerical value for living/Quaternary radiolarian species diversity, despite that the number of species (certainly <1,000) is in principle of manageable size.

Studies of living radiolarians in this time period included both study of their biogeographic and ecologic setting in ocean waters, and laboratory study of living and dead individual specimens. Particularly important was the work of Anderson, who largely took over the study from the earlier work of Hollande, Cachon, and Cachon. Between the 1970s and 2000s he extensively studied the biology of the living radiolarian cell, and used TEM images of

Figure 1.20 Jean Pierre Caulet (?–) and Catherine Nigrini (1939–2005). Caulet's early years were spent in French colonial North Africa. He completed his studies at the National Museum of Natural History in Paris and began working there on radiolarians by the beginning of the 1970s. Initial work was based mostly on deep-sea sediments collected by French research ships (Caulet had become curator of these cores at the Paris Museum) but by the 1980s he was also working with deep-sea drilling sediment materials and publishing in the drilling program reports, while working on Cenozoic radiolarian taxonomy (author or co-author of 68 species) and biostratigraphy with several colleagues around the world. In the 1990s he developed, in collaboration with Nigrini and others, the URI method for estimating paleoproductivity from radiolarian assemblages. With Nigrini he began, and continues to develop in retirement, the Radworld catalog database of radiolarian taxonomy. Cathy Nigrini grew up in Canada and completed her doctoral studies at Cambridge, UK, in 1966. She worked on radiolarians from surface and Cenozoic sediments, describing new species (42 species in the Suzuki database) and revising several genera, and participated on several deep-sea drilling legs. She also helped define and calibrate Cenozoic radiolarian event biostratigraphy and led the radiolarian taxonomic catalog development needed for major paleoceanographic research projects such as CLIMAP. Nigrini worked with others to build comprehensive taxonomic catalogs such as Radworld. Nigrini held many short-term associate appointments but mostly worked from home locations in the USA, tracking her husband's career in the oil industry. (Lyle and Moore, 2002). Caulet and Nigrini are both co-authors of the most recent prior reference book on radiolarians (De Wever et al., 2001).

microtome section cells to describe the structure of cell organelles. He also carried out basic observations of feeding, shell formation, and other aspects of radiolarian biology. Anderson's (1983) book synthesized the available biologic knowledge of radiolarians and included, albeit somewhat sketchily, an overview of their taxonomy and fossil record. Although Petrushevskaya (1981) had recently published (in Russian) a monographic treatment of nassellarians, one of the two main groups of living radiolarians, Anderson's book was the first modern volume on radiolarians in the English language since Haeckel's Challenger report and was widely read by radiolarian workers. Anderson also trained A. Matsuoka in radiolarian biologic research and Matsuoka in turn developed such studies on his return to Japan.

Rapid expansion of both plankton and marine micropaleontologic research in the late 1960s and 1970s had made clear that the biogeographic patterns of living plankton and fossil plankton were remarkably similar, which was an essential piece of information needed to use the microfossil record to study paleoceanography. But how the living plankton assemblage was transferred into marine sediments was unclear, and indeed, simple calculations of how long single cells would take to sink to the ocean bottom suggested that the patterns should not match nearly as well as they did. At the same time, geochemists were beginning to explore how biologically active elements such as carbon and nitrogen were cycled within the oceans. Both groups converged in the late 1970s on the use of newly developed sediment traps (e.g. Honjo et al., 1980) to document and quantify how surface particles (mostly plankton) sank through the water column and were altered by dissolution, consumption, and other processes, transferring elemental material to deep waters, and both organisms and their chemical contents into the deep-sea sedimentary record. This work yielded the first quantitative understanding of the ocean's role in carbon cycling (the term "carbon pump" was coined, based on this early research, by Volk and Hoffert in 1985), and explained how single cells were bundled into larger objects (fecal pellets, marine snow) which sank much more rapidly, thus preserving the biogeographic patterns of the living assemblages. Takahashi (1981) in particular studied the sinking processes of radiolarians, documenting the relatively rapid transport of radiolarian shells to great depths, which contributed to a modern understanding of how biologically active elements are cycled in the oceans. Some elements, particularly carbon, are largely recycled in the upper water column, but others, including silica, are mostly exported to the ocean floor, where dissolution in the upper sediment surface occurs (Tréguer and De La Rocha, 2013). By use of newer sediment trap models with separate holders for each month, it also became possible to study how the flux of particles varied with time. This was used by Takahashi and other plankton biologists to document for the first time with reasonably complete time series data the seasonal patterns of plankton groups, and even that of individual species, which until then had only been known very vaguely from highly irregular, time-limited plankton tow data. Takahashi also created what was at the time the most comprehensive taxonomic catalog for the living assemblage, using advances in scanning electron microscope technology to extensively illustrate the shells of numerous species. His monograph (Takahashi, 1981) became for many years a key reference for study of living or late Neogene radiolarians.

Mesozoic research increased dramatically in the 1980s and 1990s both as the use of better preparation methods (hydrofluoric [HF] acid to extract radiolarians from silicified rocks and even cherts: Pessagno and Newport, 1972) became standard, and with more understanding of concretions and other special facies where better preserved radiolarians could be found. A further impetus was the dramatic change in earth science research resulting from the development of plate tectonic theory. This new way of thinking about the earth essentially created a fundamentally new modern earth sciences, transforming geology from a semi-descriptive field to one based on quantifiable mechanisms fully based in the primary physical sciences (Oreskes and Le Grand, 2003). The theory was developed by the same style of research that grew out of WWII military management of science: international, multidisciplinary teams, frequently funded by the U.S. Navy research system, and largely based at the oceanographic institutions that also gave birth to marine micropaleontology (Oreskes and Le Grand, 2003). Plate tectonics helped explain many otherwise unexplainable empirical patterns, such as the occurrence of deep-sea sediments, particularly radiolarian-bearing cherty rocks, in specific geologic settings in mountain belts, often at considerable elevation (Figure 1.21; Bernoulli and Jenkyns, 2009a, b). Workers such as Baumgartner and De Wever, and their many students, in the 1980s and 1990s used radiolarians to decipher the history of the Alps and other areas of the "Tethyan" Mediterranean (e.g. Baumgartner et al., 1995). Another idea derived from this new paradigm of plate

Figure 1.21 Radiolarites in the Alps. Thin-bedded cherty rocks containing radiolarians and associated with relatively rare rock types (for land sections) such as sepentinites and pillow lavas are widespread in mountain belts around the world, and this association (the "Steinmann trinity"), as well as their position in other "geosynclinal" facies such as greywackes had puzzled geologists since at least the 1930s (Bailey and McCallien, 1950; Bernoulli and Jenkyns, 2009a,b). New radiolarian extraction methods such as HF permitted taxonomic and biostratigraphic study of the assemblages, which helped clarify the genesis of these rocks and how such geologic environments fit into plate tectonic theory. Source: courtesy of Daniel Bernouilli.

tectonics was that many geologic terranes (the tectonic blocks that make up large parts of the continents' mountain belts) were now recognized as wandering plate fragments, with distinct geographic and temporal origins. Mountain belt lands thus had complex, composite geologic histories, which could in many cases best be resolved with radiolarian assemblages, as these were one of the very few fossils still preserved in the often highly altered metamorphic rocks that these terranes were composed of, and could provide information on the ages, and even, via the species composition, the approximate paleolatitude (tropical, temperate, or polar) of each distinct tectonic block. New analytic methods such as the unitary association method (Guex, late 1970s, summarized in Guex, 1991) were developed to extract robust stratigraphic information from these assemblages, where preservation was often poor and, due to the highly complex tectonic setting, even basic principles of stratigraphy such as superposition of younger on top of older rocks could not be safely assumed. International cooperation on Mesozoic taxonomy also yielded at least partial catalogs of both genera and species, e.g. Baumgartner et al. (1995). One consequence of the growth of Mesozoic, and later Paleozoic studies was however a degree of decoupling of the methods, goals and research style from those developed in Cenozoic and living radiolarian research. Pre-Cenozoic workers generally worked on field geologic problems, in geology departments or geologic surveys. They had comparatively little training in biology or oceanography, or opportunity to address research questions in these fields. The taxonomy also was largely new. Higher level systematics in particular was not only different, but built in a different style than for the Cenozoic, where reference to living materials and the legacy of early biologic studies were important constraints.

Although the initial growth in radiolarian studies using these new methods was in Mesozoic work, studies of Paleozoic rocks soon followed. Early workers included Holdsworth (e.g. 1977), Nazarov (1975, and many subsequent papers; Figure 1.22), and Kozur (numerous articles beginning in the 1970s, Figure 1.23), who worked on both late Paleozoic and early Mesozoic rocks. Although at first it was uncertain if enough radiolarian material could be found to be stratigraphically useful, it soon became clear that, albeit with lower resolution, radiolarian zonations could indeed be established and used in applied geologic studies (Holdsworth and Jones, 1980).

Figure 1.22 Boris B. Nazarov. Source: courtesy of Marina Afanasieva.

While the above changes that expanded Mesozoic and Paleozoic research affected most countries, the impact on research in Japan was particularly profound (Ichikawa, 1982). The tectonic terranes where much of the research was concentrated made up only the border regions in larger continental blocks like North America, but were the dominant type of

Figure 1.23 Heinz W. Kozur (1942–2013). After starting his career in East Germany, Kozur moved to Hungary to evade political harassment, but again for political reasons lost his position there in 1984. He continued to work independently, and in cooperation with other universities, particularly in Halle. Kozur worked mostly on late Paleozoic and early Mesozoic materials and studied both conodonts and radiolarians. An extremely prolific author with approximately 600 publications, he described, usually with co-author Mostler, 884 radiolarian species, second only to Haeckel. Source: Bachmann, 2014. Reproduced with permission of Mertmann.

rock formations in Japan, which consists largely of a complex of accreted terranes and oceanic material formed by Mesozoic-Recent subduction of the margin of the Pacific Plate. With the spread of plate tectonic theory, more detailed studies of many sections were done in Japan, revealing the complex history of the rocks. Radiolarians were noted to occur frequently in these samples, and began to be used for geologic dating (Nakagawa and Nakaseko, 1977). Many formations previously thought to be Paleozoic were recognized, primarily from radiolarians, to be Mesozoic. This had such a dramatic impact on the geologic study of Japan that it is locally referred to as the "radiolarian revolution" (Takemura and Hori, 2015). Radiolarian biostratigraphic work increased substantially in Japan in the subsequent years, going from an average of just a few papers per year through most of the 1970s to nearly 100 papers per year in the 1980s (Yao et al., 2001). Japanese radiolarian workers also played an larger role beginning in the 1980s in international meetings such as the Interrad Congresses. These meetings had initially been dominated by workers from the USA, Europe, and the USSR, and had developed in the 1980s out of a merger of more regional meetings that had been held during the 1970s, such as the Eurorrad meetings, and meetings associated with Nakaseko and his Osaka home institution in Japan.

Radiolarian research began in China gradually in the 1980s to increase in intensity, with the number of papers published per decade essentially doubling each decade from ca. 30 papers in the 1980s into the 2000s with nearly 140 papers (based on data from radiolarian community literature archives). Studies of surface sediments in the seas bordering China were an early focus but with time the large majority of studies were on the biostratigraphy and paleogeography of Paleozoic and early Mesozoic faunas.

While Mesozoic, and later Paleozoic research expanded, by the mid-1980s the study of deep-sea sediments, the primary focus of Cenozoic workers, had become dominated by geochemical methods. Although at first geochemistry had complemented, it soon increasingly displaced micropaleontologic-based estimates of paleoenvironmental conditions, using stable isotope ratios of oxygen and carbon, followed by an ever-increasing suite of other geochemical proxies. The large majority of these geochemical signals were extracted from the calcite of foraminiferal shells, and as a result, carbonate sediments and the carbonate of fossils were increasingly the primary focus of deep-sea sediment research. Studies of siliceous microfossil groups such as radiolarians and diatoms became more marginal. It also did not help that the taxonomy of the highly diverse Cenozoic radiolarian assemblages was still burdened by Haeckel's inadequate nomenclature system, with the type materials largely inaccessible in East Germany behind the "Iron Curtain." Studies of Mesozoic and Paleozoic radiolarians by contrast could largely start with a clean slate, so that taxonomic progress with the older materials was much more rapid. Cenozoic research thus declined gradually but substantially in the 1990s (from ca. 60 papers to only 20 papers/year).

One of the more significant developments in radiolarian research during this period of growth was the failure in many, though not all, countries to establish a strong presence in regular university positions, as these offer the best opportunities to teach students who then go on to broaden the field and increase its importance in the research landscape. In both the USA and Europe, although a few important workers during this time period were university based and indeed trained a new generation of workers (e.g. Hays and Anderson, at Columbia University, NY; Pessagno, U. Texas, Dallas), many workers, and also the next generation of their students, worked instead in organizations such as oceanographic institutions and geologic surveys, where teaching and student training, although also done, were less important. The reasons for this are multiple. Paleontology in general was generally seen by the 1970s as a somewhat marginal specialty in both the earth sciences and in biology. These fields had become more quantitative and reductionist in style after WWII (i.e. molecular biology, which rapidly displaced organismal biology in the late twentieth century). Both the primary data – subjective observations of morphology – and the classical style of analysis that characterized much of paleontology did not fit well with the newer paradigm. Most universities hired at best only one paleontologist, who usually was either an invertebrate or vertebrate paleontologist, both specialties with a large established base (and lobby) and who had developed a more modern, analytic research program with paleobiology – a subject that micropaleontologists had largely avoided. Invertebrate and vertebrate paleontologists were also well suited to teaching general courses within the undergraduate earth science or biology curricula, and vertebrate paleontologists could provide useful publicity for their departments. When micropaleontologists were hired they were usually foraminiferal specialists who could integrate with geochemists in paleoceanographic research, or, in universities with the specialty, contribute to training programs for oil industry research needs. Establishment of substantial numbers of academic positions for radiolarian specialists only occurred in Japan, where the unique geology of the country supported a substantial number of professorships in radiolarian research, and in consequence, the training of large numbers of students. Indeed in Japan, so many students were trained that they are informally referred to in Japan according to approximate wave of study – first (1940s–1960s: Ichikawa, Nakeseko), second, and third generation. With rare exceptions (e.g. Caulet and later Lazarus) radiolarian workers also did not hold natural history museum curatorships, and large new collections of radiolarian materials were not created. Combined with the loss of older collections this has hindered developing the taxonomic foundations of the field, both by lack of reference material, and lack of dedicated taxonomist positions.

Current Period (2000-Present)

Radiolarian research in the first two decades of the twenty-first century is largely a continuation of the themes and trends established in prior decades. For example, taxonomic and biostratigraphic work still makes up the large majority of research. A major advance in radiolarian taxonomy was marked by the first comprehensive, documented synthesis of higher level radiolarian taxonomy by Dumitrica in De Wever et al. (2001). In addition to these well-established themes, there are a few new developments worth noting.

The most important is that molecular biologic studies of radiolarians are becoming increasingly common and are beginning to cover enough taxa to have a real impact on our understanding of systematics of the group (e.g. Sandin et al., 2019; 2020), and thus taxonomic practice by radiolarian specialists. Early molecular biologic studies of radiolarians, e.g. Amaral-Zettler (1997), were not aimed at understanding the taxonomy of the group as much as determining if the main higher level categories (Polycystina, Acantharia, Phaeodaria) were part of the same high-level clade, and what the relationships of these main groups were to other similar groups of protists such as the Foraminifera. This focus on the relationships *between* radiolarians and other groups of protists has been common in most molecular research on protists as the broader picture – and the possibility to discover new, high-level clades of organisms – is much more attractive to most researchers than resolving systematic details within a single clade, particularly one with already well-established, morphologically based systematics. The improved detail and breadth of molecular studies in recent years, mostly by research groups at oceanographic institutions in France and Japan, reflects improved methodologies, such as primers that specifically work with protistan genetic material, and vastly improved computing capabilities that allow both more automation, and the ability to analyze much larger data sets. In addition, it also is part of a broader trend towards understanding the vast, and until the advent of molecular methods, poorly understood diversity of the protists, and in particular, the marine plankton, as it is becoming clearer that these ecosystems are central to understanding global geochemical cycles and how the earth will respond to future climate change.

A second new development is the increasing availability of larger databases for radiolarian taxonomy and radiolarian occurrences. Several groups of workers have over the years compiled large digital catalogs of radiolarian taxonomy (each with >10,000 species), including Radworld (Caulet, Sanfilippo and Nigrini, begun in the late 1990s using Microsoft's "Access" database: Nigrini et al., 1998) and a parallel effort by Suzuki (Suzuki, 2016), also over many years, using the custom DBase Paleotax application (Löser, 2004). These databases have concentrated on capturing the original descriptions, with only a modest effort to synonymize names. Other much smaller catalogs for genus taxonomy have been made public via regular journal publishing (Paleozoic genera: Danelian et al., 2017; Mesozoic genera: O'Dogherty et al., 2009), and are thereby digitally accessible, albeit only in pdf format. However the larger, relational databases for species taxonomy are single-user, offline systems and thus have not become widely available to the research community. An exception (was) the Baumgartner et al. (1995) catalog of selected mid-Mesozic Tethyan species which was published as a web database in the 1990s and was accessible through the 2000s, but is, as of July 2019, no longer online. The community website (www.radiolaria.org: Dolven, 2001), which was started in late 2000 is another exception as the catalog section has always been an online system.

The content however has always depended on community contributions. Lack of funding, declining numbers of taxonomic specialists, and decisions by the authors of the offline database systems not to merge with the community site meant that content at radiolaria.org stalled in the late 2000s at just several hundred species, almost entirely of Cenozoic age.

Radiolarian occurrence data is also becoming available online. Major compilations of occurrence data include WoRRAD (Boltovskoy et al., 2010), currently available online (via direct sql query) to the NSB system, NSB/Neptune: (www.nsb-mfn-berlin.de: Lazarus, 1994; Spencer-Cervato, 1999; Renaudie et. al., 2020) which covers deep-sea data, mostly for the Cenozoic, and part of the Mesozoic land record in the Paleobiology Database (www.pbdb.org: Alroy et al., 2001, with much of the entry work by Kiessling and coworkers). As NSB and PBDB are still actively being developed and are widely used, more data will presumably be added to them in the future. There is also a database by Russian workers (Agarkov et al., 2019) which is not widely used and appears to hold data resolved only to geologic stages. As important as catalogs are to supporting better taxonomy, the availability of radiolarian occurrence data is perhaps even more significant. One of the striking aspects of radiolarian research in the last decades is the relative rarity of paleobiologic studies, even though paleobiology has been the single most important theme in the rest of paleontologic research in this time period (and as noted, has been important to securing university positions); and paleobiologic studies are, if not common, represented by dozens of publications for other microfossil groups such as foraminifera. Occurrence data for radiolarians allows, for the first time, larger-scale paleobiologic syntheses of patterns of radiolarian occurrence and evolutionary dynamics over geologic time. Importantly, these occurrence databases, unlike the offline species catalogs, include synonym information in the form of curated taxonomic name lists as these are essential to interpret the published data without creating substantially inflated diversity estimates. Merging these two aspects of taxonomic information is thus essential.

Work is currently underway which should soon make at least part of the broad range of offline taxonomic catalog data generally available, and link this to the existing online catalog and occurrence data. The Cenozoic taxa are being put online at mikrotax.org (Suzuki, Caulet, Lazarus, and Young, in prep.), and interest is growing in converting all the catalogs into generally accessible resources. Community developed and supported catalogs, linked to the occurrence data, should go a long way to resolving the perennial problem of inadequate radiolarian taxonomy. These developments reflect the increasing power of the internet, and increasing technical skills among radiolarian specialists. In a related development that particularly affects Cenozoic and living radiolarian taxonomy, the taxonomic legacy of Ehrenberg and Haeckel were finally addressed by a team of Japanese and German taxonomists who, making use of the new accessibility of the Ehrenberg and Haeckel collections resulting from the reunification of Germany, re-imaged hundreds of original specimens in the collections left behind by these early workers (Tanimura and Aita, 2009). This information should make possible a stable revision of the confusing species, and particularly genus level taxonomy of Cenozoic-Recent radiolarians.

A last feature of the last two decades is an overall decline in research activity, at least as measured by the annual numbers of publications (Figure 1.1). The extent of decline in activity shown in this figure is certainly exaggerated by the, normal for literature databases, failure to capture all of the most recently published literature, but the very large magnitude in decline is also too much to ascribe solely to database errors, and also reflects real-world changes in the radiolarian community. After many years of growth, or at least stability, academic positions for radiolarian workers are in decline in Japan, and fewer radiolarian workers, or micropaleontologists of any sort, are still working in other organizations such as oceanographic institutions that used to be major employers of radiolarian specialists. The one bright spot in these data is the steady increase in the numbers of papers on radiolarian biology, to the extent that in the last few years biology research output is equal to that of the more traditional, biostratigraphically oriented work on Paleozoic and Mesozoic material. These newer trends have implications for the future development of the field, but a discussion is deferred until the end of this book, where it is integrated into a summary of the main themes of the volume.

2 Biology

General Characteristics of Planktonic Protist Biology

Radiolarians are a fully planktonic group of organisms, and, although some species are found in coastal waters, the majority of the species are found only, or predominantly, in the pelagic realm of the open ocean (see Chapter 3: Ecology). Living cells, mostly in the size range of 50–500 μm, show a range of morphologies, including solitary and colonial forms (Figures 2.1, 2.2). The biologic characteristics of radiolarians reflects both unique adaptations by the group, but equally a variety of constraints that are common to protists, the pelagic environment, or both together. Constraints of this sort are not normally described in reference works on radiolarians, but are important. Many fundamental physical characteristics of the environment are very different for pelagic protists in comparison to larger benthic or terrestrial organisms, and our intuition, based on our own experience, is not always a good guide to understanding this special world. The pelagic realm is highly three-dimensional, nutrients are very dilute, and it is in constant motion with only a few stable structures, but has a complex hierarchy of temporary structures with sizes from cm to >10,000 km, with durations from minutes to decades. Water itself behaves very differently than "normal" at the scale of single-celled plankton. For this reason,

Paleobiology of the Polycystine Radiolaria, First Edition. David Lazarus, Noritoshi Suzuki, Yoshiyuki Ishitani and Kozo Takahashi.

we begin by briefly reviewing what is known, or inferred about the world of pelagic (mostly single-celled) plankton, and only then continue with the specifics of radiolarian cell biology. Most of this work is based on general principles of physics and chemistry, and is thus, in principle, applicable to all pelagic plankton. However, most work to date has concentrated the specific applications to plankton groups of established ecologic importance, e.g. the bacteria, phytoplankton or copepods, rather than radiolarians. More extensive summaries are available in books such as Fenchel (1987), Reynolds (2006), or Kiørboe (2008), and in the summary below, information not otherwise cited may be found in these volumes. This type of theoretical plankton ecology represents one of the three – largely disjunct – major approaches to the study of plankton. The other two are the classic style of studying the morphology of the organisms themselves (most radiolarian publications are of this sort); and the more recent field of molecular studies (Jürgens and Massana, 2008).

Physical Characteristics of the Pelagic Ocean

Ocean waters can be divided into different environments based on water depth, distance from coastlines, or other characteristics (Pinet, 2012). In coastal waters, winds, waves, and tides mix waters sufficiently that stable vertical structures are largely absent (except near freshwater inputs, such as estuaries), although a wide variety of dynamic temporary structures such as fronts and localized fast currents (e.g. jets) develop. Nutrient inputs from rivers, and from reworking of sediment from the bottom, are high, and the bottom is close enough to the surface to be accessible to individual organism life cycles. This distinct environment of coastal waters is often separated from those of the open ocean near the outer edge of the marine shelf by a semi-permanent shelf-break front. The waters overlying the shelf belong to the neritic zone, while those over the deeper waters beyond the shelf are in the oceanic zone. The oceanic pelagic realm can be subdivided by water depth and/or characteristics linked to depth, such as light penetration. The most important of these are the boundary between the photic and aphotic zones, above which enough light penetrates that photosynthesis can take place; and the thermocline, separating relatively well-mixed, usually warmer surface waters above from less mixed, more stably stratified, typically cooler waters below. The thermocline is usually associated with significant water density and nutrient concentration contrasts and biologic materials, due to increased production from the nutrients, tend to be concentrated in thermocline waters.

The global ocean is mixed on time scales of more than a few hundred years by surface and deep ocean currents (Brown et al., 1989). Within each major ocean basin (Pacific, Atlantic, and Indian), tropical surface currents run east to west and then branch poleward at the western margin of the basins to form counterclockwise gyres, returning between ca. 40–45° latitude N or S along the

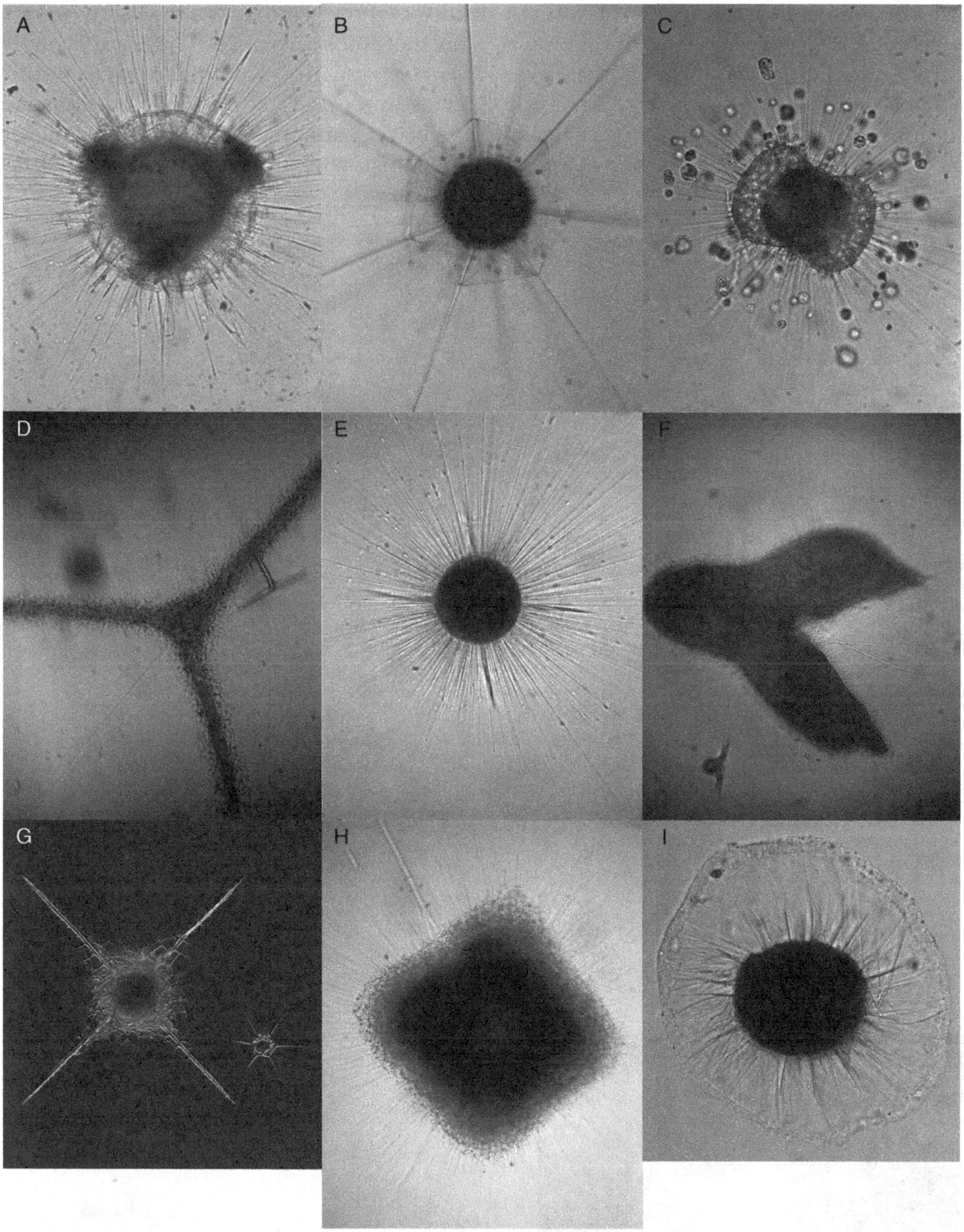

Figure 2.1 Living spumellarians. (A) *Arachnosphaera hexasphaera* Popofsky. B) *Astrosphaera hexagonalis* Haeckel. C) *Didymocyrtis tetrathalamus tetrathalamus* (Haeckel). D) *Myelastrum lobum* (Swanberg, Anderson, and Bennett). E) Hexastylidae gen. et sp. indet. F) *Myelastrum* aff. *trinibrachium* Takahashi. G) *Cleveiplegma boreale* (Cleve). H) *Spongaster tetras tetras* Ehrenberg. I) Spongodiscid with gelatinous envelope. Most images show external radiating pseudopods and actinopods; pseudoflagellum clearly visible in D, F and H; symbionts in C. Source: A–F, H courtesy Dr. A. Matsuoka; G by John Dolan/Inigo Montoya; I by Y. Ishitani. (*See color plate section for the color representation of this figure.*)

eastern margins to the tropics. Poleward of these gyres are polar waters with their own surface circulation systems – a small gyre in the North Pacific, a clockwise Antarctic Circumpolar Current in the south (marking the northern edge of the Southern Ocean), and a more complex circulation in the northernmost Atlantic. These surface currents thoroughly mix surface waters within gyres in a matter of decades, and transport significant amounts of water between even the most widely separated surface ocean

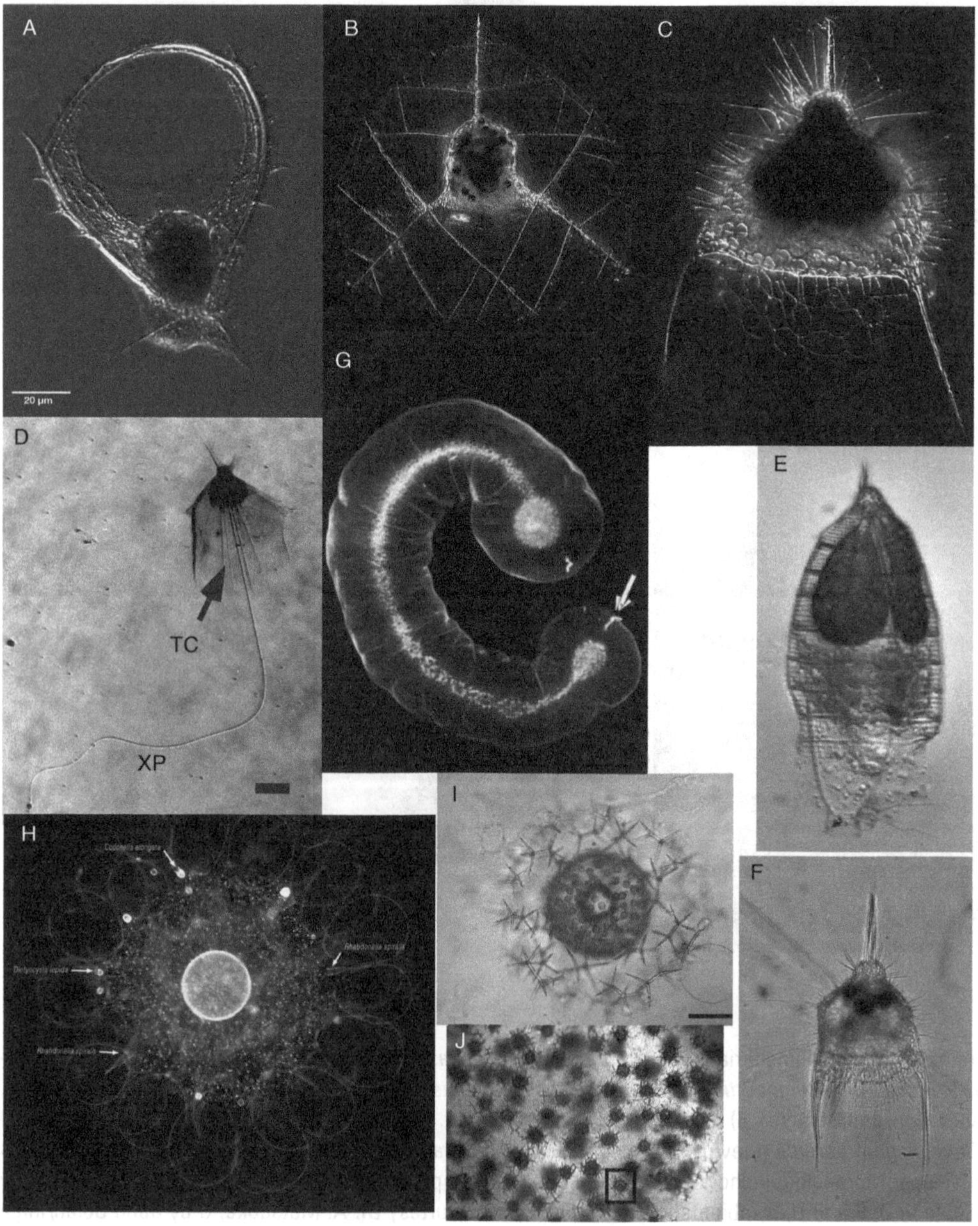

regions on the scale of just a few years (Jönsson and Watson, 2016). Below these surface currents, cold deep water currents – the thermohaline circulation – run between the polar regions. The combination of currents and mixing produces a rather small number of large bodies of water with fairly uniform physical and chemical characteristics – the ocean water masses. These more or less stable, planetary-scale features of the ocean play a dominant role in plankton biogeography and are discussed in more detail in Chapter 3: Ecology.

Numerous smaller-scale structures exist in the ocean but tend to be ephemeral, lasting only a few months to less than an hour. Seasonal variations occur in the positions of ocean fronts between water masses, and in the degree to which the surface waters above the thermocline are mixed by winds and waves. The main currents do not move as simple linear streams but meander. Frequently, eddies in various size scales break off and create, for a period of days to weeks, small patches of distinct water conditions. At still smaller scales, a variety of structures such as microfronts, thin layers of contrasting water, patches due to internal waves, or aggregations of suspended particles due to turbulence are seen, lasting for a few days to only hours. Each of these structures, some only at cm scales, can strongly affect the distribution of nutrients and suspended food particles (Prairie et al., 2012), and thus the perceived environment of small, short-lived planktonic organisms such as radiolaria and other plankton.

The apparent uniformity of the environment of plankton had once suggested to ecologists that, under standard theories of "competitive exclusion," only a few species of plankton (particularly phytoplankton) should exist. The actual diversity of plankton, with hundreds, if not thousands of species in the phytoplankton alone, was considered to be a paradox (Hutchinson, 1961). More recent works suggest that higher diversity can be expected in theory due to the complexity and chaotic behavior created by multispecies interactions (Tilman et al., 1982; Huisman and Weissing, 1999). However, it is not clear if even revised theory would predict such high observed diversities in a uniform environment. As more information is obtained about the ocean environment, it is clear that the great range in complexity in structure and its variation over time mean that the planktonic physical environment is far

Figure 2.2 Living nassellarians. A) *Zygocircus* sp. B) *Callimitra* sp. C) *Pterocanium trilobum* with numerous fine external spines supporting axopods. A–C are Rugol-stained specimens that show dense colored central capsules in central part of shell, much thinner cytoplasm with granules extending over distal skeletal elements. D) *Dictyocodon prometheus*, showing terminal cone of axopods ("TC") and long pseudoflagellum "XP." E) *Eucyrtidium* sp. showing multilobed nucleus in apex of shell and less dense cytoplasm extending to base. F) *Pterocanium praetextum praetextum*, with short terminal cone of axopods. G–J: Collodaria. G) Colony of *Collozoum caudatum*, with cells (white spots), thick gelatinous sheath, and excreted waste material (white threads at either end [arrow]). H) Solitary form *Thallassicolla pelagica* with outer layer of alveoli, symbiont cells (green dots), central capsule (blue sphere), and captured prey (labeled). I) single-cell of collodarian from a colony (J: partial view of a similar colony with single cell marked by box), showing outer envelope with spicules, yellow central capsule, and orange symbiont cells. Source: A–C, H from the online image archive of the Observatoire Océanologique de Villefranche Sur Mer (John Dolan/Inigo Montoya): (use under Creative Commons licence); D from Sugiyama et al. (2008); E by Y. Ishitani; F by N. Suzuki; G from Swanberg and Anderson (1981); I from Yuasa and Takahashi (2014). (*See color plate section for the color representation of this figure.*)

from uniform. Most of the structures are short-lived, but re-occur regularly, and thus can, in principle, be exploited as adaptive niches by different species. Together with the opportunities for diversification created by biotic interactions between species, there is no reason any longer to think that the diversity of plankton is paradoxical. This can be stated even if we as yet have almost no insight into the specific microenvironments and adaptations that are the basis for most individual taxon niches.

Plankton Taxa

All organisms have the same basic needs of reproduction, growth, feeding, and avoidance of predators. In the pelagic realm, organisms of different sizes meet these needs in very different ways, with the most important determinant usually being the organism's size. Most larger inhabitants of the pelagic realm (> ca. 1 cm) are "nekton," capable to some degree of moving through the water, which allows them to significantly control which specific local water layers they will inhabit, to actively pursue food sources, and to try to evade predators. Smaller organisms (plankton) have such limited abilities to move in the water that they are largely passive inhabitants of the water they live in, and with only limited abilities to pursue food or move to evade predation. Plankton itself can be classified in many ways, the most important being by ecologic role: phytoplankton (photosynthetic), zooplankton, or mixotrophic plankton. Mixotrophy is a common, possibly dominant behavior where organisms have a mixture of both abilities, typically the ability to ingest food particles by phytoplankton cells, or photosynthesis via photosynthetic symbionts in zooplankton hosts. Another common classification is according to size: femto- (viruses), pico- (<2μm; bacteria and protists), nano- (2-20μm; protists), micro- (20-200μm; protists, including most phytoplankton), meso- (.2-2 mm; copepods, pteropods), and mega- (those larger invertebrates that do not strongly swim and thus are not nekton, e.g. jellyfish, tunicates). Size is one of the most important constraints on how organisms behave and feed. Most taxa can feed with reasonable efficiency only on a fairly narrow range of particles. Filter feeders such as ciliates generate currents and filter small particles (<10% their own size) from the water, while "raptorial" type feeders actively capture particles, which typically are >10% their own size. These constraints dictate many aspects of cell morphology and ecosystem structure. While organisms make up the large majority of particles found in the pelagic ocean (at least by diversity), other components of the water include dead organic matter in various forms: "marine snow" – clumps of material sinking through the water column; particulate organic carbon (POC) – macroscopic individual particles of organic material; and dissolved organic carbon (DOC) – very small particles or mere clumps of organic molecules. These organic materials form a key component of marine food webs, being a source of food in particular for small plankton such as bacteria, which then are consumed by a complex web of other, larger plankton taxa. This "microbial loop" thus cycles nutrients within the water column and is dominated by zooplankton. Zooplankton taxa can thus participate in ecosystems by grazing of primary producers (phytoplankton) or by preying on other zooplankton, and many taxa do both.

Ecologic and Behavioral Constraints due to Small Body Size

One of the most significant things about plankton is that it is mostly so small (femto- to meso- size classes) that the special physical properties of water at small scales are important to understanding their biology. The primary difference is the relative dominance of viscous forces vs. inertial forces (i.e., movement following Newtonian laws) in water at small scales. This difference is measured by the Reynolds Number (Re), which is the ratio of inertial to viscous forces. For small mesoplankton this value is <1, and may be as low as 10^{-5} for bacteria. When viscous forces dominate (Re<1), water flow is laminar, not turbulent, and the surrounding medium is so thick as to resemble movement, at larger scales, of objects through a thick syrup. Many basic biological processes behave differently at this small scale. Swimming, for example, can only be done using certain types of motion, while other types (such rowing using stiff oars) produce only wiggling about in the same place (Purcell, 1977). Laminar flow means that the water flowing over an organism – and thus available for content sampling – comes, at least directly, from a thin sheet of water extending upstream from the organism. It is not a sample of water from a larger volume in the immediate region of the cell, as is true for larger organisms where turbulence is dominant. This affects the ability to detect or capture food particles, so long as these particles are not themselves motile. In particular, in order for stationary cells (like radiolaria) to have larger numbers of (non-motile) particles from the environment move across their surface, they need to inhabit water where laminar flow gradients exist: shear. Small microfronts or other locations in the water where moderate rates of shear exist (.02, Itsweire et al., 1988) may thus provide better access to food sources. Generally, the weak ability to move, and the limited ability to perceive the broader environment (since such organisms do not have vision) suggest that successfully and rapidly managing near-cell surface interactions are paramount for organisms – between them and food particles, or, for cells that can move, to chemical or water motion tracers that give the direction of food. How such constraints are expressed in the morphologic adaptations of specific taxa is still, however, only very poorly understood.

Other characteristics that are associated with small cell sizes include metabolism, growth, and biomass in ecosystems, all of which show characteristic differences when compared to more familiar macroscopic organisms. In "typical" macroscopic organisms, the majority of food consumed is used by the organism simply to maintain life (metabolic needs) and only a relatively small fraction is used for growth. In protists this is not true. Metabolic needs are usually so small that most of the food consumed is used for growth (Fenchel, 1987). One consequence of this is that there tend to be approximately equal amounts of biomass at each size class/trophic level, i.e. many fewer individuals, but larger ones, so that the approximate biomass remains nearly the same. Growth rates also tend to differ from "normal" expectations, in that phytoplankton groups tend to have much more rapid growth and turnover than the zooplankton groups feeding on them (Fenchel, 1987). Thus, it is not uncommon to find inverted biomass pyramids in ocean ecosystems, with more standing zooplankton biomass than phytoplankton biomass. This implies that zooplankton in such cases are able to graze phytoplankton rapidly enough to prevent them from expanding their biomass rapidly until they exhaust available nutrient resources. For predatory zooplankton such as radiolaria, the relatively large amounts of zooplankton biomass (similar to, or even exceeding, phytoplankton biomass) must affect feeding strategies and possibly preferred organisms for consumption, although it is not yet known if radiolarians have such preferences (see "Feeding").

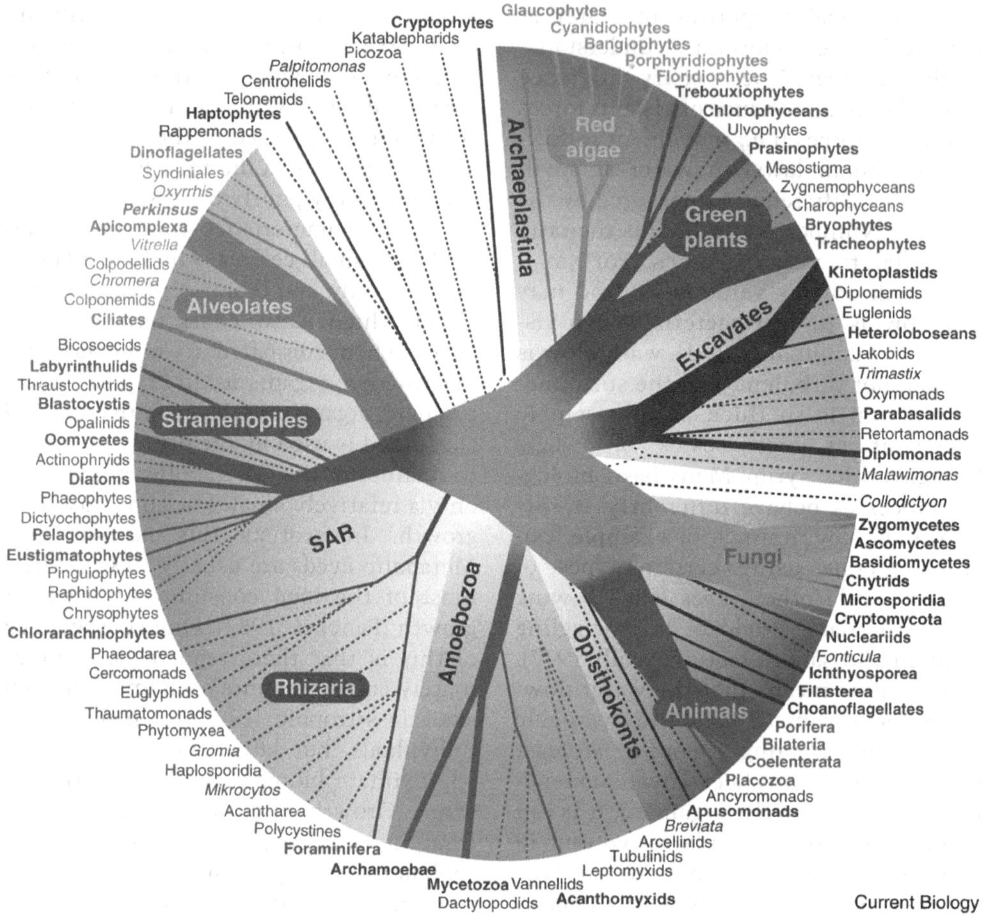

Figure 2.3 A phylogeny of eukaryotic life (e.g. without bacterial groups), showing the placement of polycystine radiolarians within the broader clade of Rhizaria. Polycystines are closely related to acantharians and foraminifera, but not to phaeodarians. The term "radiolarians" is still widely used for acantharians, polycystines, and phaeodarians, but in this book will mean only polycystines. The width of the branching lines in this figure is proportional to the amount of molecular information available for each group, which makes clear that relatively little is known yet about rhizarians in general. Source: Burki, *Current Biology*, 2004. Reproduced with permission of Elsevier. (*See color plate section for the color representation of this figure.*)

Basic Radiolarian Cellular Structure

Living polycystine radiolarians are classified within the protist super-clade Rhizaria (discussed more fully in Chapter 4: Genetics; Figure 2.3). They are taxonomically divided into Spumellaria (radially symmetric) and Nassellaria (conical geometry), with the latter having a morphologically distinct subgroup: the partially colonial Collodaria. Each of these main groups has characteristic differences in the basic morphology of the cell, although all share many characters that unite the group

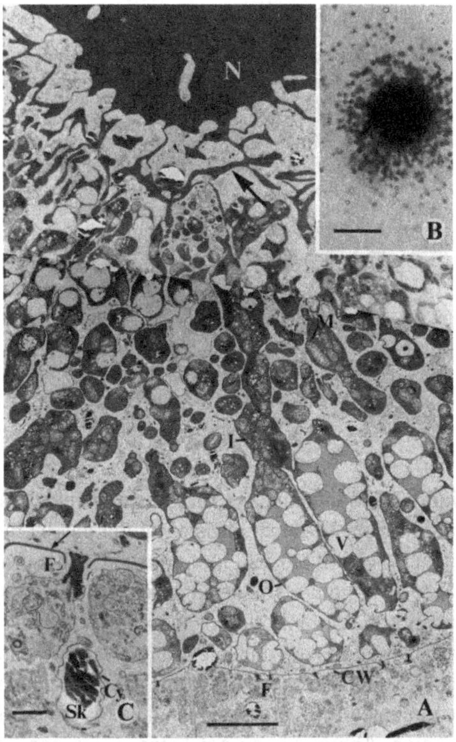

Figure 2.4 Example of details of cell morphology revealed by transmission electron microscopy (TEM). A) shows cross-section of a cell from center to just beyond capsular wall, inset B) shows living individual (transmitted light micrograph), inset C) shows transition across capsular wall (arrow) in more detail. N: nucleus; I: inner zone; O: outer zone; CW: capsular wall; F: fusule; Sk: siliceous skeletal element (fractured by thin-sectioning needed to make preparation) surrounded by Cy: cytokalymma. Scale for main image: 10 μm, for C: 1 μm. Source: Anderson (1983). Reproduced with permission of Springer.

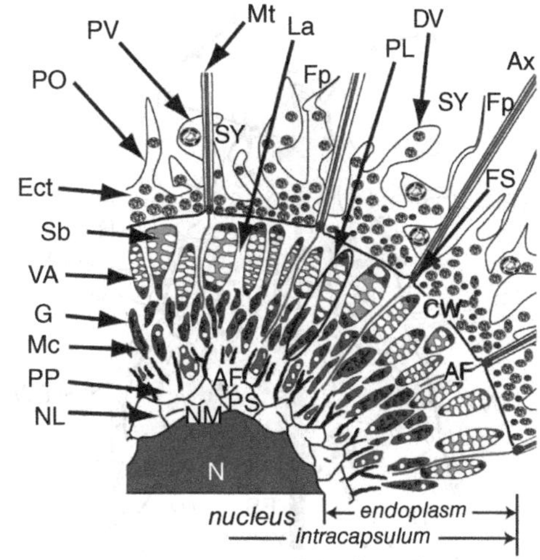

Figure 2.5 Main cellular structures of a spumellarian. The three main cellular compartments are the nucleus, endoplasm, and (external to CW – capsular or cell wall) the extracapsulum. Numerous specialized organelles are found in each compartment, and are further localized within compartments into radial zones. In many species, symbionts or parasites (SY) are also found, particularly in the extracapsulum. Other abbreviations: AF – axopodial filaments; Ax – axopodia; CW – capsular wall; DV – digestive vacuole; Ect – extracapsulum; Fp – filopodia; FS – fusule; G – golgi body; La – electron translucent lacune; Mc – mitochondria; Mt – microtubulines; NL – nuclear lobe; NM – nuclear membrane; N – nucleus; PL – protoplasmic lobe; PO – protoplasmic envelope; PP – vacuolated protoplasmic process; PS – perinuclear cisterna; PV – perialgal vacuole; Sb – granular storage substance; VA – vacuole. Source: From Suzuki and Aita (2011).

and distinguish them from other, related forms within the Rhizaria, such as Acantharia or Foraminifera. This information on cellular structures was largely obtained by extensive studies of stained thin-sections of radiolarian cells using the transmission electron microscope (TEM), although mostly from species with reduced skeletal development, as the presence of a robust opaline shell made sectioning difficult (Figure 2.4). The polycystine cell (Figures 2.5, 2.6, 2.7) consists of the usual eukaryotic complement of basic organelles: nucleus, ribosomes, mitochondria, golgi bodies, endoplasmic reticulum, and vacuoles, all embedded in the main cytoplasm of the cell. Some taxa appear to have multiple nuclei (Collodaria), although the

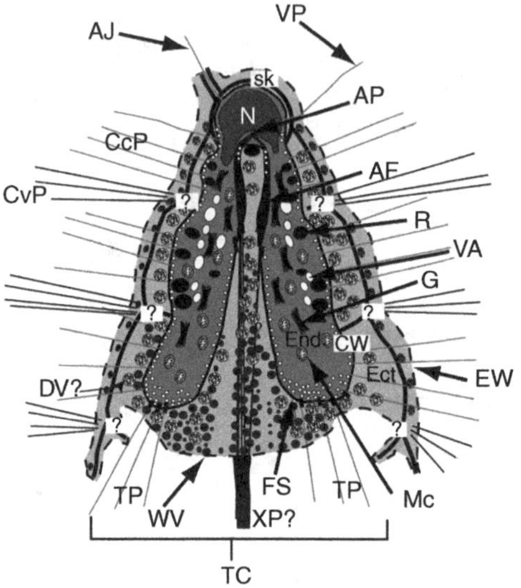

Figure 2.6 Main cellular structures of a (cyrtid) nassellarian. The conical symmetry obscures but does not truly change the primary compartmental structure of nucleus, endoplasm, and extracapsulum. The endoplasm is, however, folded into lobes along the axis of the cell with the nucleus at the tip, surrounding a well-developed axial flagellar structure (AF) that extends below the base of the cell as a pseudopodal flagellum (axial projection, XP) which is surrounded by the basket of pseudopods that together make up the terminal cone (TC) used in food capture. The podoconus is comprised of the axoplast (Ap), Axoflagellum (AF), and internal microtubule bundles (partially shown and vertical lines along the axis of the cell) that extend and connect to the basal external conical axopod structures. Other abbreviations (see also Figure 2.5): AJ – apical projections; VP – ventral projection; sk – skeleton (shell); R – electron dense reserve droplet; End – endoplasm; EW – extracapsular wall; TP – terminal projection; WV – waste vacuole; CvP –projection from convexity; CcP – projection from concavity. Source: Suzuki and Aita (2011).

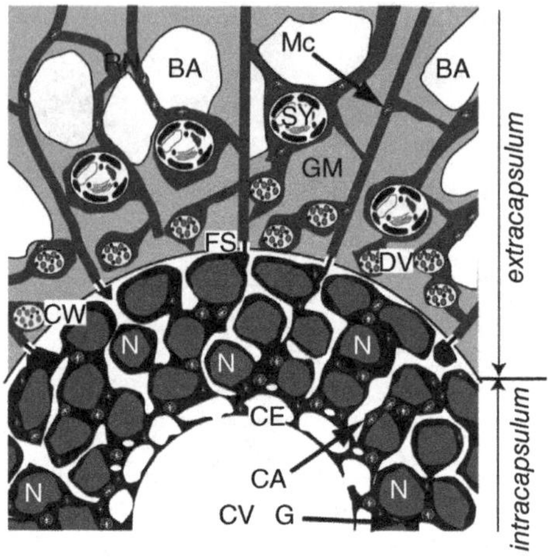

Figure 2.7 Main cellular structures of a collodarid nassellarian. The main division of the cell by the capsular wall (CW) into intracapsular and extracapsular regions is the same as in other radiolarians, but the organization of the intracapsulum is variable, often with multiple nuclei (N) or at least distinct nuclear lobes. The center of the cell may be occupied by a large lipid vacuole (CV) in some species. Abbreviations not listed in Figures 2.5 or 2.6: BA – bubble-like alveoli; CA – canal; GM – gelatinous matrix. Source: Suzuki and Aita (2011).

reported number and location within the cell vary (Swanberg and Anderson, 1981; Suzuki et al., 2009a). One of the most characteristic cellular structures of radiolarians and other rhizarians is presence of a thin spherical capsular membrane or wall. This separates the cytoplasm into two radial zones, and controls biochemical and material exchanges between the endoplasm inside the capsular membrane and the ectoplasm or extracapsulum outside it. Most polycystines show further organization of the cytoplasm within each of these regions. This structurally enforced organization of cellular activities represents the most advanced state of organization in large Rhizaria. Foraminifera show similar tendencies but lack a strict, structurally demarked division of the cell, and other rhizarians show little cellular specialization (Anderson, 1983). Numerous fine slits, plus larger, but still small (ca. .5–1 μm) pores (fusules), distributed more or less

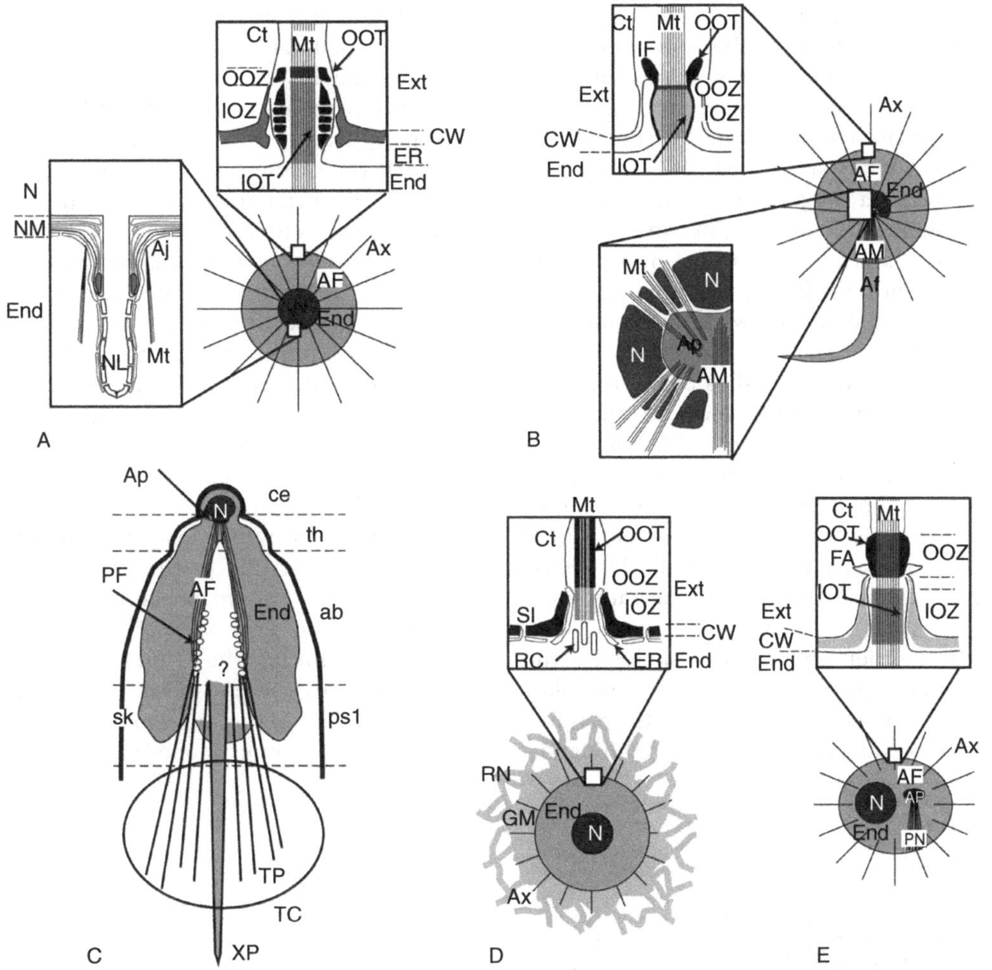

Figure 2.8 Diagrammatic views of fine structures in different groups of polycystine radiolarians. A–B: Spumellarians, C–E: Nassellarians. The relationship between the axoplast structure Ap (when present), nucleus N, and system of axopods (Ax, Am with their central mitochondrial tubule structures Mt) show considerable variation, as do the details of the fusules (complex structures –labeled parts not elaborated here – shown in inset boxes on cell outer boundaries) where the external axopods pass through the cell wall, with the microtubules extending into the interior of the cell. Additional abbreviations in "C": ce – cephalis; th – thorax; ab – abdomen; ps1 – first post-abdominal segment. Source: Suzuki and Aita (2011).

uniformly over the capsular wall, permit the exchange of cellular material between the two zones (Figure 2.8). Most of the organelles of the cell are found only in the endoplasm, which often shows a radial, lobed internal structure with alignment of vesicles and other cellular organelles. The ectoplasm by contrast is highly vacuolated and in some species is dominated by bubble-like alveoli which visually obscure the inner structures of the living cell (Figure 2.2H). Digestive vacuoles with prey, and other vacuoles with symbionts, in species possessing these, are found only in the extracapsulum. There are substantial differences in structural detail within the two main living subgroups of Polycystina. Spumellaria largely have the structures just defined (Figure 2.5), but in Nassellaria (Figure 2.6) the capsular wall is elongated-conical or multilobed, with one end at the apex of the shell, and with the fusules concentrated at the opposite, broader end. Multiple parallel bundles of microtubules extend through the fusules into the extracapsular region (Figure 2.8), they meet internally in a large conical structure called the podoconus (Figure 2.6). The extracapsular region is often not as extensive or as vacuolated as in many Spumellaria, and the siliceous skeleton is thus usually more easily seen.

The other primary non-skeletal structural element that distinguishes radiolarians from most other protists is the complex set of pseudopodia that radiate throughout the cell. These are differentiated into axopodia, rhizopodia, and the reticular network. Axopods are stiffened, rather long, straight pseudopodia, embedded in the extracapsulum and extending outwards to the very periphery of the organism, and originating at the fusules of the capsular membrane. The central stiffening structures of axopods, the axonemes, are built from complex bundles of microtubules running the axis of the axopod, and connected to each other in a set of centrally joined parallel curled laminae, which thus form a somewhat spiral shape in cross-section (Figure 2.9). Although the structural strength characteristics have not been formally analyzed, the combination suggests substantial strength via the short ribs and central tubes, but with considerable ability to flex and twist due to the curved, free outer edges of the main lamellae. The microtubules from the bases of the axopods usually pass through the fusules of the capsular wall into the endoplasm, where, in most Spumellaria, they are joined together with those from other axopods in a central

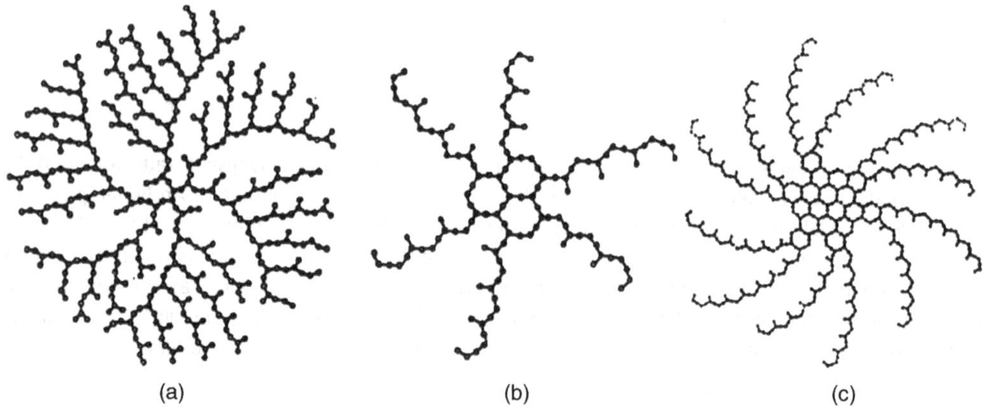

Figure 2.9 Cross-sections of axonemes consisting of axopod microtubule bundles with branching and spiral structures. Primary rays, often curved, join in often box-like hexagonal central structures, and short secondary branches project at ca. 90° from the primary rays. In three dimensions the structures are curved lamellae with short ribs, and central hexagonal tubes. Source: Anderson (1983), after Cachon and Cachon (1971 Reproduced with permission of Elsevier.

structure, the axoplast, which is located near the center of the cell, often surrounded by a fold of the nucleus. Rhizopods are long, unstiffened pseudopodia in the extracapsulum that form an elaborate network and maintain cytoplasmic continuity in this region of the cell. Even finer radiating strands of cytoplasm – filopods – are seen in the periphery of the ectoplasm.

Some spumellarian and many nassellarian taxa are known to possess specialized, extra-long pseudopodia, which at least in some cases appear to be bundles of axopodia. This specialized structure, variously called an axoflagellum, pseudoflagellum, or (in Nassellaria) axial projection is anchored within the endoplasm (the podoconus in Nassellaria) and extends further into the external environment than other cellular extensions. Nassellarians also often possess a "terminal cone" surrounding the main axoflagellum, which is made up of subparallel axopods radiating from the base of the aperture (Sugiyama and Anderson, 1997a; Sugiyama et al., 2008).

The above is meant to provide a summary of a wealth of more detailed observations on cellular structures, based primarily on the monographic studies or reviews of Hollande and Enjumet (1960), Anderson (1983), Suzuki and Aita (2011), and Suzuki and Not (2015), which should be consulted for further information.

Skeleton

Polycystina are the only planktonic rhizarian group that builds skeletons out of essentially pure opal (amorphous hydrated silica, $SiO_2.nH_2O$). The skeleton (also called test or shell) of living polycystines is characterized by one of three basic types. Isolated spicules are found only in the Collodaria; shells based on 3D radial symmetry (variations of spheres) define the Spumellaria. Lastly, shells based on conical geometry (2D radial symmetry around a linear axis with a polarity along the axis) characterize the Nassellaria. Polycystine skeletons (Figures 2.10, 2.11) are generally constructed as lattices – open meshworks of bars and pores, not solid layers. Although there is a great range of variation, most spumellarian skeletons are built from more or less concentric multiple layers, connected by radial bars or "beams"; and nassellarian shells from multiple cylindrical "segments" which together form a cone. The apical segment of a nassellarian shell at the

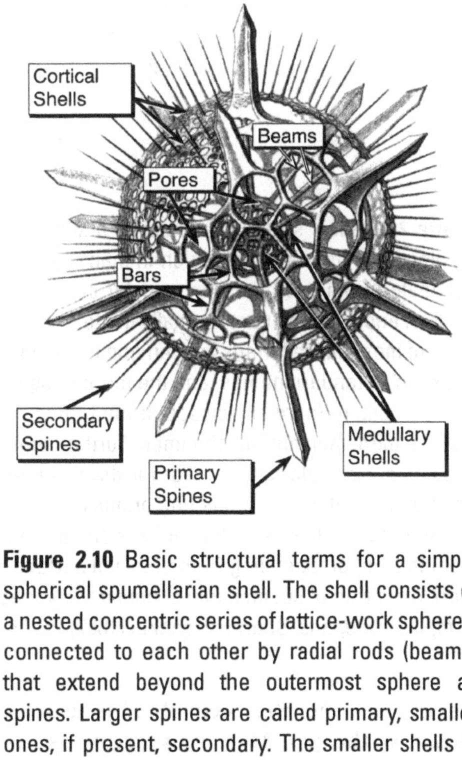

Figure 2.10 Basic structural terms for a simple spherical spumellarian shell. The shell consists of a nested concentric series of lattice-work spheres, connected to each other by radial rods (beams) that extend beyond the outermost sphere as spines. Larger spines are called primary, smaller ones, if present, secondary. The smaller shells in the interior are called medullary shells while the outermost shell (or shells, if more than one is present with nearly the same radius) are called cortical. The lattice-work elements that make up each sphere are called bars and the openings between pores. Numerous variations and additional terms exist which are covered in Chapter 5 (Taxonomy). Drawing is of *Cromyechinus icosacanthus*. Source: Haeckel (1887) p. 263, PL. 30, f. 1.

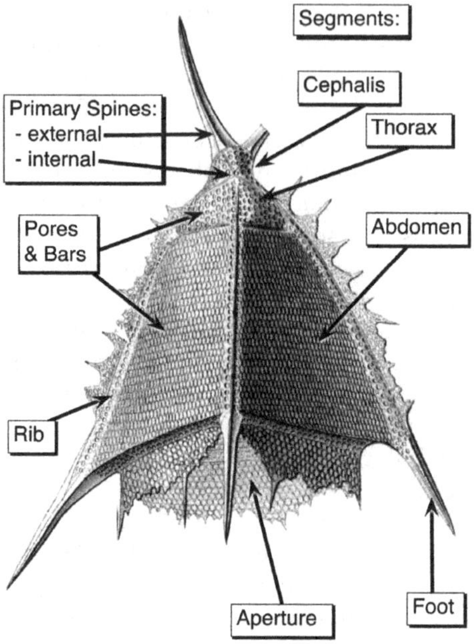

Figure 2.11 Basic structural terms for a simple three-segmented nassellarian shell. Nassellarian shell topology is that of a cone, which in many families is divided into a series of segments, and in turn are delineated from each other by annular constrictions or other visible changes in shell morphology. The walls of each segment are normally made of a lattice-work of bars that delimit openings – the pores. The base of most nassellarians is open and is called the aperture. The first three segments beginning at the apex have standard names, and the morphology of these is usually of major taxonomic importance. The apical segment (normally closed at the top) is called the cephalis, the next segment below the thorax, and the third segment the abdomen. Further segments, when present, are simply called post-abdominal segments. Nassellarians, except for the divergent subclade Collodaria are characterized by the presence of a set of internal rods (the primary spines or spicules) inside the apex of the shell, often located at the transition between the cephalis and thorax. These elements frequently extend into larger rod-like elements on, or extending from the shell surface (ribs and spines, or if extending below the open base, feet). As is true of spumellarians, numerous variations exist that are detailed in Chapter 5. Drawing is of *Artopilium elegans.* Source: Haeckel (1887) PL. 75, f. 1.

top of the cone, the "cephalis," is usually closed at the top. Most polycystine skeletons also have spines projecting from the outer shell through the ectoplasm. The characters of the skeleton are the primary basis of taxonomic classification in radiolarians, and will thus be discussed in considerably more detail in Chapter 5 (Taxonomy and Fossil Record).

Polycystine skeletons are closely embedded in the overall structure of the cell. Parts of the skeleton are normally embedded within the endoplasm, but the main ("outer") part of the shell is normally outside the central capsule in the ectoplasm, although frequently it lies near the boundary between ecto- and endoplasm. The spines are normally closely associated with the radially arranged network of actinopods in the outer part of the cell. Actinopods also extend through the pores in the outer shell(s), although detailed studies of

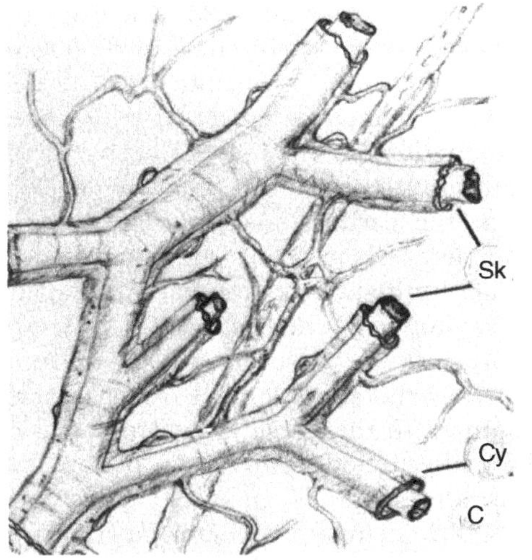

Figure 2.12 Illustration of relationship between external skeletal elements and sheath-like covering of cytoplasm (cytokalymma:Cy) which protects the skeleton (Sk) from dissolution, allows by cytoplasmic streaming movement of food particles, symbionts, etc. into or out of the main cell, and which can repair broken skeletal elements if necessary. Source: Anderson (1983). Reproduced with permission of Springer.

arrangement and variation between taxa are lacking. All parts of the skeleton in living radiolarians are located within the cytoplasm, within a cytoplasmic sheath called the cytokalymma (Figure 2.12). Although several studies have examined the positional relationships of shell and cytoplasm characters in more detail, so far no further consistent patterns have been seen, and the number of taxa examined is too small in any case to make useful generalizations.

Skeleton Formation and Growth

Given the central importance of the skeleton to (fossil) radiolarian research, the biology of how the skeleton is formed is of particular interest. Significant progress has been made by study of living specimens in the laboratory, although given difficulties in culturing radiolarians (see "Life Cycle"), our knowledge is still very incomplete. Our understanding of skeletal growth is supplemented by observations of skeletons in differing stages of completeness, and from differing environments, in both living plankton and fossil assemblages.

Early ideas on radiolarian skeleton development contrasted two possibilities: that the skeleton was developed as a biologic structure by the cell, or, that the skeleton was a more or less incidental mineral deposit created by physical processes along the boundaries of biologic structures. The latter view was promoted by D'Arcy Thompson (1942) in an influential book on biologic form. Thompson based his argument on the similarity of some radiolarian structures to those produced by simple physical processes present in interacting inorganic materials. Thompson argued that the living cell consisted of numerous bubble-like alveoli. Between these bubbles, by analogy to the curved interfaces or "fronts" that form between soap bubbles, skeletal material was deposited, presumably due to chemical processes in the external environment. The dichotomy however is false. Modern understanding of organic development, also in multicellular organisms, recognizes that biologic control in the generation of form is, in general, expressed via modulation – within the cell – of physical processes very much of the sort described by Thompson as indicative of inorganic behavior occurring extracellularly. Organisms generate biologic form via the spatial expression of genetically coded biologic chemicals, and from these, structural patterns are created by allowing physical processes such as diffusion to act, including the development of fronts between diffusion fields. Thus, biologic control of form often, particularly at small physical scales, produces structures that reflect the physical "language" that was used to generate them. The similarity of biologic form to those produced by known inorganic physical processes is thus not an indicator for or against biologic control of generation.

That polycystine skeletons are biologically controlled structures was definitely established by studies of the cellular processes involved in silica deposition (summarized by Anderson, in his 1983 book). The radiolarian siliceous skeleton is not deposited externally to the cell but internally. The site of deposition is a thickened cytokalymma called the silica deposition vacuole (SDV), which is created internally by the cell in the form that the skeletal element will subsequently have. Silica is initially deposited within the SDV in the form of very small spherules (<< 1 μm), which aggregate and eventually form the solid, transparent mass of opaline silica that characterizes the final skeleton. It is not known what biologic mechanisms are used by the cell to control the shape of the structure formed, but structures are frequently grown without being adjacent to aveoli, nor do many radiolarian shell structures actually resemble the simple forms used as analogies by Thompson. More generally, combined molecular-morphologic studies of other unicellular organisms such as yeast have found that most cellular morphology is under the control of positive adaptive selection (e.g. Ho et al., 2017). Such observations provide further support for functional and taxonomic interpretation of skeletal and other morphologic characters in radiolaria.

Radiolarian shells can grow by two fundamentally different methods: formation of new primary structures entirely from the organic SDV template, or by extending the shell from an existing structure. This latter may be by developing a new structure attached to the old (for example, subdividing a pore by an additional cross-bar), or simply by layering or thickening of a form already largely in existence. Both types of growth have been observed in the laboratory (Anderson, 1983; Anderson and Bennett, 1985; Nishimura, 1990; Ogane et al., 2009, 2010; Suzuki et al., 2013). Equally of interest, particularly for understanding taxonomy and phylogeny, is the ontogeny of the shell – how it develops from the earliest stages to the mature, completed skeleton. So far, there is only rather limited direct observational data on this. Growth of shells already at least partially formed has been studied in some detail. Laboratory studies of specimens captured in the ocean have demonstrated how silica is added and form is produced in the later stages of growth (Figures 2.13, 2.14). Unfortunately, there are no direct observations of how the very earliest parts of shells are formed, and in particular when, in the course of shell growth, the innermost parts of both spumellarian and nassellarian shells are developed. The observed youngest stages seen are, however, fairly close. Small (in the mature form, inner) shells are seen in multi-shell layered spherical spumellarians, or two or three hoops with the central sphere in spongodiscid spumellarians. Naked internal spicular systems are the smallest observed structures and are the presumed initial segment of conical nassellarians (Hollande and Enjumet, 1960; Swanberg and Bjørklund, 1987; Suzuki et al., 2009b). In Hollande and Enjumet's work, actual growth of spherical spumellarians over a period of many days was observed. Small shells in the ectoplasm of the initially small individual cell became engulfed in the endoplasm of the cell as it grew, with a sequence of new, larger shells being formed in the now-larger ectoplasm. In other studies, such as Swanberg and Bjørklund (1987) or Petrushevskaya (1962), growth is not directly observed. It is assumed that the innermost parts of the shells are always the first formed, but this assumption has yet to be confirmed by actual observations of growth in living cells.

Mature radiolarian shells do sometimes contain at least an indirect indication of

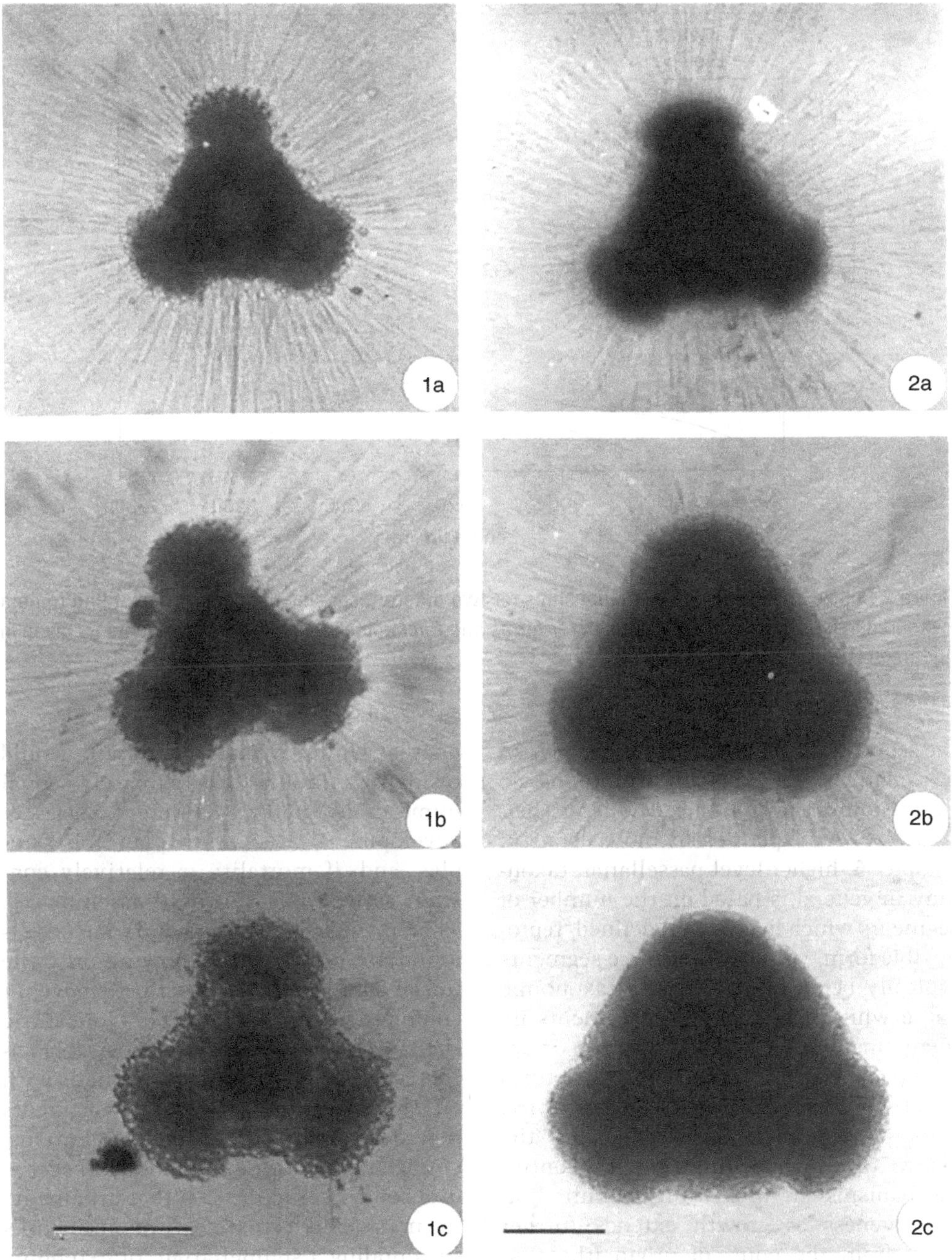

Figure 2.13 Growth observed in two individuals of the spumellarian *Dictyocoryne truncatum* over approximately three weeks' time in laboratory culture. Both individuals collected from the plankton near Barbados. The second individual began at a growth stage similar to that at which the first individual ended, providing a joint time series of growth from a small form with only stubs for arms to mature specimens with terminal arm shapes and well-developed interarm skeletal material ("patagium"). Source: Matsuoka, Mar. Micro., 1992. Reproduced with permission of Elsevier.

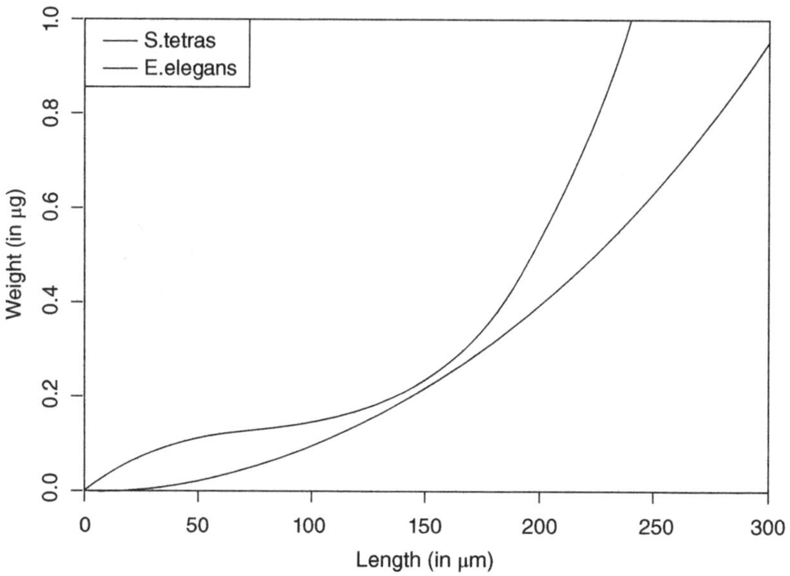

Figure 2.14 Size and shell weight relationships for two different species of spumellarians fitted to data measured in laboratory culture (*Spongaster tetras* and *Euchitonia elegans*). Source: equations given in Anderson (1996).

the sequence of shell growth in the form, in nassellarians, of changing morphologic expression of control over details of shell structure. As explained in more detail in Chapter 5, higher-level nassellarian taxonomy in general is based on the number of segments which have a well-defined, reproducible form, with the first three segments typically being individually of taxonomic value, while higher numbered segments are increasingly similar to each other, or are seen to contain few reproducible structures usable for taxonomic purposes (Figure 2.15). This pattern presumably indicates the degree to which cellular growth control mechanisms are increasingly limited in effectiveness as growth extends further away from the point of origin (the first cephalic segment).

Indirect evidence suggests that radiolarian shell growth is not uniform in rate but concentrated into relatively short phases. Uniform rates of growth would result in finding approximately equal numbers of shells in different stages of development, at least in plankton samples, and, if mortality is relatively constant as well, also in fossil assemblages. This is not so. Most radiolarian shells found are either those showing only the initial small forms as described above, or complete shells – multiple concentric shells including the outermost ("cortical") shell in spumellarians, or shells with all the characteristic conical segments that define the species in nassellarians. Ontogenetic variation is mostly observed in secondary silicification that ornaments the main shell form (fine, thin meshworks surrounding a spumellarian, or extending from the base of the opening of a conical nassellarian). This suggests that the main shell is formed fairly rapidly early in the development of the individual.

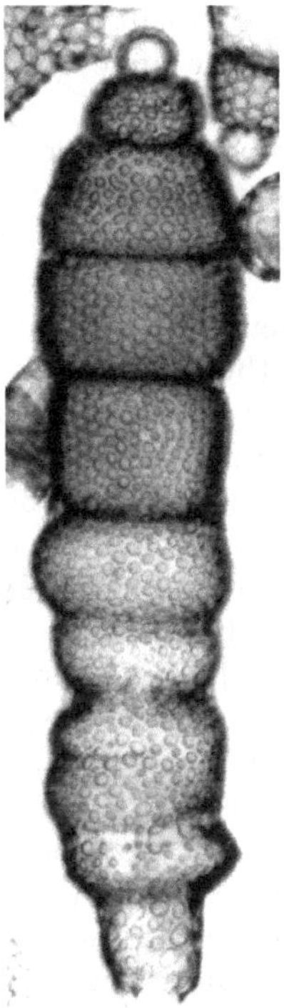

Figure 2.15 Decreasing control of fine structure expression with segment number in a specimen of a multisegmented theoperid radiolarian (*Stichocorys peregrina*). The first four to five segments have individually distinctive characteristics, are symmetric in outline, and the pores within each segment are fairly uniform in size, shape, and arrangement. Subsequent segments show gradually increasing irregularities. Specimen from Pliocene tropical Pacific sediments. Source: Photo by S. Trubovitz.

Size

Radiolarians are classified as part of the microzooplankton (20–200 μm) and are relatively large for protist plankton (Sieburth et al., 1978; Caron and Swanberg, 1990); indeed, they are referred to as the "protozoan giants of the plankton" (Fenchel, 1987). They are thus able to feed on larger prey organisms fairly high in the food chain. This size value refers only to the individual cells: colonial forms (see "Colonial Forms") are much larger and would be classified as macro- or megaplankton. Most individual species of radiolarians have relatively constant sizes, with most individuals in a population varying in dimensions by only a few percent from the average typical for each species. Variation is also very limited for most species observed in different biogeographic locations. A few species are observed to show substantial gradients in shell size with environment (Bjørklund, 1977), although it is not yet known (due to lack of genetic studies) if this reflects within species variation, or variation between pseudo-cryptic species. While most species have outer shell dimensions between ca. 70 and 300 μm, a few taxa having maximum dimensions of only 50–60 μm or >500 μm are known. These dimensions refer only to the main outer shell of the skeleton. Most living taxa have an additional external layer of cellular ectoplasm, and an even larger volume is affected by the individual extended spines, actinopods, and finer rhizopodial network. Characteristic dimensions for these parts of the living cell have however not been compiled. Some species though are large enough when alive to be visible to the naked eye and can be individually collected *in situ* within the water column by divers (Swanberg and Anderson, 1985).

Colonial Forms

Many of the species of collodarid radiolarians are colonial rather than, as is true for all other polycystines, living as solitary cells. The colonies (Figure 2.2G, J) range in size from a few mm and a few dozen individual component cells to very large colonies up to 3 m in length and ca. 10,000 individual cells (Swanberg and Anderson, 1981; Anderson, 1983). The colony consists of individual, generally widely separated collodarian cells linked together via their ectoplasmic rhizopodial networks, all of which is embedded in a sometimes very thick gelatinous sheath. The colonies can have various forms – mostly spheres or ellipsoids – but a variety of other shapes are reported, including long cylinders, flat ribbons, toroids, and even more complex forms. At least some radiolarian colonies exhibit a significant degree of coordinated behavior. Nutrients are shared between individual cells of the colony (Anderson et al., 1985), and they show a coordinated control of the distribution of algal symbionts (near the colony center when dark, distributed to the colony periphery in light). Colonies of *Collozoum caudatum* are capable of contraction and regulation of shape in response to physical disturbance. They also possess a central core along the axis of which digested waste material is apparently excreted, since the ends of the colonies typically have tail-like strands of gelatin composed primarily of debris such as empty tintinnid (ciliate) loricae and the shells of juvenile molluscs (Figure 2.2G; Swanberg and Anderson, 1981).

Life Cycle

One of the greatest limitations in understanding the biology of planktonic rhizarians is that they are generally unculturable. Although they can be captured in the wild and the individual cells maintained in the laboratory, reproduction of the next generation has not been obtained. This makes understanding the polycystine life cycle in particular very difficult, as the generality of this culturing problem means that there is not even data on closely related forms to use as analogies for missing polycystine data. At the moment the most closely related taxa whose life cycle is known are the benthic foraminifera, which show a variety of reproductive modes, but mostly an alternation of sexual and asexual phases, with the sexual phase ending in the subdivision of the cell mass into a large number of small flagellated gametes that are released into the environment and fuse to create the next generation (Bock et al., 1985). A similar pattern of flagellated swarmer production has been seen in other planktonic rhizarians in laboratory observations, including several groups of polycystine radiolarians: Actinommidae [*Plegmosphaera*] (Anderson, 1983); artiscins (Kimoto et al., 2011); Spongodiscidae (Matsuoka, 2007); and Collodaria (Anderson, 1983; Yuasa and Takahashi, 2014). During sporogenesis, axopodia and ectoplasm are first shed and the endoplasm changes color, sometimes into a milky granular appearance. These granules burst out as swimming flagellated swarmers from the test, and the skeleton becomes completely empty in the end (Anderson, 1983). The swarmers are small (a few μm in size), biflagellate, and have a crystal of celestite (strontium sulfate) in the center (Figure 2.16) – the same mineral used by Acantharia to make their skeletons. The swarmers however do not survive long, and are not (unlike for benthic foraminifera) observed to fuse, which would confirm that they are indeed gametes. Several studies have identified picoeukaryote-sized cells belonging to radiolaria in environmental molecular surveys of the deep ocean. It is thought that many or most of these may be reproductive cells released by such mass division of mature individuals, which in turn suggests that submergence to depth (several hundred meters or more) may be part of the radiolarian reproductive cycle (see summary in Yuasa and Takahashi, 2014). In earlier studies, the release of flagellated endoparasites from infected radiolarian cells was confused

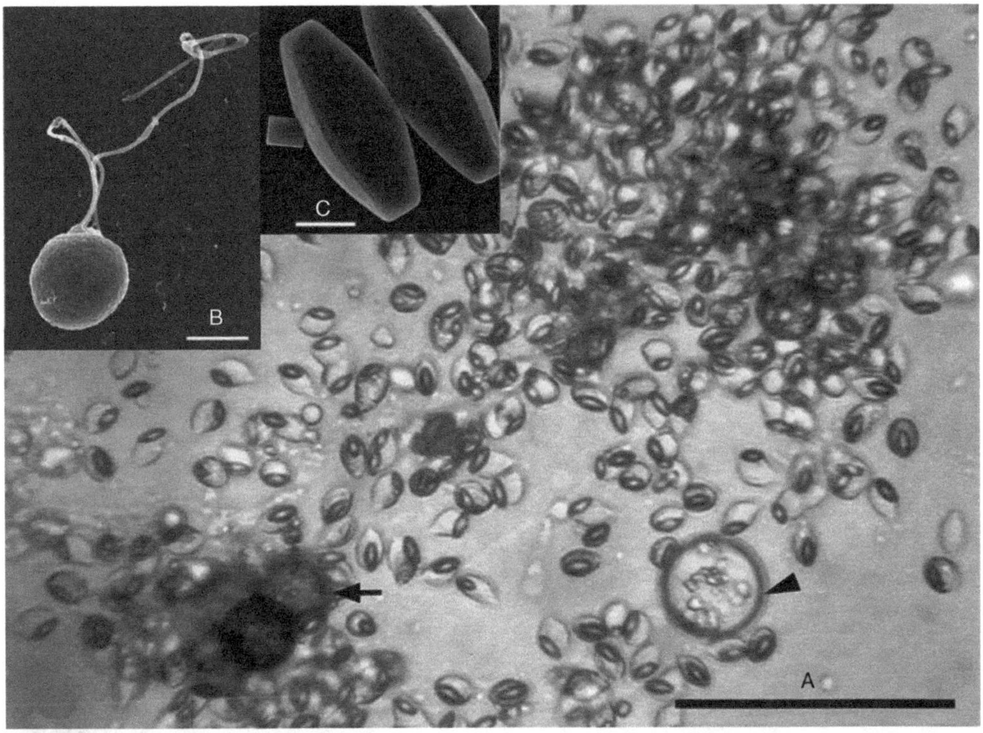

Figure 2.16 Reproductive flagellated swarmers. A) Breakup of a solitary cell of the collodarian *Sphaerozoum punctatum* (Figure 5.29.1) releases hundreds of small flagellated reproductive cells (small clear ovoids), each containing a prominent crystal of strontium sulfate (elongate grey inclusions). Symbiont cells (in life, brown) marked by black arrow. B) SEM image of a single swarmer cell showing characteristic double flagella. C) SEM image of isolated strontium sulfate crystals. Scale bars A: 100 μm; B: 4 μm; C: 2 μm. Source: Yuasa and Takahashi (2014). Reproduced with permission of Elsevier.

with the production of reproductive swarmers, but morphologic differences can be used to distinguish the two different processes (Anderson, 1983). Generally, however, most eukaryotes are originally sexual, including groups such as Rhizaria, with asexuality being a derived and relatively uncommon condition (Lahr et al., 2011). It is thus best to assume, unless evidence to the contrary is found, that polycystines are also sexual, if not restricted only to this mode of reproduction. Binary fission in polycystines, for example, is known, although it has only been reported for a few taxa, e.g. colonial collodarians, which have been observed to show binary fission of central capsules (Anderson, 1983; Anderson and Gupta, 1998, Not [pers. comm., 2018]). Skeletal formation may occur approximately simultaneously only after repeated fission of naked cells throughout a colony, or sequentially, when new cells form by budding from parent cells. In binary fission, the ectoplasm and algal symbionts are also evenly divided into two daughter cells. Binary fission has not been reported in other polycystines so far, though the rare occurrence of

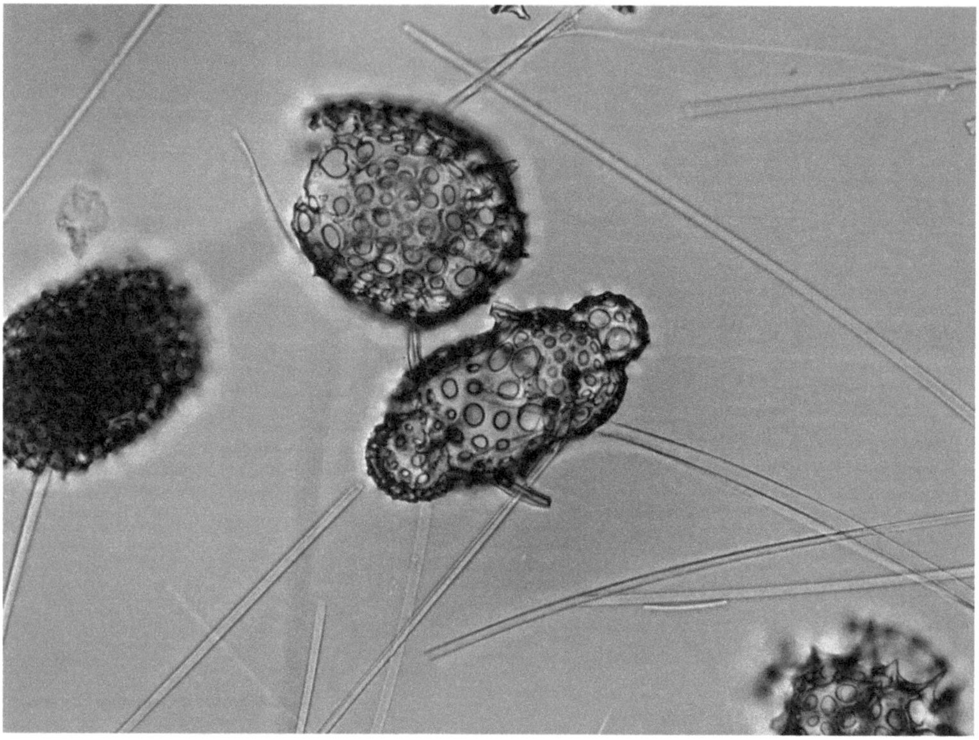

Figure 2.17 Twinned nassellarian shells (*Pseudodictyophimus* sp.) joined at their apertures. From Pleistocene Antarctic sediments (751A-1-1,98 cm, sl.2 8.2/108.7). Extended depth of focus composite image by dbl.

twinned shells (Takahashi and Honjo, 1981; Figure 2.17) suggests that it does at least occasionally occur in other radiolarians as well.

Longevity

Given the inability either to maintain radiolarians in continuous culture, or to observe them continuously in nature, only indirect estimates of longevity are possible. Radiolarian cells in culture normally live for a few days to up to five weeks (Anderson, 1983; Matsuoka and Anderson, 1992). These values are underestimates in that they represent only part of the life cycle: cells are always collected from the ocean only after they have grown to the point that they are identifiable as radiolarians. On the other hand, given the artificial conditions, they may also be overestimates of the average longevity of free-living populations. Only a few species have been studied in laboratory culture and these are relatively large spumellarians or collodarians, and thus not very representative of radiolarian diversity. Studies of cell abundances and sinking fluxes in water column can be used to estimate average residence or turnover times for entire living assemblages (Casey et al., 1971; Takahashi, 1983, Takahashi and Honjo, 1983; Welling and Pisias, 1998) and suggest that average longevities for individuals in natural populations are typically only a few days, but can be up to a few weeks.

Motility

Radiolarians do not have any substantial ability to move actively through the water except vertically, by altering their effective buoyancy (Anderson, 1983). Radiolaria are able (for example, when physically disturbed) to temporarily shed their ectoplasm, thus increasing their density so that they sink through the water. The vacuolated ectoplasm and the presence of lipid droplets offer the opportunity for the cell to regulate its density and thus maintain a consistent position in the water column. Although vertical movement would, in principle, allow radiolarians to follow chemical tracer plumes in the water column to food sources, no behavioral studies exist. The presence of diverse polycystine radiolarian DNA in deep sub-thermocline waters (see Chapter 4) suggests that radiolarians frequently do move into deeper waters, possibly as part of a regular life cycle strategy, although much of the deep water material may come from sinking dead cells (Boeuf et al., 2019). Lastly, radiolarians do have a limited ability to move over solid surfaces by means of their pseudopodia (Hollande and Enjumet, 1960), which would conceivably be of some use if feeding while attached to the surfaces of detrital matter in the water column (marine snow).

Feeding

Radiolarians are primarily classic suspension ("lie in wait") feeders, waiting passively for suspended or swimming food particles to come in contact with them (Figure 2.18). The vacuolated ectoplasm and widely radiating network of actinopods and rhizopods substantially increase the effective size of the cell with only minimal increase in cytoplasm volume. These can thus be seen as adaptations to improve food capture rates, as these depend on the effective cross-sectional area of the cell. Anderson (1983) provides an extensive summary of feeding behavior. Prey is captured by contact with axopods, the rhizopodial network, or, when present, the axoflagellum (Matsuoka, 2007) as the surface of these appears to be very sticky. Small particles are moved along the axopods to the main cell by cytoplasmic streaming, while contraction of the pseudopodia can bring larger prey into the cell for digestion. Captured prey is first enclosed in a vacuole and then digested. Rhizopods infiltrate larger prey (e.g. copepods) within the digestive vacuoles. Although selective acceptance or rejection of captured prey has been reported in some laboratory studies, rather indiscriminate feeding is suggested in other studies by recovery in radiolarians collected from ocean waters of a wide range of residual food particles, including empty shells of prey. A very wide variety of nano-, micro-, and macroplankton taxa have been reported as prey, including bacteria, small flagellates, tintinnids, diatoms, various other thecate and non-thecate algae, copepods, and other invertebrate plankton including naupli (larvae). Many radiolarians also derive nutrients from symbionts (see below). Lastly, radiolarians (and other rhizarians) are known to feed on each other (Caron and Swanberg, 1990). Various authors have suggested that certain species or clades within radiolarians have distinctive feeding behaviors (Casey et al., 1979; Matsuoka, 2007). Matsuoka (2007), in particular, suggests that food capture using long axoflagella is restricted to multisegmented Nassellaria, i.e. those species with >2–3 shell segments (Figure 2.19). This division also corresponds to a taxonomic distinction: the first three segments of nassellarian shells are developmentally

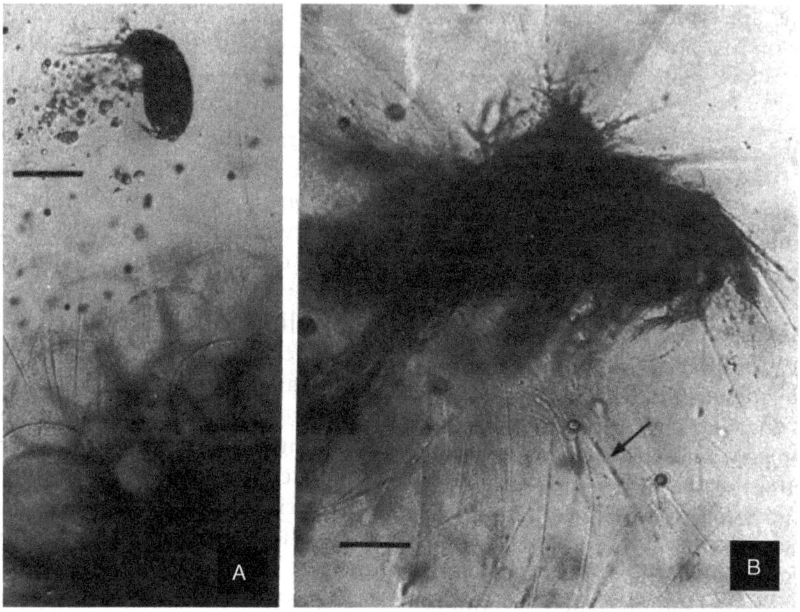

Figure 2.18 Two stages in the A) capture and B) engulfment of a copepod by a large solitary radiolarian (*Thalassicolla nucleata*). After initial capture by tips of rhizopods in A, prey is drawn closer and wrapped (close-up, B) in additional rhizopods (arrow) before being drawn within extracapsulum for digestion. Scales: A – 200 μm; B – 50 μm. Source: Anderson (1983). Reproduced with permission of Springer.

different than all subsequent segments (see Chapter 5: Taxonomy and Fossil Record). It should be noted however that these hypotheses are based on very limited data and are thus not yet very robust. Clearly though, deeper dwelling taxa cannot derive nutrition from photosynthetic symbionts, and are thus dependent on other food sources. The skeleton of radiolarians presumably also plays a role in feeding, in that it provides a strong framework to anchor the feeding network of actinopods and rhizopods, and thus allows radiolarians to successfully capture relatively large prey. No specific biomechanical analysis of this has been done for radiolaria, but the presumption is supported by the observation that predatory feeding on larger protists is confined to those species possessing spines in a closely related group – the planktonic foraminifera (Hemleben et al., 1989). Spine structures should thus be of considerable importance to the cell, though the optimum number and size must be a complex trade-off between metabolic costs of growth and shell weight (Furbish and Arnold, 1997 – for planktonic foraminifera) as well as shell strength and feeding efficiency.

Plankton biologists distinguish between suspension or filter feeding. Suspension feeders capture material not moving in the water column, e.g. by the particles swimming into the feeding organism, or by the feeding organism generating a feeding current to draw particles into the feeding structures. In flux feeding, the particles are sinking through the water column and are intercepted by the feeding organism (Jackson, 1993). Flux

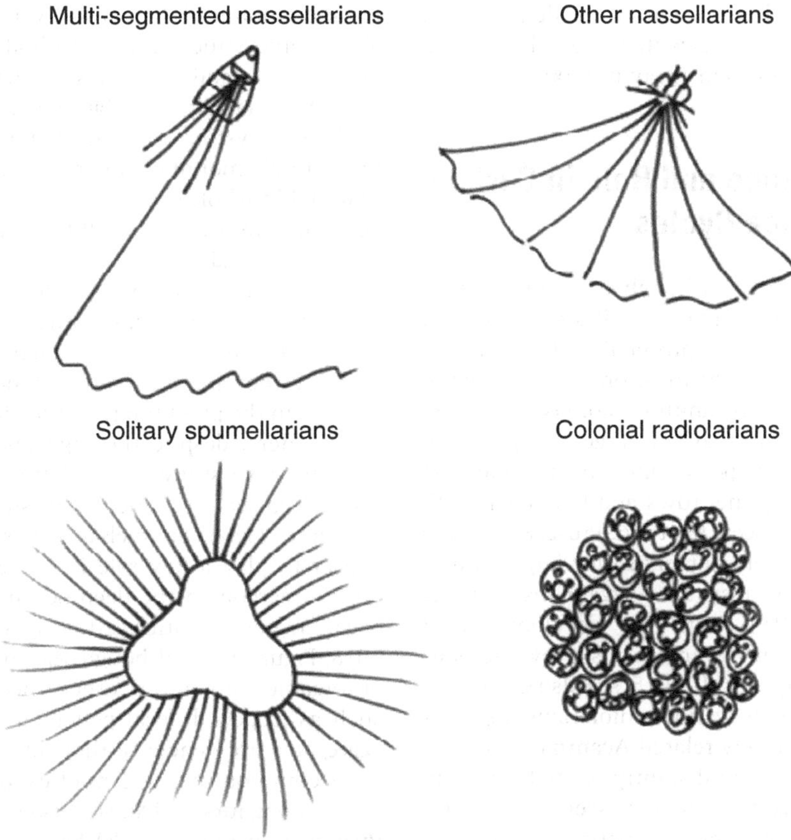

Figure 2.19 Examples of known feeding strategies. Multisegmented nassellarians create a basket (terminal cone) of axopods, and also use a long axial projection to fish for prey. Other nassellarians, particularly those with three or fewer segments use a net-like terminal cone only. Solitary spumellarians use a dispersed network of axopods to trap prey, while colonial radiolarians use primarily food created by symbiotic algae. Many species appear to use a mixture of strategies. Source: Matsuoka (2007). Reproduced with permission of Springer.

feeders may preferentially feed on larger particles such as sinking aggregates, as these, due to their faster sinking speeds, are more likely to be encountered (Jackson, 1993). Although work so far has been done only on Phaeodaria (Gowing and Bentham, 1994; Stukel et al., 2018, 2019), the similarity in biology suggests that many radiolarian species may also be flux feeders, at least for polycystine radiolarians living below the mixed layer.

Predators

Little is known about predators. Copepods, amphipods, salps, invertebrate larvae (e.g., of molluscs, lobsters) and other macrozooplankton filter feeders are reported to feed on radiolarians, along with any other food particles in the same size range that they capture (Haeckel, 1887; Caron and Swanberg, 1990; Wang and Jeffs, 2014). Possibly due to relatively low distribution densities of radiolarians in seawater, specialised feeders on radiolarians may not exist.

Abundance and Role in Carbon and Silica Cycles

Until recently, although occasionally suggested as important contributors to export of carbon (e.g. Lampitt et al., 2009), radiolarians were thought to be only a minor component of zooplankton biomass, as most data concentrated on larger zooplankton such as copepods, and most collection methods (e.g. net-tows and long-term sediment traps, see Chapter 6: Preservation and Methods) were optimized for larger and/or dissolution resistant organisms (Stukel et al., 2018). With the development of molecular environmental survey methods (e.g. de Vargas et al., 2015; see also Chapter 4: Genetics) polycystine radiolarians, together with the closely related Acantharia and the somewhat more distantly related rhizarian group Phaeodaria, are now seen to be a dominant component in marine ecosystems (Figure 2.20), although the number of studies and amount of data is still very limited. The polycystine groups Collodaria and Spumellaria and to a lesser degree the Acantharia (a rhizarian clade closely related to the Polycystina) make up a large fraction of the total amount of eukaryotic molecular material in whole ocean environmental surveys (de Vargas et al., 2015, Gutierrez-Rodriguez et al., 2018) and are, except for metazoans (particularly copepods), the most common component of the micro- to meso-sized zooplankton in molecular surveys. Notably, they are much more abundant (and diverse) in the plankton than the planktonic foraminifera, despite the high abundances and intensive study of this latter group in the deep-sea sediment fossil record (Kennett, 1982). Spumellaria in particular appear to play a key role in the export of organic carbon to the deep ocean in some ocean regions (Gutierrez-Rodriguez et al., 2018, Figure 2.20), although seasonal export of organic carbon by other mechanisms such as (copepod) fecal pellets and diatom aggregates are generally thought to be the most common mechanism of export in the ocean (Longhurst, 1998, see also Chapter 6: Preservation and Methods). Large rhizarians, in general, are able to intercept a substantial

Figure 2.20 Abundances of radiolarians (Spumellaria, Nassellaria/Collodaria) and other groups of large Rhizaria (Acantharia) in the upper water column of the California Current System, in a) water samples from the photic zone, b) sediment traps fixed with formalin deployed for three days just below the photic zone, and c) sediment traps without formalin. Each column C1–C5 refers to a separate sampling station in a transect running from near the coast (C1) to open ocean (C5). In each panel the contribution of all groups together to the total protist biomass is shown by white dots. "Radiolarians" in this figure refers both to Polycystina and Acantharia. Collodarians and acantharians dominate the photic zone composition of radiolarian groups and make up 20–40% of the total protist biomass (the rest is mostly dinoflagellates). Export of carbon below the photic zone (panel b) is however dominated, in both absolute and relative measures, by spumellarians. The decrease in abundance of spumellarians in traps without formalin (panel c) shows how rapidly the spumellarian signal decays when organisms can consume trap contents and skeletal material dissolves. The relatively high abundance of Acantharia among "radiolarians" in panel c is due to the presence in all traps of added strontium that prevents dissolution of Acantharian material. Source: Gutierrez-Rodriguez et al, ISME Journal, 2018. Reproduced with permission of Nature group.

fraction of the sinking material exported from the photic zone (Stukel et al., 2018) in deeper ocean water layers, and thus also play an important role in the remineralization (consumption) of organic carbon in the water column. The new awareness of the abundance of radiolarians (both Polycystina and Phaeodaria) in the oceans is also leading to new estimates of their importance in ocean silica cycling. They may be responsible for ca 10% of primary silica production (Llopis-Monferrer et al., 2020), and due to their dissolution resistance, an even larger fraction of silica export to sediments.

Symbiosis

Many species of radiolarians harbor algal symbionts ("zooxanthellae"). The taxonomic affinity of the symbionts has only been studied in relatively few species but

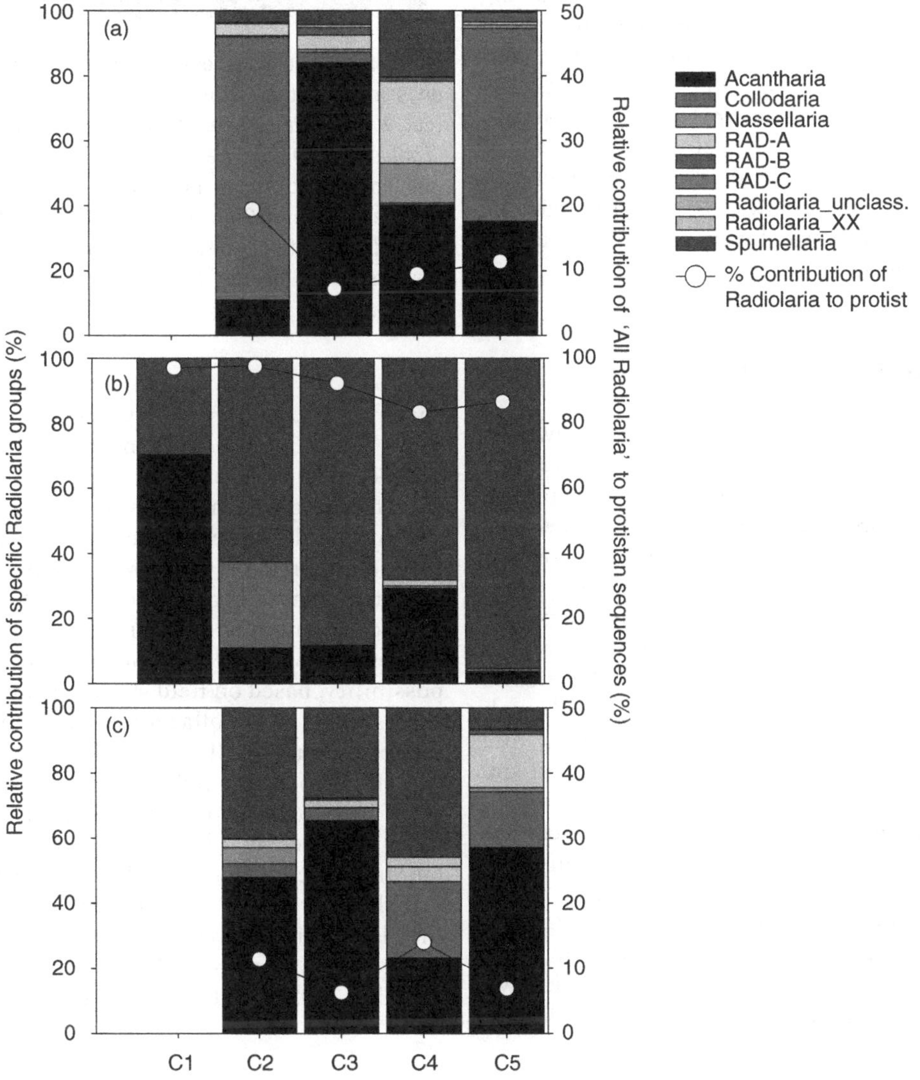

these have so far mostly been found to be dinoflagellates or prasinophytes, although bacterial symbionts have been reported for a few species (Anderson, 1983; Matsuoka, 1993; Sugiyama and Anderson, 1997; Probert et al., 2014; DeCelle et al., 2015). The symbiont cells are found within vacuoles, usually in the extracapsulum. These radiolarians are mixotrophs, as, in addition to food captured by predation, they obtain a part of their nutrients from their symbionts. This has been shown by C^{14} radioactive tracer studies showing the transfer of newly fixed carbon from symbionts to the radiolarian host cell (Anderson, 1978; 1983). Mixotrophs significantly affect pelagic ecosystems, enhancing overall cell abundance, cell size, biomass, and export productivity (Ward and Follows, 2016), although the role of mixotrophy in radiolarians is not yet clear.

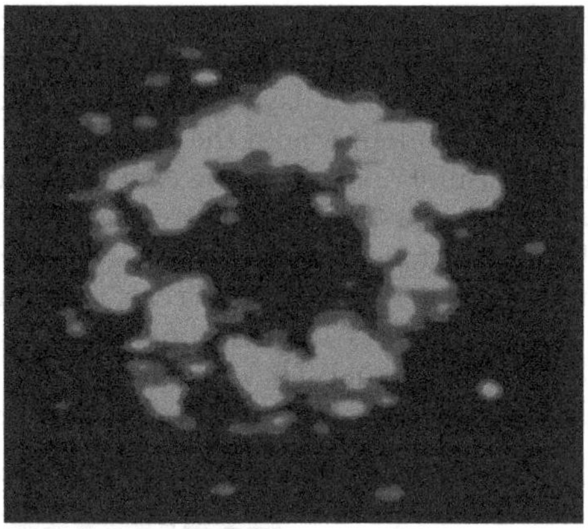

Figure 2.21 Bioluminescence – Image of blue light emission from the extracapsular layer of a large (ca. 2 mm total diameter) solitary collodarian (*Thalassicola nucleata* – Figure 2.2H). Light production is invoked by mechanical stimulation. Source: Widder (2010), after Latz et al. (1991). Reproduced with permission of Oxford. *See color plate section for the color representation of this figure.*

Bioluminescence

Many marine organisms are known to emit light, including many invertebrate taxa and also some species of protist plankton, and are thought to have evolved (or possibly incorporated by some form of external gene capture) numerous times (Haddock et al., 2010). Among the planktonic rhizarians, bioluminescence is known to occur regularly in Phaeodaria and in one group of polycystines – the collodarians, including both solitary and colonial forms (Figure 2.21; Campbell and Herring, 1990; Widder, 2002; Haddock et al. 2010). The chemical system used is one only known from radiolarians and Cnidarians (jellyfish and the like; Campbell and Herring, 1990), and given this specificity, reproducibility in field collections, and the absence of known bioluminescence in other planktonic rhizarians (e.g. other polycystines, planktonic foraminifera) it seems unlikely to be a coincidental artefact from recently ingested food, but an evolved, conserved system. The biologic function is unknown. Collodaria, so far as is known, lack photoreceptors, nor are they motile, and this excludes many functions known for other groups of bioluminescent organisms. One possibility, based on field studies of bioluminescence in dinoflagellates and copepod behavior, is a "burglar alarm" effect, e.g. that producing light, but only when disturbed (such as by an attack on the light producer by a lower-level predator), has an indirect defensive effect – the light signals to other, higher-level, visual, motile predators where their natural prey – the lower-level predator – is (Widder, 2002; Haddock et al., 2010).

Summary

Radiolarian biology, although still very poorly known in detail, does suggest some general conclusions. Radiolarians are, for protists, quite large, and are able to feed on a remarkably large range of prey sizes and types, mostly other protists. Compared to some other protist plankton they are not particularly numerically abundant (See Chapter 3: Ecology) but are ubiquitous, inhabiting most pelagic environments. Their general strategy of passive suspension feeding is augmented not only by their large size but by their extended network of pseudopodia so that the cell is, in effect, sitting in the middle of a biologic web made of sticky, thread-like material. These characteristics suggest that radiolarians are similar to spiders in terrestrial ecosystems. Spiders, too, are lie-in-wait predators of most other insects, capable of ingesting a wide variety of prey, and are a successful, if numerically not very abundant, part of most terrestrial ecosystems. This observation applies to solitary forms of radiolaria. Colonial forms of radiolaria are different in that, by changing their primary nutrition to photosynthetic symbiont algae, they have shifted both their typical size and ecologic behavior to become somewhat similar to the holopelagic macrophyte species of *Sargassum*, and have even been considered as pelagic analogs to benthic coral reefs (Swanberg, 1979). Collodaria are also a significant component of overall protist biomass in many ocean environments (Biard et al., 2017). These analogies should not be taken too closely given the enormous differences in ecology and environment of the taxa being compared, but nonetheless provide a simple summary image or framework to use when thinking about radiolarian biology in other contexts, including ecologic and evolutionary behavior as discussed later in this book.

3 Ecology

Introduction

Radiolarian biogeography is mostly derived from information gathered from more general studies of the biogeography of plankton. This subject has a long history in marine biology, but growth in understanding has been very gradual and is still quite limited. The reason can be appreciated by considering the unique challenges this subject presents. Unlike studies of biogeography on land, the ocean surface, as viewed from a ship, is essentially featureless and devoid of biogeographically relevant information, while visibility in the water is extremely limited. Further, the organisms of interest are, due to their small size, mostly invisible "in the field." All of our knowledge of pelagic biogeography was therefore, until rather recently, obtained by laboratory examination of scattered samples of ocean water, and extrapolating patterns from the extremely limited number of data points thus obtained in space and time. In the last few decades, this limited view has been supplemented by satellite observations of ocean surface water structures, and biogeographically relevant distribution patterns of ocean color created by the pigments of ocean phytoplankton (Longhurst, 1998). Although the link between ocean color and plankton biogeography is indirect and imperfect, satellite

Paleobiology of the Polycystine Radiolaria, First Edition. David Lazarus, Noritoshi Suzuki, Yoshiyuki Ishitani and Kozo Takahashi.

data has brought for the first time a broad spatial, "sighted" view of ocean biotic distributions, compared to the essentially "blind" sampling that preceded it.

The ocean can be visualized as consisting of two main parts. A fairly thin (ca. 100–200 m) lower-density warm surface layer extending from the cool temperate regions through the tropics floats on top of a much larger deep ocean made of colder, denser water, which outcrops at the surface in polar regions, and lies below the warm lid waters in the tropics (Figure 3.1). The deep ocean circulation is controlled by water density, due to temperature and salinity, and geophysical factors affecting fluid flow on planetary scales. The details are mostly not important for radiolarian biology except for two things. First, the way the deep ocean stores, redistributes, and returns dissolved nutrients to the surface ocean affects radiolarian distributions significantly, as nutrient availability controls the growth of phytoplankton and thus ultimately food for radiolarians. The second is the tendency of water to flow very long distances along the lines of equal density. In the oceans, density increases nonlinearly with depth, but in a fairly uniform way with depth over very broad areas of the ocean. Large areas of the ocean can thus be seen as composed of a series of thin water layers, each with a fairly constant composition areally. The layers are not completely flat however, but curve upwards with increasing latitude, so that they eventually outcrop at the surface. Water thus flows from higher latitude surface regions equatorward along thin, extended layers in the subsurface over very long distances. This means that temperate, or polar regions north and south of the equator, which at the surface are widely separated, can be nonetheless connected by this "laminar" or sheet-like style of subsurface water flow (note that there is turbulence via small (meter scale) eddies within each water mass;, thus flow is not laminar at smaller physical scales, and is not to be confused with laminar flow at millimeter scales, e.g. Chapter 2: Biology). This has implications for biogeography, as discussed later in this chapter. Surface ocean characteristics (i.e. the upper ca. 200 m) are by contrast of great importance in radiolarian biogeography and ecology. As pointed out in Chapter 2, the basic patterns of pelagic plankton biogeography are controlled by the circulation of basin-wide gyres in the surface water layer – tropical-subtropical, and polar. These, in turn, are driven by the global pattern of atmospheric circulation and the interaction of ocean currents as they encounter the continental margins. These major circulation features determine the primary surface water masses, within which water conditions tend to show similar characteristics in temperature, salinity and nutrients. Areas between these regions – oceanic fronts – show high gradients in water characteristics, and often locally strengthened current flows. All of these fronts are very dynamic and many are rather transitory, but a few, notably the Antarctic Polar Front system, are persistent, if also dynamic features of ocean circulation. The primary patterns of ocean currents are shown in Figure 3.2, which shows the main gyre systems and ocean basins. The pattern shown in Figure 3.2 is highly simplified, ignoring many smaller-scale features. It is also a long-term (ca. decadal) average of current flow patterns and at any given moment the actual flow pattern is usually much more complex, with numerous eddies and meanders seen at various spatial scales, most of which however only last from a few days to a few weeks. An example from satellite imagery is shown in Figure 3.3. Lastly, circulation in the Indian Ocean is complicated by seasonal reversal in current flow (not shown) due to the monsoon winds.

The main surface-water mass characteristics are determined to some degree by local factors such as net precipitation or evaporation from the surface, or wind strength. Mostly though there is a broad pattern where tropical surface waters are fairly stable in characteristics throughout the year, while in increasingly higher latitudes there is a seasonal cycle of stability. Heating, cooling, and changing wind strength create a strong seasonal cycle outside the tropics which once or twice per year mixes deeper waters into the surface. These mixing events control the availability of dissolved nutrients and thus the growth of phytoplankton. The vertical transition zone between the warm surface and cold deep water layers is the thermocline, which is also normally also the center of the zone of high water density contrast – the pycnocline. The thermocline depth, and thus the thickness of the surface mixed layer, varies according to latitude, season, and strength of wind mixing (Figure 3.4). In a few areas of the ocean, converging winds, or wind interaction with basin edges, create diverging patterns of ocean surface water flow. In these areas, deeper, colder, nutrient-rich water upwells, creating a shallow thermocline and enhanced surface productivity. Upwelling areas of the oceans are located in the open ocean in the tropics, and along the eastern margins of ocean basins in mid-latitudes. By contrast, in the central regions of the low-latitude gyres, the net direction of vertical water flow from the gyre circulation is downwards, the thermocline is deeper, and little mixing of nutrient-rich deep water into the surface waters occurs. These areas of the ocean are much less productive (oligotrophic) than the more productive (eutrophic) upwelling regions. The high-latitude gyres (North Pacific and the Southern Ocean) are areas of net upwelling of water and are thus at least seasonally more eutrophic (Figure 3.5).

The reason why deep waters are more nutrient rich than surface waters comes from the combined effect of biologic productivity and the barrier that the pycnocline places on vertical mixing of ocean waters. Although the large majority of organic material created by phytoplankton is consumed soon after formation by grazing in the surface water layer, some fraction of this material sinks below the thermocline, where, either in the deep-water column or on the sea floor, it is mostly consumed and degraded back into dissolved nutrients again (a process called remineralization). Because of the density contrast of waters, very little of this deep water is normally mixed back into the surface, and deep waters thus become traps that accumulate and become enriched in nutrients. It is only where waters upwell (upwelling regions, polar regions, seasonal wind-driven mixing in mid latitudes) that these nutrients are restored to the surface. These restored nutrients, along with the input of new nutrients to the ocean, e.g. in coastal regions from rivers, allow new growth of plankton. In addition to this primary pattern, in polar regions the oceans are covered, seasonally or year-round, with sea ice. This ice reduces light penetration into the ocean and shields the waters from wind-driven mixing, both of which limit ocean productivity.

All of these factors taken together, along with shipboard and satellite data on plankton distribution, form the basis for the modern classification of plankton biogeography. This has been most extensively developed by Longhurst (1998), who, in a general synthesis of marine biogeography (coastal as well as open ocean) divided the modern ocean into just four primary biomes (Polar, Westerlies [subtropical regions], Trades [tropical], and Coastal) and a larger number of more gradational provinces (ca. 30 open ocean, plus ca. 20 coastal). These provinces reflect the splitting of biomes into a dozen or so sub-biomes in individual ocean basins (e.g., Atlantic, Pacific, Indian), spatial gradients within these regions in basins, and special conditions near frontal regions. The boundaries between such biogeographic divisions in the ocean are of course (literally) fluid, and vary somewhat with time. Alternate attempts to classify the ocean however yield very similar patterns, despite minor differences in the locations of boundaries (Fay and McKinley, 2014; Reygondeau and Donn, 2018; Figure 3.6). Although radiolarians as such played a minimal role in these analyses, radiolarian distributions, as we shall see, do largely conform to these biogeographic concepts.

Biogeography

Water column sampling of radiolarian cells, or in more recent years, their environmental DNA (Biard et al., 2017) has been fairly infrequent, with many geographic gaps in coverage. The number of specimens collected in any given sample in cell-based studies is often relatively low, and samples, whether cell or molecular, are usually from a single very short time interval or small range of water depths. The details of the geographic distribution of species over the global ocean are thus often difficult to discern. Plankton-based biographic studies of radiolarians are also rather labor intensive and until recently have been restricted to regions, rather than basin wide (e.g. Renz, 1976; Boltovskoy and Riedel, 1980; 1987; Boltovskoy and Jankilevich, 1985; Boltovskoy et al., 1996; Nimmergut and Abelmann, 2002; Yamashita et al., 2002). Radiolarian species distributions can also be studied using radiolarian shells from the surface sediments of the ocean floor (e.g. in regional studies: North Pacific: Nigrini, 1970; Atlantic: Goll and Bjørklund, 1971, 1974; Indian: Johnson and Nigrini, 1980, 1982; Arctic: Bjørklund and Kruglikova, 2003; Global: Lombari and Boden, 1985). The shells found in sediments are not a perfect, unbiased representation of the living radiolarian faunas – there are several biases due to dissolution of shells and other factors that are described in Chapter 6. The patterns seen however are for most polycystine species reasonably good approximations of the pattern seen in the water column, but have the major advantage of being based on much more extensive information. Currently, there are many times more samples of radiolarians from surface sediment from all regions of

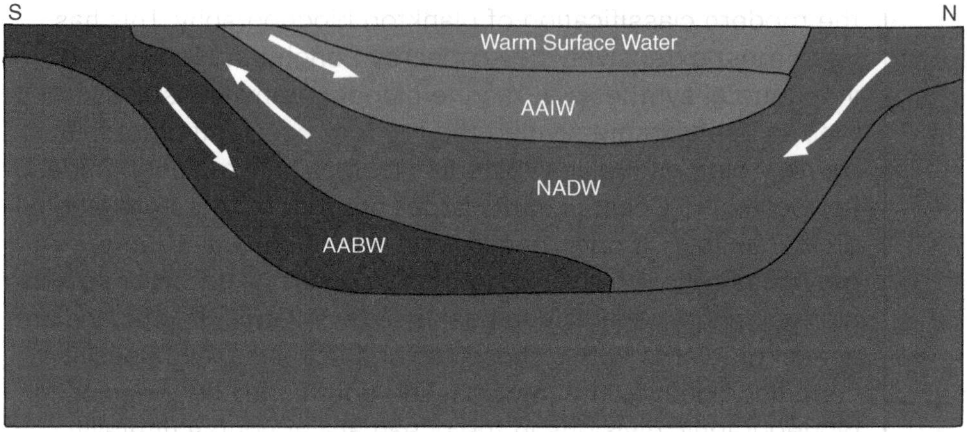

Figure 3.1 Atlantic Ocean deep-water mass circulation in latitudinal cross-section. Cold Antarctic Bottom Water (AABW) forms near the Antarctic continent in winter and spreads northward along the ocean bottom, displacing the slightly less dense North Atlantic Deep Water (NADW) that forms in the Norwegian-Greenland Sea and flows south. NADW flows upward and mixes into the surface in the Southern Ocean. Near the Antarctic Polar Front cold, but less dense surface water submerges and flows as Antarctic Intermediate Water (AAIW) north between the top of NADW and the warm surface water "lid" of low- to mid-latitudes. A similar, if less complete pattern (due to less than full pole-to-pole extent) is seen in other ocean basins. Note diagram is extremely vertically exaggerated (>1000X); in reality the main water masses at the scale of this printed diagram would be thinner than the paper of the page they are printed on. (*See color plate section for the color representation of this figure.*)

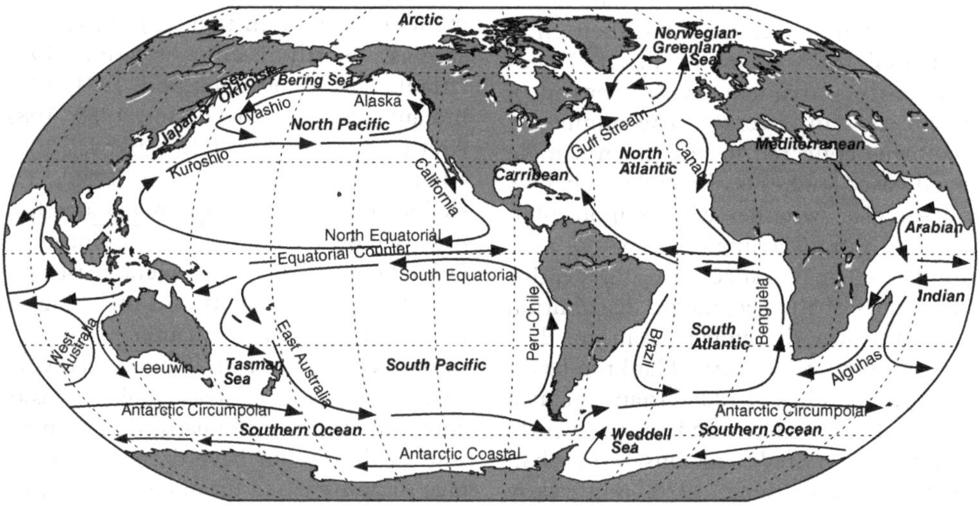

Figure 3.2 Major ocean surface currents (plain font) and ocean names (italic). Simplified from various literature sources, basemap created with Matplotlib.

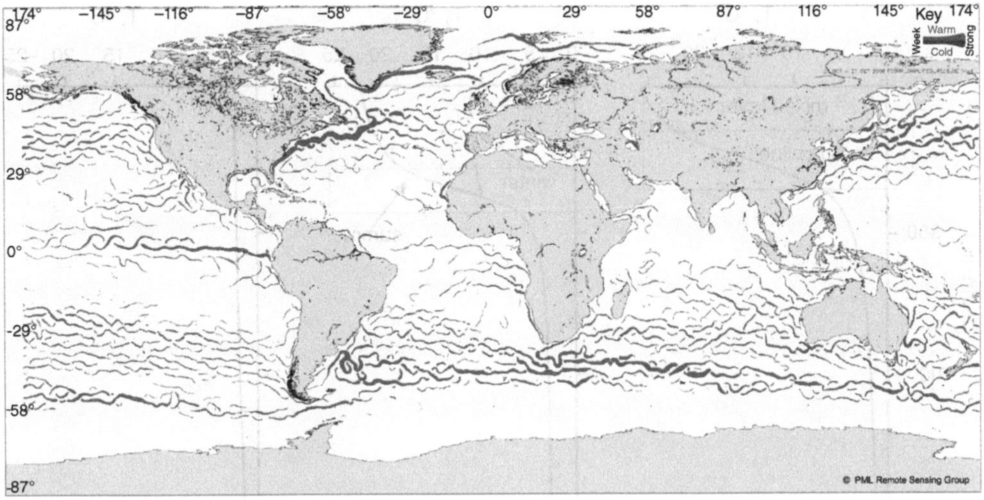

Figure 3.3 Global pattern of open ocean surface fronts during a single week (21–27 October 2008, from analysis of satellite data), illustrating the highly complex, dynamic nature of ocean surface boundaries. Note meanders and discontinous lateral extent of individual regions of contrast, as well as conformity of mean position of regions of stronger frontal development with classically defined ocean boundaries (gyres, fronts, biotic provinces), and stable central gyre regions of little contrast. Source: Image courtesy P. Miller, Plymouth Marine labs, from Miller (in prep). See also Miller et al. (2013).

the ocean floor (many thousands) than are available from the water column. Further, these samples generally contain much larger numbers of radiolarian shells, representing an average over hundreds, or thousands of years of sedimentation of radiolarians at the sample site. Together, these characteristics allow clearer patterns of distribution to be constructed. We will use both types of data in the following summary of radiolarian biogeography.

Radiolarian distributions in both the water column and surface sediments were most recently, and for the first time, comprehensively synthesized for the entire ocean by Boltovskoy et al. (2010), and this publication provides most of the information summarized here. For many species, distributional information is still very incomplete, with presence or quantitative abundance information only available from a relatively small number of localities, and particularly for water column data. Many taxa, however, are recorded quite frequently and the distributions of these can be used to illustrate the main patterns of radiolarian biogeography.

Tropical species distributions are fairly common among radiolarians (Figure 3.7). Species in this category are most common within ca. 20° latitude from the equator, although they can be found in lesser abundance also in the subtropics and mid-latitudes (ca. 40° N/S). Warm-water cosmopolitan species (Figure 3.8A) are found throughout the warm waters of the oceans, being absent only in polar regions, but are less common in the tropics, with maximum abundances in middle latitudes (ca. 20–40° N/S). Cool temperate species (Figure 3.8B) are most common in the cooler subtropics to subpolar regions (40–60° N/S), although they are present in very low numbers also in tropical regions, mostly in deeper waters below the mixed layer, and more abundantly in upwelling regions

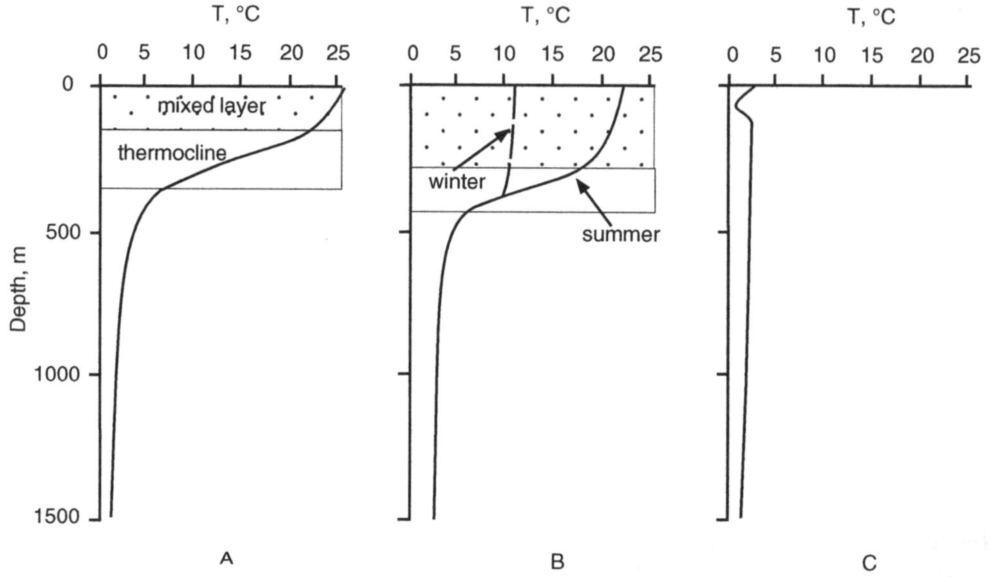

Figure 3.4 Representative temperature profiles vs. water depth for different regions of the ocean. In tropical regions (A) relatively warm (25°C or even more) water in the upper 100–200 m is mixed by winds. A transition region, the thermocline, separates the warm mixed layer from the cold deep ocean below. This structure is stable year-round. In temperate regions (B) the temperature contrast between surface and deep ocean in summer is somewhat less than in the tropics and the depth of the mixed layer is deeper. This is in part due to substantial cooling during the winter which reduces the thermal contrast and permits deeper mixing, aided by winter storm winds. In the high polar regions (C, e.g. the Southern Ocean) surface-water temperatures are only seasonally slightly higher at the surface and for much of the year there is little thermal contrast. Ocean waters can be mixed very deeply. Redrawn and modified from profiles published online by the Bigelow Laboratory (www.bigelow.org).

along the margins of ocean basins. Although most such species are bipolar, a few are found only in one hemisphere, or even (e.g. *Ceratospyris borealis*, North Pacific) in a single ocean basin, which is unusual as most radiolarian species (and most plankton species in general) are found in all ocean basins in the same latitudinal oceanic biome. Lastly, there are polar species (Figure 3.8C), some endemic to either the Arctic or Antarctic, that are mostly found poleward of the polar fronts at 50–60° N/S. These species, too, may occasionally be found in lower latitudes, although they are generally rare and found only at or below the thermocline in low latitude regions.

In addition to these main patterns there are the distributions of species in low- to mid-latitude upwelling regions, which contain a mixture of species characteristic for the latitude, cool to cold water taxa that are latitudinally displaced from their main latitudinal location, and as well a few species that appear to be endemic to upwelling locations (Nigrini and Caulet, 1992). Upwelling biogeography is discussed in more detail in Chapter 7.

Lastly, there are a few species which are found commonly only in a very limited area of the oceans, even if they are not endemic to the area. A good example is *Cycladophora davisiana*, which, because

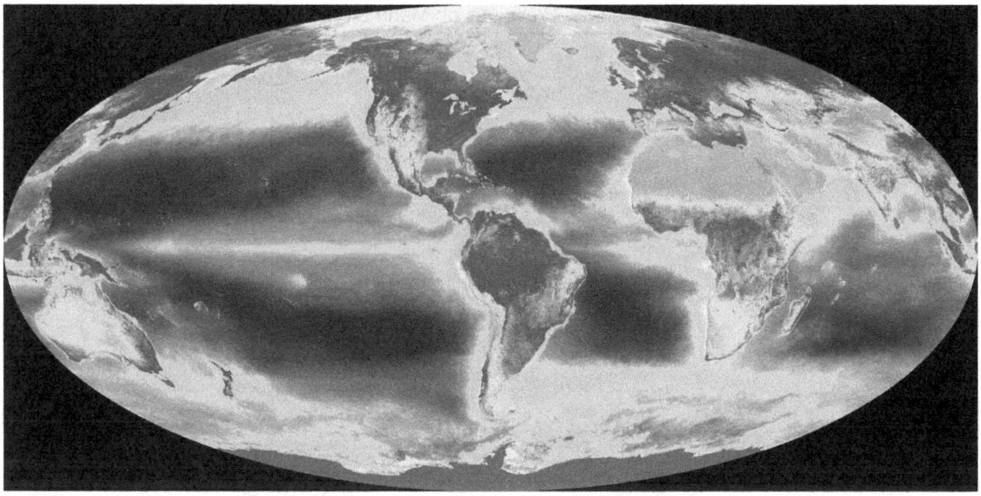

Figure 3.5 Map of global chlorophyll concentration in the oceans (and on land), averaged over several years of satellite imaging. Chlorophyll concentration is a reasonable proxy for primary ocean productivity (the total amount of organic carbon fixed in the near surface waters), although not necessarily for net (aka new) productivity, which measures how much organic material is exported into deeper water layers vs. being consumed and recycled in the surface water. False color image – highest concentrations in red, grading through orange, yellow light blue, dark blue and (lowest concentration) purple. Concentrations vary substantially seasonally and yearly. Source: NASA/SeaWifs (public domain). (*See color plate section for the color representation of this figure.*)

of its being much more common in large areas of the ocean in Pleistocene glacial periods (see Chapters 7 and 8), has been a target for ecologic research. This taxon has a rather unique distribution in the modern ocean, being abundant only in the Okhotsk Sea where its relative abundance in total radiolarians exceeds 30%, whereas elsewhere in the modern ocean it is typically present but almost always <10% in surface sediment assemblages, and usually <1% (Hays et al., 2013). With more than three decades of work since the 1980s (e.g., Morley and Hays, 1983; Morley, 1983), the ecologic association of this taxon with seasonal sea-ice presence or ice-margin conditions has been clearly characterized. This taxon is not abundant in areas where perennial sea-ice cover occurs. This situation is analogous to that of a typical sea-ice-related diatom, *Thalassiosira antarctica* (resting spores), which also are correlated to seasonal sea-ice and ice margin conditions rather than perennial ice cover. In the perennial ice-covered environment, neither *T. antarctica* nor *C. davisiana* can produce sufficient numbers of dissolution-resistant specimens to be preserved in fossil record.

Most data on radiolarian taxa distributions has been done at the level of morphologically defined species. No significant data yet exists on the genetic variation within radiolarian morphospecies (see Chapter 4, Genetics). In other groups of marine protist plankton such as foraminifera, most morphospecies have been shown by molecular studies to consist of several genetically distinguishable species. Some of these, but not all, are also definable by minor differences in

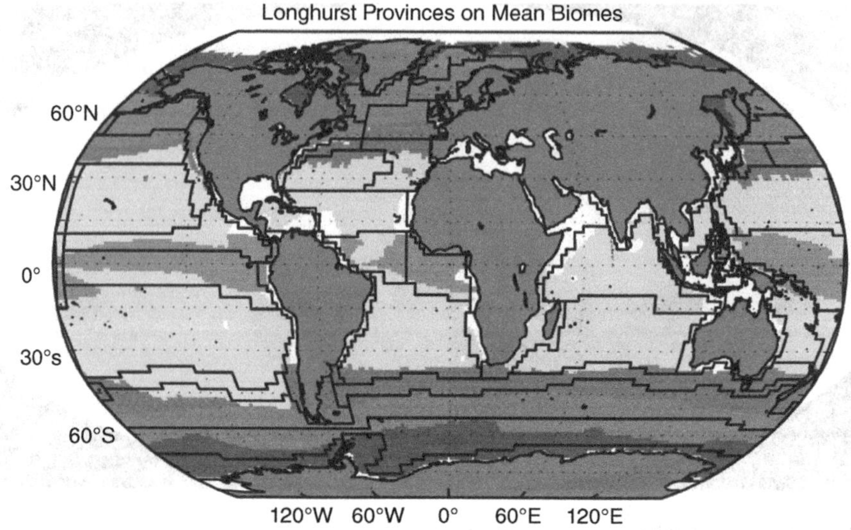

Figure 3.6 Global map of major pelagic biogeographic regions, showing two different classification schemes, both based on physical and chemical ocean characteristics. Black outlined areas are ocean provinces of Longhurst (1998), colored regions ocean biomes from Fay and McKinley (2014). Orange: equatorial; yellow: subtropical permanently stratified; green: subtropical seasonally stratified; light blue: subpolar seasonally stratified; dark blue: (sea) ice. White areas are unclassified and mostly correspond to upwelling regions, coastal areas and marginal seas. Longhurst provinces are subdivisions of his coarser scale biomes, which more closely match those of Fay and McKinley. Note similarity in structure, despite minor differences in placement of boundaries. Source: Earth System Science Data. Licensed under CC BY 4.0. (*See color plate section for the color representation of this figure.*)

morphology that until now had been treated as being not taxonomically significant even at the subspecies level. The individual molecularly identified (pseudo) cryptic species in these groups tend to be geographically less widely distributed than the morphospecies as a whole, typically being common only within one side of a biome in any given ocean basin (i.e. being found primarily within individual provinces in the Longhurst scheme). Morphologic variation within individual

→

Figure 3.7 Distribution of a tropical species of radiolarian (*Euchitonia elegans*), based on a synthesis of virtually all available published data (Boltovskoy et al., 2010). Shown are relative abundances in color and size coded dots (key in upper panel lower left corner) representing percent categories for surface water (above 150 m), deep waters (below 150 m), and in surface sediments. Simple presence/absence shown by crosses and open circles, respectively. Absence only shown for locations where based on source publication taxa was looked for but not found. Data for distribution in the water column is much less complete than from surface sediments. *E. elegans* is most common within ca. 20° of the equator. It is also present in lower abundances throughout the subtropics (ca. 40°N/S), but only rarely and sporadically in cold-water regions near the poles. Source: Boltovskoy et al. (2010). Reproduced with permission of The Palaeontological Association. (*See color plate section for the color representation of this figure.*)

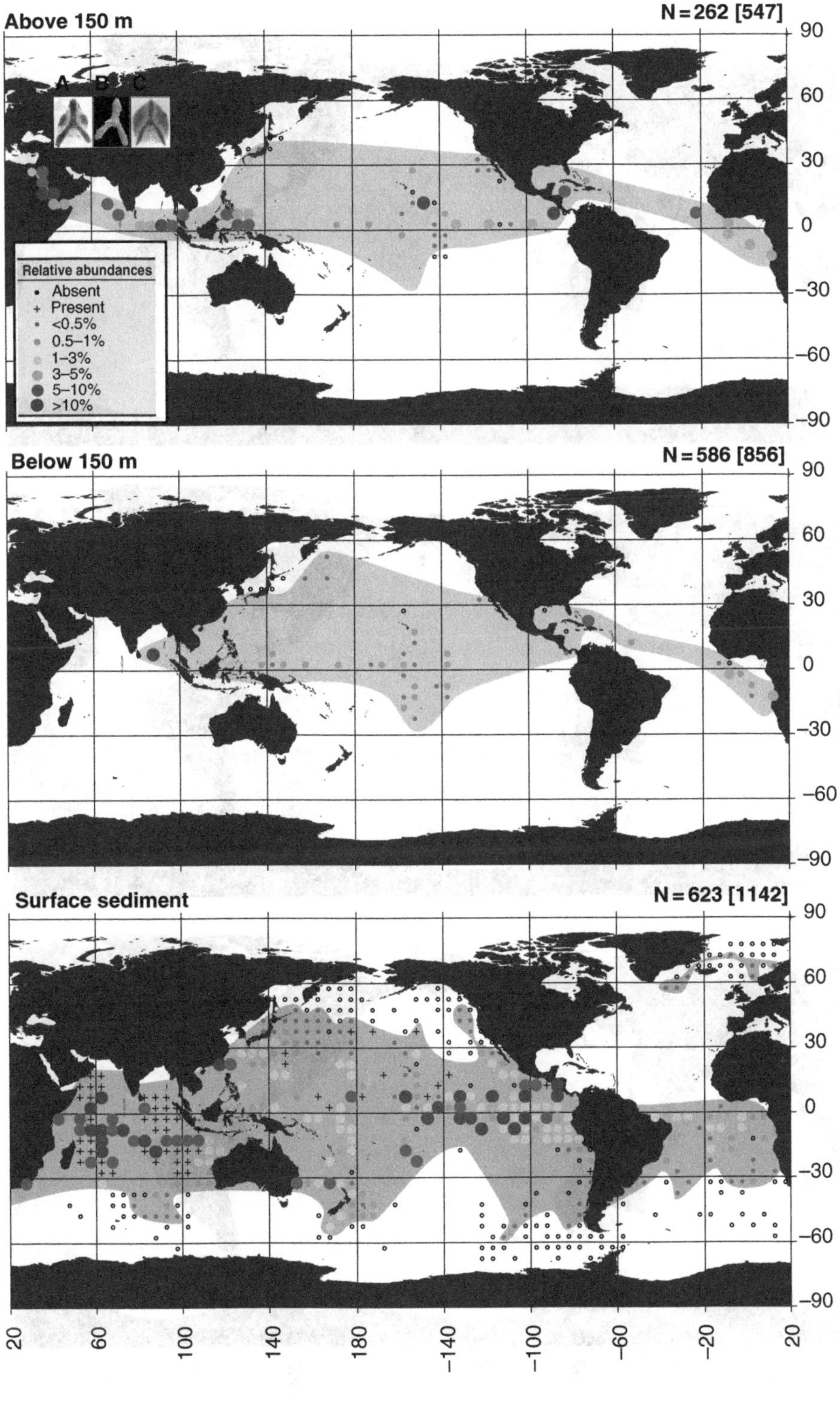

Above 150 m
N=262 [547]
A B C
Relative abundances
Absent
Present
<0.5%
0.5–1%
1–3%
3–5%
5–10%
>10%
Below 150 m
N=586 [856]
Surface sediment
N=623 [1142]
90
60
30
0
–30
–60
–90
20
60
100
140
180
–140
–100
–60
–20
20

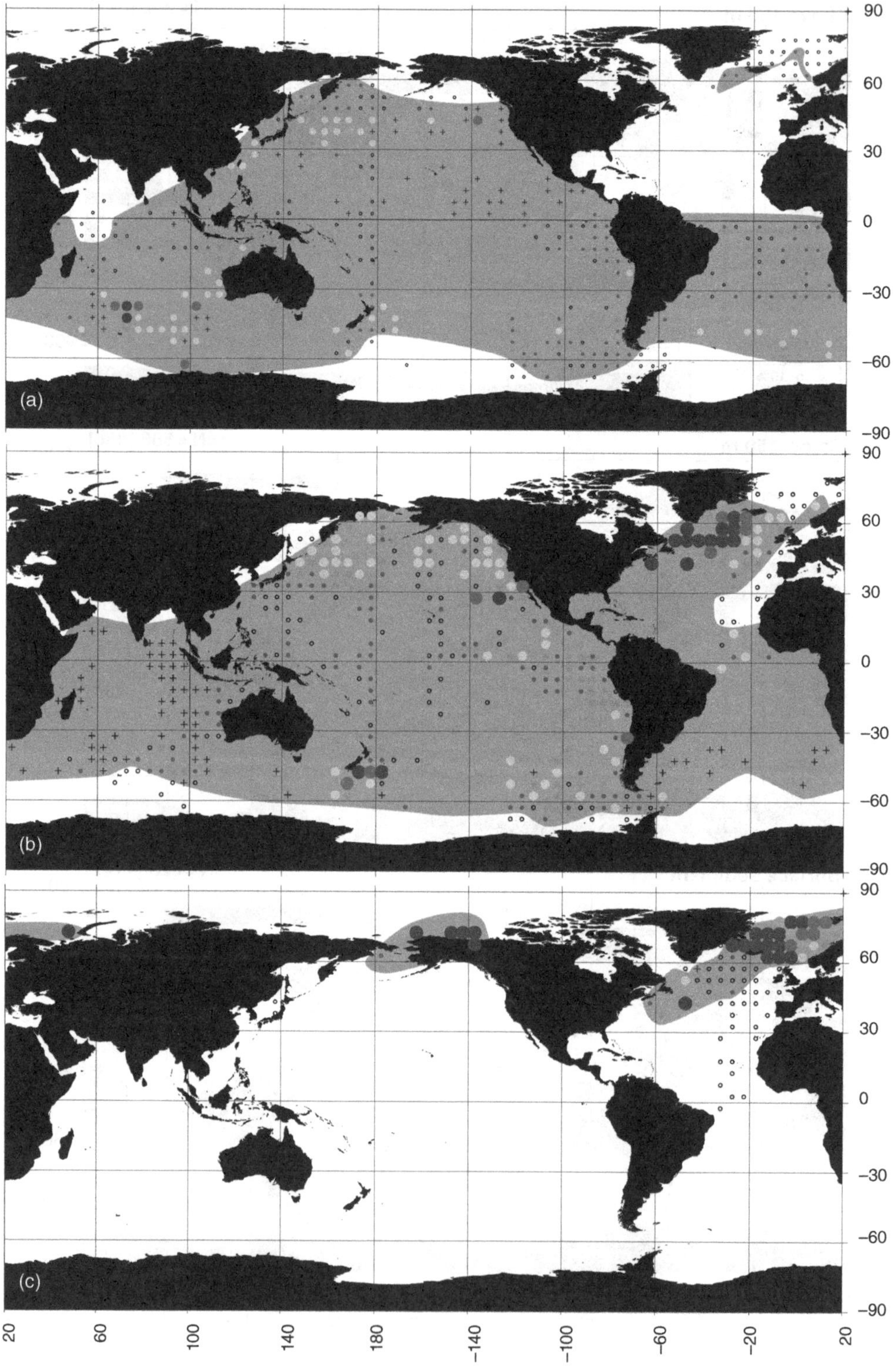

90
60
30
0
−30
−60
−90
(a)
(b)
(c)
20
60
100
140
180
−140
−100
−60
−20
20

radiolarian morphospecies is quite common and in some cases this variation has been examined geographically (Bjorklund, 1977; Johnson and Nigrini, 1980). These studies have found that subspecies level variation is often associated with specific geographic/environmental conditions (Figures 3.9, 3.10). Although the available data is too limited to draw conclusions, it may be that subspecies variation in radiolarians is also associated with genetic variation, or even represents individual genetic species with, relative to the enclosing morphospecies, more narrowly ranging ecologic preferences and geographic habitats.

From all of the above: knowledge of distributions for many species in the water column, and the faunal distributions of even more from sediment samples, radiolarian biogeographic patterns can be generalized in the form of a biogeographic map (Figure 3.11). Radiolarian species are generally found throughout one or more of the biogeographic regions shown. These patterns are compatible with those seen in other groups of plankton, and the biome and province boundaries determined by plankton ecologists using physical ocean parameters (e.g., Longhurst, Figure 3.6 above). However, radiolarian distributions tend to show more dispersion than in some other groups of plankton. In particular, it is common for subtropical, temperate, or even polar species to be present, albeit in relatively low numbers, also in lower-latitude regions. This characteristic of radiolarian distribution can only be understood by examining the vertical distribution of species in the water column, and how these distribution profiles vary geographically.

Vertical Distribution

Radiolarians *sensu lato* have long been recognized to occupy different depth zones in the ocean. Haeckel (1887) defined several depth zones on the basis of radiolarian taxa, although he primarily considered higher-level clades, e.g. that Phaeodarians were characteristic of very deep waters. Depth habitats of individual species of polycystine radiolarians have only been studied in detail relatively recently (1970s or later), although a few studies exist that date back to the 1950s: Reshetnyak (1955) for example, which although primarily aimed at characterizing the depth zonations of phaeodarians did mention a few polycystines. Petrushevskaya (1971a) was one of the first detailed studies of vertical distributional patterns of polycystine radiolarians, including live and empty shells down to ca. 5 km in the water column at two stations in the central Equatorial Pacific. This study first

Figure 3.8 Plotting conventions as in Figure 3.7. All data shown is from surface sediments only. A – Distribution of the actinommid spumellarian *Actinomma medianum*. Although present over a broad latitudinal range, this typical subtropical cosmopolitan species is most abundant in mid latitudes, ca. 20–40° N/S, and is much less common both in the tropics or in more polar regions. B – Distribution of *Botryostrobus aquilonaris*, an artostrobid nassellarian. This species is primarily found in subpolar regions between 40–60° N/S but also to some extent in low- to mid-latitude areas of upwelling such as the Peru–Chile current region and the eastern Equatorial Pacific. C – An example of a polar endemic species, the cannobotryid nassellarian *Amphimelissa setosa*. This species is a northern polar endemic, mostly in the Arctic Ocean but also found where currents such as the East Greenland Current transport cold water southwards along the eastern margin of ocean basins. Source: All figures from Boltovskoy et al. (2010). Reproduced with permission of The Palaeontological Association. (*See color plate section for the color representation of this figure.*)

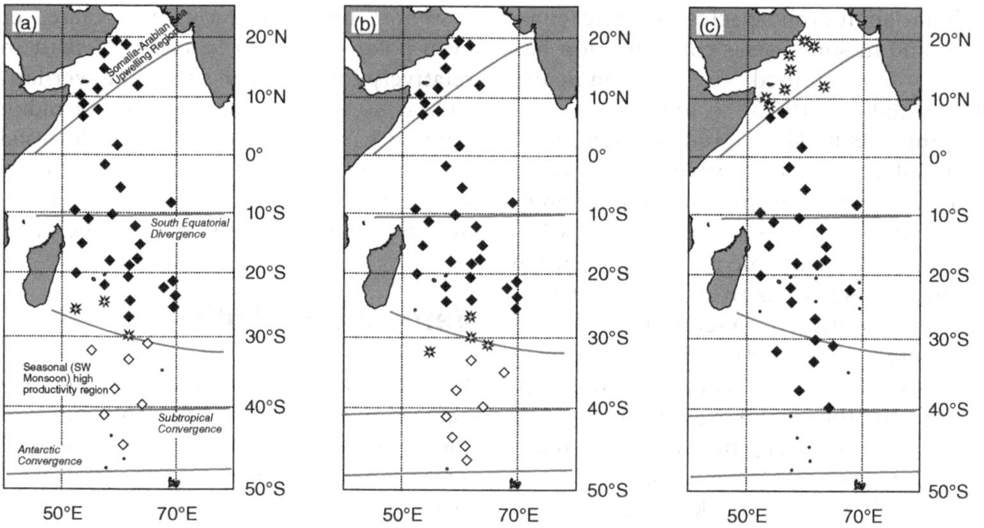

Figure 3.9 Distribution of selected radiolarian subspecies pairs in surface sediments of the Indian Ocean. Diamonds – present, small dot – absent, stars – both subspecies present. Boundaries of ocean regions approximate, based on literature. Northern boundary of seasonal high productivity SE of Madagascar and extent of Somali–Arabian Upwelling Region from SeaWifs Imagery (http://oceancolor.gsfc.nasa.gov/cms/). Region between Subtropical and Antarctic Convergences approximately equal to the seasonally stratified substropical biome of Fay and McKinley (2014), Somali–Arabian Upwelling region to their "unclassified" region. A – black: *Spongaster tetras tetras*, white: *S. t. irregularis*; B – black: *Pterocanium praetextum praetextum*, white: *P. p. eucolpum*; C – black: *Lamprocyclas maritalis polypora*, starred – *L. m. polypora* and *L. m. ventricosa*. Radiolarian distribution data from Johnson and Nigrini (1980).

established that some species are surface dwellers (0–100 m, e.g. *Pseudocubus obeliscus*) and others are subsurface dwellers (e.g. 75–100 m and deeper, *Litharachnium tentorium*). Lastly, she observed that the subsurface dwellers were found most abundantly at a depth of 300 m, but the populations densities of these taxa were much less than those of the surface dwellers. She further noted that all of the taxa encountered in the plankton were also found in surface sediments, but the percentages of individual taxa are very different from those of plankton assemblage. This observation was confirmed in the study of Renz (1976), which was the first to report quantitative distribution patterns in a transect of stations in the equatorial Pacific crossing the equator. She employed pumps and 35 μm sieves to ensure inclusion of the smallest radiolarians. By documenting the vertical distribution patterns of 23 polycystine taxa from the middle latitude eastern North Pacific for the first time, Kling (1979) became a key study in radiolarian vertical distribution research. This study included some of the important taxa in paleoceanography (e.g. *Cycladophora davisiana* with maximum abundance at 200 m). Four separate depth zones were characterized: (1) surface zone ca. 0–50 m; (2) upper subsurface maximum zone, maximum at 50–100 m; (3) lower subsurface maximum zone, maximum at

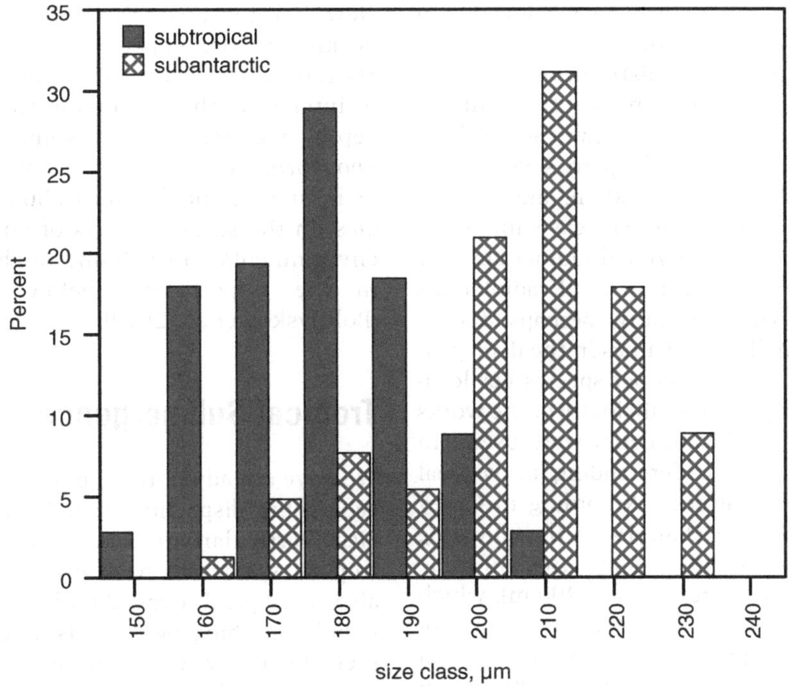

Figure 3.10 Variation in cortical (outer) shell size vs. latitude in the species *Actinomma haysi* summed across several surface sediment sample pairs north (subtropical) and south (subantarctic) of the Subtropical Convergence in the South Atlantic (ca. 40°S). Replotted from data presented in Bjørklund (1977).

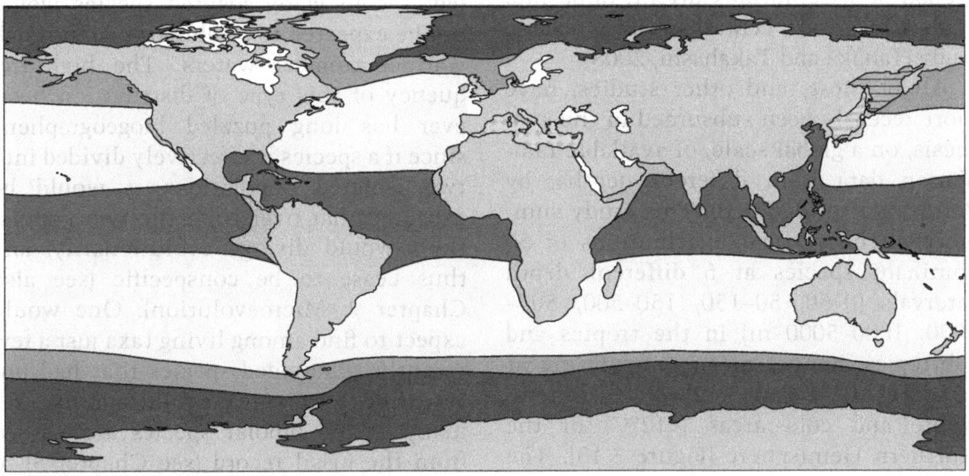

Figure 3.11 Approximate biogeographic provinces of modern radiolaria, modified and updated from distributions proposed by Goll (1976). Medium grey – Tropical; white – Subtropical; light grey – Subpolar; dark grey – Polar; light grey – low- to mid-latitude Upwelling. Patterns mark special faunas in marginal seas (Mediterranean, Japan, and Okhotsk). Source: Lazarus, Geol. Soc. London, 2011. Reproduced with permission of Geol. Soc.

100–200 m; and (4) deep zone below 200 m (Kling, 1979). McMillen and Casey (1978) reported the depth distribution of many polycystine radiolarians in the Gulf of Mexico and the Caribbean Sea. These authors discriminated up to seven depth zones (e.g. 0–50, 50–100 m, and several deeper) at four stations, and gave numerical information for individual radiolarian taxa as well. Only values for total radiolarians were shown graphically, perhaps due to mostly small count values in the data (generally <10 specimens per species per location and depth interval). Many of the works of Casey and his students were useful in advancing our understanding of vertical distribution patterns. According to Casey (1977) and Casey et al. (1979), the following faunas were found off southern California: (1) a Shallow Water Fauna (<200 m), which fluctuates seasonally, with invasions of Subarctic, Equatorial, and North Central Shallow water faunas; (2) at 200–500 m depth a Transition Central Fauna is encountered; and (3) at 500–1000 m Subarctic Intermediate Fauna. Other primary studies include Kling and Boltovskoy (1995) across the southern California Current, Yamashita et al. (2002), and a central subarctic Pacific study (Tanaka and Takahashi, 2008).

All of these, and other studies, have more recently been subsumed in the synthesis, on a global scale, of available radiolarian data from different depths, by Boltovskoy et al. (2010). This study summarized the vertical distributions of 50 dominant species at 5 different depth intervals (0–50, 50–150, 150–500, 500–1000, 1000–5000 m) in the tropics and subtropics (annual mean temperature at 10 m >12° C) (Figure 3.12) as well as temperate and cold areas (<12°C) of the Northern Hemisphere (Figure 3.13). The vertical distribution patterns are shown as percentages of individual taxon's contribution to overall radiolarian assemblage. Vertical abundance profiles of the same five depth intervals for selected taxa show (Boltovskoy et al., 2010): 7 taxa peaking in the upper 50 m; 31 taxa at 50–150 m; and 5 taxa at 150–500 m. Also of interest is the geographic variation in depth preferences for some species. *Spongotrochus glacialis* is most abundant, and one of the most abundant species, in the surface waters of north polar environments, but is found in abundance in the tropics only below 1000 m (Boltovskoy et al., 2010).

Tropical Submergence

We have already noted the relatively high geographic dispersion of radiolarian species, particularly the tendency of non-tropical species to have low abundances also in tropical areas. Another feature of radiolarian biogeography is one that is seen in many other groups of plankton – the bipolar distribution of many cool water species, with essentially no presence in the tropical surface waters in-between. Such distribution makes sense in terms of an individual species' ecology – cold water-adapted species would not be expected to be able to live in much warmer tropical waters. The high frequency of this type of distribution however has long puzzled biogeographers, since if a species is effectively divided into two isolated populations it would be expected that over time the two populations would diverge evolutionarily, and thus cease to be conspecific (see also Chapter 9: Microevolution). One would expect to find among living taxa just a few recently subdivided species that had not yet time to diverge evolutionarily. Yet many of the bipolar species are known from the fossil record (see Chapter 5) to have had bipolar distributions for many millions of years. How can this be? One hypothesis long in the literature is that bipolar species are connected by "stepping-stone" small populations in the

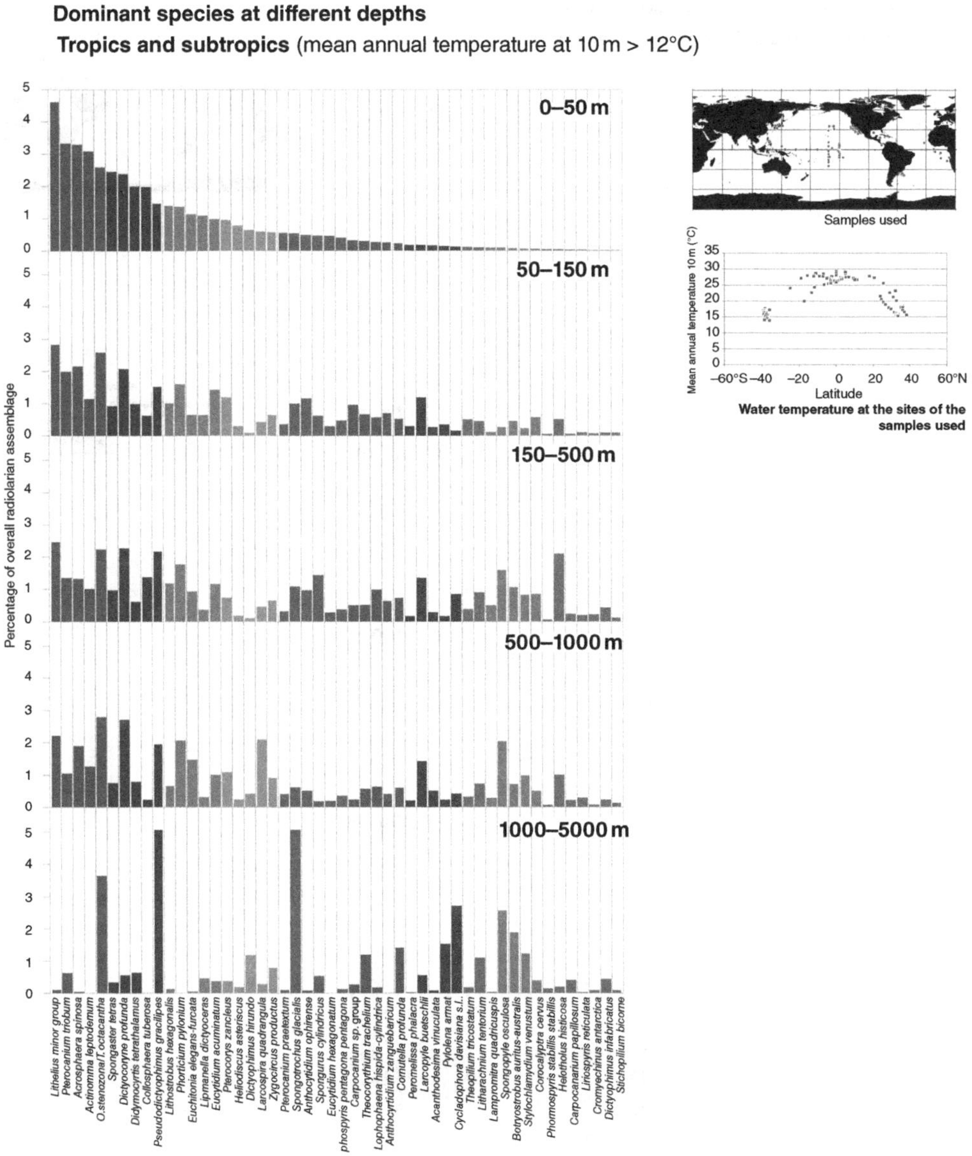

Figure 3.12 Vertical depth characteristics of 50 most abundant polycystine taxa in the tropics and subtropics (mean annual temperature >12°C) Source: Boltovskoy et al., 2010. Reproduced with permission of The Palaeontological Association.

upwelling regions on the eastern margins of ocean basins in mid- to low latitudes (Johnson and Brinton, 1963). Not all bipolar species however appear to have such lower latitude upwelling populations, and in any case, the current systems in the eastern tropics do not flow directly poleward to the next "stepping-stone"

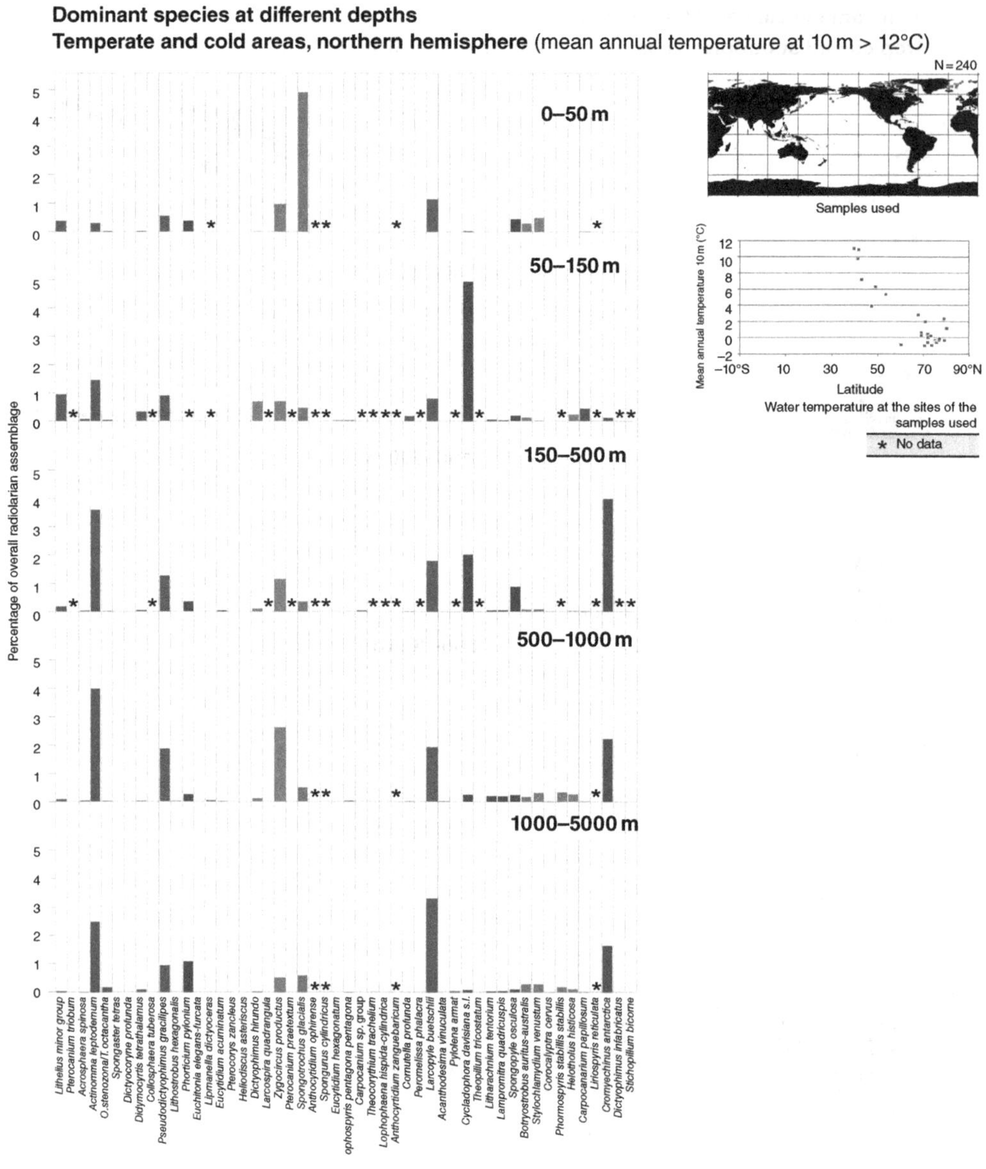

Figure 3.13 Vertical depth characteristics of 50 most abundant polycystine taxa in the temperate and cold areas of the northern Hemisphere (mean annual temperature <12°C) Source: Boltovskoy et al., 2010. Reproduced with permission of The Palaeontological Association.

upwelling region, but instead flow westward into the maximum warmth western tropical waters (Figure 3.2), where cold water forms would not be expected to survive.

Although not a solution that necessarily can be applied to all plankton, for those species which can survive below the photic zone (e.g. zooplankton, mixotrophs functioning solely in zooplanktonic mode, or phytoplankton with resting spore stages), one solution may be via tropical submergence. This was first proposed for radiolarians by Casey (1971), and elaborated on in numerous subsequent papers (e.g. Casey et al., 1982); several other studies are summarized in Boltovskoy (2017). Casey (1971) primarily focused on the vertical distribution of radiolarians, summarizing what was at that time known, including the fact that some species (like *S. glacialis*, cited above) seemed to be found in the surface water layers in high latitudes but in deeper water layers in the tropics. He proposed that radiolarian species occurrences in ocean waters were largely controlled by water characteristics, and not that much by water depth. This has an important consequence. As noted at the beginning of this chapter, the upper part of the main cold-water column shows a strong tendency to laminar flow, with surface waters in the high latitudes flowing into low latitudes in the subsurface, following levels of constant water density (Figure 3.1). Thus, polar plankton species that primarily track water masses would be entrained and be transported into low latitude regions, but within cold waters that they are better adapted to. In this way, the polar faunas can be considered in contact and not actually isolated into fully separate populations, and thus capable of showing long-term, stable patterns of surface bipolar distribution. Similarly, species with bimodal latitudinal subtropical or cool temperate distributions could be linked by subsurface water flow. The extent of this exchange has not been quantified in a comprehensive way but must, at least for polar species, be fairly low, as otherwise polar species would, like temperate species, also be common in sediments from tropical regions.

As important as this hypothesis is, there is still relatively little hard data to back it up. In Casey's work, most references to changing water depth habits for species was only given in anecdotal or generic form, without actual comparative data for species. This information gap has subsequently been partially filled in later studies, confirming at least in some cases the shift of polar surface water species to greater tropical depths, e.g. Ishitani and Takahashi (2007), Ishitani et al. (2008). These authors concluded that tropical submergence is only true for certain species, e.g. those species that favor cold waters but show no correlations to other environmental factors such as salinity, productivity, and dissolved oxygen (e.g. Ishitani and Takahashi, 2007), suggesting that they have a high tolerance to most environmental variables except for temperature. Boltovskoy (2017) has provided the most complete synthesis, based on the still very patchy available published plankton data from Boltovskoy et al. (2010). He estimates that 20–30% of the deep-dwelling taxa in low latitudes are advected populations from more polar regions. Tropical submergence patterns are observed in several other plankton groups (non-polycystine "radiolaria" clades, Foraminifera, and Pteropoda) (Boltovskoy 1988, 1994, 2017).

Longitudinal Gradients and Upwelling Assemblages

Due to the physical oceanographic mechanisms that create it, coastal upwelling is present normally along only the eastern limbs of basin-wide circulation gyres, and this, along with other ocean circulation factors, leads to a longitudinal gradient in primary productivity and other water characteristics within the gyre. While radiolarian species are generally present throughout a gyre, they may be more common in the eastern or western part (e.g. Kamikuri et al., 2007). There are also occasionally differences in the inter-basinal distribution of radiolarian taxa. According to Boltovskoy et al. (2010), there are taxa that are absent or very scarce in either the Atlantic or the Indian Ocean, and quite abundant in the other two major oceanic basins. There are eleven taxa belonging to this category: *Actinomma delicatulum*, *Cycladophora cosma cosma* (as *C. davisiana cornutoides*), *Dictyophimus infabricatus*, *Druppatractus irregularis*, *Lamprocyrtis nigriniae*, *Larcospira quadrangula*, *Peripyramis circumtexta*, *Tholospyris scaphipes* (as *Phormospyris stabilis scaphipes*), *Pterocanium korotnevi*, *Larcopyle buetschlii*, and *Stylochlamydium venustum* (for details see Boltovskoy et al., 2010). However, interestingly enough none of the species listed are scarce in the Pacific but abundant in either the Atlantic or the Indian oceans. Furthermore, based on the radiolarian distribution of the Arctic Ocean and adjacent areas, Bjørklund and Kruglikova (2003) stressed that the modern radiolarian fauna of the Arctic, which is dominated by Nassellaria over Spumellaria, is recruited from the North Atlantic, but not from the North Pacific.

In addition to these general basinal patterns, there are a few species that appear to be restricted to upwelling regions along the margins of ocean basins in low to mid latitudes. These are described in Chapter 7.

Latitudinal Gradients

As is true for most other groups of organisms, there is a gradient of diversity from high values in the tropics to lower values at the poles, which has been noted since the beginnings of radiolarian research (e.g. Haeckel, 1887): "*In the case of the Radiolaria as well as of many other classes of animals, the law holds good that the richest development of forms and the greatest number of species occurs between the tropics, whilst the frigid zones (both Arctic and Antarctic) exhibit great masses of individuals, but relatively few genera and species.*" Subsequent work has confirmed this with quantitative data (e.g. Boltovskoy et al., 2010; Boltovskoy and Correa, 2016; Figure 3.14). Individual plankton sample diversities can reach 120 taxa in the tropics whereas a much lower diversity of <20 taxa per sample is found in high latitudes. This trend is true for all three types of samples: plankton, sediment traps, and sediments. Samples collected by sediment traps tend to show higher species number than those from plankton and sediments. This can be attributed to better preservation of samples by traps than in sediments as well as better and longer coverage of time span represented in sediment traps than in plankton tows, plus the cumulative effect of settling of species from different water layers into the traps. It should be noted that the *absolute* diversity values in individual samples are much lower than the total diversity in a latitude band as sample sizes are too low to fully recover diversity (see also Chapter 9 for discussion of sampling), but the relative patterns are valid, and are seen also in global compilations of diversity vs. latitude (Boltovskoy and Correa, 2016).

Coastal Gradients

Although occurrences of radiolarians in very shallow water neritic environments appear to be rare, they are also known. A monospecific population of a single nassellarian radiolarian taxon – *Lophophaena rioplatensis* was well documented with salinity as low as 15.4 psu in la Plata estuary off Buenos Aires, Argentina (Boltovskoy et al., 2003). Neritic radiolarian populations, also of low diversity, have been

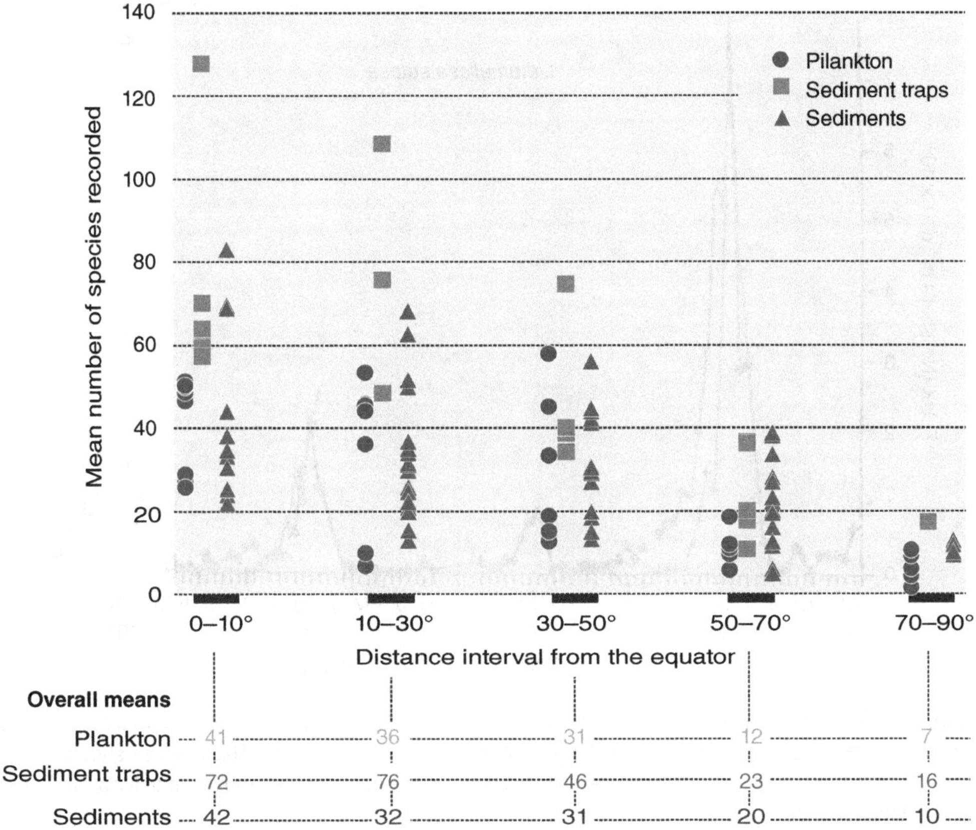

Figure 3.14 Latitudinal distribution of number of species reported per sample in three different types of samples: plankton, sediment traps, and sediments Source: Boltovskoy et al., 2010. Reproduced with permission of The Palaeontological Association.

reported from the Norwegian fjords (Bjorklund and Swanberg, 1987). Storms can also move more diverse open ocean populations temporarily into shallow waters. Thus, it is clear that radiolarians are not entirely restricted to hemipelagic to pelagic realms even though most taxa are.

Seasonal Variability

Seasonal distribution patterns of radiolarians were not well known until time-series sediment trapping became more of a standard tool in investigating radiolarian seasonality. Nevertheless, a handful of studies were published (Renz, 1976; Casey, 1977; 1979) with ship plankton tow sampling at different times of year. However, not much seasonal variability was seen in the region studied – the central Equatorial Pacific. Sediment trap studies (Honjo et al., 1980) have since shown that many radiolarian species are primarily present only at certain seasons of the year in regions outside the tropics with a strong seasonal cycle in ocean conditions. Examples from the Gulf

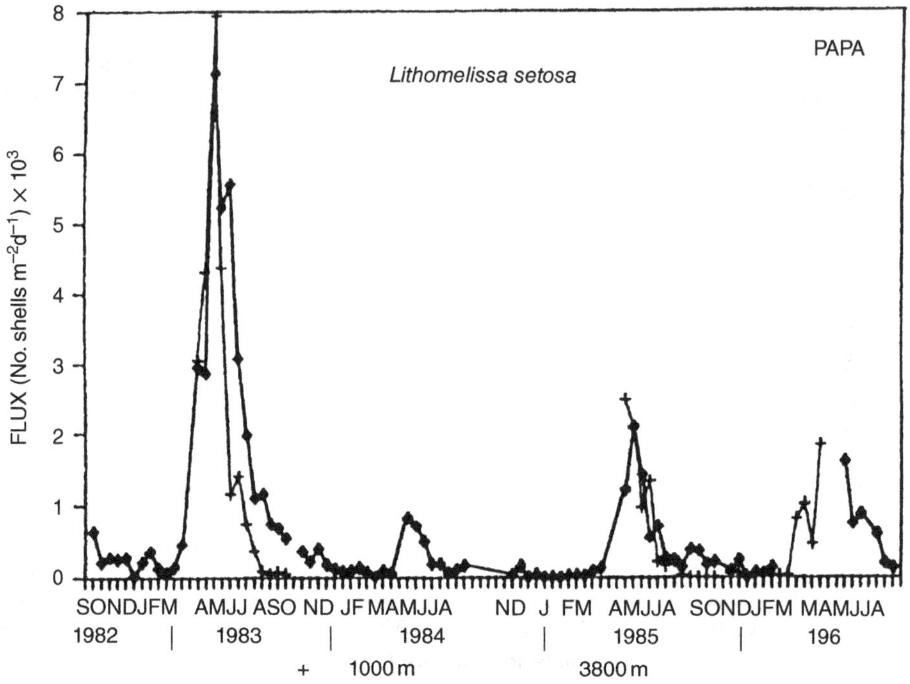

Figure 3.15 Time series of specimen flux of the small nassellarian species *Lithomelissa setosa* in two deep sediment traps at the North Pacific long-term monitoring station PAPA. This species shows a consistent single spring maximum. Source: Takahashi (1997). Reproduced with permission of News of Osaka Micropaleontologists.

of Alaska (Takahashi, 1987, 1997) include *Lithomelissa setosa* with a spring flux maximum (Figure 3.15), and *Ceratospyris borealis* (Figure 3.16) whose flux mimics the increases of diatoms during both spring and fall. In more polar regions with seasonal sea ice, these patterns can shift. *C. borealis*, for example, is, like many other species such as *Rhizoplegma boreale*, most common in the Sea of Okhotsk during the fall, as sea ice still covers this region in the springtime. In sea-ice covered regions, most of the radiolarian annual production occurs only during the sea-ice free time interval, both in the North Pacific (Okazaki et al., 2003) and Southern Ocean (Abelmann and Gersonde, 1991; Abelmann, 1992) (Figure 3.17). However, in sea-ice covered regions not all taxa are equally affected. Deeper-dwelling forms such as *Cycladophora davisiana* (centered at 200 m or below in the Sea of Okhotsk (Nimmergut and Abelmann, 2002) tend to be present even when sea ice is present, while surface water species such as *Spongotrochus glacialis, Stylochlamydium venustum,* and *Stylodictya validispina* (Hulsemann, 1963; Kling, 1979; Morley and Stepien, 1985; Nimmergut and Abelmann, 2002) are sufficiently inhibited by sea ice that their relative percentages can be used as a sea-ice tracer (Morley and Hays,1983; Morley, 1983).

In a few cases, ecologic partitioning between closely related species via seasonal

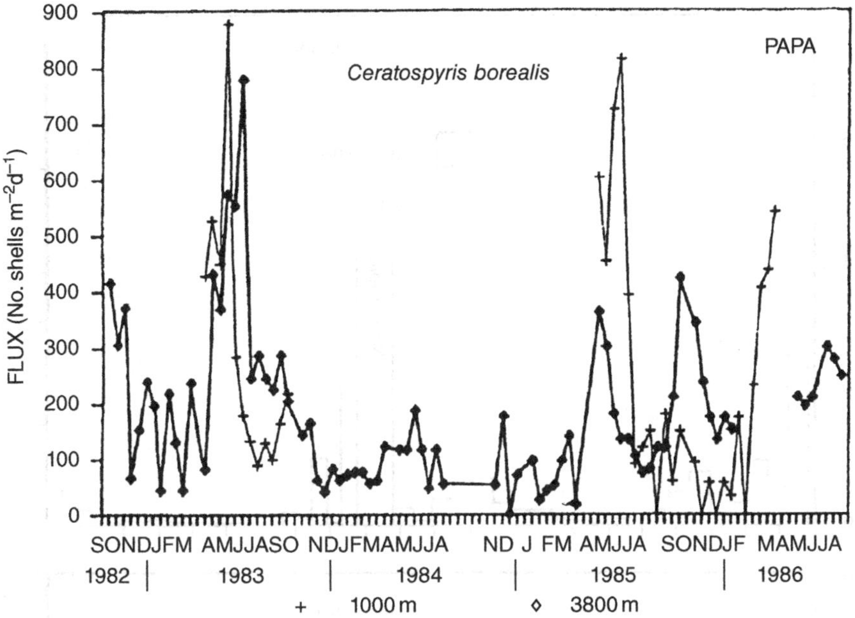

Figure 3.16 Time series flux in PAPA sediment traps of another nassellarian, *Ceratospyris borealis*, showing dual peaks in spring and fall. Source: Takahashi (1997). Reproduced with permission of News of Osaka Micropaleontologists.

shifts has been suggested. *Pterocanium korotonevi* and *Pterocanium diplotriaena*, two morphologically similar species, both show a double-peaked spring/summer pattern in the Gulf of Alaska, but *P. korotnevi* is relatively more common in the spring peak while *P. diplotriaena* is more common in the summer. This may be due to different adaptations to the changing composition of phytoplankton between the two seasonal intervals (Takahashi, 1987).

In addition to species-level patterns, differences have also been noted in adult and juvenile forms by season, again using sediment traps. In the same data above, *Pterocanium korotonevi* and *Pterocanium diplotriaena* adults peaked in the summer (July–August) whereas juvenile forms showed more constant levels over the year, hinting at differences in reproduction and growth on a seasonal basis.

Interannual Variability

Radiolarian fluxes also change over longer periods of time, at least in high-latitude regions (Takahashi et al., 1989; Takahashi, 1997; Ikenoue et al., 2012). In the North Pacific and Bering Sea, lower fluxes of both siliceous phytoplankton and radiolarians are observed in years with relatively deep pycnoclines, suggesting the influence via water mixing of nutrients on radiolarian productivity on an interannual basis in surface waters. Pycnocline depth can in turn be related to known interannual changes in ocean circulation, such as El Niño and the Arctic Oscillation.

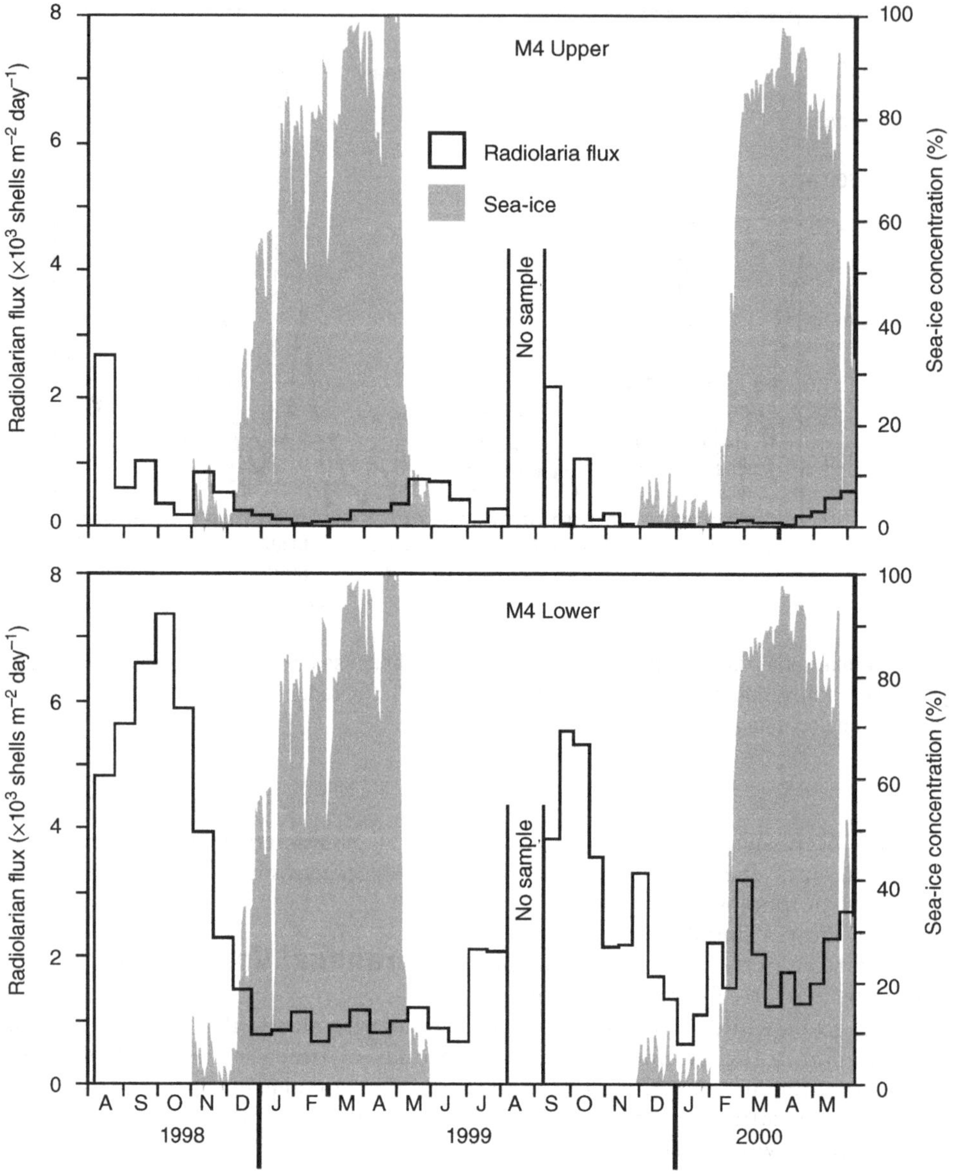

Figure 3.17 Annual production in polar environments and its control by seasonal sea ice. Two-year (1998–2000) fluxes of total radiolarians and seasonal sea-ice cover at Station M4 of the Okhotsk Sea Source: Okazaki et al., 2003. Reproduced with permission of Elsevier.

4 Genetics

Introduction

Genetics deals with traits of inheritance from parents to offspring, and is an important source of information for taxonomy and phylogenetics.

Most polycystine radiolarians have siliceous shells which can be preserved as fossils. Morphologic and stratigraphic studies of radiolarians have been until now the primary source of taxonomic information for the group. However, these studies do not include shell-less groups (i.e. naked or highly soluble shell-forming specimens, including many possible sister taxa to the radiolaria). Moreover, phylogenetic relationships at the suprageneric level are not always easily deduced from the shell morphology for a variety of reasons, e.g. lack of intermediary forms (particularly at higher taxonomic levels), convergence of shell forms i.e. iterative evolution. Molecular phylogeny can provide helpful information to resolve these problems. Beginning with Amaral Zettler et al. (1997), molecular analyses of radiolaria have begun to provide clues to the phylogeny and early evolution of radiolaria. This chapter gives an overview of macro- and microevolutionary relationships of radiolaria based on molecular phylogeny. Studies of evolutionary processes that are based at least in part on molecular data are given in Chapter 9: Evolution.

Paleobiology of the Polycystine Radiolaria, First Edition. David Lazarus, Noritoshi Suzuki, Yoshiyuki Ishitani and Kozo Takahashi.

It should also be noted here that one can use the term "Radiolaria" as a formal taxonomic grouping that includes not only the Polycystina, but also other groups that are related to, but not part of, the Polycystina. When we do so in this chapter, we capitalize the term (i.e. "Radiolaria") to make this meaning clear, and in such instances explicitly use Polycystina or "polycystine radiolarians" to refer to just the fossil-forming radiolarians that are the subject of the rest of this book.

Molecular Phylogenetic Position of "Radiolarians" within Eukaryotes

There are several groups of protists that are at least occasionally referred to in the literature as "radiolarians," or which are thought to be at least closely related to them. Until now, in this book, we have only defined these in detail for the living polycystine radiolarians: Nassellaria, Spumellaria, and Collodaria (Chapter 2: Biology). The other groups have been mentioned only in the briefest of terms, but in this chapter as we are interested in the phylogenetic relationships between these higher taxa a more detailed definition of the groups and their distinguishing characteristics is warranted.

In the traditional literature of radiolarian biologic studies, two other groups of protists have been included together with the polycystines as the formal taxon Radiolaria: the Acantharia and the Phaeodaria. This broad definition of Radiolaria originates in the taxonomic work of early taxonomists such as Ehrenberg and Haeckel.

Acantharia (Suzuki and Not, 2015) are large solitary plankton (ca. 0.05–5 mm) with strontium sulfate shells based on 20 radial spines arranged in a uniform geometric arrangement called Müller's rule. Although the composition and geometry of the shells are different from polycystine radiolarians, common cytoplasmic features exist as well, such as the existence of a central capsule, and the arrangement of microtubules in the axopodida.

Phaeodaria were traditionally grouped in the radiolarians due to their similar feature of having a central capsule and possessing hollow silica shells. But they lack bridge structures between the microtubules in their pseudopodia. The most typical feature of this group is the existence of phaeodium which is an aggregate of their wastes, in the cell. To date, molecular phylogeny excludes them from the Polycystina, but places them as close relatives of cercozoans (Suzuki and Not, 2015).

In addition to these radiolarian groups, early molecular phylogenies placed one other group of protists placed into close relative of Acantharia: the Foraminifera (Cavalier-Smith, 2003). Foraminifera are large benthic or planktonic protists with external pseudopods. However, they differ in many respects from other "radiolarian" groups, e.g. they possess strong non- or only microporous shells made of calcium carbonate and lack many of the cytoplasmic features that characterize the Polycystina or related forms. Their phylogenetic relationship to Radiolaria is still unclear.

Lastly, there is the unusual taxonomic group Taxopodia, which consists of just a single genus: *Sticholonche*. *Sticholonche* was traditionally described as heliozoan organism; that is, it possesses motile siliceous spines, which it uses to move through the water with a complex rowing motion. It lacks a central capsule (the endoplasm is delimited by the capsular membrane inside the cell), which is the most common feature within other radiolarians (Cachon and Cachon, 1978).

Molecular Studies of Radiolarian's Position within Eukaryotes

Ribosomes are tiny organelles that exist in large numbers within cells, and play a central role in protein synthesis. They are made of a smaller and a larger part, or "subunits." The smaller subunit reads messenger RNA coming from the cell nucleus DNA, while the larger subunit uses the information from the RNA to assemble amino acids into proteins. Ribosomes are themselves made out of proteins and RNA. Small subunit ribosomal RNA (SSU or 18S rRNA), one of three units of ribosomal RNA, is one of the genes most used to clarify organismal phylogeny due to the availability of data from many groups of eukaryotes and a useful balance between variation and conservation in gene sequence. Through SSU rRNA analyses including radiolarians, Cavalier-Smith (2003) defined the eukaryotic super group "Rhizaria" including Filosa, Endomyxa, Foraminifera, and Radiolaria. He further grouped Radiolaria and Foraminifera into "Retaria" following their morphological traits, e.g. reticulopodia and multiple porous walls. The rhizarian group has one unique molecular character vs. all other eukaryotes: insertions between ubiquitin monomers. Ubiquitin is, as the name implies, a ubiquitous protein which is highly conserved among eukaryotes and plays an essential role in cellular processes: apoptosis, signal transduction, endocytosis, and cell cycle regulation. Because of the high degree of conservation seen in this molecule and its pivotal functional role, this unique pattern of insertions between ubiquitin monomers of Rhizaria is thought to be critical to solve rhizarian phylogeny. Bass et al. (2005) investigated this polyubiquitin-insertion, and determined that radiolarians lack insertions, foraminifers and endomyxans have one, and filosans have one or two. This suggests that Radiolaria branched in the most basal position in Rhizaria, and successively branched into Foraminifera, Endomyxa, and Filosa (Figure 4.1). Even though this hypothesis is compatible with the well-established rhizarian phylogenetic relationship, the relationship suggested above is only supported by insertion of ubiquitin monomers – other molecular data give different results. Moreira et al. (2007) explored phylogenetic relationships within rhizarians using SSU and large subunit (LSU) rRNAs (another subunit of ribosomal RNA), and grouped Radiolaria and Foraminifera out of Endomyxa and Filosa. This phylogenetic relationship is also supported by a separate 189-gene phylogeny (Burki et al., 2010). The close relationship between Radiolaria and Foraminifera was strongly supported in their phylogeny: in fact, Radiolaria were nested within Foraminifera because of too few taxonomic samples from Radiolaria. Ishitani et al. (2011) collected SSU and LSU rRNAs from a diverse group of Radiolaria, and confirmed the sister relationship between Radiolaria and Foraminifera (Figure 4.2). They also identified several variants of radiolarian actin, alpha-tubulin, and beta-tubulin, which are large molecules involved in the construction of the microfilaments and microtubules that play a major role in cell structure and shape. They constructed a phylogeny using this information (Figure 4.3) which suggests that divergent foraminiferal beta-tubulin molecules (Habura et al., 2005, Bassen et al., 2016) originated from the common ancestor of Retaria. Radiolarians also have a similar structure called the "stereoplasmic rod" (Cachon and Cachon, 1971), suggesting

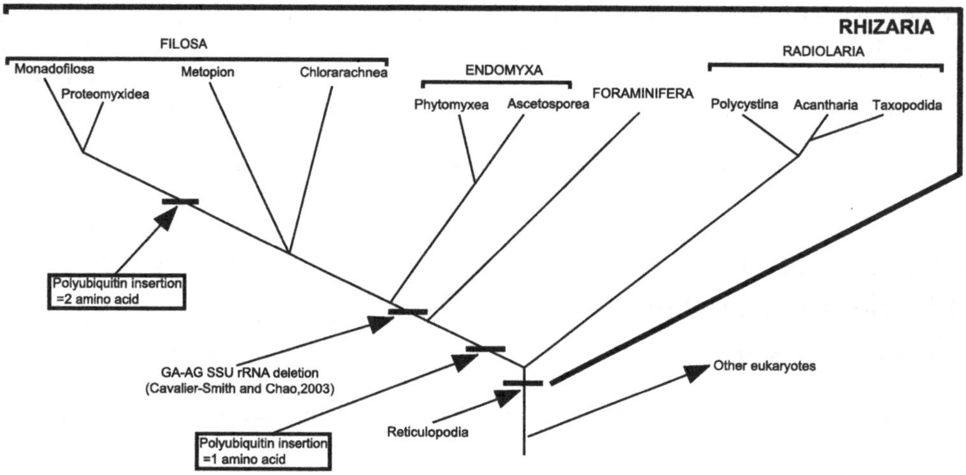

Figure 4.1 Schematic tree showing relative phylogenetic positions of the taxa under study as jointly informed by SSU rRNA phylogenies and the partitioning suggested by the distribution of presence/absence, and the size, of the polyubiquitin insertion. simplified after Bass et al., 2005. Reproduced with permission of Elsevier.

a synapomorphy (shared character that characterizes a group) of Retaria (Ishitani et al., 2011; Hou et al., 2013). To date, the monophyly of Retaria is strongly supported by various multigene studies. However, the phylogenetic relationship between Radiolaria and Foraminifera within the Retaria is still debated (Foraminifera are derived from Radiolaria: Sierra et al., 2015; Foraminifera are sister to Radiolaria: Krabberød et al., 2017) because of incomplete gene sampling coverage or still developing phylogenetic methodology. Further multigene phylogenies with completely sampled genes will presumably resolve the phylogenetic relationships within Retaria.

Relationships of Radiolarian Clades

Since the time of Haeckel (1866), Radiolaria have been considered to be composed of Acantharia, Polycystina (Spumellaria and Nassellaria), and Phaeodaria based on their common morphology of radial pseudopodia and mineralized shells. However, Acantharia are the only group possessing celestite shells (others having silica shells) and they also have a specific cell feature called the myoneme (a contractile organelle surrounding spines). These morphological differences were considered to be sufficient to exclude Acantharia from Radiolaria in more recent taxonomic systems (e.g. Riedel, 1967; Anderson, 1983). Because of limited available information from the cell structure of these organisms, it had not been possible to robustly determine the phylogenetic relationships of Radiolaria, and many scientists explored this issue using radiolarian SSU rRNA, e.g. Amaral-Zettler et al. (1997) and others. These SSU rRNA studies indicated that Acantharia is in fact within the clade Radiolaria (Lopez-Garcia et al., 2002) but that Phaeodaria is external to it (being at most within the common higher clade of Filosa) (Polet et al., 2004). Intriguingly, Nikolaev et al. (2004) suggested that *Sticholonche* (Taxopodida) also belongs to

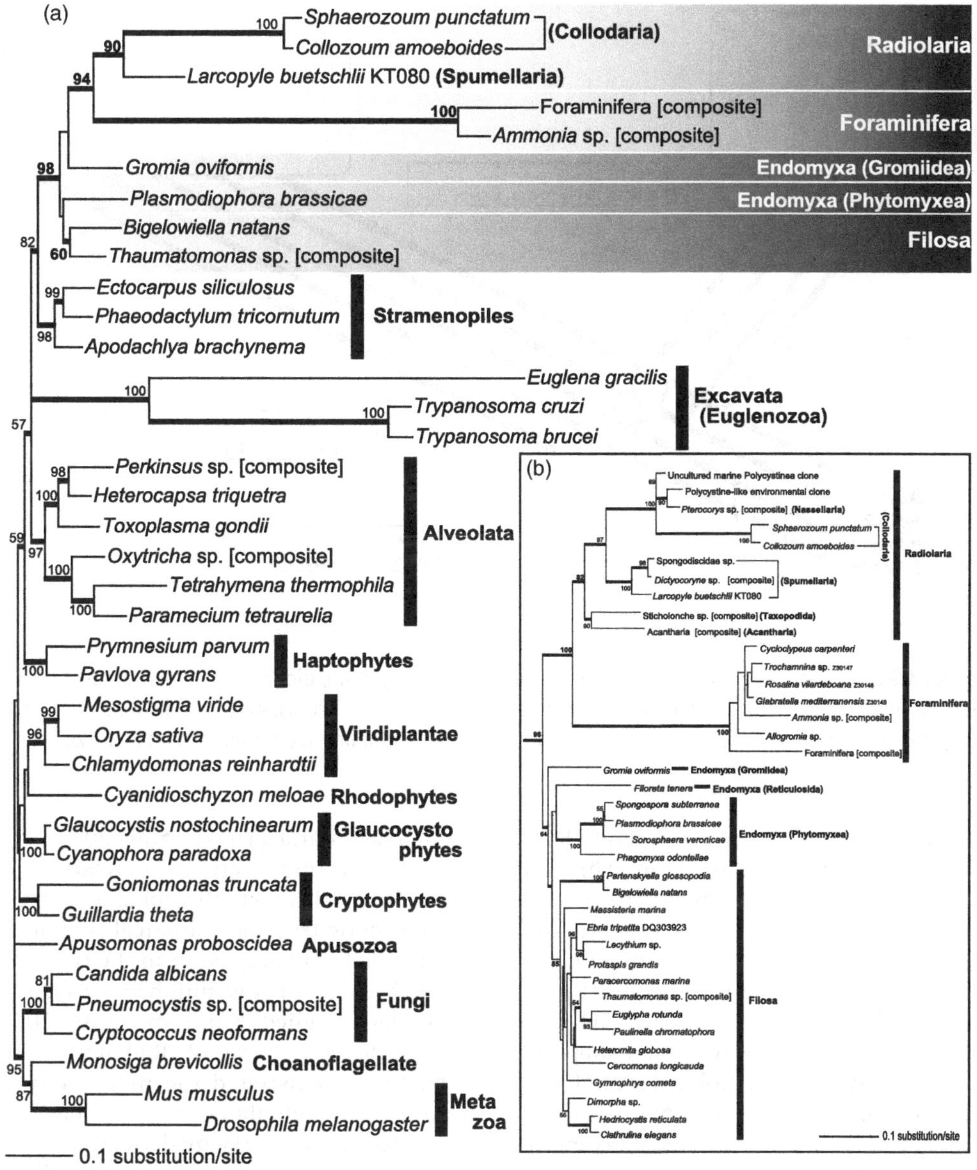

Figure 4.2 (a) Maximum-likelihood (ML) phylogeny of a broad range of eukaryotes based on concatenation of SSU rRNA and LSU rRNA (38 taxa, 3050 sites) reconstructed under separate model conditions. Rhizarian members are shaded. (b) The rhizarian clade in the maximum-likelihood phylogeny of a broad range of eukaryotes based on concatenation of SSU rRNA and LSU rRNA (68 taxa, 3050 sites) under concatenate model conditions. Bootstrap values (BV) over 50% are shown. Thick branches represent the bipartitions supported by over 0.90 Bayesian posterior probabilities (PP) Source: Ishitani et al., Mar. Micro., 2011. Reproduced with permission of Elsevier.

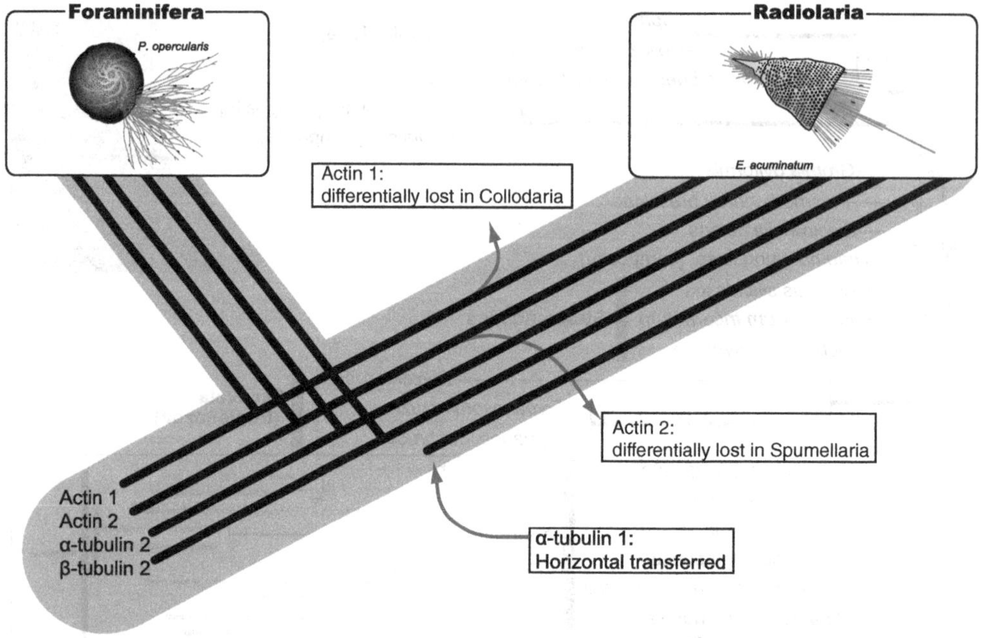

Figure 4.3 Schematic evolutionary patterns of genes encoding actin, alpha-tubulin, and beta-tubulin of Retaria, inferring from the phylogenies of each gene in Ishitani et al., (2011). Arrows indicate that horizontal gene transfer and differential loss arose in radiolarian lineage (inferred from Ishitani et al., 2011).

Radiolaria based on the phylogeny of SSU rRNA and gene-encoding actin. This is in contrast to Cavalier-Smith (1993) who grouped Phaeodaria with Acantharia, because of shared common morphological feature of myoneme and defined the group so created as the Spasmaria. The SSU rRNA phylogenies have however been supported in later studies of the molecular phylogeny of Radiolaria (Suzuki and Not, 2015). In modern community consensus taxonomic systems Radiolaria is composed of Acantharia, Polycystina, and Taxopodida (Adl et al., 2012).

Phylogenetic relationships between the Radiolarian orders is still not entirely clear. Takahashi et al. (2004) were the first to report on the SSU rRNA of shell-bearing Spumellaria, and suggested that they were closely related with Acantharia, but not with Collodaria. SSU rRNA information from Nassellaria (Yuasa et al. 2005) suggested that they were close relatives of Collodaria, although the topology of the phylogenetic tree and statistical support of each node (branching point in the tree) was weak because of low numbers of taxa in each clade studied. Kunitomo et al. (2006) added 19 more taxa of Radiolaria, and robustly demonstrated the monophyly of Collodaria, Spumellaria, and Acantharia. In their study, Radiolaria are divided into two subclades: one consisting of Nassellaria and Collodaria, and another consisting of Acantharia plus the subclade Spumellaria and Taxopodida. However, the node consisting of Acantharia, Spumellaria, and Taxopodida was weakly supported in the statistical analyses (Figure 4.4). Even though taxonomic sampling has been further increased (e.g., Ishitani et al., 2012), the phylogenetic positions of Acantharia,

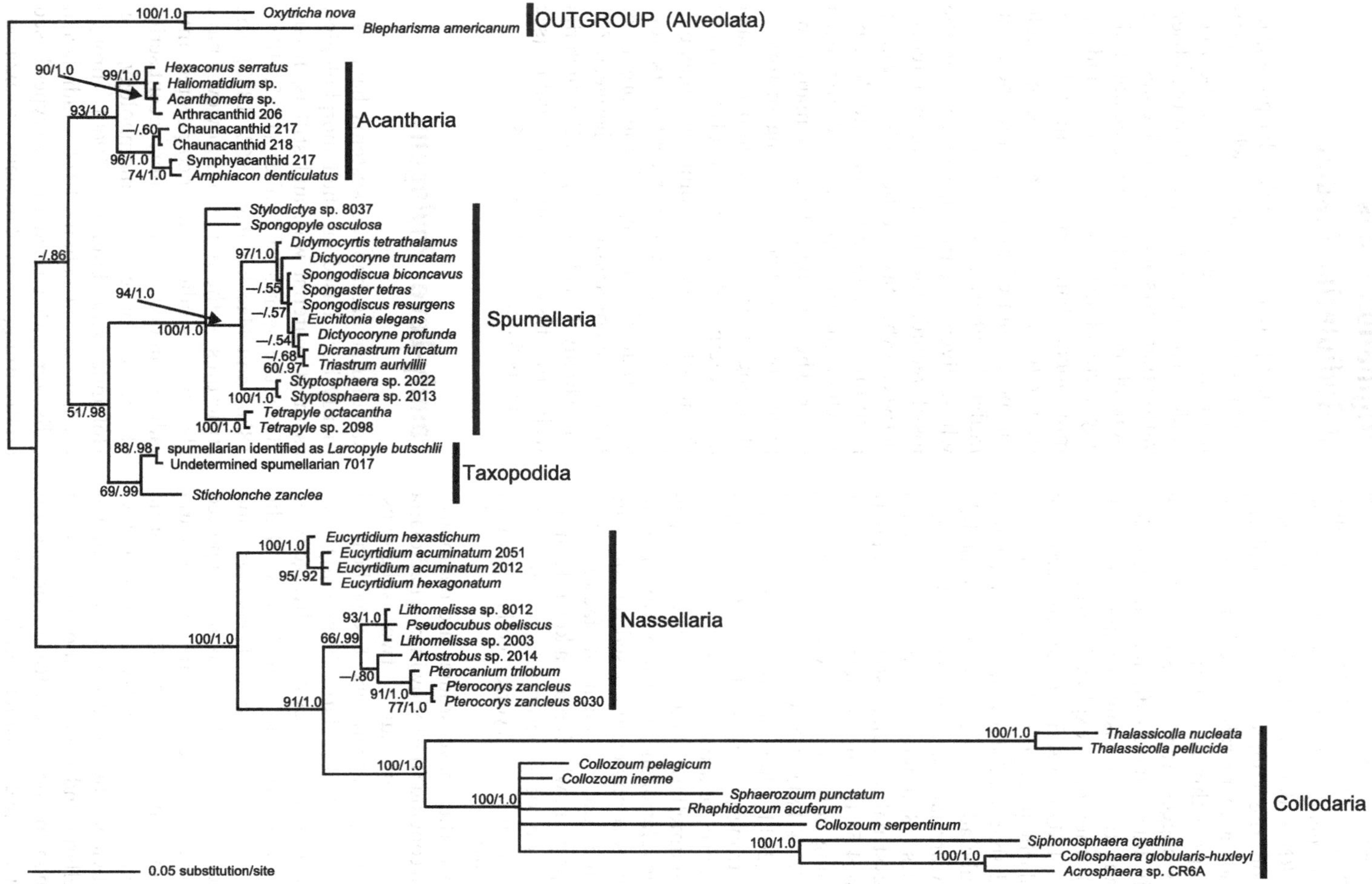

Figure 4.4 Bayesian phylogeny of radiolarians based on SSU rRNA (49 taxa, 963 sites). Numbers at nodes present PP over 0.50 and BV over 50% from ML analyses. Source: simplified after Kunitimo et al., 2006. Reproduced with permission of Elsevier.

Taxopodida, and Spumellaria were still uncertain in SSU rRNA-based phylogenies.

Thus, even with a large number of taxa, SSU rRNA data, although able to shed light on the composition of Radiolaria, was not able to reliably determine relationships within Radiolaria. These results clearly suggest that the SSU rRNA gene simply did not have sufficient information relevant to this issue. Krabberød et al. (2011) therefore explored using other genes – LSU rRNA from a diverse group of Radiolaria (except Collodaria) –and suggested that Radiolaria were divided into two subclades: one consists of Taxopodida and Acantharia (Spasmaria); and another includes Spumellaria and Nassellaria (Polycystina) with high statistical support (Figure 4.5). In spite of their phylogeny not including collodarians, by concatenating the SSU + LSU rRNA data from Ishitani et al. (2011) they obtained a phylogeny that strongly suggested that Collodaria were closer relatives of Nassellaria rather than Spumellaria (Figure 4.2). This phylogenetic relationship was also strongly supported by a phylogeny based on the gene encoding the "elongation factor like" (EFL) protein, which plays a key role in the translation of genetic information into proteins in eukaryotic cells (Ishitani et al., 2012).

When the above studies are combined, we can conclude the following: (1) Radiolaria is a clade including two subgroups: Spasmaria (Acantharia and Taxopodida) and Polycystina (Spumellaria, Nassellaria, and Collodaria); (2) Polycystine phylogeny is that Spumellaria branch at the base of the clade, with monophyly of Nassellaria and Collodaria. This phylogeny is also supported by morphological characteristics: Spasmaria have spasmin-like myonemes; Polycystina have siliceous shells and two cytoplasm layers divided by a capsular membrane; Nassellaria and (some) Collodaria share a common morphological feature of triangular spicules (e.g. Cavalier-Smith, 1993).

Origination Times of Radiolarian Clades

Another potential of molecular phylogeny is molecular clock analysis, which can estimate the timing of divergence between clades. The method is based on the observation that parts of certain molecules – sections of molecules thought to be not under any form of natural selection – undergo mutational change at approximately constant rates. If the rate for a given molecule can be calibrated by comparing the accumulated differences between two taxa whose divergence time is well dated in the fossil record, the cumulative differences between other taxa in the molecular composition can be used to calculate the time at which the two taxa diverged from each other. The method is beset by many complications and there is a long history of changing estimates for a given pair of taxa as the methodologies have evolved, and as the fossil calibration points are updated. With this as a caution, recent work suggests that the timing of the origin of Polycystina or Spasmaria was between 700 and 540 my (Decelle et al., 2012), which fits well to fossil occurrence data (see Chapters 5 and 9 and Figure 4.6).

Family-Level Phylogeny

Radiolarian taxonomic systems have primarily been based on shell morphologies, as most material was, and still is, fossil. Since the works of Müller (1855, 1857, 1858) and Haeckel (e.g. 1862, 1887), the structures of the living cell however have also been studied and incorporated into radiolarian taxonomy. In more recent series of studies of cellular morphology, e.g. Hollande and Enjumet (1960) through Cachon and Cachon (1972), radiolarian cells were categorized into five types based on the arrangement of nucleus and axoplast

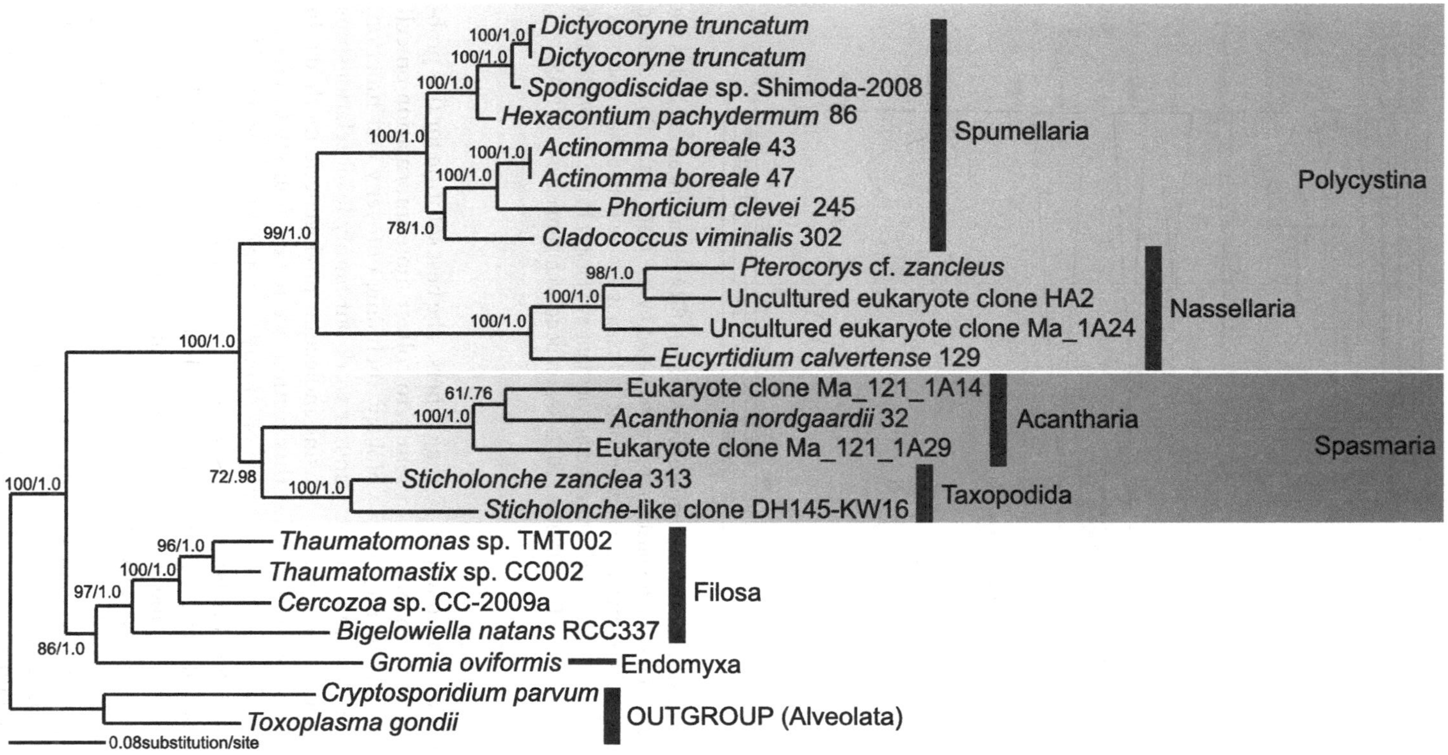

Figure 4.5 ML phylogeny of Radiolaria inferred from an SSU + LSU rDNA (24 taxa, 4163 sites). Values at nodes represent BV from ML analysis and Bayesian PP. Source: simplified after Kraberrød et al., 2011; Licensed under CC BY 4.0. Radiolarian members are shaded.

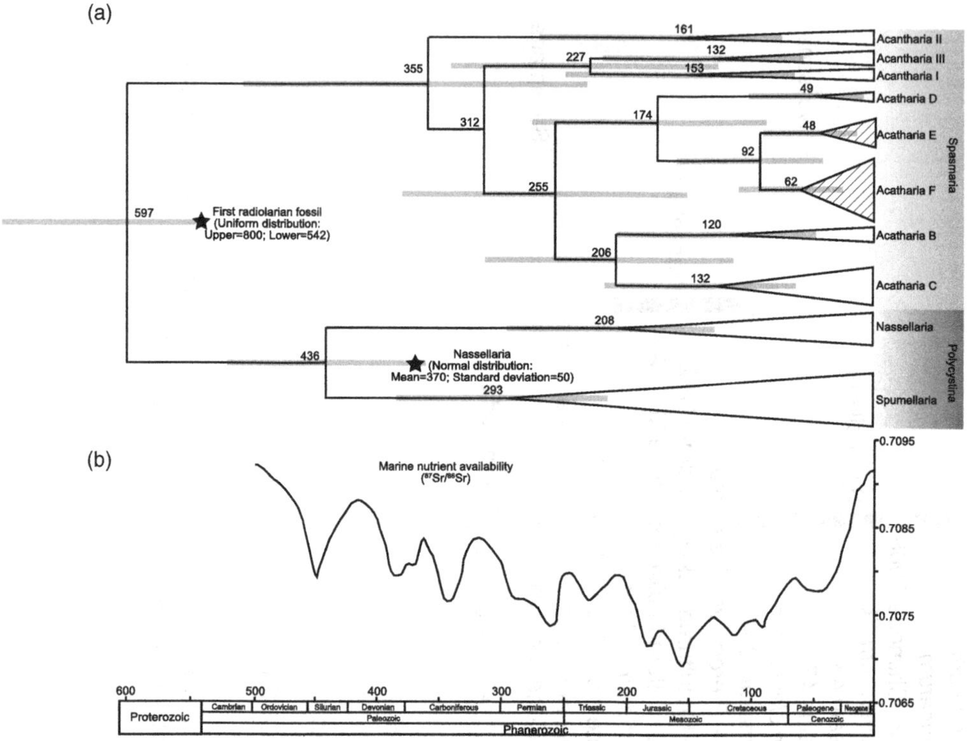

Figure 4.6 (a) Divergence time estimation of Radiolaria based on SSU rRNA. Numbers on the nodes are indicated estimated divergence time. Time constraints are indicated by the black stars. Dark-shaded bars around nodes are indicated 95% credible intervals. Hatched clades are indicated the symbionts-bearing acantharians. (b) Ratio values of $^{87}Sr/^{86}Sr$, reflecting nutrient availability in the ocean, through geological time. Source: simplified after Decelle et al., 2012; Reproduced with permission of PNAS.

(the central meeting point of axopodial filaments within the cell, see Chapter 2: Biology). Based on these cellular morphological differences and initial spicule structures, De Wever et al. (2001) proposed a family level taxonomy based on characters of the innermost shell (in Spumellaria), which tends, in the limited data known so far, to change morphology depending on nucleus – and axoplast – shapes and locations. However, as discussed more fully in Chapter 5 (Taxonomy), current understanding of radiolarian biology is insufficient to *a priori* choose particular characters as more important for taxonomic purposes than others. The best solution for this problem is to collect more information on cell biology, including ontogeny and function, and/or work out family level relationships using molecular phylogenetic methods. In this section, we review four studies focusing on family level taxonomy: Spumellaria (Ishitani et al., 2012), Collodaria (Biard et al., 2015), Nassellaria (Sandin et al., 2019) and, as suitable studies of Polycystina are so rare, a recent study of Acantharia (Decelle et al., 2012). These studies are first attempts to clarify family level taxonomy and phylogeny of Radiolaria, but still do not cover more than a small fraction of the diversity of Radiolaria. Further accumulation of radiolarian molecular and cellular information in future studies is still urgently needed to improve phylogenetic understanding.

Spumellaria (Shell-Bearing Radiolarians)

Since the work of Takahashi et al. (2004) several spumellarian SSU rRNA studies have been performed, attempting to clarify phylogenetic relationships within Spumellaria. Yuasa et al. (2009) collected SSU rRNA information from Hexalonchidae, which is classified quite differently by two different morphological phylogenies, one proposed by Haeckel (1881) and Riedel (1967) and the other by De Wever et al. (2001). Both Haeckel and Riedel placed Hexalonchidae into Spumellaria, based on characteristics of the outer shells. However, De Wever et al. (2001) used the inner shell characters to place Hexalonchidae into Entactinaria, which is a clade sister to, but not in Spumellaria. The SSU rRNA phylogeny clearly suggests that Hexalonchidae is nested within other spumellarians, and rejects the "Entactinaria" hypothesis. In a later study, Ishitani et al. (2012) collected SSU rRNA data mainly from Spongodiscidae. This family is defined almost entirely on the presence of a spongy external shell which, being optically opaque, has prevented any study of the group's internal structures. They suggested Spongodiscidae can be divided into two subclades (Figure 4.7). Moreover, they also inferred based on their results that Spumellaria as a whole can be divided into two subclades (Spumellaria I & II). Spumellaria I were mostly cold-water forms while Spumellaria II were predominantly warm-water, which, if confirmed by additional data, might suggest ecologic preferences for higher taxa groups. Although SSU rRNA data from seven spumellarian families has been collected to date, this may still only be a fraction of the living families of Spumellaria (7 familes in this book, but 27 in Anderson (1983) or 21 in De Wever et al. (2001)). In the most recent study (Sandin et al., 2020 preprint) the subdivision of Spumellaria into new groups, not based on either the Riedel (1967) or De Wever (2001) family concepts is supported. Data is however still too limited to propose a robust new classification (see also Chapter 5: Taxonomy and Fossil Record).

Collodaria (Colonial or Naked Radiolarians)

The taxonomy of collodarians has long been a challenge. While most described species are in the subclade Collosphaeridae and have shells, these shells are relatively simple in structure and have no obvious shared characters with other polycystine radiolarians. Other collodarians do not even possess a real shell but instead have only loose, dispersed, morphologically simple triaxonic spicules, which look only somewhat similar to, but are not easily homologized with nassellarian primary spicules. The collodarians have, since Haeckel, been grouped together into a single clade, and placed within the Spumellaria, mostly because of their spherical cell shape and the lack of any characters that would either assign them to Nassellaria or distinguish them as a sister clade to either Nassellaria or Spumellaria. Because of the limited information available from shells, collodarians need to be classified at least in part based on cell structure. However, current cell structure-based studies are still not adequate to cover most species of collodarians (e.g. Anderson, 1976). Therefore, Collodaria represent one of the least analyzed groups in radiolarian morphological phylogeny.

Amaral-Zettler et al. (1999), in a pioneering study using SSU rRNA, proposed that Collodaria should be divided into three subclades (Thalassicollidae, Sphaerozoidae, and Collosphaeridae), and suggested that the colonial lifestyle arose after the divergence of Thalassicollidae. Ishitani et al. (2012) suggested collodarians could be divided into four families (Thalassicollidae, Sphaerozoidae, Collosphaeridae, and

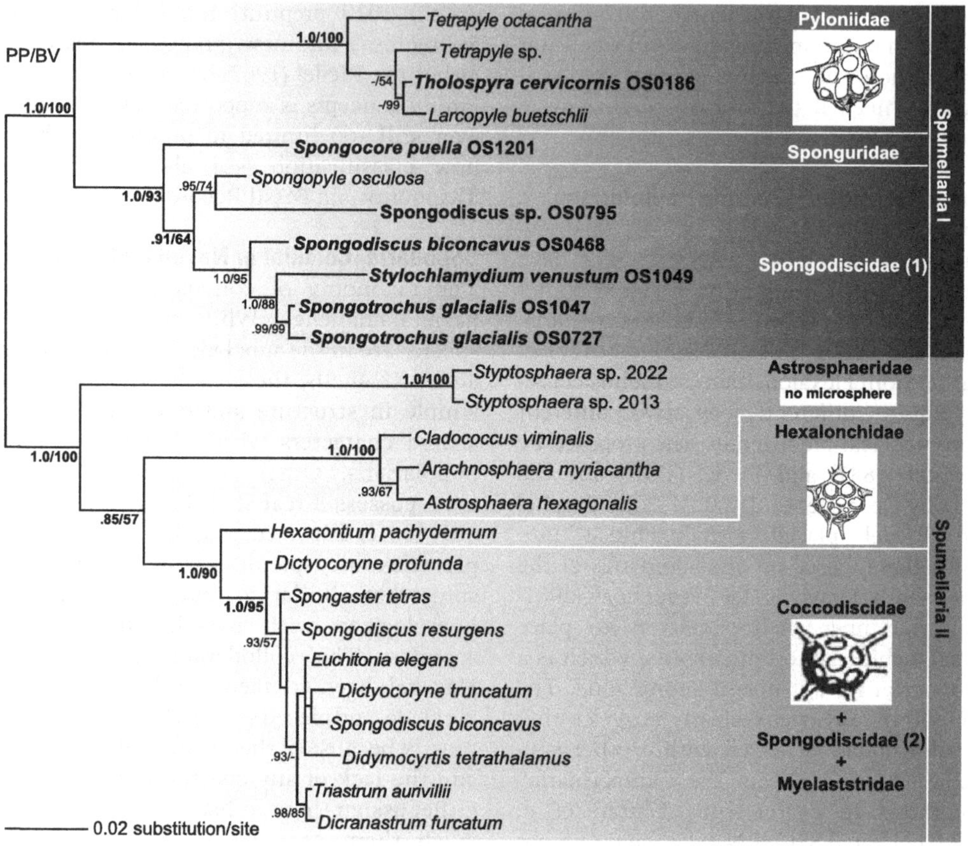

Figure 4.7 Bayesian phylogeny of spumellarians based on SSU rDNA (26 taxa, 1365 sites). A schematic inner structure simplified from De Wever et al. (2001) is below the family name. Numbers at nodes present Bayesian PP over 0.50 and BV from the ML analysis over 50%. Source: simplified after Ishitani et al., Deep Sea Res., 2012. Reproduced with permission of Elsevier.

Collophidae) based on a SSU rRNA phylogeny combined with a consensus of morphological traits from the nucleus and central capsule (e.g. Anderson et al., 1999). However, increasing taxonomic sampling and molecular data (including partial LSU rRNA) suggest Thalassicollidae is not a coherent clade, with individual specimens assignable to all three of the other groups (Biard et al., 2015). Collodarian systematics is thus still unclear; in particular, it is still hampered by the difficulty in identifying taxonomically informative characters from their morphological traits.

Nassellaria

Of all the major living clades of Radiolaria, the Nassellaria have by far the most detailed taxonomic framework, as their morphologies are rich in characters, yet geometrically easier to interpret than is true for many Spumellaria. Family level classification of Nassellaria was first really established by Haeckel (1881; 1887) and revised by Riedel (1967), both workers referring to living and Cenozoic fossil forms. Although Haeckel proposed (purely speculative) higher-level phylogenies and evolutionary scenarios for all forms of Radiolaria, including

Nassellaria, Riedel largely confined the taxonomic treatment to definition of families, with relationships between them left unstated. Mesozoic workers defined a large number of new families for older morphologies, again using the primacy of inner spicule structures, and in De Wever et al (2001) this alternate classification was extended to Cenozoic and living forms (see Chapter 5). Sandin et al. (2019) have provided the first reasonably large-scale molecular systematic analysis of the Nassellarians (Figure 4.8). They find that the older classification of Haeckel and Riedel, which is based on a variety of external and internal characters, largely fits the pattern seen in the molecular results. Relationships between families are however documented for the first time, and include some surprises, such as the inclusion of Cannobotryidae in the Plagiacanthidae. Other results are less surprising but nonetheless not predictable by prior morphologic analyses alone, and as noted by the authors, also not by internal characters. The large clade of Theoperidae in the Riedel classification was defined by possessing a uniform set of simple internal spicular structures, a small, simple cephalis and three, usually more conical, lattice shell segments. Riedel noted that the family was probably polyphyletic, and indeed it is revealed as consisting of at least three, partially quite distantly related clades in the molecular analysis.

Acantharia

The taxonomy of Acantharia has been very little studied, as the strontium sulfate shells of the group do not fossilize, and thus they have not been of any interest to paleontologists, nor have biologists shown much interest in them. Until recently, the last major study of Acantharia and their taxonomy, based on shell characteristics, was published by Schewiakoff (1926). Decelle et al. (2012) renewed study of the group and collected many acantharian SSU and LSU rRNAs. They refined the application of morphologic characters important for higher taxonomy, i.e. the central junction of the spicules, degree of outer shell development, and the development of lateral apophyses on the main spicules (Figure 4.9).They suggested an evolutionary history of Acantharia in which acantharian ancestors had loose central junctions, which became tighter in descendant forms, and in parallel, the development of apophesea or shells. Also, the divergence time estimation for the evolutionary association with symbiotic algae suggested that it might be related to unusually low marine nutrient availability (Decelle et al., 2012; Figure 4.6). It should be noted however that in addition to the high age uncertainty in this study, the Jurassic-mid Cretaceous interval was thought in general to be an interval with gradually increasing, not decreasing, rates of nutrient input vs. the earlier Triassic period (Gibbs et al., 1999). Further, warm global temperatures, low ocean stratification, and low wind stress would have led to mostly mesotrophic conditions in upper ocean waters, not oligotrophic ones (Hay, 2008). This conclusion is supported by widespread deposition of organic carbon-rich sediments, particularly in oceanic anoxic event intervals (OAEs), which are interpreted as indicators of high export productivity (Meyers, 2006).

Microevolution of Radiolaria

Molecular data can also be used to study the genetic structure of single morphospecies and/or gene flow between populations. As most studies of radiolarians are built on the assumption that morphospecies are biologic species, understanding the true relationship between these two types of observational data is an urgent task that as yet is not well studied. Studies began with Amaral-Zettler et al. (1998), who examined the cryptic diversity of *Thalassicolla* (Collodaria). They

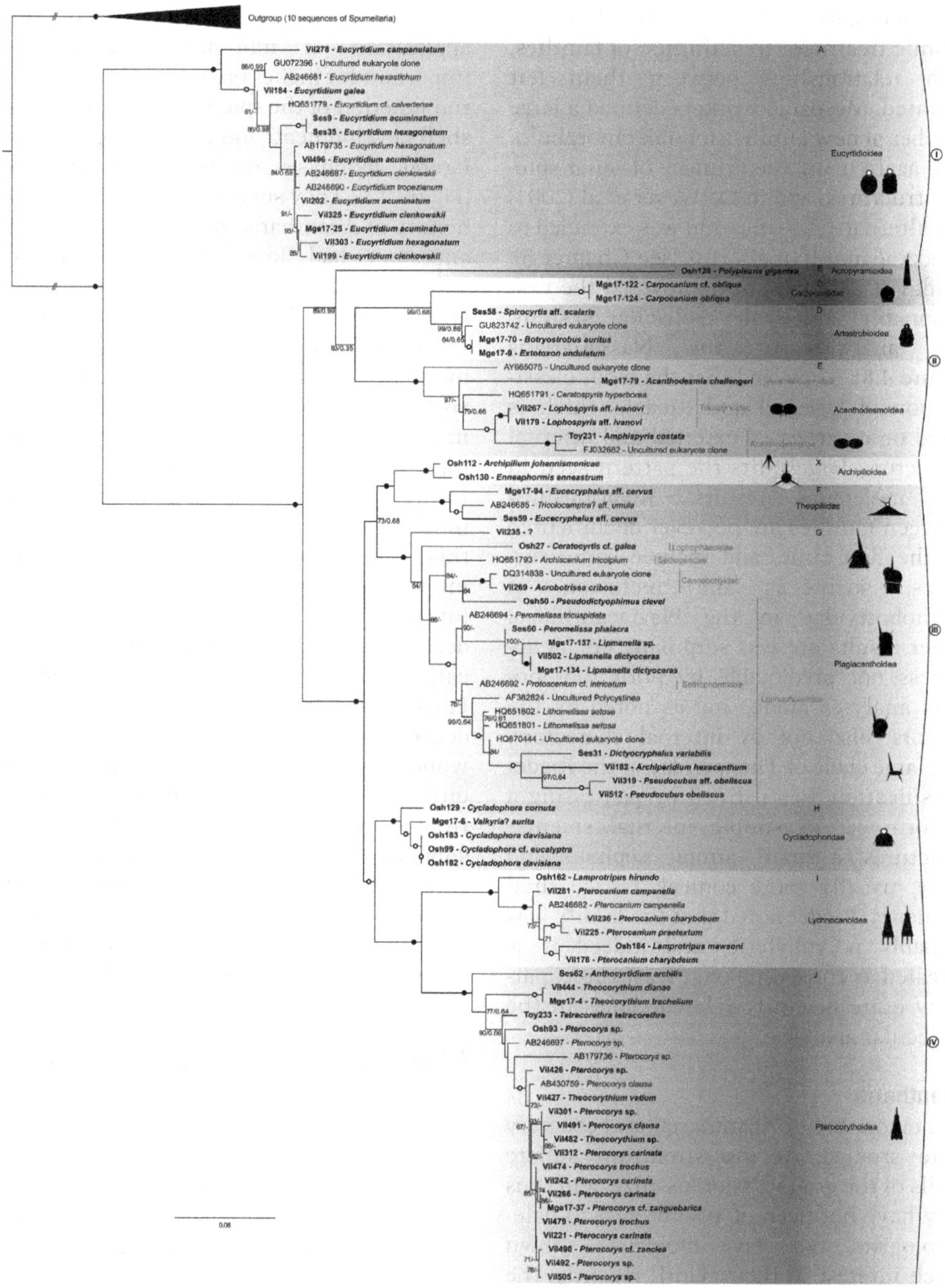

Figure 4.8 Molecular classification of living Nassellaria, based on complete 18S and partial 28S rDNA. Nodes with filled black dots are more robustly supported than those with open white dots. Bold names are sequences from study, others from prior literature. Source: Sandin et al., Protist, 2019. Reproduced with permission of Elsevier.

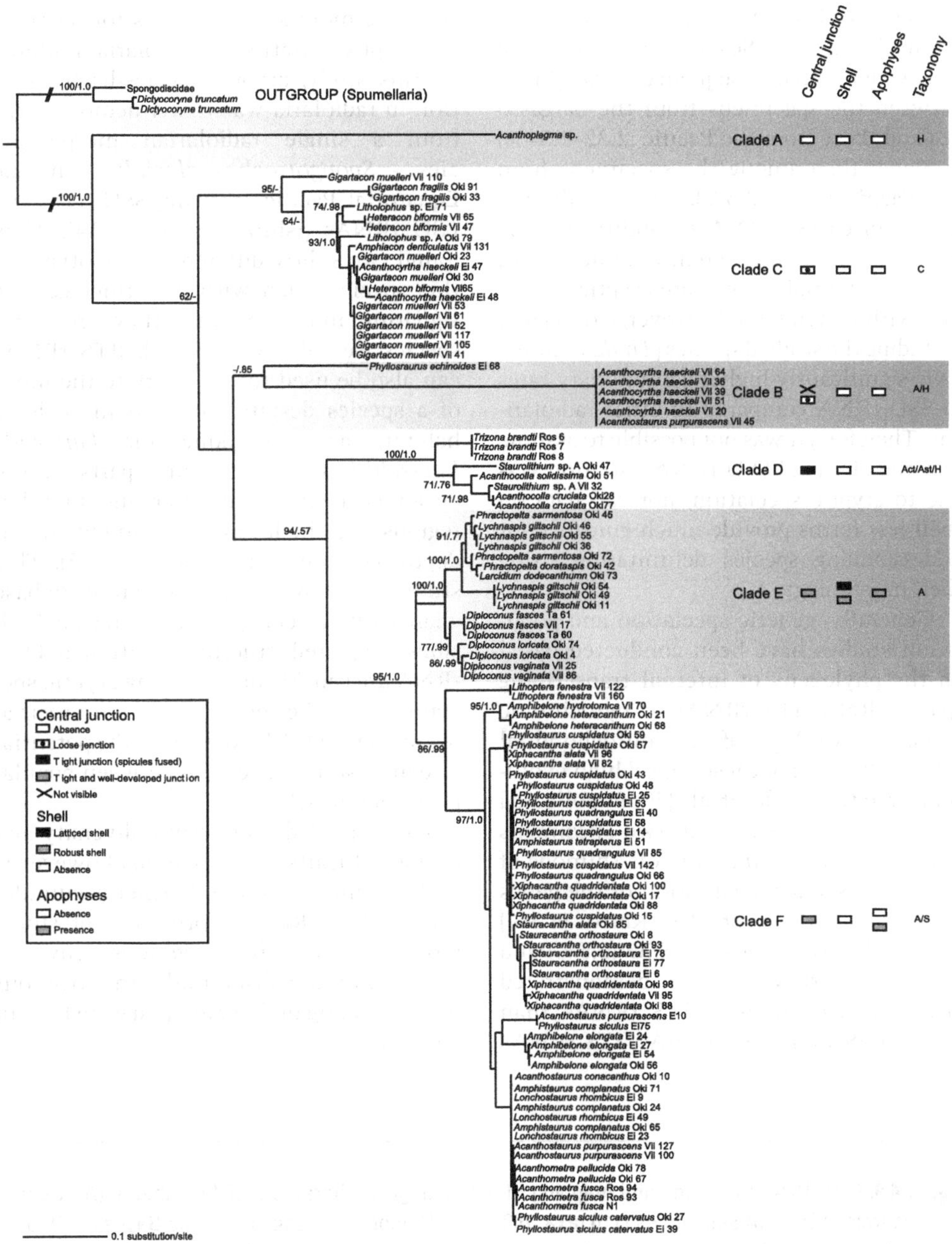

Central junction
Shell
Apophyses
Taxonomy
100/1.0
Spongodiscidae
Dictyocoryne truncatum
Dictyocoryne truncatum
OUTGROUP (Spumellaria)
Acanthoplegma sp.
Clade A
H
100/1.0
95/-
Gigartacon muelleri Vil 110
Gigartacon fragilis Oki 91
Gigartacon fragilis Oki 33
74/.98
Litholophus sp. Ei 71
Heteracon biformis Vil 65
Heteracon biformis Vil 47
64/-
93/1.0
Litholophus sp. A Oki 79
Amphiacon denticulatus Vil 131
Gigartacon muelleri Oki 23
Acanthocyrtha haeckeli Ei 47
Gigartacon muelleri Oki 30
Heteracon biformis VII65
Acanthocyrtha haeckeli Ei 48
Gigartacon muelleri Vil 53
Gigartacon muelleri Vil 61
Gigartacon muelleri Vil 52
Gigartacon muelleri Vil 117
Gigartacon muelleri Vil 105
Gigartacon muelleri Vil 41
Clade C
C
62/-
-/.85
Phyllosraurus echinoides Ei 68
Acanthocyrtha haeckeli Vil 64
Acanthocyrtha haeckeli Vil 36
Acanthocyrtha haeckeli Vil 39
Acanthocyrtha haeckeli Vil 51
Acanthocyrtha haeckeli Vil 20
Acanthostaurus purpurascens Vil 45
Clade B
A/H
100/1.0
Trizona brandti Ros 6
Trizona brandti Ros 7
Trizona brandti Ros 8
Staurolithium sp. A Oki 47
Acanthocolla solidissima Oki 51
71/.76
Staurolithium sp. A Vil 32
71/.98
Acanthocolla cruciata Oki28
Acanthocolla cruciata Oki77
Clade D
Act/Ast/H
94/.57
Phractopelta sarmentosa Oki 45
Lychnaspis giltschii Oki 46
91/.77
Lychnaspis giltschii Oki 55
Lychnaspis giltschii Oki 36
100/1.0
Phractopelta sarmentosa Oki 72
Phractopelta dorataspis Oki 42
Larcidium dodecanthumn Oki 73
100/1.0
Lychnaspis giltschii Oki 54
Lychnaspis giltschii Oki 49
Lychnaspis giltschii Oki 11
Diploconus fasces Ta 61
Diploconus fasces Vil 17
Diploconus fasces Ta 60
77/.99
Diploconus loricata Oki 74
Diploconus loricata Oki 4
86/.99
Diploconus vaginata Vil 25
Diploconus vaginata Vil 86
Clade E
A
95/1.0
Lithoptera fenestra Vil 122
Lithoptera fenestra Vil 160
95/1.0
Amphibelone hydrotomica Vil 70
Amphibelone heteracanthum Oki 21
Amphibelone heteracanthum Oki 68
86/.99
Phyllostaurus cuspidatus Oki 59
Phyllostaurus cuspidatus Oki 57
Xiphacantha alata Vil 96
Xiphacantha alata Vil 82
Phyllostaurus cuspidatus Oki 43
Phyllostaurus cuspidatus Oki 48
Phyllostaurus cuspidatus Ei 25
Phyllostaurus cuspidatus Ei 53
97/1.0
Phyllostaurus quadrangulus Ei 40
Phyllostaurus cuspidatus Ei 58
Phyllostaurus cuspidatus Ei 14
Amphistaurus tetrapterus Ei 51
Phyllostaurus quadrangulus Vil 85
Phyllostaurus cuspidatus Vil 142
Phyllostaurus quadrangulus Oki 66
Xiphacantha quadridentata Oki 100
Xiphacantha quadridentata Oki 17
Xiphacantha quadridentata Oki 88
Phyllostaurus cuspidatus Oki 15
Stauracantha alata Oki 85
Stauracantha orthostaura Oki 8
Stauracantha orthostaura Oki 93
Stauracantha orthostaura Ei 78
Stauracantha orthostaura Ei 77
Stauracantha orthostaura Ei 6
Xiphacantha quadridentata Oki 98
Xiphacantha quadridentata Vil 95
Xiphacantha quadridentata Oki 88
Acanthostaurus purpurascens E10
Phyllostaurus siculus Ei75
Amphibelone elongata Ei 24
Amphibelone elongata Ei 27
Amphibelone elongata Ei 54
Amphibelone elongata Oki 56
Acanthostaurus conacanthus Oki 10
Amphistaurus complanatus Oki 71
Lonchostaurus rhombicus Ei 9
Amphistaurus complanatus Oki 24
Lonchostaurus rhombicus Ei 49
Amphistaurus complanatus Oki 65
Lonchostaurus rhombicus Ei 23
Acanthostaurus purpurascens Vil 127
Acanthostaurus purpurascens Vil 100
Acanthometra pellucida Oki 78
Acanthometra pellucida Oki 67
Acanthometra fusca Ros 94
Acanthometra fusca Ros 93
Acanthometra fusca N1
Phyllostaurus siculus catervatus Oki 27
Phyllostaurus siculus catervatus Ei 39
Clade F
A/S
Central junction
Absence
Loose jenction
T ight junction (spicules fused)
T ight and well-developed junct ion
Not visible
Shell
Latticed shell
Robust shell
Absence
Apophyses
Absence
Presence
0.1 substitution/site

collected SSU rRNA of *Thalassicolla nucleata* in the Sargasso Sea and the Pacific, and discovered that sequence variability between the specimens from the Sargasso Sea and those from the Pacific (2.32–2.54%) is higher than among the specimens from the Pacific (0.45–2.03%). This result, the first reported SSU rRNA variability of single morpho-species in different oceanic basins, has the potential to evaluate cryptic speciation within radiolaria. However, Collodaria, including the studied species (*Thalassicolla*), have significantly higher evolutionary rates in SSU rRNA compared to other radiolarians. Therefore, it was not possible to assume that the observed SSU rRNA variation was due to cryptic speciation, nor can data on shell-less forms provide much guidance for understanding species delimitation using shell morphology.

Generally, genetic speciation and population studies have been conducted based on the phylogeny of internal transcribing spacer rRNA (ITS rRNA) or cytochrome oxidase I (COX I) of the mitochondrial genome because of these genes' high evolutionary rates. Ando et al. (2009) collected ITS rRNA sequences from two specimens of *Dictyocoryne truncatum* off Shimoda of Japan, and sequence variation of the genes showed that those of ITS rRNAs (4.3%) were higher than those of SSU (0.5%) and LSU (0.5%) rRNAs. They demonstrated faster evolving rates of ITS rRNAs than those of SSU and LSU rRNAs. ITS rRNAs could be more suitable genes for identifying cryptic variation of radiolaria. Building on this study, cryptic species-level variation in radiolaria was firstly demonstrated from a single radiolarian morphospecies – *Spongotrochus glacialis* – in the Equatorial Pacific by using SSU, ITS, and LSU rRNAs (Ishitani et al., 2014). They found two subtly different morphotypes in this species, each with a distinct genetic signature indicating that they are sister species (see also Chapter 9). IITS rRNAs can also be used to demonstrate the unity of a species despite variation in ecologic habitat, as was done for *Larcopyle buetschlii*, which in some parts of the ocean is a surface dweller and in other regions (e.g. the Japan Sea) is found only at depth (Ishitani and Takishita, 2015). This study also showed that compensatory base changes (base changes occurring on both sides of a paired structural position) in ITS2 rRNA are capable of describing cryptic speciation, and the mosaic variation of variable regions in ITS1 rRNA have the potential to demonstrate gene flow between populations (Figure 4.10).

Our knowledge of cryptic diversity and ecological traits of radiolarians however is still extremely limited. Further accumulation of ITS rRNA information of single morpho-species from various environments would substantially improve our understanding of diversity and gene flow in radiolaria.

Figure 4.9. LSU rDNA phylogeny and mapping of morphological characters of Acantharia. (A) ML tree inferred from LSU rDNA sequences (110 taxa, 702 sites). Numbers at nodes indicated Bayesian PP over 60% and BV over 0.8 from ML analysis. Branches with double barred symbol are reduced to half length for clarity. (B) Skeleton traits identified for each clade mapped onto the molecular phylogeny: 1) junction of spicules at the cell center, 2) type of the shell, and 3) presence/absence of apophyses, which are small lateral extensions on each spicule. The four columns on the right indicate the associated order for each taxa according to the traditional taxonomy of Schewiakoff (1926): H = Holacanthida; C = Chaunacanthida; A = Arthracanthida; S = Symphiacanthida; and Act/Ast = Actinastra/Astrolophi. Source: Decelle et al., 2012 Reproduced with permission of PNAS.

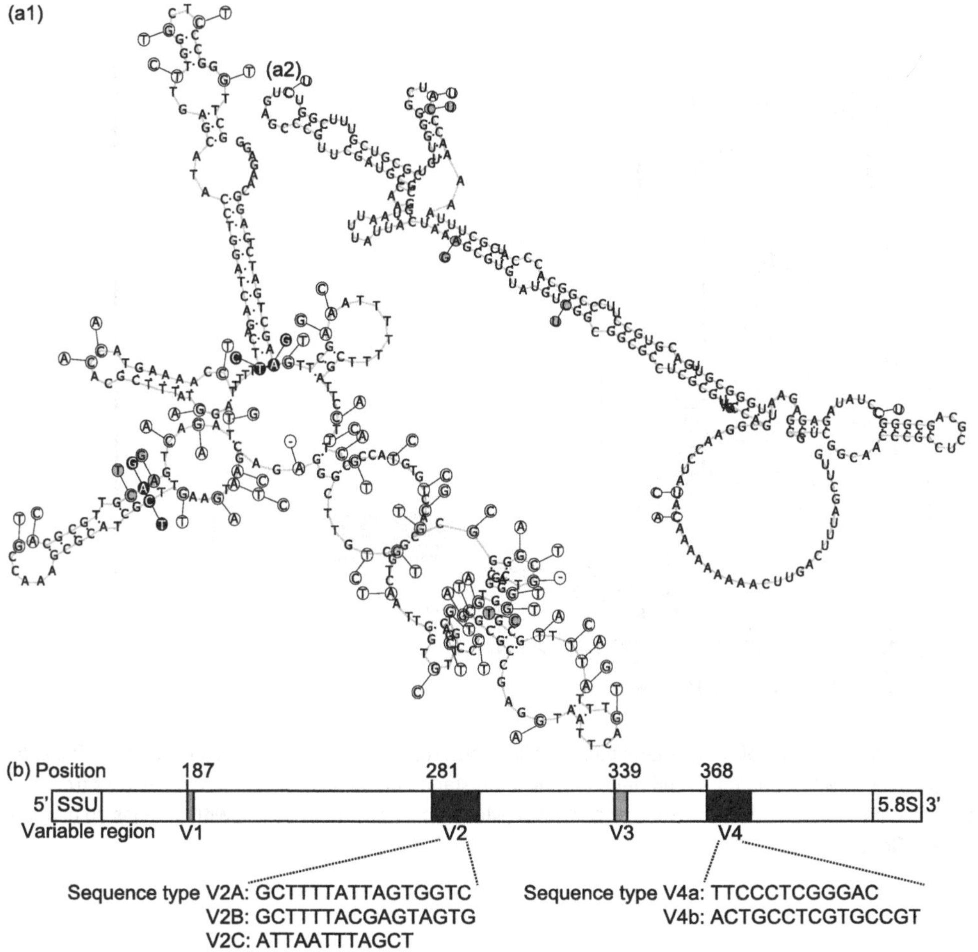

Figure 4.10 (a) Secondary-structure model of ITS2 in (a1) *Spongotrochus glacialis* KH1004_893 and (a2) *Larcopyle buetschlii* KT1006_007_Cl07. (a1) Black circles indicate compensatory base changes in comparison between two sibling species of *S. glacialis.* Gray circles indicate hemi-compensatory base changes, and white circles indicate single base changes. (a2) Gray circles indicate hemi-compensatory base changes in comparison with other individuals of *L. buetschlii*, and white circles indicate single base changes. (b) Schematic of the variation of ITS1 sequences in *Larcopyle buetschlii*. Variable regions are designated V1, V2, V3, and V4. V2 and V4, with more than 12 bp, have sequence types V2A/V2B/V2C and V4a/V4b, respectively. Source: Ishitani, Y. and Takishita, K., 2015. Reproduced with permission of Oxford University Press.

Diversity of Pico-Radiolarian Material

Since the work of Díez et al. (2001), an unexpectedly high diversity of pico- and nannoplankton-sized eukaryotes have been revealed by SSU rRNA surveys from environmental samples from water or soils. Not et al. (2007) collected SSU rRNA from pico-sized organisms (<5 μm) in seawater samples (environmental

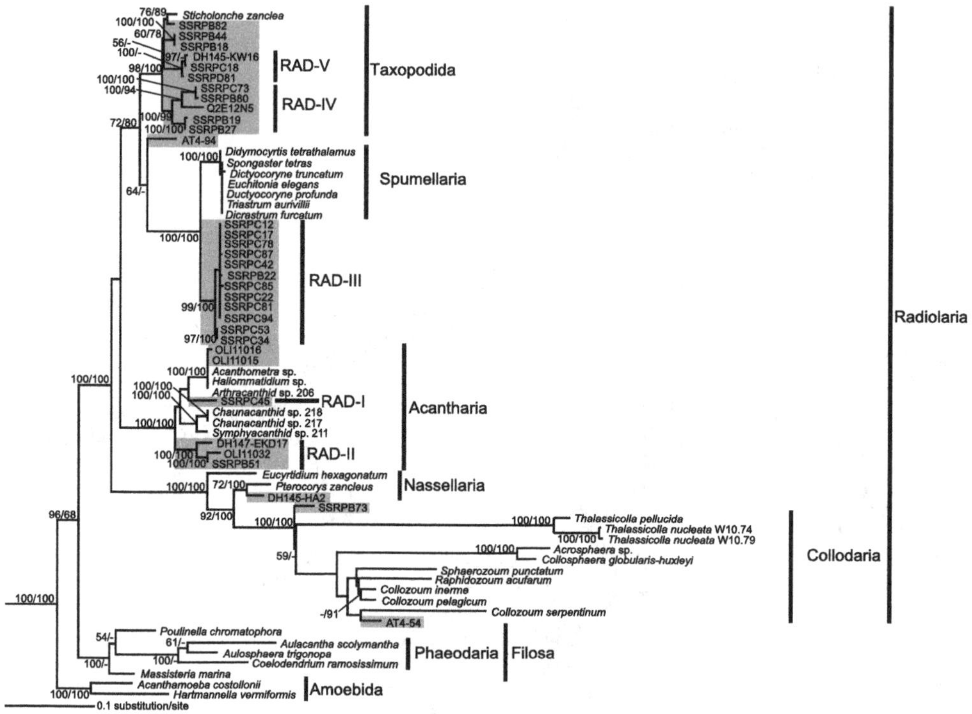

Figure 4.11 ML phylogeny of radiolarians based on SSU rDNA (69 taxa, 1369 sites). Numbers at nodes indicated BV over 50% from both ML and neighbor-joining distance methods. Environmental samples from the water or sediments were shaded. Source: Not, Envir. Microbiol., 2007. Reproduced with permission of John Wiley and Sons.

samples) rather than discrete particulate samples, and explored an unexpected diversity of radiolarians. They identified a new clade unrecognized from molecular studies of larger radiolarians (Figure 4.11). The molecular signature of these organisms are so far known only from SSU rRNA of environmental samples, and it is still uncertain if they are distinct pico-sized organisms or swarmers, or even just free rDNA from dead specimens of larger radiolarians which to date have not been studied by molecular methods and are thus "molecularly unknown." Gilg et al. (2010) indicated that these SSU rRNAs may come from unicellular flagellated organisms (suggesting swarmers) by using the "fluorescence in situ hybridization" (FISH) technique which can yield fluorescence from labeled DNA (Figure 4.12). Further work on the molecular characteristics of isolated radiolarians and FISH analyses of environmental samples will be needed to understand the meaning of such small-size fraction environmental molecular data.

Transcriptomics of Radiolaria

Transcriptomics is the analysis of all RNA in a cell. The RNA is expressed from DNA and is used by the cell later by traduction

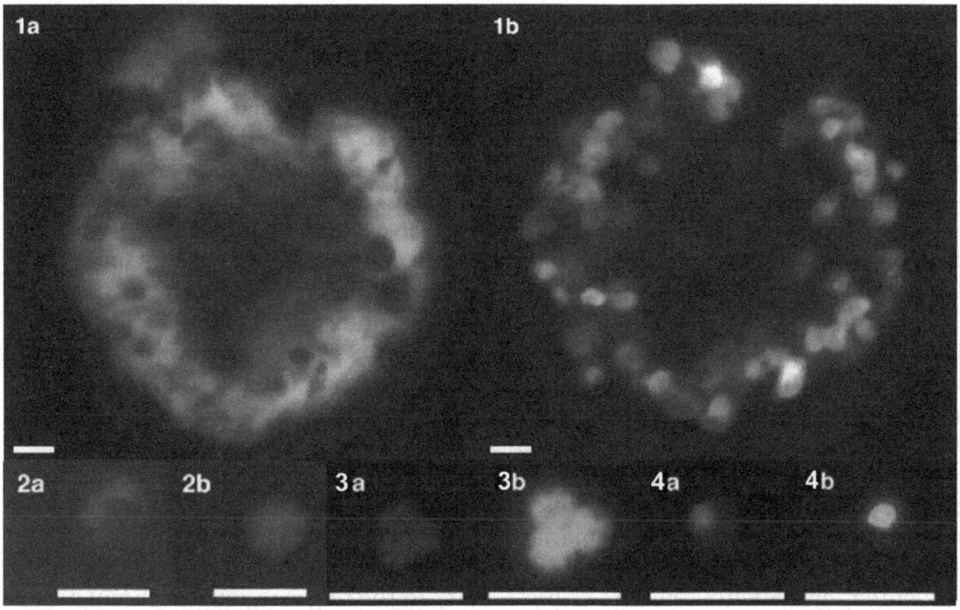

Figure 4.12 Photomicrographs of acantharia and relatives. All scale bars = 10 um. Each "a" image shows a hybridized cell and each "b" image shows the corresponding DAPI image. (1) hybridized with acantharia-specific probe a497. All but one known species of acantharian is polynucleated in the vegetative stage. Skeleton disassociation was observed because Sr was not added to the fixation medium. (2) An acantharia swarmer, targeted with a497. Note the single nucleus at this stage. (3–4) Cells targeted by UC-1-189, a probe specific for unidentified acantharian. Source: Gilg et al., Protist, 2010. Reproduced with permission of Elsevier.

to amino acids to create the cell's proteins. Transcriptomes, a snapshot of expressed DNA content, are useful for understanding characteristic metabolic or evolutionary traits. The first transcriptomic analysis of radiolaria (Balzano et al., 2015) revealed that the genes encoding the protein c-type lectin, which plays various roles in cell functioning, were found only from symbiont-bearing lineages, suggesting that c-type lectin is likely involved in the host/symbiont specific recognition . A transcriptomic analysis of Rhizaria revealed that genetic innovation in cytoskelton genes (alpha-tubulin, beta-tubulin, actin, and actin-related proteins) have shaped cytoskeletal structure in the group (Kraberrød et al., 2017). Transcriptomics has the potential to reveal links between radiolarian morphology and metabolism, and how this has changed through their evolutionary history.

Methodology

Molecular analysis of radiolarians is conducted by basic molecular methods, consisting of four steps: DNA extraction, polymerase chain reaction (PCR), cloning, and sequencing. Radiolaria often have numerous symbionts and generally all have captured organisms as prey (i.e. dinoflagellates, diatoms, and tintinnids) inside the cell (Anderson, 1983). This complex mixed-source composition of the protoplasm makes it difficult to extract only a targeted radiolarian DNA. In this section, we describe three techniques to avoid contamination by symbionts and/or prey organisms: two methods before the DNA extraction and another one to amplify only radiolarian DNA in the PCR step.

DNA Extraction

Radiolarian cells include RNA, DNA, protein, lipid, polysaccharide membrane, and other components. Some substances, e.g. proteins and polysaccharides, influence the later procedures for genetic studies. For the DNA extraction from radiolarian cells, chemical buffers (e.g. Chelex resin, GITC buffer, and alkaline buffer) or proteinase buffers are used to break/remove membranes, proteins, and lipids. These buffers allow extracting DNA from radiolarian cells, including not only radiolarians but also symbionts and prey organisms. Pioneering studies of radiolarian genetics have tried to eliminate these contaminant DNA sources before the extraction step.

Reproductive Cell Method

Many radiolarians either consume or expel their endosymbiotic algae just before swarmer cell formation. In order to reduce the potential of amplifying contaminants, pioneering studies of radiolarian genetics used radiolarian reproductive cells just before the swarmer cell release (Amaral-Zettler et al., 1997). They collected targeted DNA by using this method from Acantharia and Collodaria.

Dissecting Cell Method

Radiolarian cells have two primary cytoplasmic layers bounded by the capsular membrane: the ectoplasm (outside the capsular membrane) including symbionts and digestive vacuoles and the endoplasm (inside the capsular membrane) including the radiolarian nucleus (see Chapter 2: Biology). Amaral-Zettler et al. (1997) dissected the cells of the radiolarian species *Thalassicolla nucleata*, and physically divided the endoplasm and ectoplasm to extract only radiolarian DNA from endoplasm (Figure 4.13). They successfully extracted pure radiolarian DNA from the endoplasm of radiolarians. This technique obtained SSU rDNA information from Collodaria, Spumellaria and Nassellaria.

PCR

The PCR method, one of the most common techniques in biology, can amplify target regions of DNA by using primers (containing 10–30 nucleotides complementary to the target region). Accumulation of SSU rDNA data from radiolarians has allowed the design of specific primers which amplify only targeted sequences. Most of the molecular studies of radiolarians now use this simple and easy method to amplify the targeted radiolarian DNA (Table 4.1).

Summary

The studies summarized in this chapter clearly indicate the enormous potential to improve radiolarian taxonomy, and to provide the foundational understanding of how radiolarian shell morphology is related to key biological concepts such as species. Despite the remarkable advances in recent years, it is also clear that so far only the most preliminary information has been obtained, and that our understanding of the genetics of radiolarians is still in a very early stage of development. It can only be hoped that the next decades bring continued rapid advances in this key area of radiolarian research.

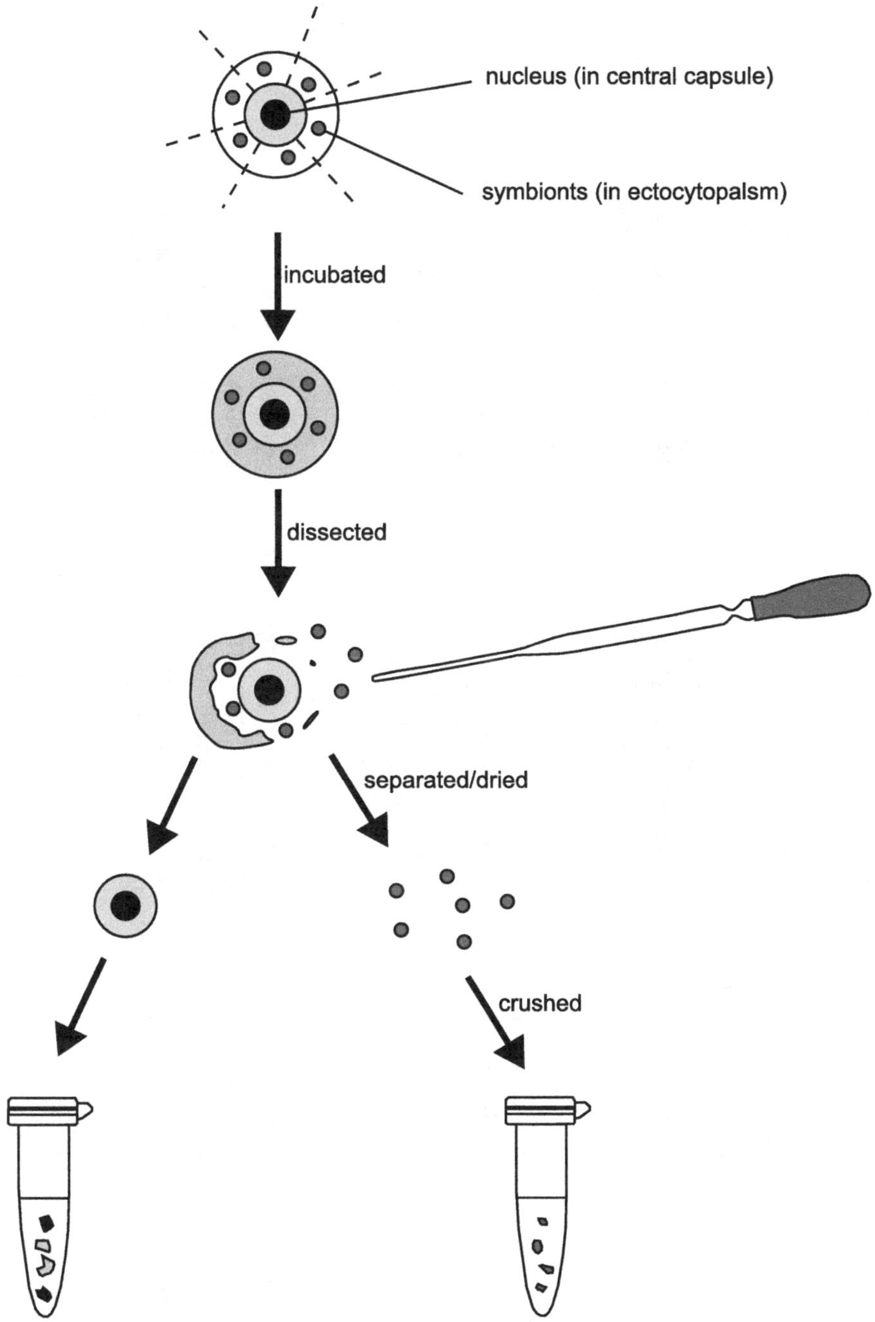

Figure 4.13 Flowchart of the DNA extraction from a single radiolarian cell by using dissecting method
Source: Yuasa et al., 2009.

Table 4.1 List of specific primers for radiolarian taxa.

	Primer Name	Sense	Alignment	Gene	Specificity
Amaral-Zettler et al., 1999	R906	F	TAT TAG TAT TTT RTC GTT	SSU rDNA	Collodaria
	R1451bio	R	TAT TGT AGC CCG TGC GCT	SSU rDNA	Collodaria
Amaral-Zettler and Caron., 2000	A497	F	GTG AGT TGA TTG GAA TGA	SSU rDNA	Acantharia
	A899	R	TCG TCA TAC AAA GGT CCA	SSU rDNA	Acantharia
Polet et al., 2004	s12pc	F	AGR TAT TAR TAT TTT RKC G	SSU rDNA	Polycystina
	s12.2pcr	R	CCT AGC TGA TAT AGT TTA CAG	SSU rDNA	*Collozoum inerme*
	s15.1pc	R	CAK CCA CAA AAT CAG GAC AGG	SSU rDNA	*Thalassicolla pellucida*
	s17.1pc	R	AAT GTA CTA TTT AGT AAG TGA TAG	SSU rDNA	*Thalassophysa pelagica*
Yuasa et al., 2004	Spu-191F	F	GCG ACT YAC GAA GCC CTG TA	SSU rDNA	Spumellaria
	Spu 1731R	R	ACT TCG RGC CTC CCG TT	SSU rDNA	Spumellaria
Ando et al., 2009	18S-Spumellaria	F	AAT ACR TGC ACK AAG GGG CGA	SSU rDNA	Spumellaria
	18S-Nassellaria	F	GCG TAT CAT TCA AGT TTC TGA	SSU rDNA	Nassellaria
	ITS-a3	R	TCA CCA TCT TTC GGG TCC CAA CA	LSU rDNA	Cercozoa
	ITS-18S-RadI	F	ACC GCC CGT CGC TCC TAC CG	SSU rDNA	Radiolaria
	28S-Rad2	F	TAA GCG GAG GAA AAG AAA	LSU rDNA	Radiolaria
	ITS-28S-RadI	R	CCC TCA CGG TAC TTG TTC GC	LSU rDNA	Radiolaria
	18S-riv-RadI	R	CCA CCA ACT AAG AAC GGC CA	SSU rDNA	Radiolaria
	18S-5.8-Rad2	R	GCG TTC TTC ATC GTT GCG	5.8S rDNA	Radiolaria
Krabberød et al. 2011	3180R	R	GGG TAA AAC TAA CCT GTC TCA CGA CGG TC	LSU rDNA	Rad?
	28S_Rad3_R	R	CGG TCT TCA AAG TTT TCA TTT G	LSU rDNA	Rad?
	Sticho1	F	TAC ATG CAC GAA GGT CCA AC	LSU rDNA	Sticho?
Ishitani et al., 2012a	S32TAS	F	CCA GCT CCA ATA GCG TAT RC	SSU rDNA	Taxopodida+Acantharia+Spumellaria
	S51TAS	R	YAA GAA TTT CAC CTC TCG CTT	SSU rDNA	Taxopodida+Acantharia+Spumellaria
Ishitani et al., 2012b	S50NC	F	GGA AGG GCA CCA CA	SSU rDNA	Nassellaria+Collodaria
	S81NC	R	TCA CAG ACC TGT TAT TAT TGC HW	SSU rDNA	Nassellaria+Collodaria
Decelle et al., 2012	S32 J	F	CCA GCT CCA ATA GCG TAT AC	SSU rDNA	Radiolaria
	S879	R	CCA ACT GTC CCT ATC AAT CAT	SSU rDNA	Radiolaria

5 Taxonomy and Fossil Record

Introduction

In this chapter we cover two major subjects which are conveniently presented together: in Part 1 the taxonomy of radiolarians, and in Part 2 the basic patterns of their occurrence over time in the fossil record.

Taxonomy is a branch of biologic science concerned with the formal description of the evolutionary structure of living things above the level of the individual. Our understanding of radiolarian diversity, stratigraphy, geographic distribution of species, and almost all other subjects is built on the basis of radiolarian taxonomy, which defines both species and higher level taxa. The goal of this chapter is to present an overview of the theoretical principles and major features of radiolarian taxonomy in enough detail that the reader can read and understand the primary taxonomic literature. It is not meant to be an exhaustive review of radiolarian taxonomy at any particular taxonomic level. To some extent recent monographic publications exist, although coverage is incomplete, and the content still in part very provisional (e.g. all Phanerozoic families – De Wever et al., 2001; O'Dogherty et al., 2009 and Caridroit et al., 2017 for Mesozoic and Paleozoic genera, and various smaller catalogs for species). Although most readers are assumed to have had at least some prior knowledge

Paleobiology of the Polycystine Radiolaria, First Edition. David Lazarus, Noritoshi Suzuki, Yoshiyuki Ishitani and Kozo Takahashi.

of biologic taxonomy, short summaries of essential principles are given. The descriptive section provides a short summary of the morphology and characteristics of a selection of species from Cenozoic families of radiolarians, together with a smaller number of Mesozoic and Paleozoic forms. A focus on Cenozoic taxa reflects not only the interests of the majority of the authors, but also the reality that the majority of paleobiologic studies of radiolarians have been done on Cenozoic materials. Only shell morphology is considered, as this is the only source of information available for fossil material; to date, there has been only one, very restricted, attempt to use protoplasmic structures in radiolarian taxonomy (Hollande and Enjumet, 1960) which has not been widely adopted by subsequent workers.

The distribution of radiolarians over geologic time is summarized in the last part of this chapter. Together, the taxonomy and geologic distribution of radiolara provides the observational data needed to understand how radiolarians have been used to understand the biologic and geologic processes discussed in the last third of this volume (paleoceanography, stratigraphy, and evolution).

PART 1 - Radiolarian Taxonomy

Principles of Species-Level Taxonomy

The large majority of living things belong to natural entities of cross-breeding populations, separated from other such populations. These cross-breeding groups are *species*, and delimiting and describing these is the realm of *descriptive* taxonomy. Through repeated events of *cladogenetic speciation* – the creation of new species – a range of descendent species are created over geologic time. These increasingly diverge in characteristics from the original form, eventually creating organisms so different that their differences are noted by assigning them to different higher-level taxa (genera, families, or other groups). This simple definition of taxonomic research hides many complexities. Species, in this so-called "biologic species" definition, belong to cross-breeding populations, and are distinguished from other species by lack of breeding. In reality, only for a very few organisms has breeding behavior been examined in order to determine their species assignment – such data is normally not available even for living species, and clearly impossible for fossil material. Lack of breeding may also exist in nature within living populations due to the presence of (for taxonomic purposes) purely temporary physical barriers that prevent populations from meeting, requiring the definition of a species to be modified to include "actual or potentially cross-breeding," with all the ambiguity that comes from the concept of "potentially." In practice, descriptive taxonomy is based on the concept of clustering in biologic characters (morphologic, molecular or other).

Within species, variation in characters is usually continuous and somewhat limited in scope. Character differences between species are larger, and marked by the absence of significant intermediate states. This pattern comes from the exchange of genetic data between individuals in a cross-breeding population ("*gene-flow*") that to some degree homogenizes the variation between individuals in characters, together with the lack of such genetic exchange between different species. How precisely character variation is determined and evaluated depends on the group of organisms being studied. For the most part in the practice of descriptive taxonomy using morphology (genetic methods differ, see Chapter 4: Genetics), both observation and evaluation of character variation is done subjectively, by examination of characters in numerous individuals belonging to the putative species (singular or plural), and assigning individuals into perceived clusters. Many methods exist to make this process at least partially quantitative, but most are very time-consuming and are limited to only certain types of morphologic characters, and thus are mostly employed only after initial subjective determination of clusters, to validate the patterns already inferred. Of central importance is the availability of a sufficient number of individuals to be able to see the full nature of variation, and the relative frequencies of different morphologies. If this is lacking, morphologic gaps or seeming peaks in variation may be seen simply due to inadequate sample sizes. If these unrepresentative patterns of variation are used, species may be defined that are only minor variations within a single biologic species. In radiolarian studies this has frequently been a problem. Many early descriptive taxonomic analyses of radiolarians were based on very small numbers of individuals from plankton samples, and in more modern research, radiolarians from older geologic time intervals where preservation is poor may yield only a few individuals for study, making robust descriptive taxonomy difficult.

More importantly for the taxonomy of protist groups, many protists are not known to be sexual, and the very concept of cross-breeding is problematic. This is specifically true for the polycystine radiolaria (see Chapter 4: Genetics). Yet, because of the ubiquity of species in most of nature, scientists have chosen to use the concept of species in taxonomy even for groups of organisms where the biologic basis of cross-breeding has not been established. All that is required is that individuals in the group studied show the same pattern of clustering in taxonomic characteristics for groups of similar individuals, together with gaps in variation between different clusters. This appears to be generally true of radiolarians, although the clarity of pattern can vary substantially depending on which particular group of radiolarians is studied. Another biologic issue for descriptive radiolarian taxonomy is the potential existence of cryptic species – genetically distinct biologic species which show no consistent differences in morphology, particularly skeletal morphology, and thus are assigned to the same morphologically defined species. Despite being a common phenomenon in many other groups of protists, the existence or frequency of cryptic species in radiolarians has not yet been significantly explored (see Chapter 4: Genetics). The existence of cryptic species within morphospecies does not invalidate the utility of morphospecies definitions for many types of research, but does affect e.g. the way models of speciation are applied to patterns of morphologic change in studies of speciation (see Chapter 9: Evolution).

Descriptive taxonomy done for fossil material is different than taxonomic work on only living material, even if the analyses make use of only same types of characters, e.g. skeletal morphology. In the definition given above (determining patterns of variation within vs. between clusters of morphology) there is no concept of an additional dimension: morphology within biologic species as it exists over geologic time. Species as they exist today, or at any other moment in geologic time, are in reality only momentary snapshots of a continuous series of populations. The entire time series of populations for a single distinct cluster, from the origin of the species to its extinction, is called a *lineage*. If a lineage has evolved significantly over time – a process called *anagenesis* – different morphologies can exist at different points in time within a single lineage which are as different from each other as the differences which typically separate species of different lineages from each other at any one point in time. Anagenesis appears to be very common in radiolarians, and in all other groups of marine protists with good fossil records where it has been studied (see Chapter 9: Evolution). The different morphologies created by anagenesis have been given individual species names by taxonomists, with some lineages showing enough evolutionary change to have created a series of several distinct anagenetic morphospecies. This practice is practical: many morphospecies transitions within lineages are used as geologic time markers (see Chapter 8: Biostratigraphy), and giving names to the morphologies created by anagenesis provides, in the names, a measure of how much anagenesis has contributed to species-level evolution. Since the transition from one morphologic species to another is continuous, a practical problem arises in how best to mark the boundary between one named anagenetic species and the next. There is no universal agreement among paleontologists in general, or among radiolarian workers, on how to do this. In one approach (a morphotypic definition of the range), the youngest and oldest specimens matching the morphologic definition of an anagenetic species found in a series of samples taken from a geologic section may be used to mark the stratigraphic range of the morphospecies. In a gradually evolving population, this results in an overlap in geologic time between the disappearance of the older and the appearance of the younger anagenetic morphospecies within an evolving lineage. Alternatively, in an "evolutionary" range definition, the point in time where the descendant morphology becomes more common than the ancestral can be used to define the transition between two anagenetic morphospecies. Neither method is ideal. Morphotypic definition uses the occurrences of rare specimens at the ends of the range and finding such specimens is partially dependent on sample size. Evolutionary definitions require accurate assessment of relative frequencies of each morphotype, and assume that the transition of dominance occurs only once, or rapidly enough that the transition interval of nearly equal abundances is short. In practice most workers tend to mix the two definitions, subjectively using both types of information to determine the ranges of anagenetic morphospecies.

Anagenetic change is not however the same pattern, or process, as *cladogenesis*: the splitting of one ancestral species into two (or more) descendant species. The distinction is primarily of importance in studies of biologic evolution, where the differences between the two types of evolution often need to be separated in order to study a specific type of evolutionary process accurately (see Chapter 9: Evolution and Ezard et al., 2012). In particular, the first or last appearance of an anagenetic morphospecies in geologic time, while recording evolutionary change, may not necessarily correspond to the origin or extinction of a biologic species, if in fact the morphospecies is only an intermediate evolutionary state in a continuously evolving lineage whose origin via cladogenesis precedes, and extinction as a distinct biologic population postdates, the range of the morphospecies itself.

Species definitions must not only meet theoretical criteria; as the basis for identification of specimens in daily applied work they must also be practical. Definitions that require rarely preserved or difficult-to-see characters, or examination with relatively time-consuming methods mean in practice that the definition will rarely be used, and specimens will be assigned instead to a broader catch-all category, such as the genus of the species: <genus-name> spp. Such data is of relatively little use, and impractical species taxonomy that helps create it should be avoided whenever possible.

Rules for Describing and Naming Species

Early in development of the biologic sciences it became clear that some set of rules were needed for describing new species, both to ensure that the descriptions were sound and useful, and to resolve conflicts in the use of species names that arose when two or more people independently described what subsequently was realized to be the same biologic species. Separate Rules of Nomenclature were developed for animals, plants, bacteria, and viruses, with the Zoological Code applicable to radiolarians (the current code can be accessed at the International Commission on Zoological Nomenclature's website: www.iczn.org). The Code is very complex (nearly 100 pages in length) and written in a dense legalistic style; most new students, and probably most working scientists, find it rather intimidating. The basic rules covering the name for a new species however are not that complicated (a short summary is given at the ICZN site). Names for species are separated into two parts (binomial): a genus name and a species (or "trivial") name, both of which should be in Latin grammatical form, although the word sources need not be Latin in origin. Each binomial name should be, within the scope of the relevant Code (Zoologic, Botanic, etc.), unique. The genus name is the first name in the hierarchy of higher-level classification, and thus is meant to group together a set of (evolutionarily) related species. This system has some positive aspects. By allowing words to be used over and over again so long as they are in different combinations, it is possible to create unique name combinations from a relatively small, manageable pool of unique words. A similar system underlies names for people in most human cultures. Second, the generic half of the name gives some information on the relationship of the species to other forms. The downside of the system is also the same last point. When the classification is not artificial, the generic name is actually an embedded hypothesis about evolutionary relationships, and thus can change whenever the scientific hypothesis changes. This creates a major amount of instability in the names of biologic species, and has driven more than one taxonomic database administrator to despair. An ideal system would strictly separate all unstable information from a stable, unique identification code for the object. Indeed, taxonomic database administrators are trying to develop a global system of unique, stable identification codes to supplement the classic biologic binomial system, primarily for use in computer-to-computer exchange of data.

More important than the name itself is ensuring that the new species is really new. This requires a careful, complete search of all possibly relevant taxonomic literature before deciding to publish. For many groups of organisms, comprehensive catalogs exist, making this a relatively easy task. For radiolarians, catalogs are very incomplete, so that a separate literature search is unavoidable. Fortunately, much of the primary taxonomic literature is now available in electronically searchable form, so that most relevant papers can be found by searching large archives of digital publications (e.g. libraries of PDF files). It is also very important that the description and illustrations are detailed and clear and include a differential diagnosis: how the new species can be distinguished from other, previously named similar species. Multiple images are generally needed to illustrate both overall morphology and key structural details, together with the range of variation between individuals. Many earlier descriptions were lacking in this regard, due in large part to cost restrictions on the publication of expensive plates in printed journals. With the advent of electronic publication this is no longer an issue, and there is no reason anymore not to fully illustrate new species descriptions. Lastly, the *holotype* specimen must actually be deposited and accessible in a archival institution (e.g. a major natural history museum) for each new named species. This has often not been done properly for radiolarian species: type specimens have frequently been held back by the author for later deposit, the deposit obligation then forgotten, and the type material eventually lost upon the author's retirement, as the universities where most radiolarian workers are employed rarely have the ability to ensure long-term archiving of biologic or fossil materials, and the importance, even the existence, of type specimens is easily overlooked in inconspicuous boxes of microscope slides. Radiolarian taxonomy has also suffered in that many older type specimens properly deposited in museums were lost, particularly in Germany, due to destruction of collections in World War II (Chapter 1: History; Lazarus, 2014).

The holotype is essential to maintaining taxonomic nomenclature. It is a partial solution to the problem of how to assign the use of an existing species name when the concept of what the species constitutes changes (due to presumably better quality data and more research). As illustrated in Figure 5.1, if later study shows that what had been thought to be one species should be divided into two (or more) species, the existing species name continues to be used, but only for the newly created species which contains the holotype specimen. For the other new species, a new name must be published – with description – in the literature. This publication should normally also include a revised description of the older name, e.g. the reduced scope of the meaning of the name vs. older literature. The Code of Nomenclature also regulates the use of species names when the opposite situation occurs. When new research shows that what were previously described as two (or more) species really belong to a single species, the oldest of the published names is retained for further use, and the other names are considered to be synonyms of the "senior" name. This rule of using the oldest name provides an objective way to decide which name gets used, and which is discarded. Without this, taxonomists could (and did) argue about "priority" based on personal considerations.

This system of priority is not without problems. Most importantly names, including many radiolarian names, published in the early history of biologic research often have descriptions and illustrations so poor

Figure 5.1 Use of type specimens to provide continuity in use of names during taxonomic revision. Upper panel: specimen observations (dots), taxa definitions (ellipses) and specimens chosen to be types for each species (large dots). Middle panel: additional data reveals one species ("B") to be two separate species (disjunct clusters of dots). Lower panel: existing name "B" is retained and assigned to the new species definition (larger ellipse) that contains the type specimen for the older species. A new type specimen and new name "C" is created for the other part of the original species.

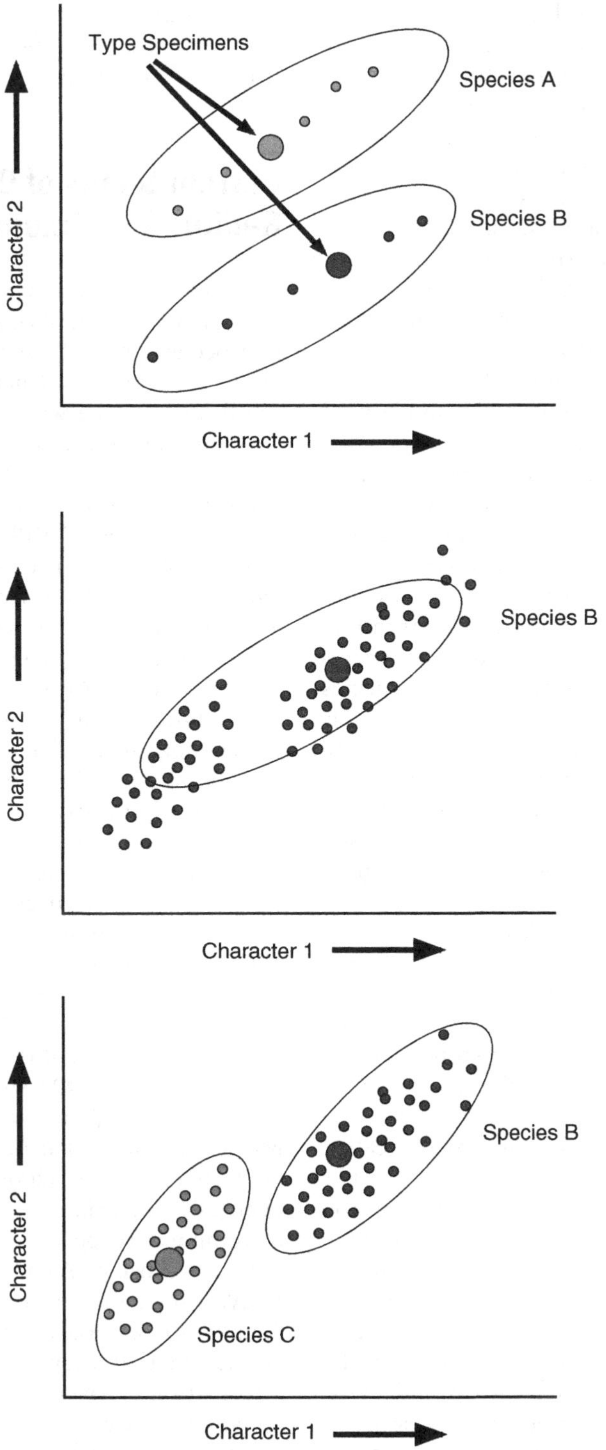

Type Specimens
Species A
Species B
Character 2
Character 1
Species B
Character 2
Character 1
Species B
Species C
Character 2
Character 1

that in practice it is difficult or impossible to determine to which modern species concept the name corresponds to. This may be due to limitations on printing in older times, particularly for good illustrations; or that the species seen many years ago was the only one of its kind and due to its morphologic distinctness, needed only a short description to determine it. Meanwhile, since the original publication, many new species have been discovered, all much more similar to each other and to the old species, so that the original description is no longer adequate to distinguish between a series of relatively similar forms. For these problems, the holotype specimen normally plays a critical role. According to the Code, the holotype should be re-examined, and the name that it stands for should be used for whatever modern species concept it belongs to. Should this re-examination discover that two or more older species names belong to a single modern species concept, the rule of priority is used, as described above, to decide which to retain, and which to place in synonymy. However, for many older species names for radiolarians, no holotype specimen was ever designated by the author, or the original specimen has since been lost (see Chapter 1: History). Again, the Code has a logical rule: a new specimen should be chosen to serve as the type of the species name, based, if possible, on the original material studied by the author. If there was no original holotype designated, the newly chosen specimen is called a lectotype; if there was one but it was lost, the new specimen is called a neotype. Yet even this rule is often not sufficient, if no original material from the author has been preserved, and the author's original description was not sufficiently clear as to where the original material came from to make possible the collection of new material that is unquestionably of the same sort as that used in the original study. In last resort, it has often been necessary to declare an old radiolarian species name a *nomen dubium* – a name of unclarifiable meaning, which simply must be rejected from further scientific use.

Current Status of Descriptive Radiolarian Taxonomy

The large majority of living Polycystina are thought to be described (around 600 morphospecies: WoRMS database; Appeltans et al., 2012; see also Chapters 2–4), even if the differences between many species are still unclear and a significant number of species must be lumped together for practical work (to around 300 practical units, Boltovskoy et al., 2010), and a significant number of small species may have been overlooked in this estimate (see below). The majority of fossil species of polycystine radiolarians, by contrast, have not yet been described. Although no comprehensive catalogs exist, several individual partial catalogs are available: an online catalog of >600 species (descriptions, images, and synonymy), mostly of Neogene–Recent forms at www.radiolaria.org; a fairly comprehensive taxonomically evaluated list of ca. 1,200 valid Cenozoic species names, together with numerous synonyms, accessible at the NSB marine microfossil database at *www.nsb-mfn-berlin.de* (Lazarus et al., 2015); a Late Jurassic to Early Cretaceous catalog (Baumgartner et al., 1995) with 424 species; and lists of published, though not evaluated, names for Mesozoic and Paleozoic species in O'Dogherty et al. (2009) and Aitchison et al. (2017). The largest databases (in terms of temporal scope and sheer numbers of taxa) are those of Agarkov (bio-world.org), RadWorld (Caulet et al, unpublished), and Suzuki (2016, not available online). All are similar in that they primarily give basic information on over 15,000 species names (of which only ca. 10,000 are thought to be validly published according to the rules of nomenclature) but only very limited information on synonyms. In addition, the Caulet gives detailed original descriptions for genera and species. The Suzuki database gives all known papers citing the species names, and in most cases a species illustration. As the geologic age of the samples (to Epoch or Stage) is also noted, these databases can provide age-specific overviews of taxonomic effort. Publication effort can be summarized (Figure 5.2), showing roughly equal numbers of taxonomic papers for Cenozoic and Mesozoic, but much fewer for the Paleozoic. A simple summary of reported taxonomic diversity vs. age (Figure 5.3) similarly shows that there are many more published Mesozoic and Cenozoic species than Paleozoic ones.

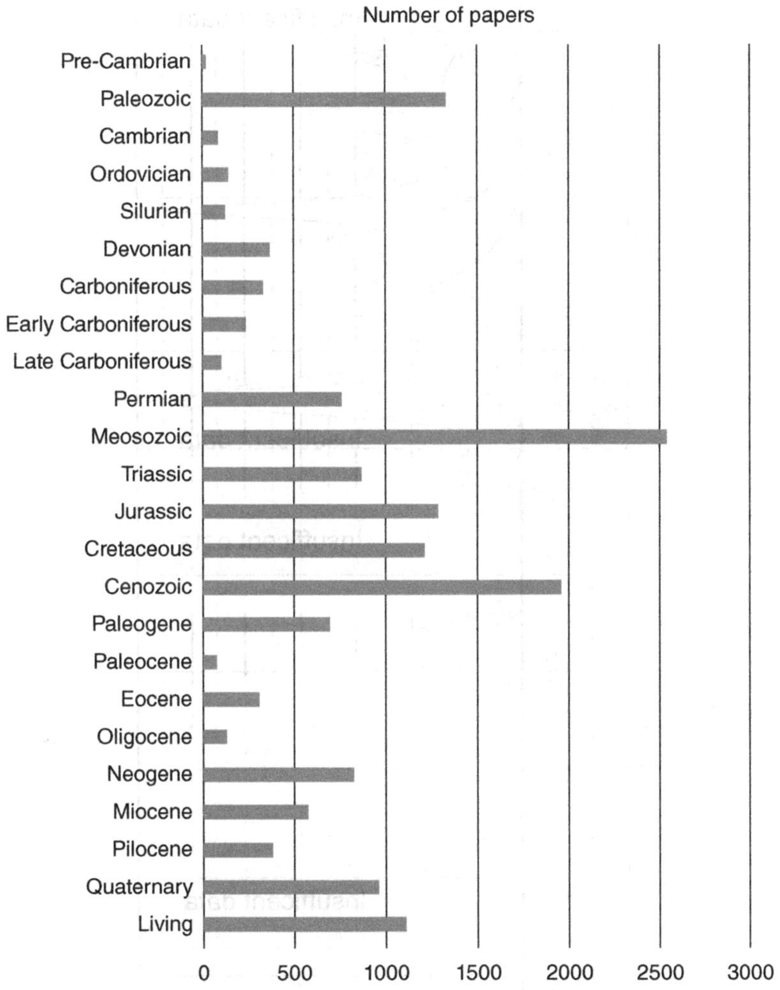

Figure 5.2 Numbers of taxonomic papers by geologic time interval, from the Suzuki database. Papers with descriptions from more than one time interval were counted in all relevant intervals, or, if less clear, in the next higher time category. For these reasons, the sub-interval values do not sum to the higher level ones.

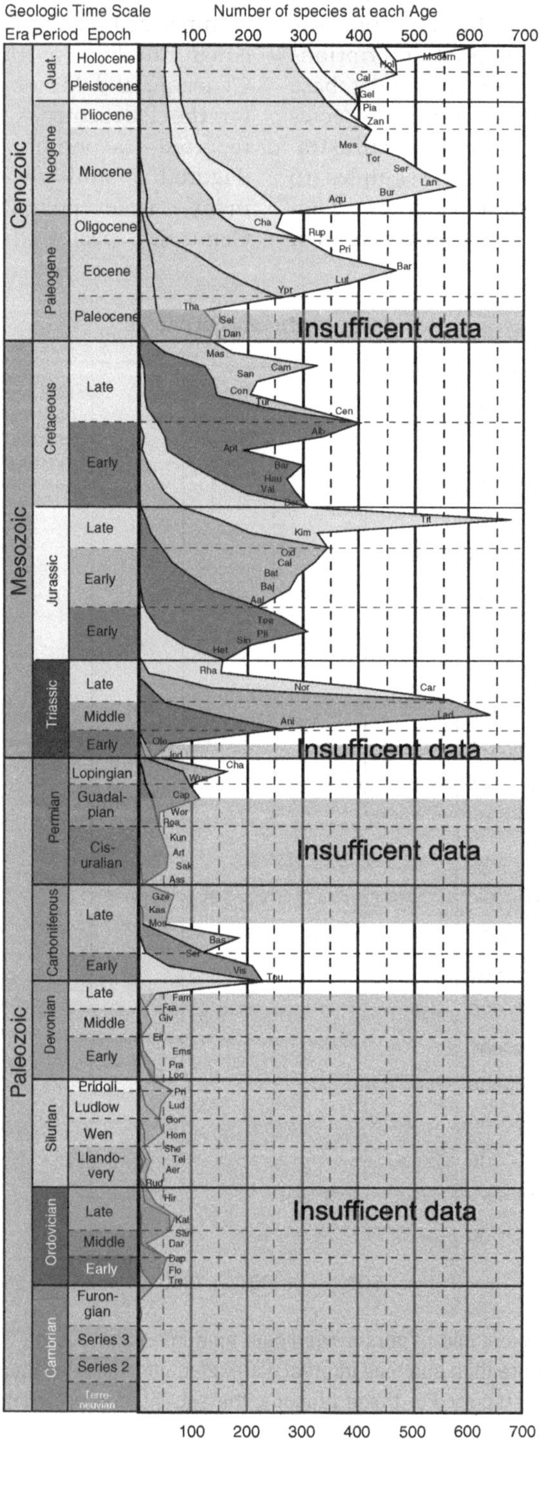
Geologic Time Scale
Number of species at each Age
Era Period Epoch
100 200 300 400 500 600 700
Cenozoic
Mesozoic
Paleozoic
Quat.
Neogene
Paleogene
Cretaceous
Jurassic
Triassic
Permian
Carboniferous
Devonian
Silurian
Ordovician
Cambrian
Holocene
Pleistocene
Pliocene
Miocene
Oligocene
Eocene
Paleocene
Late
Early
Late
Early
Early
Late
Middle
Early
Lopingian
Guadal-pian
Cis-uralian
Late
Early
Late
Middle
Early
Pridoli
Ludlow
Wen
Llando-very
Late
Middle
Early
Furon-gian
Series 3
Series 2
Terre-neuvian
Insufficent data
Insufficent data
Insufficent data
Insufficent data

However, it is not possible to interpret this figure as a true biologic diversity curve (Suzuki, 2016; see also Chapter 9). Various problems including biased sampling and taxonomic effort have to be accounted for. Also, as the Cenozoic list (Lazarus et al., 2015), which does evaluate names for synonyms, includes nearly twice as many invalid names (synonyms, nomina dubia, etc.) as valid, we must assume that the number of distinct valid Phanerozoic species is similarly affected, e.g. only ca. 3,000–4,000 distinct, valid species so far published.

Due to the unusually high quality of preservation of diversity in the fossil record in the Cenozoic, we have a fairly good idea of approximately how many total Cenozoic fossil species exist, in comparison to the number that have been actually described. At the time of writing, perhaps only one third of the Cenozoic fossil morphospecies have been described. The numerous undescribed Cenozoic forms, visible in the many microslides made from deep-sea sediments but ignored, simply reflect the limits of manpower – many more Cenozoic radiolarian fossils have been collected in the last decades than can be described by the small number of radiolarian taxonomists that work on this material. Descriptive work has concentrated on those forms most useful for biostratigraphy or paleoceanography. Less common species, species with indistinct or clearly diachronous first or last occurrence patterns in geologic time, and forms with complex morphologies have been studied relatively little. For older time intervals the picture is more speculative, as preservation is by no means as complete, or as consistent. However, if we assume that diversity in the Mesozoic was similar or slightly higher than today's; that (more debatable) the large majority of species have been preserved in the fossil record, that Paleozoic diversity was half modern values (see diversity estimates in Chapter 9); and assume a mean longevity for species of ca. 8 my (a value based on various studies of microfossil ranges in the literature), we can roughly estimate how many species may have originally been present: about 20,000. A similar value comes from extrapolating the above cited catalog name counts using completeness estimates.

The numbers of species described by different authors also differs dramatically. As noted in Chapter 1, early work was dominated not only by descriptions of new species, many new species were published per paper, so that a relatively small number of authors (in addition to Haeckel) contributed many species descriptions to the literature. More recent work has required not only more detailed documentation of new species, but also the need to check the ever-increasing literature for possible prior publication. Without comprehensive, searchable catalogs, this has made it ever more difficult to produce good new taxonomic descriptions. Nonetheless, many more recent authors have also published numerous new species – a list of the top 100 authors is given in Table 5.1 as an aid to identifying key literature sources for taxonomic referencing.

←

Figure 5.3 Total number of available species names in the Suzuki (2016) database (9,870), with the diversity history of each geologic epoch, as determined by the recorded total range of each species, shown as a colored cohort. Species only reported as living are marked separately as "Modern." Note that available names are simply those formally published according to the Rules of Zoologic Nomenclature. Many of these are known to be synonyms, and there are many other biases such as unequal preservation or recording effort by time interval, so this figure is not a direct estimate of biologic diversity. Source: Suzuki (2016). Reproduced with permission of Paleontological Society of Japan. (*See color plate section for the color representation of this figure.*)

Table 5.1 Most prolific authors of new radiolarian species.

Name	As First Author	As Second Author	Total
Haeckel	2970		2970
Kozur	884		884
Mostler	28	757	785
Ehrenberg	533		533
Pessagno	482	31	513
Dumitrica	461		461
Campbell	160	201	361
Clark	200	160	360
Won	265	24	289
Vinassa	288		288
Nazarov	209	52	261
Squinabol	245		245
Hinde	231		231
Yeh	228		228
Carter	184	29	213
Tan	179	22	201
Foreman	197		197
Popofsky	187		187
Blome	109	67	176
Kozlova	159	16	175
Cheng	146	24	170
Lipman	141		141
Tekin	138		138
Bragin	120	11	131
Hull	116	13	129
Ormiston	16	110	126
Riedel	63	63	126
Sanfilippo	84	39	123
Lazarus	15	105	120
De Wever	82	36	118
O'Dogherty	110		110
Tikhomirova	110		110
Sugiyama	96		96
Bragina	95		95
Feng	95		95
Petrushevskaya	65	29	94
Nakaseko	93		93
Renaudie	93		93
Chen	76	15	91
Wang	66	24	90
Wu	78	11	89
Yang	73	10	83
Afanasieva	82		82
Carnevale	82		82
Sashida	61	17	78
Nishimura	44	33	77
Parona	77		77
Noble	54	22	76
St	73		73
Li	58	14	72
Vishnevskaya	62	10	72
Gorbovets	1	70	71
Borisenko	70		70
Aliev	69		69
Whalen	22	47	69
Caulet	50	18	68
Baumgartner	67		67
Dreyer	67		67
Principi	66		66
Jud	65		65
Takemura	65		65
Mast	64		64
Neviani	64		64
Goll	63		63
Holdsworth	10	50	60
O'Connor	58		58
Steiger	46	10	56
Aitchison	53		53
Iams	0	50	50
Isakova	50		50
Schwartzapfel	50		50
Gorican	48		48
Deflandre	47		47
Amon	25	20	45
Moksyakova	45		45
Ichikawa	44		44
Lucchese	44		44
Zhang	44		44
Bjørklund	19	24	43
Kiessling	43		43
Krasheninnikov	42		42
Yao	31	10	41
Nestell	24	15	39
MacDonald	25		25
Nigrini	38		38
Renz	37		37
Cachon (and as Enjumet)	3	33	36
Hollande	36		36

Table 5.1 (Continued)

Name	As First Author	As Second Author	Total
Cleve	33		33
Pantanelli	33		33
Ruedemann	33		33
Wilson	0	33	33
Cordey	32		32
Su	11	21	32
Frizzell	31		31
Lahm	31		31
Liu	12	19	31
Middour	0	31	31
Aita	30		30
Mamedov	29		29

100 most prolific authors of new radiolarian species, publishing either as first or second author. Source: Suzuki taxonomic database.

Principles of Higher-Level Taxonomy

Determining and describing the divergent branches of natural relatedness due to evolution among organisms is the domain of higher-level taxonomy, also called *systematics*. Not all higher-level taxonomy is actually systematics. Systems of taxonomy (also called classification) can be devised that do not attempt to reflect the pattern of evolutionary relatedness. Sometimes these are very practical, e.g. a field guide to the identification of wildflowers that classifies flowers first by color (a character that has no higher-level natural taxonomic significance in plant evolution), then by other non-systematic criteria, etc., to allow rapid location in the book of the description of a particular species. Artificial, top-down classifications however have limits even as identification guides, since they will usually artificially divide natural (evolutionarily related) lower-level groups into two or more partitions, requiring duplicate entries in the classification to accommodate the divided groups (Figure 5.4). At the extreme, when rigorously applied down to the level of individuals, this can produce absurdity. For example, if all mammals were to be classified first by hair color, then people and cats with black hair would belong in one higher-level category, while white-haired cats and (older) people would be assigned to another. Within both the "black hair" and "white hair" higher-level taxa one would have to create duplicate sets of lower level taxa for "catlike" and "people-like" body forms. Also in this example, individuals with both hair color types (e.g. late-middle-aged people) might even be considered as (evolutionary) links between the higher-level categories. This occurs because the characteristic chosen for the higher-level classification (hair color) actually varies within individual species, and even within single individuals over their lifetime. This example is not purely didactic, as one of the main early sources of radiolarian taxonomic names (Haeckel, 1887) makes use of a highly artificial higher-level classification, which resulted in the creation of many biologically meaningless lower-level names, both for genera and species (Lazarus, 2014).

Haeckel and the Beginnings of Higher-Level Radiolarian Taxonomy

Haeckel was one of the most influential workers on radiolarians, particularly through the publication of his great radiolarian monograph based primarily on the materials collected by the H.M.S. Challenger expedition (Haeckel, 1887). As has been noted in Chapter 1, he was also an influential figure in the development of evolution research, being a very prominent promotor of Darwin's theories. Haeckel in particular stressed the importance of reconstructing phylogeny as evidence for

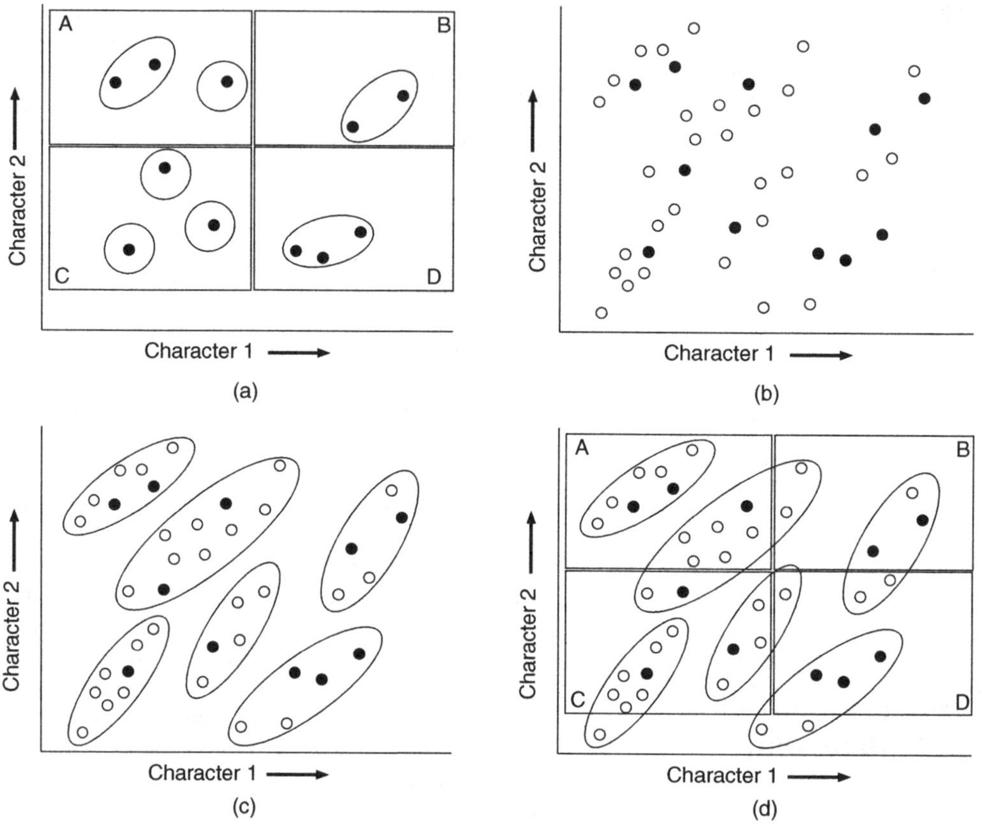

Figure 5.4 Effect of artificial classification criteria. A) Initial observations (black dots), initial lower level taxa (circles and ellipses, e.g. species) and artificial higher-level classification (boxes, e.g. genera). B) Additional data (open circles). C) New, "natural" lower-level taxa. Note how largest ellipse now combines two taxa previously in separate higher taxa. D) Conflict with higher-level classification. If higher taxa definitions are retained, many natural taxa must be split into multiple lower-level taxa, e.g. as in upper left panel "A." Largest ellipse must be split into three (one new artificial "species" to hold new data discovered in higher taxon box "B"). Other problems exist as well: observations exist that do not fit boundaries of higher taxa concepts (ellipses extending below box boundaries). In reality, taxa data are not two-dimensional but multidimensional and the extent of problem multiplies as additional criteria and higher taxa definitions are included.

evolution, and published many evolutionary trees, for biologic diversity as a whole, and also for individual groups such as radiolarians. A recent analysis of Haeckel's taxonomy in relation to that of other early workers (Lazarus, 2014) suggests that Haeckel developed, relative to more conservative earlier workers such as Ehrenberg, a very complex hierarchy of higher-level taxa (the formal definition of branches in a phylogenetic tree) in his study of radiolarians. This was not only to provide a practical means to manage the great diversity of forms encountered in his study. At least in part he wished to use the patterns of connectedness and change in these higher-level taxa as evidence for the existence of evolution in his broader program of promoting evolution and evolutionary theory to society. However, because at that time virtually nothing was known about radiolarian biology, development, or even variation within species, there was little objective basis for constructing a natural classification. Haeckel therefore created his complex higher-level taxonomy based on rather arbitrary sets of characteristics, including gross geometry, symmetry, and speculations on the inferred evolutionary history (more on this below). This mixed, but partially very artificial classification, acknowledged as such by him, had the inevitable effect of artificially splitting species, natural groups of biologically related species (and higher taxa) into two or more artificial categories, with replication of characteristics and thus also the creation of many artificial taxonomic names. The great majority (ca. 90%) of Haeckel's species and the large majority of his genera are probably not valid (Figure 5.5, Lazarus, 2014); Caulet et al. (2017) report that only ca. 400 of the over 1600 Cenozoic genera, mostly from Haeckel, are probably valid. Most likely, a large proportion of his families, subfamilies, and other higher-level categories are also biologically invalid (Lazarus, 2014). Clearing away these names via synonymization remains a major task and barrier to the development of a robust, useful taxonomy for the radiolarian groups. A second negative consequence of Haeckel's artificial taxonomy was the inclusion of biologically rather distantly related forms in the same

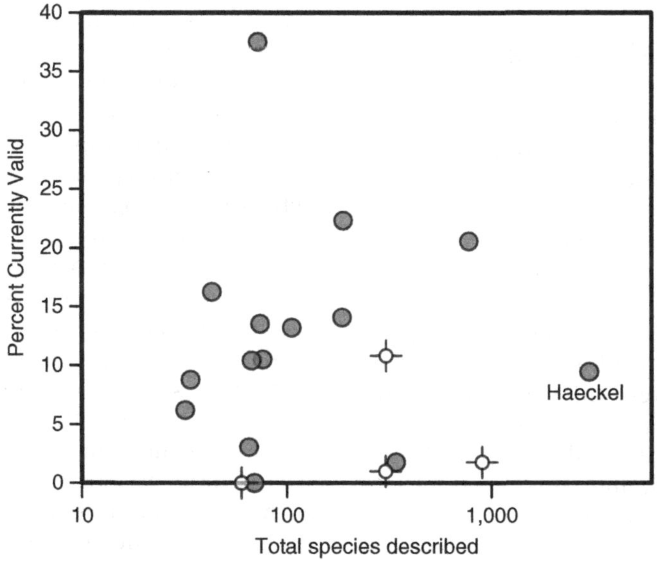

Figure 5.5 Current use of polycystine radiolarian taxonomic names published during the first century of radiolarian research, by author (only early German language workers analyzed). Open circles with crosses data known to be biased by methods used in study, and thus not suited to analysis; solid circles – number of species described by an author vs. number still considered to be valid, and in use in modern literature. Despite general trend for more prolific authors to have higher percent usage (possibly a "convenience" effect where modern workers consult only a few larger monographs to find matching early species descriptions), Haeckel's species are among the least used (<10%). Note log scale on x-axis. Source: Lazarus J. Micropal. (2014). Licensed under CC BY 3.0.

formal, low-level taxon, e.g. genus. Haeckelian genera frequently lumped together living species and unrelated species from much older geologic time intervals, giving the false impression that radiolarians had not evolved significantly over the Phanerozoic, and were thus not useful for stratigraphy (Moore et al., 1952; see also Chapter 8: Biostratigraphy). Equally problematic is that the speculative "style" of taxonomic research established by Haeckel became a template for many later workers as well, particularly for higher-level taxonomic names, although some early workers recognized the problems of Haeckel's taxonomy and attempted to reform it (Lazarus, 2014). Few of these reforms, regardless of intent, are however based on any significant biologic data and as such are little more than speculation.

Biologic Systematics

Higher-level taxonomies that are based on natural evolutionary relationships avoid the problems described above of artificial classification. Natural (also called biologic) classifications also provide insight into evolutionary processes from analyses of the structure of their contained taxa, and thus are an important resource for many types of evolutionary study. How does one create a biologic higher-level taxonomy that reflects evolutionary relationships (taxa defined by shared descent from a common ancestor), since an arbitrary, artificial taxonomy clearly is only of comparatively limited use? This problem has engaged biologists for centuries and many different approaches have been developed, each using a different primary type of data, or distinct method for evaluating data.

Higher-level taxonomic methods can be divided into three principal types. The first two are based on analyses of characters in organisms: those based on overall similarity, and those based on a nested hierarchy of characteristics. The two categories are not mutually exclusive in practice. People have been using both methods since the origins of humans to classify the world into usable categories, generally using overall similarity first, and hierarchical or single character definitions as a secondary refinement. This is what has given us the categories primarily based on overall similarity, "cats" and "people," with black-haired cats and older people being secondary refinements within each. The second major approach to classification is the exclusive use of nested sets of single characters to define a hierarchy of groups. The criteria can be almost anything, and although they can be very arbitrary and artificial, a variety of practical classification schemes can also exist. For example, predatory vs. non-predatory animals, fast vs. slow predators, etc., can be very useful nested classification systems if you are potential prey. Many words in human language still reflect this utilitarian approach: "crops" being all plants that share the characteristic of being cultivated, even though not all crops are edible (e.g. cotton), and the evolutionary origins of the species included are extremely heterogenous (a few, such as mushrooms, are not even actually plants); while "edible crops" define the subcategory of those we use as food. In biologic taxonomy however the goal has been to reconstruct the sequence of evolution: the phylogenetic pattern of branching descent with modification that has created the diversity of different species. The third approach, logically, is to trace out the evolutionary history of a group of organisms using the fossil record. This uses both character and geologic information: similarity of morphology to connect populations representing single species or higher taxa that are found in geologic time-series. Although theoretically robust, this method, referred to as *stratophenetics*, requires a degree of completeness in the fossil record, and our documentation of it, that is rarely available even for more recently originated, lower hierarchical-level taxa even in groups that have a substantial fossil record, and essentially unknown for higher-level taxa that originated in older geologic time intervals. It is routinely used in radiolarian research only for species, or at most genus-level taxonomy: to connect gradually changing populations within an evolving lineage together, and less commonly, to document the process of cladogenetic speciation (see Chapter 9: Evolution). This is particularly true for numerous stratophenetically defined genera such as those used extensively in Cenozoic taxonomy (Sanfilippo et al., 1985; Chapter 8: Stratigraphy). These actually correspond to a single anagenetically evolving lineage and are based on detailed observation of continuous morphologic series, and thus are not the same as genera concepts for groups of distinct lineages, but hypothesized, on the basis of character analysis, to be closely related.

Although character-based methods are the only practical ones for most higher-level taxonomy beyond the genus level, even with the relatively well-fossilized radiolarians, fossil data is still useful, albeit in a more restricted way. Since the classification is meant to reflect evolutionary history, the nested set of taxa also implies a time ordering of each group's evolutionary origin. This sequence of first appearances should in principle match that seen in the fossil record, and discrepancies can be used to test the reliability of the character-based analysis. In practice there are many uncertainties on how best to evaluate and compare to each other the relative reliability/uncertainty of both character grouping and geologic time of origin data, which has made this approach difficult to employ without controversy. This problem has not yet been evaluated in radiolarian studies.

The above methods, and many variations and complications, are areas of major concern in biologic research, and the subject is represented by a large, complex, and often disputatious literature extending back many decades. Many individual approaches have coalesced into "schools" of taxonomic methodology, such as "phenetics" (quantitative, generally computer based, similarity classification) or "cladistics" (also quantitative and computer based, but rigorously focused on hierarchical classification). At the moment, cladistic classification is the most common method used in biologic systematics, both for classical morphologic, as well as molecular, data. In reality, no single method is always superior, and even superiority requires a personal decision as to the relative importance of phylogeny vs. similarity in classification. Hierarchical methods (e.g. cladistics) assume, at least as an initial default, that all characters that appear in evolution are retained by all descendants in a group; that no character appears more than once, that all characters have been accurately identified and recorded, and (usually) that all characters are equal in weight and uncorrelated to each other. Almost all these assumptions are to some degree false in real-world studies, and although the discrepancies are not necessarily fatal, they do very frequently have a substantial negative effect on the accuracy and robustness of the results. Classification by overall similarity, for example, also has limits. It typically groups together taxa that are near the joint base of a section of phylogeny, but on different adjacent evolutionary branches, since these "early" taxa, unlike later forms, still share more similar characters than they do divergent ones. Depending on the importance of always reflecting phylogeny in one's view of classification, this may or may not be a significant limitation of the methodology.

Higher-Level Taxonomy in Radiolaria

Traditionally, higher (above the genus level) taxa in radiolarians have been defined using both methods of overall similarity, and by single characteristics thought to reflect high-level evolutionary relationships. In marked contrast to the systematic research in many other groups of organisms, this work has proceeded in more recent years without almost any explicit reference to the formal theory and/or quantitative methods of modern systematics (summarized above). There are, for example, no substantial morphology-based studies of polycystine systematics either using formal phenetic or cladistic methods, and cladistic analyses using molecular data have until very recently focused on the relationship of polycystines to other groups; and as yet have included too few family-level groups to provide an adequate basis for the systematic structure of polycystines (see Chapter 4: Genetics). Existing classification systems are still almost entirely based on the morphology of the shell, and are done using earlier methodologic approaches to systematic analysis. Early classifications of course were created before systematic theory developed. Ehrenberg largely declined to create higher-level categories within radiolarians, and his generic definitions were also relatively few. Haeckel was the first to create an extensive higher-level systematics, and the first to do so based on theoretical principles (evolution and phylogeny). His Challenger report (1887) included a great number of higher-level taxonomic ranks (Legion, Sublegion, Order, Family, Subfamily, plus several informal lower groupings) and an equally impressive number of instances: e.g. 20 Orders and 85 Families, and many more numerous lower-level groups above the genus level (Lazarus, 2014). As noted above however, his classification was not based on any significant amount of biologic data that would have permitted an evolutionary classification, and his higher-level taxa are in fact often quite artificial. Some of his taxa however are still used today. This reflects a special aspect of radiolarians which is very important in their classification.

Radiolarians show an extreme degree of variation in basic constructional principles. It has been noted earlier (Chapter 2: Biology) that living polycystine radiolarians can be divided into two higher-level taxa based on fundamentally different structural organization (*Baupläne* – plural of *Bauplan*, the German word for construction plan): the Spumellaria, with a spherical-radial symmetry; and Nassellaria, with a conical form (axial gradient-radial symmetry). Within these two broad geometric categories many other variations exist, including other groups ranked at the same level, and many lower-ranked categories within them, which share basic symmetry principles but differ in other structural details. Many of these "details" are, geometrically seen, more dramatically different between these subgroups than those typically seen between different phyla of invertebrates. For example, within the Spumellaria there are shells in the form of simple or nested sets of spheres, flat discs of nested rings, bipolar spears or three-dimensional hexagonal stars, shells that look like ringed planets (a central sphere surrounded by a much larger largely free-standing orbital ring), spirals (single or double, in-plane, or twisted, 3D coiling), nested 3D sets of bands, and many others. These dramatically different geometries have two important effects on the classification of radiolarians. One, each distinct geometric topology provides an obvious basis for grouping species together at the level of fundamental geometric Bauplan. Such groupings are the basis for most family-level taxa in radiolarian classification, and most of those of Haeckel's higher-level taxa erected using these topologic, Baupläne criteria (in contrast to those based on other, simpler characteristics such as spherical vs. oblate spheroid shape) are still in use. Second, the very distinctiveness of the geometries means that there are relatively few characters that describe higher-level morphologic differences and yet are still found among multiple family-level groups. Most of the characters that define taxa with orbital rings simply don't exist for taxa that have spiral shells, and vice-versa. It has thus been more difficult to find shell morphology characters suited to determining higher-level relationships between families than is true for many other groups of organisms.

Bauplan-based higher level taxa, typically formally as families, have been elaborated for radiolarians by many authors. Traditionally, Cenozoic–Recent, Mesozoic, and Paleozoic workers tended to work fairly independently of each other (see Chapter 1: History) and the higher-level systematics developed by these three research communities had little to do with each other (De Wever et al., 2001; Afanasieva et al., 2005).

One approach that has been very popular in creating radiolarian higher-level taxonomy is the use of characters that are thought to appear early in individual shell growth as the basis for classification. This is based on the practical aspect that these characters, typically found in the central part of radiolarian shells are to some degree found across otherwise topologically very different family-level groups. They thus provide one of the few sources of characters widespread enough to allow higher-level taxonomic analyses. Second, it is inferred that these characters, being located in the most central part of the shell, are also those first secreted as the cell began to grow. These characters are thus inferred to be ontogenetically early (*ontogeny* – a termed coined by Haeckel in 1866 in his Generelle Morphologie, is the study of the development of the form of a biologic individual over its lifetime). Lastly, as inferred early ontogenetic characters, it is implied that they are most likely to reveal patterns of early (i.e. deep, high-level branching) phylogeny. This last is based on the theory that ontogeny gives phylogenetic history, an idea that dates back to before Haeckel, but which Haeckel promoted in a very strong form, and which he called the "biogenetic law" (Gould, 1977; Mayr, 1982). In Haeckel's version, the stages of ontogenetic development of the individual reliably "recapitulate" the phylogenetic history of the (adult stages of the) organism. Higher-level taxonomy based on implied ontogenetically early characteristics was a common feature of many early studies of radiolarian taxonomy (e.g. Dreyer, Mast, and Popofsky; see Lazarus, 2014), building on the tradition established by Haeckel. This tradition continued into modern times, particularly in Soviet and former Soviet countries (e.g. Petrushevskaya 1971, 1986), and is also used as the basis for ca. 25% of the relatively recently created higher level taxa of Dumitrica, as synthesized in De Wever et al. (2001; see also Chapter 1: History). This latter work also provides, for the first time, a description of the central structures for most families, including the majority where the defining characters are based more on the general Bauplan. It is the most comprehensive synthesis of radiolarian taxonomy since Haeckel's Challenger monograph from 1887 and is a landmark study that will influence radiolarian taxonomists for many decades to come. De Wever et al. (2001) lists and gives succinct yet reasonably detailed descriptions plus illustrations for 131 radiolarian families, plus a large number of subfamily categories (De Wever et al., 2003). The taxonomy covers the entire Phanerozoic record and thus is much larger than the almost exclusively Cenozoic-Recent material covered by Haeckel (1887). Yet, at the family level at least the classification is less elaborate than Haeckel's (53 Cenozoic-Recent families vs. Haeckel's 85), despite the growth of material available for study in the last century. This illustrates well the over-elaborated nature of many of Haeckel's higher taxa. A parallel, if less detailed synthesis by Afanasieva et al. (2005) lists 132 families, and in content is very similar to that of De Wever et al. (only 19 family-level taxa are given in Afanasieva et al. that do not appear in the index of De Wever et al.). This similarity is encouraging, suggesting that the large majority of family-level taxa in these publications are based on broadly accepted criteria. The De Wever et al. (2001) family-level taxonomy is indeed for the most part based on distinct Baupläne, and taxa long established in the primary literature. Those family-level taxa based on analyses of central shell structures are mostly more recent (several being introduced for the first time in the De Wever volume). These are more speculative in nature, being based on several not yet verified assumptions. The ontogenetic stage of central shell characters has not yet been established, as almost no data on the relevant stages in the early ontogeny of radiolarians actually exists, despite claims to the contrary in De Wever et al., 2001 and several other authors, e.g. Afanasieva et al. (2005) (Chapter 2: Biology). The objective reproducibility of subjective mapping to formal geometric elements, particularly the primary spicules, from the small, often irregularly shaped bars and spines seen in the central shell has yet to be demonstrated. Nor has it been demonstrated that these structures are truly homologous between groups. Lastly, the biogenetic law itself has largely been discredited (Gould, 1977; Mayr, 1982). Although ontogeny does often provide information on phylogeny, there are numerous occasions where it does not, and it can only be used as one source of possible information on phylogeny and higher-level systematics. Thus, it is not certain, even if all the preceding concerns are satisfied, that these structures actually contain a clear signal of phylogenetic history. Using central shell characters to define families will eventually be validated, or not, at least for living taxa, as more molecular data on relationships become available. The few studies so far published have not supported the De Wever et al. (2001) scheme, and provide only partial support for the earlier Riedel (1967) concepts. De Wever et al. for example place the genus *Hexacontium* and its family Hexalonchidae in the order Entactinaria, while more traditional classifications based on the Bauplan and totality of characters place these in the Spumellaria. Molecular data on *H. pachydermum* (Yuasa et al., 2009), a species whose morphology is well known (e.g. Cortese and Bjorklund, 1998) place *Hexacontium* firmly within the Spumellaria. As entactinarians are mostly known as fossils and no other, morphologically more typical living Entactinaria have been analyzed with molecular methods, this result is somewhat ambiguous, but clearly the primary spicules of *Hexacontium*, despite having an "Entactinarian" structure, are not a reliable character for determining to which order the taxon should be assigned. More comprehensive, if still very preliminary studies by Sandin et al. 2019, 2020) have also largely confirmed the importance of all characters, in particular those of the outer shell, in higher level taxonomy, even though several classical Riedel (1967) groups in these analyses are poly- or paraphyletic.

In addition to these general syntheses, individual studies have significantly altered family-level classifications for some time intervals. This is particularly true for the lower Paleozoic, where additional material has been found in recent years that has significantly improved our understanding of early radiolarian diversity.

The Observational Basis of Taxonomy: Structures of the Radiolarian Shell

Although many radiolarian shell structures are specific to individual Baupläne and are described in the family in which they occur, some structures are found throughout most radiolarian taxa and are introduced here.

Spicules – These are the most primitive construction elements of radiolarian shells. One form in particular is of special importance (Figure 5.6). It consists of two clusters of radiating spines (one cluster of five, the other of four spines), one spine of each being shared as a linking "median" bar, with the remaining 4/3 spines projecting away from the ends of this bar. This construct, the "primary," or "initial" spicules, is found in only some groups of radiolarians, and in others is absent or highly altered in form. When present however it is located in the most central part of the shell and thus is a major focus for ontogeny-based methods of systematics. Each individual spine, as well as secondary spines that may grow from them are named, and their development in individual taxa is normally part of the taxon description.

Lattice wall – Most radiolarian shells are primarily built out of porous planar meshworks that are shaped into different geometric forms (typically curved into the surface of spheres, cylinders, or other shapes) (Figure 5.7A–H). This meshwork ("lattice wall") consists of a mesh of rod-like "bars," that join to each other at "nodes," and thereby create a two-dimensional pattern of open "pores." Bars may be simple rod-like structures, but usually are more triangular in cross-section, or bear ridges on the outward-facing side ("crested," thus creating "framed" pores). The pore shapes may be nearly circular to highly polygonal, the pores uniform or irregular in size, closely or widely spaced (corresponding respectively to thin or wide bars), and the arrangement of the pores can be a regular hexagonal packing, irregular, in rows, or show size gradients.

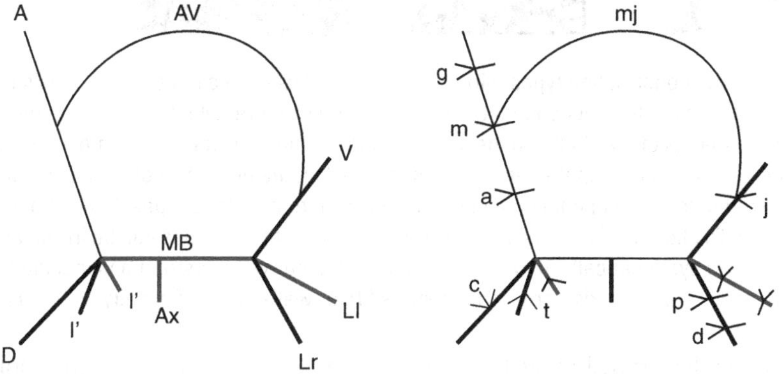

Figure 5.6 Nassellarian shells all (other than Collodaria) possess a set of primary spicules (spines). These are located inside the main lattice-wall shell near its apex. The main spicular elements are constant. At each end of a median bar (MB) radiate a cluster of three or four spines. At one end the apical spine (A) points towards the shell apex and the dorsal spine (D) extends distally, while two other spines (left and right secondary laterals l') extend to each side. At the other end there is a similar arrangement of an apical directed vertical spine (V) and two primary left and right laterals (Ll and Lr) but no fourth spine. A downward-directed spine from the middle of the median bar is the axobate and is not always present. Many derived structures exist in some taxa, including a wide range of arches (AV and mj in this figure) and secondary spines attached to the primary spines. Secondary spines and arches can be described by name according to location of attachment (g, galear; m, mitral; a, anterior; t, tergal; c, cervical; p, pectoral; j, jugal; d, unnamed additional locations). Note: In this drawing medium grey lines are in the plane of the page, thick black lines extend out of the plane of the page towards the viewer, thin grey lines out of the page away from the reader. Source: Renaudie and Lazarus, 2012. Licensed under CC BY 3.0.

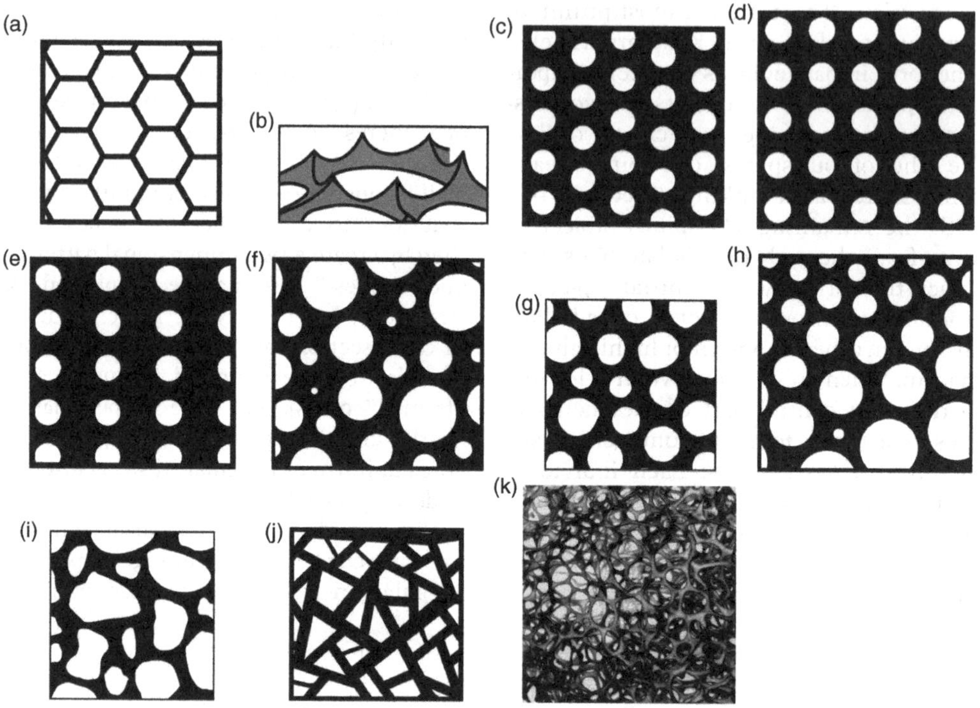

Figure 5.7 Shell wall construction types. Most radiolarian shells have a lattice wall, consisting of a regular, repeating pattern of elements connected into a planar structure, which is curved in the third dimension to form the shapes of the shells. A) the simplest form is thin bars connected into a regular polygonal mesh; B) frames – typically cuspate ridges, may be formed on the outer side of the meshwork and with a flat annulus around the pore opening convert many pores into circular shapes. Pore arrangements may be C) staggered; D) aligned; E) in rows. Pore sizes may vary (F), their shapes can be irregular (G), and the pore size may vary systematically in gradients (H). Non-lattice wall constructions include hyaline walls with pores (I), walls made of criss-crossed flat strips (J), or walls made of spongy material (K).

Shell walls may be formed by at least two other types of construction: (porous) hyaline sheets (Figure 5.7I), and irregular meshworks formed by randomly criss-crossing strips or bands of silica (Figure 5.7J). Hyaline [clear, uniform] walls are found in many nassellarian cephalises, and are the typical structure in certain groups such as collosphaerids. Pores in hyaline walls can be of any shape or size but are often irregular, and normally do not have frames. Walls made of criss-crossing strips are characteristic of orosphaerids and a few other groups. There is no real regularity of pattern and thus they are not true lattice-wall constructs.

Spongy material – a commonly occurring element in radiolarian shells, it is, as the name suggests, an irregular, three-dimensional network of thin interconnected bars and intervening open space (Figure 5.7K). Spongy material is often formed into sheets and these may be used instead of lattice wall to form the walls of shells in some species.

Spines – long rods attached at one end to the shell, generally radiating outwards and tapering towards a point distally. Spines may be cylindrical (circular to elliptical in cross-section) or bladed (three flat, or thickened bands joined at the center of the spine), straight or curved, occasionally even twisted. Spines extending from the middle part of a conical shell and connected to it by a sheet of lattice wall are called wings; spines extending from the base of shell (in those where a base is defined) are called feet.

Beams – rods (cylindrical or bladed) that are joined at both ends to other shell structures. Beams typically connect shell layers together in radiolarians with multilayered shells.

Higher-Level Taxonomy in this Book

Given the existence of multiple, recent, comprehensive syntheses of higher-level taxonomy in radiolaria (e.g. De Wever et al., 2001; Afanasieva et al., 2005) we do not attempt to reproduce here a similar complete compendium. Instead we describe and illustrate a representative subset of radiolarian diversity. Given our own interests and expertise (mostly in late Cenozoic or living species) and their relative importance to studies of paleobiology, we mostly illustrate and describe Cenozoic families, but cover these completely. We use in this book the more traditional classification for these Cenozoic forms of Riedel (1967), rather than that given by De Wever et al. (2001), as, in addition to concerns about the biologic underpinnings given above, the vast majority of existing Cenozoic literature makes use of the Riedel classification and any new student of Cenozoic radiolarians must be fully aware of this traditional system. Differences between the two classifications for Cenozoic forms are given in Table 5.2, and, if different, the equivalents in the De Wever classification are given in [brackets] in the detailed listing below. Using this more "traditional" definition, the vast majority of Cenozoic species belong to only two orders: Spumellaria and Nassellaria, with only two (in the Cenozoic) low diversity families (Orosphaeridae and Saturnalidae) in the order Entactinaria. There are several additional Orders known only from the Mesozoic or Paleozoic, and a very high diversity of families (reflecting to some degree the above mentioned differences in taxonomic practice by workers specializing in different geologic time intervals). For pre-Cenozoic taxa we show only a few common and/or practically important family-level taxa, choosing preferentially those taxa that are robustly based on distinctive Baupläne. For each Cenozoic family the key morphologic characteristics are briefly described (of the skeleton only – see Chapter 2 for cellular anatomy), illustrated, the geologic range given, and important individual species or genera noted as appropriate. Total known Cenozoic species diversity for each higher taxon is given from the compilation of Cenozoic names by Lazarus et al. (2015). With only a few exceptions we do not give any other high level taxa for Polycystina other than the Order, or sub-family level taxa other than the Genus, as we feel that the stability of such intermediate ranked taxa is currently mostly still inadequate. We do note informal groupings as appropriate, though it should be noted that in some families many taxa are not included in any defined subgroup and thus one should assume that in all families with subgroup names there is also the group "other." We also provide guidance as to how important, in a practical sense, each Cenozoic family is by using the NSB database to estimate how commonly each family is reported in the deep-sea drilling literature (Table 5.3). Although undoubtedly an imperfect metric, the deep-sea drilling program reports are certainly the most important single body of literature for Cenozoic radiolarian research and thus the commonness of reported records in the database gives a rough estimate as to what extent taxa in each family are used in applied research. No estimates are given for pre-Cenozoic families as the NSB database has too few occurrence records from older literature to be useful and no other databases provide

Table 5.2 Comparison between higher level taxonomic names for Cenozoic radiolarians according to Riedel (1967) and as used in this book; and in De Wever et al. (2001). Primary differences between this book and the De Wever classification are in the latter elevation of many informal subfamily concepts to family rank, use of different names, and occasionally only partially overlapping core concepts for units.

Taxon Name	Taxon Rank	Family	Order	Typical Genera	Name(s) in De Wever book	Differences in Concept
NASSELLARIA	**order**					
Collozoidae	family				Collodaria (order)	molecular studies show this clade a subgroup within Nassellaria, thus demotion to family rank within Nassellaria order
Sphaerozoidae	subfamily	Collozoidae	Nassellaria	*Sphaerozoum*	family	
Thalassicollidae	subfamily	Collozoidae	Nassellaria	*Thalassicolla*	Thalassosphaeridae (family)	
Collosphaeridae	subfamily	Collozoidae	Nassellaria	*Acrosphaera, Collosphaera*	family	
Plagiacanthidae	family		Nassellaria		---	name used in DW for what is a separate subfamily concept in this book; no equivalent larger taxon concept given in DW
(plagonids)	subfamily	Plagiacanthidae	Nassellaria	*Neosemantis*	Plagiacanthidae (family)	
(lophophaenids)	subfamily	Plagiacanthidae	Nassellaria	*Peridium*	Lophophaenidae (family, partim)	
(sethoperids)	subfamily	Plagiacanthidae	Nassellaria	*Callimitra, Cladoscenium*	Sethoperidae (family)	
(sethophormids)	subfamily	Plagiacanthidae	Nassellaria	*Enneaphormis, Corocalyptra*	Sethophormididae (family), Theopiliidae (family, partim)	
(lithomelissids)	subfamily	Plagiacanthidae	Nassellaria	*Antarctissa, Lithomelissa*	Lophophaenidae (family, partim)	
Theoperidae	family		Nassellaria			
(cycladophorids)	subfamily	Theoperidae	Nassellaria	*Cycladophora*	Theopiliidae (family, partim)	
(pterocanids)	subfamily	Theoperidae	Nassellaria	*Pterocanium, Lychnocanoma*	Ultranaporidae (family)	

(eucyrtinids)	subfamily	Theoperidae	Nassellaria	*Eucyrtidium*	Eucyrtidiidae (family), Theocotylidae (family), Theoperidae (family)	
(lophocyrtids)	subfamily	Theoperidae	Nassellaria	*Lophocyrtis*	Lophocyrtiidae (family)	
(cornutellids)	subfamily	Theoperidae	Nassellaria	*Cornutella*	Acropyramididae (family)	
Artostrobiidae	family		Nassellaria	*Botryostrobus, Siphocampe*	-same-	
Trissocyclidae	family		Nassellaria	*Tholospyris, Dendrospyris*	Triospyrididae, Acanthodismiidae	
(stephanids)	subfamily	Trissocyclidae	Nassellaria	*Zygocircus*	Stephaniidae	
Pterocorythidae	family		Nassellaria	*Lamprocyclas*	-same-	
Carpocaniidae	family		Nassellaria	*Carpocanium*	Carpocaniinae (subfamily)	DW uses Carpocaniidae (family) to include Mesozoic forms with feet, etc. as subfamily Diacanthocapsinae
Cannobotryidae	family		Nassellaria	*Saccospyris*	-same-	
Nassellaria-other			Nassellaria	minor families		
SPUMELLARIA	**order**					
Actinommidae	family		Spumellaria	*Actinomma*	Actinommidae (family), Entapiidae (Paleocene–Eocene family), Rhizosphaeridae (family in order Entactinaria)	
(stylosphaerids)	subfamily	Actinommidae	Spumellaria	*Stylatractus*	Stylosphaeridae (family)	
(hexastylids)	subfamily	Actinommidae	Spumellaria	*Hexastylus*	Hexastylidae (family in order Entactinaria)	
(astrosphaerids)	subfamily	Actinommidae	Spumellaria	*Astrosphaera*	Astrosphaeridae (family)	
Coccodiscidae	family		Spumellaria		-same-	
(coccodiscids)	subfamily	Coccodiscidae	Spumellaria	*Lithocyclia*		
(artiscins)	subfamily	Coccodiscidae	Spumellaria	*Diartus*	Artiscinae (subfamily)	
Heliodiscidae	family		Spumellaria	*Heliodiscus*	-same-	

(Continued)

Table 5.2 (Continued)

Taxon Name	Taxon Rank	Family	Order	Typical Genera	Name(s) in De Wever book	Differences in Concept
Litheliidae	family		Spumellaria	*Larcopyle*	-same-	in DW placed in distinct superfamily together with Sponguridae, but DW notes this concept artificial; no real difference to other spongodiscids
Pyloniidae	family		Spumellaria	*Tetrapyle*	-same-	
Spongodiscidae	family		Spumellaria		-same-	DW separates most subfamilies as used by us into separate families
(spongodiscids)	subfamily	Spongodiscidae	Spumellaria	*Spongodiscus, Euchitonia*		
(stylodictyids)	subfamily	Spongodiscidae	Spumellaria	*Stylodictya, Stylochlamydium*		
(myelastrids)	subfamily	Spongodiscidae	Spumellaria	*Myelastrum*	Myelastridae (family)	
(spongurids)	subfamily	Spongodiscidae	Spumellaria	*Spongurus*	Sponguridae (family)	
(porodiscids)	subfamily	Spongodiscidae	Spumellaria	*Circodiscus*	Larnacillidae (family)	
Tholoniidae	family		Spumellaria		-same-	
Spumellaria-other			Spumellaria	minor families	Suttoniidae, Patulibracchiidae (Trimanicula, Plioc., Dumitrica1991RevMicropal), Pseudoaulophacidae (Pentapylonium implicatum, Recent, Dumitrica1991RevMicropal)	
ENTACTINARIA	**order**					
Orosphaeridae	family		Entactinaria	*Orosphaera*	-same-	in order Entactinaria
Saturnalidae	family		Entactinaria	*Saturnalis*		

Comparison between higher-level taxonomic names for Cenozoic radiolarians according to Riedel (1967), as used in this book; and in De Wever et al. (2001). Primary differences in the De Wever classification are elevation of many informal subfamily concepts to family rank; use of different names; and occasionally only partially overlapping core concepts for units.

Table 5.3 Relative frequency of higher-level taxa in species occurrence data in the NSB deep-sea microfossil database. Three-quarters of all reported species occurrences are for Nassellaria, only 25% for Spumellaria. Two to three families in turn make up ca. 75% of all occurrences for each order: Actinommidae and Coccodiscidae for Spumellaria; Theoperidae, Pterocorythidae and Artostrobiidae for Nassellaria. These frequencies reflect the relative utility of different taxa in applied studies, particularly biostratigraphy, and differ substantially from the true relative ratios of diversity or abundance in the full fossil assemblages.

Family	Order	NSB count	% of Order	% of all Radiolaria	
Actinommidae	S	11,129	40.7	10.4	
Coccodiscidae	S	9,421	34.5	8.8	
Spongodiscidae	S	5,275	19.3	4.9	
Litheliidae	S	731	2.7	0.7	
Pyloniidae	S	483	1.8	0.5	
Heliodiscidae	S	274	1.0	0.3	
		total Spumellaria:			**27,313**
					25.5%
Theoperidae	N	40,571	50.8	37.9	
Pterocorythidae	N	10,878	13.6	10.2	
Artostrobiidae	N	10,101	12.7	9.4	
Trissocyclidae	N	10,008	12.5	9.3	
Plagiacanthidae	N	3,644	4.6	3.4	
Carpocaniidae	N	1,735	2.2	1.6	
Cannobotryidae	N	1,596	2.0	1.5	
Collosphaeridae	N	1,252	1.6	1.2	
Amphipyndacidae	N	6	0.0	0.0	
Unknown	---	8	0.0	0.0	
		total Nassellaria:			**79,799**
					74.5%
				Total Occurrences:	**107,112**

equivalent coverage (radiolarian data in the PBDB or GBDB are databases still limited to a few time intervals and regions).

Formal Classification of Polycystina

Cenozoic Taxa

Order Spumellaria Ehrenberg 1876

(Figures 5.8–5.14)

Definition: Polycystina with diverse variations on spherical point symmetry, often with multiple concentric shells connected by radiating beams, and lacking an initial spicule in the interior of the shell.

Geologic range: Paleozoic to Recent

Subtaxa: described in separate entries below

Remarks:

Described species of Cenozoic Spumellaria make up only about one third the total number of described Cenozoic Polycystina (Lazarus et al., 2015), and only about a quarter of published occurrence records (Table 5.3). Despite substantial progress in spumellarian morphologic and systematic studies, particularly by Dumitrica and co-workers (summarized in De Wever et al., 2001), our understanding of the taxonomy and systematics of this order remain (relative to the Nassellaria) poor. The complex, multi-layered shells of

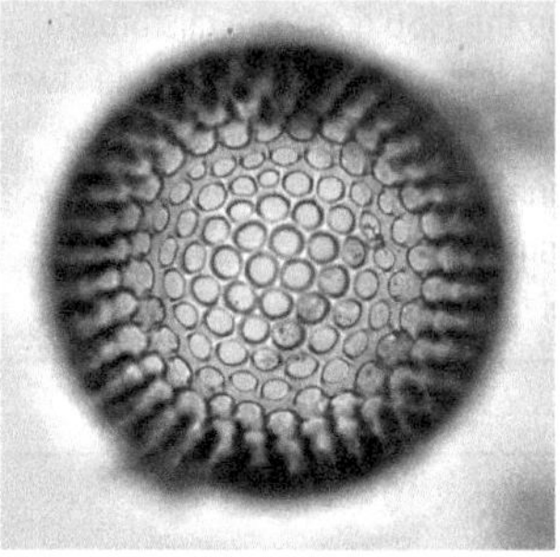

1. Cenosphaera elysia?

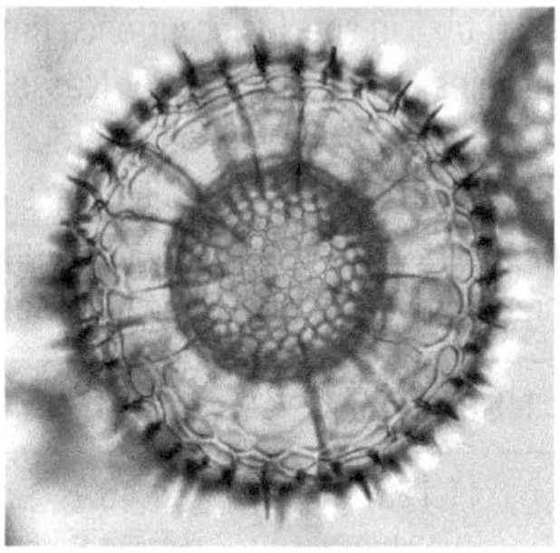

2. Liosphaera antarctica

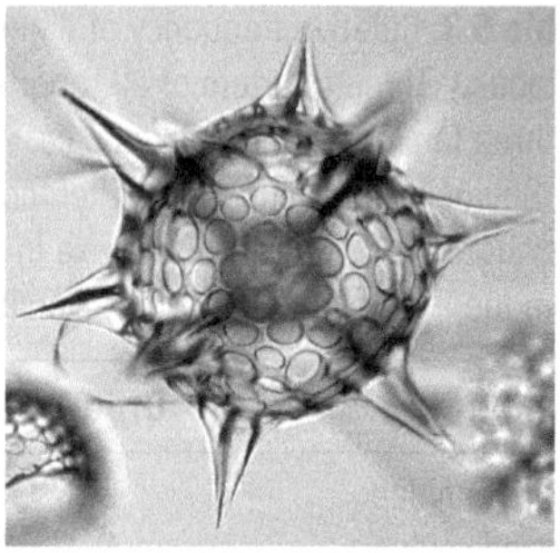

3. Actinomma golownini

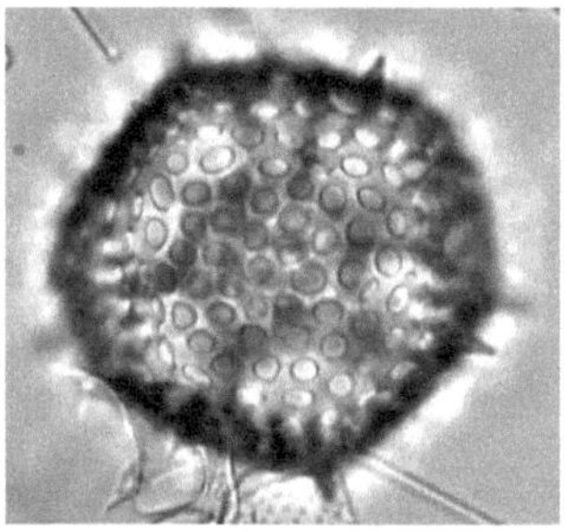

4. Haliomma miocenica

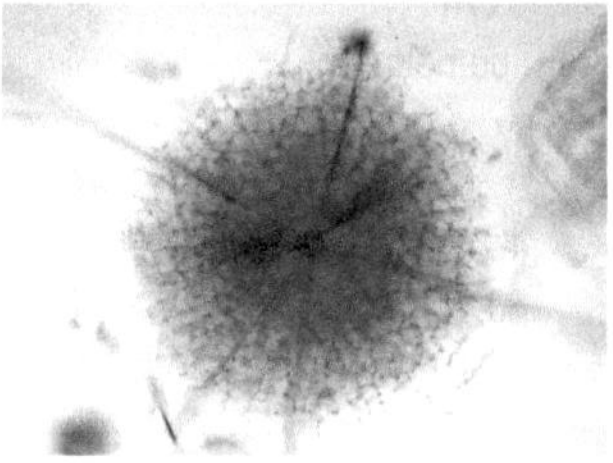

5. Centrocubus octostylus

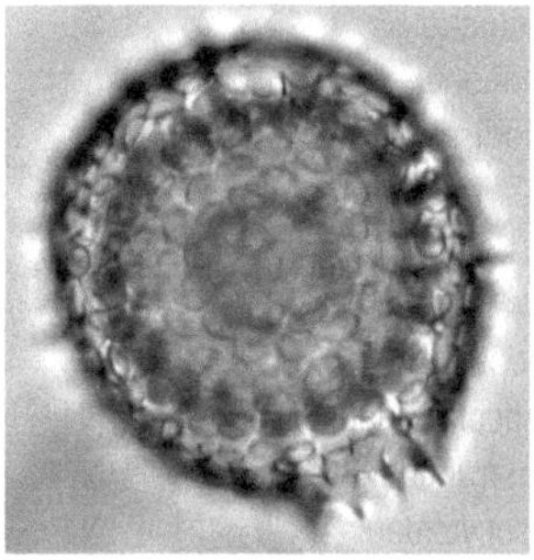

6. Sphaeropyle antarctica

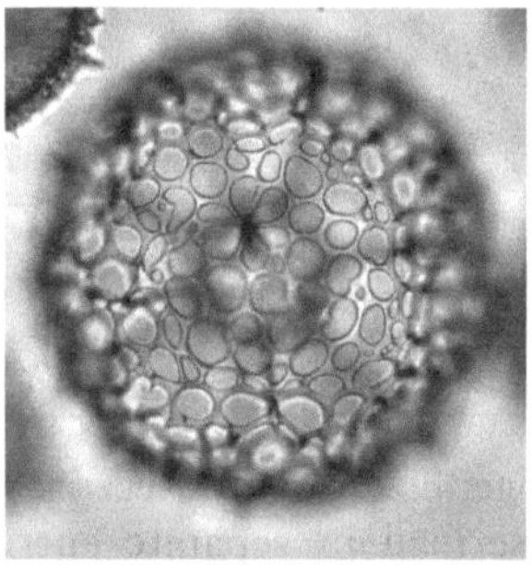

7. Haliomma macrodoras

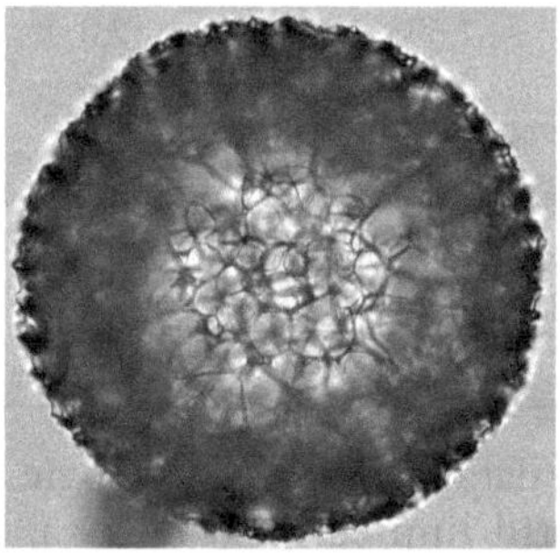

8. Rhizosphaera antarctica?

9. Arachnosphaera dichotoma

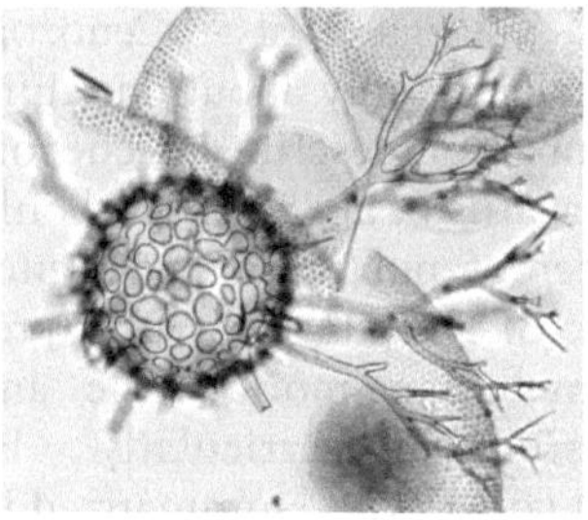

10. Cladococcus sp.

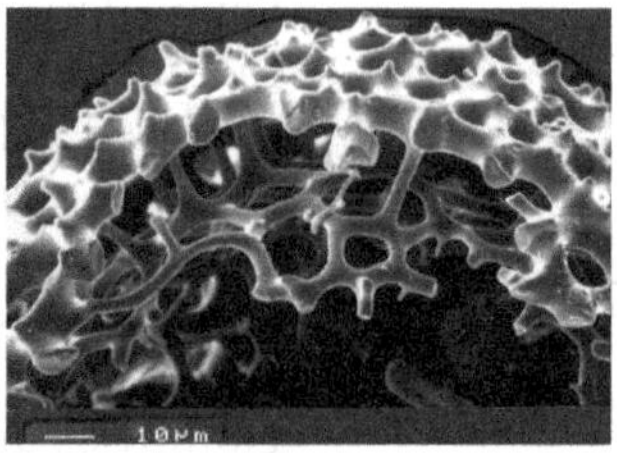

11. Carposphaera Q mellitosphaera

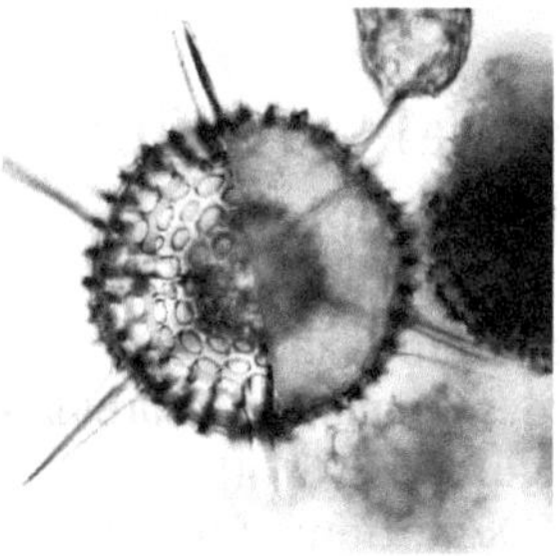

12. Actinomma popofski

many forms make both observation of structure and assignment of specimens to described taxa in routine studies often difficult. As noted, there are also unresolved differences in opinion as to the utility of characters. Lastly, although the majority of radiolarian cytologic work has been done on living Spumellaria (summarized in Anderson, 1983), integration of cytology and skeletal taxonomic characters is still largely lacking.

Family Actinommidae *Haeckel 1862 (Figures 5.8–5.10, 5.15)*

Definition: Spherical to ellipsoidal radiolarians with one or more mostly concentric lattice shells, connected by radial beams, sometimes with radial external spines (Figure 5.15). A diverse, probably polyphyletic group (e.g. Sandin et al., 2020). Name as used in De Wever et al. (2001) is much narrower in scope due to elevation of many subgroups to family status.

Explanation of plate captions

Species name: short description identifying key characters and illustrating meaning of morphologic terms (composite if image made by merging multiple focal plane images of same specimen); sample name or other source information - NN abbreviation of image library or person who took image. Key: JR Johan Renaudie (Antarctic Neogene); DL David Lazarus [MM_DL - DL Antarctic mid Miocene]; PG Patrick Grünert; vdP van de Paverd Ph.D. Thesis 1995 (surface sediments, tropical Indian and Pacific Oceans). Other sources given in full e.g. Haeckel, 1887.

Figure 5.8 Explanation of plate captions

1 – Actinommidae. *Cenosphaera elysia?*: Simple actinommid with single cortical shell, medium sized, regular, close-packed circular framed pores; 690B-7H-2 27/29cm JR. 2 – *Liosphaera antarctica*: Large medullary shell with small regular close-packed circular pores and strongly contrasting thin, large polygonal pored cortical shell, numerous connecting beams, thorny shell surface; 747A-7-6, 45, 16.2/137.4 b MM_DL. 3 – *Actinomma golownini*: Large pored, subcubic shaped cortical shell, 14 external large, tapered bladed spines, 6 only extend as beams to medullary shells (composite); Antarctic Miocene DL. 4 – *Haliomma miocenica*: Irregularly shaped thin-walled cortical shell with medium-small subcircular, irregularly arranged pores, numerous short external spines; 701C-10-6, 056 cm DL. 5 – *Centrocubus octostylus*: Small polygonal medullary shell, very thick spongy cortical layer and extremely long, serrated external spines; Challenger Plankton, Station 352 DL. 6 – *Sphaeropyle antarctica*: Two spherical medullary shells, outer very large, ellipsoidal cortical shell with prominent pylome, short external spines; 745B-22-4, 053 cm DL. 7 – *Haliomma macrodoras*: Near spherical thin-walled cortical shell with medium but variable sized, irregularly shaped pores, medium-small medullary shell, attachment points of beams to cortical shell depressed, pores at attachment points altered to petallate form; 751A-13H-CC JR. 8 – *Rhizosphaera antarctica?*: Polygonal inner medullary shell, surrounded by medullary meshwork, attached via numerous thin beams to thin-walled spongy cortical shell; 747A-1H-3 45/47cm JR. 9 – *Arachnosphaera dichotoma*: Similar to *Centrocubus* but large spherical polygonal pored medullary shell; 747A-2H-1 45/47cm #1 a JR. 10 – *Cladococcus* sp.: Subspherical thin-walled cortical shell with subrounded polygonal pores, rough surface, numerous long, distally anatamosing external spines; Antarctic DL. 11 – *Carposphaera? mellitosphaera*: Double cortical shell, outer moderately thick, small pored, rough surface, inner thin, irregular polygonal pores, numerous connecting thin cylindrical connecting beams (SEM); vdP Pl. 16 f5b. 12 – *Actinomma popofski*: Common two medullary + one cortical shell actinommid shell type, ca. 8–12 external bladed spines extending internally as bladed beams, cortical shell subspherical with flattened faces, pores medium-sized subrounded, framed (composite); RC9-132-855cm MM_DL.

Geologic range: early Mesozoic to Recent

Subtaxa: Several informal subgroups are distinguished (with matching names in De Wever (2001)):

Actinommids (Actinommidae, Entapiidae, Rhizosphaeridae [in Entactinaria]) – One or more spherical shells and either no well-developed external spines, or numerous

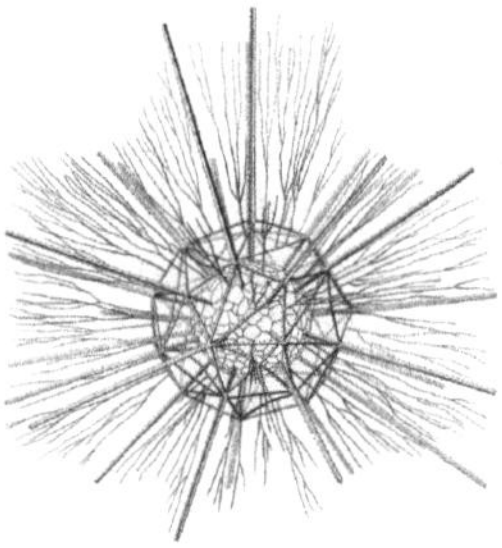

1 Drymosphaera dendrophora

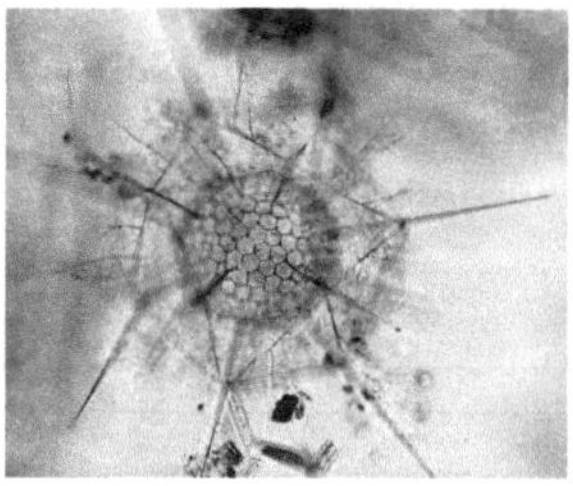

2 Astrosphaera hexagonalis

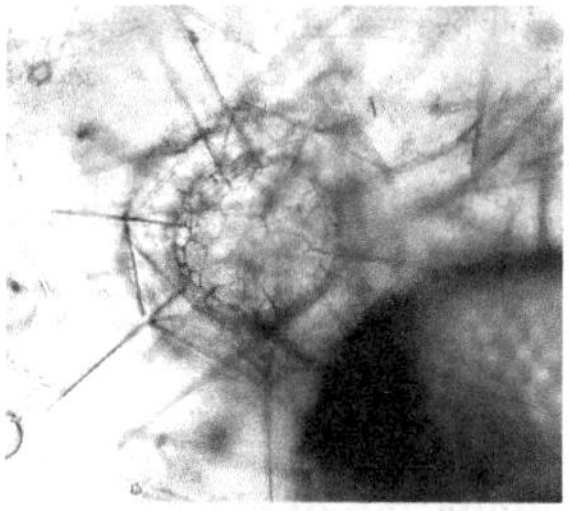

3 Drymospaera dendrophora?

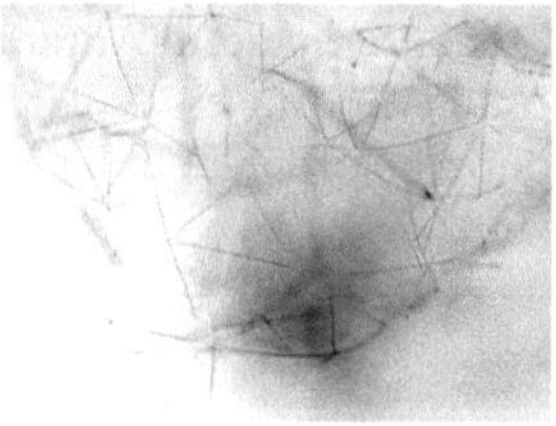

4 astrosphaerid? Challenger plankton

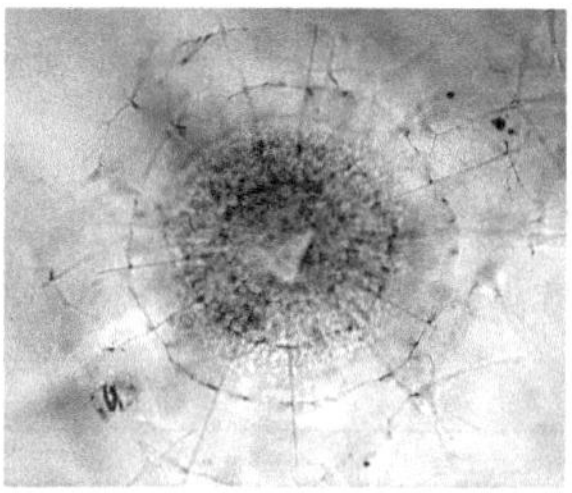

5 Arachnosphaera sp.

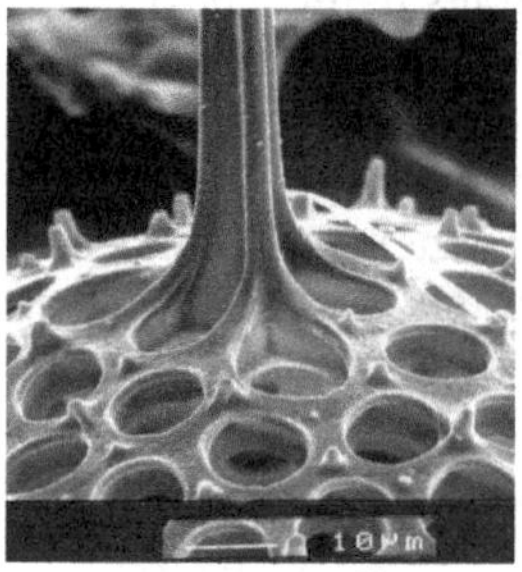

6 Hexastylus hexagonium

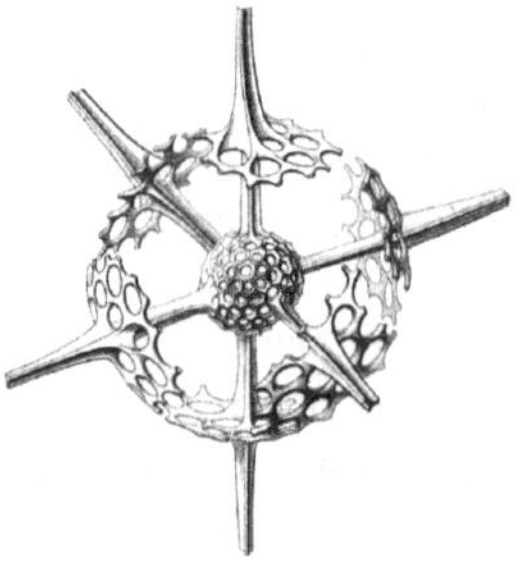

7 Hexacontium axotrias

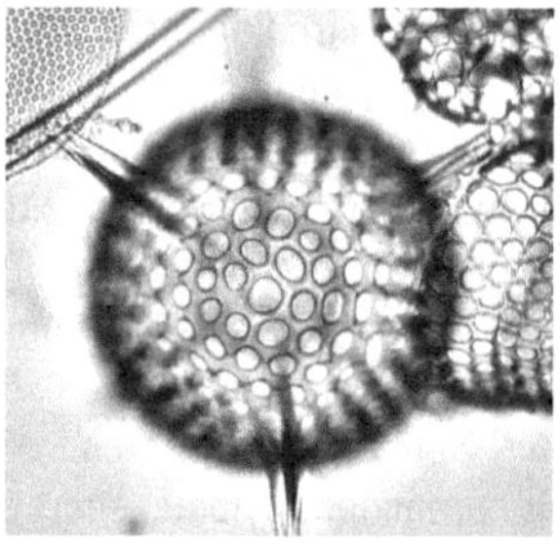

8 Hexacontium pachydermum

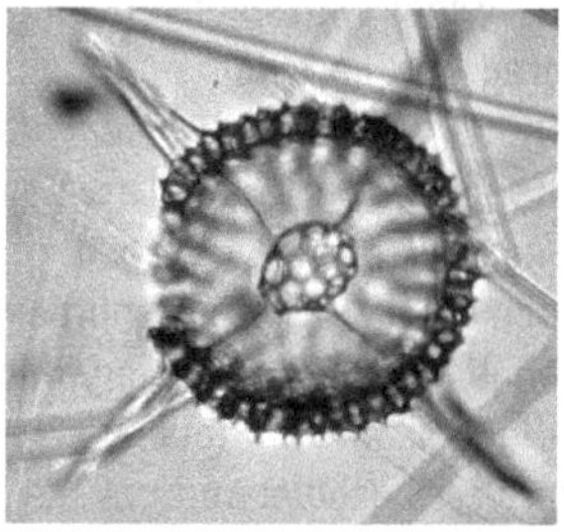

9 Hexacontium pachydermum

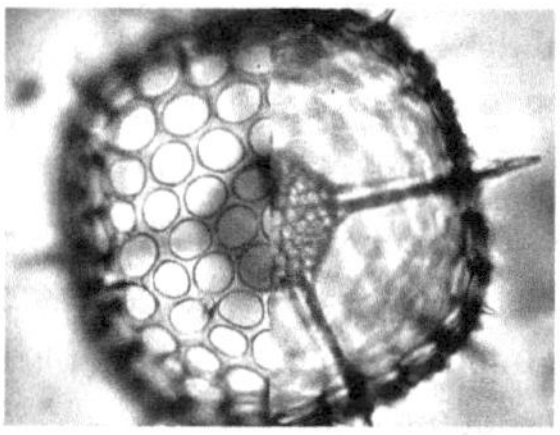

10 Hexalonche aristarchi composite

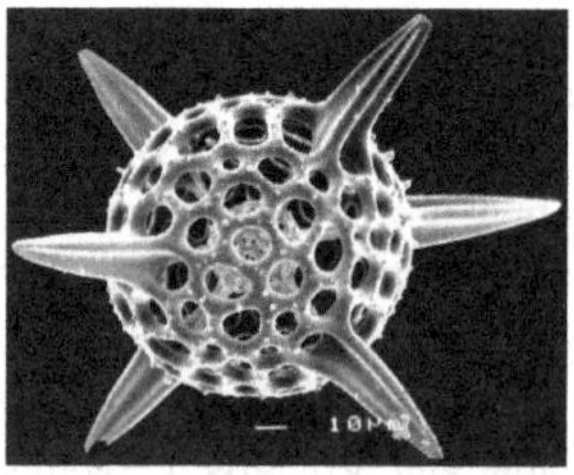

11 Hexacontium pythagoreas

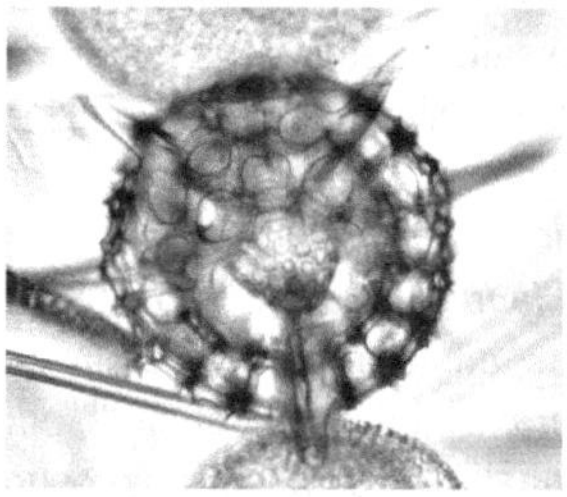

12 Hexastylid cf. Actinomma boreale composite

(more than 6), generally not showing any clear geometric alignment. Most species have a double medullary shell. Included here however are genera such as *Cenosphaera*, which by definition have only a single cortical shell, although at least some species in the genus may in fact possess a very delicate, normally not preserved medullary shell.

Stylosphaerids (Stylosphaeridae) (Figure 5.16A) – Spherical or ellipsoidal shelled Actinommidae with usually two primary external spines aligned co-axially at either (ellipsoidal) end of the shell.

Hexastylids [Hexastylidae (in Entactinaria)] – Actinommidae with six external primary spines in cubic orientation (perpendicular to the faces of an enclosing cube).

Astrosphaerids [Astrosphaeridae] – Acti nommidae with large numbers of concentric shells, often with outermost shells as open hexagonal meshes. Usually delicately constructed and thus, except possibly for the (unidentifiable) inner cortical shell only infrequently encountered as fossils.

Remarks:

Also included here in Actinommidae are multi-shelled spherical radiolarians with a variety of non-spherical medullary shells including those with spongy, poorly understood medullary structures and taxa with simple "spicular" structures (e.g. *Rhizosphaera*, Figure 5.8, f. 8), which in the De Wever et al. (2001) classification are mostly placed in the order Entactinaria.

Actinommidae are fairly common in most Cenozoic assemblages and, with 209 valid species in the Lazarus et al. (2015) compilation of Cenozoic names, are, by a factor of more than two, the most diverse group of Cenozoic spumellarians. Despite having relatively simple, more or less geometric forms, species can be defined on many characters: relative and absolute shell sizes, relative size, shape and ornamentation of pores and bars, and external spines, among others. Many of these characters are only moderately different between different described species and can be difficult to adequately illustrate in a

Figure 5.9 Actinommidae (astrosphaerids and hexastylids). 1 – *Drymosphaera dendrophora*: Drawing by Haeckel 1887 (Plate 20, Figure 1) showing typical astrosphaerid structure of hexagonal mesh cortical shell, serrated main spines. 2 – *Astrosphaera hexagonalis*: Species similar to *D. dendrophora* but larger, regular pored medullary shell; Station 213 Challenger plankton PG. 3 – *Drymospaera dendrophora?*: Compare to figures 1 and 2; Challenger plankton Station 206 PG. 4 – phaeodarian?: Not all spherical hexagonal latticework shells are Polycystina; Challenger plankton Station 352 DL. 5 – *Arachnosphaera* sp.: Delicate, multiple concentric shells, here seen in optical cross-section, cytoplasm still present in center; Challenger plankton Station 352 DL. 6 – *Hexastylus hexagonium*: SEM of double cortical shell, outer with uniform circular unframed pores, numerous small thorns, radial partitions connecting shells forming hexagonal box frames around aligned pores, base of three-bladed external spine; vdP Pl. 18 fig. 2b. 7 – *Hexacontium axotrias*: drawing by Haeckel (1887, Plate 24, Figure 3) of typical hexastylid medullary and cortical shells connected by bladed beams extending as external spines. 8 – *Hexacontium* sp.: Specimen with focus on three (of six) bladed external spines (SEM); relatively thin cortical shell with subrounded, medium-sized irregularly shaped pores; 737B-7R-1 JR. 9 – *Hexacontium pachydermum*: Specimen with thick cortical shell and strongly framed pores, focus on interior thin beams and ellipsoidal, polygonal medullary shell; 748B-5H-5, 45 cm JR. 10 – *Hexalonche aristarchi*: Subcubic thin-walled cortical shell with large uniform, subcircular closely packed pores and fine pored, subpolygonal shaped medullary shell (composite); 744A-10H-4, 60 cm JR. 11 – *Hexacontium pythagoreas*: Typical hexastylid form with strong, bladed external spines; vdP Pl. 34. 12 (SEM) – hexastylid *cf. Actinomma boreale?*: Double, large pored cortical shells with numerous connecting beams, small pored, pear-shaped medullary shell (composite); 748B-5H-4, 45 cm JR.

limited number of published images. This, together with a frequent lack of differential diagnoses and comparative studies to similar species, has led to considerable uncertainty in assignment of specimens to published descriptions. Geographic and

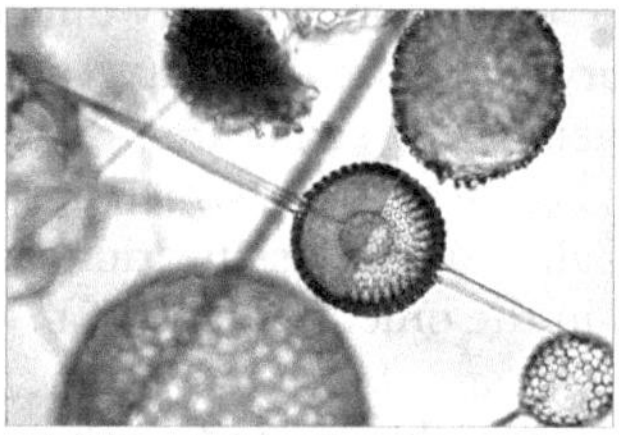

1. Stylosphaera minor composite

2. Stylatractid sp. composite

3. Stylatractus universus composite

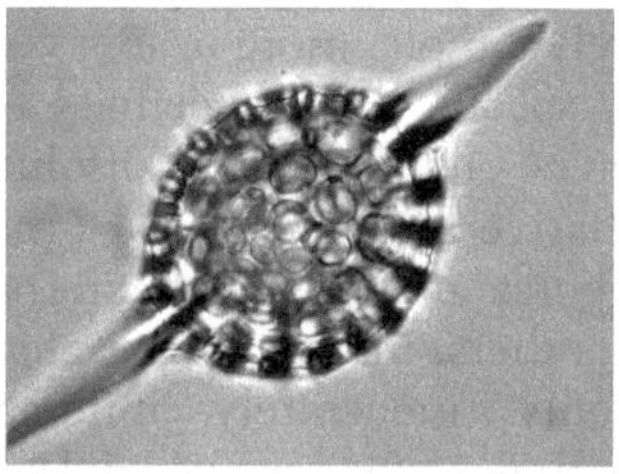

4. Stylatractus neptunus composite

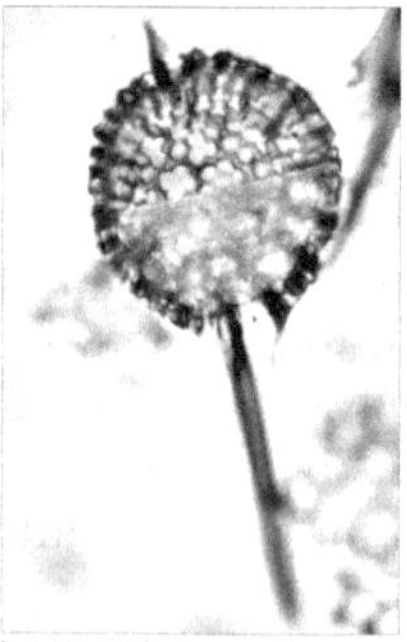

5. Sphaerostylus rosetta composite

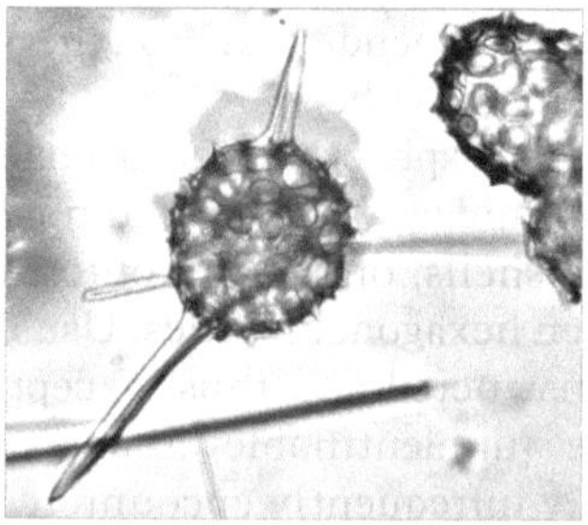

6. Druppatractus irregularis composite

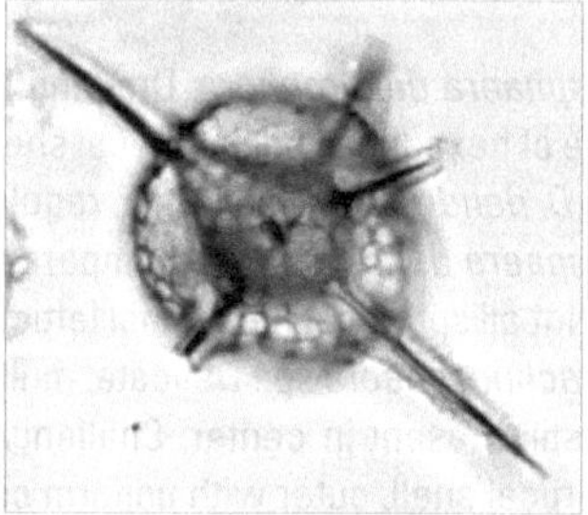

7. stylatractid sp. composite

8. spongy stylatractid composite

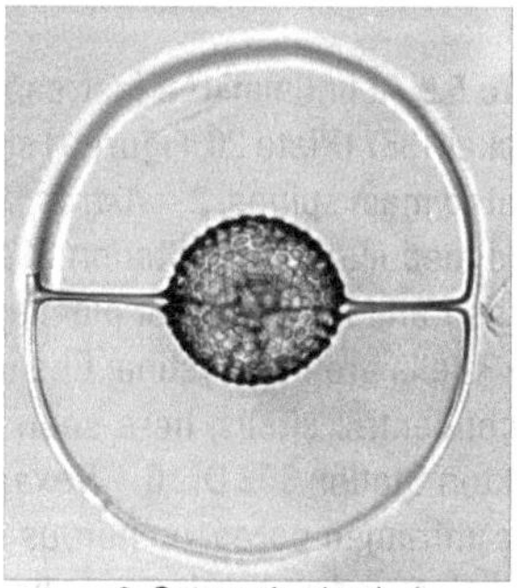

9. Saturnais circularis

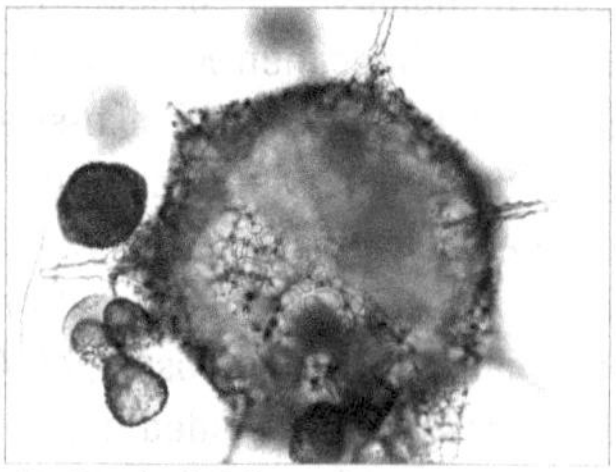

10. Orodapis? ferrealuma composite

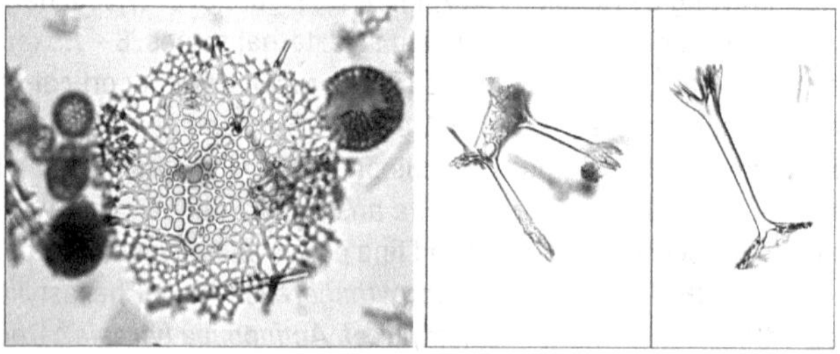

11. Oroscena carolae

ontogenetic variation in the group also seems to be, in comparison to many other radiolarian groups, fairly large, but is poorly understood (e.g. only rare studies such as Bjørklund, 1977, on *Actinomma haysi*), compounding the taxonomic uncertainties. Blueford (1982) gives a useful taxonomic analysis of tropical Miocene Actinommidae, although her review of prior literature was limited by the synthesis tools then available, and thus some of her newly described species are junior synonyms.

Actinommidae are the most commonly reported family of Spumellarians in deep-sea literature (41% of occurrence records in NSB: Table 5.3), though only 10% of all data in NSB.

Some species of stylosphaerids characteristically have more than two primary external spines (e.g. *Lithotractus timmsi*) and some specimens of these may have as many as five or six, projecting at angles nearly qualifying them as hexastylids; while evolutionary trends in some actinnomids towards reductions in spine numbers and regular, hexastylid-like cubic orientation (e.g. the evolution of *Actinomma golownini* [Figure 5.8, F3] from an unnamed ancestor – Lazarus, 1992) suggests that all these subgroups are closely related and also that they may not be monophyletic.

Family Heliodiscidae *Haeckel 1881 (Figure 5.11)*

Definition: Spumellarians with a double medullary shell, with the innermost located eccentrically. Cortical shell, in the original definition by Haeckel, was the key character – lenticular or discoidal, usually with radiating spines in equatorial plane. This definition was modified by the work of Hollande and Enjumet (1960), Dumitrica (1978), and others to be based on the (discovered by them) eccentric innermost medullary shell, and now includes genera with spherical or ellipsoidal cortical shells

Figure 5.10 Actinommidae (stylosphaerids, orosphaerids and Saturnalis). 1 – *Stylosphaera minor?*: Small, regular circular pores on cortical shell, thin-walled spherical medullary shell, long slender cylindrical spines (composite); 744A-7H-3 53/55cm JR. 2 – stylatractid sp.: Medium-large framed cortical pores, thick cylindrical unequal length tapered polar spines bladed at base (composite); 747A-10-1, 45 cm MM_DL. 3 – *Stylatractus universus*: Thick-walled, small pored ellipsoidal cortical shell, cylindrical polar spines nearly equal in length (composite); 747A-1H-5 45/47cm #2 a JR. 4 – *Stylatractus neptunus*: Large pored, thick-walled cortical shell, two ellipsoidal medullary shells, stout conical polar spines (composite); 744A-7H-5 53/55cm (111,4|2,6)b JR. 5 – *Lithatractus floridus*: Ellipsoidal cortical shell has large pores divided into rosettes of small pores, unequal bladed polar spines. Medullary shell absent in this specimen (composite); 744A-8H-3 53/55cm JR. 6 – *Druppatractus irregularis*: Bladed polar spines not colinear, large cortical pores, pear-shaped medullary shell (composite); 744A-2H-1, 53 cm JR. 7 – stylatractid*?* sp.: Incomplete specimen lacking cortical shell, with pear-shaped inner, irregular ellipsoidal outer medullary shell, bladed equatorial beams and polar spines (composite); 751A-11-4,98 cm DL. 8 – *Spongatractus pachystylus*: Spongy outer layer obscures cortical and medullary shells, strongly tapered bladed polar spines (composite); 605-22-1, 102 cm DL. 9 – *Saturnalis circularis*: Spherical, thin-walled cortical shell, tips of polar spines joined to bladed ring encircling shell; vdP Pl. 30 f3. 10 – *Orodapis? ferrealuma*: Very large polyhedral cortical shell with distinctive entactinarian lattice wall of fused, thin, randomly oriented silica strips forming irregular polygonal pores, external spines with buttressed bases extend internally as thin beams to small polygonal medullary shell (composite); 747A-2H-5, 47 cm JR. 11 1–3 – *Oroscena carolae*: Fragments showing entactinarian lattice wall and edges of polyhedron, distally thickening, terminally branched and forked cylindrical external spines; 744A cores 8-10 JR.

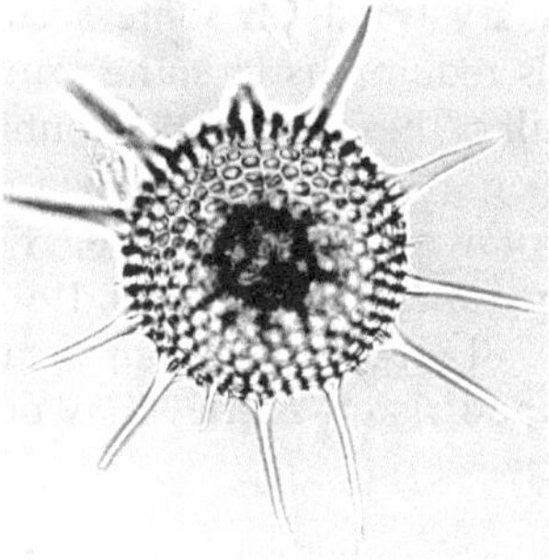

1. Heliodiscus asteriscus

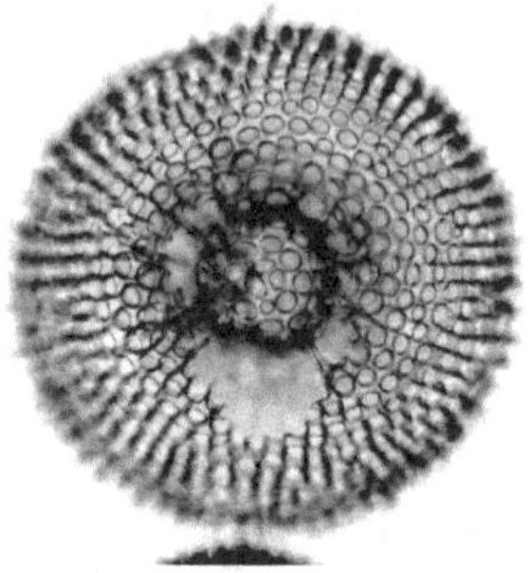

2. Heliodiscus sp. composite

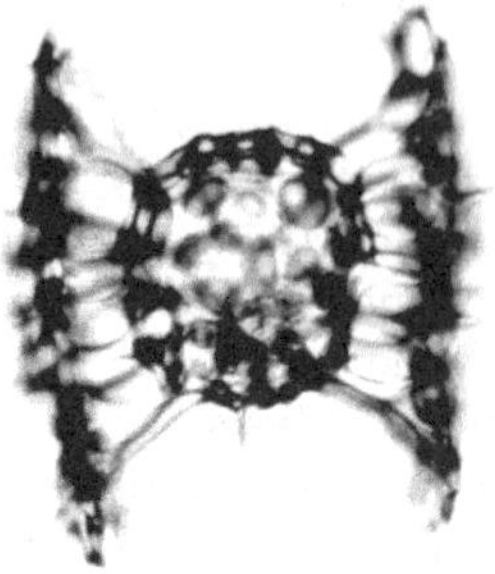

3. Heliodiscus sp.

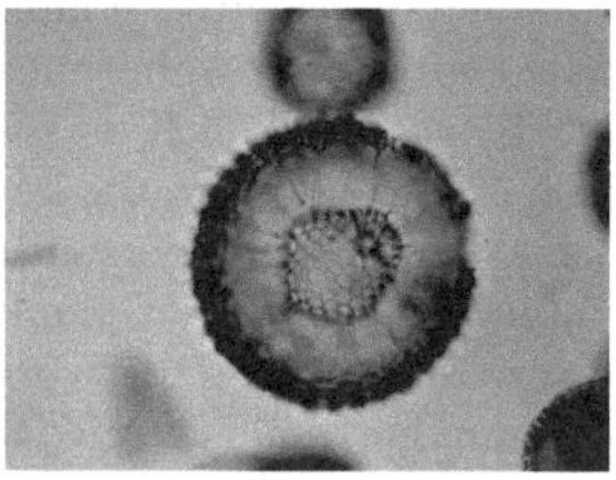

4. Excentrodiscus japonicus

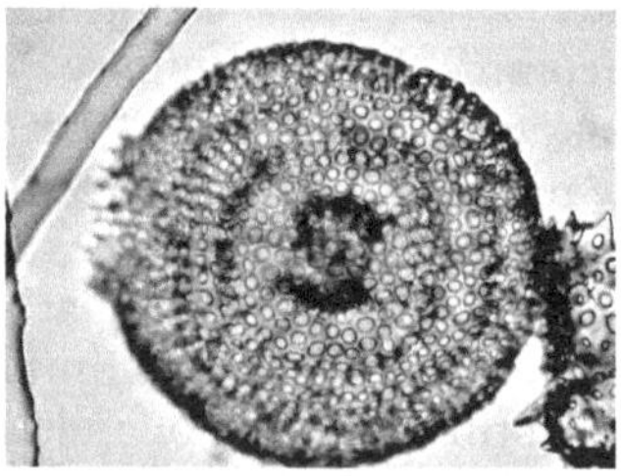

5. Circodiscus ellipticus

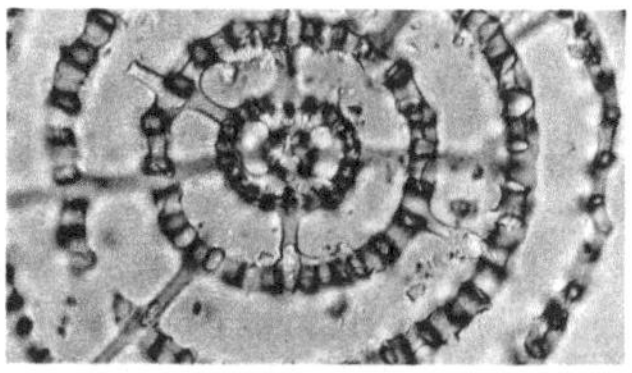

6. Circodiscus microporus

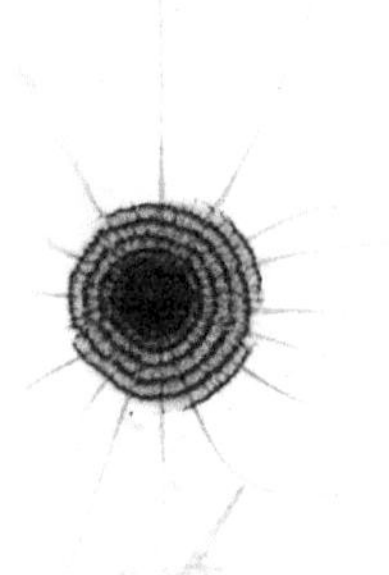

7. Stylodictya sp.

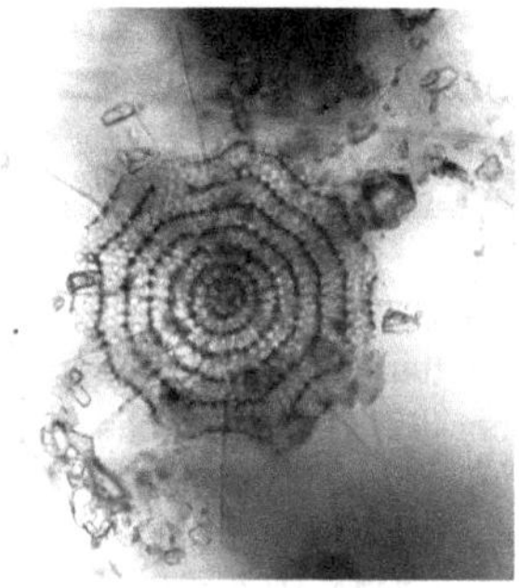

8. Stylochlamydium venustum

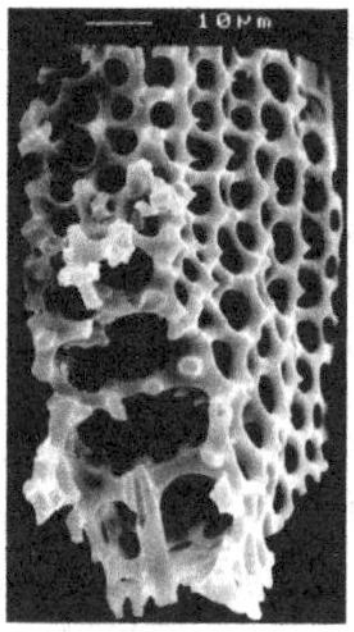

9. Stylodictya sp.

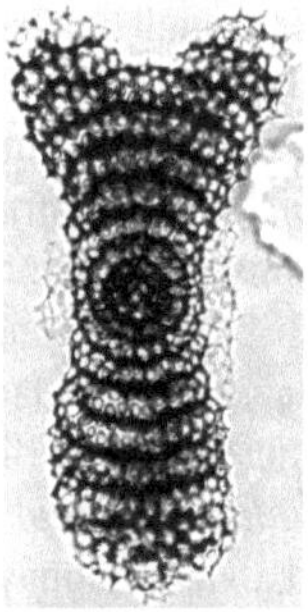

10. Amphirhopalum ypsilon

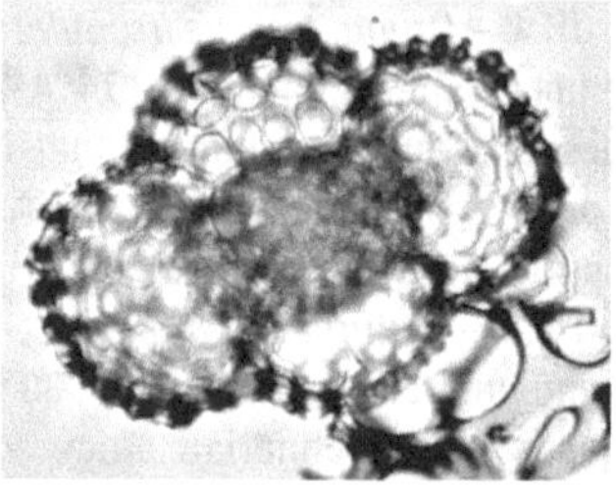

11. Cubotholus spp

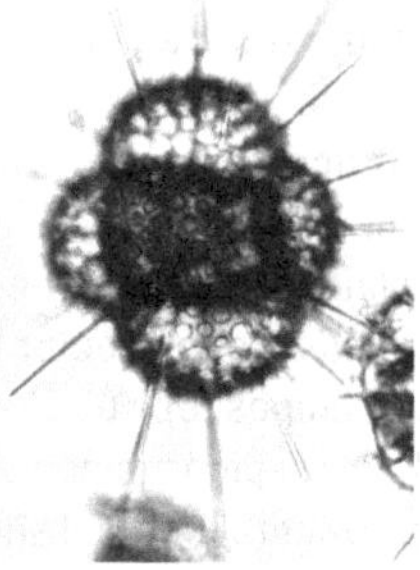

12. Amphitholus metallasson

more typical of Actinommidae, e.g. *Excentrodiscus* and *Excentrosphaerella*.

Geologic range: Paleocene to Recent

Subgroups: none

Remarks:

It is not uncommon to see in transmitted light observation other spherical spumellarians, e.g. presumed Actinommidae, with eccentrically positioned inner medullary shells. These have until now been assumed to be simply loose shells broken free from their beam attachments and thus lying against the wall of an enclosing shell, but more detailed examination may discover other actinommids that have a true eccentric medullary shell and thus belong to Heliodiscidae.

Heliodiscids, at least as currently known, are very low diversity and normally only one or two species, at low abundances, are found in a given assemblage. They are rarely reported on in the literature (0.3% of NSB data).

Family **Coccodiscidae** *Haeckel 1862, emend. Sanfilippo and Riedel 1980 (Figure 5.12)*

Definition: Spumellarians with non-spherical cortical shells, typically discoid or bipolar, and possessing a double medullary shell.

Geologic range: Early Eocene to Recent

Subgroups:

Coccodiscinae Haeckel 1862 - cortical shell discoidal or divided into sometimes spongy radial arms arranged in the equatorial plane (Eocene–Recent).

Artiscinae Haeckel 1881, emend. Riedel 1967 - cortical shell typically bilobed polar extending perpendicular to the

Figure 5.11 Heliodiscidae, stylodictyids, Tholoniidae. 1 – *Heliodiscus asteriscus*: Lenticular thin-walled cortical shell, long, cylindrical spines in equatorial plane, two spherical medullary shells, inner eccentrically located on wall of outer; vdP Pl. 36 f2f. 2 – *Heliodiscus?* sp.: Broken specimen missing spines, central area of cortical shell, exposing medullary shells; 747A-10-2, 45 cm MM_DL. 3 – *Heliodiscus* sp.: Incomplete specimen in side view consisting of medullary shells and beams connecting to rudimentary cortical shell; vdP Pl. 37 f6. 4 – *Excentrodiscus japonicus*: Broken spherical cortical shell exposing irregular shaped outer medullary shell and small, spherical inner eccentrically located medullary shell 689B-3R-5, 136 cm JR. 5 – *Circodiscus ellipticus*: Discoidal, thin, small circular pored cortical shell with large pylome, two internal rings plus central oblong larnacilla structure; 744A-10H-2 JR. 6 – *Circodiscus microporus*: Dissected cross-section of shell in equatorial plane with radial cylindrical beams and central larnacilla structure; Dumitrica 1989, Pl. 9 f7. 7 – *Stylodictya* sp.: Flat, discoidal cortical shell, small circular pores, with numerous internal rings (septa), spongy central area, long, delicate, partically curved spines radiating from disc edge in equatorial plane; Station 271 Challenger plankton PG. 8 – *Stylochlamydium venustum*: Flat, discoidal cortical shell, small circular pores, internal rings increasingly irregular with distance from center, partially discontinuous, concave outwards, thin radiating equatorial spines extending internally as beams; St 206. 9 – *Stylodictya* sp.: SEM of broken section of discoidal shell with small cortical pores, pored internal rings fragmentary radial beam; vdP Pl.52 f1bf [as *Perichlamydium arachium*]. 10 – *Amphirhopalum ypsilon*: Flattened shell in form of two opposed radially chambered strips, one bifurcating distally, center with spherical medullary shells, cortical shell thin, surface with small circular pores, delicate veil surrounding central part of shell; vdP Pl. 51 f4 [as *Perichlamydium wirchowii*]. 11 – *Cubotholus* sp.: Flattened hemispherical lobes along dimensive axes, rounded polygonal, uniform medium closely packed pores in cortical shell; 744A-8H-4, 62 cm JR. 12 – *Amphitholus metallasson*: Similar to *Cubotholus* sp. (f. 11) but cortical pores smaller, round, shell with numerous thin cylindrical radiating external spines; vdP Pl. 63 f9f.

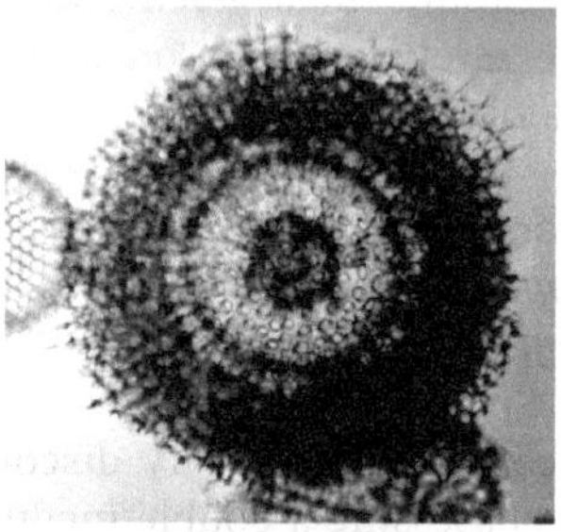

1. Lithocyclia ocellus composite

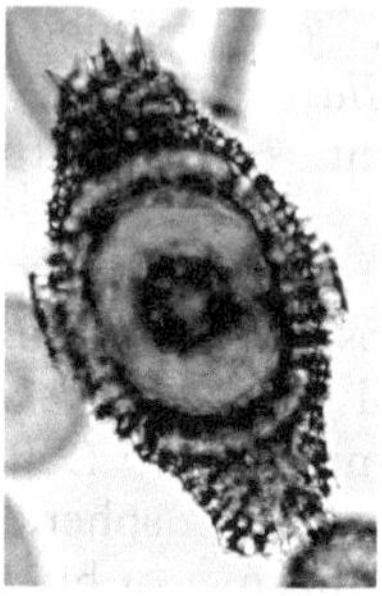

2. Didymocyrtis cf. laticonus

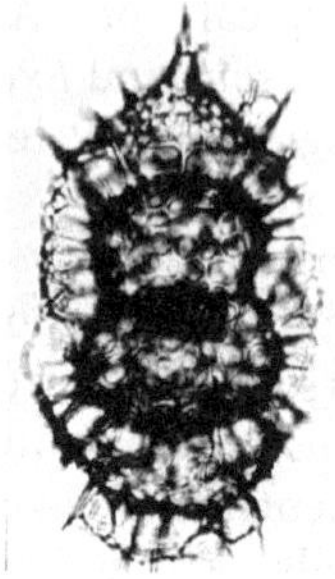

3. Didymocyrtis tetrathalamus

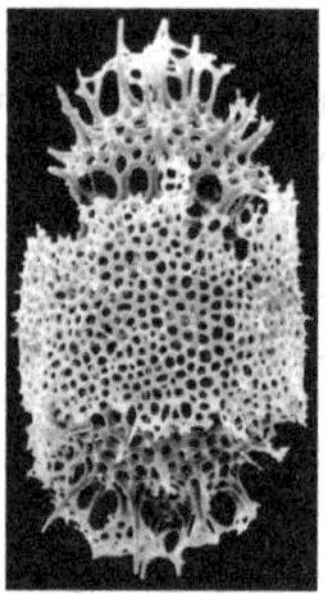

4. Didymocyrtis tetrathalamus

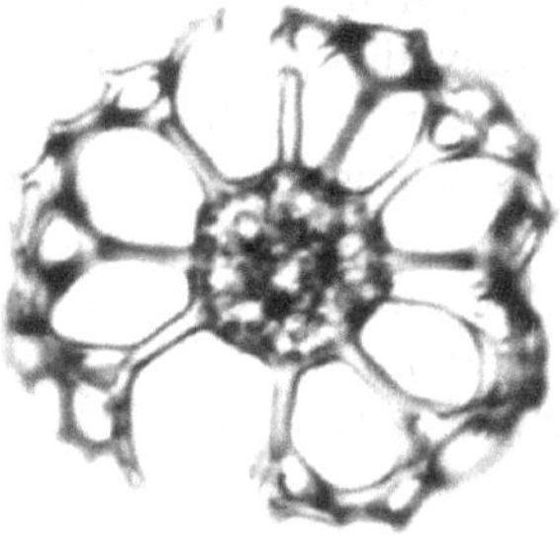

5. Didymocyrtis medullary shell

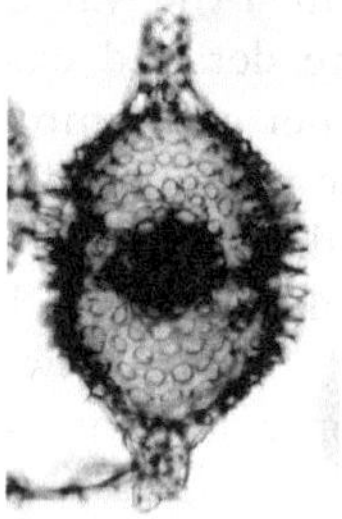

6. Didymocyrtis violina composite

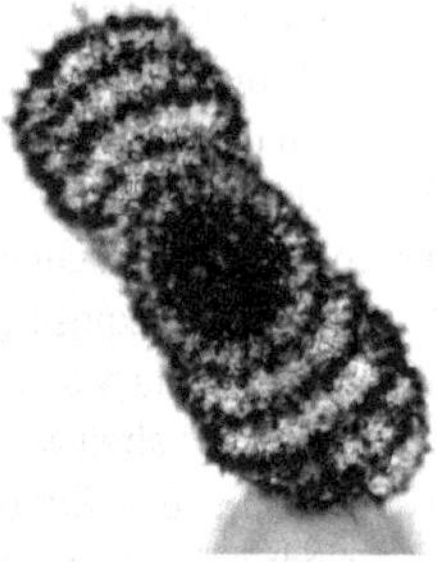

7. Amphymenium challengerae

8. Amphymenium amphystylium composite

9. Spongocore puella

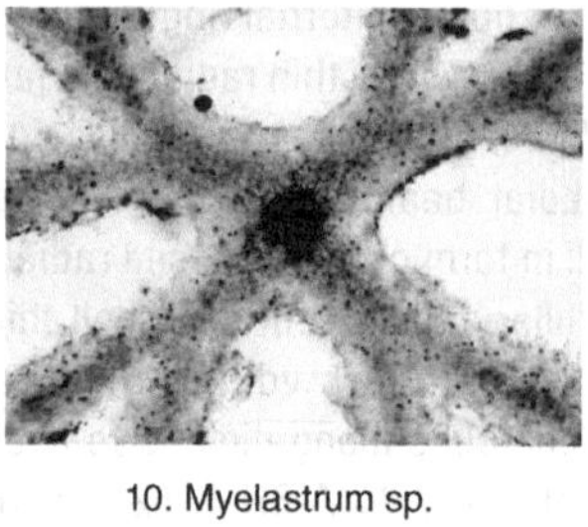

10. Myelastrum sp.

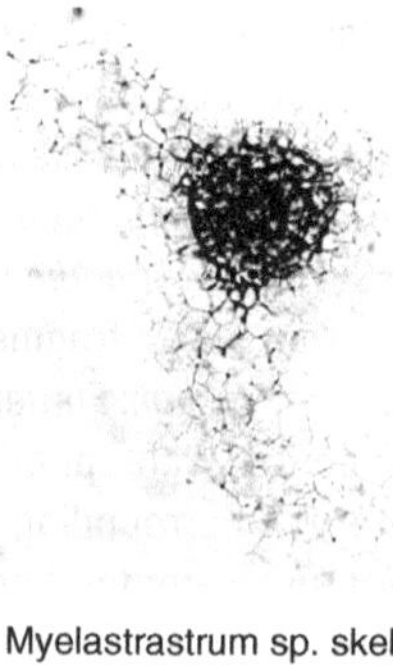

11. Myelastrastrum sp. skeleton

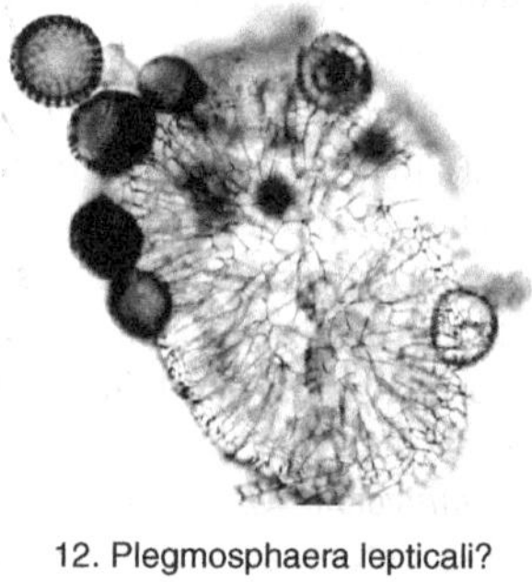

12. Plegmosphaera lepticali?

equatorial axis defined by the beams connecting the cortical and medullary shells (Oligocene–Recent).

Remarks:

Artiscinae is comprised of only a very small number of gradually evolving lineages whose origin and evolutionary relationships have been well studied, and that have been subdivided into anagenetic chronospecies for low- to mid-latitude biostratigraphy, particularly the genera *Didymocyrtis* and *Diartus* (see Chapter 8: Stratigraphy). Species are distinguished by the shape of the cortical lobes, development of chambered polar arms ("caps"), and the development in some forms of small ridges ("plica") on the cortical shell lobes. The group originated from the Coccodiscinae in the late Oligocene (Sanfilipppo and Riedel, 1980). Some species of Coccodiscinae are also stratigraphically important in the Paleogene.

Coccodiscids are, next to the Actinommidae, the most commonly reported spumellarian group (35% of occurrences/9% of all radiolarian data), primarily due to the use in both Paleogene and Neogene of species in standard biostratigraphic zonations. There are only 33 valid Cenozoic species and thus they not very diverse.

Family Pyloniidae *Haeckel 1881 (Figures 5.13, 5.16B)*

Definition: Spumellarians with a *larnacilla* structure: lattice shells made of a nested set of ellipsoidal bands, larger bands attaching at a constant (for the taxon) angle to the outer tip of the preceding inner band and successively oriented on different axes of three-dimensional symmetry. The remaining open regions between the inner and outer enclosing bands of shell are called *gates*.

Geologic range: late Cretaceous?/ Paleocene to Recent

Subgroups: many proposed by different authors but not included here. Two

Figure 5.12 Coccodiscidae, Spongodiscidae. 1 – *Lithocyclia ocellus*: Lenticular cortical shell, small circular pores, spherical medullary shells, circular ring at mid-radial distance, spongy shell interior in outer part of shell (composite); 605-22-1, 102 cm DL. 2 – *Didymocyrtis cf. laticonus*: Ellipsoidal (double?) cortical shell with equatorial constriction, broad, short spongy polar caps; 746A-6H-2, 53 cm JR. 3–5 *Didymocyrtis tetrathalamus*: Three different specimens of the same species, illustrating within-species variablity and differences in imaging. 3: Transmitted light, specimen with partial external veil and many spines; vdP Pl. 39 f10. 4: Similar morphology but SEM – only external features visible; vdP Pl. 39 f9. 5: Incomplete specimen, axial view of medullary shell; vdP Pl. 39 f1. 6: Plankton specimen, complete veil, no external spines; Station 271 Challenger PG. 6 – *Didymocyrtis violina*: Ellipsoidal cortical shell with equatorial constriction, thin, spongy tubular polar caps (composite); 573B-14-5, 120 cm DL. 7 – *Amphymenium challengerae*: Ellipsoidal cortical shell, large spongy? ellipsoidal medullary shell, bipolar chambered arms broader than cortical shell; 737B-3R-3 53/55cm JR. 8 – *Amphymenium amphystylium*: Long, narrow chambered bipolar arms expanding in width distally from small indistinct center, tubular veil around central, narrow portion of shell, pylome-penetrating chambers towards center at tip of one arm (composite); RC9-132, 855 cm MM_DL. 9 – *Spongocore puella*: Bipolar spongy arms with indistinct chambers, substantial width expansion distally, rounded tips, pylome with spines at one end, double veils around central part of shell; 744A-8H-1 JR. 10 – *Myelastrum* sp.: Very large shell of spongy bifurcating arms in equatorial plane, only central shell strongly silicified, specimen with numerous symbiont cells (dark dots); Station 271 Challenger plankton PG. 11 – *Myelastrum* sp.: Isolated skeletal material – central medullary region and weakly silicified arm; vdP Pl. 51 f10. 12 – *Plegmosphaera lepticali*: Very large ellipsoidal shell of compact, thin, anatamosing radial beams surrounding nearly hollow center; 748B-6-2, 45 cm JR.

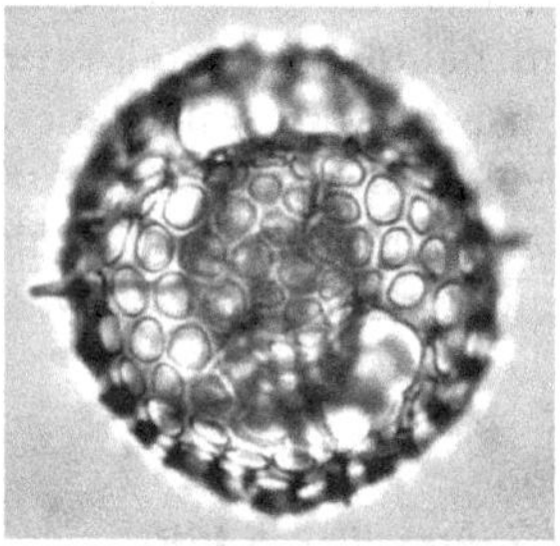

1 - Larcospira quadrangula?

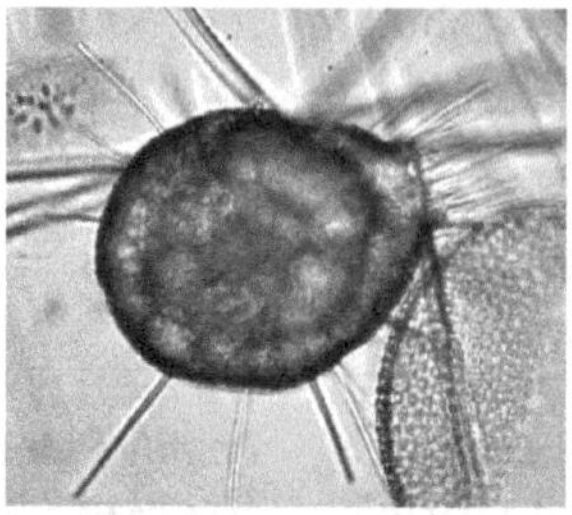

2 Larcopyle labyrinthusa composite

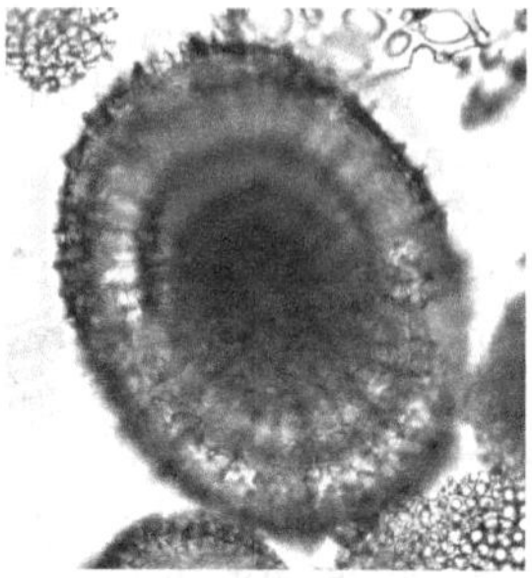

3 Larcopyle polyacantha composite

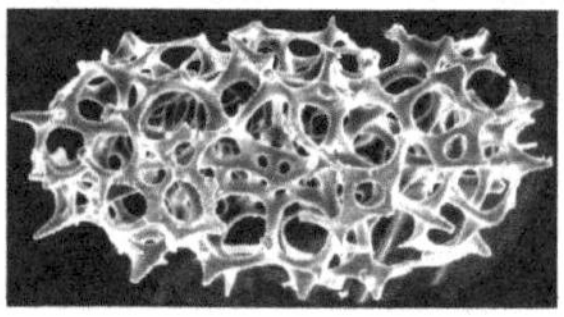

4 - Larcopyle buetschli

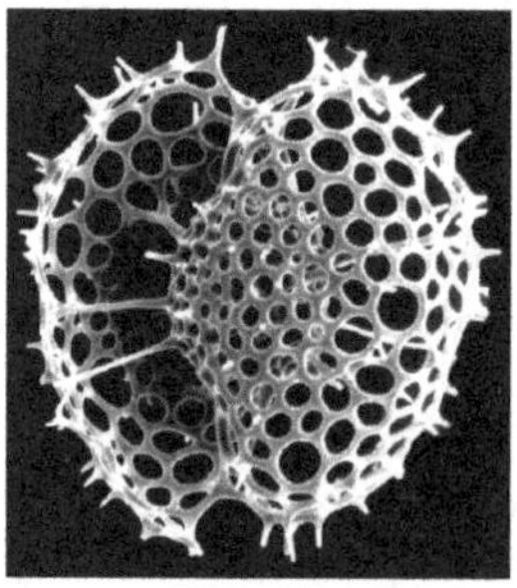

5 - Larcopyle quadrangula

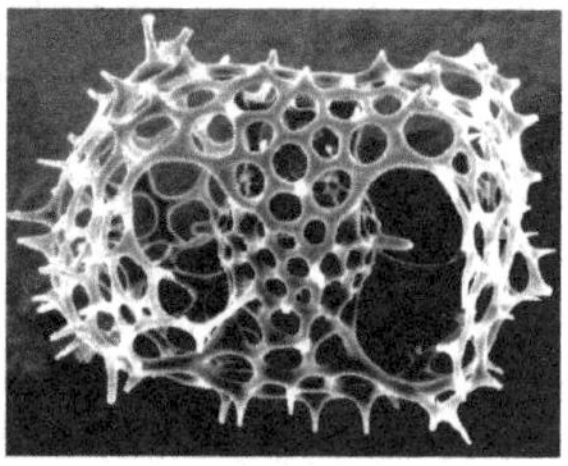

6 - Tetrapyle octacantha

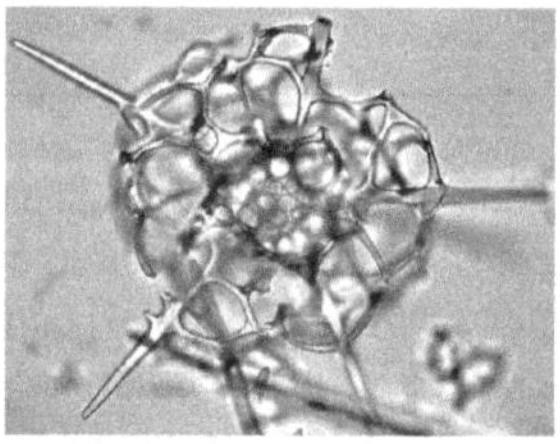

7 Pentapylonium implicatum composite

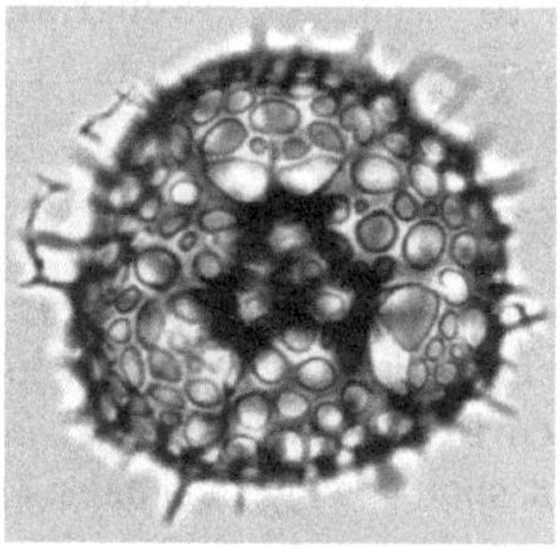

8 - Hexapyle

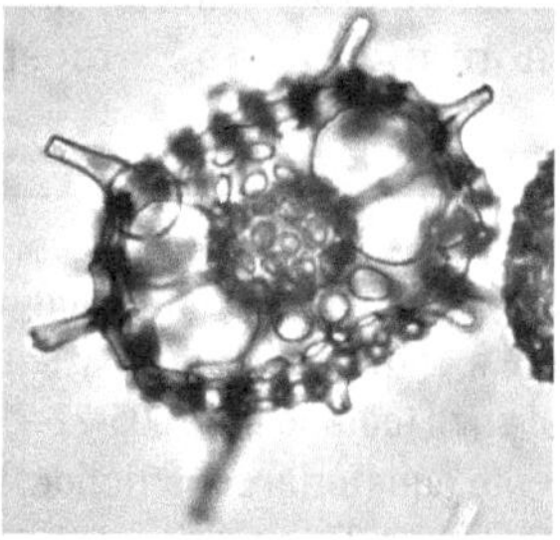

9 - Octopyle stenozona

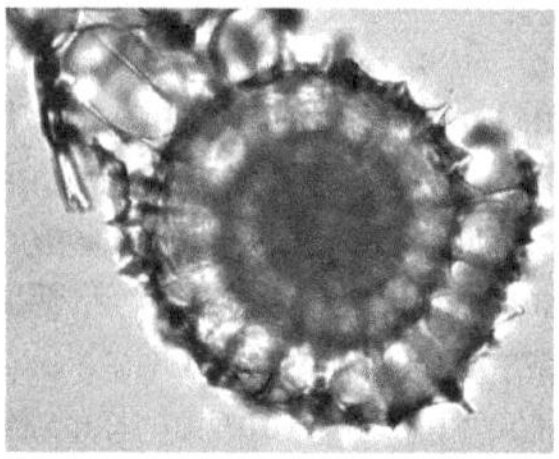

10 - Lithelius nautiloides

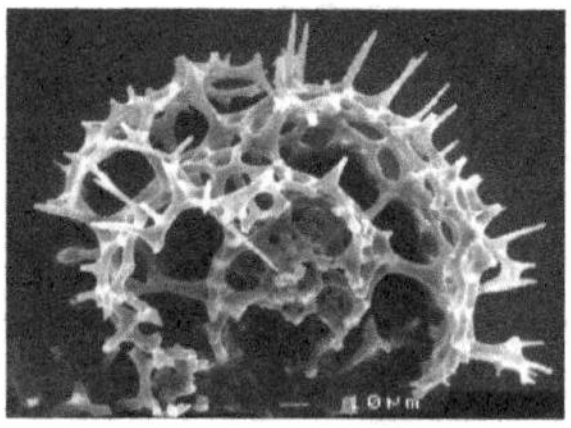

11 lithelid SEM showing merged spiral

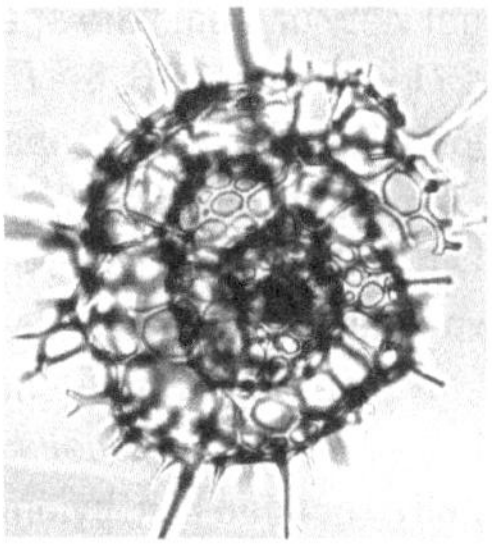

12 Phortium inner structure w. banded lattice development

geometric types can be distinguished. *Tetrapyle* and similar forms have 90° angles between bands; *Hexapyle* and related forms have 120° angles.

Remarks:

Pyloniids are the simplest of several groups of spumellarians possessing, at least in some members, a larnacilla structure, e.g. many Lithelidae. The optical appearance of larnacilla structures varies according to the angle at which they are viewed and can appear in some orientations to be similar to the coiled structures typical for lithelids, although lithelids can possess true larnacilla structures as well. Lastly, the spiral shells in lithelids can, via the irregular development of lattice wall, sometimes give the appearance of having gates. Determining the true nature of structures is thus difficult and despite much improvement in the last decades (several studies by Dumitrica e.g. 1988, 1989, 2020; summary in De Wever et al., 2001), the higher level systematics of all these forms remains a matter of debate. The stratigraphic range is also, depending on how widely the group is defined, a bit uncertain.

Pyloniids as defined here are, with 17 valid Cenozoic species in Lazarus et al. (2015), only moderately diverse and are only rarely reported in the deep-sea radiolarian literature (2% of spumellarian occurrences, ca .5% of all radiolarian data).

Zhang and Suzuki (2017) also recently proposed a major revision of living pyloniids, lithelids, and related forms based on shell structures, with 10 genera and 34 species or subspecies recognized just for the Recent time interval.

Family Lithelidae *Haeckel 1862 (Figures 5.13, 5.16C)*

Definition: Spumellarians with shells consisting of single or multiple spirals, generally in a single plane, emanating from a simple spherical or larnacilla medullary shell, and having either an open termination or ending in an enclosing, usually ellipsoidal cortical shell. The cortical shell may possess a pylome, an opening, typically at one pole of the ellipsoidal shell, frequently surrounded by distinctly differentiated lattice wall and/or radiating spines.

Geologic range: basal Mesozoic? to Recent

Figure 5.13 Pyloniidae, Lithelidae. 1 – *Larcospira* sp.: Double spiral merging at outer surface to form ellipsoidal shell, thin wall, medium-sized pores, thin bars, smooth surface; 746A-8H-CC, 22 cm JR. 2 – *Larcopyle labyrinthusa*: Thin-walled cortical shell with large, spine-surrounded pylome, widely spaced inner whorls, small number of internal beams (composite); 751A-11H-2, 98 cm JR. 3 – *Larcopyle polyacantha polyacantha*: Ellipsoidal shell of multiple, tightly spaced spirals with numerous thin beams penetrating as external spines, small irregularly distributed cortical pores, rough surface, pylome moderately developed (composite); 747A-4H-1, 45 cm JR. 4 – *Larcopyle?* sp.: SEM of incomplete specimen showing small pored central structure, coarse lattice, and penetrating spines; vdP Pl.54 f6. 5 – *Larcopyle quadrangula*: Ellipsoidal, rapidly expanding double spiral, thin-walled shell (SEM); vdP Pl.55 f4. 6 – *Tetrapyle octacantha*: Rounded subquadratic shape formed by larnacilla bands of thin lattice wall (SEM); vdP Pl. 56 f5. 7 – *Pentapylonium implicatum*: Strong cylindrical radial spines from triangular pylonid central structure, shell of complex fenestrated, coarse lattice-work girdle bands in changing orientations (composite); 744A-3H-1, 53 cm JR. 8 – *Hexapyle*: Pylonid structure of nested bands, but with a triangular symmetry; vdP Pl. 55 f9. 9 – *Octopyle stenozona*: Larnacilla structure, similar to *Tetrapyle* but gate openings divided by beams; 747A-2H-7, 45 cm JR. 10 – *Lithelius nautiloides*: Single (sometimes double) moderately evolute coiled discoidal shell, moderate numbers of strong radial spines in coiling plane and rough, thorny surface; 744A-3H-1, 53 cm JR. 11 – lithelid: SEM showing merged spiral; vdP Pl.58 f2. 12 – *Phortium*: inner spirals w. banded (intermittent) lattice development; vdP Pl.57 f3.

Subgroups: several proposed by various authors but (see remarks below) none considered here to be robust enough to use.

Remarks:

In the De Wever et al. (2001) classification, family assignment is based solely on the interpreted central shell structure, and thus many classic lithelid species and genera are assigned by them to other families, with only a very few taxa remaining in Lithelidae. As noted elsewhere, despite the clear need for systematic revision, we retain the more classic family concept given the uncertain state of understanding of morphologic characters and still nascent development of radiolarian molecular systematics.

Cenozoic lithelids are moderately diverse with 24 accepted species, but are only infrequently reported in the literature (3% of spumellarian deep-sea occurrences/ 0.7% of all data).

Family Tholonidae *Haeckel 1887 (Figure 5.11)*

Definition: Spumellarians with nested lobed shells arranged in mutually perpendicular axes.

Geologic range: Oligocene? to Recent

Subgroups: none

Remarks:

Very low diversity and generally rare, the group is little studied and has not been used for applied research, e.g. stratigraphy (e.g. there are no occurrence records for tholonids in the NSB database). Consequently, even the stratigraphic range of the group is not well determined. De Wever et al. (2001) interpret the shell structure as a closed variant of the larnacillid structure. It is not clear, if so, if the group is then really a separate family or only a derived subset of another larnacillid-bearing spumellarian group.

Family Spongodiscidae *Haeckel 1862 (Figures 5.14, 5.17)*

Definition: flattened spumellarians with lenticular or radially arranged arms extending from a small medullary shell, and usually with at least partially spongy shell structure (Figure 5.17). Shell volume divided in many forms by concentric rings or spiraling interior walls, and often with radial spines embedded in and approximately in the equatorial plane of the shell.

Figure 5.14 Spongodiscidae. 1 – *Spongodiscus craticulatus*: Spongy disc, thick central area, radial bands with different thickness, numerous short,fine equatorial spines, pylome; 744A-7H-2 JR. 2 – *Spongodiscus osculosus*: Uniform spongy disc tapering in thickness towards perimeter, fine pored enclosing cortical shell, large pylome with spines; 744A-8H-1 JR. 3 – *Spongaster tetras*: Quadrate, flattened spongy shell divided into four larger lobes meeting central spongy area, thinner spongy layers between lobes; Station 271 Challenger plankton PG. 4 – *Spongodiscus biconcavus?*: Spongy disc (this specimen somewhat elliptical), pylome forms notch in outline; 737A-26X-1 JR. 5 – *Spongotrochus glacialis*: Spongy disc, thicker at center, 8–12 cylindrical beams extending as spines in equatorial plane; 747A-6H-2, 45 cm JR. 6 – *Dictyocoryne profunda*: Triagonal flattened shell with distally lobed spongy arms and thin bands of spongy webbing ("patagium") between arms, attached to lobes, central area showing closely spaced rings; 737A-27X-2 JR. 7 – *Schizodiscus favus*: Spongy disc with thickened center (side view); 744A-5H-1, 53 cm JR. 8 – *Euchitonia*: Three narrow spongy arms in a plane with angle smaller between two of them, irregularly developed patagium, central thickening (SEM); vdP Pl. 50 f5. 9 – *Euchitonia*: SEM close-up of spongy arm and patagial webbing; vdP Pl. 59 f2. 10 – spongodiscid: SEM of spongy, smooth-surfaced disc, central thickening and larger pore, presumably pylome (edge view); vdP Pl. 41 f1b. 11 – *Dictyocoryne truncatum?*: Similar to *D. profunda* but shorter arms, flattened arm tips (SEM); vdP Pl. 43 f5a. 12 – spongodiscid: Close-up of spongy arm and thin covering veil (SEM); vdP Pl. 43 f7.

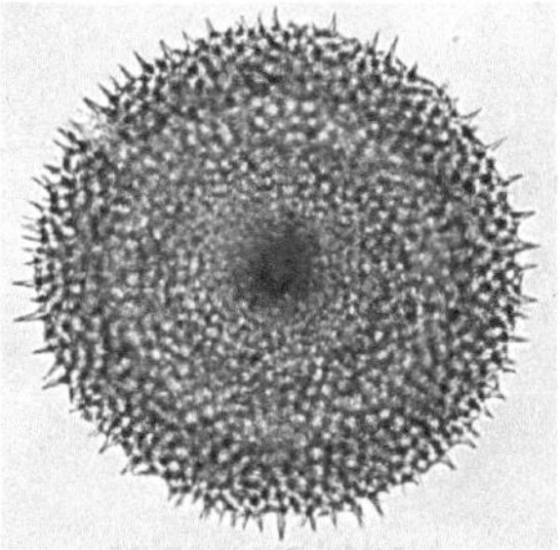

1 Spongodiscus craticulatus

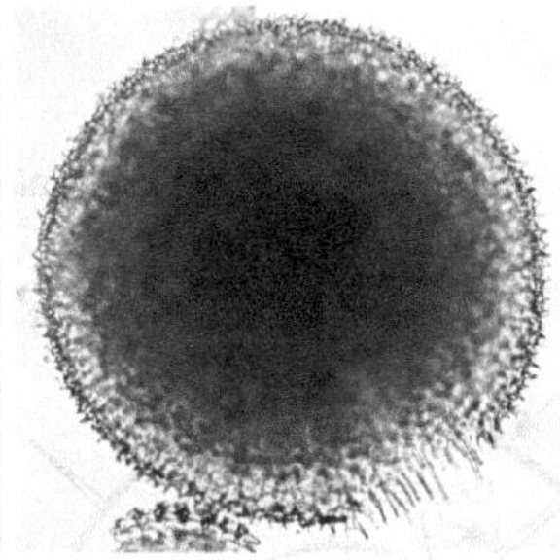

2 - Spongodiscus osculosus

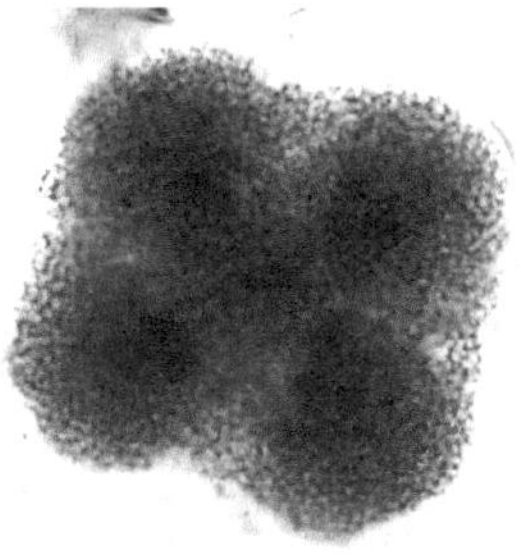

3 Spongaster tetras composite

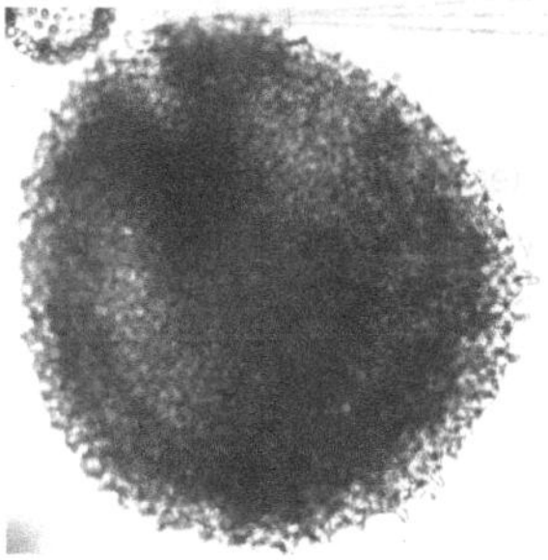

4 - Spongaster biconcavus?

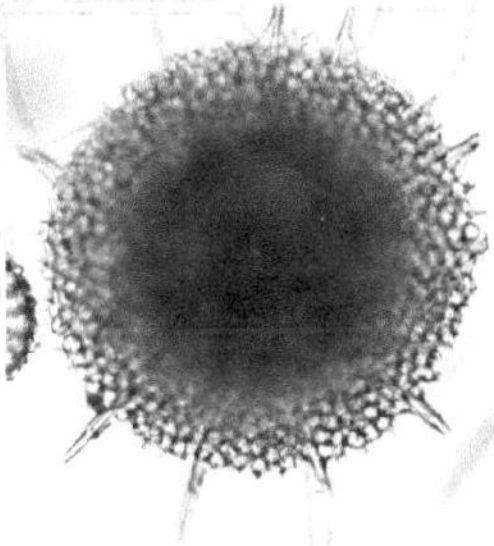

5 - Spongotrochus glacialis

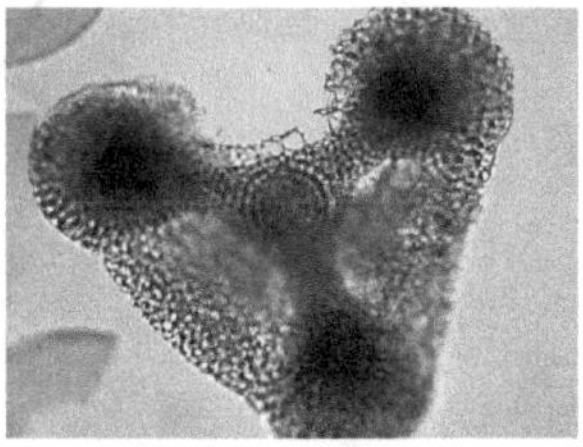

6 - Dictyocoryne profunda

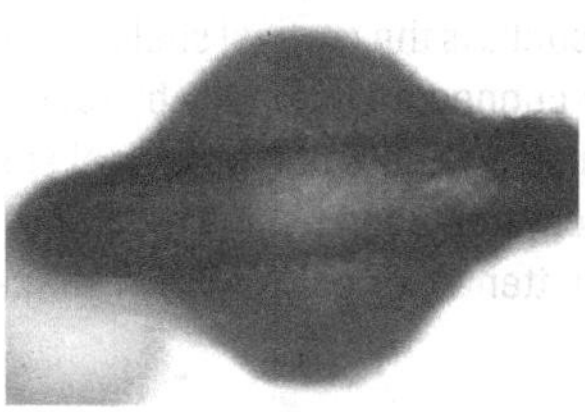

7 - Schizodiscus favus

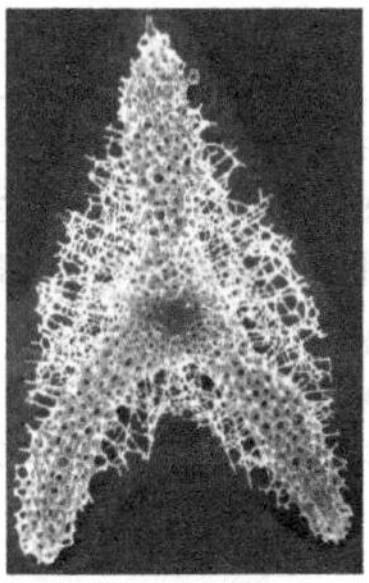

8 Euchitonia

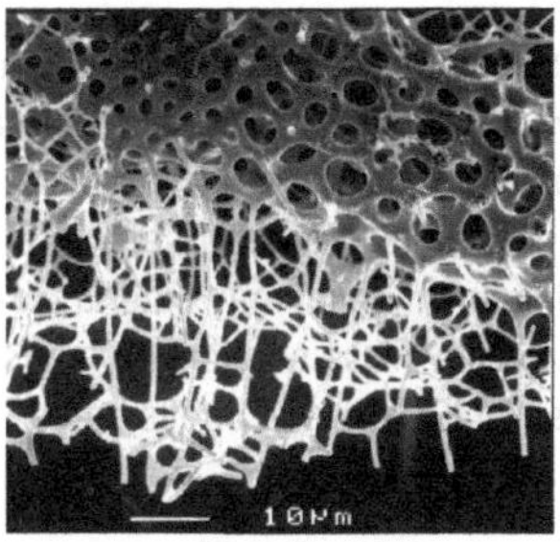

9 Euchitonia

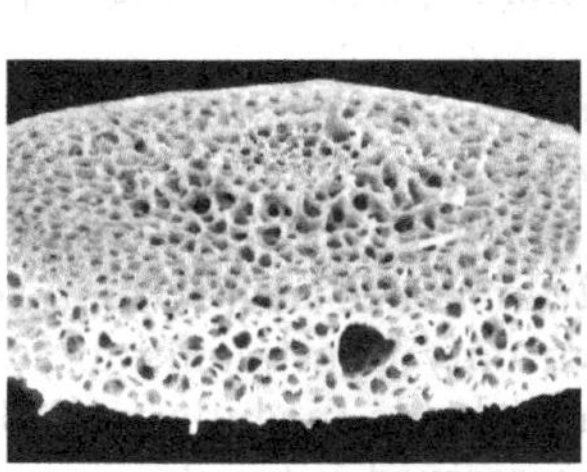

10 spongodiscid

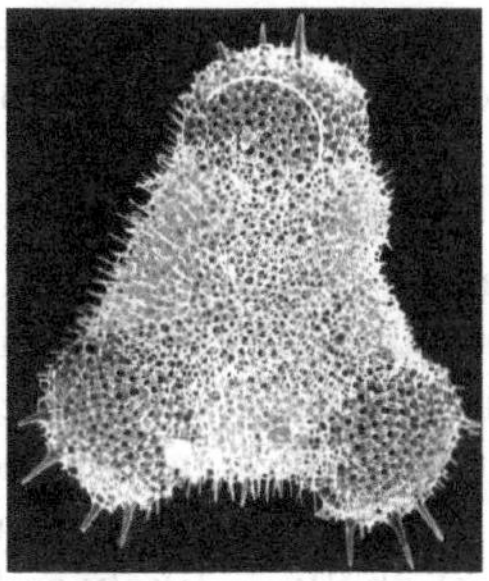

11 Dictyocoryne truncatum

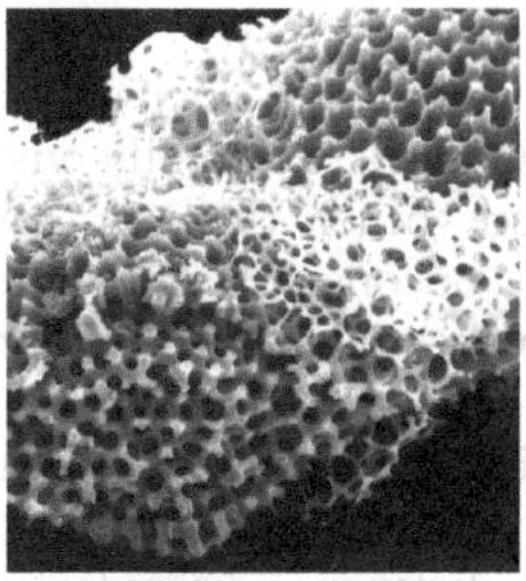

12 spongodiscid

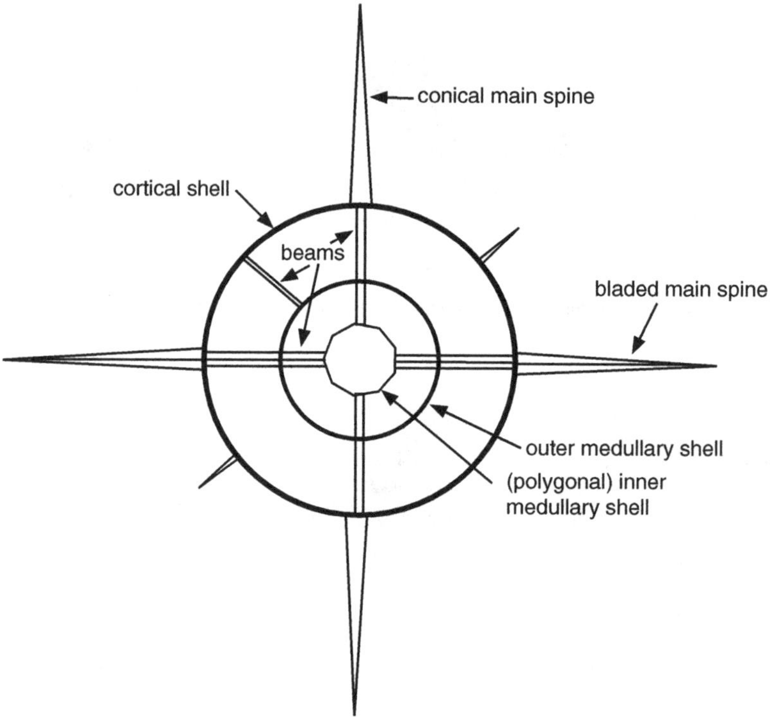

Figure 5.15 Main structural components of spumellarian shells. The simplest structural type is actinommid (shown here); most other spumellarian shells are variants, or have at least elements of this form. In the actinommid form the outermost in the series of concentric spherical shells is the cortical shell; the remaining, inner shells (if present) are usually called medullary. Shells are connected internally by radial struts called beams, which may extend externally as spines. Spines may also originate at the cortical shell and have various cross-sectional shapes – conical (circular cross-section) or bladed (typically 3 flat or lobed blades at 120° angles). Shells can be fully spherical or be variously flattened, or polyhedral in shape.

Geologic range: Triassic?/Late Cretaceous to Recent

Subgroups:

Spongodiscids – shells are discoidal and typically lack externally visible internal structure other than in some forms a pylome, or consist of chambered arms lying in a plane, plus one or more medullary shells. Some taxa with planar radial beams extending as spines. Shell structure partially or mostly spongy.

Stylodictyids – discoidal cortical shells with small circular pores, divided into a series of concentric annular chambers by internal walls; in some forms the wall is spiral rather than a series of concentric rings. Planar to sub-planar radial beams extending as spines common. Spongy shell wall is commonly present in at least part of the shell.

Myelastrids – Spongy-armed Spongodiscidae with well-silicified medullary shells and much more weakly silicified, often distally branched arms. Usually quite large. Known only from plankton as only the indistinguishable medullary shells are likely to be preserved as fossils.

Spongurids – bipolar spongy and/or chambered arms extending from an ellipsoidal central shell with one or more medullary

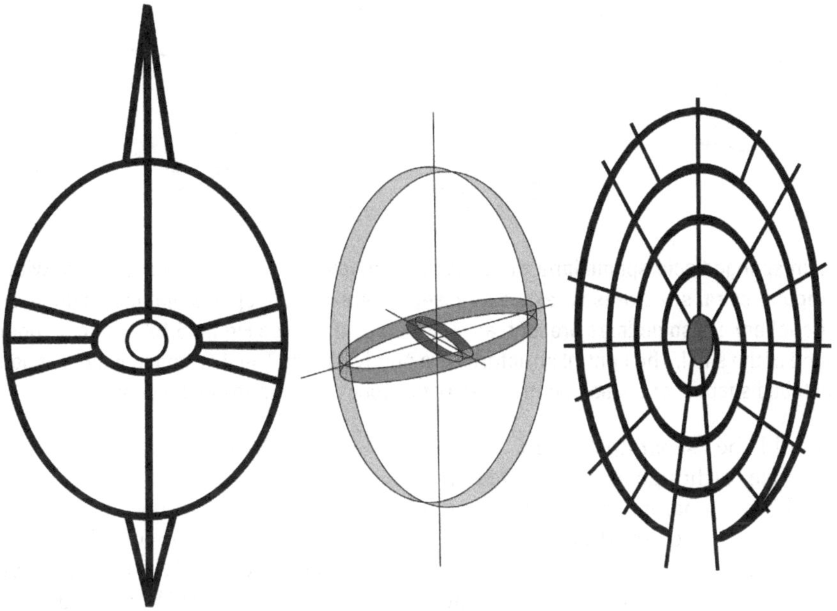

Figure 5.16 A) Stylosphaerid shell structure. In this minor variant from actinommid form, the cortical shell is often ellipsoidal, external spines are reduced, usually to just two, polar opposite, spines, usually of unequal length. Other than the beams connecting medullary shells to the polar spines additional beams may be present, often restricted to a region near the equatorial plane. B) The larnacilla structure is characteristic of the Pyloniidae and is present in some lithelids as well. It consists of three ringlike bands of lattice wall nested successively inside each other with the plane of each band set in one of each of the three spatial dimensions. C) Lithelid structure is based on spirals, either single (as here), double, or multiple. The external shell shape can be spherical but usually is ellipsoidal. Depending on orientation the spiral(s) are clearly visible or appear as irregularly nested ellipses. Numerous radial beams connect layers and some may extend as short, usually simple, needle-like external spines. The exterior may be open – simply the last spiral, or closed into an ellipsoidal outer shell. A pylome, an opening at one end extending into the interior is present in some species. The central core of the shell may be a simple polyhedral shape or a small larnacilla structure like 5.16B.

shells. Short spines and/or a pylome may exist at the ends of the spongy arms.

Porodiscids – Discoidal, circular to ellipsoidal shells with small circular pores, small medullary shells, and a small number of widely spaced internal concentric walls dividing the shell into circular chambers. Typically possess a pylome.

Remarks:

The Spongodiscidae is most likely a polyphyletic grouping (Sandin et al., 2020) based more on a combination of similar external characters and a lack of information on internal ones; however current knowledge is insufficient in our opinion to formally reclassify them.

Although most features of the porodiscid shell resemble stylodictyids, specimens of porodiscids have been shown to have a larnacilla medullary shell (De Wever et al., 2001). This, together with differences in shell structure, make it likely that these

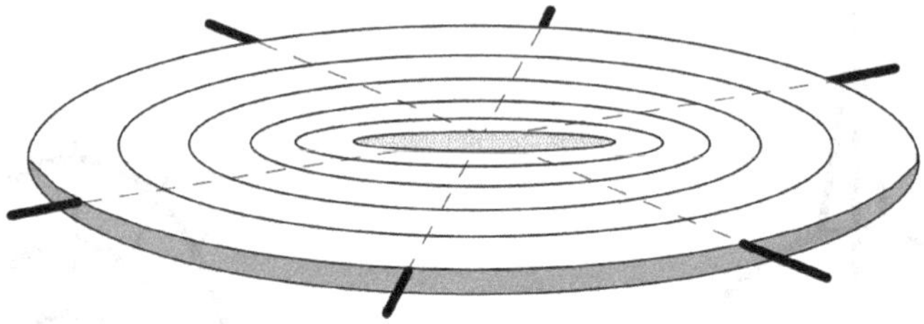

Figure 5.17 In spongodiscid spumellaria the shell is a spongy disc (or variations thereof), with internal rings or irregular spiral structures weakly expressed. Internal radial spines that may extend externally along the periphery are sometimes present, as may be a pylome (see Figure 5.16C) at some point on the external edge of the shell. The central structures are poorly known. The disc shape may be modified into a variety of lobed shapes, or a small central disc with spongy radial arms in the plane.

forms should be separated from other Spongodiscidae, although it is not clear if they should have separate family rank status or be assigned to another clade, e.g. pylonid or lithelid.

With 33 valid Cenozoic species the Spongodiscidae is moderately diverse and also commonly reported in the literature (19% of all spumellarian data/9% of all data in the NSB database).

Order Nassellaria Ehrenberg 1876

(Figures 5.18, 5.19; 5.20–5.29)

Definition: Polycystina mostly with axial symmetry, generally polar. General morphologic features and terms are given in Figure 5.18. Most taxa possess a constant set of primary spicules (Figure 5.19A), consisting of a median bar terminated by on one end a cluster of three spines (one vertical and two primary laterals), and at the other end a cluster of four spines (apical [pointing up], dorsal [sideward or down], and secondary left and right lateral). In most species the primary spicules are situated at the base of a partially closed sub-spherical or lobed, cap-like lattice wall, the cephalis (examples of cephalic structures shown in Figure 5.19B–F). Additional cylindric segments may successively append to the base of the cephalic lattice wall, and bear external structures such as wings or feet.

Geologic Range: Basal Triassic (possibly late Paleozoic) to Recent

Subtaxa: listed below

Remarks:

Described species of Cenozoic Nassellaria make up two-thirds the total number of described Cenozoic Polycystina (Lazarus et al., 2015), and three-quarters of published occurrence records (Table 5.3). Their dominance in the published literature is not necessarily due to dominance in numbers of specimens in sediments, as in many cases spumellarian species (particularly spongodiscids and lithelids) are the most common forms. Nassellarians are simply easier to understand morphologically and to identify than many spumellarian forms, and thus have been the major focus of taxonomic research and use in Cenozoic biostratigraphy, paleoceanography, and evolutionary analyses. Nassellarians make up a slight majority of published (raw, unevaluated) Mesozoic species as well. Paleozoic occurrences are based on the similarity in spicular structure of at least some specimens of the Paleozoic families Archocyrtiidae and Popofskyellidae (together, 111 named species, but only 6% of all Paleozoic species names). There is however a very substantial gap (>60 my) between the youngest (Carboniferous) Paleozoic forms and the nassellarians from the early Triassic (Noble et al., 2017), and molecular studies date the nassellarian origin to the early Triassic (Sandin

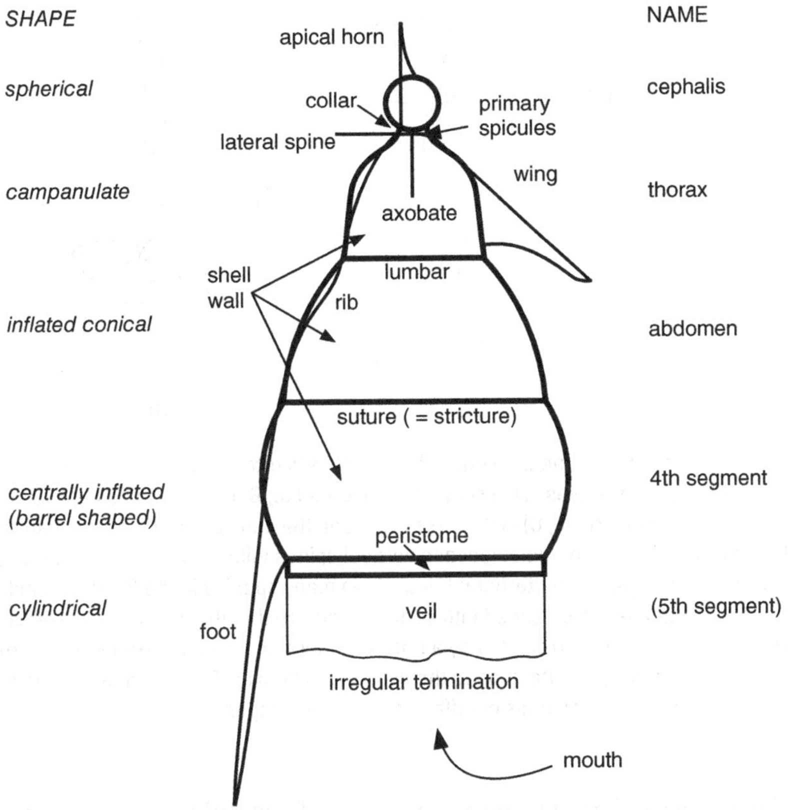

Figure 5.18 Main structural components of nassellarian shells. The conical geometry is divided into segments. The first three have standard names, as shown on the right. Further segments are simply referred to by number. A peristome – a short ring of distinct shell material, typically poreless, at the base of the shell – may be present. The base of the shell, if open, is called the mouth or basal aperture. In some shells a thin veil of irregularly porous shell material may be attached below the last numbered segment (a veil normally is not considered a segment by itself). Segments are characterized in part by their shape in outline. Some of the most common shapes, and on the left their descriptions, are shown. The juncture between segments is a suture or stricture. That between the first and second segments (cephalis to thorax) is the collar, between thorax and abdomen lumbar. These may be marked in addition internally by a thin annulus of shell material – a septum (not shown). Spines often extend from the internal primary spines (Figure 5.6) near the base of the cephalis (approximately at the cephalic-thorax suture). Examples shown here include a spine vertically through and beyond the apex of the cephalis, forming an apical horn; a free-standing lateral spine; a downward laterally directed spine connected by lattice wall to the thorax thus forming a wing; a long spine, mostly embedded in the lattice wall of the segments, thus forming a rib, and extending to, and projecting beyond the base of the shell, forming a foot. Lastly an axobate is shown – a spine, or cluster of spines, that may extend distally downwards in the middle of the thorax.

et al., 2019). The Paleozoic forms are thus most likely an example of convergent evolution.

Molecular studies suggest that the diverse, and in modern oceans, very abundant group of collozoids are phylogenetically

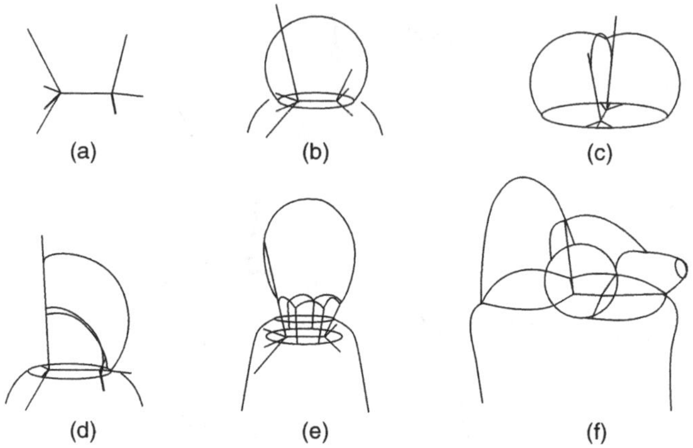

Figure 5.19 The cephalis (first segment) of nassellarian shells is often strongly modified and is typical for several families. A) primary spicules, as in some Plagiacanthidae; B) location at base of simple spherical cephalis, e.g. as in the Theoperidae; C) in the Trissocycliidae the cephalis is subdivided into two lobes by a "D ring" formed by the joining of the apical and vertical spines with an arch AV; D) paired lobes at the base of the cephalis distinguish the Pterocorythidae; E) in many species in the Plagiacanthidae the lowermost pores of the cephalis are extended into vertical struts so that the median bar is offset below the main cephalic wall and is often surrounded by a ring. The thoracic wall may extend only partially up the struts or enclose them entirely by attaching to the main cephalic wall; F) in Cannobotryidae the cephalis is subdivided into a complex set of lobes of different sizes and shapes.

nested within the Nassellaria, but are based on their diversity and unique adaptations they were previously given Order rank. The Nassellaria would thus be paraphyletic. Here the collozoids are treated as a superfamily within the order Nassellaria.

Family Plagiacanthidae *Hertwig 1879 (Figure 5.19A, B, E; 5.20–5.22)*

Definition: Nassellaria usually with, in addition to the primary spicules, well-developed additional arches and secondary spines, but reduced development of external lattice wall shell, having no, one, or at most two segmented shells; and an undivided (unlobed) cephalis.

Geologic range: early Triassic to Recent

Subgroups:

Plagonids (Plagiacanthidae) – lacking a well-developed lattice shell, consist only of spicules and structures such as arches connected to them.

Lophophaenids (Lophophaenidae, partim.) – possessing a well-developed cephalis but at most only a rudimentary thorax.

Sethoperids (Sethoperidae) – cephalis and well-developed external spines, joined distally by bands of lattice wall.

Sethophormids (Sethophormidae; Theopiliidae, partim.) – small, usually cap-like cephalis with, if present, usually cylindrical horns and flat, wide angled thorax with regular termination.

Lithomelissids (Lophophaenidae, partim.) – Two segmented nassellaria with relatively large cephalis, thorax, irregularly terminated; feet, if present, generally arising from within thorax.

Remarks:

This is a large and probably polyphyletic group which has been subdivided by numerous previous authors into various partially incompatible clades, with a rich confusion of nomenclature for higher taxa, including, in addition to the taxa used in De Wever et al. (2001), Plagoniidae, Plectoidea, and many others. Our choice of Plagiacanthidae is rather arbitrary but allows a match in name, if not scope, to the classification of De Wever et al. (2001), which elevates several subgroups into families. With 145 recognized valid species, plagiacanthids are the fourth most diverse family of radiolarians and the third most diverse group of Nassellaria. Some plagiacanthids are difficult to distinguish from incomplete (juvenile) specimens of other nassellarians. Many species also are smaller than the 63 μm mesh size used in many studies. For both these reasons they are underrepresented in the deep-sea drilling literature – only 5% of nassellarian occurrences (3% of all radiolarian occurrence data). Trubovitz et al. (2019) for example report many more undescribed species even in well studied late Neogene tropical sediments.

Family Trissocyclidae *(Haeckel) Goll 1968 [superfamily Acanthodesmiacea] (Figures 5.19C; 5.23, 5.24)*

Definition: Nassellaria which possess a saggital ring ("D-ring") structure, which divides the cephalic lattice wall (present in most species) into left and right lateral lobes. The saggital ring is formed by the primary nassellarian spicules of median bar at the base, a straight upward-pointing apical spine, a lateral upward-pointing vertical spine, and a curved arch joining them near the apex of the skeleton.

Geologic range: Paleocene to Recent

Subgroups:

Stephanids [Stephaniidae] – forms lacking a cephalic wall and thus consist only of the D-ring and ornamentation.

The remainder of Trissocyclidae is not here divided, although De Wever et al. (2001) divides them into two families (Triospyrididae and Acanthodesmiidae) primarily on the presence or not of well-developed feet.

Remarks

A very diverse group with 156 species, the second highest diversity in the Nassellaria. Trissocyclids also make up a substantial fraction of NSB data with 12.5% of nassellarian occurrences and 9.3% of all radiolarian occurrence data. Goll (1968 and many subsequent papers) has provided the most comprehensive taxonomic revision and description of species.

Family Theoperidae *Haeckel 1881 (Figures 5.19B, 5.24–5.26)*

(Eucyrtidiidae, Theocotylidae, Theoperidae, Lophocyrtiidae,

Acropyramididae, Theopiliidae [partim.], Ultranaporidae [partim.]?)

Definition: Nassellarians with a simple (neither divided or lobed), usually relatively small cephalis and at least one, generally several post-cephalic segments, usually also with internal septa at segment sutures.

Geologic range: Triassic to Recent

Subgroups:

Eucyrtinids (Eucyrtidiidae) – Multisegmented (generally >3 segments) theoperids with small, subspherical cephalises and lacking terminal feet or substantial lateral wings.

Cornutellids [Acropyramididae] – Two segmented nassellaria with small, hyaline cephalises and elongated, conical thoraxes, irregularly terminated, lacking external wings or feet. Pores in thorax aligned in horizontal rows between vertical thoracic ribs, elliptical to rectangular in shape, bars circular in cross section.

Lophocyrtids [Lophocyrtiidae] – Three segmented forms with approximately cylindrical shell shapes and in many species a relatively elongated abdomen. Two arches connect the apical and primary lateral spines within the cephalis. Taxonomic review by Sanfilippo (1990) and Sanfilippo and Caulet (1998).

Cycladophorids [Theopiliidae (partim.)] – Two segmented forms with a small, subspherical cephalis bearing bladed thorns

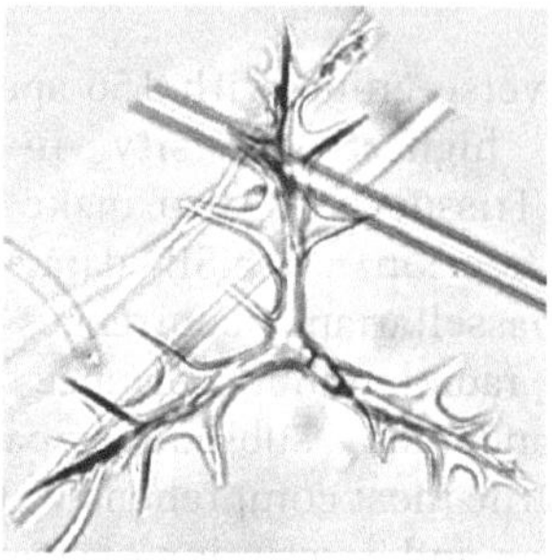

1 - Neosemantis bjoerklundi mimicus

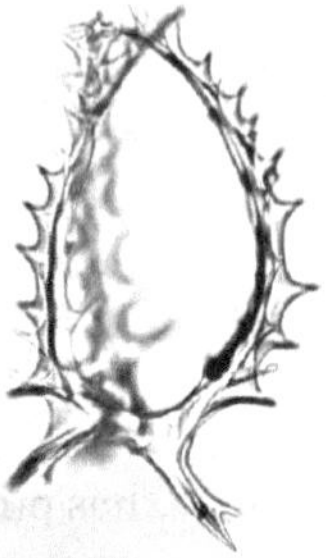

2- Neosemantis distephanus

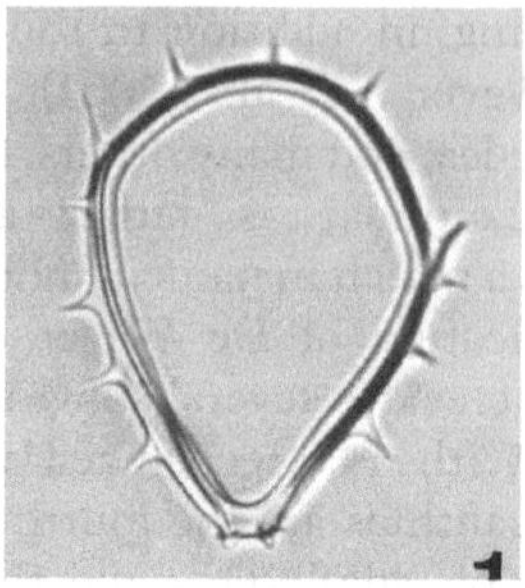

3 - Zygocircus productus

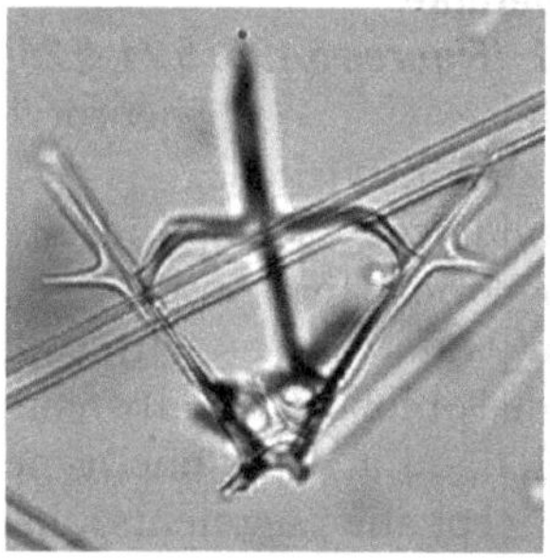

4 - Pseudocubus obeliscus

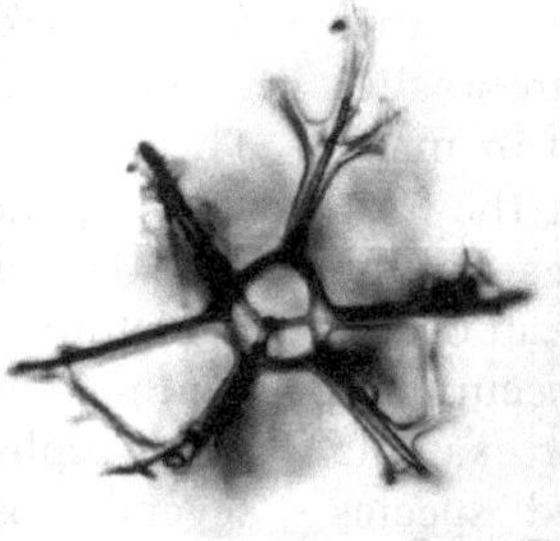

5 - Pseudocubus obeliscus

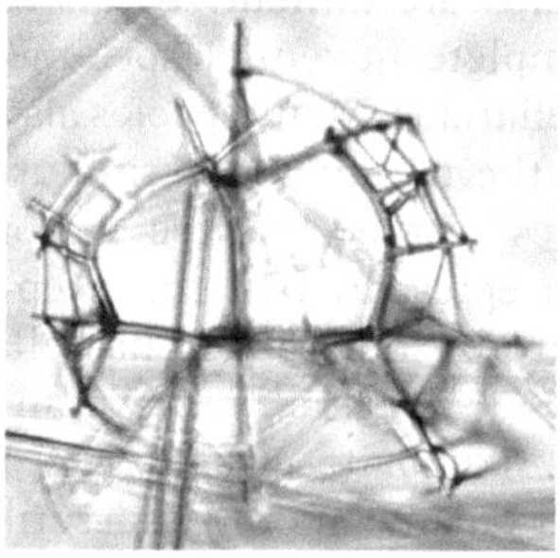

6 - Protoscenium? sp.

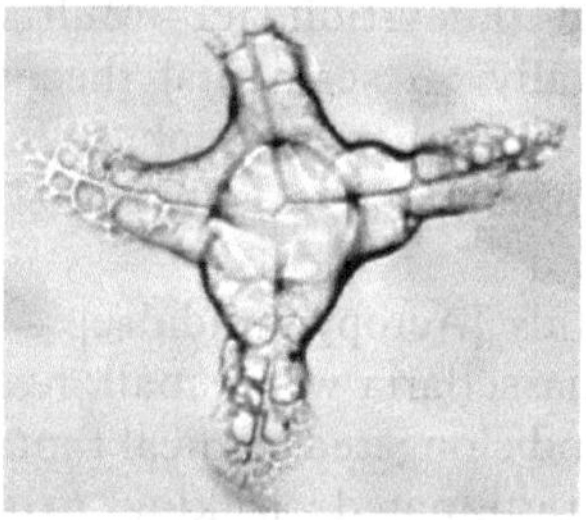

7 - Plectagenidium deflandrei

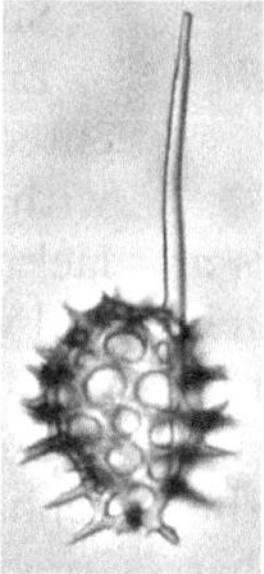

8 - Peridium longispinum

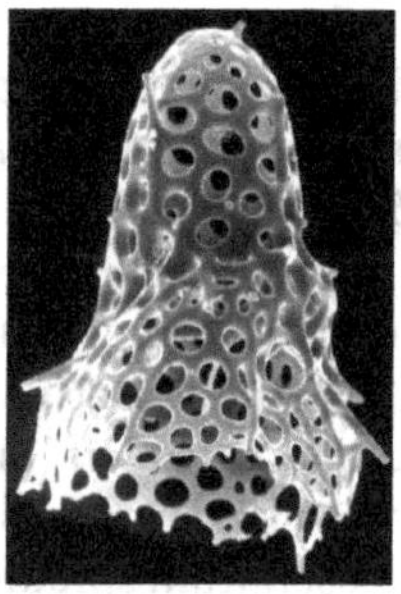

9 - Lophophaena variablis

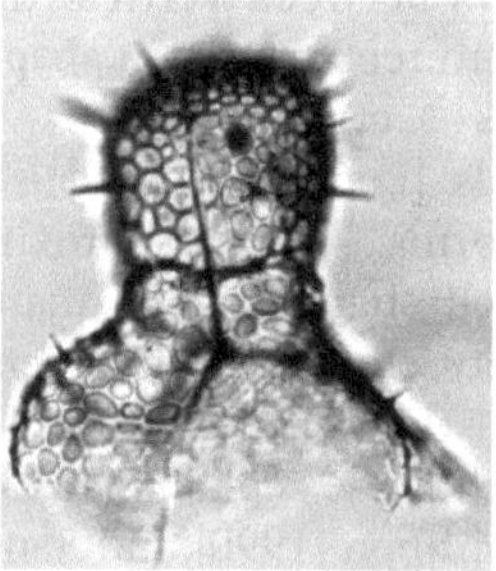

10 - lophophaenid sp.

11 - Peromelissa thoracites

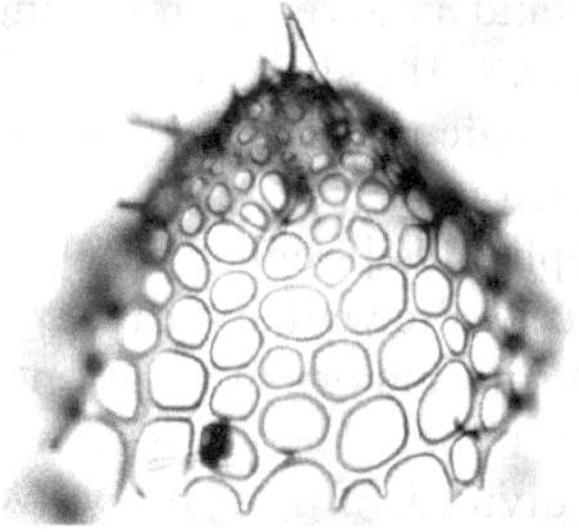

12 - Ceratocyrtis galeus

and conical to campanulate thorax, in most species strongly flared. Most species with a distinct septum at the base of the thorax. Abdomen reduced to small, irregularly developed fringe. Taxonomic review by Lombari and Lazarus (1988).

Pterocanids [Ultranaporidae (partim.?)] – Two segmented nassellarians with a small, subspherical cephalis, large campanulate to truncate pyramidal thorax, with three well-developed ribs extending as prominent downward flared bladed feet.

Remarks:

The Theoperidae have long been assumed to be a polyphyletic group (Riedel, 1967), as indeed initial molecular results suggest (Sandin et al., 2019). Several of the families suggested by De Wever et al. (2001) however do not seem to be well supported by the initial results from molecular work (Sandin et al., 2019), nor is the latter yet detailed enough to suggest a robust new family structure. Theoperids, with 347 accepted species, are by far the most diverse family of Cenozoic radiolaria. They make up more than half of all NSB occurrence data for nassellarians, and over a third of all radiolarian data.

The molecular results of Sandin et al. (2019) suggest that the cycladophorids and pterocanids are a separate group than the rest of the theoperids, and are part of a larger clade including the Pterocorythidae. The reported stratigraphic range of both the cycladophorids and Pterocorythidae are Cenozoic and younger (De Wever et al., 2001), as are traditionally defined pterocanids, including, e.g. *Pterocanium*. The Utranaporidae however, as defined by De Wever et al. (2001), includes not only Cenozoic genera such as *Pterocanium*, but also many Mesozoic taxa such as *Ultranapora*, and ranges back to the early Triassic. This is in conflict with the stratigraphic ranges for the other taxa in the molecularly defined clade. There is in addition a substantial (nearly 30 my) stratigraphic gap between the Mesozoic and Cenozoic members of Ultranaporidae in the late Cretaceous (O'Dogherty et al., 2011). This may represent another example

Figure 5.20 Plagiacanthidae (1). 1 – *Neosemantis bjoerklundi mimicus*: Skeleton of only reduced set of primary spicules arranged in approximate triangle, spicules curved, bladed, thorny (basal view). 2 – *Neosemantis distephanus*: Side view showing curved spicules, not joined at tips; 747A-7-2, 45-7 cm MM_DL. 3 – *Zygocircus productus*: Small median bar at base with strongly reduced lateral spines, much longer, bladed apical and shorter vertical spines joined distally by arch to form sagittal ring; van der Paverd, Pl. 62 f14. 4, 5 – *Pseudocubus obeliscus*: Small median bar forms tripod inside collar ring, 3–5 bladed columns (2 are apical and vertical spines) expand upward, joined distally by convex arches, are branched (verticils); 4: (side view) 744A-6H-4 JR; 5: (basal view) vdP Pl. 60 f5. 6 – *Protoscenium?* sp.: One of many plagiacanthid species with elaborate cephalic spine elements but no lattice wall; 751A-2H-3, 98 cm JR. 7 – *Plectagonidium deflandrei*: Approximately quadratic arrangement of primary spicules from small median bar, increasingly branched verticils distally attaching at tips to hyaline veil; 751A-7H-2, 98 cm JR. 8 – *Peridium longispinum*: Typical shell form of tapered ovoid cephalis with small median bar at base, strong cylindrical apical spine projecting vertically, numerous thorns on lattice wall; 747A-2H-2, 45 cm JR. 9 – *Lophophaena variablis*: Elongate cephalis, skirt-like, incomplete thorax, spines of primary spicules project laterally from upper thorax (SEM); van der Paverd, Pl. 65 f11. 10 – lophophenid sp.: Specimen with arches extending from apical spine above median bar in lower cephalis; flattened top; van der Paverd, Pl. 66 f6 [as *Tripospyris tricostata*]. 11 – *Lithomelissa thoracites*: Somewhat irregular, egg-shaped cephalis and similar-sized, irregularly terminated thorax with medium-sized, irregular pores, thin, smooth shell, spines generally prominent, directed apically, laterally, and ventrally van der Paverd, Pl. 66 f9 [as *Peromelissa t.*]. 12 – *Ceratocyrtis galeus*: Small cap-shaped cephalis with medium-sized apical spine, shoulder on conical thorax with large, rounded polygonal pores; 747A-2H-5, 45 cm JR.

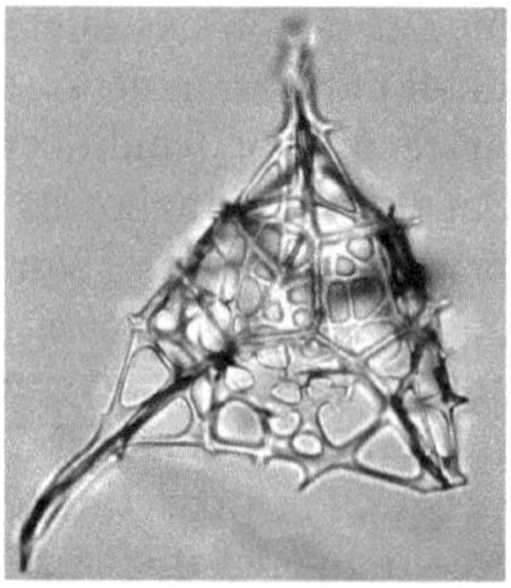

1 Archiscenium quadrispinum composite

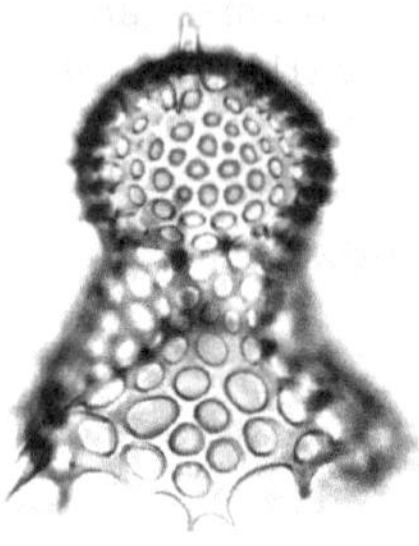

2 Lophophaena sp.

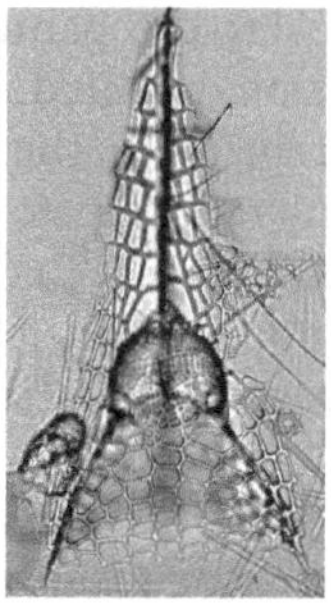

3 Callimitra solocribrata

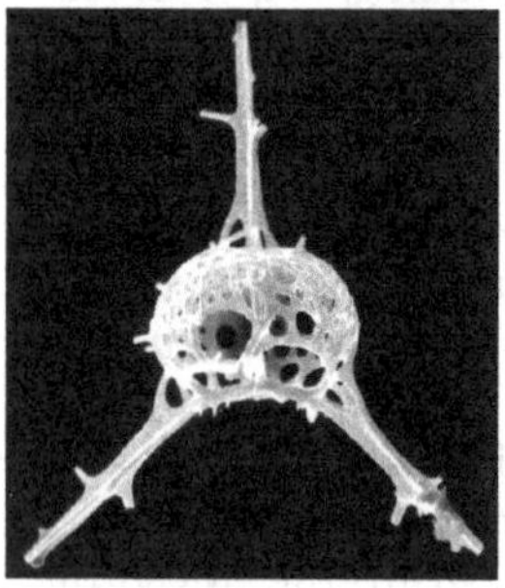

4 Clathrocanium coarctatum

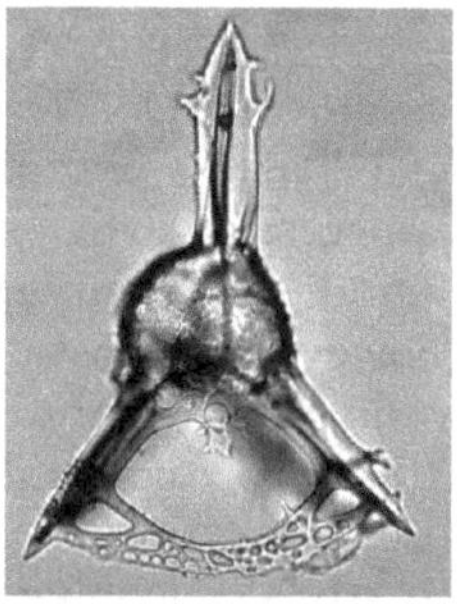

5 Clathrocanium coarctatum

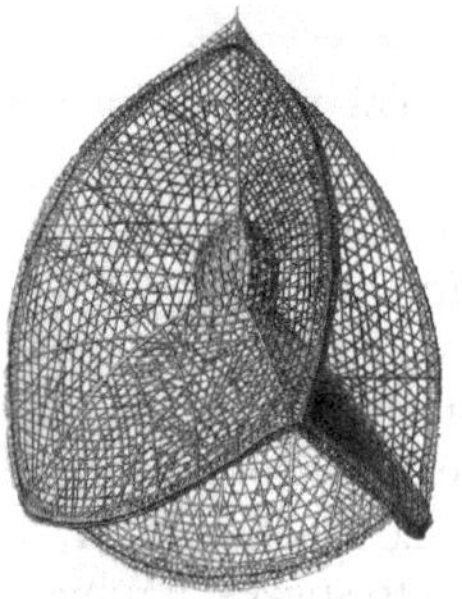

6 Callimitra emmae

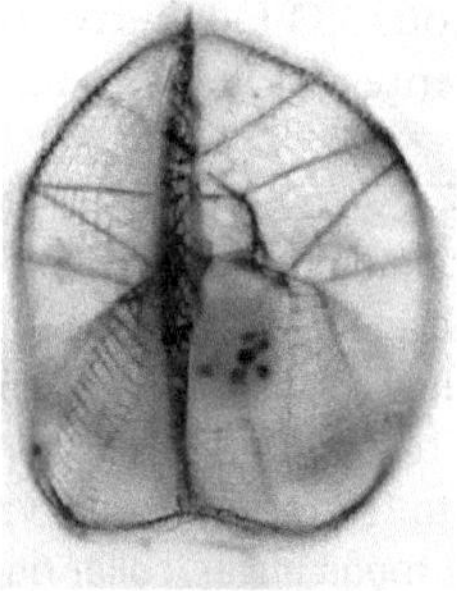

7 Callimitra sp.

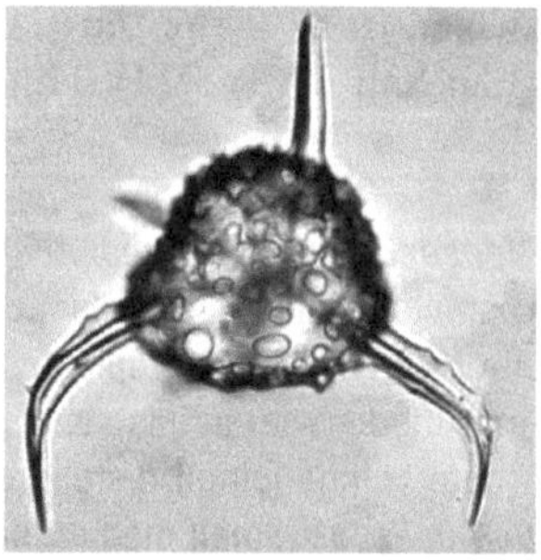

8 Corythomelissa horrida

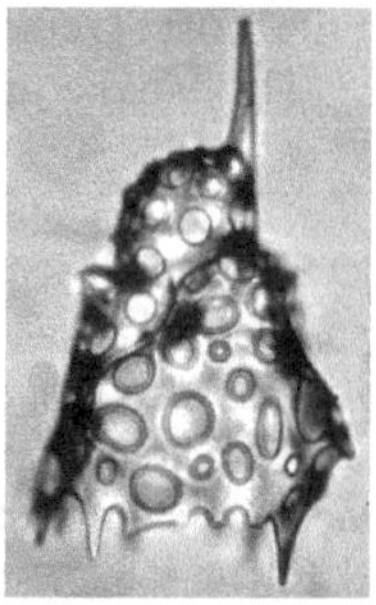

9 Lithomelissa stigi

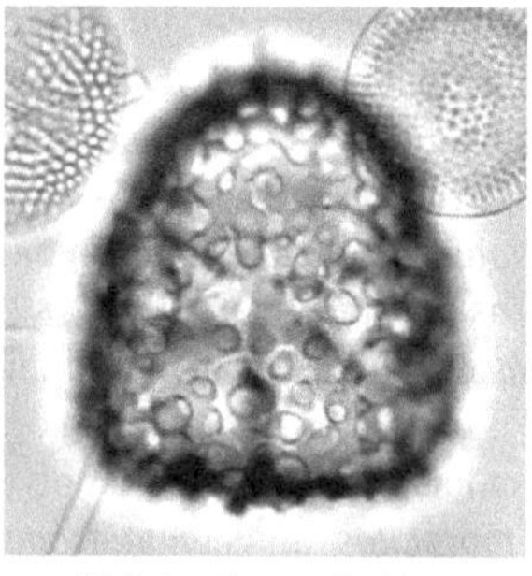

10 Antarctissa cylindria

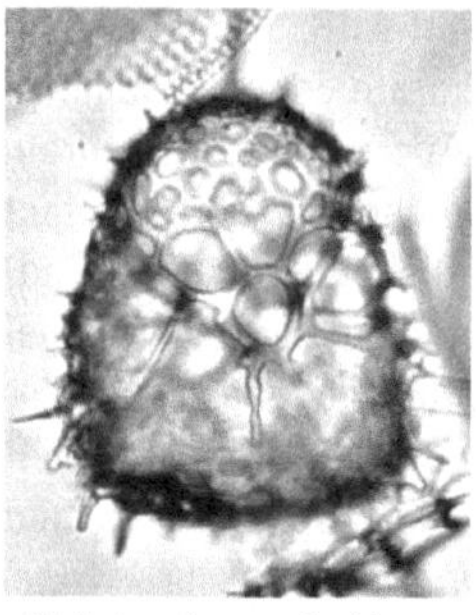

11 Antarctissa cylindrica

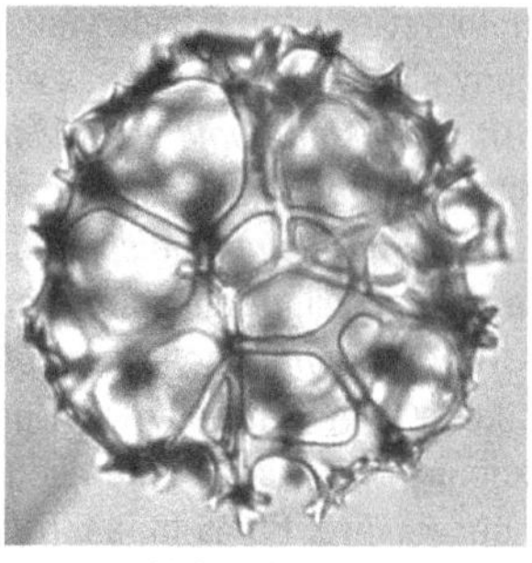

12 Helotholus haysi

of convergent evolution, with the Mesozoic and Cenozoic species belonging to two different clades.

Family Artostrobiidae *Riedel 1967 (Figure 5.27)*

Definition: Multisegmented nassellarians with a large pore in the cephalic lattice wall, often extended into an external tube, surrounding the vertical spine. Pores, particularly in post-thoracic segments, aligned in horizontal rows, frequently with horizontal hyaline bands. No basal feet or internal septa.

Geologic range: Jurassic to Recent

Subgroups: none

Remarks:

Species morphology in this group is unusually plastic and assignment of individuals to taxa is sometimes difficult. The main taxonomic reference is Nigrini (1977). Published Cenozoic artostrobid diversity is only moderately diverse with 64 species but is well reported in deep-sea occurrences (13% of nassellarian, 9% of all radiolarian data).

Family Pterocorythidae *(Haeckel) Moore 1972 (Figures 5.19D, 5.28)*

Definition: Two or three segmented nassellarians with an elongated cephalis which has two lateral lobes at the base, internally

Figure 5.21 Plagiacanthidae (2). 1 – *Archiscenium quadrispinum*: three curved, bladed feet and apical spine, tent-like pyramidal cephalis of loose, irregular bars, polygonal pores (composite); 744A-5H-2 JR. 2 – *Lophophaena* sp.: Slightly flattened, large spherical cephalis, small uniform sized subrounded pores, thick bars, with apical spine, free inside cephalis. Thorax short, conical, irregularly terminated, medium pores, distinct neck at cephalic suture; RC9-132-855 cm MM_DL. 3 – *Callimitra solocribrata*: Large, hemispherical thin-walled cephalis with irregular meshwork bars, polygonal pores, three long divergent feet and even longer apical spine joined by strong, coarse polygonal pored meshwork webbing; 747A-1H-2, 82 cm JR. 4, 5 – *Clathrocanium coarctatum*: Compressed hemispherical thin-walled cephalis, small irregular pores, strong bladed apical spine with distal thorns, three divergent feet distally joined by arcuate, convex downwards bands of lattice wall; 4: Incomplete specimen, one foot missing (SEM); van der Paverd, Pl. 64 f1; 5: Complete specimen with more silicification (cephalic wall nearly hyaline, feet and apical spine more heavily silicified); 737B-7R-1 JR. 6, 7 – *Callimitra emmae*: Hemispherical cephalis, at center of large webbed structure made of flat sheets of webbing connecting the three long, indistinct divergent feet and apical spine. Webbing lattice formed by three sets of thin parallel threads crossing at angles to form roughly triangular pores, with thicker, widely spaced primary and more numerous smaller secondary threads. Outer edges of webbing planes convex, delimited by additional thickened latticework; 6: Haeckel, 1887, Plate 63, Figure 3 (drawing); 7: Specimen with cytoplasm below cephalis (composite); Station 271 Challenger plankton PG. 8 – *Corythomelissa horrida*: Cap-like cephalis, indistinct suture to short, divergent thorax. lattice wall rough with small irregular pores. Strong apical horn and divergent, curved three-bladed feet; 690B-6H-4, 22 cm JR. 9 – *Lithomelissa stigi*: Cephalis medium-sized, large-pored, thick, rough wall, hemisphere vertically flattened at attachment to strong vertical apical horn, thorax with circular to elliptical large pores, rough surface, irregular or toothed termination; 690B-6H-2, 22 cm JR. 10, 11 – *Antarctissa cylindrica*: Large hemispherical cephalis merging smoothly without noticable change in outline to short cylindrical thorax, typically closed by basal plate. Shell surface rough with small irregularly distributed subcircular pores. Interior structure of spicules and additional arches and spines complex; 10: Specimen with focus on shell exterior; 701C-10-6, 56 cm DL; 11: Specimen with side broken away showing small tilted median bar, axobate and complex structure of spicules and arches; 745B-19H-5, 46 cm JR. 12 – *Helotholus haysi*: Species with internal structures similar to *A. cylindrica*, specimen view upwards from thorax aperture, median bar in center and above it ring and spoke structure of arches and internal bars; 746A-8H-3, 53 cm JR.

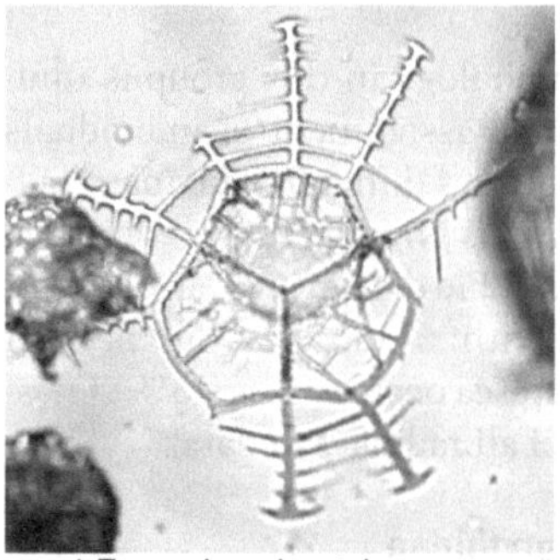

1 Enneaphormis rotula

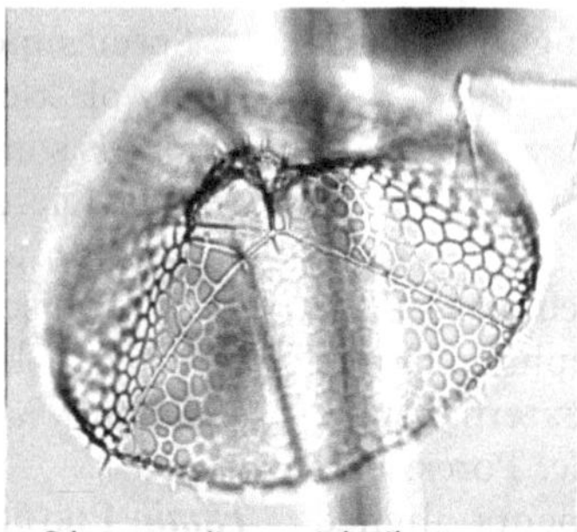

2 Lampromitra pentalactis

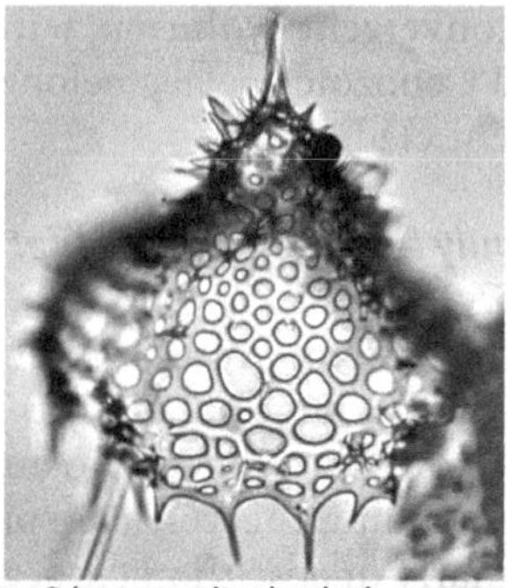

3 Lampromitra huxleyi

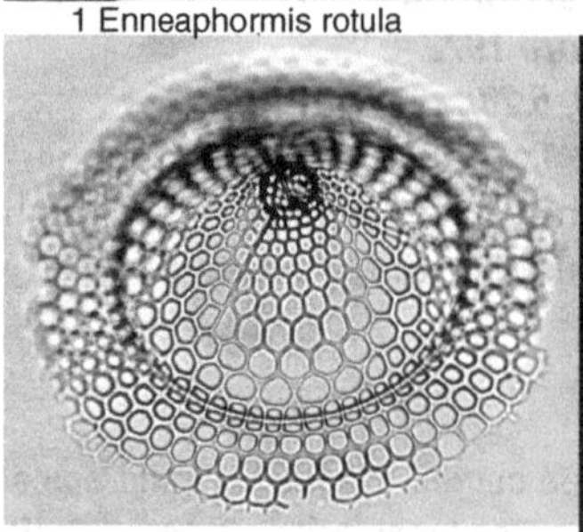

4 Eucecryphalus tricostatus

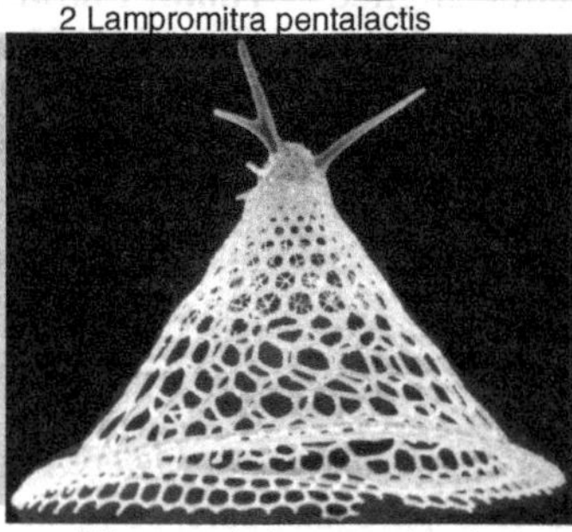

5 Eucecryphalus gegenbauri

6 Litharachnium tentorium

Peripyramis circumtexta

8 Cornutella clathrata

9 Artostrobus annulatus

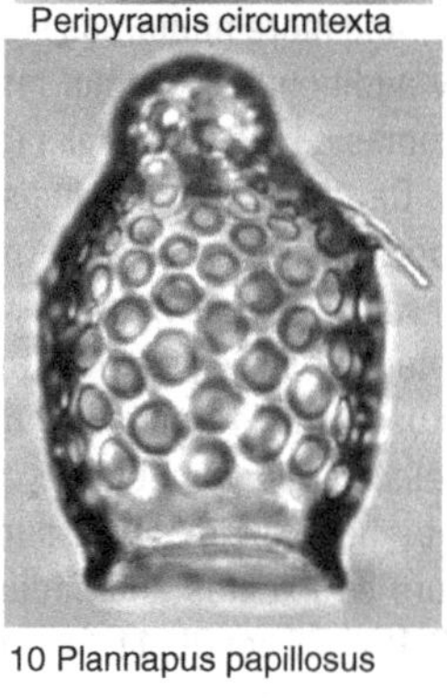

10 Plannapus papillosus

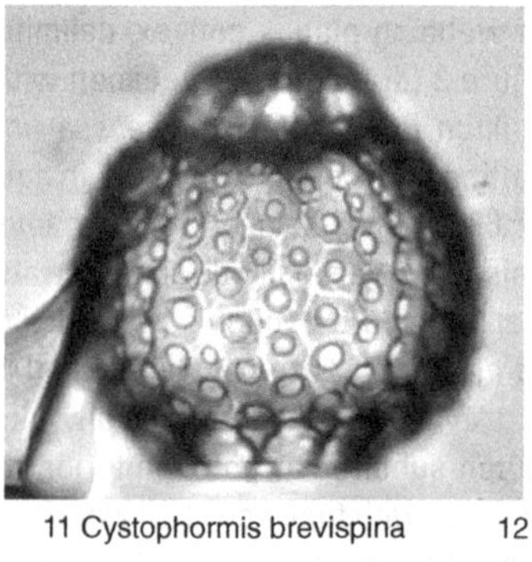

11 Cystophormis brevispina

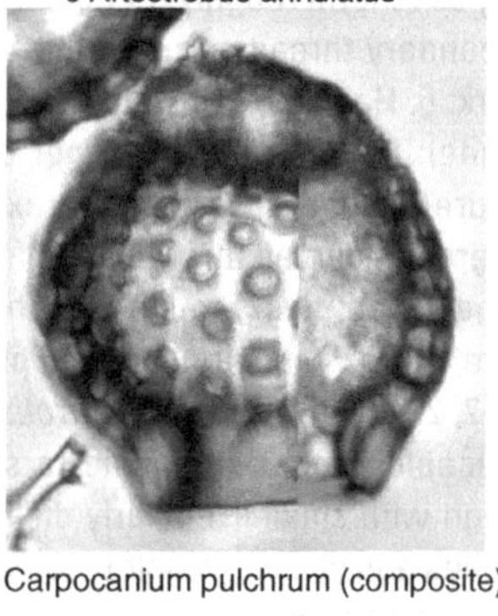

12 - Carpocanium pulchrum (composite)

marked by arches from the apical spine. Cephalis usually with a single vertical bladed horn.

Geologic range: basal Eocene to Recent

Subgroups: none

Remarks:

De Wever et al. (2001) note that the origin of this family appears to be from a late Paleocene theoperid. Several genera have been subject to taxonomic review, including *Calocycletta* (Moore, 1972), *Anthocyrtidium* (Nigrini and Caulet, 1988), and *Pterocorys* (Caulet and Nigrini, 1988), plus many species in other genera such as *Podocyrtis* that are parts of evolving Cenozoic lineages used for biostratigraphy (Sanfilippo et al., 1985). The lobes at the base of the cephalis are not always visible depending on orientation but the overall shape of the cephalis, and the frequent

Figure 5.22 Plagiacanthidae, Theoperidae (cornutellids), Carpocaniidae, 1 – *Enneaphormis rotula*: Very flat cephalis with irregular meshwork wall, polygonal pores, unequally hexagonal basal (collar) ring, three primary spicules within ring. Thorax resembles a spider web with 8–10 near horizontal radial spines connected by multiple parallel straight circumferential segments; 748B-8H-4, 45 cm JR. 2 – *Lampromitra pentalactis*: Flattened cephalis merging imperceptibly into equally flattened conical thorax, both with thin-shelled, medium-sized polygonal pores. Three primary spicules at base of cephalis extend into, and are joined by, additional thoracic ribs. Fine teeth on base of thorax (peristome); van de Paverd, Pl. 67 f2. 3 – *Lampromitra huxleyi*: Cephalis small, hemi- to subspherical, small pores, thick walls, short conical horn. Thorax flattened conical, polygonal pores, peristome with prominent vertically downward pointing teeth and smaller row approximately horizontal; 748B-6H-3, 45 cm JR. 4 – *Eucecryphalus tricostatus*: Small hemispherical cephalis, flattened conical thorax, broad horizontal abdominal brim, both thorax and brim with regular polygonal pores, very thin shell, three weakly developed thoracic ribs extending from primary spicule tripod at base of cephalis; van de Paverd, Pl. 69 f5. 5 – *Eucecryphalus gegenbauri*: Cephalis small, hemispherical, with two cylindrical horns, thorax conical, upper thorax pores small, regular polygonal, with fine mesh, lower thorax pores significantly larger, less regular, peristome with thin horizontal brim of small polygonal pores (SEM); van de Paverd (as *Sethoconus gegenbaueri*), Pl. 69 f2. 6 – *Litharachnium tentorium*: Extremely small spherical cephalis, thorax starting as narrow, curving gradually (concave upward) to flat conical shape or even recurving upward, lattice-wall formed by regular rectangular rows of pores, a dozen or more thicker thoracic ribs, small pored thickened peristome in complete specimens (SEM); (as *L. tentatorium*), van de Paverd, Pl. 71 f1b. Theoperidae: 7 – *Peripyramis circumtexta*: Small, hyaline subspherical cephalis, narrow conical thorax with 8–10 strong thoracic ribs, large subrectangular pores in staggered rows, short spines and secondary meshwork on lower thorax in some specimens; 745B-19-5, 46 cm DL. 8 – *Cornutella clathrata*: very small hyaline nearly spherical cephalis with thin conical spine, thorax narrow cone slightly curved to one side, pores rounded subrectangular-ellipsodal, wall thick, surface smooth; 747A-5H-3, 45 cm JR. 9 – *Artostrobus annulatus*: Small hemispherical hyaline cephalis with prominent apical horn and two characteristic horizontal spines extending from top of thorax. Thorax thick walled, smooth, narrow conical with rows of small subrounded pores; 744A-6H-4 JR. 10 – *Plannapus papillosus*: Flattened hemispherical cephalis with scattered small circular pores, distinctly larger barrel shaped thorax with large rounded polygonal pores in staggered vertical rows, thin lateral spine at shallow distal angle from upper thorax, hyaline cylindrical peristome; 744A-7H-2 JR. 11 – *Cystophormis petrushevskayae*: Flattened, multichambered carpocanid cephalis, short barrel-shaped thorax with staggered rows of small framed pores, short hyaline peristome with ridges; JR. 12 – *Carpocanium pulchrum*: Very similar to *C. brevispina* but longer peristome (in many specimens with teeth), somewhat larger thorax pores, small apical thorn on cephalis; 747A-6H-2, 45 cm JR.

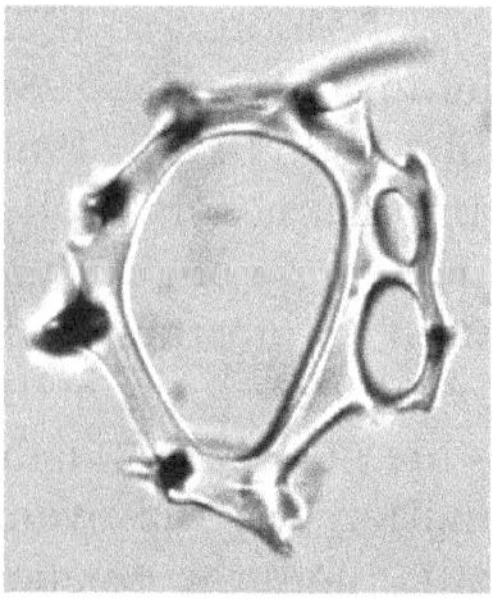

1 Unidentified trissecyclid

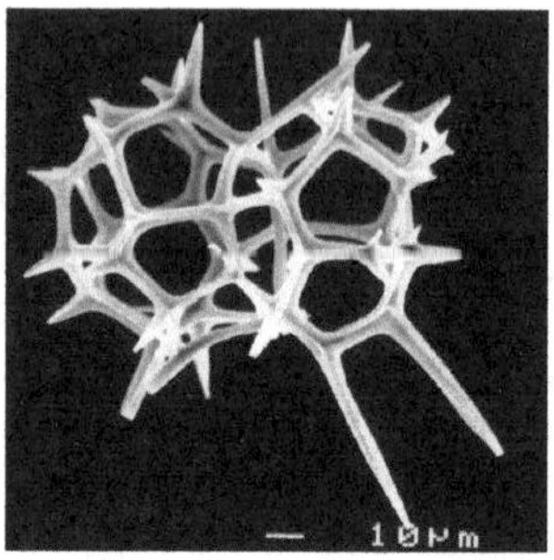

2 - Lophospyris pentagona pentagona

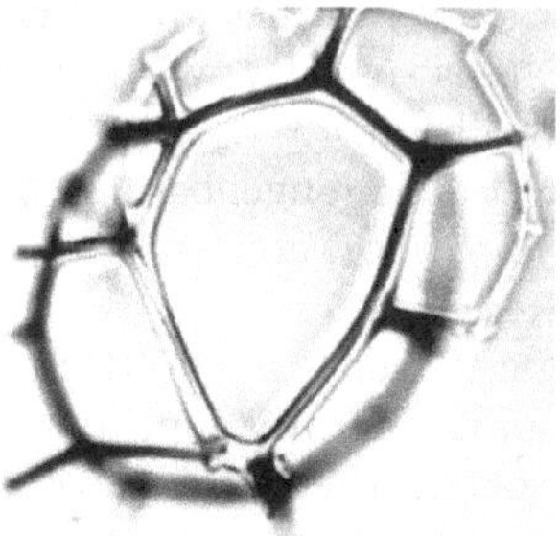

3 - Lophospyris pentagona pentagona

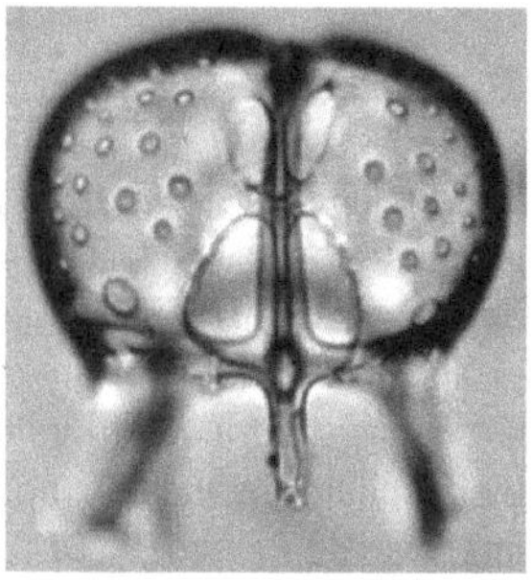

4 - Dendrospyris stabilis scaphipes

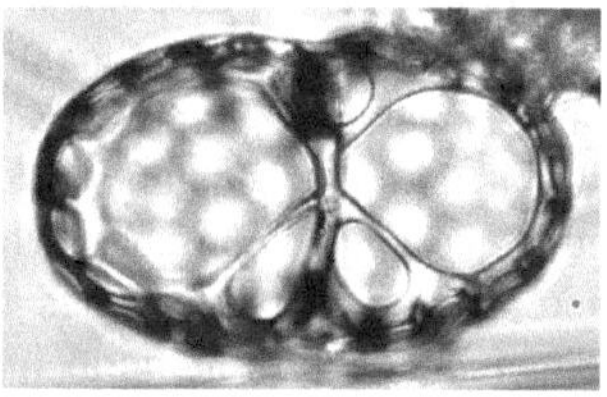

5 - Dendrospyris stabilis?

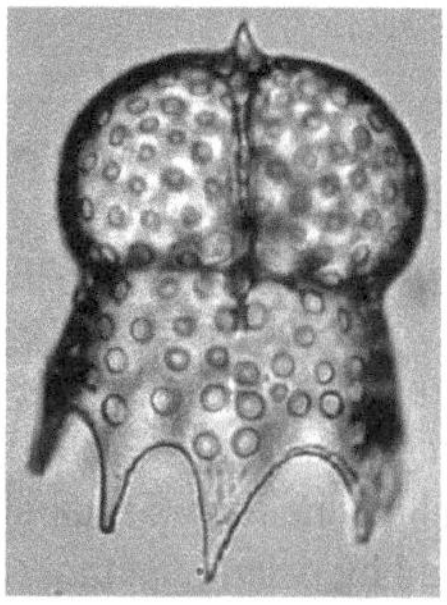

6 - Rhodospyris pulchra

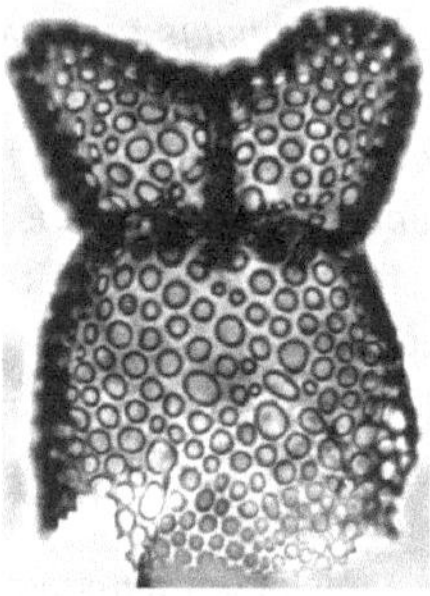

7 - Dendrospyris stabilis stabilis [?]

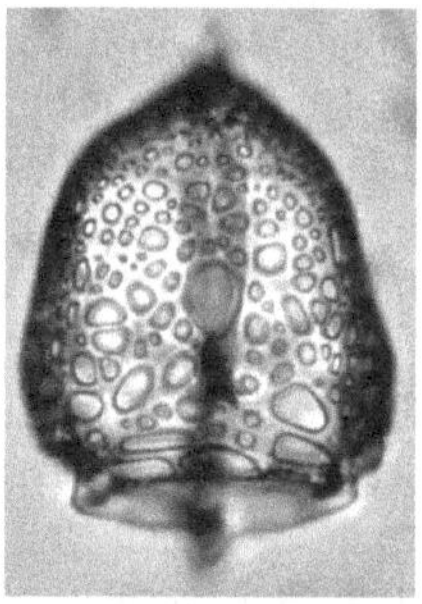

8 - Platybursa harpoi

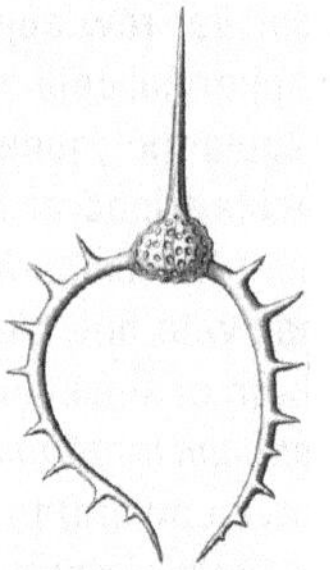

9 - Dorcadospyris dentata

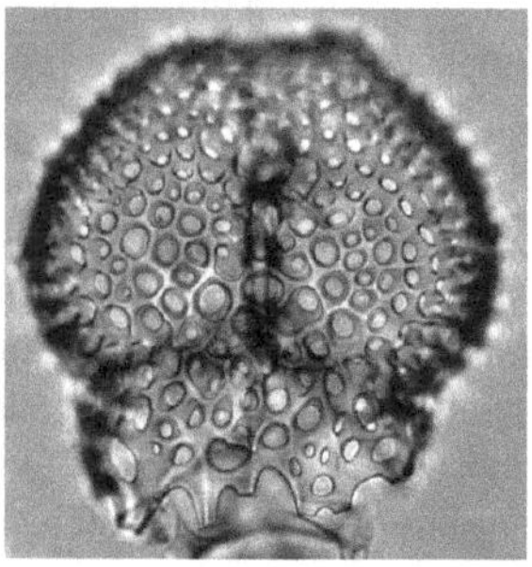

10 - Dendrospyris megalocephalis

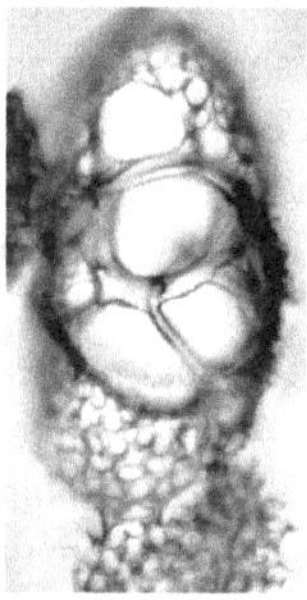

11 Dendrospyris megalocephalis (composite)

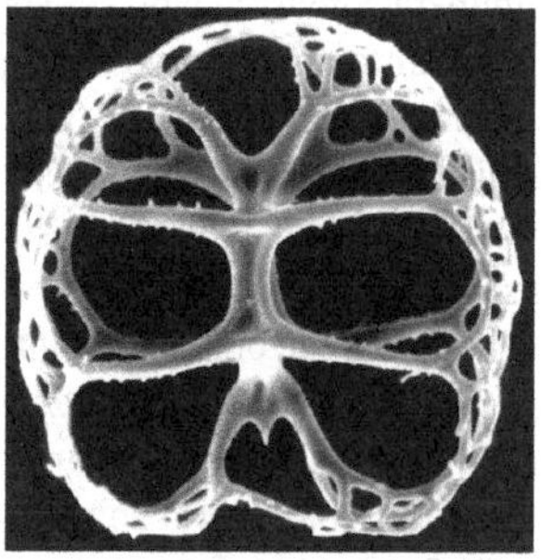

12 - Amphispyris reticulata

presence of a "neck" or narrowing at the cephalic base makes identification of this family fairly easy. Species determination however often depends on characters near the base of the shell and thus many broken specimens cannot be determined to species level. Because of their extensive use in biostratigraphy, pterocorythids are the second most common type of occurrence data in the Cenozoic (14% of Nassellaria, 10% of all radiolarian data). They are moderately diverse (78 accepted Cenozoic species).

Family* Carpocaniidae *(Haeckel) Riedel, 1967 [Carpocaniinae] (Figure 5.22)

Definition: Generally, two segmented forms with a reduced cephalis that is nearly indistinguishable in external outline from the thorax. Cephalis usually laterally elongated and subdivided into three or more chambers by internal bars or septa.

Geologic range: Eocene (Late Jurassic?) to Recent

Subgroups: as defined here, none

Remarks:

De Wever et al. (2001) use a broader definition than Riedel that includes, as subfamily, O'Dogherty's (1994) Diacanthocapsidae. This family contain many mid–late Mesozoic forms with an encased cephalis but with a variety of characters not seen in Cenozoic materials, and thus may not be directly phylogenetically related. Although a few species of carpocanids possess distinctive characters (usually near the base of the shell) they are morphologically rather monotone and not much studied, with only

Figure 5.23 Trissocyclidae. 1 – unidentified incomplete trissocylid: Side view showing "D-ring" structure with median bar at base, primary spicules A and V (apical and vertical) connected at ends by arch. Rudiments of cephalic lattice wall also present; 738B-1H-1, 53 cm JR. 2, 3 – *Lophospyris pentagona pentagona*: Bilobed cephalis with small number of extremely large polygonal pores and bladed bars, long, bladed feet; 2: van der Paverd Pl. 63 f3 (SEM); 3: side view of broken specimen showing D-ring in central ("sagittal") plane of shell; van der Paverd Pl. 63 f8. 4 – *Dendrospyris stabilis scaphipes*: Highly variable species with several subspecies, rank, and generic assignment uncertain. Large paired "sagittal" pores next to D-ring are typical of many trissocyclids. *D. s. scaphipes* has smooth cephalic wall, scattered small circular pores, three strong bladed feet; 747A-9H-8, 45 cm JR. 5 – *Dendrospyris stabilis?*: Basal view showing common form of trissocyclid lateral spine connections to lattice wall, forming large, rounded/polygonal basal pores; 744A-6H-4 JR. 6 – *Rhodospyris pulchra*: Cephalis lacks sagittal pores, thorax well developed, lattice wall of both smooth, scattered small circular pores, thorax terminated by large teeth; 744A-10H-2 JR. 7 – *Dendrospyris stabilis stabilis?*: Bilocular cephalis, well-developed thorax, both with small similar-sized pores and smooth shell, clear sagittal and collar stricture, no sagittal pores or basal feet. Specimen shown with compressed subtriangular cephalic lobes is one of many known morphologic variants; van der Paverd Pl. 63 f13 (as *Lophospyris s. s.*). 8 – *Platybursa harpoi*: Laterally compressed, apically elongated cephalic lobes, small, irregularly sized lattice wall pores, smooth shell, no thorax, feet absent or reduced to small spines; 747A-2H-3, 45 cm JR. 9 – *Dorcadospyris dentata*: Small, thick-walled ellipsoidal cephalis with indistinct sagittal constriction, strong apical horn, no thorax, two extremely large curved cylindrical feet bearing outward pointing thorns (drawing); Haeckel, 1887, Plate 85, Figure 6. 10, 11 – *Dendrospyris megalocephalis* 10: Large, flattened inflated cephalis extending apically well above D-ring, with thin shell, small subrounded framed pores, small thorax with irregular or toothed termination 737A-27X-2 JR; 11: Side view showing D-ring in center of shell and additional arches above and below (composite); 747A-6H-3, 45 cm JR. 12 – *Amphispyris reticulata*: Strongly vertically compressed shell, small D-ring in center connected by arches and lateral bar pairs at base and top of shell that anatamose to form approximately cyclindrical band of irregularly pored lattice wall, open at top and base of shell (SEM, apical view); van der Paverd Pl. 61 f6.

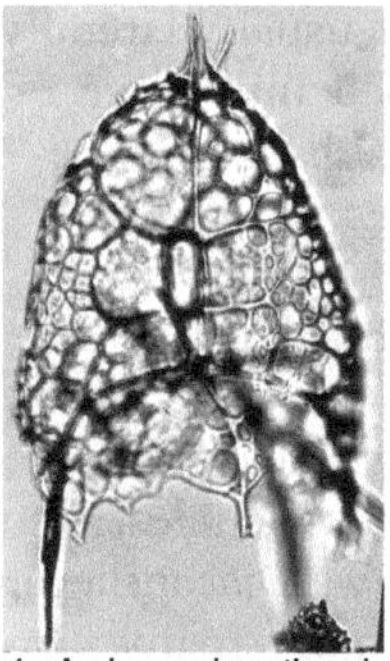

1 - Androspyris anthropiscus

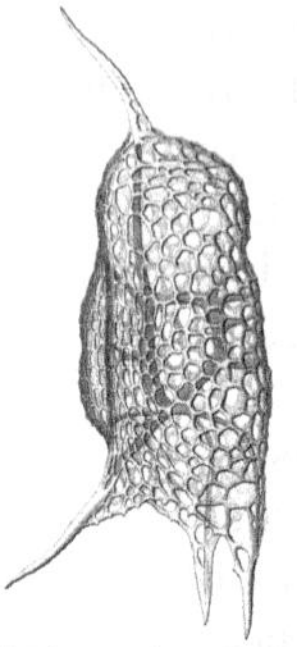

2 - Androspyris anthropiscus

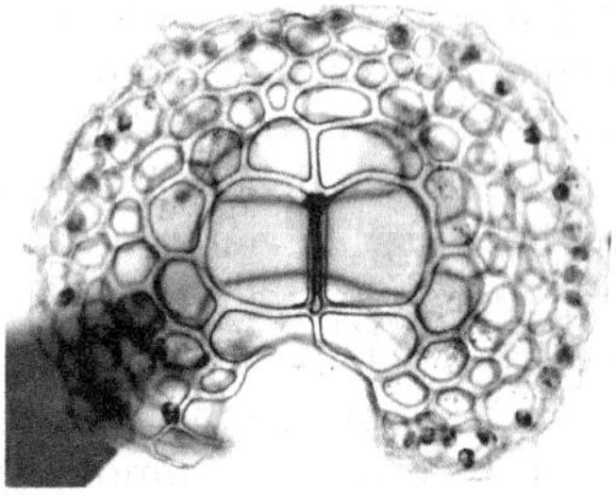

3 - Nephrospyris renilla

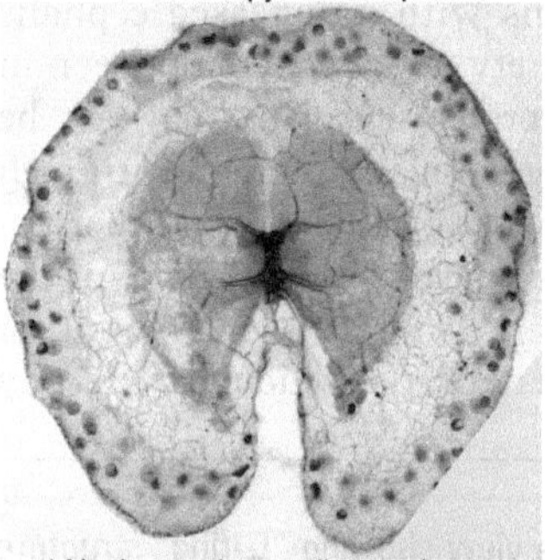

4 Nephrospyris paradictyum (composite)

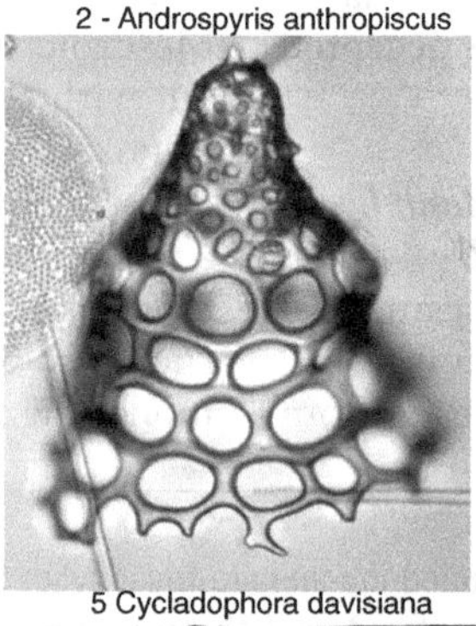

5 Cycladophora davisiana

6 Cycladophora cosma

7 Cycladophora pliocenica (composite)

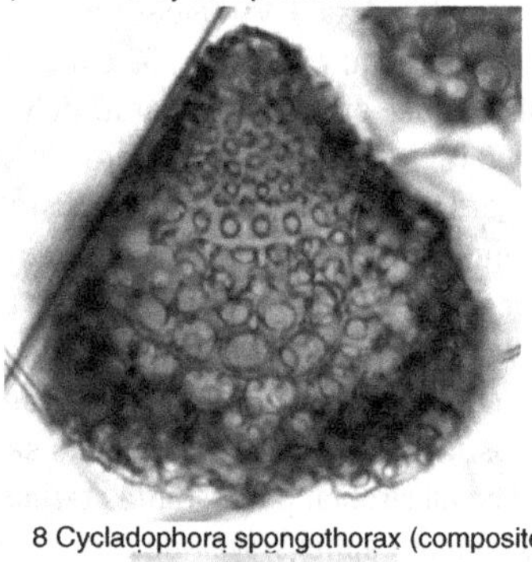

8 Cycladophora spongothorax (composite)

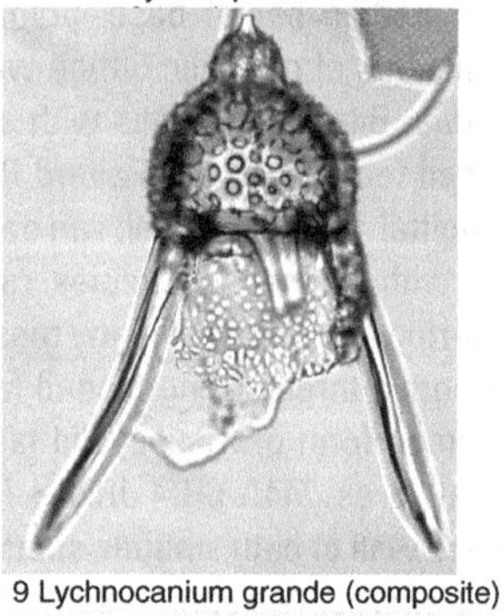

9 Lychnocanium grande (composite)

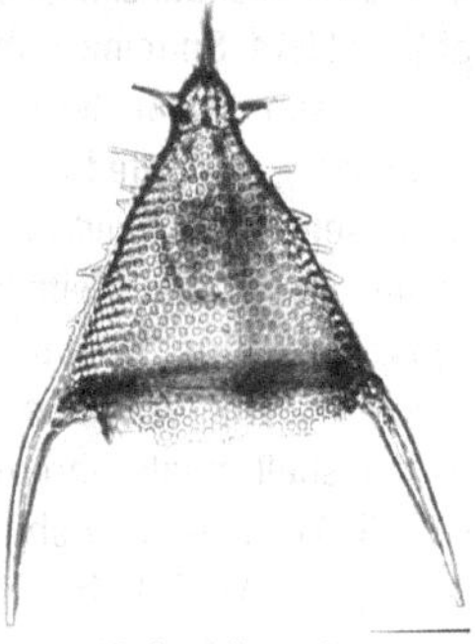

10 Artopilium elegans

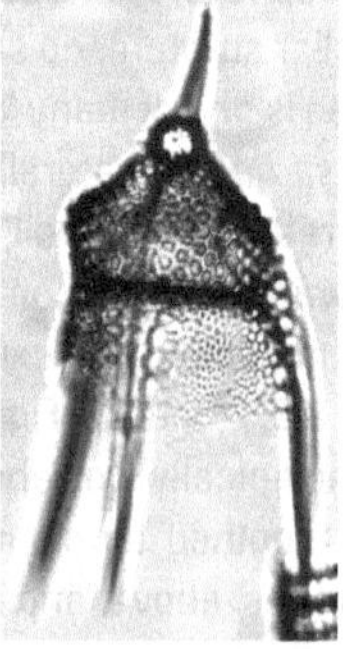

11 Pterocanium praetextum

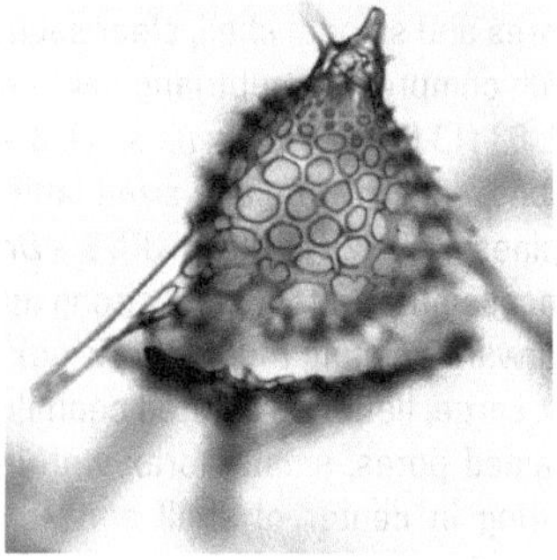

12 Dictyophimus larus

21 accepted species and ca. 2% of reported occurrence data.

Family Cannobotryidae *Haeckel, 1881 (Figures 5.19F, 5.28)*

Definition: Nassellarians with a multi-lobed cephalis (three or more chambers, usually of different sizes) and often with irregular shapes including tubular extensions.

Geologic range: mid-Jurassic (O'Dogherty et al., 2011) to Recent

Subgroups: none

Remarks:

With 28 Cenozoic species, the family is not very diverse, and is only rarely reported in the literature (2% or less of NSB records).

Superfamily Collodaria

Definition: Colonial (rarely solitary) radiolarians with reduced or absent shells, or

Figure 5.24 Trissocyclidae, Theoperidae (cycladophorids, pterocanids). 1, 2 – *Androspyris anthropiscus*: Shell strongly compressed along dorso-ventral axis, large, thin-walled, irregular polygonal pores, D-ring occupies only central half of cephalis, laterally connected to lattice wall, cephalis merges to irregularly terminated thorax without change in shell outline or wall characteristics, three basal feet and tapered apical spine; 1: Oblique view with D-ring visible as dark outline in middle of shell; van der Paverd Pl. 62 f5 (as *Tripospyris a.*); 2: Oblique view showing internal D-ring and orientation of external spines; Haeckel, 1887, Plate 83, Figure 8 (drawing). 3 – *Nephrospyris renilla*: Vertically compressed trissocyclid similar to *A. reticulata* (Plate 11) but much better developed lattice shell of polygonal pores, strong notch in outline gives kidney shape, and thus name. Specimen from plankton has peripheral ring of symbionts; Haeckel 1887 Station 271 Challenger plankton PG. 4 – *Nephrospyris paradictyum*: Similar to *N. renilla*, but only central D-ring well silicified, thus not preserved in identifiable form in sediments (see analogous *Myelastrum* in Spumellaria – Plate 5). Plankton specimen with preserved cytoplasm and symbionts; Haeckel 1887 Station 271 Challenger plankton PG. Theoperidae: 5 – *Cycladophora davisiana*: Dicyrtid (two-segmented) shell, small subspherical cephalis, thick, smooth barred thorax with small upper thorax pores, large lower thorax. Upper thorax conical, lower thorax variable but usually partially nearly cylindrical thus forming "shoulder" to upper thorax. Termination irregular; 751A-1-2, 7 cm JR.– *Cycladophora cosma*: Similar to *C. davisiana* but thorax simple conical, thinner walled, often with basal downwards-lateral directed teeth; 744A-6H-1, 53 cm JR. 7 – *Cycladophora pliocenica*: Small subspherical cephalis with stout vertical and upward-lateral bladed horns, strongly campanulate thorax with large, polygonal framed lower thorax pores, distinct terminal septum, and basal teeth (some specimens with rudimentary abdomen); 747A-2H-7, 45 cm JR. 8 – *Cycladophora spongothorax*: Shell shape similar to *C. pliocenica* but somewhat stepped outline due to external annular ridges on thorax, covered by loose spongy layer; 1138A-17-2, 105 cm DL. 9 – *Lychnocanium grande*: Dicyrtid with three long, stout gradually divergent bladed feet, occasionally with incomplete, delicate irregularly pored webbing, thorax hemispherical to subspherical (convergent towards base), small circular pores in heavy, framed bars, short, stout cyclindrical apical thorn on cephalis (composite); 746A-5H-1, 53 cm JR. 10 – *Artopilium elegans*: Pterocanid form with two angled three-bladed vertical spines on cephalis (here seen edge-on), conical thorax with very uniform rows of small circular pores, three divergent bladed feet, variably developed free abdominal lattice wall, often with lateral thorns on thoracic ribs; vdP Pl. 67 f1. 11 – *Pterocanium praetextum*: Small spherical thick-walled hyaline cephalis with cylindrical apical horn, "biretta" shaped thorax formed by pyramidal inflated lattice wall between sunken thoracic ribs, long parallel bladed feet with porous bases, veil connecting adjacent feet; vdP Pl. 70 f10. 12 – *Dictyophimus larus*: Small hemispherical cephalis with one vertical, one divergent cyclindrical apical horns, large polygonal pored rough surfaced conical thorax extending inperceptably to irregularly terminated abdomen, heavy cylindrical thoracic ribs extending from lower thorax as prominent wings with latticed butresses; 747A-10-4, 45 cm MM_DL.

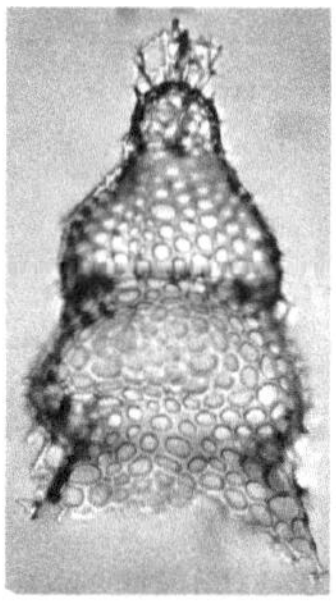
1 Artophormis cf. gracilis (composite)

2 Anthocyrtella callopisma

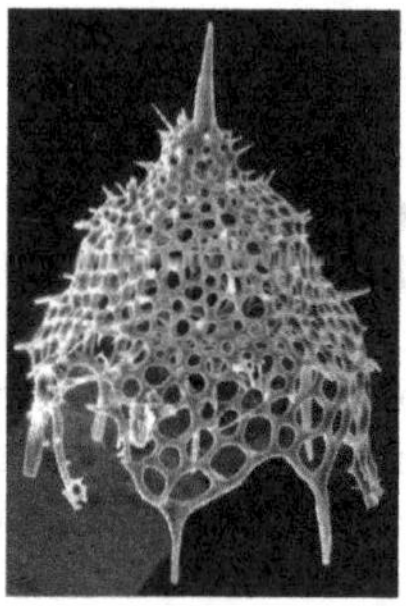
3 Calocyclas monumentum

4 Lithostrobus cf. hexagonalis

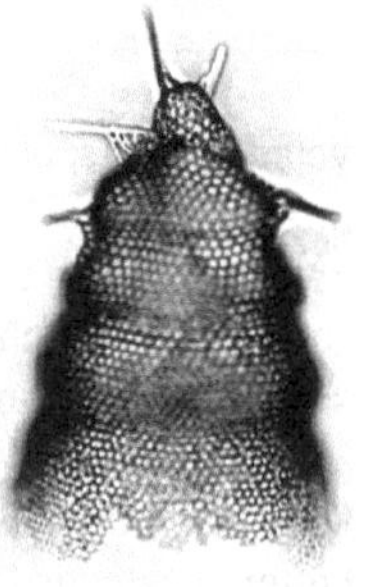
5 Stichopilium bicorne

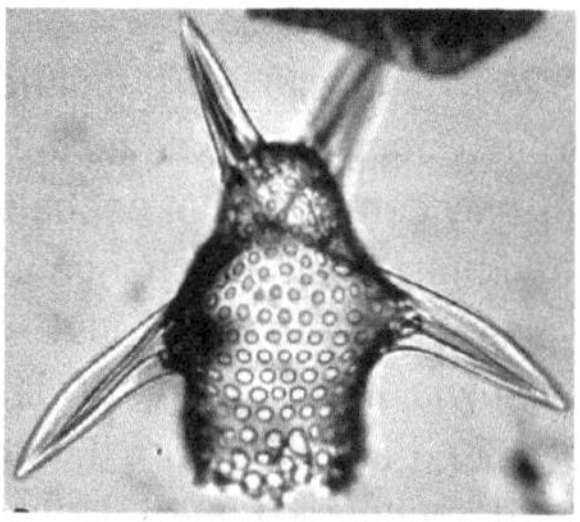
6 Stichopilium cf. bicorne

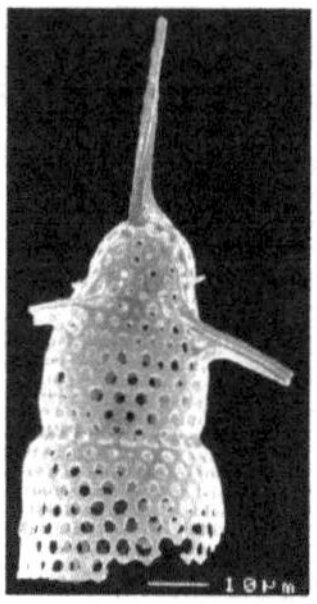
7 Stichopilum kruegeri

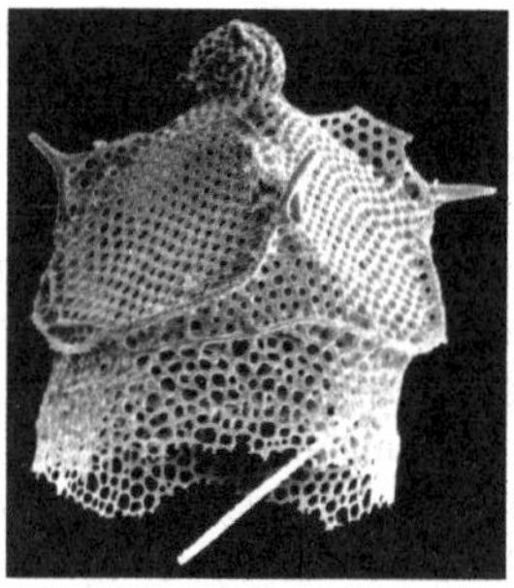
8 Lipmanella bombus

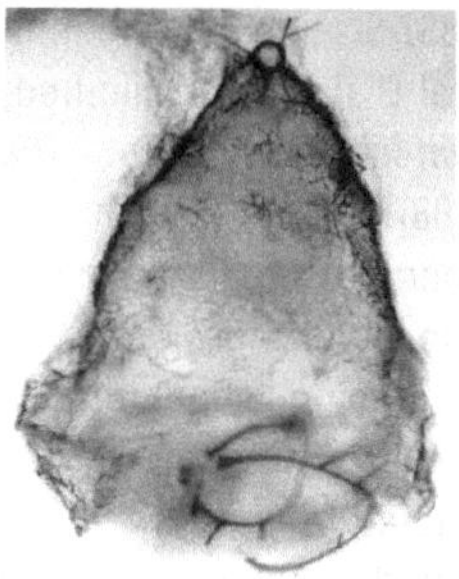
9 Sethoconus venosus (composite)

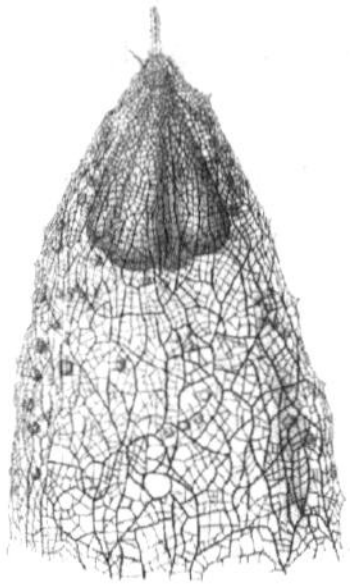
10 Sethoconus venosus

11 Sethoconus tabulatus?

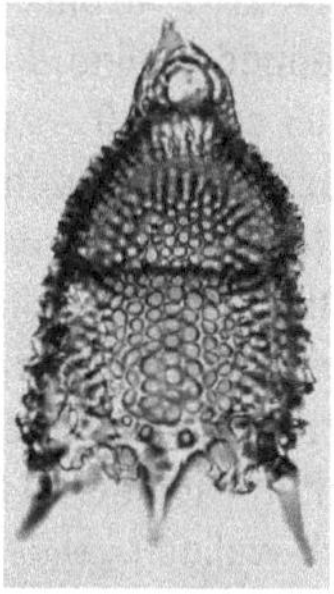
12 Thyrsocyrtis clausa (composite)

possessing isolated double spicules, and generally spherical symmetry of cell structures.

Geologic range: (Eocene?) Miocene to Recent

Subgroups:

Family* Collosphaeridae *Müller, 1858
(Figure 5.29)

Collodaria with a single, more or less spherical lattice shell, made of a flat sheet of hyaline material and bearing pores of various sizes and shapes. In some species pore edges are variously enlarged into rims, spines or tubes.

Geologic range: (Eocene?) Miocene to Recent

Family* Sphaerozoidae *Haeckel, 1862
(Figure 5.29)

Collodaria without a lattice shell but bearing isolated spicules embedded in the cytoplasm. Spicules have a central bar and branched ends. These are mostly three, occasionally four, or even more rayed, but not normally three plus four as in other nassellarian primary spicules. It is unknown if the spicules of sphaerozoids are derived but altered from the nassellarian

Figure 5.25 Theoperidae. 1 – *Artophormis cf. gracilis*: Hemispherical cephalis with stout apical horn surrounded by corona of spines with connecting meshes, truncate campanulate thin-walled, moderate pored thorax and abdomen, fourth segment with multiple ribs, termination irregular; 748B-8H-4, 45 cm JR. 2 – *Anthocyrtella callopisma*: Four segmented theoperid with thin-walled, truncate campanulate thorax and abdomen, fourth segment with noticeably larger pores and several ribs, generic assignment uncertain; 746A-13X-1, 53 cm JR. 3 – *Calocyclas monumentum*: Broad conical three-segmented shell, thin-walled, medium sized, uniform close packed polygonal pores, cephalis subspherical with strong apical horn, abdomen terminated by serrated lattice wall and/or spines (SEM); vdP Pl. 75 f10 (as *Lamprocyclas aegles*). 4 – *Lithostrobus cf. hexagonalis*: Broad, cap-shaped cephalis merging indistinctly into thorax, large spine extending from primary spicules as wing at thorax base, multiple post thoracic segments indistinct, marked only by step-like changes in outline, all segments including cephalis with thin wall, small circular pores in horizontal rows; 744A-8H-2, 60 cm JR. 5 – *Stichopilium bicorne*: cap-shaped thin-walled cephalis with two apical horns, broad thorax with wings, multiple post-thoracic thin-walled, fine pored segments distinguishable by changes in shell outline but no internal septa, irregular termination vdP Pl. 75 f1. 6 – *Stichopilium cf. bicorne*: Form with very robust, bladed cephalic horns and lateral spines, more widely spaced lattice wall pores, narrow abdomen; 748B-6H-1, 45 cm JR. 7 – *Stichopilium kruegeri*: Species marked by small size and elongate, barrel-shaped post-cephalic segments (SEM); vdP Pl. 75 f5. 8 – *Lipmanella bombus*: Spherical cephalis, pyramidal thorax with small, uniform, hexagonally arranged pores, flat or slightly concave sides, external raised keels, adjacent keels curve and joined together at thorax base. Lateral external spines extend from thoracic ribs. Short subcylindrical abdomen, irregular termination (SEM); vdP Pl. 71 f 8 (as *Eucyrtidium anomalum forma bombus*). 9, 10 – *Sethoconus venosus*: Small, hyaline spherical cephalis with apical and lateral horns, very large irregularly conical, thin-walled thorax which folds inwards at termination. Thorax weakly silicified, composed of thin strip-like bars and small polygonal pores. Wall surface bent into shallow angles; 9: Challenger Plankton Station 352 DL; 10: (drawing) Haeckel, 1887, Plate 55, Figure 2. 11 – *Sethoconus tabulatus?*: Cap-like cephalis, long, irregularly conical thorax with large polygonal-hexagonal pores in staggered vertical rows, bars cylindrical, moderately thick. This specimen has unusually large, branched apical horn; 747A-5H-3, 45 cm JR. 12 – *Thyrsocyrtis clausa*: Three-segmented theoperid with very thick-walled, hyaline spherical cephalis with small apical horn, campanulate thorax, distinct lumbar constriction, and internal septum. Thorax walls near vertical but indented between straight sections leading to three terminal tapering conical feet. Abdomen and thorax lattice wall similar, small-medium close-packed subcircular pores; 747A-10-2, 45 cm MM_DL.

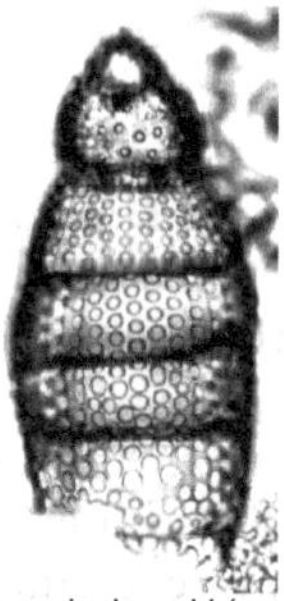

1 Eucyrtidium cienkowski (composite)

2 Eucyrtidium hexagonatum

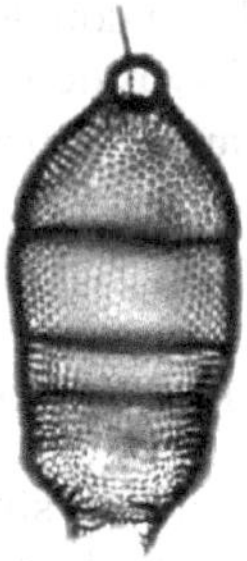

3 Eucyrtidium anomalum

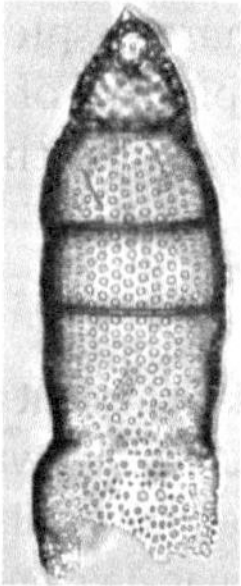

4 Eucyrtidium acuminatum? (composite)

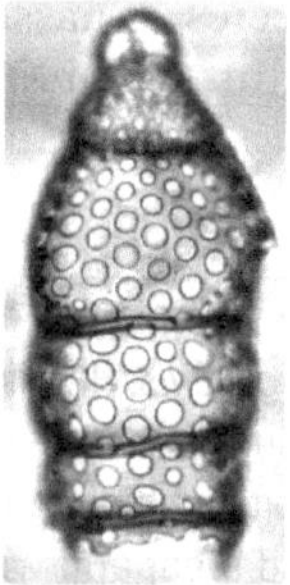

5 Eucyrtidium teuscheri

6 Eucyrtidium teuscheri

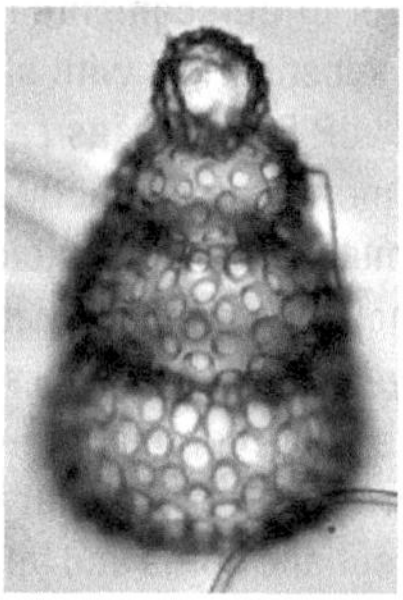

7 Stichocorys cf. peregrina

8 Cyrtocapsella cornuta (composite)

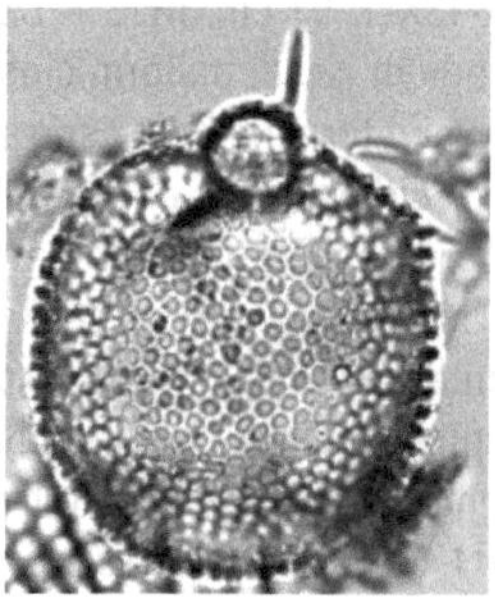

9 - Lithopera bacca (composite)

10 Clinorhabdus robusta

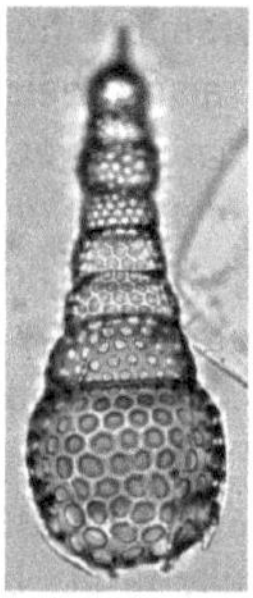

11 Cyrtopera laguncula (composite)

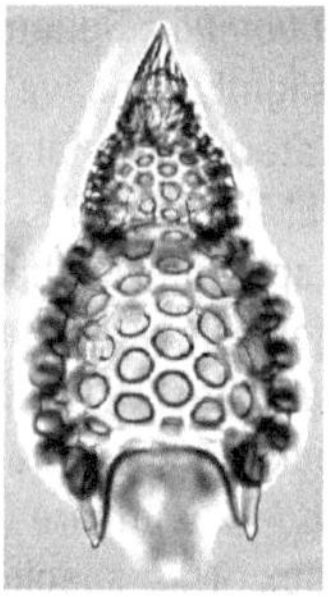

12 Podocyrtis trachodes (composite)

primary spicule, or are convergent structures of differing origin.

Family* Collophidiidae *Biard and Suzuki, in Biard et al., 2015

This clade lacks any form of skeleton, even spicules. The formal definition by Biard and Suzuki (2015) caps earlier morphologic (Anderson et al., 1999) and molecular studies (Amaral-Zettler et al., 1999; Ishitani et al., 2012) that established the distinctiveness of the group. A detailed morphologic description of the group as a whole has not been given yet.

Remarks:

The systematics of the Collodaria are undergoing rapid revisions as new molecular data become available. Their original assignment to Spumellaria in classical taxonomy (e.g. Haeckel, 1887) was even then seen as provisional as the morphology of the cells and skeletons were not very

Figure 5.26 Theoperidae (eucyrtinids). 1 – *Eucyrtidium cienkowski*: *Eucyrtidium* are multisegmented theoperids with approximately cylindrical shapes, small cephalises, approximately conical thoraxes, mostly uniform post-thoracic segments, with internal septa. *E. cienkowski* is distinguished by strongly silicified cephalis and thorax, conical abdomen, regular circular pores in staggered vertical rows alternating with low longitudinal ridges; 747A-8-6, 45 cm MM_DL. 2 – *Eucyrtidium hexagonatum*: Species with numerous short gradually expanding sub-cylindrical segments, all segment sutures smooth – no change in outline, lattice wall thin, pores uniform, small circular, arranged in rows; vdP Pl. 73 f6 (as *E. tropezianum*). 3 – *Eucyrtidium anomalum*: Very long, conical-campanulate thorax, often also elongate abdomen, very small hexagonally packed circular pores, complete specimens with constricted base; vdP Pl. 73 f1. 4 – *Eucyrtidium acuminatum?*: Specimen with heavily silicified cephalis and thorax; thinner-walled post thoracic segments rather long, but varying length, have small, widely spaced circular pores, bars flat, nearly as broad as pore diameters; 737B-5R-2 JR. 5, 6 – *Eucyrtidium teuscheri*: Thin, cap-shaped cephalis, small conical thorax, large abdomen, prominent lateral wing, post-thoracic segments with large circular pores and broad flat smooth bars, septa between segments irregularly developed; 6: 745B-18H-2, 53 cm JR; 6: Individual with non-horizontal, incomplete septa below abdomen; 747A-1H-1, 45 cm JR. 7 – *Stichocorys cf. peregrina*: Small, spherical, thick-walled cephalis with inconspicuous apical thorn, series of centrally inflated, short post cephalic segments with internal septa, small circular framed pores in thick, rough surfaced shell; 747A-5H-3, 45 cm JR. 8 – *Cyrtocapsella cornuta*: Four-segmented theoperid with small thick-walled cephalis, medium-sized inflated thorax, very large conical-campanulate abdomen and hemispherical fourth segment with only small circular mouth at base of shell; Site 1137, photo by Sarah Trubovitz. 9 – *Lithopera bacca*: Small spherical cephalis with well-developed apical horn at side, large, inflated ovoid thorax, nearly closed at base, with furrows on top marking ribs extending from base of thorax into lateral spines, lattice wall of small irregular closely packed pores; photo by Sarah Trubovitz. 10 – *Clinorhabdus robusta*: Small spherical cephalis, long thorax, conical at top, inflated cylindrical at base, with two offset, opposite thin lateral spines, distinct lumbar suture and septum, long subcylindrical abdomen partially closed at base, pores uniform to slightly expanding distally, subcircular, irregularly packed or in staggered rows; 744A-8H-1 JR. 11 – *Cyrtopera laguncula*: Theoperid with ca. 7–10 segments. These are initially narrow, gradually expanding in width, mostly shorter than wide, with small circular pores in horizontal rows. In complete specimens the apical horn is long and curved, and the last segment is inflated, sub-spherical, and encloses the base of the shell; 744A-3H-1, 53 cm JR. 12 – *Podocyrtis trachodes*: Small hemispherical cephalis with stout bladed apical horn, campanulate thorax, distinct lumbar stricture and internal septum, very large conical-campanulate abdomen with maximum width near base, strong basal peristome with three broad, tapered bladed feet. Pores of thorax and abdomen subcircular, medium-large (increasing in size distally), framed, shell surface rough; Indian Ocean Eocene DL.

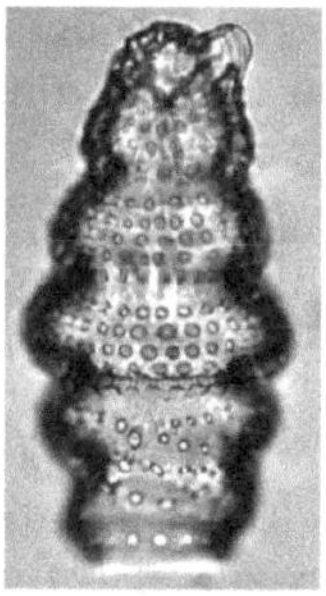

1 Botryostrobus bramlettei

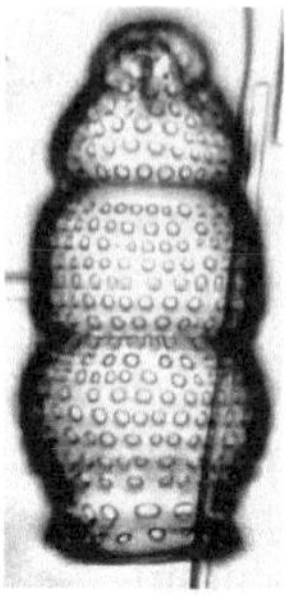

2 Phormostichoartus corbula?

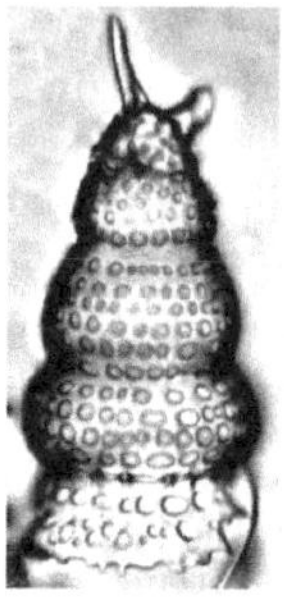

3 Siphostichartus jahnae

4 Spyrocyrtis scalaris

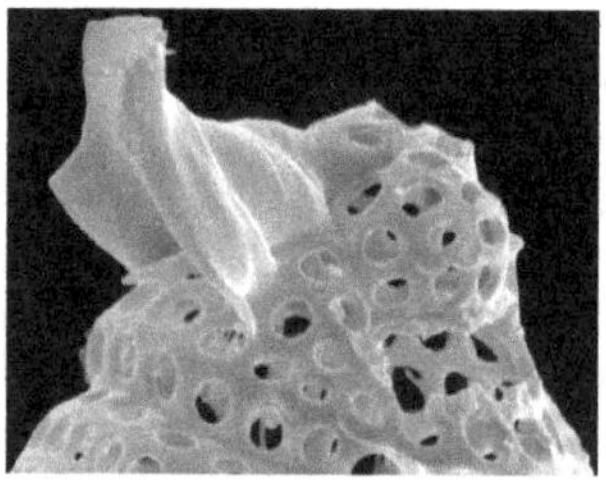

5 Spyrocyrtis scalaris

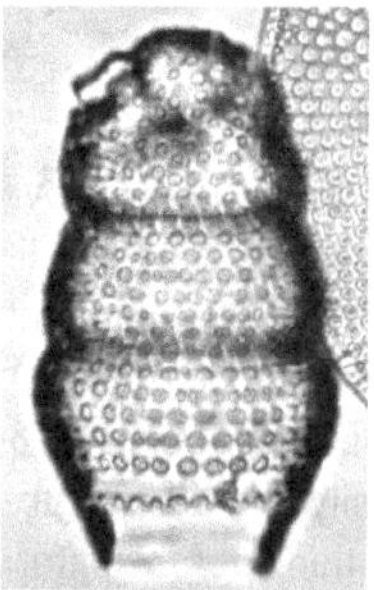

6 Phormostichoartus corbula

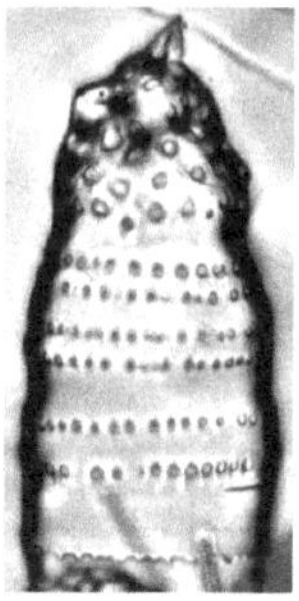

unnamed artostrobid

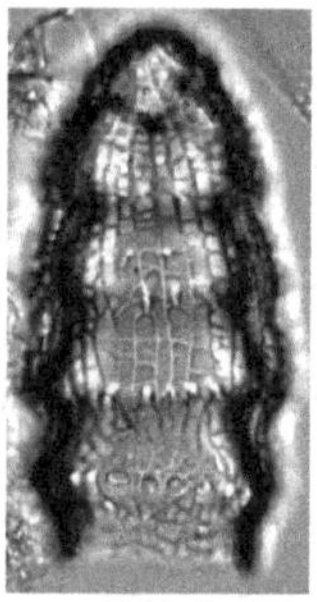

8 - Siphocampe arachnea (composite)

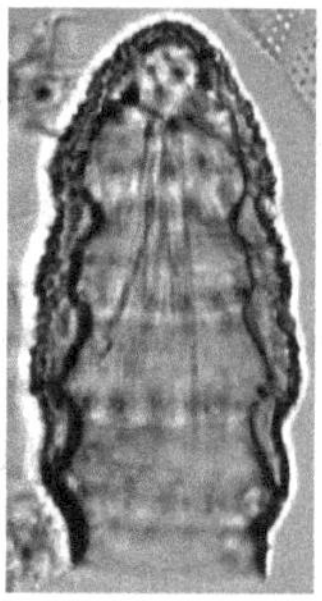

9 Siphocampe arachnea

10 Anthocyrtidium sp. (composite)

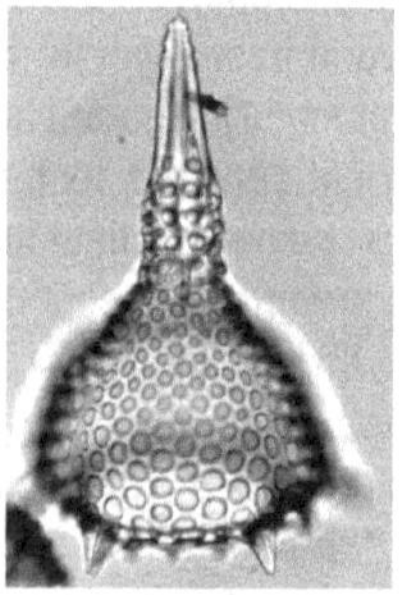

11 Anthocyrtidium ophirense

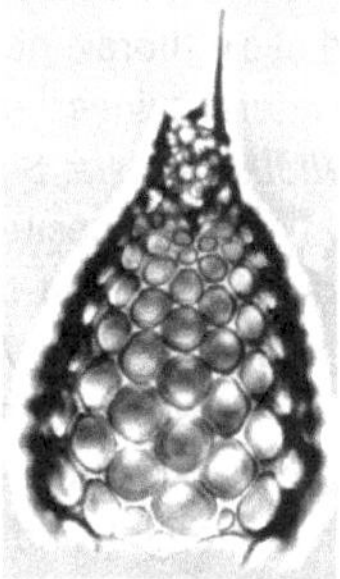

12 Lamprocyrtis nigriniae

similar with other Spumellaria, sharing essentially only a spherical symmetry. Thalassicollidae – large solitary species lacking any skeletal structure – were long given family status but are not a monophyletic group (Biard et al., 2015), and appear instead to be a morphologic stage in life cycle of some Sphaerozoidae.

The geologic range of the group is also not well established. The isolated spicules of Sphaerozoidae have no known biostratigraphic or paleoceanographic value and are rarely reported in the literature (there are no occurrence records in the NSB database), and it is also unclear if such spicules can reliably be differentiated from similar simple spicules created by other groups of organisms e.g. sponges. Collosphaeridae, with 45 accepted species, is not very diverse, nor are they commonly reported in the fossil literature (1–2% of NSB data). Collosphaerids are robustly represented as

Figure 5.27 Artostrobiidae, Pterocorythidae. 1 – *Botryostrobus bramlettei*: Small cephalis, large obliquely upwards vertical tube, fairly small, conical-inflated thorax, indistinct collar stricture. Abdomen and fourth segment similar, increasing in width, fifth segment narrower, all three short, centrally inflated, each with four to five rows of small circular pores and bars nearly equal in width, sutures lacking pores but no internal septa; 745B-18H-2, 53 cm JR. 2, 6 – *Phormostichoartus corbula*: Cap-shaped cephalis, lateral ventral tube lies on shoulder of thorax, conical-campanulate thorax with ca. 4 rows medium-sized subcircular pores, post-thoracic segments flattened barrel-shaped, nearly equal width, significantly longer than thorax, with 6-10 rows of small circular pores, distinct constriction at suture, with internal septum; 2: specimen with cephalic tube pointing directly at viewer; 751A-12H-3, 98 cm JR; 6: specimen with tube 90° to viewer; 737A-27X-2 JR. 3 – *Siphostichartus jahnae*: Rounded truncate conical cephalis with flared, lateral upward ("duck-billed") vertical tube and prominent apical horn; post-cephalic segments conical-campanulate, 4–6 rows small circular pores, sutures constricted but no internal septa, fifth segment narrower, irregular termination; JR. 4, 5 – *Spirocyrtis scalaris*: Small subspherical cephalis with strong duck-billed lateral vertical tube lying on thorax, short conical-campanulate thorax, numerous short conical post thoracic segments with several rows small subrounded pores, distinct steps in profile due to abrupt increases in width at segment sutures; 4: vdP Pl. 77 f1 5: vdP Pl. 77 f4b SEM (as *Botryostrobus s.*). 7 – unnamed artostrobid: Small, nearly hyaline cephalis with stout apical horn, lateral vertical tube, cap-shaped thorax with widely spaced, scattered large subrounded pores, post thoracic segments approximately equal in width, 2–4 widely spaced rows of small pores, sutures marked by slight constriction only; 751A-12H-1, 98 cm JR. 8, 9 – *Siphocampe arachnea*: Hemispherical cephalis, several post-cephalic segments nearly same width, conical-campanulate, indented sutures, isolated single rows of pores at maximum width of each segment, surface of shell covered by reticulate network of ridges (Figure 8), shell thick-walled, prominent multiple axobates in middle of shell (Figure 9); Antarctic Neogene DL. Pterocorythidae: 10 – *Anthocyrtidium sp.*: Small conical cephalis with bilobed pterocorythid base and strong conical apical horn, barrel-shaped thorax with widely spaced staggered vertical rows of small circular pores and weak longitudinal ridges, hyaline peristome elongated at base into irregular row of teeth; 751A-14H-CC JR. 11 – *Anthocyrtidium ophirense*: Thick-walled, elongate cephalis with large subcircular pores, paired pores on either side of apical spine embedded in cephalic wall, strong bladed conical apical horn, collar suture indistinct, conical-campanulate thorax with uniform, hexagonally packed subrounded medium-sized pores, row of stout, bladed, pointed spines at base of thorax, thin peristome with downwards pointing small teeth; 532-20-2 B. Bittniok. 12 – *Lamprocyrtis nigriniae*: Cylindrical cephalis, open at top with prominent bladed apical spine at side, indistinct basal cephalic lobes and collar suture, thorax elongate conical-campanulate with large, hexagonally packed subcircular pores and narrow bars, indeterminate termination; vdP Pl. 75 f6 (as *Lamprocyclas n.*).

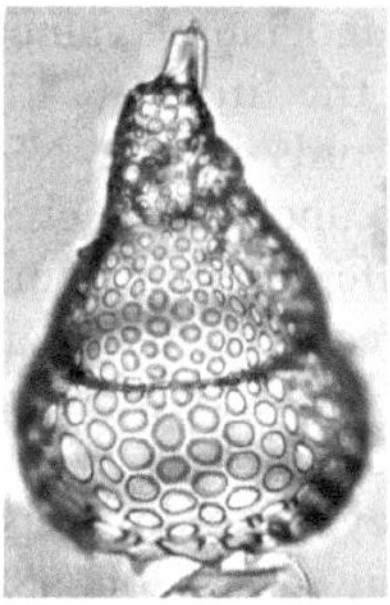

1 Lamprocyclas hannai

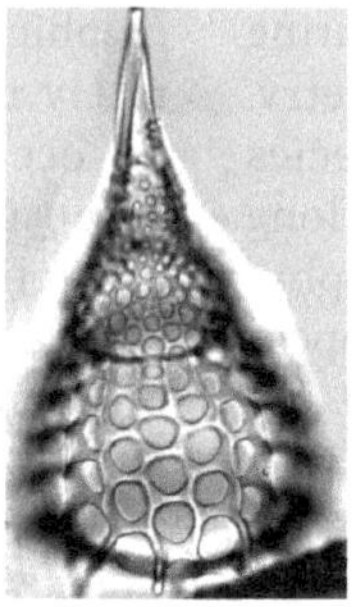

2 - Lamprocyclas cf. maritalis (composite)

3 Pterocorys hertwigii

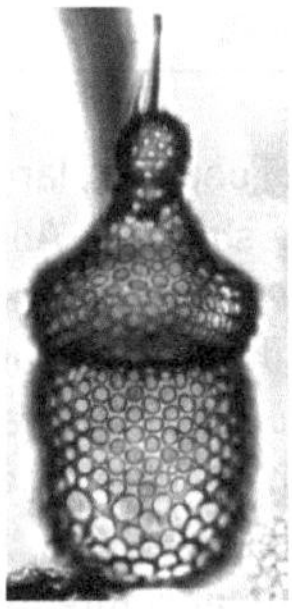

4 Theocorythium trachelium

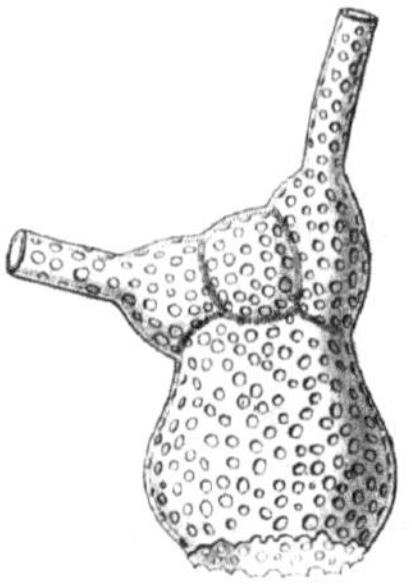

5 Acrobotrys disolenia

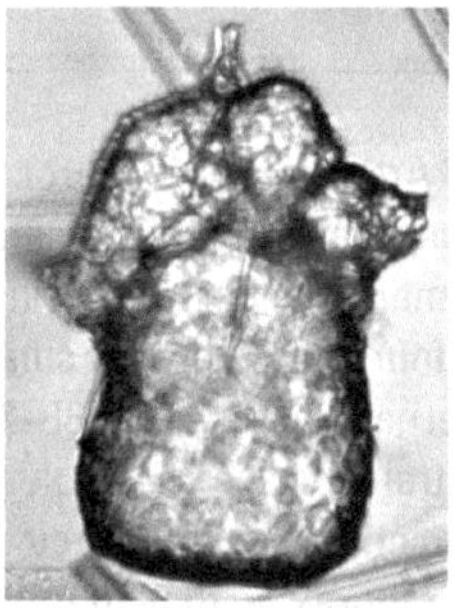

unnamed cannobotryid

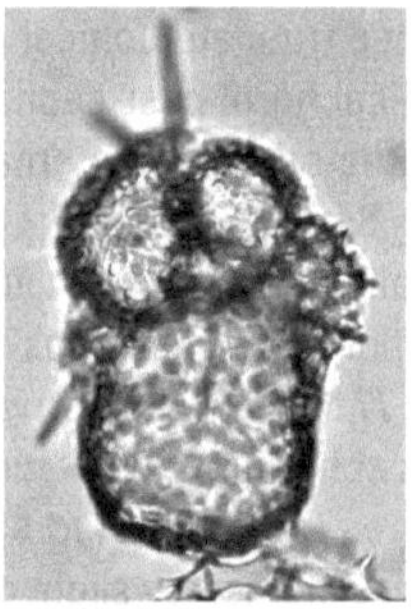

7 Saccospyris praeantarctica

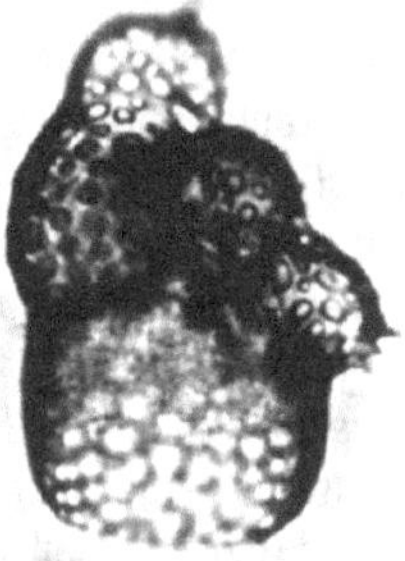

8 Botryopyle sp.

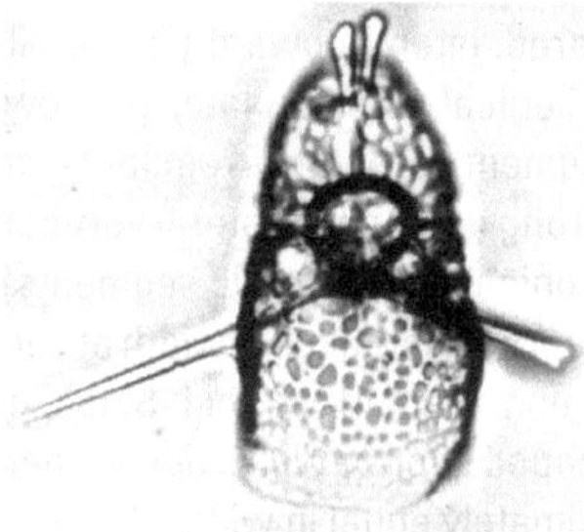

9 Botryocella multicellaris

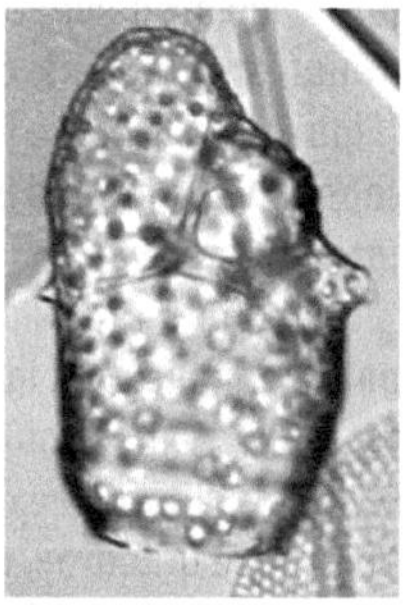

10 Saccospyris sp.

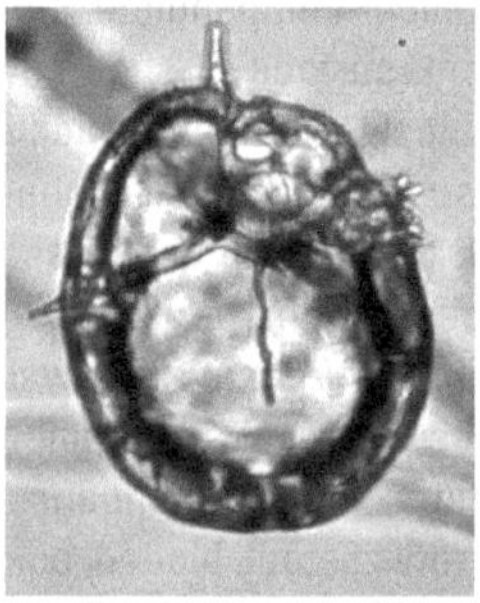

11 Saccospyris conithorax

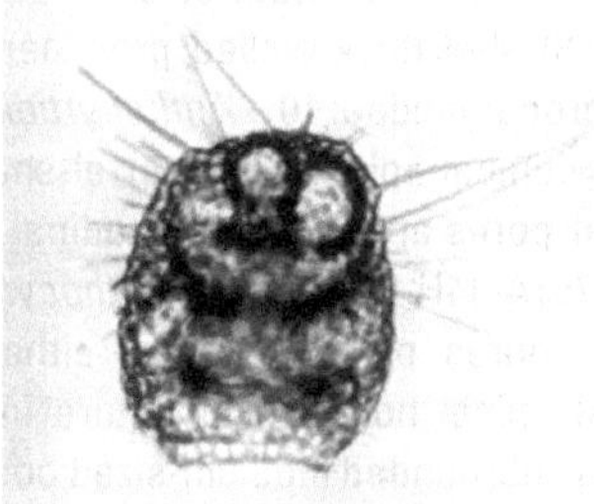

12 -Botryocyrtis scutum

fossils back to the middle Miocene, with scarcer occurrences to the basal Miocene (Bjørklund and Goll, 1979), but older records and identifications are debatable. The oldest well-documented material is from the Eocene (Goll and Caulet, 1985) but the morphology of the shells is distinctive enough that assignment to Collosphaeridae while possible, is uncertain. Molecular clock estimates (Ishitani et al., 2012) place the origin of the Collodaria in the mid-Cenozoic (ca. 33 Ma, error range of 40–25 Ma).

The collodarians are, based on recent environmental molecular surveys, very diverse, and are among the most abundant larger zooplanktonic protists in the modern oceans (de Vargas et al., 2015; Biard

Figure 5.28 Pterocorythidae, Cannobotryidae. 1 – *Lamprocyclas hannai*: Medium-sized, porous, elongate cephalis with two basal lobes, bladed apical spine, conical thorax with small close-set irregular pores, distinct lumbar suture and internal septum, short barrel shaped abdomen approximately same length as thorax with medium-sized irregular pores, mouth constricted with small denticles at base; 737B-8R-2 JR. 2 – *Lamprocyclas cf. maritalis*: Cephalis similar to *L. hannai*, lumbar stricture less distinct, abdomen conical, distally constricted, pores of abdomen medium-large, bars thick, strong row of teeth at base of abdomen wall and second row from distinct basal septum (peristome) at mouth; 1084-56X B. Bittniok. 3 – *Pterocorys hertwigii*: Typical pterocorythid cephalic structures, campanulate thorax with small regular pores, broad barrel-shaped abdomen with small regular circular pores, no distinct termination, several longitudinal ridges run down shell from ca. mid thorax to near mouth (SEM); vdP Pl. 74 f4 (as *P. zancleus forma h.*). 4 – *Theocorythium trachelium*: Globular to ellipsoidal cephalis with bladed apical spine, constricted at base where lateral lobes are placed creating "neck," upper thorax conical, changing contour to attach to larger, short subcylindrical lower thorax. Lumbar stricture pronounced, internal septum, Abdomen elongate cylindrical, same diameter or slightly narrower than thorax, peristome with downwards pointing teeth, Pores of thorax and abdomen similar – medium-small, subcircular, in close-packed staggered vertical rows; vdP Pl. 74 f17. Cannobotryidae: 5 – *Acrobotrys disolenia*: Three-lobed cephalis with tubes extending from left and right lobes, sac-like thorax with irregular termination, entire lattice wall similar with small irregular pores (drawing); Haeckel, 1887, Plate 96, Figure 10. 6 – unnamed cannobotyrid: Three-lobed cephalis with one flat-lying large chamber extending down one side of upper thorax, narrow tubes or spines extending from left and right lobe tips, apical spine at suture of large and middle lobe, subcyclindrical thorax, closed at base, axobate present; 751A-2H-3, 98 cm JR. 7 – *Saccospyris praeantarctica*: similar to unnamed species above but larger cephalic lobe approximately hemispherical, lateral spines more downwardly directed, thorax significantly smaller in relation to cephalis; 744A-4H-1, 53 cm JR. 8 – *Botryopyle* sp.: Cephalis with three main lobes of unequal sizes, the largest extending vertically on one side of the shell and itself subdivided into two lobes, short cylindrial thorax; vdP Pl. 77 f17. 9 – *Botryocella multicellaris*: View from side with smallest lobe of cephalis showing prominent lateral spines and a pair of spines penetrating the large lobe at apex of the cephalis; vdP Pl. 77 f14 (as *Botryopyle m.*). 10 – *Saccospyris* sp.: Specimen with only two distinct lobes, largely fused into each other and marked only by interior spine from median bar; 745B-17-4, 53 cm DL. 11 – *Saccospyris conithorax*: Small, thick-walled shell, fully closed at base, with only a few scattered, indistinct pores, cephalic lobes fused, marked only by location of primary spicules. Axobate present. 738B-1H-3, 53 cm JR. 12 – *Botryocyrtis scutum*: Three main lobes of cephalis approximately globular and similar in size, one one side displaced distally to lie partially under the middle lobe. Poorly developed additional structure at base of cephalis below main lobes. Short cylindrical post-cephalic segments indistinctly visible inside thick mantle surrounding lower cephalis and rest of shell. Outer shell shows numerous small, close packed pores, many fine spines radiate from upper thorax and cephalis; vdP Pl. 77 f6 (as *Botryostrobus s.*).

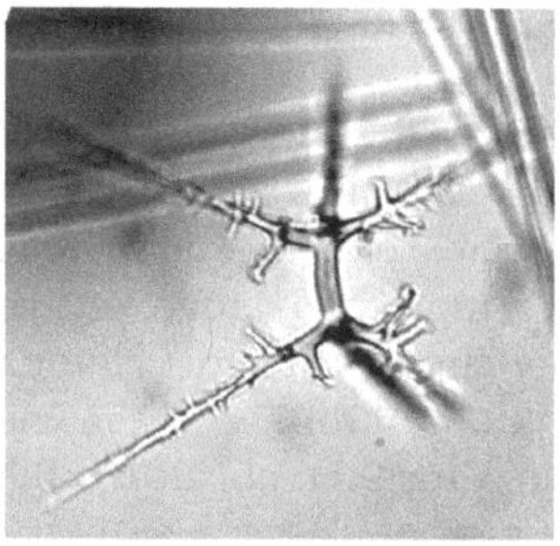

1 Sphaerozoum punctatum

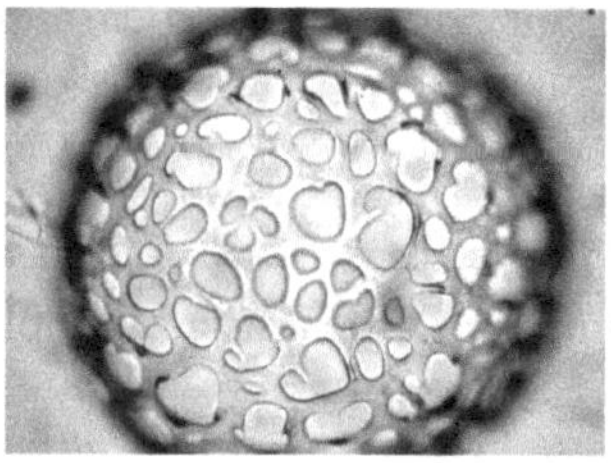

2 Acrosphaera cyrtodon?

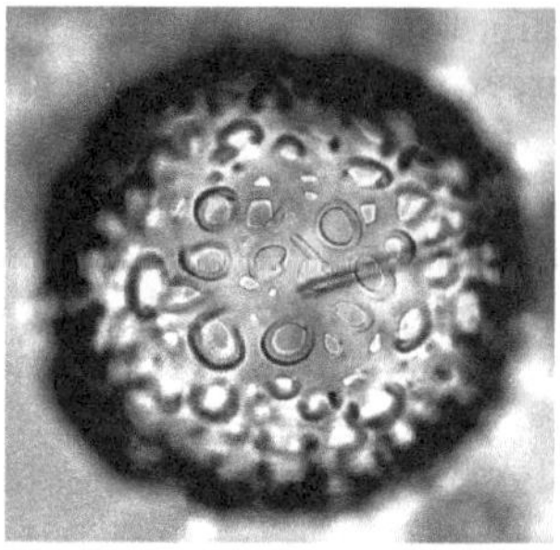

3 Acrosphaera labrata

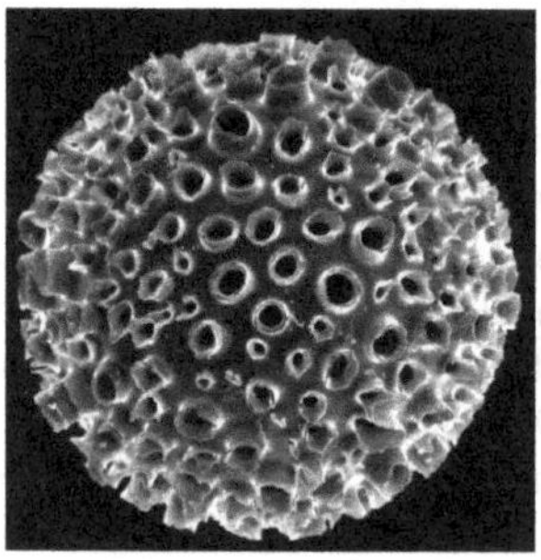

4 Siphonosphaera tubulosa

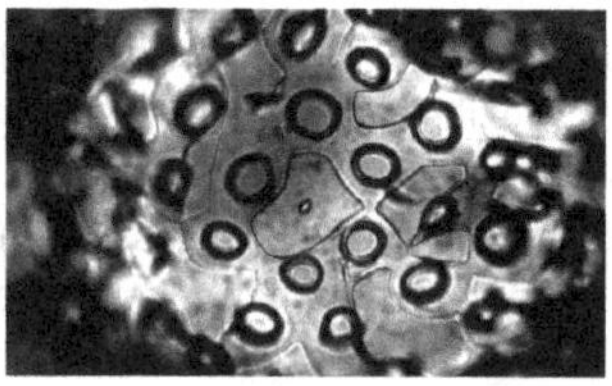

5 Acrosphaera australis

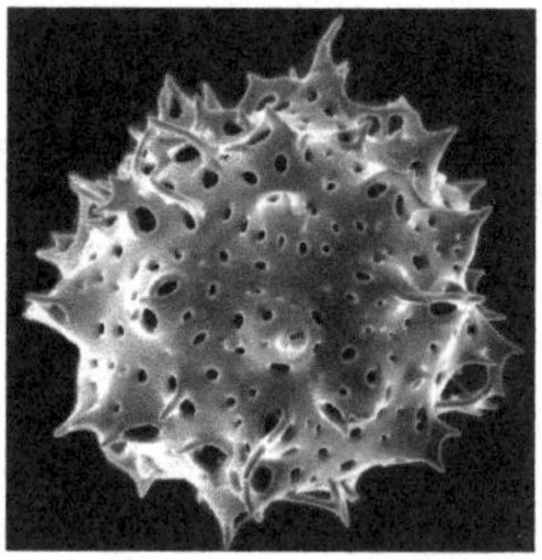

6 Acrosphaera spinosa

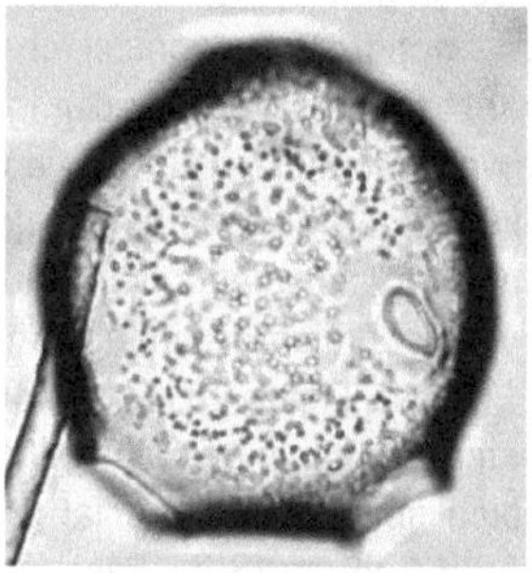

7 Siphonosphaera magnisphaera

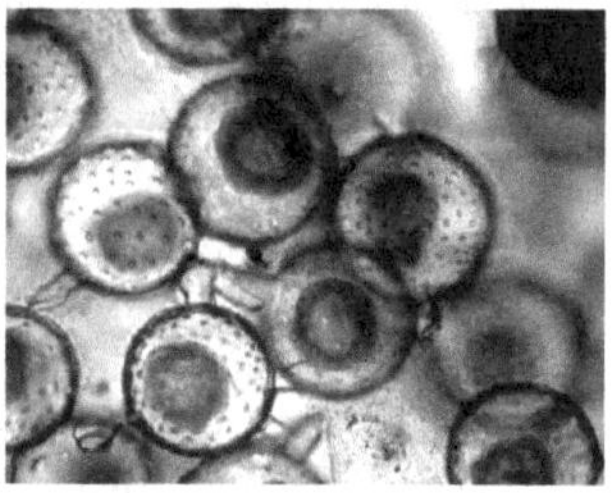

8 Siphonosphaera socialis colony

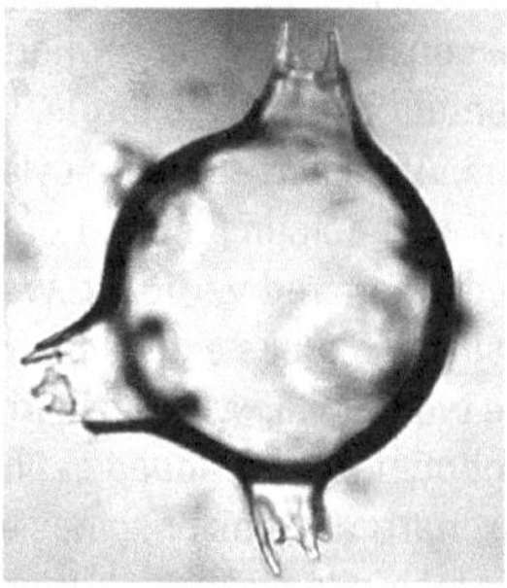

9 Siphonosphaera abelmannae

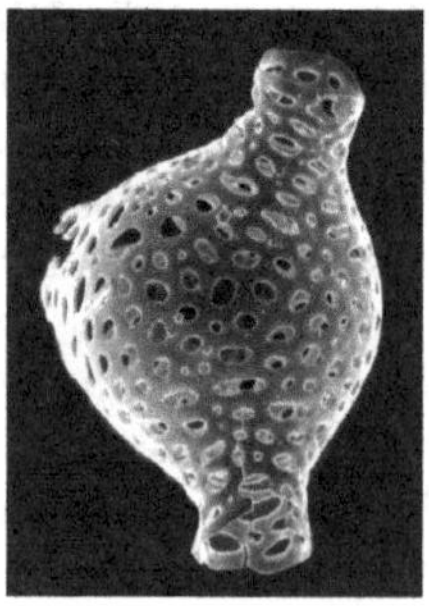

10 Solenosphaera zanguebarica

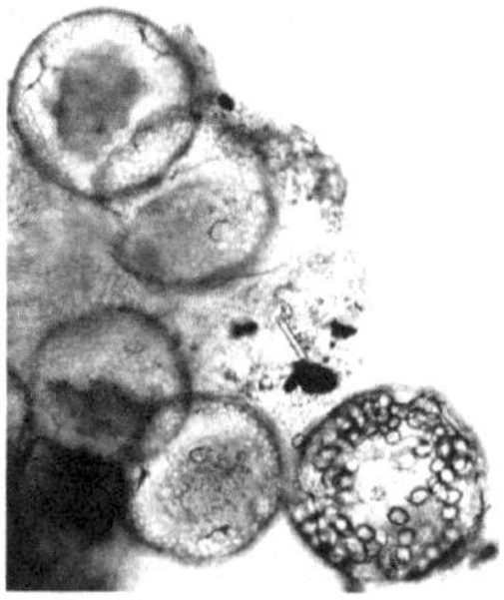

11 Buccinosphaera invaginata colony

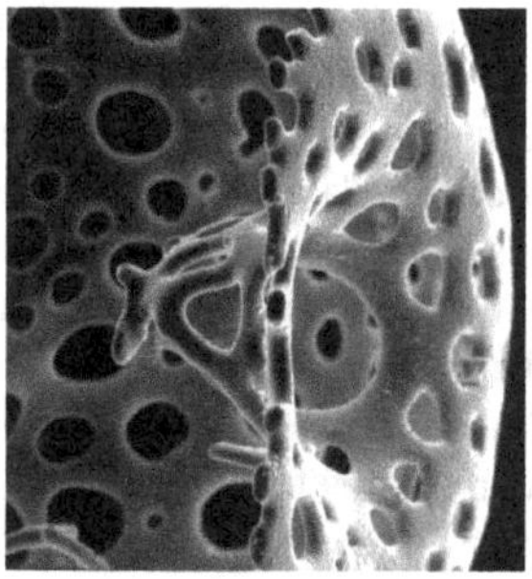

12 Buccinosphaera invaginata

et al., 2017). Taxonomic understanding of this diversity however is almost completely lacking, except for the shelled Collophaeridae.

Order Entactinaria
Family* Orosphaeridae *Haeckel, 1887
(Figure 5.10)

Definition: Large to very large polyhedral to subspherical shells composed usually of two types of irregular lattice bands, thicker and thinner, criss-crossing to form irregular polygonal pores. Thick, circular in cross-section external spines attach to thicker lattice bands and arranged at varying degrees from fully radial to tangential to the shell surface, and often have external buttresses. Inner structure reduced to a few thin beams connected via terminal branches to a small median bar.

Geologic range: Eocene to Recent

Subgroups: none

Remarks:

De Wever et al. (2001) hypothesize that the orosphaerids derived from Polyentactiniidae, a Cretaceous family with generally similar shells, by modification of the external (in Polyentactiniidae, three-bladed) spine structure. While plausible, no stratigraphically or morphologically intermediate forms are known. Orosphaerids are normally found only as fragments in the fossil record and have no known stratigraphic or paleoceanographic application. There are no records of them in the NSB database and they are usually only mentioned briefly, if seen, in the general text of reports. Because of their massive structure they are often the only type of radiolarian material left (along with sponge spicules) in highly dissolved biosiliceous fossil assemblages and thus their abundant presence is sometimes used as a dissolution indicator.

Figure 5.29 Collosphaeridae. 1 – *Sphaerozoum punctatum*: Simple spicule with central bar and three spines radiating from the central bar ends, spines with bristle-like side branches; 744A-6H-4 JR. 2 – *Acrosphaera cyrtodon?*: Typical collosphaerid spherical shell with flat, broad hyaline bars, but slightly rough surface from pore edges, with variably sized, irregular polygonal pores, often with one or more tooth-like projections partially subdividing them; 747A-8H-5, 45 cm JR. 3 – *Acrosphaera labrata*: Spherical collosphaerid where pore rims are thickened into rounded lip-like structures; Lazarus ODP Leg 120. 4 – *Siphonosphaera tubulosa*: Spherical collosphaerid with medium sized pores, pore rims extended to short tubes, less than or equal to pore diameters (SEM); vdP Pl. 9 f2. 5 – *Acrosphaera australis*: Spherical collosphaerid with moderately widely spaced, medium-sized tubed pores, distal ends of pores connected to layer of flat hyaline bands forming incomplete secondary outer cortical shell; Lazarus ODP Leg 120. 6 – *Acrosphaera spinosa*: Surface of shell with scattered small subcircular pores, shell surface deformed into numerous low conical protrubences, most terminating in a thorn (SEM); vdP Pl. 5 f6. 7 – *Siphonosphaera magnisphaera*: Spherical shell with numerous scattered small circular pores, a few very large subcircular pores with short tapered hyaline tubular rims; 751A-13H-2 98 cm JR. 8 – *Siphonosphaera socialis*: Spherical shell with a few scattered small circular pores, a few long tubes of medium diameter with irregular terminations (colony); Challenger Plankton Station 293 DL. 9 – *Siphonosphaera abelmannae*: Almost completely hyaline spherical shell with a few short tapering tubes with irregularly toothed rims; 744A-10H-6 60 cm JR. 10 – *Solenosphaera zanguebarica*: Spherical shell with a few large short cylindrical tubes, regularly terminated. Both shell and tubes with medium sized subrounded to ellipsoidal pores (SEM); vdP Pl. 13 f3 (as *Trisolenia megalactis*). 11, 12 – *Buccinosphaera invaginata*: Somewhat irregularly spherical shell with medium sized, subrounded pores. Irregular shape due to invaginated shell wall around larger pores; 11: colony (one specimen with celestite crystals indicating formation of reproductive swarmers), Challenger Plankton, PG; 12: SEM of inward indented pore structure; vdP Pl. 4 f4.

The order Entactinaria is much more diverse in the Mesozoic and Paleozoic, as detailed in De Wever et al. (2001).

Family Saturnalidae *Deflandre 1953 (Figure 5.10)*

Definition: Polycystines with a spherical, lattice-wall, or spongy cortical shell connected to an equatorial ring or set of rings by two polar spines, and extending internally as beams to a small mesh-like medullary shell. The ring(s) may be circular or elliptical, smooth or bladed, and bear radially arranged co-planar flattened spines.

Geologic range: Triassic to Recent

Subgroups: several given in De Wever et al. (2001) not included here. Axopruninae sensu De Wever et al. (2001) we retain in the Actinommidae.

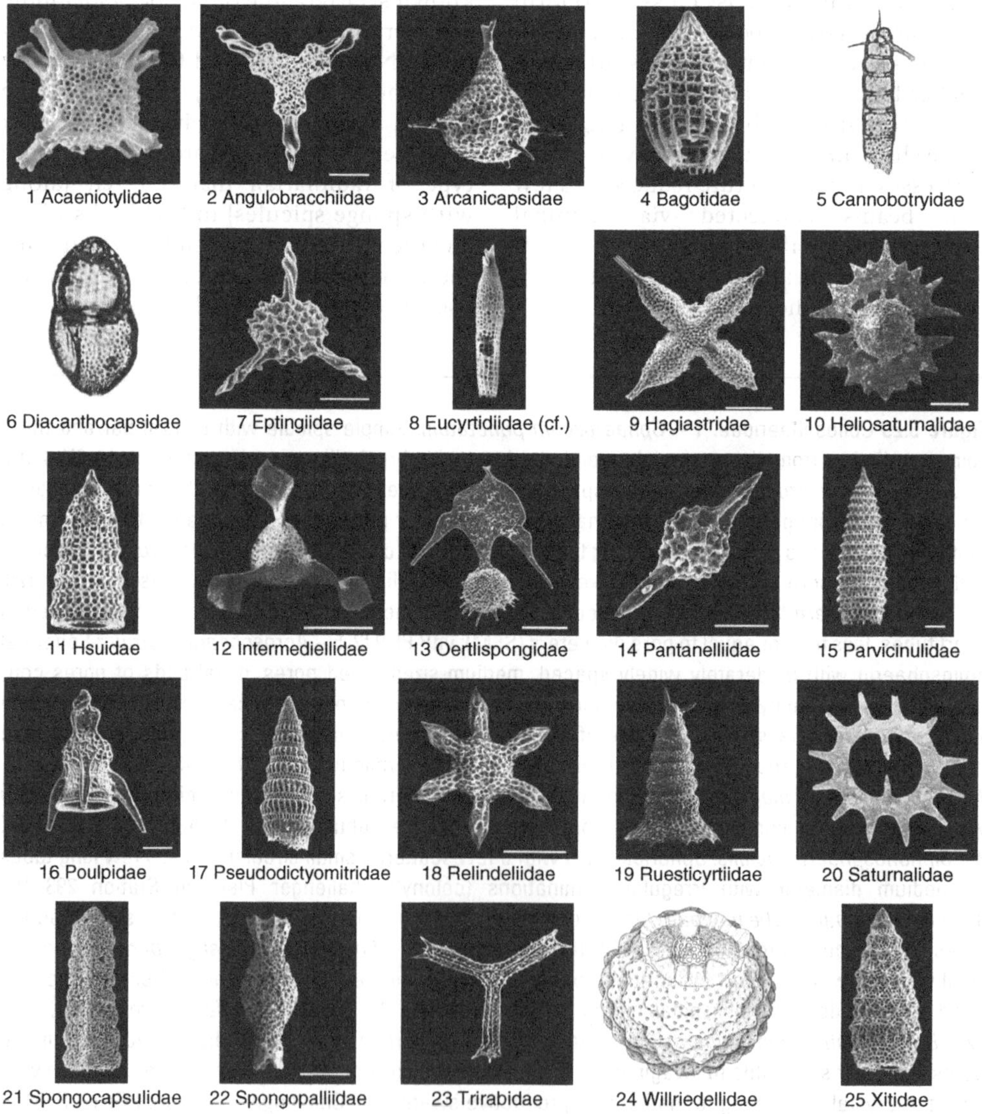

Remarks:

Placement of this group into Entactinaria follows De Wever et al. (2001) although most other aspects of the shell morphology are typical of the Spumellaria. The distinctive ring, however, is unlike any other structure in Spumellaria, so the purely morphologic affinities to either order are ambiguous.

The only known Cenozoic species is *Saturnalis circularis* with a reported range of early Eocene to Recent (NSB database search 15.10.17), but the group was much more diverse in the Mesozoic.

Mesozoic and Paleozoic Taxa

(Figures 5.30, 5.31)

Mesozoic and Paleozoic radiolarians are collectively (vs. the Cenozoic) very diverse, and contain the majority of all described species and families of radiolarians. As these are extensively documented in De Wever et al. (2001), O'Dogherty et al. (2009), and Caradroit et al. (2017), we illustrate only a few forms in Plates 18 and 19 to indicate the great range of forms. A full listing of family taxa (as of 2018) is given as Table 5.4. This list includes more than a dozen families not given in De Wever et al. (2001), which reflects the rapidly evolving state of higher-level taxa concepts in pre-Cenozoic radiolaria. The Paleozoic also contains three orders – Albaillellaria, Latentifistularia, and Archaeospicularia – not found in the Mesozoic or younger. The reported species diversity of Mesozoic and Paleozoic Orders is given in Table 5.5, along with, for comparison, Cenozoic species diversity by Order from Lazarus et al. (2015). It should be noted however that simple comparisons of diversity numbers between Eras is not possible due to, in addition to preservation, taxonomic practice, and primary diversity reporting standards that differ by Era; differing geologic age durations of Eras, valid species only (Cenozoic) vs. all published names in older time intervals, as well as the degree of comprehensiveness in the source publications.

Species-Level Variation in Radiolaria

Until now we have described higher-level taxonomy in radiolarians, e.g. orders and families, and via the plates, given numerous

Figure 5.30 Mesozoic radiolaria. All SEM images except as noted. 1 – *Tappanella novacubica*, Hull 1997, pl. 21 fig. 7; 2 – *Risella tledoensis*, Carter 1993, pl. 9 fig. 10; 3 – *Religa globosa*, Whalen & Carter 2002, pl. 15 fig. 2; 4 – *Canutus tipperi*, Pessagno & Whalen 1982, pl. 4 fig. 7; 5 – *Bisphaerocephalina? apimelos*, Foreman 1968, pl. 6 fig. 1 (transmitted light); 6 – *Myllocercion acineton*, Foreman 1968, pl. 5 fig. 11b (transmitted light); 7 – *Xenorum largum*, Blome 1984, fig. 3; 8 – *Foremania sandilandsensis*, Carter et al. 1998, pl. 24 fig. 20; 9 – *Crucella messinae*, Pessagno 1971, pl. 6 fig. 1; 10 – *Spongosaturnalis bipartitus*, Kozur & Mostler 1972, pl. 3 fig. 17; 11 – *Crubus chengi*, Yeh 1987, pl. 18 fig. 15; 12 – *Zhamojdasphaera latispinosa*, Kozur & Mostler 1979, pl. 12 fig. 5; 13 – *Scutispongus tortlispinus*, Kozur & Mostler 1996, pl. 1 fig. 11; 14 – *Pantanellium riedeli*, Pessagno 1977, pl. 6 fig. 5; 15 – *Proparvicingula moniliformis*, Carter 1993, pl. 18 fig. 15; 16 – *Annulohaeckelella longipedis*, Kozur & Mostler 2006, pl. 2 fig. 4; 17 – *Pseudodictyomitra pentacolaenis*, Pessagno 1977, pl. 8 fig. 3; 18 – *Hexacatoma elegantissima*, Tekin & Mostler 2005, pl. 5 fig. 3; 19 – *Pararuesticyrtium densiporatum*, Kozur & Mostler 1981, pl. 16 fig. 1; 20 – *Palaeosaturnalis levis*, Donofrio & Mostler 1978, pl. 2 fig. 1; 21 – *Anachoreta sagitta*, O'Dogherty 1994, pl. 15 fig. 26; 22 – *Bitubopyle coronata*, Kozur & Mostler 2006, pl. 5 fig. 12; 23 – *Paronaella? casmaliaensis*, Pessagno 1977, pl. 1 fig. 6; 24 – *Holocryptocanium tuberculatum*, Dumitrica 1970, pl. 16 fig. 102 (drawing); 25 – *Xitus plenus*, Pessagno 1977, pl. 9 fig. 15.

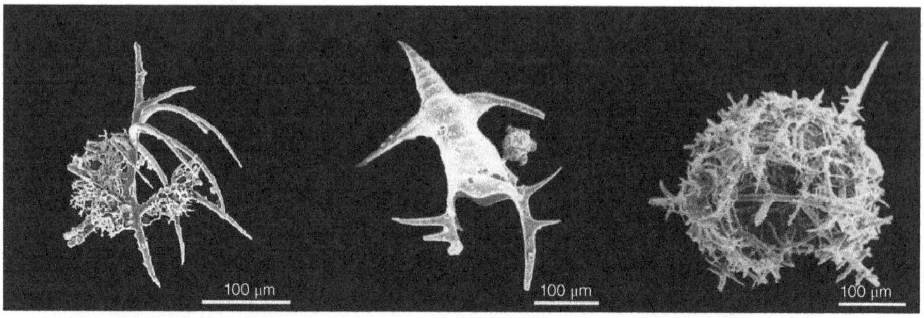

1 Ceratoikiscidae 2 Follicucullidae 3 Echidninidae

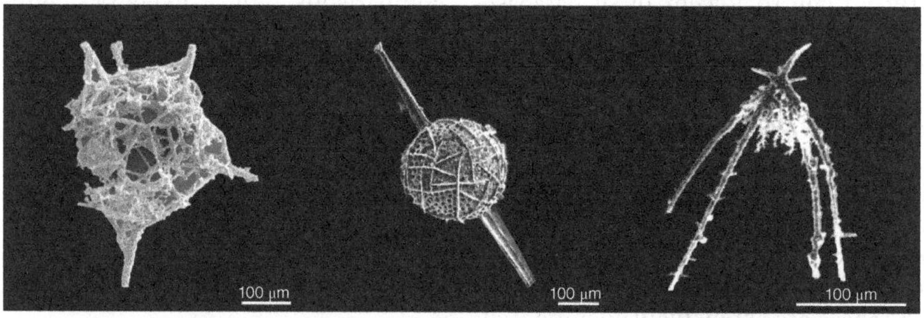

4 Entactiniidae 5 Entactiniidae 6 Palaeoscenidiidae

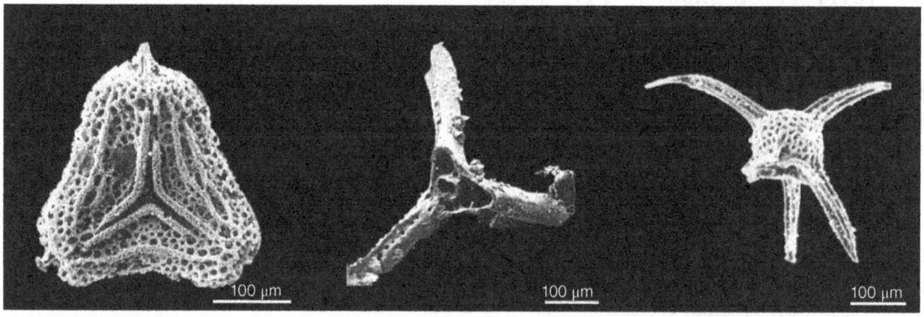

7 Cauletellidae 8 Ormistonellidae 9 Archocyrtiidae

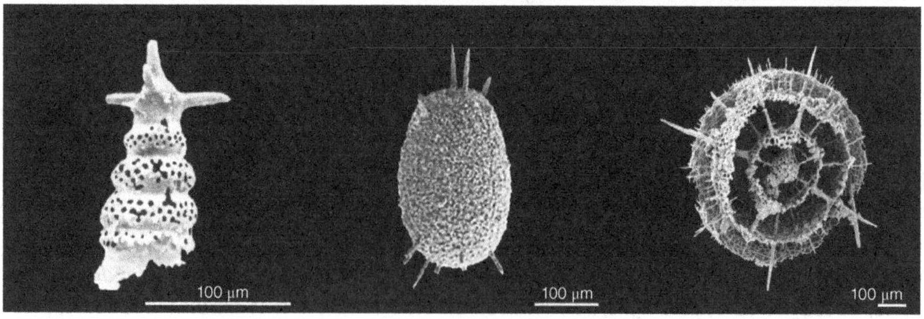

10 Popofskyellidae 11 Copycintridae 12 Inaniguttidae

Table 5.4 List of Paleozoic and Mesozoic families. Paleozoic from Caridroit et al. (2017), Mesozoic updated from O'Dogherty et al. (2009) by O'Dogherty (pers. comm., March 2018).

Paleozoic higher taxa

Order Albaillellaria Deflandre, 1953

Albaillellidae (Deflandre, 1953) Holdsworth (1977)
Ceratoikiscidae Holdsworth, 1969
Follicucullidae Ormiston & Babcock, 1979

Order Archaeospicularia Dumitrica, Caridroit & De Wever, 2000

Archeoentactiniidae Won & Below, 1999
Echidninidae Kozur, Mostler & Repetski, 1996
Protoentactiniidae Kozur, Mostler & Repetski, 1996
Palaeospiculidae Won in Won & Below, 1999
Secuicollactidae Nazarov & Ormiston, 1984

Order Entactinaria Kozur & Mostler, 1982

Entactiniidae Riedel, 1967
Haplentactiniidae Nazarov in Nazarov & Popov, 1980
Palaeolithocycliidae Kozur & Mostler, 1989
Palaeoscenidiidae Riedel, 1967
Pylentonemidae Deflandre, 1963
Tetrentactiniidae Kozur & Mostler, 1979

Order Latentifistularia Caridroit, De Wever & Dumitrica, 1999

Cauletellidae Caridroit, De Wever & Dumitrica, 1999
Latentifistulidae Nazarov & Ormiston, 1983
Ormistonellidae (De Wever & Caridroit, 1984) Dumitrica in De Wever et al. (2001)
Pseudolitheliidae (Kozur & Mostler, 1989) De Wever et al. (2001)
Tetratormentidae Nestell & Nestell, 2010
Ruzhencevispongidae Kozur, 1980

Order Spumellaria Ehrenberg, 1876

Antygoporidae Maletz & Bruton, 2007
Copicyntridae Kozur & Mostler, 1989
Haplotaeniatidae Won, Blodgett & Nestor, 2002
Inaniguttidae (Nazarov & Ormiston, 1984) Danelian & Popov (2003)

?Order Nassellaria Ehrenberg, 1876

Archocyrtiidae (Kozur & Mostler, 1981) Cheng (1986)
Popofskyellidae (Kozur & Mostler, 1981) Cheng (1986)

Order Incertae Sedis

Corythoecidae Nazarov in Nazarov & Rudenko, 1981
Proventocitidae Aitchison, 1998
Aspiculidae Won, Iams & Reed, 2005

Mesozoic higher taxa

Order Albaillellaria Deflandre, 1953

Albaillellidae Deflandre, (Deflandre, 1953) Holdsworth (1977)
Follicucullidae Ormiston & Babcock, 1979

(continued)

Figure 5.31 Paleozoic radiolaria (all images SEM). 1 – *Kappaforma insecta*, Noble & Lenz 2007, fig. 8.5. 2 – *Holdsworthella permica*, Kozur 1981, fig. 2. 3 – *Grosmorneus globosus*, Won & Iams 2002, fig. 4.18. 4 – *Costaentactinia foremanae*, Seo & Won 2009, pl. 2 fig. 13. 5 – *Entactinosphaera strangulata*, Nazarov & Ormiston 1985, pl. 2 fig. 1. 6 – *Palaeoscenidium flammatum*, Goodbody 1986, pl. 1 fig. 4. 7 – *Triplanospongos musashiensis*, Sashida & Tonishi 1988, fig. 9.7. 8 – *Latentifistula valdeinepta*, Nazarov & Ormiston 1985, pl. 3 fig. 9. 9 – *Mostlerium unicum*, Cheng 1986, pl. 4 fig. 8. 10 – *Kantollum pittsburgense*, Cheng 1986, pl. 2 fig. 4. 11 – *Tamonella multispinosa*, Dumitricia et al. 1980, pl. 10 fig. 8. 12 – *Plussatispila magnilimax*, MacDonald 2006, fig. 3.1.

Table 5.4 (Continued)

Order Entactinaria Kozur & Mostler, 1982

Centrocubidae Hollande & Enjumet, 1960
Entactiniidae Riedel, 1967
Eptingiidae Dumitrica, 1978
Heliosaturnalidae Kozur & Mostler, 1972
Heptacladidae Dumitrica, Kozur & Mostler, 1980
Hexalonchidae Haeckel, 1881
Hexapylomellidae Kozur & Mostler, 1979
Hexasaturnalidae Kozur & Mostler, 1983
Hexastylidae Haeckel, 1881
Hollandocachoniidae Dumitrica, 2013
Isometractiniidae Dumitrica, 2013
Italosaturnalidae Mostler & Krainer, 1994
Kungalariidae Dumitrica & Carter, 1999
Multiarcusellidae Kozur & Mostler, 1979
Odoghertitidae Dumitrica, 2013
Paracuboctostylidae Li & Sashida, 2012
Parentactiniidae Kozur & Mostler, 1981
Pentactinocarpidae Dumitrica, 1978B
Polyentactiniidae Nazarov, 1975
Quinquecapsulariidae Dumitrica, 1995
Rhizosphaeridae Haeckel, 1881
Saturnalidae Deflandre, 1953
Spongosaturnaloididae Kozur & Mostler, 1983
Thalassothamnidae Haecker, 1906
Triumbraculidae Dumitrica, 2013

Order Latentifistularia Caridroit, De Wever & Dumitrica, 1999

Cauletellidae Caridroit, De Wever & Dumitrica, 1999
Latentifistulidae Nazarov & Ormiston, 1983
Ormistonellidae De Wever & Caridroit, 1984
Pseudolitheliidae Kozur & Mostler, 1989

Order Spumellaria Ehrenberg, 1876

Acaeniotylidae Yang, 1993
Actinommidae Haeckel, 1862
Angulobracchiidae Baumgartner, 1980
Archaeospongoprunidae Pessagno, 1973
Astrosphaeridae Haeckel, 1881
Bernoulliidae Pessagno, Blome & Hull In Pessagno et al., 1993
Bolenidae Hull, 1997
Capnuchosphaeridae De Wever in De Wever et al., 1979
Catenopylidae Dumitrica, 1989
Cavaspongiidae Pessagno, 1973
Conocaryommidae Lipman, 1969
Dactyliosphaeridae Squinabol, 1904
Domuzdagiidae Dumitrica & Zügel, 2002
Emiluviidae Dumitrica, 1995
Gomberellidae Kozur & Mostler, 1981
Hagiastridae Riedel, 1971
Heliodiscidae Haeckel, 1881
Hexaporobrachiidae Kozur & Mostler, 1979
Intermediellidae Lahm, 1984
Larnacillidae Haeckel, 1887
Lithelidae Haeckel, 1862
Mendacastridae Dumitrica & Zügel, 2002
Miropylidae Dumitrica, 1988
Oertlispongidae Kozur & Mostler in Dumitrica et al., 1980
Pantanelliidae Pessagno, 1977
Paratriassoastridae Kozur & Mostler, 1981
Patruliidae Dumitrica, 1989
Patulibracchiidae Pessagno, 1971
Phaseliformidae Pessagno, 1972
Pseudoaulophacidae Riedel, 1967
Pyramispongiidae Kozur & Mostler, 1978
Relindellidae Kozur & Mostler in Dumitrica et al., 1980
Sphaerostylidae Haeckel, 1881
Spongodiscidae Haeckel, 1862
Spongopalliidae Kozur, Krainer & Mostler, 1996
Spongotortilispinidae Kozur & Mostler in Moix et al., 2007
Sponguridae Haeckel, 1862
Trimeridianellidae Dumitrica and Tekin, 2014
Tritrabidae Baumgartner, 1980
Veghicycliidae Kozur & Mostler, 1972
Xiphostylidae Haeckel, 1881

Order Nassellaria Ehrenberg, 1876

Acropyramididae Haeckel, 1881
Amphipyndacidae Riedel, 1967
Anisicyrtiidae Kozur & Mostler, 1981
Arcanicapsidae Takemura, 1986

(continued)

Table 5.4 (Continued)

Archaeodictyomitridae Pessagno, 1976
Archaeosemantidae Kozur & Mostler, 1981
Archocyrtiidae Kozur and Mostler, 1981
Artostrobiidae Riedel, 1967
Bagotidae Pessagno & Whalen, 1982
Bulbocyrtiidae Kozur & Mostler, 1981
Cannobotryidae Haeckel, 1881
Canoptidae Pessagno in Pessagno et al., 1979
Cuniculiformidae De Wever, 1982
Deflandrecyrtiidae Kozur & Mostler, 1979
Diacanthocapsidae O'Dogherty, 1994
Echinocampidae Bragin, 2009
Eucyrtidiellidae Takemura, 1986
Eucyrtidiidae Ehrenberg, 1847
Foremanellinidae Dumitrica, 1982
Goestlingellidae Dumitrica, 2017
Gongylothoracidae Bak, 1999
Gradinariidae Dumitrica, 2017
Hexapylocapsidae Dumitrica & Zügel, 1998
Hsuidae Pessagno & Whalen, 1982
Japonocapsidae Kozur, 1984
Livarellidae Kozur & Mostler, 1981
Lophophaenidae Haeckel, 1881
Minocapsidae O'Dogherty, Goričan and Gawlick, 2017
Nabolellidae Kozur & Mostler, 1979
Nakasekoellidae Kozur, 1984
Neosciadiocapsidae Pessagno, 1969
Obeliscoitidae O'Dogherty, 1994
Parvicingulidae Pessagno, 1977
Plagiacanthidae Hertwig, 1879
Planispinocyrtiidae Kozur & Mostler, 1981
Poulpidae De Wever, 1981
Pseudodictyomitridae Pessagno, 1977
Pseudosaturniformidae Kozur & Mostler, 1979
Rotaformidae Pessagno, 1970
Ruesticyrtiidae Kozur & Mostler, 1979
Sanfilippoellidae Kozur & Mostler, 1979
Sethoperidae Haeckel, 1881
Spinolobocyrtiidae Kozur & Mostler, 2006
Spinomersinellidae Kozur, Moix & Ozsvárt,2007
Spongocapsulidae Pessagno, 1977
Spongolophophaenidae Kozur & Mostler, 1994
Spongosilicarmigeridae Dumitrica in De Wever et al., 2001
Stichocapsidae Haeckel, 1881
Tertoniidae Dumitrica & Zügel, 2003
Tetraspinocyrtiidae Kozur & Mostler, 1994
Theopiliidae Haeckel, 1881Tripedurnulidae Dumitrica, 1991
Ultranaporidae Pessagno, 1977
Unumidae Kozur, 1984
Williriedellidae Dumitrica, 1970
Xiphothecaellidae Kozur & Mostler, 1981
Xitidae Pessagno, 1977
Xitomitridae O'Dogherty, Goričan and Gawlick, 2016

examples of species descriptions. Important though these subjects are, for most radiolarian workers, and in particular for students beginning their studies, the most difficult problem faced in daily work is understanding the variation in the morphology of radiolarian shells *within* species, and how to distinguish this variation from variation *between* closely related species. Although this has been introduced previously (in the discussion of type specimens), how this actually appears in practice has not. While to some extent this can only be understood by long experience of identifying individual specimens, its importance is such that a few comments should be made. Most radiolarian species show a fairly constant basic set of shell characters, with (in comparison to other protists, or even many multicellular organisms) only moderate amounts of within-species variation. As noted in Chapter 2: Biology, ontogenetic and size variation also tends to be fairly modest for most species. For a wide variety of characters in Spumellaria and Nassellaria, Lazarus and Pallant (1989) noted a coefficient of variation (the standard deviation divided by the mean) of mostly between 10–20%. While this is more than is typical for vertebrate skeletal characters (5–10% for cranial measurements in carnivores: Dayan et al. [2002]), it is actually equal to, or even less than is

Table 5.5 Species diversity of radiolarian orders by geologic era. Sources: Cenozoic: NSB database, only valid species counted; Mesozoic: O'Dogherty et al. (2009) pp. 357–369; Paleozoic: Aitchison et al. (2017) SOM spreadsheet col. B (sorted, counted). In Paleozoic data names with only a blank or a dash were not included in totals: these are mostly n. dub; invalid, or genus header lines.

Era	Order	N species	Percent
Cenozoic	Spumellaria	369	30.93
	Nassellaria	820	68.73
	Entactinaria	1	0.08
	other	3	0.25
	total species	1193	
Mesozoic	Spumellaria	2714	43.1
	Nassellaria	3185	50.6
	Entactinaria	386	6.1
	other	5	0.1
	total species	6290	
Paleozoic	Entactinaria	726	37.6
	Albaillellaria	364	18.8
	Spumellaria	207	10.7
	Latentifistularia	194	10.0
	incert. cedis	177	9.2
	Archaeospicularia	154	8.0
	?Nassellaria?	111	5.7
	blank or dash	*893*	
	[spong spic; not rad]	*44*	
	total radiolarian species	1933	

* Note that the Mesozoic and Paleozoic values are from unfiltered lists of published names and thus include numerous synonyms; also, differences in sampling effort and total amount of geologic time represented by each interval mean that numbers do not represent relative true diversity between Eras.

found in many "typical" macroinvertebrate groups: 10–20% in a bivalve (Araujo et al., 1993), or 15–30%, and more, in shell characters of a gastropod (Sundberg 1988).

The challenge comes from the nature of the characters that are typically used to distinguish between species, which are usually fairly small differences in shape, or relative sizes of shell elements (Figures 5.32, 5.33). Close attention to detail and use of multiple characters allow most specimens within a single sample to be unambiguously assigned to species, although a few specimens may show transitional combinations of characters that make assignment difficult or even impossible. In addition, there is the problem of species assignment within anagenetically evolving lineages, due to the continuous change in character states as one defined species evolves into another over evolutionary time. For this problem only a fairly arbitrary set of rules can be constructed, unique for each species pair discrimination.

Ideally rules to discriminate between closely related species should be part of the initial species' descriptions but frequently are not, if for no other reason than that many of the other species in pairwise comparisons were not yet described when a species was initially published. Such "differential diagnoses" thus often need to be added as an emended diagnosis. As they are sometimes included in newer species descriptions, these can sometimes be consulted to better understand the concept of related older species. For these and other reasons simply reporting a species name is often not enough to document precisely what was seen and assigned to that name in a particular study. Radiolarian publications therefore still tend to include plates showing the species reported in the paper, with notes on how problematic determinations were handled. Lastly, for many species, identification requires preservation of at least few key characters. Specimens which are ontogenetically very incomplete,

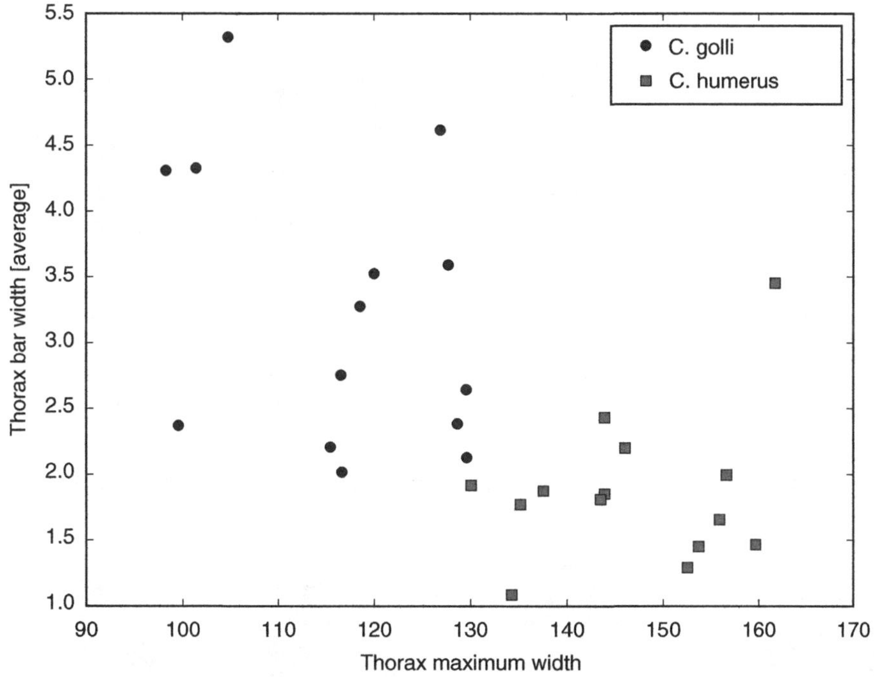

Figure 5.32 Typical patterns of variation in characters used to distinguish sister species, here in the radiolarian genus *Cycladophora*. Species pairs such as *C. golli* and *C. humerus* (illustrated in Figure 5.33) show continuous but largely non-overlapping ranges in characters. Using both (or better, multiple) characters allows clear assignment of almost all specimens. Compare to Figures A and B of this chapter. Source: From data given in Lombari and Lazarus (1988).

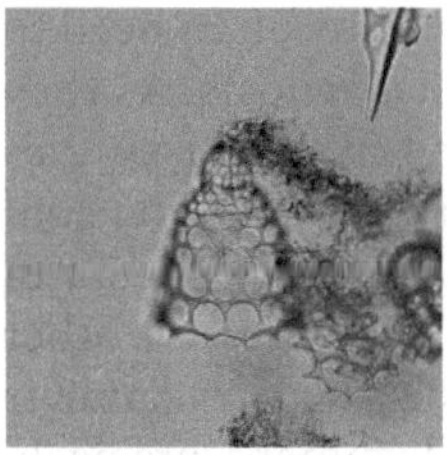
Cycladophora conica
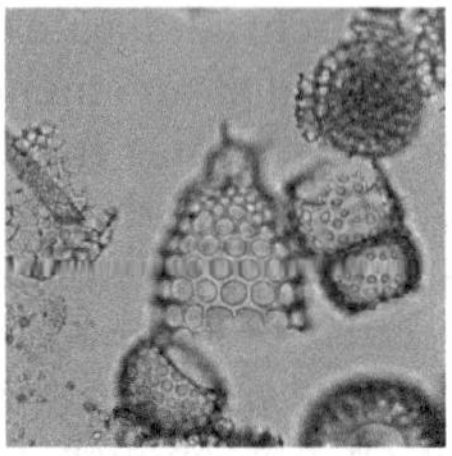
Cycladophora conica
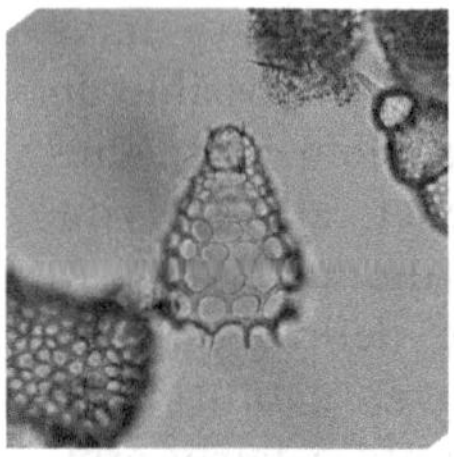
Cycladophora conica
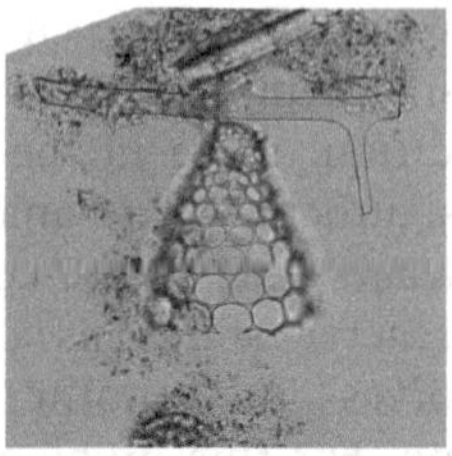
Cycladophora conica

Cycladophora cosma

Cycladophora cosma
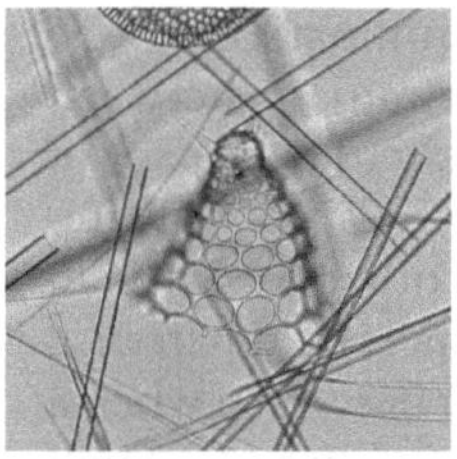
Cycladophora cosma
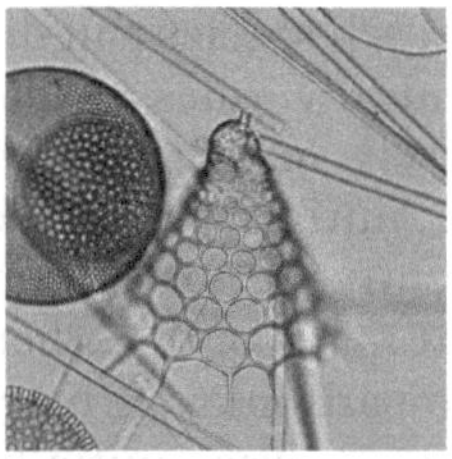
Cycladophora cosma
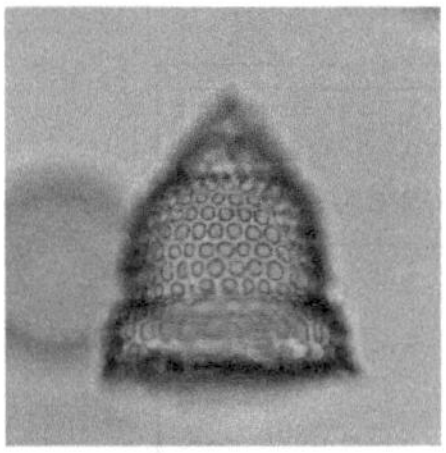
Cycladophora golli

Cycladophora golli

Cycladophora golli
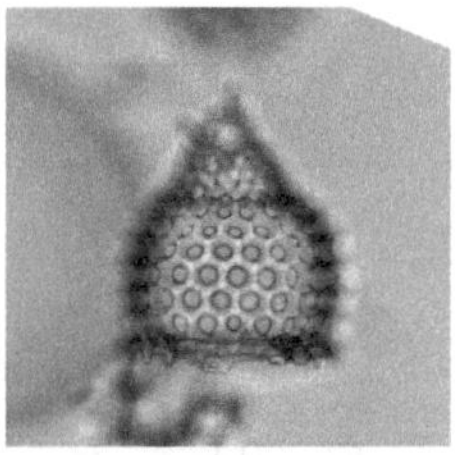
Cycladophora golli
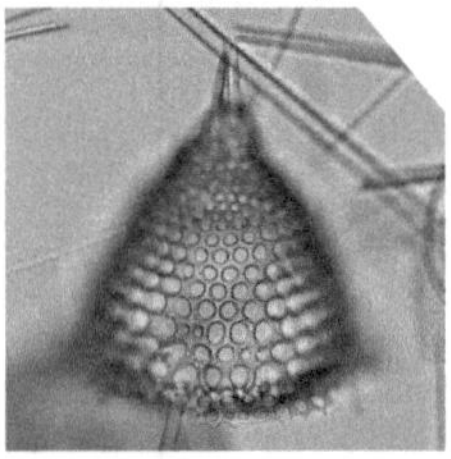
Cycladophora humerus
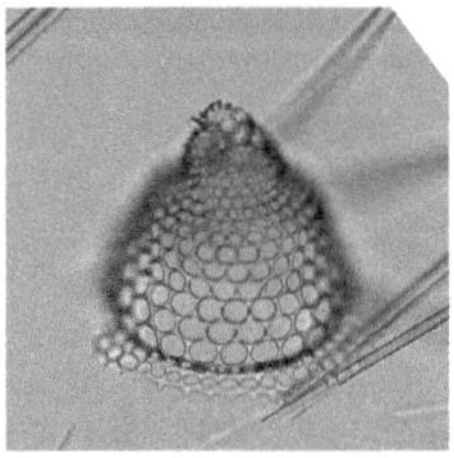
Cycladophora humerus

Cycladophora humerus
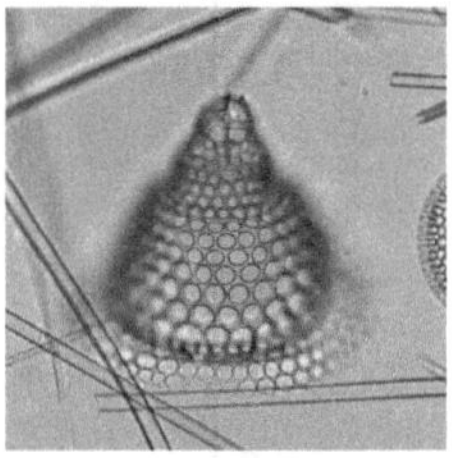
Cycladophora humerus
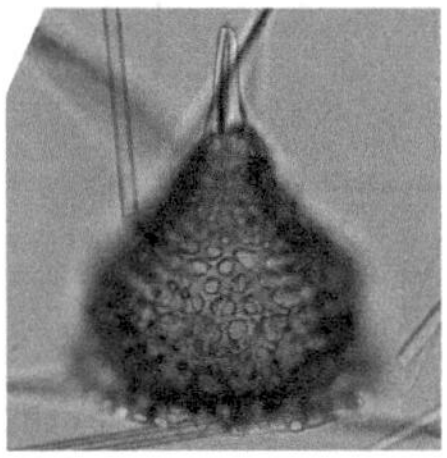
Cycladophora spongothorax
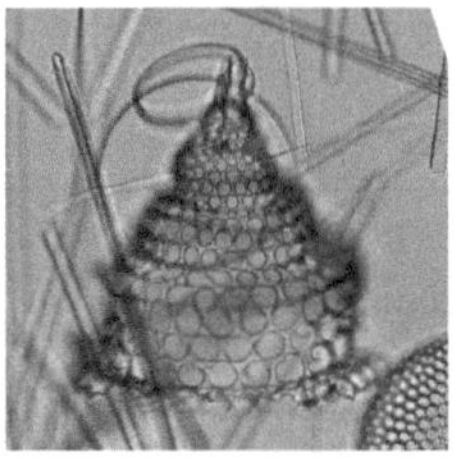
Cycladophora spongothorax
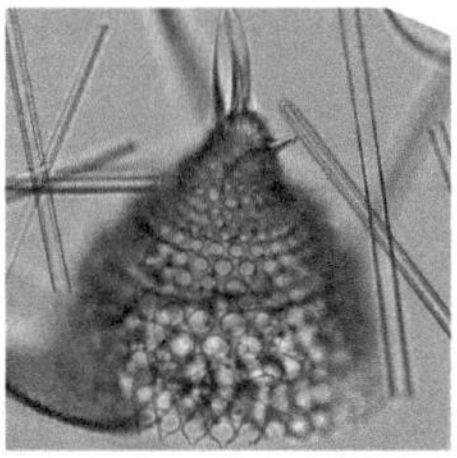
Cycladophora spongothorax
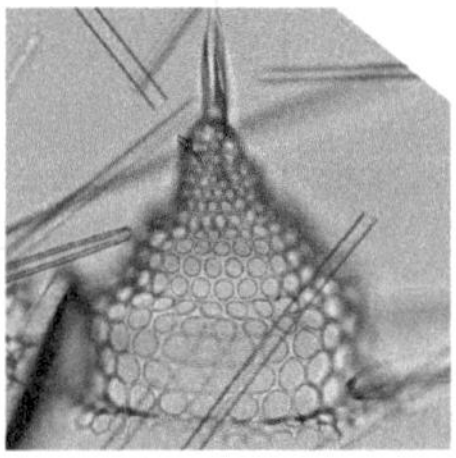
Cycladophora spongothorax

broken, oriented on the slide at an inopportune angle, or which are in part obscured by other material on the slide (sediment, diatoms, or other radiolarians) are often not identifiable to species level. These problems can result in a significant fraction of the individuals on a slide being only identifiable to genus, or even family level.

PART 2 - Summary of the Radiolarian Fossil Record

Phanerozoic radiolaria have been documented in hundreds, if not thousands, of papers. The quality of the fossil record itself, though, is very uneven (amount of rock record, preservation, and other factors; see Chapter 6), and must always be considered when summarizing the literature. Permian radiolarians for example are found today in diverse regions worldwide, including the marginal regions of the present west Pacific (e.g. China, Far-East Russia, Japan, Thailand), North America, Urals, New Zealand, and Chile. However, almost all these regions, except those of the Urals, are actually Tethyan deposits (a tropical ocean region that ran from the current Mediterranean to the Indian Ocean) after the reconstruction of plate movement, and consequently our understanding of radiolarian faunal changes in the Permian is only regional, not global.

Preservation can also strongly affect diversity, and differences in preservation between time intervals or regions can thus bias our perception of comparative diversity. A typical case is known from the Devonian and Carboniferous. The Upper Devonian and Lower Carboniferous (=Mississippian) has the most diversified fauna in the Paleozoic, exceeding 510 species. By contrast, the known species in the Lower to Middle Devonian and the Upper Carboniferous are only 30 species and 90 species, respectively. This is most likely primarily a preservation artifact: The Upper Devonian to Lower Carboniferous radiolarians are largely reported from phosphatic or carbonate nodules/concretions, whereas the Lower to Middle Devonian and the Upper Carboniferous radiolarians are extracted from chert, tuffaceous chert, or limestone. Phosphatic, carbonate, or manganese nodules/concretions preserve radiolarians far better than do cherts, tuffs,

Figure 5.33 Within and between (sister) species variation in radiolarians. Shown are 4 specimens each from 5 selected species in the same genus *Cycladophora* (out of a total of 13 valid species in the genus). Additional single specimens of *C. cosma* and *C. spongothorax*, together with two other species: *C. bicornis* and *C. davisiana*, are also illustrated in Figure 5.24. Specimens here were chosen for completeness and suitable orientation to see morphology, but otherwise were not pre-selected to be "typical" for the species, and thus represent the variation normally encountered in actual populations. All specimens are at the same magnification. *C. conica* is distinguished by the small size, simple, narrow, slightly convex thorax, moderately large thorax pores, and lack of any abdomen. *C. cosma* is similar to *C. conica* but slightly larger and the thorax expands more broadly at the base (often slightly concave in outline), with larger basal pores. *C. golli* has relatively small thorax pores, thicker thorax bars, a campanulate thorax with a distinct, rounded step between the upper and lower thorax, and in most specimens a short abdomen. In some specimens, i.e. the rightmost image (formally named as subspecies *C. g. regipileus*) the base of the thorax is noticeably recurved, so the lower thorax is barrel shaped. *C. humerus* has a flat-sided, small-pored conical lower thorax, a distinct rounded step or shoulder to the upper thorax, and a short, broadly conical basal skirt in most specimens. *C. spongothorax* has two step-like changes in the thorax profile and in most specimens at least some external spongy material covering the lower thorax.

or limestones; indeed, the quality of preservation in such nodules and concretions can be nearly as good as in Holocene radiolarians found in sea-bottom surface sediments. Radiolarians in tuffs, cherts, and limestones by contrast are much more poorly preserved, and the preservation quality is further reduced by the strong acids used to prepare these samples (see Chapter 6). Relatively diverse radiolarian faunas are also known in the Silurian, Anisian to Ladinian of the Triassic, Upper Jurassic, and the Santonian–Campanian of the Cretaceous, where radiolarian species are largely described from nodules/concretions.

Family diversities can be used as they are less likely to biased by these preservational and sampling issues. Most syntheses of family level change are based on the De Wever et al. (2001) family classification and thus this classification is also used here, even if some individual taxa are, as noted above, still somewhat speculative in nature. While substantial changes in higher-level diversity, mostly at the family level, are also noted in the following review, an exhaustive treatment of taxic changes is given in De Wever et al. (2001), which should be consulted for additional details. Species diversity change and evolutionary interpretation of the fossil record are treated separately in Chapter 9: Evolution.

Lastly, it should also be noted that in most of the individual studies cited (at least, prior to the Cenozoic), sample sizes (numbers of specimens observed) is generally only a few hundred individuals, with many species being recorded at very low abundances and only very sporadically in samples. The observed diversity and position of evolutionary turnover events in individual sections is thus at least in part obscured by limits of sample size.

Cambrian and Ordovician

Molecular phylogenetic studies suggest the emergence of radiolarians in the latest Precambrian (e.g. Decelle et al., 2012b). The most plausible of uncertain oldest fossils are recorded from the Terreneuvian (basal Cambrian; Cao et al., 2014) – geologic time scale figures with the ages of named intervals are given in Chapter 8: Stratigraphy, although only one specimen is known so far. The oldest widely accepted radiolarian occurrences are from the base of the middle Cambrian Series 3 (Maletz, 2011). Additional ambiguous specimens that could possibly be radiolarians have been reported from the Terreneuvian (Braun et al., 2007) and the lower Cambrian Series 2 (Obut and Iwata, 2000, Pouille et al., 2011, Korovnikov et al., 2013). If the beginning of skeletalized radiolarians is in the Terreveuvian, radiolarian skeletalization developed gradually in parallel with the "Cambrian Explosion" in biodiversity. The oldest radiolarians from the lower Cambrian Series 2 (if indeed radiolarians) belong to the Order Archaeospicularia (*Palaeospiculum* and *Archeoentactinia*), possessing a simple spicule geometrically similar to the hexactine spicule of sponges. The hexactine spicular-like primitive radiolarians evolved in the Furongian (late Cambrian) into spherical radiolarians like *Polyentactinia hunanensis*, made of a three-dimensional mesh with a high number of outer spines and an internal hollow space. Cambrian radiolarians, including questionable reports (Maletz, 2011), have been found in bioclastic grainstone (White, 1986), subtidal micritic limestone (Dong et al., 1997), interbedded fine-grained limestone and shale sequences (Won and Iams, 2002), and carbonate concretions in shales (Won and Below, 1999; Braun et al., 2007), suggesting that Cambrian radiolarians lived in subtidal to continental slope water environments (Dong et al., 1997; Won and Iams, 2002) at least. Due to limited age resolution and numbers of localities, the causes, timing, and environmental context of initial silica mineralization in radiolarians have not been determined.

Ordovician radiolarians are relatively well dated with co-occurring graptolites and conodonts (Maletz, 2011). Although the known Ordovician radiolarian faunas are based on isolated spot collections, a significant faunal turnover occurred in Tremadocian–Florian of the Early Ordovician. The dominant families since the late Cambrian – Palaeospiculumidae, Parachaeoentactinidae, Protoentactiniidae, and Echidninidae – were largely replaced by labyrinthine-type Haplotaeniatumidae (Maletz, 2007). The next major faunal turnover occurred in the latter mid-Ordovician (Darriwilian): Antygoporidae completely disappeared prior to the Darriwilian and were followed by Inaniguttidae with long spines.

Ordovician radiolarians are also found from argillaceous limestone which is thought to have deposited from outermost shelf or slope environments (Dunham and Murphy, 1976; Kozur et al., 1996b), but are more commonly reported from siliceous limestone or thin chert beds intercalated in limestone from continental shelf settings such as in Kazakhstan (Nazarov and Popov, 1980) and North China (Cui et al., 2000). As Ordovician radiolarians are often associated with graptolites in offshore (deep-water) black shales (e.g. Eisenack, 1971; Maletz and Bruton, 2005, 2007), the distribution of radiolarians had expanded further offshore, vs. their Cambrian habitats, by the Middle Ordovician. Deep-water radiolarian ribbon chert, a key lithologic facies for much of the fossil record of radiolarians, first appeares in the late Tremadocian to the latest Darriwilian in the Lachlan Orogen, Australia (Umeda et al., 1992; Percival, 2011), in the late Darriwilian in the Northern Belt of the Southern Uplands (Danelian and Clarkson, 1998), and in the Late Ordovician of the Normanskill cherts of New York State (Ruedemann and Wilson, 1936). The appearance of radiolarian ribbon chert was temporally close to the faunal turnover from Antygoporidae to Inaniguttidae with long spines. The Ordovician radiolarian cherts in the Burubaital Formation in Kazakhstan (Tolmacheva et al., 2004) and the Lachlan Orogen in Australia (Percival, 2011) are commonly associated with lingulate brachiopods, sponges, and other metazoan fossils, suggesting that Ordovician radiolarians inhabited high-fertility, oxygenized open oceans on the continental shelf.

Although our understanding of radiolarians in the Cambro-Ordovician is still very sketchy and undergoing constant revision, it is clear that the standing diversity, of species or families, was relatively low. In the De Wever et al. (2001) compilation, Ordovician standing family diversity was only ca. 5 families, a small fraction of the diversity seen in the same compilation later in the fossil record.

Silurian to the Lower Carboniferous

Radiolarians show throughout their Phanerozoic evolutionary history repeated patterns of morphological convergence among different phylogenetic groups toward similar general external forms: spherical, elliptical, conical, discoidal, and others. However, as the Cambrian and Ordovician radiolarians are comprised only of spherical and hexactine spicular-like forms (Maletz, 2011), several of the basic forms appeared for the first time in the Late Ordovician to Devonian. The first appearances of bilateral morphotypes, for example, are the genus *Protoceratoikiscum* (Ceratoikiscidae, Albaillellaria) in the Darriwilian of the Late Ordovician (Maletz, 2011) and *Ceratoikiscum* (Ceratoikiscidae, Albaillellaria) in the Wenlockian of the middle Silurian (Renz, 1988; Noble and Lenz, 2007). Significant evolutionary events in Silurian Albaillellaria occur during the *lundgreni* Extinction Event in the middle Silurian. This event is marked by its global impact on the plankton community including graptolites, rather than primarily on shallow-water benthos, and is associated with a positive carbon isotope excursion peak between the early and late Homerian (late Wenlockian) (Lenz et al., 2006; Noble et al., 2012). Radiolarian species diversity in local sections drops from 28 species to 6 species with only 2 species persisting across the event (Noble et al., 2012). After this crisis the diversity and abundance quickly recovered in the Canadian Arctic, as well as in sections from Japan.

New radiolarian morphologies subsequently appeared between the Frasnian of the Late Devonian and the Visean of the Early Carboniferous. The first radiolarian with a pore-less test evolved as the genus *Protoholoteciscus* (Albaillellidae, Albaillellaria) in the Frasnian (Aitchison, 1993). A conical nassellarian-like form with segments or segment-like undulation belonging to Popofskyellidae and a monocyrtid nassellarian-like form with three feet belonging to Archocyrtiidae first appeared in the Famennian. A triangular plate-like morphotype with three spongy arms similar to later genera such as *Latentifistula, Ouaka, Pseudotormentus, Scharfenbergia,* and *Trilacertus* appeared in the Visean (Won, 1983; Cheng, 1986; Schwartzapfel and Holdsworth, 1996).

Mass extinction events in invertebrate metazoans in the pre-Devonian are recognized at the Cambrian–Ordovician boundary and the Ordovician–Silurian boundary, but no significant responses have been detected in radiolarians. Several other extinction levels are known in the Devonian to have affected radiolarians. Radiolarian behavior across the late Devonian Frasnian–Famennian mass extinction event has been examined in pelagic chert sections in China (Zhang et al., 2008). The Frasnian–Famennian boundary is placed within the radiolarian *Helenifore robustum* Zone. Members of Ceratoikiscidae temporarily disappeared until at least the level of the upper Kellwasser extinction. Some Entactiniidae persist, although some, such as *Trilonche,* disappear only to gradually reappear above the upper Kellwasser extinction. The diversity of radiolarian faunas gradually diversified after the Nehden extinction in the middle Famennian.

Late Paleozoic to Late Mesozoic Siliceous Sedimentation

Sedimentary rocks with high radiolarian content are found throughout the geological record, but they are more common in the time interval from roughly mid–late Carboniferous to early Cretaceous. These deposits are 90% or more SiO_2, primarily from radiolarians and siliceous sponge spicules, and are called radiolarite or chert. Radiolarite and chert rarely contain much carbonate or terrigenous matter so that the sedimentation rate is low: ca 1.8–2.0 mm/ky between the Bajocian and Tithonian (excluding Oxfordian) in the western Tethys (De Wever and Cordey, 1986), and 1–3 mm/ky in the Triassic in the Panthalassa (Hori et al., 1993). Note: in this chapter, the term Panthalassa is applied for the wide oceans around the supercontinent Pangea, while the "V" shaped embayment between Africa–India and Asia is called the Tethys. After the formation of the Atlantic Ocean by the complete breakup of the west Gondwana into South America and Africa, the term proto-Pacific is applied for the oceans between the America–Africa continent and the Eurasia.

This long interval has frequently been referred to by radiolarian workers as the Radiolarian Optimum Event, or interval (ROE) (De Wever et al., 1986). The Permian Chert Event named by Murchey and Jones (1992) is part of this ROE pattern. The onset of radiolarian-rich sedimentation is seen in the Moscovian (the middle-late Carboniferous e.g. middle Pennsylvanian) in the Panthalassa, based on the recent age determination of radiolarian assemblages (Nestell et al., 2012) and expanded over much of the low-latitude Panthalassic ocean in the middle-late Carboniferous to the Early Cretaceous (Baumgartner, 2013). Radiolarian-rich sediments declined in the Tithonian of the Late Jurassic in the Tethys while it continued up to the early Aptian in the Panthalassic ocean.

This pattern of silica-rich sedimentation is due to gradually changing patterns of sedimentary deposition and global geochemical cycling over the Phanerozoic, particularly of the elements silicon and carbon. In the late pre-Cambrian and early Paleozoic, radiolarians were still apparently relatively uncommon in the plankton, and exported (relative to later time intervals) little biogenic silica into sediments (Maliva et al., 1989; Kidder and Irwin, 2001). Most silica reaching the ocean from continental weathering was removed either by inorganic precipitation as sedimentary minerals or by siliceous sponges in shallow-water environments. Carbon, input into the environment from volcanic activity, was mostly captured and removed from the system as biogenic carbonate in the form of invertebrate shells in shallow benthic environments (Katz et al., 2004; Ridgwell, 2005). These environments were widespread in much of the Paleozoic through late Mesozoic as sea level was mostly relatively high and extensive parts of the continents were typically covered by shallow seas. By the mid-Paleozoic radiolarians had become abundant enough that a significant amount of the total silica from weathering was deposited by them in more offshore environments. In waters deep enough to be below the carbonate compensation depth (the water depth in oceans where rates of carbonate dissolution balance rates of carbonate input, resulting in no net carbonate deposition) radiolarian-rich biogenic silica deposits began to accumulate (Ridgwell, 2005). This pattern continued until the mid-Mesozoic, when the evolution of carbonate-secreting phytoplankton (e.g. the coccolithophores and discoasters) shifted the main locus of carbonate deposition to deep-water environments, and the high flux of carbonate into the sediments from these primary producers, compared to the fluxes of silica from radiolarian zooplankton, changed the pattern of primary sediment deposition from silica-rich to carbonate-rich (chalks, limestones, and similar rocks; Hüneke and Henrich, 2011). Falling sea level and the reduction of shallow sea areas also contributed to this shift, particularly in the Cenozoic. These changes in sedimentation and element cycling also had significant impacts on biotic evolution, which are described in Chapter 9: Evolution.

Mass Extinctions at the End of the Paleozoic Era

The mass extinction that defines the end of the Paleozoic (e.g. Raup, 1979) has in recent years been shown to be a complex event. Several million years prior to the main end-Permian Extinction (Clapham et al., 2009; Wignall et al., 2012; Payne and Clapham, 2012) there was a less severe Capitanian Mass Extinction (also called the end-Guadalupian crisis). This Capitanian Mass Extinction is likely to be related with the Kamura cooling event that was marked by high positive $\delta^{13}C_{carb}$ values, reaching up to + 7.0‰ (Isozaki et al., 2007; Isozaki, 2009). Further, the recovery of biotas after the main extinction event in the early Triassic was also marked by repeated environmental instability.

Radiolarian biostratigraphic changes from the Capitanian to the Wuchiapingian have been examined in the pelagic Panthalassa ocean (Mitsumura and Kamata, 2009), shallower continental shelf in South China (Sun and Xia, 2006), and in shelf limestone beds associated with fusulinids in North America (Noble et al., 2011; Jin et al., 2012). The general trends of radiolarian faunal characteristics are similar to each other: *Follicucullus* (Follicucullidae, Albaillellaria) was dominant below the Capitanian–Wuchiapingian boundary (i.e. the Guadalupian–Lopingian boundary), but became rare above the boundary. No signs of an extinction or faunal turnover in radiolarians however were detected either in the Capitanian or across the Capitanian–Wuchiapingian boundary. In the Capitanian Lamar Limestone and lower part of the Reef Trail Member of the Bell Canyon Formations in North America, a close correspondence between lithology and radiolarian fauna was recognized in that spumellarian-rich intervals are composed of silty carbonate rocks and the *Follicucullus*-rich horizons are found in purer limestone beds (Noble et al., 2011). The spumellarian-rich interval corresponded to periods of increased terrigenous input and elevated marine productivity while the *Follicucullus*-rich intervals represent periods of lower productivity, as evidenced by sedimentary organic carbon concentrations, $\delta^{18}O_{carb}$ values, Mg/Ca ratios, and lithology (Jin et al., 2012). Across the Capitanian–Wuchiapingian boundary, a diversification event likely occurred in the radiolarians of the Panthalassa ocean (Mitsumura and Kamata, 2009).

The end-Permian Extinction Event (end-Changhsingian) is the most severe extinction in the Phanerozoic, with pronounced terrestrial ecosystem turnover and elimination of between 70 and 80% of all marine invertebrate genera (Clapham et al., 2009; Schneebeli-Hermann et al., 2013). This event coincided with a negative stable carbon-isotope excursion (CIE) of 3–6‰ in marine carbonate rocks and 3–10‰ in bulk sedimentary organic matter (OM), and the cessation of chert deposition (the "Chert Gap"). It is closely associated with massive volcanic activity, particularly the Siberian Traps, and is thought to ultimately be due to the effects that carbon and other volcanic substance injections had on the atmosphere and oceans. Many lines of evidence suggest that temperatures on land and in the oceans had risen to high levels while oceans had become acidified and partially euxinic (Payne and Clapham, 2012; Takahashi et al., 2014). A complex, and not yet well-understood set of interactions led to reductions in primary productivity and the abundances of many marine organisms, culminating in ecosystem collapse and mass extinction.

The Permian–Triassic boundary is formally defined by the first occurrence of the conodont *Hindeodus parvus* in the global stratigraphic reference section in China (Yin et al., 2001; Jiang et al., 2007), but the exact horizon of the Permian-Triassic boundary in a section varies according to the local paleontologic and/or chemostratigraphic criteria used. Despite such uncertainties, a fairly clear global pattern of radiolarian evolutionary change is seen. In the low-latitude pelagic Panthalassa ocean, a decrease in radiolarian abundance coincided with an increase in TOC content near the top of the Permian, implying that radiolarian extinction occurred at the end of the Permian, coinciding with oceanic anoxia but with undiminished primary productivity (Takahashi et al., 2009a, b). Radiolarians suffered extinctions in the end-Permian at the ordinal level (De Wever et al., 2007, Figure 5.34). Paleozoic-type radiolarians belonging to Orders Albaillellaria and Latentifistularia largely went extinct at the end-Permian, and Order Entactinaria

Figure 5.34 Paleozoic–Mesozoic radiolarian family ranges showing extent of evolutionary turnover across the Permo-Triassic boundary and early Triassic. The turnover, both extinctions and originations, extended over a large part of the Triassic, with only three families clearly becoming extinct at the Permo-Triassic boundary itself. Source: From De Wever et al., Eclogae Geologicae Helvetiae (2007), which uses the taxonomic concepts of De Wever et al. (2001). Reproduced with permission of Springer.

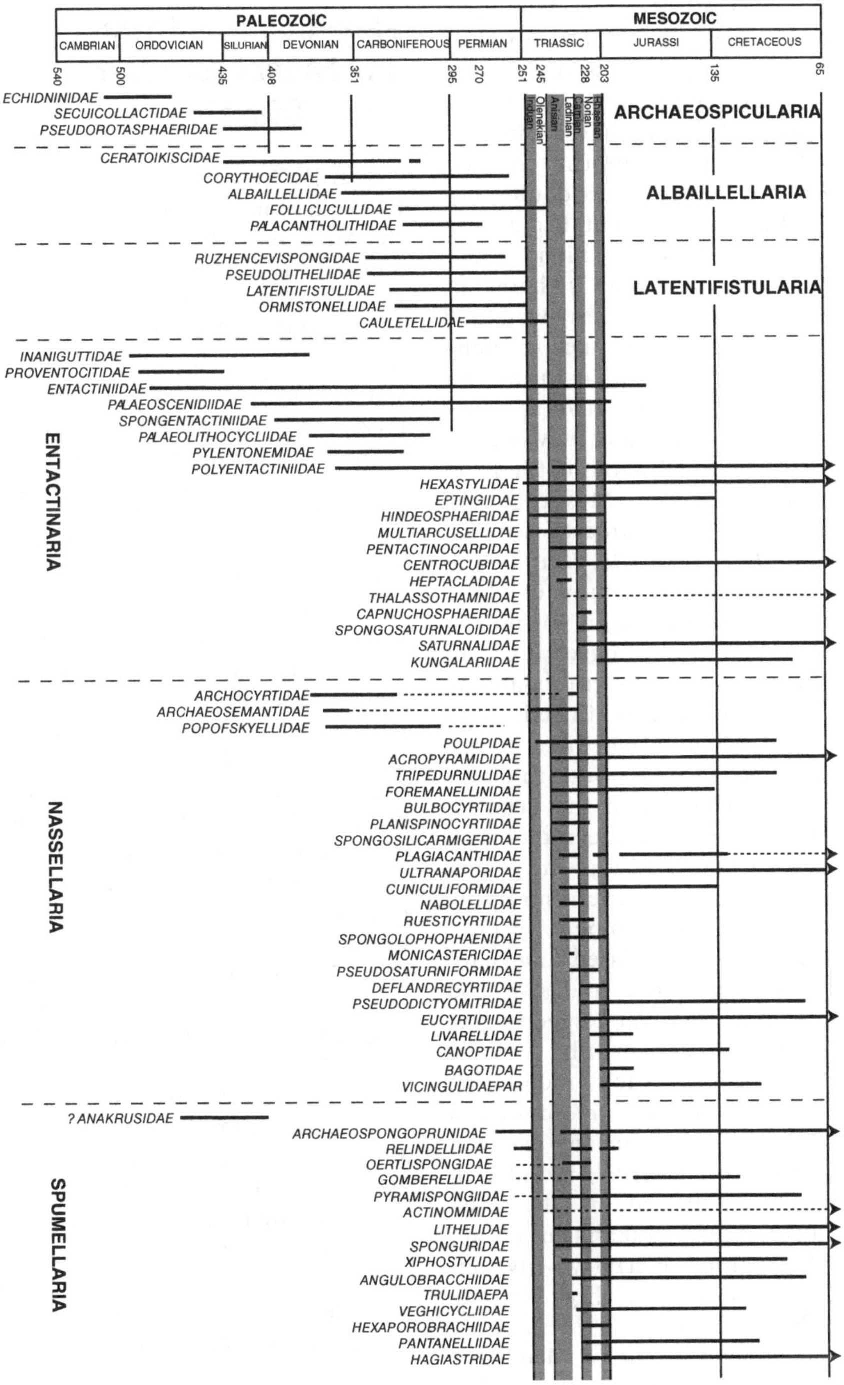

PALEOZOIC
MESOZOIC
CAMBRIAN
ORDOVICIAN
SILURIAN
DEVONIAN
CARBONIFEROUS
PERMIAN
TRIASSIC
JURASSI
CRETACEOUS
540
500
435
408
351
295
270
251
245
228
203
135
65
Indian
Olenekian
Anisian
Ladinian
Carnian
Norian
Rhaetian
ARCHAEOSPICULARIA
ALBAILLELLARIA
LATENTIFISTULARIA
ENTACTINARIA
NASSELLARIA
SPUMELLARIA
ECHIDNINIDAE
SECUICOLLACTIDAE
PSEUDOROTASPHAERIDAE
CERATOIKISCIDAE
CORYTHOECIDAE
ALBAILLELLIDAE
FOLLICUCULLIDAE
PALACANTHOLITHIDAE
RUZHENCEVISPONGIDAE
PSEUDOLITHELIIDAE
LATENTIFISTULIDAE
ORMISTONELLIDAE
CAULETELLIDAE
INANIGUTTIDAE
PROVENTOCITIDAE
ENTACTINIIDAE
PALAEOSCENIDIIDAE
SPONGENTACTINIIDAE
PALAEOLITHOCYCLIIDAE
PYLENTONEMIDAE
POLYENTACTINIIDAE
HEXASTYLIDAE
EPTINGIIDAE
HINDEOSPHAERIDAE
MULTIARCUSELLIDAE
PENTACTINOCARPIDAE
CENTROCUBIDAE
HEPTACLADIDAE
THALASSOTHAMNIDAE
CAPNUCHOSPHAERIDAE
SPONGOSATURNALOIDIDAE
SATURNALIDAE
KUNGALARIIDAE
ARCHOCYRTIDAE
ARCHAEOSEMANTIDAE
POPOFSKYELLIDAE
POULPIDAE
ACROPYRAMIDIDAE
TRIPEDURNULIDAE
FOREMANELLINIDAE
BULBOCYRTIIDAE
PLANISPINOCYRTIIDAE
SPONGOSILICARMIGERIDAE
PLAGIACANTHIDAE
ULTRANAPORIDAE
CUNICULIFORMIDAE
NABOLELLIDAE
RUESTICYRTIIDAE
SPONGOLOPHOPHAENIDAE
MONICASTERICIDAE
PSEUDOSATURNIFORMIDAE
DEFLANDRECYRTIIDAE
PSEUDODICTYOMITRIDAE
EUCYRTIDIIDAE
LIVARELLIDAE
CANOPTIDAE
BAGOTIDAE
VICINGULIDAEPAR
? ANAKRUSIDAE
ARCHAEOSPONGOPRUNIDAE
RELINDELLIIDAE
OERTLISPONGIDAE
GOMBERELLIDAE
PYRAMISPONGIIDAE
ACTINOMMIDAE
LITHELIDAE
SPONGURIDAE
XIPHOSTYLIDAE
ANGULOBRACCHIIDAE
TRULIIDAEPA
VEGHICYCLIIDAE
HEXAPOROBRACHIIDAE
PANTANELLIIDAE
HAGIASTRIDAE

significantly decreased in diversity across the Permian–Triassic boundary. Instead of these formerly dominant Paleozoic-type radiolarians, new orders, e.g. Nassellaria and Spumellaria, diversified by the Middle Triassic. The extinction process of radiolarians in the end-Permian is well documented as double crises in shallower continental platform sediments in South China (Feng et al., 2007). The first crisis is characterized by extinction at species levels and second one is by disappearance of many genera. Prior to the first crisis in the late Changhsingian, rapid evolutionary events occurred, with many new species and genera appearing, including the Paleozoic-type taxa and some Mesozoic-type progenitors such as *Paroertilspongus, Tetrapaurinella, Paurinella, Tamonella,* and *Tetraspongodiscus*. Many of these species are very short-ranged within the late Changhsingian, suggesting an attempt to accommodate to changing environments. After this initial rapid evolutionary event, the first crisis brought a sudden decrease both in diversity and productivity, and the destruction of the majority of the radiolarian community. A total of 22% of the genera disappeared and 51% of genera surviving were reduced to just a single species. This first crisis is also related to a sharp reduction of foraminifer and ostracod diversity at both genus and species levels. In the second crisis, only a single species survived in each radiolarian genus. These radiolarians are small and have simple forms without any external spines. During this interval, radiolarian chert deposition halted in the low-latitude Panthalassa (Takahashi et al., 2009a, b). High-productivity environments persisted in the high-latitude southern hemisphere, as evidenced by ribbon chert continuously deposited from the Late Permian to the Middle Triassic (Suzuki et al., 2007).

Basal Mesozoic Scarcity of Radiolarian Fossils and Faunal Turnover (Early Triassic)

The Early Triassic (Induan and overlying Olenekian) are the first few million years of the Mesozoic. This interval is marked by an extreme scarcity of siliceous microfossils and chert (Racki, 2000; Beauchamp and Baud, 2002; Takahashi et al., 2009a, b) with cherts known only from the Oruatemanu Formation in New Zealand (Suzuki et al., 2007) and the upper Dalong Formation in southwest China (Yang et al., 2012). This phenomenon is called the "Chert Gap." The last Permian-type radiolarians evolved, in parallel with the appearance of the first Mesozoic-type monocyrtid nassellarian species during this interval in the early Induan (Kamata, 2007; Hori et al., 2011); while the turnover from dominantly Permian- to Mesozoic-type radiolarians started within the late Induan (Hori et al., 2007; De Wever et al., 2007; Figure 5.35). Biogenic siliceous sediments gradually began to be deposited again, e.g. siliceous claystone deposition in the basal Olenekian (Takahashi et al., 2009a, b); discontinuous chert deposition in the latest Olenekian (Suzuki et al., 2002; Takahashi et al., 2009b); and the full recovery of chert deposition in the basal Anisian, the lowermost unit of the middle Triassic (Takahashi et al., 2009b). Using the Triassic geologic timescale of Ogg (2012), the restoration of biosiliceous sedimentation took ca. 5 Myr. The overall pattern of radiolarian turnover during the Triassic is unclear in detail due to the limited number of well-preserved assemblages but appears to have been extended, rather than truly abrupt. No less than nine Paleozoic families survive into the early Triassic. Of these, six become extinct within the Triassic (De Wever et al., 2007).

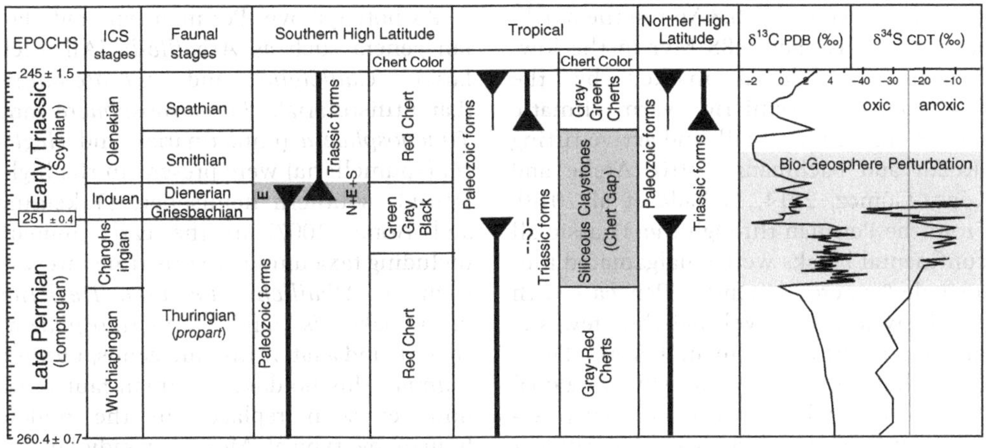

Figure 5.35 Geographic differences in preserved patterns of radiolarian evolutionary turnover across the Permo-Triassic boundary and early Triassic. The only stratigraphically continuous records – in southern high latitudes – show Paleozoic Entactinaria-dominated faunas gradually being replaced over a ca. 1 my long interval by new Mesozoic forms rich in nassellarians and spumellarians. In tropical and north high latitude regions a 3 my gap in the record exists between latest Paleozoic time and the oldest Triassic radiolarians. Paleozoic forms however are still common in these later time intervals. Source: De Wever et al., Eclogae Geologicae Helvetiae (2007). Reproduced with permission of Springer.

The Chert Gap interval of approximate 5 Myr occurred during a more general pattern of delayed recovery in other biotic groups, which is attributed to the effects of extreme environmental conditions in the Early Triassic (Tong et al., 2007). During the Induan and Olenekian, multiple extinction events are recognized in ammonoids and conodonts in association with three Early Triassic global negative carbon isotopic excursions (Stanley, 2009), relatively high $\delta^{34}S$ sulfide values (Takahashi et al., 2013), and other indications of anoxia and a disturbed, unstable carbon cycle (Takahashi et al., 2009b). However, a total of 204 species radiolarian species were already present in the late Olenekian (Suzuki et al., 2002), and 140 species are known from the Anisian (Suzuki et al., 2002). Ribbon chert deposition completely recovered by the middle Anisian (Sugiyama, 1997).

Triassic

Mesozoic oceans are in general better understood than those of the Paleozoic, particularly for the later Mesozoic, where substantial amounts of unmetamorphosed rock, or even only partially lithified sediment, is still available for detailed paleoceanographic studies. Early Mesozoic oceans recovered from the end-Permian extinction event over much of the Triassic Period, with episodes of widespread anoxia in the basal Triassic, and a general cooling trend throughout the period, starting however from the extremely warm conditions associated with the end-Permian extinction event (Ogg et al., 2004; Preto et al. 2010; Stanley 2009). Other than a poorly understood mid-Triassic climatic event (the "Pluvial Event" in the Carnian: Preto et al. 2010), no broad, global climate patterns are known. The Carnian pluvial event (CPE) in the middle Carnian is a short-lived event (0.7–0.8 Myr in the northwestern Tethys and 0.88 Myr in the low-latitude Panthalassa) marked by the development of semiarid, warm climatic conditions, sea-level fall, and active rifting (Kozur and Bachmann, 2010, Arche and López-Gómez, 2014, Nakada et al 2014). From the Permian through the Triassic all continental blocks were amalgamated into a single "supercontinent" – Pangea – with relatively low sea level and thus few significant shallow continental seas. There was thus also only a single global ocean – Panthalassa, with a partially isolated (by island arc chains) westernmost arm called the Proto-Tethys. Ocean models suggest that the Panthalassic ocean broadly resembled the Pacific Ocean of today, with subtropical and subpolar gyres in both hemispheres, and strong equatorial upwelling (Winguth et al. 2002). Latitudinal temperature gradients and general ocean overturning were however much reduced compared to the modern ocean, and there were thus presumably relatively few, weakly delimited biogeographic provinces.

One of the most important events in the Phanerozoic history of the plankton began in the Triassic, and accelerated in the Jurassic: the origin of major "red" plastid containing phytoplankton groups, e.g. the coccolithophores and dinoflagellates, which, along with diatoms, dominate modern plankton (Falkowski et al. 2004). Why these groups originated and radiated is still a matter for speculation, but their spread in the plankton must have changed the nature of food sources for radiolarians and other phytoplankton grazers. Coccolithophores also exported for the first time substantial amounts of carbonate into the deep ocean, gradually shifting the locus of carbonate sedimentation over the late Triassic and Jurassic from the shelves to the deep sea, and thereby adding a new mechanism (carbonate compensation depth feedbacks) to the regulation of atmospheric CO_2 and global climate (Ridgwell and Zeebe 2005).

As noted above, Permian-type radiolarian genera such as *Albaillella* (Albaillellaria), *Cauletella* and *Oruatemanua* (Latentifistularia), *Entactinosphaera* and *Triaenosphaera* (Entactinaria), and *Hegleria* (Spumellaria) were present in the high-latitude southern hemisphere (Takemura and Aono, 2007) in the early Induan, including taxa unique to this time interval, such as *Albaillella aotearoa, Hegleria? arrowrockensis* and *Entactinosphaera? spoerlii* and a latentifistulid genus, *Oruatemanua*. This holdover Permian fauna was however soon replaced in the middle Induan by typical Mesozoic radiolarians, particularly nassellarian taxa. Except for the questionable Devonian "nassellarians" belonging to Popofskyellidae and Archocyrtiidae (Cheng, 1986), the oldest indisputable nassellarian is *Tripedocorbis? blackae*, a monocyrtid nassellarian from the lower Induan (Hori et al., 2011). These earliest nassellarians include the monocyrtid nassellarians *Hozmadia* cf. *ozawai, Tripedocorbis kozuri, Tripedocassis oruatemanuensis, Triassospongocyrtis conica, Triassospongocyrtis* cf. *longispinosa* and *Triassospongocyrtis gracilis*, and a precursor morphotype of multisegmented nassellarians – *Zevius* sp. in the late Induan (Kamata, 2007).

More than 1,700 species belonging to 282 genera have been formally described for the Triassic (O'Dogherty et al., 2010). Although these numbers probably include numerous synonyms, clearly, Triassic diversity was much higher than in the Permian. This is also seen in family diversity, which increased from fewer than 20 families in the late Paleozoic to over 50 families in the Triassic (De Wever et al., 2001; Figure 5.36). A rapid explosion of new morphotypes is seen in the middle to late Anisian, maximum diversification at the genus level was attained during the early Carnian (approximately mid-Triassic), and progressive decline of diversity was observed in the later Triassic (the Norian to Rhaetian;

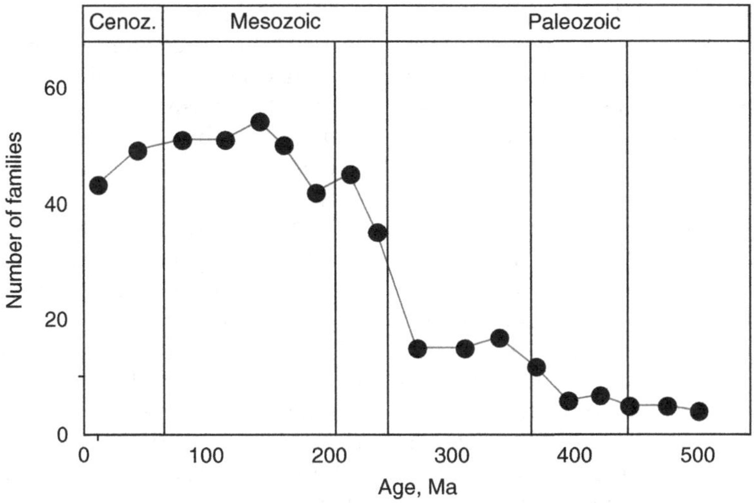

Figure 5.36 Phanerozoic family diversity of radiolaria, showing a series of monotonically increasing steps from low early Paleozoic levels of just a few families, to ca. 15 families through the late Paleozoic, and a very large increase through the lower Mesozoic (mostly in the Triassic) to ca. 45–50 families per time interval for the remainder of the Mesozoic and Cenozoic. Vertical lines mark the "big five" mass extinction events in the history of life. Source: From data compiled from De Wever et al. (2001).

O'Dogherty et al., 2010). The progressive decline of diversity from the Norian to Rhaetian mainly occurred between the late middle and early late Norian by disappearance of all Capnuchosphaeridae (except for the genus *Icrioma*) and Capnodocinae (Carter and Orchard, 2007). All the species belonging to these two families possess very characteristic hollow spines (like uncooked penne-type pasta noodles) and were dominant in the late Carnian to middle Norian. Finally, any radiolarians with twisted spines disappeared by the Triassic-Jurassic boundary. Faunal changes started by the latest Rhaetian when radiolarians reduced in size and multicyrtid nassellarians became to have lesser number of segments (Longridge et al., 2007).

Radiolarian fossils are sufficiently common and widespread by the Triassic that biogeographic patterns can be distinguished for the first time. The oldest-known radiolarian biogeographic provinces can be seen in the late Olenekian, with a western Panthalassa fauna of 143 species and the eastern Tethyan fauna with 71 species (Suzuki et al., 2002). These two biogeographic provinces are very distinctive with only 11 species are common. An anti-tropical distribution pattern at the genus level is recognized in Anisian by the presence of the genus *Glomeropyle* in both the high-latitude southern hemisphere of New Zealand and the high-latitude northern hemisphere of northeastern Siberia (Aita and Bragin, 1999). Both observed species diversity and biogeographic patterns can be strongly influenced by lithology and paleoceanographic conditions, and patterns similar to those of radiolarians are also known for other Triassic plankton such as the conodonts (Narkiewicz and Szulc, 2004), where the patterns are interpreted as being caused by the extent of epipieric seas (e.g. the Muschelkalk of the Tethys region), and by migration regulation by two ocean gateways: the East Carpathian and Silesian-Moravian Gates.

The diversification and endemism of radiolarians from the early Anisian to the early Ladinian has not been examined, but these the mechanisms proposed for conodonts might have affected radiolarian faunas as well.

Although most studies of Triassic oceans have concentrated on the boundaries to the underlying Permian or overlying Jurassic, some global events are known within the Triassic. Radiolarians responded to the Carnian event primarily via changing biogeography, with a significant penetration of tropical-subtropical Tethyan radiolarians into the boreal regions (Bragin et al., 2012). Other geologic events, such as the Manicouagan impact event at 215.5 Ma in the late middle Norian, may have triggered regional extinctions of marine metazoan ammonoids, bivalves, conodonts, and of terrestrial herbivorous dicynodonts and floras, but no effect has been recognized for radiolarian fauna (Onoue et al., 2012). Radiolarian faunal turnover occurred instead in the late middle Norian, ~1 Myr after the impact event.

Triassic–Jurassic Boundary Mass Extinction

The Triassic–Jurassic boundary (Rhaetian–Hettangian boundary) is a period of extinction of marine and non-marine biota, and is considered to be one of the big five mass extinction events in the history of life (Sepkoski, 1996). A negative C-isotope pulse is seen just below the Triassic-Jurassic boundary, coinciding with the disappearance of the end-Triassic benthic faunas and the drowning of widespread Rhaetian carbonate platforms (Galli et al., 2005). The negative C-isotopic pulse may have resulted from the sudden release of gas hydrates, and was followed by a positive C-isotope excursion starting at the Triassic-Jurassic transition, which records a global change in organic carbon burial rates. These indicators of changes in the carbon cycle are probably in response to elevated atmospheric gases, particularly CO_2, caused by one of the largest known extrusive magmatic events in earth history – the Central Atlantic Magmatic Province (CAMP) volcanism. CAMP basalts are distributed throughout the margins of the Atlantic Ocean basins and are thought to be associated with the break-up of Pangea and the origin of the Atlantic Ocean (Marzoli et al., 1999). Although the timing of this magmatic activity vs. the extinctions, and the specific environmental impacts are still debated, it seems likely that, as true for the Permo-Triassic extinction, perturbations to the carbon cycle caused by volcanic activity created adverse environmental conditions (warming, ocean acidification, etc.) that led to the extinctions.

As true also for ammonoids and conodonts, extinctions of radiolarians occurred just at the Triassic-Jurassic boundary (Pálfy et al., 2000). The magnitude however differs significantly depending on the data used and how it is evaluated. In local sections the magnitude of change can be quite large and can be shown to be closely associated with a negative shift of $\delta^{13}C_{org}$ (Ward et al., 2001). Radiolarian faunas in the Queen Charlotte Islands and Japan abruptly changed across the Triassic–Jurassic boundary (Carter and Hori, 2005). Rhaetian radiolarians belonging to 20 genera and 130 species completely disappeared near the Triassic-Jurassic boundary. Index taxa such as *Bettracium*, *Risella*, *Globolaxtorum tozeri*, *Livarella valida*, and *Pseudohagiastrum giganteum* disappeared. Post-extinction faunas in these sections are characterized by a few species like small, primitive spumellarians with spongy or irregularly latticed meshwork and rod-like spines. These early post-Triassic-Jurassic boundary species also completely disappeared by the end of the Hettangian, the first stage of the

Jurassic. Just after the Triassic-Jurassic boundary, new genera such as *Charlottea*, *Udalia*, and *Parahsuum* s. l. appeared both in the Queen Charlotte Islands and Japan. However, quantitative examination of radiolarian species diversity from global compilations of occurrence data shows that, while significantly higher than prior background extinction rates, the magnitude of the end-Triassic extinction in radiolaria is more modest than seen in local sections, and less than the (extended) extinction that occurred at the Permo-Triassic boundary (Kocsis et al., 2014). The differences between local sections and a global compilation are presumably due in part to the incompleteness of diversity in local sections, and the confounding effects of changing local preservation, or depositional conditions, on the diversity observed in individual sections. However, the presence of distinctive faunas, mostly of small forms in the earliest post-extinction level assemblages suggests that at least locally a major perturbation to the radiolarians took place that is not fully captured in the larger scale compilations.

Jurassic

The Jurassic was a warm and humid climate period characterized by high CO_2 and low O_2 atmospheric levels, with several brief high-latitude glaciations in the Pliensbachian, Bajocian-Bathonian, late Callovian-early Oxfordian, and Tithonian. The Jurassic Period marks the beginning of a more "modern" ocean system, with the break-up of Pangea creating a new low-latitude seaway – the central Atlantic – between North America and Eurasia, and probably also between North and South America by the Late Jurassic. Together with the western tropics of the Panthalassic Ocean (the Tethys – from between Africa and Eurasia to what is now the eastern edge of the Indian Ocean), this low-latitude ocean is the source of many of the Mesozoic marine rocks, now exposed on land in European mountain belts, whose faunas have formed the basis of much classical radiolarian research. Sea levels rose during the Jurassic and again created extensive shallow seas over continental shelves and epicontinental seas over large areas of continent interiors.

While the Jurassic is more famous as the beginning of the "age of dinosaurs" on land, the Jurassic was also the time period when the evolution of key phytoplankton groups fundamentally altered the functioning of the global carbon cycle and distribution of nutrients in the oceans. Modern dinoflagellates displaced the earlier, presumably related acritarchs in the late Triassic and early Jurassic (Fensome et al., 1996). Coccolithophores in particular, although having first evolved in the late Triassic, first became diverse in the mid Jurassic, shifting the earth's carbonate deposition from shelf to deep-sea environments (Bown et al., 2004; Ridgwell, 2005). Deep-sea sediments were still largely carbonate free at the beginning of the Jurassic but by the end of the Period increasing pelagic carbonate production by rapidly evolving coccolithophores, and also the first planktonic foraminifera began to displace the until-then common mixture of clays and radiolarian-dominated biosiliceous sediments (now preserved as radiolarian cherts) with calcareous oozes (Hüneke and Henrich 2011). This permanently affected the style of radiolarian preservation in all subsequent geologic time intervals (see Chapter 6).

Early and Middle Jurassic Radiolaria

After a short (<1 my) period of low post-extinction diversity in the early Hettangian, radiolarian species diversity gradually recovered through the middle Hettangian and Sinemurian, with the diversification of forms with triradiate spines and evolution of several new genera (e.g. *Protokatroma*, *Saitoum*) (Longridge et al., 2007). Most survivors of typical Rhaetian radiolarians disappeared by the end of the Hettangian (Carter and Hori, 2005), but a few conservative forms such as *Canoptum*, *Crucella*, *Orbiculiformella*, *Pantanellium*, and *Paronaella* survived and gradually diversified in the Hettangian at the genus level, while their descendants radiated in the Jurassic and Cretaceous (Longridge et al., 2007). Taxa like *Bistarkum* and *Cornutella* with significantly discontinuous stratigraphic ranges (reflecting either morphologic convergence of unrelated forms, or major gaps in fossil recovery) are reported in the Pliensbachian/Toarcian and later. Radiolarians diversified in species number during the earliest Pliensbachian, reaching an Early Jurassic maximum in the mid-Pliensbachian (Goričan et al., 2013). Overall, Jurassic radiolarian diversity, whether measured at the species, genus, or family level, was on average similar to that of the later Triassic.

Several oceanic anoxic events (OAEs) are known in the Jurassic, particularly the "T-OAE," associated with the Pliensbachian-Toarcian boundary (late Early Jurassic). This is recorded as multiple events in some sections, e.g. in radiolarian-bedded chert in Japan (Ikeda and Hori, 2014). The T-OAE is marked by carbon isotope excursions (~5‰ $\delta^{13}C$ excursion in the exogenic carbon reservoirs), and is thought to have been associated with a 1,000 ppm rise in atmospheric CO_2 and a brief interval of 6–7 °C global warming. These led to anoxia, organic carbon deposition, and biotic extinctions (Little and Benton, 1995; Jenkyns and Clayton, 1997; Beerling & Brentnall, 2007). The T-OAE also strongly affected primary producers such as coccolithophores and dinoflagellates in the western Tethys (Fraguas et al., 2012), causing a biocalcification crisis in calcareous phytoplankton and the disappearance of dissolution susceptible benthos (Tremolada et al., 2005). The T-OAE may have been caused by the volcanic activity of the Karoo-Ferrar large igneous province (LIP) (Ikeda and Hori, 2014). Significant radiolarian faunal turnovers and reductions in diversity are recorded at this time, and multisegmented nassellarians became smaller (Hori et al., 1997). Early post-extinction radiolarian faunas are characterized by *Praeparvicingula*-species (Parvicingulidae, Nassellaria) and the appearance of *H. hexagonus*, although spongy spumellarians also survived in this extinction interval (Goričan et al., 2013). Radiolarian species diversity recovered in the middle to late Toarcian.

In Europe, radiolarian-rich sediments became more common starting in the early Bajocian (Middle Jurassic), reaching a local abundance maximum by the late Callovian (Baumgartner, 2013), are correlated to a positive shift in local carbon isotope curves indicating eutrophication. Species diversity was as high as 300 species in the low-to mid-latitude Panthalassa in the Middle Jurassic (Toarcian-Bajocian (Yao, 1997). Yao (1997) documented extinction rates in local sections from Japan that became gradually higher from the Toarcian to the Bajocian by the disappearance of the short-ranging species in the low- to mid-latitude Panthalassa. Danelian and Johnson (2001) however noted an interval of high first-occurrence rates in the western Tethys during the Aalenian–Oxfordian. As neither of these studies compensated for sample-size effects (see Chapter 9: Evolution), the differences could be a statistical artifact. Alternately, the different patterns could be due to differing evolutionary patterns in biogeographic provinces: the mid-Jurassic species composition between the Panthalassa region (Yao, 1997) and the western Tethys (Danelian and Johnson, 2001) are quite different with only a few common species. In the western Tethys, radiolarian diversity changes are associated with a $\delta^{13}C$ shift (Bartolini et al., 1999), again implicating changes in the carbon cycle as a cause of radiolarian diversity change.

Jurassic radiolarian biogeographic provinces are in general a controversial issue. A classification was originally proposed by Pessagno and Blome (1986); provinces were both defined and analyzed in a study by Kiessling (1999; 2002; Figure 5.37, see also Chapter 9). A variety of alternatives have however been suggested and debated (see summary for the mid-Jurassic interval in Goričan et al., 2013). According to Goričan et al. (2013), in the Pliensbachian–Toarcian interval, 16 genera (e.g. *Bagotum, Canoptum, Crucella, Hagiastrum, Hsuum, Katroma, Napora, Orbiculiformella, Parahsuum, Paronaella, Praeconocaryomma, Praeparvicingula*) are common in both low- and mid-latitudes, but some of these genera show biogeographic differences at the species level. The dominant genera in the Tethys and the low-latitude Panthalassa are *Ares, Bistarkum, Eucyrtidiellum, Foremania, Gigi, Pleesus, Pseudoeucyrtis*, spherical nassellarians such as *Minocapsa*, and most saturnalids. The mid-latitudes in the Panthalassa are by contrast characterized by common occurrences of *Bipedis, Canutus, Haeckelicyrtium, Wrangellium, Perispyridium*, and *Turanta*.

Jurassic radiolarians are characterized by a variety of morphologies in the surface textures in multisegmented nassellarians (e.g. parvicingulids) and by the diversification of spherical nassellarians (e.g. *Striatojaponocapsa, Williriedellum*, and *Zhamoidellum*). Although spherical nassellarians range from the Late Triassic to the Maastrichtian (latest Cretaceous), they are only diverse between the Aalenian (earliest Middle

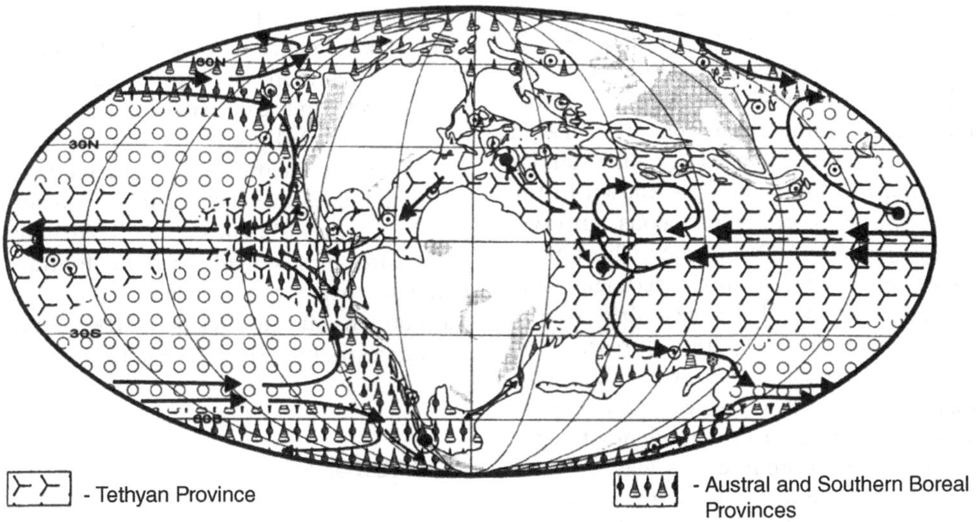

Figure 5.37 Biogeographic provinces of radiolaria in the latest Jurassic (Tithonian), with hypothetical surface currents as proposed by Parrish (1992). Open circles with solid dot centers mark locations of samples (large dots: author's own data; small dots: literature data). Source: From Kiessling (1999). Reproduced with permission of Micropress.org.

Jurassic) to the earliest Cretaceous. While large, robust radiolarians are known from several time intervals in the Phanerozoic (Ordovician, Jurassic–Cretaceous, and the middle to upper Eocene), possibly the largest radiolarian tests by internal volume of the entire Phanerozoic are those in the genus *Mirifusus*. *Mirifusus* species appeared in the Aalenian (the Middle Jurassic) (e.g. *Mirifusus proavus*). In contrast to the many other, normally sized species of the mid-Jurassic, species in this genus increased in size significantly and their surface ornamentation became more elaborate in the Bajocian (the late Middle Jurassic), and reached their maximum size in the Callovian. (e.g. 600 μm diameter in *Mirifusus guadalupensi*; Guex et al. 2012).

Late Jurassic–Early Cretaceous

The mid/late Jurassic to Early Cretaceous interval has long been studied, particularly in Tethyan sections (e.g. Goričan, 1994; Jud, 1994; and many other papers; summarized in Baumgartner et al., 1995). Significant global changes in radiolarian (or other marine) biotas are not known in the Late Jurassic–Early Cretaceous interval (Pessagno et al., 2009), but the highest number of species for a single geologic time interval (865) has been described in the Tithonian/Berriasian (latest Jurassic and early Cretaceous; Kiessling, 2002). This high species diversity is partially a taxonomic artifact, due to the tendency to split Mesozoic taxa more finely, at least vs. later Cenozoic and Recent species (Kiessling, 2002). However, as this practice is common throughout the Mesozoic, diversity for this time interval is indeed high vs. other intervals in the Mesozoic, and can in part be explained by a significant latitudinal diversity gradient in the Tithonian–Berriasian, with significantly differentiated faunas, both between poles and tropic, and also between northern and southern polar faunas (Kiessling, 2002). Similar gradients have been noted for gastropods and bivalves, but not for other nektonic macro-invertebrates (Hallam, 1994; Crame, 1996). As the latitudinal thermal gradient in the Tithonian–Berriasian is much gentler than that of the present day, and there was no prominent development of polar ice, salinity, and productivity variation might have contributed to the radiolarian diversity gradient. Why this interval in time should be higher than both preceding and succeeding time intervals however has not been studied.

With the radiation of the coccolithophores and deposition of deep-sea phytoplankton carbonate, widespread radiolarite deposition came to an end, in the Tithonian in the Tethys and somewhat later, the early Aptian, in the Panthalassa. Radiolarian-bearing sedimentary rocks from this point on are dominantly other lithologies, e.g. siliceous mudstone or mudstone in accretionary complexes in Japan and Russia (Kasai et al., 2000; Nakae, 2000a, 2000b; Yamato Omine Research Group, 2005, 2007; Ishida, 2009). In the deep ocean, e.g. Site 801 of the Ocean Drilling Program in the western equatorial Pacific, alternating red radiolarite and claystone beds formed instead of just radiolarite (Matsuoka, 1992). Although radiolarites occasionally can be found in younger rocks, the ocean's sedimentary mechanisms changed permanently, if gradually, across the Jurassic–Cretaceous boundary.

Cretaceous

The Cretaceous period is known as one of the warmest climatic periods in the Phanerozoic, in association with typically high Mesozoic atmospheric CO_2 levels of around 2000 ppm (approximately 8 times modern values – Royer et al., 2004) and reduced latitudinal gradients in sea surface temperature. Cretaceous warmth is inferred not only from a variety of biotic and sedimentologic indicators (Hay and Floegel, 2012) but also from the absolute values of oxygen isotopes, which, in Cretaceous and younger sediments only, can be used to calculate absolute temperature values (there is a systematic, unexplained shift in geologically older values – Veizer and Prokoph, 2015). In addition, the global heat flow was high due to the increase in global oceanic ridge length that resulted from the breakup of Pangea (Müller et al., 2013). The South Atlantic opened as did the Indian and Atlantic sectors of the Southern Ocean. Many large igneous provinces (LIPs) formed, such as Caribbean, Kerguelen, Ontong-Java, and Shatsky plateaus (Bryan and Ferrari, 2013). The surface oceans were warm and (relative to today) rather saline. Deep-ocean waters, although cooler than the surface waters, were still much warmer than in the modern ocean and thus the vertical density contrast in the oceans was much lower than the modern condition (Otto-Bliesner et al. 2002). This probably affected ocean mixing of nutrients, but details are unknown, since mixing is also affected by ocean surface wind stress, and in the warm, equable climates of the Cretaceous these were lower than today. Coccolithophores reached the highest level of diversity in their evolutionary history (Bown et al. 2004) and formed widespread chalk deposits both in shallow seas and in deep ocean basins. The Cretaceous is also marked by a few global and several more regional organic black shales which were formed by oxygen anoxic events (OAEs), which in turn are linked to tectonically driven rises in atmospheric pCO_2 (Jenkyns, 2010). Black shales are primarily seen in the low-latitude Atlantic and Tethys, although some horizons of black shales are recognized in the Shatsky Plateau in the proto-Pacific. Black shales are poorly developed or not known in the middle- to high-latitude proto-Pacific and the Russian boreal regions. Paleoceanographers have been debating the causes of OAEs for decades, stressing increased export productivity, decreased oxygen supply from reduced circulation, or other factors. Regardless of cause, OAEs significantly affected the evolution of Cretaceous plankton. Cretaceous marine plankton show several episodes of enhanced taxonomic turnover, particularly during OAEs (e.g. Leckie et al., 2002). Radiolarians were also strongly affected by OAEs (Erbacher, 1994). At the family level, radiolarian diversity and turnover remained similar to that of the Jurassic, except at the end of the Cretaceous, where a major extinction and turnover took place (De Wever et al., 2001).

Although latitudinal sea-surface temperature gradients were reduced in the Cretaceous, radiolarian provincialism or latitudinal zonal provinces are known from at least the late Aptian through Maastrichtian (Empson-Morin, 1984, O'Dogherty, 1994, Takahashi, 1999). These biogeographic faunal differences however are much less pronounced than faunal changes within a biogeographic region over the course of the Cretaceous.

The earliest Cretaceous interval (Berriasian–Barremian) is considered to be a warm, stable climate period with sea-surface temperature exceeding 32°C at 15°–20°N, and an average of 26°C at ~53°S (Littler et al., 2011). The earliest Berriasian radiolarian faunas show a close similarity to those of the preceeding latest Jurassic. Species that appeared in the Middle–Late Jurassic such as *Mirifusus dianae*, *Zhamoidellum ovum*, and *Williriedellum ruesti* persisted into the Berriasian. Cosmopolitan species such as *Pantanellium riedeli*, *Pseudodictyomitra excellens*, *Tethysetta boesii*, *Thanarla brouweri*, and *Archaeodictyomitra lacrimula* rangedthoughthisinterval,intotheBarremian–Aptian.

Radiolarian faunas changed gradually between the Berriasian and the lower Aptian (Jud, 1994, Dumitrica et al., 1997; Ishii et al., 2009, Kurilov and Vishnevskaya,

2011, Pessagno, 1977c). Peaks in some taxa such as *Pantanellium* are associated with a Valanginian OAE (Erba et al., 1999; Bartolini, 2003).

A more widespread OAE in the lower Cretaceous, OAE1, has been resolved by numerous studies into a pair of OAEs – 1a and 1b. OAE1a lasted for approximately 1 my in the early Aptian, and radiolarian-bearing deposits have been studied at various stages over the development of this event both in the Tethys and in the paleo-low latitude central Pacific Shatsky Rise (Premoli Silva et al., 1999, Bralower et al., 2006, Danelian et al., 2002, Michalík et al., 2008). Radiolarian chert in Shikoku, Japan (Shimanto Belt), also records this event, although black shale was not deposited (Ihoriya et al., 2009). However, the accumulation V, U, and S in the Shimanto chert is a signature of anoxic conditions (Ihoriya et al., 2009). This suggests OAE1a conditions developed above the lower bathypelagic zone in the low-latitude proto-Pacific but did not reach more deeper and shallower water depths. The response of radiolarians to OAE1a is well documented in the west Tethys. According to Erbacher et al., (1996), 41% of radiolarian species went extinct and 26% of the faunal diversity originated across the OAE1a interval in the Umbria-Marche Basin, Italy. High levels of faunal turnover were also seen in Slovakia (Koňhora Formation, the Pieniny Klippen Belt units of the Outer Western Carpathians), where extinctions and originations both exceed 50% (Michalík et al., 2008). Biogeographic differences exist between locations, despite similarity in turnover magnitude. For example, faunas of the Carpathians are characterized by the dominant, high diversified Williriedellidae, a few *Crucella*, *Paronaella*, and *Angulobracchia*, and the absence of Pantanelliidae (Michalík et al., 2008), while similarly aged faunas in Greece are dominated by diversified, abundant Archaeodictyomitridae (Danelian et al., 2002).

The range of many common species such as *Pseudodictyomitra lodogaensis*, *Archaeodictyomitra sliteri*, *Holocryptocanium barubui*, *Archaeodictyomitra vulgaris*, *Xitus spicularis*, *Spongocapsula? zamorensis*, *Thanarla praeveneta*, *Archaeodictyomitra simplex*, *Pseudodictyomitra macrocephala*, and *Stichomitra communis* is so long that in early studies it was represented by a single zone, the *Dictyomitra veneta* Zone (O'Dogherty, 1994). Refined zonations were developed by examining changes in abundance and the biostratigraphic occurrences of other, less common species. Major faunal changes can be seen in these less-common species (Taketani, 1982, O'Dogherty, 1994; Bragina, 2004, Palechek et al., 2010).

The late Aptian to ca. middle Albian is a several million year-long interval in the Cretaceous with a complex history of oceanic anoxia. The interval is collectively called OAE1b, with several shorter, stronger, regional, or global anoxic events or "horizons" (named variously as the Jacob, Kilian, Paquier, Urbino, and Leenhald). The Paquier event in the basal Albian (ca. 115 Ma; Ogg and Hinnov, 2012) is perhaps the most intense. Substantial evolutionary turnover in planktonic foraminifera faunas occurs at the species level at this horizon, and there is a globally developed hiatus in many carbonate sections, indicating dissolution and perturbation of the global carbon cycle (Fraass et al., 2015). Radiolarian faunal responses to OAE1b were examined in the Dercourt Member of the Ionian Zone, Greece (Danelian et al., 2004), the upper shaley siliceous interval in the Sopoti section in Albania (Danelian et al., 2007, Danelian, 2008), and the Allam Member of the Fahdene Formation, northern Tunisia (Ben Fadhel et al., 2010, 2014). Abundant, diverse cryptocephalic nassellarians and Archaeodictyomitridae were found from these regions and have been interpreted as a response to regional paleoceanographic changes within the Mediterranean Tethys.

Substantial evolutionary turnover in radiolarians, with 42% species extinction and 36% origination, was estimated by Erbacher et al. (1996).

The remainder of the Albian and the subsequent Cenomanian show only relatively gradual change in radiolarian faunas. The transition between the Cenomanian and overlying Turonian is marked however by the OAE2 event (also called the Bonarelli Horizon). Black shales, often with radiolarians, are found in globally distributed sections, including many in epicontinental shelf environments, as well as in deeper-water sections. Similar to OAE1a, the evidence of trace metals suggests enhanced marine productivity in OAE2 (Leckie et al., 2002), and is thought to be driven by warm climates and high rates of weathering due to enhanced volcanic CO_2 production in association with formation of the Caribbean Large Igneous Province. Detailed studies have shown that this event is complex, and includes a brief but substantial cooling episode (the "Plenus Event") immediately preceding the main warming interval (van Helmond et al., 2016). Substantial turnover is seen across the OAE2 in calcareous nannofossils, dinoflagellates, and planktonic foraminifera (Erba, 2004; Pearce et al., 2009; Fraass et al., 2015).

In contrast to the dominance of long-ranging radiolarian species in the mid-Cretaceous, many short-ranging species appeared and disappeared from the late Cenomanian to the early Turonian in the North Atlantic and in the northern margin of the Mediterranean Tethys (Thurow, 1988, Bąk, 1999). The abrupt appearance events of many radiolarian species are recognized near OAE2 in the northwest proto-Pacific, in a tectonic terrane now accreted to Japan (Taketani, 1982; correlative stage revised by O'Dogherty, 1994). Radiolarian-bearing sediments equivalent to OAE2 are reported from many locations: Southern Alps (Salvini and Marcucci Passerini, 1998, Bąk, 2011), Guibbio, Italy (Musavu-Moussavou et al., 2007), the Outer Carpathian Basin (Bąk, 2000, 2004, 2006, 2007), DSDP Hole 398D west of the Iberian Peninsula (Thurow, 1988), the Demerara Rise in the northern proto-Atlantic (Musavu-Moussavou and Danelian, 2006), and from the Western Caucasus (Bragina et al., 2007). Radiolarian response to this event varies however between locations. Radiolarian turnover was substantial in individual sections (extinction levels of 58% in Erbacher et al., 1996, 41% in Musavu-Moussavou et al., 2007, and only slightly lower origination rates (49% in Erbacher and Thurow, 1997, 35% in Musavu-Moussavou et al., 2007).

There are reasons however to question the patterns reported from local sections. Bąk (2011), in a detailed study of central European sections, noted that many species involved in the turnover event appeared before the OAE2 level (i.e. below the carbon shift and organic rich sediment facies), and further, the disappearance of many species at the end of OAE2 was only temporary, with most reappearing ca. 1 my afterwards. A similar pattern of more gradual change in radiolarian faunas before and after the OAE level was also reported from European alpine sections by Salvini and Passerini (1998). In the North Pacific (sections from Sakhalin and Japan: Taketani, 1982, Takashima et al., 2004, Yazykova et al., 2004), most radiolarian extinctions occurred within the upper Cenomanian, below black mudstones that are the local expression of the OAE, and the resultant lower diversity fauna showed no further change in the overlying lower Turonian. This suggests that the turnover levels reported in other studies may be at least in part only local, temporary ecologic responses, rather than irreversible evolutionary changes. The fauna composition across the OAE2 also varies between sections: equal ratios between nassellarians and "spumellarians and entactinarians" at Demerara Rise (Musavu-Mossavou et al.,

2007); a highly endemic fauna with *Holocryptocanium barbui* and *Cryptamphorella conara* in the Outer Carpathians (Bąk, 2011); and two dominant multisegmented nassellarians (*Archaeodictyomitra sliteri* and *Stichomitra communis*) plus the absence of staurnalids in an Italian section (Musavu-Moussavou et al., 2007).

In the Campanian, three $\delta^{13}C$ excursions are recognized in the classic, intensively studied pelagic section near El Kef, northern Tunisia, and are linked to rapid sea-level fall (Jarvis et al., 2002). Global cooling is also recognized in the Maastrichtian (MacLeod et al., 1997). Late Cretaceous radiolarians are characterised by significant though gradual turnover in the Turonian to Campanian, with many of the species being relatively short-ranging. Extinctions appear to have been more common than originations and thus diversity tended to decline. Maastrichtian faunas are uncommon and poorly known, but diversity in the Maastrichtian appears to have been lower than in the earlier Cretaceous and include a substantial number of longer-ranging species or genera from earlier Cretaceous time intervals (Foreman, 1968, Empson-Morin, 1981, Thurow, 1988, Taketani, 1995, Hollis, 1997, Takahashi, 1999, Bubík et al., 1999, Hollis and Kimura, 2001, Yazykova et al., 2004, Bandini et al., 2006, Bandini et al., 2008, Aumond et al., 2009, Ishida et al., 2010, Bragina, 2012, Benyamovskiy et al., 2012, Kochhann et al., 2013).

The K/T Extinction Event and Early Paleocene

The Mesozoic came to an end with the end of the Cretaceous. The boundary to the overlying Cenozoic is marked by one of the largest mass extinction events in earth history (the "K/T" or K–Pg), in which famously the dinosaurs died out and were replaced by mammals as the dominant terrestrial vertebrates. After a long controversy, most scientists now accept that, although latest Cretaceous volcanism and/or gradual late Cretaceous cooling climate may have indirectly contributed to this extinction event, the main cause of the K/T extinction was a large meteorite impact at the Yucatan Peninsula in Mexico (which today, although buried under more recent sediments, is an impressive 150 km in diameter and 20 km in depth)(Schulte et al. 2010). This impact created a massive, if short-term (months to years) perturbation of the environment, from tsunamis to global darkness and cooling. These led to near global ecosystem collapse and extinctions on both land and sea. It is thought that ca. 75% percent of all species and nearly 50% percent of all genera of life perished at this time (Jablonksi, 1994). The most persuasive geological evidence linking the impact event to mass extinctions is the presence, in precisely the same sediment layers, of unique geologic markers of impact debris and the last appearances of numerous fossil lineages. This has been most clearly demonstrated with fossil marine plankton, where the great abundances of specimens and continuous sedimentation characteristics of pelagic sediments show that extinctions and impact must have occurred simultaneously, or at least within a few thousand years of each other. Marine plankton in these sections was severely affected, although the degree differs significantly according to taxonomic group. The overwhelming majority of both planktonic foraminifera and calcareous nannofossils (coccolithophores and discoasters) went extinct. By contrast, diatoms and dinoflagellates appear not to have been directly affected, although the number of sections and preservation quality is generally rather poor for both.

Radiolarians are, as noted above, uncommon in Maastrichtian sections. In the early Paleocene biosiliceous deposits are extremely rare (Hollis, 1997) – a pattern very similar to the "chert gap" seen in the basal Mesozoic after the Permo-Triassic boundary mass extinction event. This makes study of radiolarians at the K/T difficult. Fortunately, there are a few sections where reasonably well preserved radiolarians are found more or less continuously through the late Maastrichtian and early Paleocene. These sections, particularly in New Zealand, document that there were no significant extinctions of radiolarians directly at the K/T boundary (Hollis, 1997, Figure 5.38). However, by the mid-Paleocene, when radiolarians again become more common in the rock record, a dramatic change in assemblages had occurred, with numerous family-level taxa having become extinct, and several new families appearing, or first becoming common including important Cenozoic clades such as the Trissocyclidae and Plagiacanthidae. Additional families are first seen only a bit later in the early Eocene, although their true timing of origin may be earlier due to limited data. Precise values for change at the family level between the Cenozoic and Mesozoic are difficult to give given the substantial differences in family level taxonomy between these two geologic Eras. The details of this transition have been documented in several papers by Hollis and coworkers (e.g. Hollis et al., 1995; Hollis, 1997, 2003), but the underlying causes of radiolarian change, or the effects

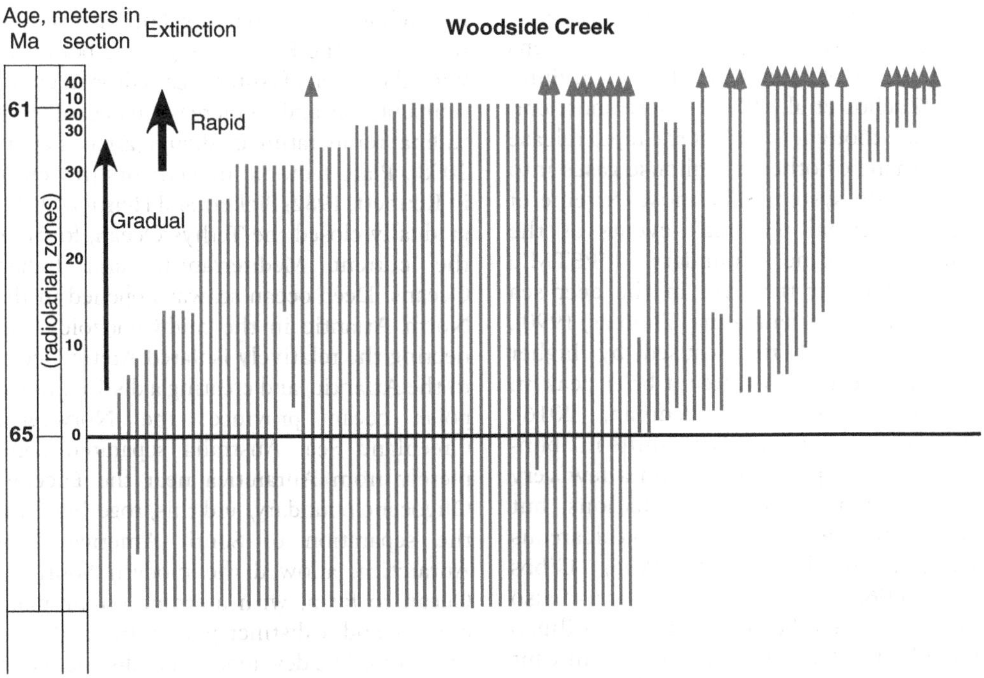

Figure 5.38 Stratigraphic ranges of species in the latest Cretaceous (Maastrichtian) and early Paleocene in a typical K/T section from New Zealand (Woodside Creek). The K/T boundary (horizontal black line) is robustly defined on multiple other microfossil and geochemical criteria. Ages in the Paleocene are based on multiple microfossil groups and although only approximate are accurate to ca. 1 my. Extinction rates (summarized by black arrows) are initially moderate but increase around 62–61 Ma. Source: From data in Hollis (1997).

on macroevolutionary patterns of higher-level taxa have not yet, however, been fully studied (see also Chapter 9).

Cenozoic

Cenozoic oceans, like those of the Mesozoic, began in a recovery mode from the mass extinction, with initially low export productivity and extremely low diversity in carbonate plankton groups. The early Cenozoic was a period of general warming, although the peak temperatures reached in the early Eocene were still much below that of peak Cretaceous warmth. This interval of time is marked by relatively low background rates of climate change, punctuated by several short intervals of rapid warming, generally a few tens to hundreds of thousands of years in duration (Zachos et al., 2008). These short episodes are thought to reflect the rapid release of methane or other greenhouse gases into the climate system. The most extreme of these "thermal maxima" events at the Paleocene–Eocene boundary ("PETM") caused a major turnover in the deep-sea benthic foraminifera fauna (Thomas, 1990). Enhanced weathering caused by higher atmospheric pCO_2 created a brief peak in oceanic silica input (Penman, 2016). Plankton was affected, with shifts in abundances, geographic ranges, and a few very short-lived new species originations, but the net effect was very minor on calcareous plankton (Kelly et al. 1996; Gibbs et al. 2006). The radiolarian response also appears to have been minimal. Sanfilippo and Blome (2001) reported no significant evolutionary turnover in the tropical Atlantic. Hollis (2006) reported 13 species first occurrences in New Zealand sections (from a prior diversity of ca. 70 species) within the PETM interval (ca. 200 kyr). However, he notes that all but three of these species are known in older intervals of sections elsewhere in the world, suggesting that migration, possibly due to the warming, is largely responsible for the observed local first appearances.

By the late Paleocene, siliceous sedimentation had resumed in many ocean basins, and the fossil record of radiolarians from this point on to the Recent is extremely good, with numerous sections deep-sea sediment sections obtained by the DSDP, ODP, and IODP drilling programs available for study. The amount of material is in fact overwhelming, with only a very small fraction of the available material having been studied in detail. Early to mid-Eocene sediments in many ocean basins are rich in silica, and when not converted by diagenesis into chert (Muttoni and Kent, 2007) are rich in radiolarians (Sanfilippo et al., 1985). Cooling throughout the remaining Eocene was terminated at the Eocene–Oligocene boundary with glaciation of Antarctica, a drop in global temperatures and an increase in oceanic latitudinal temperature gradients (Zachos et al., 2008). Plate tectonic movements (reviewed in Kennett, 1982; Bickert and Henrich, 2011) gradually closed the Tethys Ocean, forming the current Mediterranean and Indian Oceans. Deep ocean seaways opened in the North Atlantic in the mid-Cenozoic, connecting the relatively isolated Arctic Ocean to the Atlantic, and creating a distinct north polar ocean province: the Norwegian Greenland Sea. Australia separated completely from Antarctica near the Eocene–Oligocene boundary, and this, together with the separation of South America from Antarctica, allowed the modern Southern Ocean to form, with a strong circumpolar current and a distinct planktonic and benthic biota. The development of distinct polar ocean biogeographic provinces in the southern hemisphere occurred in the late Eocene. Around 35 Ma, shortly before the glaciation event that marks the Eocene–Oligocene boundary, a cool temperate, cosmopolitan radiolarian fauna was replaced by a more endemic Southern Ocean fauna (Lazarus and

Caulet, 1994; Lazarus et al., 2008; Figure 5.39). The Antarctic ice cap appeared at the Eocene–Oligocene boundary, and polar ocean deep waters cooled dramatically, reorganizing deep-ocean circulation and establishing a strong vertical density gradient over most of the global ocean (Zachos et al., 2001). In the tropics, diverse, rapidly

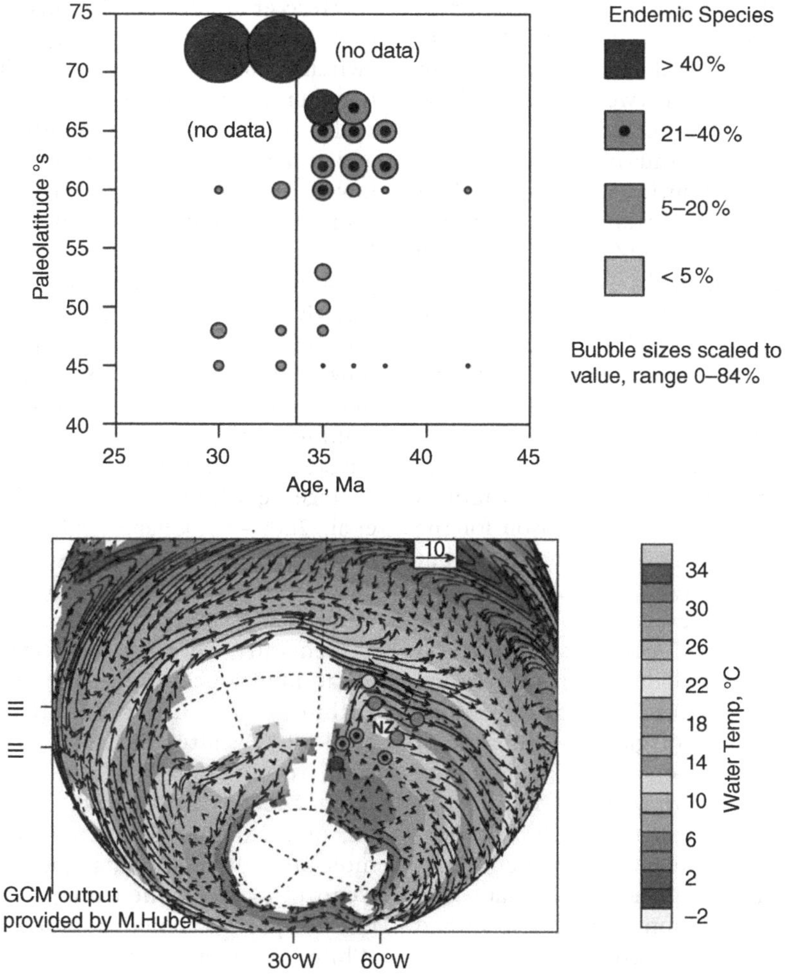

Figure 5.39 Gradual development of endemic austral radiolarian faunas in the late Eocene to early Oligocene in the southwest Pacific sector of the Southern Ocean and their correlation to the physical environment. On left, the degree of endemism is strongly related to latitude and (although geographic coverage in the SW sector is incomplete) increases over time. Increasing endemism in the Southern Ocean across the Eocene–Oligocene boundary is documented in other ocean sectors as well (Lazarus and Caulet, 1994). On right, data (colored dots off eastern Australia–New Zealand region) from a single time interval (ca. 35 Ma) are overlain on an ocean temperature and circulation reconstruction to show close correspondence of endemism and oceanographic conditions. Source: From Lazarus et al. (2008), ocean circulation/temperature reconstruction by Huber et al. (2004). Reproduced with permission of Micropress.org. (*See color plate section for the color representation of this figure.*)

evolving Eocene faunas were replaced by more slowly evolving faunas in the Oligocene (Sanfilippo et al., 1985). Cooling in the mid-Miocene stabilized the – until then, fairly dynamic – East Antarctic ice cap and increased glaciation of West Antarctica (Kennett, 1982; Bickert and Henrich, 2011). Further cooling and deepening of basement topographic barriers between Greenland and Northern Europe, together with shallowing of the still-open Isthmus of Panama gateway between the Atlantic and Pacific (Lear et al., 2003) initiated strong deep-water flow between the Norwegian-Greenland Sea and the Southern Ocean by the late Miocene. Further cooling in the late Cenozoic led to the development of northern hemisphere ice sheets and numerous strong glacial-interglacial cycles which have characterized the last few million years.

The reasons for cooling are still under debate. Atmospheric CO_2 concentrations appear to be the proximate reason for the cooling (DeConto and Pollard, 2003), but the mechanisms controlling Cenozoic CO_2 in the atmosphere are still unclear. The plate motions that reorganized the ocean systems also raised mountain ranges, and thus weathering rates. The long-term effect of increased rates of global weathering of mineral nutrients on ocean productivity, and thus sequestration of carbon in ocean sediments, is thought to have been of central importance (Misra and Froelich, 2012; Renaudie, 2016). Cenozoic climate and deep ocean circulation changes also shifted the patterns of nutrient supply and upwelling in the oceans. The stronger equator-to-pole and surface-to-deep temperature gradients and associated ocean density structure created a more strongly partitioned ocean, with localized regions of high productivity, driven by upwelling of nutrients from deeper waters, and broad regions of oligotrophy outside these regions, where strong vertical ocean stratification limited nutrient renewal and confined most primary production to recycled nutrients above the pycnocline (Lazarus et al., 2009). Biosiliceous sedimentation followed a complex pattern of shifts over the Cenozoic (Renaudie, 2016). In the late Neogene alone (Cortese et al., 2004), siliceous sedimentation in the Atlantic largely ceased and switched instead to the North Pacific, together with increased rates of biosilica sediment export in the Southern Ocean, and in newly established regions of mid- to low-latitude coastal upwelling (Benguela, California, and Peru-Chile systems). Diatoms became a major component of the phytoplankon, beginning in the mid-Cenozoic (Renaudie, 2016), in association with the establishment of strong regional upwelling. Diatoms are extremely efficient in removing dissolved silica from the water column, and reduced silica availability affected radiolarians, which produced on average significantly thinner shells (Lazarus et al., 2009 – see Chapter 9: Evolution). It is not known however if this phenomenon had any effect on rates of radiolarian evolution or taxonomic turnover.

Generally, very little change in radiolarian faunas occurred at the family level within the Cenozoic (De Wever et al., 2003), after the substantial family level turnover in the mid-Paleocene to Early Eocene. One notable exception is the origin of the Collodaria, which both molecular and fossil data place within the Cenozoic, even if the timing is uncertain (see Chapters 4: Genetics and 9: Evolution). While some families appear to become more diverse in the later Cenozoic (e.g. Plagiacanthidae), little objective data on these changes has yet been obtained. Most of the evolutionary changes appear to have occurred at the level of genus and species, and are tied to the changes in ocean environments such as the development of polar faunas. These changes and their causes are discussed further in Chapter 9: Evolution.

6 Preservation and Methods

Introduction

Until now we have described living radiolarians and their ecology; in most of the remainder of this book we will describe the fossil record of radiolarians and the various thematic studies made with it. In between the biology and paleontology of radiolarians lie the important subjects of how the living radiolarian biota is transformed by preservation processes into the fossil record, and how this fossil record is recovered for study. Both of these issues strongly affect the type and quality of science that can be done with the fossil record of radiolarians. Further, the processes of preservation, and the methods used to recover the fossils and prepare them for study, are quite different than those in most other areas of paleontology and need to be understood in order to understand much of the rest of this book.

Paleobiology of the Polycystine Radiolaria, First Edition. David Lazarus, Noritoshi Suzuki, Yoshiyuki Ishitani and Kozo Takahashi.

Preservation

Most of the geologic record is made up of small, discontinuously deposited parcels of sediment or rock, which only sporadically preserve within them fossils of organisms. This is due to the complexity of both the fossilization process, which for most organisms is rare, and also the episodic nature of sedimentation on land, or in shallow epicontinental seas. Deposits of sediments are only intermittently laid down during periods of high sea level which cover the land surface with marine waters. Over geologic time, most of the deposits are then lost due to repeated episodes of exposure to erosion during low sea level stands. Radiolarians and other microfossils have also been preserved in such shallower water sediments, in the geologic past when shallow marine seas extended over large areas of the continents. The fossil record of radiolarians from such deposits is similar in quality to those for most other types of fossil material. One of the most important characteristics of the *deep-sea* marine microfossil record for paleobiology, however, is the uniquely continuous process of fossil sedimentation. This happens because 1) fossilization for groups such as radiolarians is comparatively common; 2) there are enormous numbers of specimens involved in the process, so that even a small percentage of preservation means nearly continuous fossilization; 3) very little other material other than microfossils is deposited over most of the ocean floor; and 4) the oceans overlying deep-sea sediments are continuously present. Having a permanently present source of particles results in continuous sedimentation of material, and elimination of subsequent erosion by exposure, as occurs on land. The processes are complex, and the result is not really complete, but it is qualitatively better than any other type of fossil record.

Marine microfossil preservation patterns are strongly influenced by the chemical behavior of their shells. Carbonate-shelled marine microfossils such as planktonic foraminifera and coccolithophores share shell chemistry with the large majority of other types of fossils – molluscs, brachiopods, and the like. Carbonate preservation is strongly depth related due to the presence in the oceans of the lysocline – a depth in the water column above which shells do not dissolve, and below which shells dissolve at an increasing rate, until at sufficient depth no carbonate is preserved – the "carbonate compensation depth" or CCD.

For siliceous microfossils, the patterns of preservation are different, as the nature of the controlling chemistry is fundamentally different. In particular, the entire water column is undersaturated in silica (Figure 6.1), and in general the factors affecting dissolution are not so simple as for carbonate. Silica dissolves at any depth in the water column but the extent of dissolution depends on conditions such as temperature, level of saturation, pH, and at lesser effect pressure (Hurd, 1973). Most significant dissolution takes place above ca. 250 m in the water column, due in part to higher sea-water temperature in the upper layers (Hurd, 1972; Figure 6.2). Siliceous microfossil chemistry is also different in that silica is not sensitive to normal ranges of environmental acidity and thus, unlike carbonate microfossils, do not dissolve from the acidic conditions that occur in the presence of organic matter in sediments. Lastly, biogenic silica undergoes diagenesis quite differently than carbonate. All these differences mean that the processes and patterns of preservation of siliceous microfossils are quite different, as described below, creating a distinctly different fossil record.

Creation of the deep-sea fossil record begins with the production of fossilizable skeletons of plankton by cells living in the water column. Radiolarians are generally continuously present in most ocean waters, although they vary dramatically in absolute

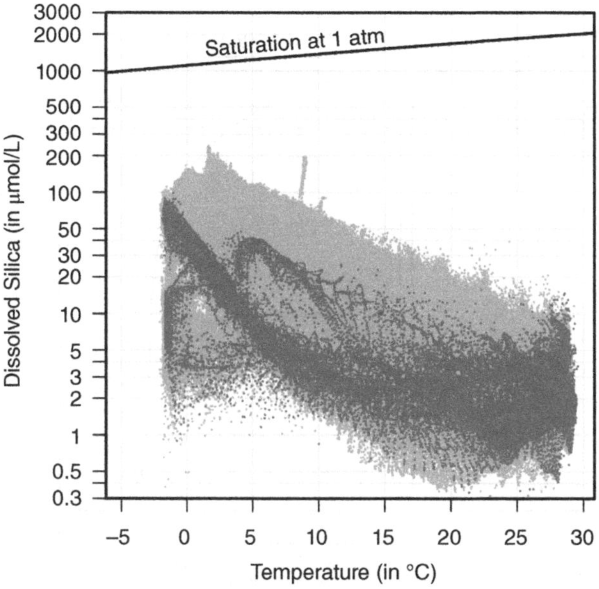

Figure 6.1 Dissolved silica concentrations in global ocean waters in relation to temperature and full saturation value. Note log scale on vertical axis. Darker values are surface waters, lighter shading are subsurface waters. Source: Renaudie (2016), Licensed under CC BY 3.0.

abundance by region (see Chapter 3, Ecology) and also, at least in some regions, by season. Generally, seasonality in productivity is correlated to latitude, which in turn is linked to seasonal supply of nutrients, as described in Chapter 3. Organic particles, including dead, skeleton-bearing radiolarian shells, must first sink below the mixed surface layers as they begin their descent to the bottom. Most organic matter in the oceans is found as particles smaller than 100 µm and is referred to as "suspended particulate matter" and assumed not to sink at significant rates. Only particles >100 µm – called "net plankton" – are thought to form particles that sink significantly (Michaels and Silver, 1988). Individual radiolarian or other protist cells are however mostly less than, or not much bigger than this 100 µm size definition, so that the time it would take to sink is so long (months to more than one year: Takahashi and Honjo, 1983) that they would almost certainly be completely consumed long before reaching the bottom several km below – organic matter, by bacteria, and the silica shells by the undersaturated ocean water (Moore, 1969: See also Chapter 1: History). Most of the marine deep-sea microfossil record is therefore created by material that has sunk much faster as it has been aggregated into much larger, more rapidly sinking clumps of material. These aggregates include fecal pellets, created by larger multicellular grazing plankton or nekton; "marine snow," which are just clumps of organic material in the mm to cm size range; and even larger (10 cm plus) "mats" of sinking organic material created by aggregating of smaller clumps during periods of particularly intense export of material in times of very high productivity (Takahashi, 1981; Takahashi, 1987; Kemp et al., 2000; Hüneke and Henrich, 2011). Such aggregates sink at the rate on the order of 100–200 m d^{-1} and

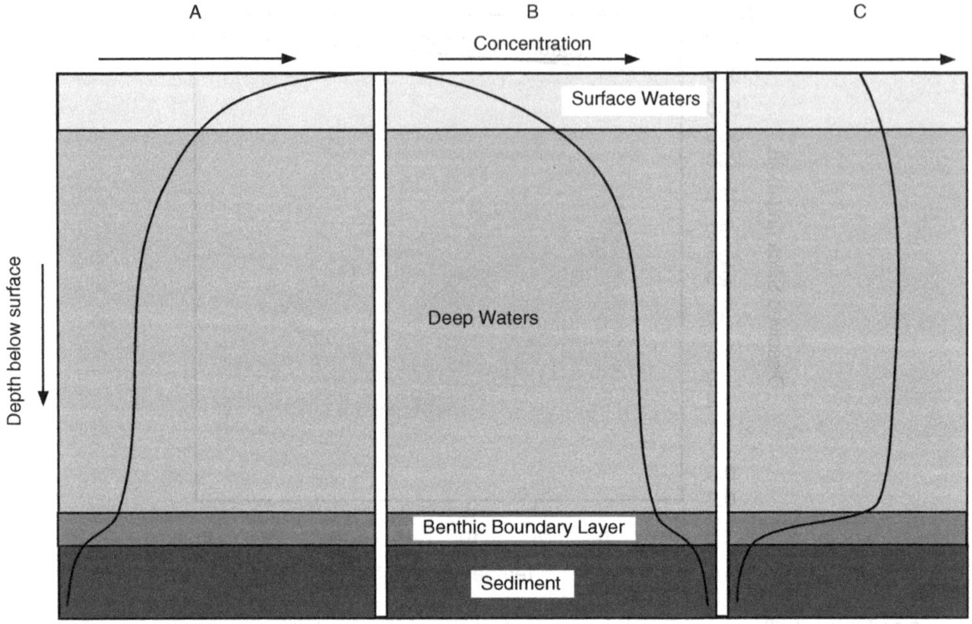

Figure 6.2 Concentrations of key biologically active elements vs. water depth in the ocean. A) Biologic source materials produced in the surface water layer decline in abundance with increasing water depth as they are degraded by biologic activity, chemical dissolution, and other processes. The pattern is non-linear, with much higher rates of decline in upper waters than deep. Relatively intense decline also occurs in the thin benthic boundary layer (nepheloid layer and mixed layer of surface sediment), with only much slower rates of change in the deeper layers of sediment below. Examples of such biologic particles include particulate organic carbon and poorly silicified shells of phytoplankton. B) Dissolved products of degradation of biologic materials in "A" show the inverse pattern of increase. Dissolved silica is an example. C) Refractory components such as the shells of many species of radiolarians show little decay with depth in the water column and may even show slight increase if some production occurs only in deep waters. Most of the decline in these components occurs within the benthic boundary layer, e.g. prolonged exposure of radiolarian shells to bottom waters undersaturated in silica. (*See color plate section for the color representation of this figure.*)

transport even small particles such as coccoliths and diatoms (Takahashi, 1986; Takahashi et al., 1989, 1997).

Due to the limited number of studies plus uncertainties in measurement methods, the contribution of radiolarians to total export flux of biologic material is still very poorly known. Radiolarians make up a significant fraction of the total export from surface water layers, although the amount varies dramatically by region, season, and other factors. In areas of high productivity, polycystine radiolarians can be one of the dominant components of the export particulate flux (Lampitt et al., 2009), and sometimes equaling or even exceeding that of phytoplankton groups (Gutierrez-Rodriguez et al., 2018). However, in most regions phytoplankton dominates, and by the time abyssal water layers are reached a large fraction of the total original flux has been remineralized – much of the organic carbon has been biologically consumed by bacteria and other deep ocean organisms (Michaels and Silver, 1988; Hüneke and Henrich, 2011). In oligotrophic environments radiolarian organic carbon fluxes are important in the upper bathyal region (Guidi et al., 2016) but export into the deepest water layers may only be a few percent of the total deep carbon composition (Boeuf et al., 2019).

Nonetheless, a significant fraction of the total original living material reaches the sea floor, although altered in composition and consisting of mostly "refractory" substances such as resistant organic compounds and skeletal material. In particular, in some regions a very large fraction of the original production of opaline radiolarian skeletal material reaches the ocean floor (Figure 6.2). The opal skeletons have also been altered in complex ways that affect further dissolution: protective surface organic layers have been removed by decomposition, increasing solubility, but adsorption of elements such as aluminum onto the surface have reduced solubility (Hurd, 1972). Radiolarian skeletons become more porous as they partially dissolve but this process seems to be limited, possibly by ion coatings that limit further dissolution (e.g. Fe, Al, Mn: Takahashi et al., 1983).

The biogenic material, having reached the ocean bottom, enters a three-layered transition zone in the preservation process: the nepheloid layer, the mixed-surface sediment layer, and the upper ca. 1 m of sediment below the mixed layer. It is in this transition zone that most of the loss of opaline skeletal material occurs, and the primary characteristics of the fossil record are created. The nepheloid layer is typically a couple hundred meters of relatively turbid water next to the ocean bottom, containing suspended particles falling from above and a variety of material being moved laterally near the bottom due to resuspension of loose surface materials by (generally quite sluggish) bottom currents and minor slumping of these surface materials due to bottom topography. This surficial reworking does not seem to normally erode older material but does affect rates of local sedimentation, an effect known as sediment focusing (Hüneke and Henrich, 2011). In some regions however, stronger bottom currents or more rugged topography (continental slopes, sea mounts) can cause significant erosion and reworking of older sediments into younger ones (Moore et al., 2012). Particle residence times in the nepheloid layer are very poorly known but evidence for at least locally long horizontal transport (e.g. Booth and Burckle, 1976) suggest that particles can spend weeks or even months being moved slowly about horizontally (thus, to some degree, blurring geographic information on scales of < ca. 100 km) before finally settling permanently onto the sea floor. Here they become part of the uppermost mixed-sediment layer, which is actively stirred together ("bioturbated") by the activity of burrowing organisms. Individual particles such as radiolarian skeletons presumably exit from this mixed layer more or less at random when by chance they are left behind in the lowermost layer of sediment and are no longer disturbed by bioturbation. How long this takes thus will vary enormously but the average – a few kyr – can be estimated from the sedimentation rate, typically between 1 and 10 cm/kyr in deep-sea environments, and the thickness of the mixed layer, typically around 10 cm. During this long time interval, particles are still exposed to dissolution by bottom water, which is mixed into and circulates through the mixed layer. This dissolution process is thus critical to forming the fossil record, and is a function of rate of sedimentation or "rain rate," e.g. flux of particles onto the sea floor (higher rates mean less time exposed to dissolution), corrosiveness of the bottom water, and concentration of opal in the sediment itself (more opal particles, particularly easily dissolved ones such as thin diatom shells, act as a buffer that protects other, more robust shells such as radiolarians from dissolving so quickly). This last effect becomes more important in the last of the three regions of preservation – the uppermost sediment below the mixed layer. Here the water in the sediment no longer mixes so freely with the bottom water of the ocean – it is "pore" water, which slowly equilibrates with the sediment particles around it. If there is abundant opal in the sediment the pore waters reach silica saturation, dissolution ceases, and the remaining opal particles have become permanent fossils. If there is very little opal in the sediment to begin with, all of it may still dissolve before the pore waters reach saturation, and no radiolarian fossils will be left. It is thus not unusual to find well-preserved radiolarians in the sediment mixed layer but to rapidly lose all preserved opal material below it. Such conditions indicate that the combination of factors needed to preserve opaline fossils was insufficient. At the other extreme the preserved fossil record can consist largely of opaline sediment (diatom or radiolarian ooze), with as many as 10^6 radiolarian fossils per cm^3. Importantly, the factors that control the creation of this record are well known and based on more or less continuous oceanographic processes. The preservation of radiolarians as fossils is thus not a rare, chance event as is true for many other types of fossil (e.g. vertebrates) but is a process with predictable behavior. This can be used to understand the probabilities of preservation in past oceans, so long as the primary causal factors (sedimentation rates, particle composition, etc.) are known or can be reasonably estimated.

Geographic Variation in Preservation

Deep-ocean waters are not uniform in the concentration of dissolved silica. As deep waters flow from one ocean basin to the next, they gradually accumulate increasing amounts of dissolved silica, and other tracers of biologic export productivity, from the rain of material coming from the surface waters above. In the modern ocean deep waters flow from the North Atlantic to the Southern Ocean, are there remixed, then flow northwards through the Pacific basin to upwell regionally in the North Pacific. Here, deep waters are the oldest and richest in silica (Figure 6.3), and the rate of dissolution of biogenic opaline fossils is much lower than in the silica-poor deep waters of the North Atlantic. These patterns of deep water flow affect the preservation of silica in ocean sediments, as do rates of input of silica from surface productivity. Concentrations of radiolarians in sediment (Figure 6.4) are also affected by dilution from other sedimentary components – carbonate fossils, diatoms, and clastic grains.

Diagenesis

Newly deposited fossils of biogenic silica – diatoms, radiolarians, sponge spicules, or other – are all chemically similar, being composed of amorphous opal: $SiO_2 \bullet nH_2O$, also called opal A. In this substance the silicon and oxygen atoms form a three-dimensional lattice and the water molecules are only loosely bound to it, occupying spaces within the lattice framework. Amorphous silica is not however very stable, and even rather modest increases in temperature and pressure cause it to alter (undergo diagenesis) into a different intermediate chemical form: opal CT or "porcellanite" (DeMaster, 2003). The CT refers to the presence of a mixture of tiny crystals of the minerals cristobalite and tridymite, typically forming minute spheres ("lepispheres"). This transformation is accompanied by loss of water and an increase in density. Radiolarians that have undergone this change lose their transparency, but the visible structure of thc shells is not obviously altered. As diagenesis continues the silica changes

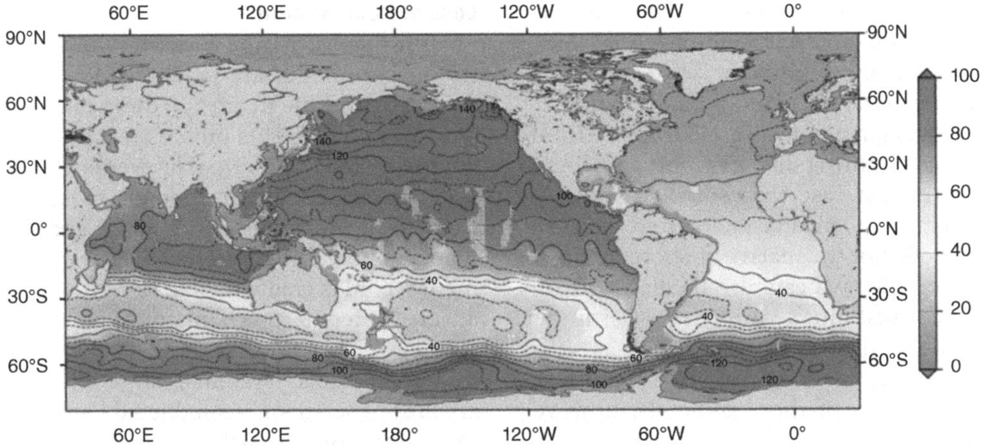

Figure 6.3 Dissolved silica concentration in ocean water at 1 km water depth, showing gradient from North Atlantic to North Pacific due to accumulation along path of deep water flow, and influence of export of biogenic silica into deep water from areas of high export productivity such as the Southern Ocean and equatorial regions. Source: World Ocean Atlas 2018 map. (*See color plate section for the color representation of this figure.*

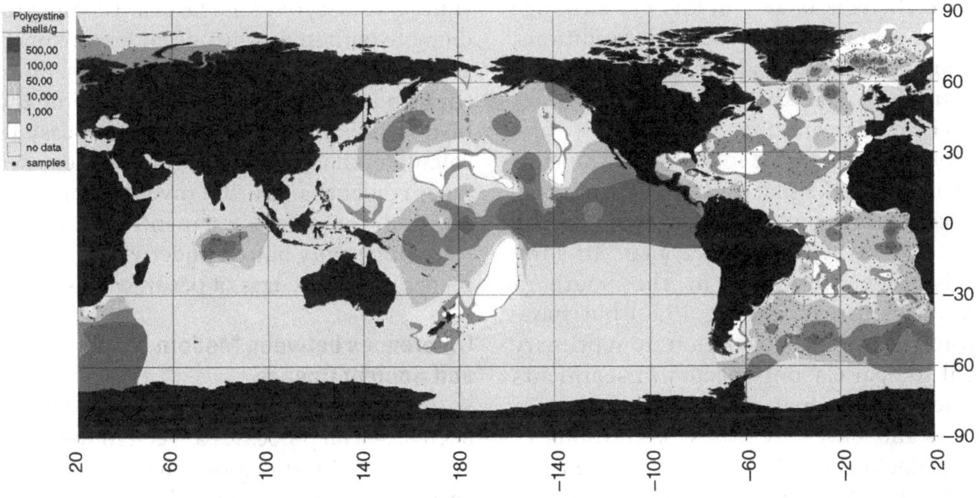

Figure 6.4 Radiolarian concentration in ocean surface sediments on a carbonate free basis. Pattern reflects input flux from surface waters, dissolution in water column and surface layer, and dilution by diatoms, and in some coastal regions, clastics. Source: Boltovskoy et al. (2010). Reproduced with permission of The Palaeontological Association. (*See color plate section for the color representation of this figure.*)

further into the stable mineral quartz. Such microfossil shells are typically composed of a microcrystalline form of quartz, and thus the general morphology of the shell tends to be preserved, although the surface becomes rougher and the crystal surfaces bend light so much that the shells are no longer easily viewed in transmitted light in the microscope. At the same time, dissolution and reprecipitation of silica occurs in the sediments, with the formation of massive layers of chert. While the large majority of microfossil shells are destroyed during this diagenetic process, it is often the case that some of the more robust shells, mostly radiolarians, survive, embedded in the matrix of chert.

Biogenic opal can also be diagenetically transformed into a variety of other minerals including clay minerals such as smectite, resulting in the destruction of the shells. Some of these minerals, however, also preserve, to at least some degree, the morphology of the original shells. The most common of these is pyrite, which tends to destroy fine structures and, in many cases, leaves only a fairly rough cast of the original shell form. The chemical processes that control pyritization of radiolarian shells is largely unknown, although it is associated with reduced conditions, organic matter, and hydrogen sulfide (Afanasieva et al., 2005).

The temperatures, pressures, and amount of time needed for these transformations to take place varies enormously. Porcellanite layers have been found only a few meters below the sea floor in Plio-Pleistocene sediments in the Southern Ocean (Bohrmann et al., 1992) but more typically, the transformation from primary opal to opal CT only occurs in sediments at several hundred meters below the sea floor, and only after tens of millions of years (DeMaster, 2003). Thus, for the most part, the Neogene, and most of the Oligocene deep-sea fossil record consists of primary opal skeletons. Eocene and older sediments show more and more diagenetic imprints, while Mesozoic and older radiolarians have usually been transformed into quartz. These differences have, as discussed below, a profound effect on the quality of the fossil record of radiolarians and the methods used to study the skeletons as well.

Loss of Rock Record

Shallow-water marine fossiliferous deposits on continents are gradually lost over geologic time due to erosion during periods of low sea level. This process results in an approximately log-normal (power law) relative abundance distribution of preserved rocks vs. geologic age (Willenbring and von Blanckenburg, 2010). The deep-sea sediment record is gradually lost as well, but by a very different mechanism: subduction. As new ocean floor is created along the mid-ocean ridge systems, an equivalent amount of sea floor is subducted back into the interior of the earth, in the last ca. 100 my, mostly at the edges of continents around the Pacific and eastern Indian Ocean. Although global rates of sea floor spreading have varied slightly over geologic time, the magnitude is fairly small so that this loss of record is approximately constant, and also, over the last ca. 100 my predictable both in magnitude and in the main geographic locations where it occurs. From this, and more detailed knowledge of the history of plate tectonics, it is also easy to calculate that very little of the deep-sea record older than ca. 200 my is preserved at all, and these few fragments of oceanic crust are deeply buried under younger sediments near the edges of ocean basins.

Differences between Modern and Ancient Oceans

So far, our discussion of preservation has been based on processes as seen in the modern ocean. In the more distant geologic past, oceans were somewhat different in character, in ways that may have affected preservation. Most importantly, prior to the late Cretaceous, or even middle Eocene, marine diatom abundances were much less than is true today (Renaudie, 2016), and thus the removal rates of silica were also less (Racki and Cordey, 2000; Lazarus et al., 2009), resulting in much higher concentrations of dissolved silicon in sea water. Rates of opal dissolution were much lower, and a much higher fraction of the silica removal in the silica cycle was carried out by radiolarians. Ocean-bottom waters were also often much warmer (ca. 10°C vs. ca. 2°C today), and rates of ocean turnover slower. All these probably did not fundamentally alter the basic patterns of opal preservation (see discussion in Chapter 7) but almost certainly would have affected absolute values – percent preservation, degree of bias in preservation, etc. Analogies to modern oceans should be made with caution for early Mesozoic or Paleozoic materials.

Quality of Radiolarian Fossil Record

The incompleteness of the fossil record has been long known (even Darwin acknowledged it), but just how incomplete is essential knowledge if we are to effectively use it. The fossil record of radiolarians of course is only of their skeletons (and that only for the polycystines that are the subject of this book): the majority of biologic information – cellular structures, genetic material, behavior – are all absent. Further, and unlike the situation for other groups of organisms, we know so little about the function of radiolarian skeletons that we cannot even indirectly infer much about the biology of fossil forms from the skeletal structures preserved. Despite these weaknesses, the fossil record of radiolarians also has significant strengths, which have been mentioned above, but which are gathered together here. The main points are (see also Lazarus, 2011 and Chapter 9):

Large numbers of specimens: Although restricted to unlithified (mostly deep-sea) sediments and a subset of other rocks, within these, it is easy to collect virtually unlimited numbers of specimens for study. There are many km of exposed sections for study in any given geologic interval, from which thousands of samples can be taken, and from each of these a vast number of specimens extracted. A pure radiolarian ooze may contain $>10^6$ specimens/cm^3, and more typical diatom oozes $10^4/cm^3$. Even lithified rocks from Mesozoic epicontinental seas often have $>10^3$ specimens/cm^3 (Empson-Morin, 1984). This does not mean that unlimited numbers of specimens are actually available for every research question. Many species are relatively rare (see also discussion in Chapter 9, Macroevolution). Further there are currently practical limits on how many specimens a single researcher can examine (up to ca. 500,000/year, working full time on observation: e.g. Renaudie and Lazarus, 2013), although this may change with automated systems (discussed below). Still, the abundance of material vastly exceeds that available for most macrofossils, and particularly the material available to the study of larger organisms such as vertebrates.

Continuous preservation in local sections: This is primarily true for the deep-sea record but is also largely true for deeper-water sections from epicontinental seas, where a continuous, if seasonally varying, input of shells into the sediment occurs. This contrasts to the much more episodic, and hard to determine, frequency at which many other types of fossil accumulate. The major limitation however is the blurring effects of bioturbation in the sediment mixed layer, together with lateral transport and reworking of surficial material in the nepheloid layer. Unlike macrofossils which are often deposited in beds which, if not actual life assemblages, are accumulations representing generally only a few decades (Behrensmeyer et al., 2000), marine microfossils, and radiolarians, are generally mixed on time scales of hundreds or even thousands of years. This has been extensively studied by paleoceanographers, who are also deeply interested in the resolution of these records. Using such things as dispersed ash layers found in sediments, which were originally very short-term (ca. 1 year) events, it has been shown that the original signal is smeared by deep-sea sedimentation over a period of thousands of years, in sediments with typical sedimentation rates of ca. 3 cm/kyr (Peng et al., 1979). There is thus, in paleoceanography, a constant search for sections with relatively high rates of sedimentation (10 cm/kyr or more). This is to improve temporal resolution, and thus study processes at time scales of hundreds of years, most importantly to understand future climate change. There is, in addition, blurring of the record over longer time scales due to reworking of specimens from older layers into younger ones. This happens simply as a consequence of having such an extremely dense fossil record with vast numbers of specimens – reworking of sediments and rocks occurs of course also in other areas of paleontology, but it is very rare for this to capture, preserve, and redeposit a fossil from an older into a younger time interval. In microfossil-rich sediments it is common, and in some regions seriously affects the stratigraphic integrity of the record (Moore et al., 2012), though the very large majority of deep-sea sediment samples (>95%) have no detectable reworking in them (Lazarus, 2011).

Good preservation of original diversity: The fossil record preserves biologic diversity quite imperfectly, particularly at the species level. It is clear that simply comparing the numbers of described living species (at least a few million) vs. fossil species of all life (e.g. ca. 400,000 taxa, all ranks, in the community Phanerozoic fossil database PBDB, over >500 my) that only a very small fraction of now-extinct species have been preserved and recovered for scientific study. Further, there is extreme bias in preservation, e.g. only a few groups of plankton have any fossil record at all (Lazarus, 2011). Further, among the plankton that do have fossil records, some preserve only a fraction – generally less than 50% – of the species, based on comparisons of the living assemblages to those preserved just below the mixed layer in deep-sea surface sediments. Using the same comparative procedure, the large majority of radiolarian species (>90%) that are known to live in the plankton are also found in surface sediments, making the radiolarian record of species diversity remarkably complete. This statement must be qualified by noting that the percent estimate is only a rough one, as the species taxonomy of the living assemblage is still not fully known (see Chapter 5), and that some species, while preserved, are so fragile that only rarely are specimens found that are so complete that they possess the morphologic details needed to identify them. Further, diagenesis and other factors tend to reduce species diversity in sections from epicontinental seas. Empson-Morin (1984) for example, found only a fraction of the species diversity in most land sections from non-pelagic sediments, which she attributed to both small numbers of individuals and fundamentally lower preserved or primary diversity in the land section samples. Still, the completeness of species diversity preservation is, at least for deep-sea sediments, an order of magnitude better than the average for the fossil record. A last caution here is that however good the actual preservation of diversity, it must have been accurately recorded before it can be analyzed. As noted in Chapter 5, radiolarian taxonomy is quite incomplete, and much of the available data only reports a subset of species, e.g. for paleoceanographic study. These observational imperfections complicate using the radiolarian fossil record for things like paleobiodiversity research (Lazarus, 2011; see also Chapter 9, Macroevolution).

Good recovery of primary biogeographic diversity: One of the main reasons for poor preservation of species diversity in most groups of organisms is that their biogeography is complex, with many distinct geographic ranges on land or along coastlines, only a few of which are, at any given time, located in regions suited to preserving specimens as fossils. With marine plankton the situation is quite different. As noted in Chapter 3, there are remarkably few biogeographic distribution patterns in radiolarians, each tied to the vast patterns of open ocean surface circulation. It is thus possible for a relatively small number of geographic locations to provide at least one geologic section with, for any given time in the geologic past, at least one sample of the species that were present from each biogeographic range (Lazarus, 2011). This completeness declines with increasing geologic age, and in e.g. the early Paleozoic it is likely that geographic coverage is extremely incomplete, adding to other problems of diagenesis and loss of sections by erosion in such old material. However, in younger time intervals where significant deep-sea sediment exists, geographic preservation is generally excellent. There is nonetheless a substantial bias towards younger material even in the Cenozoic deep-sea record. This is due to the (in the deep-sea, weak) tendency of older sediments to be eroded, gradual loss by subduction, but most significantly by the sampling process, which has to penetrate younger material to reach older, and thus recovers relatively little older material (see Methods below). All these factors together result in a typical exponential decrease in relative abundances of samples with increasing geologic age (Figure 6.5: Lazarus et al., 2014). A similar pattern exists for other types of geologic sections, e.g. Phanerozoic shallow marine, albeit with different constants in the best fitting equation (Willenbring and von Blanckenberg, 2010).

Preservation of relative abundances: In those areas of the ocean with seasonal patterns of productivity and export, the sedimentary signal is biased towards times of high export and other times of the year are underrepresented in the sediment. In general it can be concluded that species that are usually rare or quite common in the fossil record were also relatively rare or common in the plankton, but less dramatic differences in relative abundance in fossil data are not a reliable guide to the original relative abundances of species in the plankton, as shown by differential dissolution in modern ocean studies (Figure 6.6). The differences between living and fossil abundances increase as preservation declines: in areas of the modern ocean where radiolarians are very poorly preserved the diversity of species found in the sediments is much lower than in the living assemblage in the overlying ocean waters, and is composed of species with robust, dissolution-resistant shells. Diagenesis further preferentially removes more delicate forms, so that Cenozoic assemblages from deep-sea sediments are generally significantly better, if imperfect, recorders of relative species abundances than older assemblages from lithified rocks.

Methods

The study of radiolarians makes use of many methods, from initial observation or collection of material, to sample preparation, imaging, data collection and analysis. Analytic methods are covered in the relevant chapters later in this book, e.g.

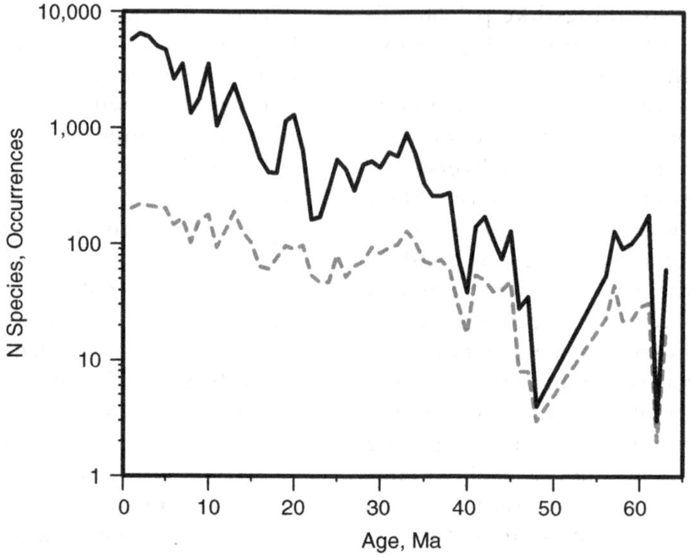

Figure 6.5 Number of diatom occurrence records (solid line) and number of species (dashed line) in the NSB deep-sea drilling marine microfossil database, as of fall 2013. Pattern of declining data density with age is similar for all fossil groups including radiolarians. Note log vertical axis and more rapid decline of occurrence data with time than species diversity; also, as is typical for raw diversity data, close correlation to number of occurrence data records. Source: Lazarus et al. 2014. Licensed under CC BY 4.0.

transfer functions and factor analysis in Chapter 7 - Paleoceanography, subsampling procedures in Chapter 9 - Evolution. Here we cover all methods from initial collection of material through the various steps needed to make data collection possible. We follow a roughly chronologic (from the point of view of the radiolarians) organization, from collecting living material through sedimented material, radiolarians from older sediments and rocks, and then the various procedures in the laboratory needed to prepare and image specimens.

Collecting Material from the Water Column

Field observation, without collection, of living radiolarians in the water column has until now not been possible due to their extremely small size, and the difficulty in identifying species using skeletal characters when these are embedded in and partially obscured by living cell protoplasm. Many promising methods have been developed in recent years to survey plankton in ocean waters (Lombard et al., 2019) but, with the exception of environmental barcoding of DNA/RNA (see Chapter 3) these have yet to be applied in the study of polycystine radiolarians in any substantial way.

There are three standard ways of sampling radiolarians from the water column: plankton tows, bottle casting, and sediment traps. The first requires use of plankton nets (Figure 6.7). These are towed behind a slowly moving ship, or even, particularly to recover living individuals with minimum damage, left to collect water via currents moving through a net attached to a stationary ship (Matsuoka et al., 2017). Because of the small size of some of the radiolarian taxa a standard plankton net for phytoplankton (i.e. 100 μm in mesh size) is

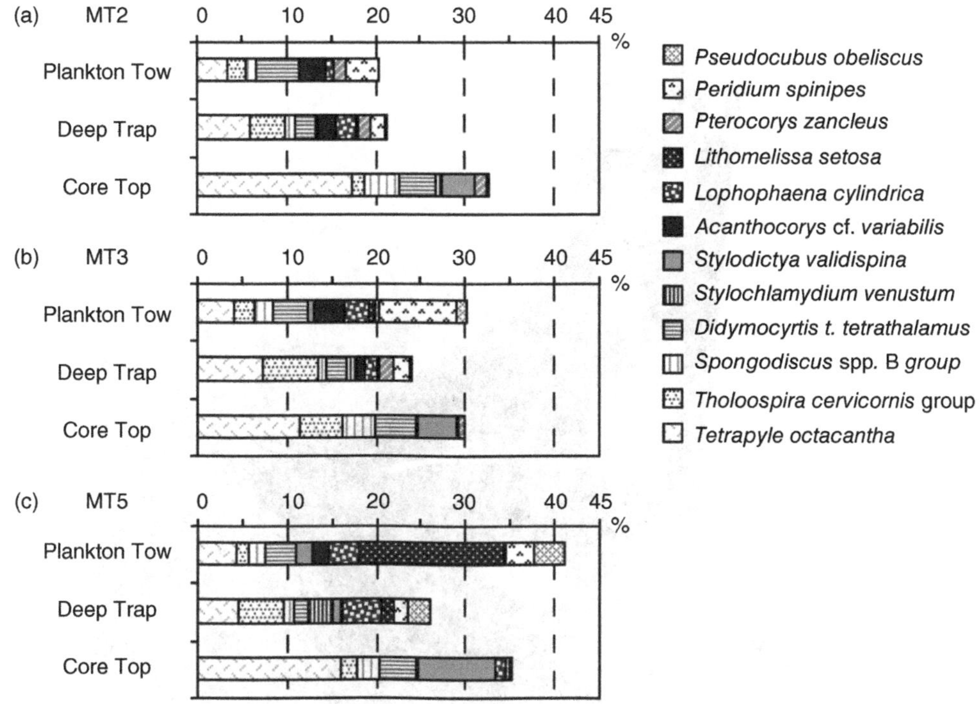

Figure 6.6 Change in relative abundance of selected radiolarian species between surface waters (plankton tow), deep ocean waters (deep trap), and in the surface sediments of the ocean floor (core top), from three different locations in the equatorial Pacific. Source: Takahashi and Yamashita, 2004. Reproduced with permission.

insufficient for radiolarians. Historically, most radiolarian workers used a mesh size of ca. 63 μm for sampling radiolarians (e.g. Kling, 1979; Kling and Boltovskoy, 1995; Boltovskoy and Riedel, 1980; Yamashita et al., 2002; Tanaka and Takahashi, 2008; Ikenoue et al, 2015). This is because that 63 μm was a standard mesh size to separate silt and sand and thus sieves of this mesh size were readily available in geoscience. Another reason for the still conservative use of 63 μm mesh size today is collaboration with other scientists who use this mesh size for other biota. Tanaka and Takahashi (2008), for instance, employed a vertical multiple plankton sampler (VMPS: mouth diameter 1 m; mesh size 63 μm) and sampled 15 intervals from 0–25 m down to 2500–3000 m. However, as living radiolarian studies have evolved, the need for further detail suggests using a mesh size of ca. 45 μm in order to recover the smallest-sized radiolarians, including juvenile forms (e.g. *Amphimelissa, Lithomelissa,* and *Antarctissa;* Nimmergut and Abelmann, 2002). Earlier, Renz (1976) used 35 μm mesh size sieves for pumped water samples.

The second main method to obtain radiolarians is bottle casting. Typically, one can obtain a parcel of water employing Niskin^R bottles or similar samplers. These are essentially cylinders which can be capped, initially left open at both ends, attached to a cable towed in the water column. They are closed by the spring-loaded

Figure 6.7 Plankton net sampling device, in this case featuring multiple nets that can be sequentially opened and closed ("MOCNESS"). (Source: LTER Network. Licensed under CC BY-SA 4.0.)

end caps, generally by sending a small weight (a "messenger") sliding down the cable which trips a lever and causes the caps to close. Because particles collected by bottles tend to settle towards the bottom of the sampler due to gravity, immediate sub-sampling from the bottle onboard is a standard way to ascertain that one gets a representative portion of a sample. The sub-sample must also be either sieved or membrane filtered on board immediately so that appropriately small amounts of the final sample are obtained for transportation to a shore laboratory. The filtered and oven-dried sample at ca. 40–50°C is good in preserving specimens since dissolution does not proceed in air. This is in contrast with the situation for liquid samples where one must take steps to preserve the integrity of samples (either or both of skeletons/organic matter depending on purpose).

Use of a pump is another way of obtaining radiolarians in lieu of bottle casting followed by filtration on board. Renz (1976), as mentioned above for example, pumped waters from 0–250 m depths and sieved through 35 μm mesh on board.

Lastly, sediment traps (Honjo and Doherty, 1988) are used to capture senescent or dead radiolarians and other material as it settles through the water column towards the sea floor (Figure 6.8). These are relatively complex devices requiring substantial ship technology to deploy, and

Figure 6.8 Sediment trap (model). Large cone-shaped trap ends in opening leading to a rotating set of collection bottles at the base of the device. Floats attached to the sides provide near-neutral buoyancy. Source: Wikipedia.

can either be free floating or moored to the ocean bottom. Moored traps, which can be set out in a vertical series, are normally, to avoid problems with surface currents and waves, placed well below the surface layer. To avoid contamination from suspended particles in the nepheloid layer near the ocean bottom they are generally are at least a couple hundred meters above the sea floor. Mechanisms can be attached that rotate a series of collecting bottles at the bottom opening of the trap, e.g. monthly, providing a time series of radiolarian and other particle fluxes into the deep-sea.

Collecting sediments

Once radiolarians reach the ocean bottom and, escaping dissolution, become part of the sediment they are considered to be fossils and at this point normally only skeletal material can be recovered. Because radiolarians in modern oceans are relatively rare in shallow shelf waters, and tend to dissolve in sediments with high non-biogenic opal overall composition, most radiolarian-bearing sediments are found only in relatively deep ocean waters – hundreds of meters to several kilometers water depth. Recovering sediment from such great ocean depths is challenging and is a major factor limiting the scope of research possible on fossil radiolarians from the deep-sea sediment record. Early ocean expeditions were able only to recover small volumes (cm^3's) of material from the surficial layer of sediment using simple metal probes, or by simply scraping mud adhering to cables or other equipment that had been lowered and recovered from the ocean bottom (Figure 6.9). This type of ocean sedimentary material was mostly all that was available to early researchers such as Haeckel. With the availability of onboard ship electrical power, it became possible to more effectively lower and recover much heavier types of bottom samplers and the much heavier, stronger cables needed for this. Larger volumes of surface sediment were possible using a variety of surface sediment samplers, although this had little consequence for radiolarian sampling as only a few cm^3 volume is needed in any case. More important were coring devices – hollow metal pipes with weights attached to the upper end which could be dropped into the sediment to recover a core – a thin (only a few cm in diameter) cross-section of the sediment penetrated. Coring technology advanced with the development of piston corers, particularly by Kullenberg, which compensated for pressure buildup inside the coring tube and could take significantly

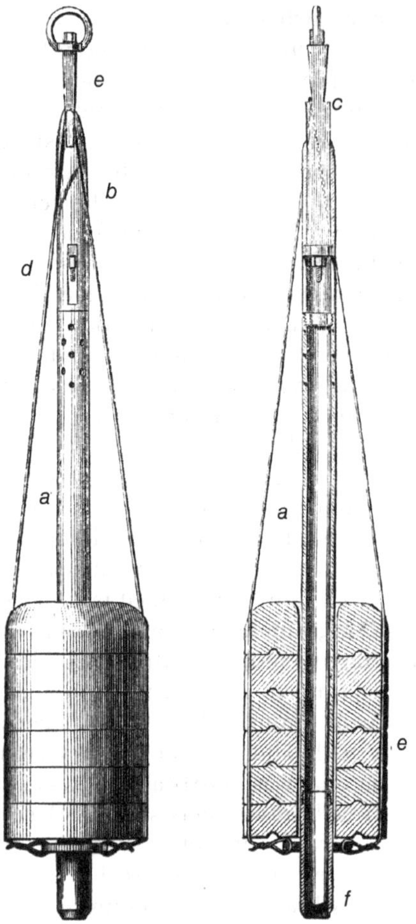

Figure 6.9 Typical sediment sampler used by the H.M.S. Challenger expedition, being a short gravity core combined with a weight and used both for depth sounding and sampling. Public domain. Source: Thomson and Murray, 1885.

longer, less disturbed cores (Kennett, 1982; Rothwell and Rack, 2006; Figure 6.10). Taking a piston core from a ship requires some time (usually an hour or more), as the corer has to be winched to near the ocean bottom prior to release, then winched back to the surface. The ship must also be nearly stationary. These time and motion constraints are often in conflict with other ocean-going science needs – slowly moving ships for plankton sampling, or to take geophysical measurements, and in more recent years piston coring has become less common as a result. Nonetheless, routine sampling with piston corers, particularly in the early decades of ocean exploration, has resulted in a global inventory of thousands of piston cores, held in marine institution sample repositories (Figure 6.11), from which an exceedingly detailed, geographically extensive picture of fossil radiolarian occurrences can be extracted. Even so, pipe-type corers are usually limited to cores of ca. 20 m in length (a few special systems manage up to 60 m) and thus can recover only the uppermost layers of the sediment column of the ocean bottom – the last few hundred thousand to (rarely) few million years of sedimentation.

The majority of the geologic time held in the deep-sea sediment record has been recovered instead by a different technology: deep-sea drilling. This method makes use of more or less standard oil-drilling technologies, which have been slightly adapted to work from a ship in the deep ocean. The most important changes are the use in most cases of a single drill pipe without a surrounding hole (as would be the case on land) or a secondary enclosing pipe which connects the hole in the ocean bottom with the ship at the surface (a "riser" pipe), either of which would normally be used to control the fluids circulating at the bottom of the hole where the drill bit is cutting into the sediments. Such "open hole" drilling makes it much more difficult to accurately control the drilling operation and thus makes coring more difficult and prone to technical problems. The other is the addition of additional motors to the ship (thrusters) which provide positional control, so the ship remains stationary above the drill hole, regardless of wind, wave, or current conditions. This is needed to prevent the long length of pipe connecting the ship to the top of the hole (the drill string) from bending or

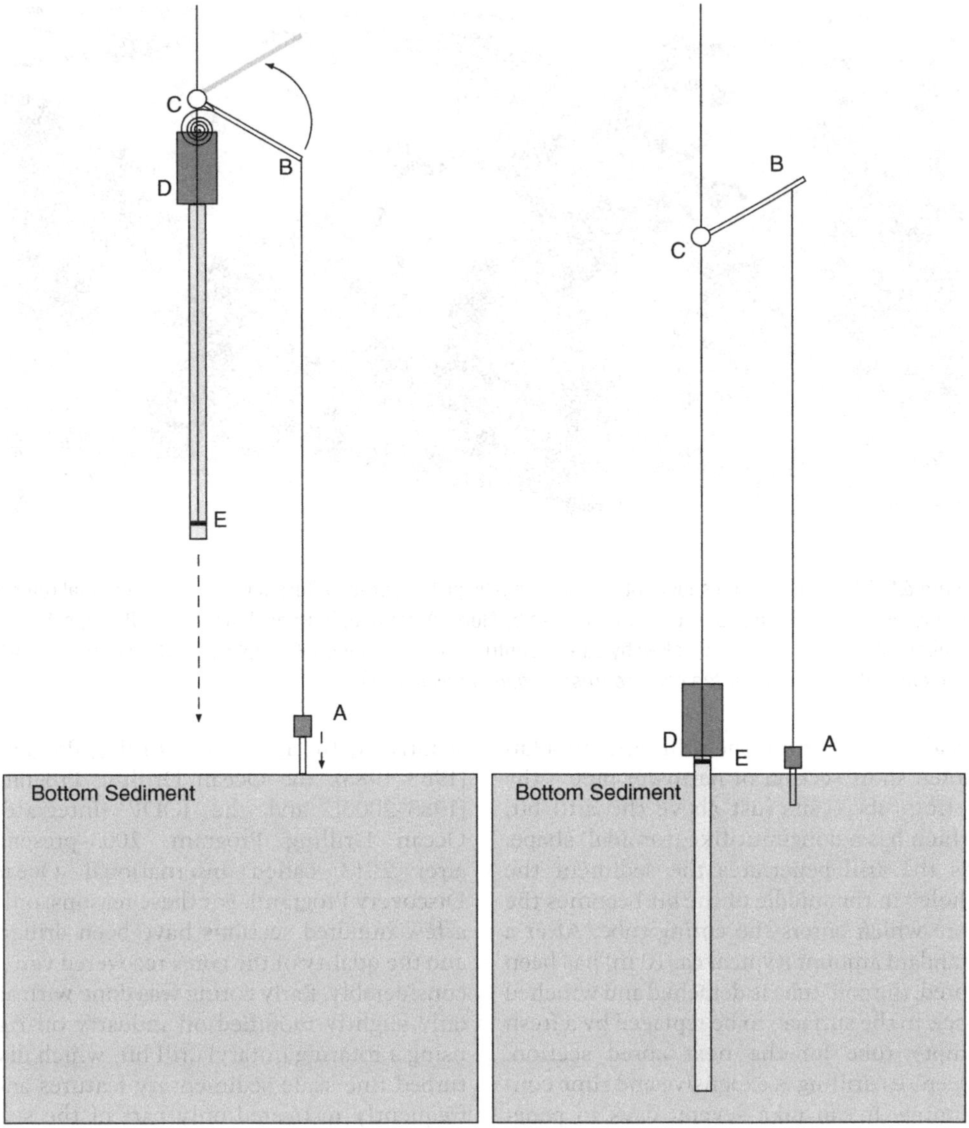

Figure 6.10 Piston corer. Left: Corer at moment of reaching bottom. Cable attached to small trigger weight A (with simple core tube attached below it) slackens, allowing spring-loaded lever B to rise and release catch mechanism C. Piston corer with heavy weight D falls free to bottom. Right: carefully measured (vs. trigger cable) coiled main corer cable (C) tautens as piston E reaches bottom. Corer falls "around" fixed piston/cable into sediment. On retrieval, sediment inside corer can't easily fall out as top of corer sealed by piston E and any downward movement of sediment would create vacuum between top of sediment and piston.

Figure 6.11 Map of deep-sea sediment samples, mostly piston cores. Different color dots in original refer to curation institution. Map shows only cores taken by North American, German, France and UK ships. Similar numbers of samples have been taken by other countries. Source: National Geophysical Data Center, NOAA. (*See color plate section for the color representation of this figure.*)

breaking off. Within the drill pipe, an additional short section of narrower pipe – the coring tube – sits just above the drill bit, which has a doughnut-like (toroidal) shape. As the drill penetrates the sediment the "hole" in the middle of the bit becomes the core which enters the coring tube. After a standard amount (typical ca. 10 m) has been cored, the core tube is detached and winched back to the surface, to be replaced by a fresh empty tube for the next cored section. Deep-sea drilling is expensive and time consuming. It can take several days to penetrate several hundred meters into the sediment to recover a single section, and the ships adapted to scientific needs are very special – currently only two exist: the Joides Resolution, and the Chikyu. Both ships are successors to the Glomar Challenger (see Chapter 1: History). Each cost several hundred million $US to build, and cost millions per year to maintain, by a long-running international consortium of countries, under the management of, consecutively, the Deep Sea Drilling Program (1968–1983), the Ocean Drilling Program (1983–2003), and the IODP (Integrated Ocean Drilling Program, 2003–present, after 2013 called International Ocean Discovery Program). For these reasons, only a few hundred sections have been drilled, and the quality of the cores recovered varies considerably. Early coring was done with an only slightly modified oil industry oil rig, using a rotating (rotary) drill bit, which disturbed fine scale sedimentary features and frequently recovered only part of the section of each core. Major improvements included the development of improved "heave compensators" which made it easier to drill continuously in sediments, instead of having the drill pipe churn up and down the hole as surface waves moved the ship. An even more important improvement was the development of the hydraulic piston corer (Figure 6.12), which took sediment cores similar in style and quality (Figure 6.13) to those obtained by the classic

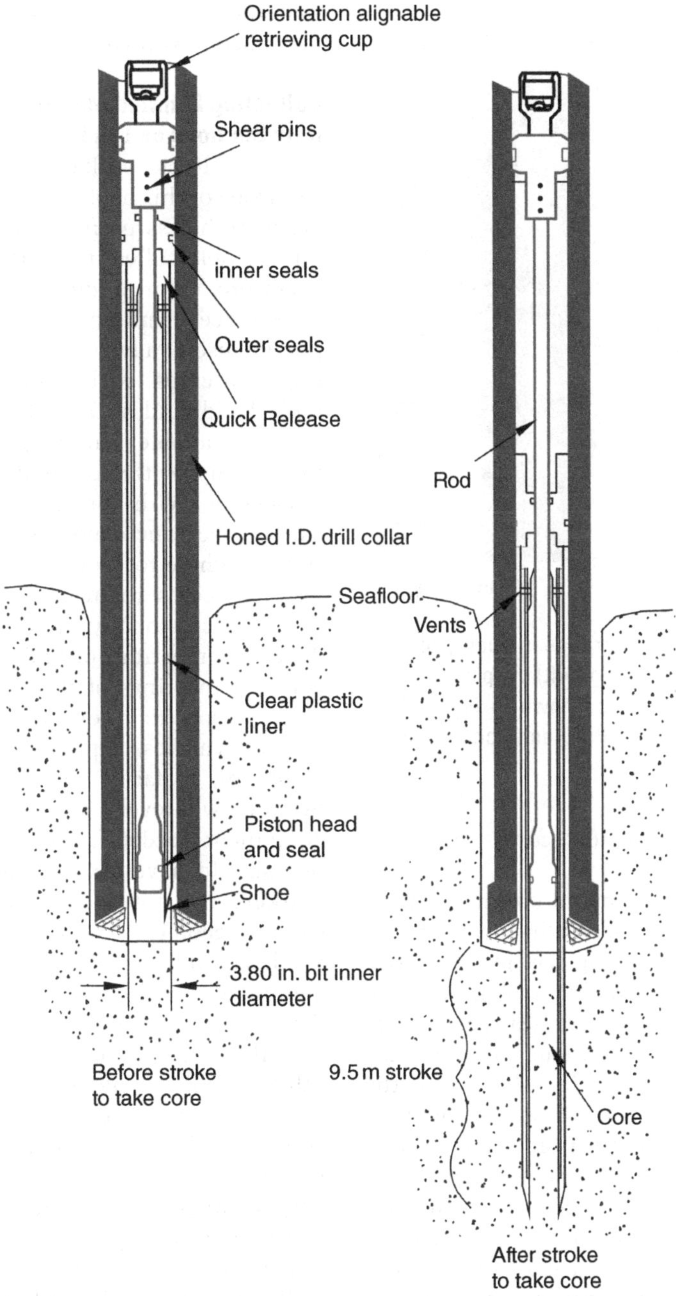

Figure 6.12 Hydraulically driven piston corer for use inside drilling pipe assembly. Although driven by water pressure instead of a heavy weight, the piston principle for coring and recovery is the same as for classic corer of Figure 6.10. (Source: Ocean Drilling Program. Licensed under CC BY 4.0.)

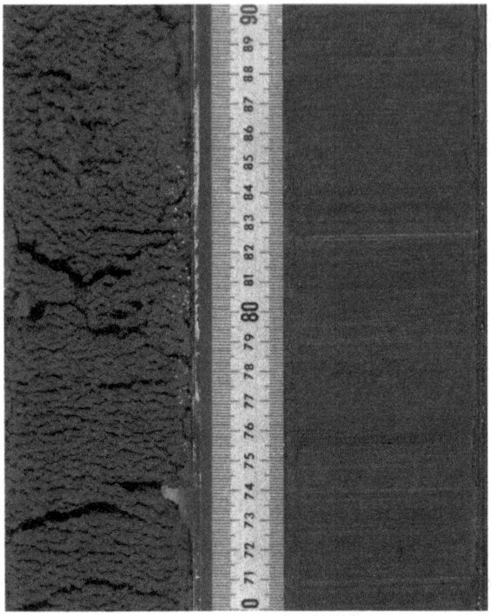

Figure 6.13 Typical difference in quality of sediment recovery by drilling (left) vs. piston coring (right) using the device shown in Figure 6.12. (Source: Ocean Drilling Program. Licensed under CC BY 4.0.)

Kullenberg-style piston corer shown in Figure 6.10.

Despite limitations on coverage and in earlier sections on quality, collectively deep-sea ocean drilling has provided reasonably good geographic and geochronologic coverage of the deep-sea sedimentary record for most of the Cenozoic (0–65 Ma), and at least some sections back into the mid- or even early Mesozoic. Equally important, radiolarians from deep-sea sediments are, for the most part, much better preserved than those recovered from lithified rock sections on land (see Diagenesis and Loss of Rock Record above), so that not only the geographic-temporal coverage, but also the accuracy of diversity and other characteristics of the assemblages are unusually good. The deep-sea drilling archives thus form one of the primary sources of material for many types of research, as described in subsequent chapters.

Collecting Lithified Material from Sections on Land

Geological field work is a subject outside of the scope of this book, but a few comments are nonetheless useful. The starting point for any field collecting work is usually the literature. Prior studies of radiolarians or other siliceous microfossils from the region provide useful advice on where to find geologic outcrops with radiolarian bearing rock. Field guides to the local geology are often available from geologic societies – many written originally for field trips associated with professional meetings. These give much more specific information on where rock outcrops can be found, how to access them, and other practical details. Occasionally these were even produced for meetings of radiolarian workers (such as meetings of the Interrad society) and give specific information on the radiolarian localities. Equipment for geologic collecting is fairly simple (Figure 6.14). Beyond whatever special clothing and equipment needed for the type of terrane, one should have sample bags, waterproof markers and notebooks, measuring stick, camera, geologic hammer, hand magnifying lens, and a sturdy pack to carry both equipment and samples. In addition, depending on the region and time required to collect all material desired, a variety of other supplies may be needed: while in most areas of the European Alps one can stay in hotels, and even enjoy excellent restaurant lunches while out collecting each day, in other regions camping equipment and complex transport logistics are required as part of the sampling effort (Figure 6.15). Collecting samples in the field also requires placing the samples in the proper geologic context, e.g. the location, geologic formation, and position within the rock exposure. Detailed information is provided in

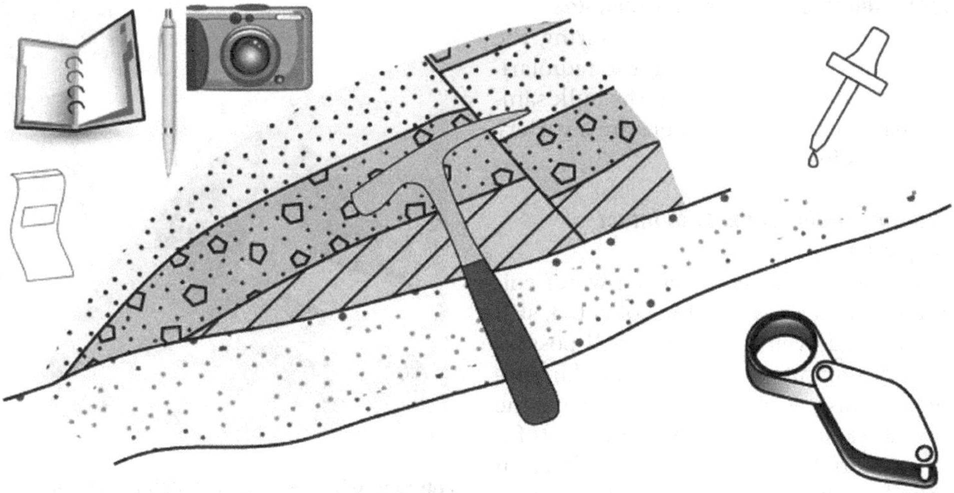

Figure 6.14 Equipment for collecting rocks in the field is fairly simple: usually a rock hammer and a simple hand magnifier to help determine the lithology is all that is required; plus equipment to etch rock surfaces, revealing dissolution resistant radiolarians as tiny bumps.

Figure 6.15 Camping is often part of collecting radiolarians from land sections. B. Nazarov on a collecting trip. Source: Courtesy of M. Afanasieva.

general guides to fieldwork, e.g. Coe (2010). Selecting specific lithologies is very dependent on the area and only a few general suggestions are possible (for details see Blome et al., 1995, De Wever et al., 2001). However, in many sedimentary rock formations carbonate or phosphatic concretions can be found. These are thought to have formed early in sediment diagenesis and often contain better preserved radiolarians than found in the general rock of the rest of the formation (e.g. Blome, 1985; Kiessling, 1999). Etching rock surfaces in the field substantially increases the chances of selecting samples with useful radiolarians. A small bottle of weak acid (e.g. 15–20% HCl) with a dropper is useful for etching the surfaces to see if they show the tiny bumps that may indicate the presence of radiolarians, as is a hand magnifier. Although hydroflouric acid in principle does a better job than HCl in etching surfaces (De Wever et al., 2001), it is too dangerous to use in field work – even a small drop, if not treated immediately can cause serious injury.

Recovering Radiolarians from samples

After collecting but before mounting the radiolarians for examination, the radiolarians need to be extracted from the bulk samples that contain them. There are a variety of methods for this.

Extracting radiolarians with intact protoplasm

Radiolarians collected from the water column may, at least when first collected, still be alive, and even if not may still possess an intact protoplasm with all cellular contents, not just, as for fossil radiolarians, the siliceous skeleton. Preparation methods for such material depends on the collection method and also the intended use. Radiolarians that are meant for live studies must be collected using the gentlest of possible methods as the external cellular structures – axopods and others – are easily damaged. Specimens collected by hand into jars is the gentlest method but this normally requires diving equipment, and only a few larger solitary species or colonial forms can be seen and thus sampled this way (Swanberg, 1979; Swanberg and Harbison, 1980). More generally, material from plankton sets or bottles are the source samples. The relatively large volume of water from the sample needs to be examined for specimens and these extracted into different containers for study, as the mixture of material in the bulk plankton sample includes highly active nekton (copepods) and other organisms whose movements would interfere with observational work. Simple reflected light stereo "picking" microscopes, or biologic inverted microscopes, are best suited – the latter provide greater magnification and are more suited to smaller species, e.g. many nassellarians. Radiolarians are selected using a pipette, either with a rubber bulb, or a standard biologic micropipette system with a holder (Figure 6.16). Specimens are transferred to trays with small volume chambers filled with fresh seawater. Specimens may be re-selected and transferred again to insure that the specimen is not contaminated by extraneous material. A detailed guide is given by Matsuoka et al. (2017).

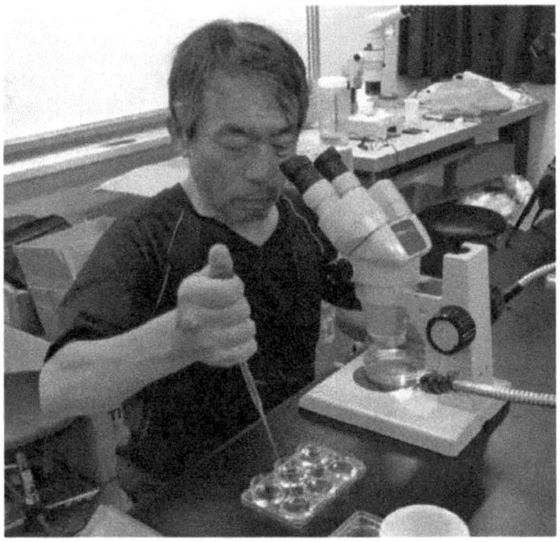

Figure 6.16 Concentrated plankton material from a net tow is examined using a stereo microscope and radiolarians seen are removed using a pipette into separate culture chambers for detailed study. Source: Matsuoka et al. (2017).

Specimens meant for study of cellular structure where the living state is not required, e.g. thin section studies using transmission electron microscopes (TEM) or DNA analysis, are treated differently. These are generally picked, as above, but then fixed with pure ethanol or other reagents and stored for further processing.

Extracting radiolarian skeletons

A large body of knowledge has been developed on how best to separate radiolarian shells from the matrix of other materials they may be embedded in. The literature documenting the methods is scattered and incomplete, with many "recipes" being transferred between workers informally. We provide a simple flow chart that illustrates some of the possibilities (Figure 6.17), and describe additional procedures below, but even this is far from exhaustive.

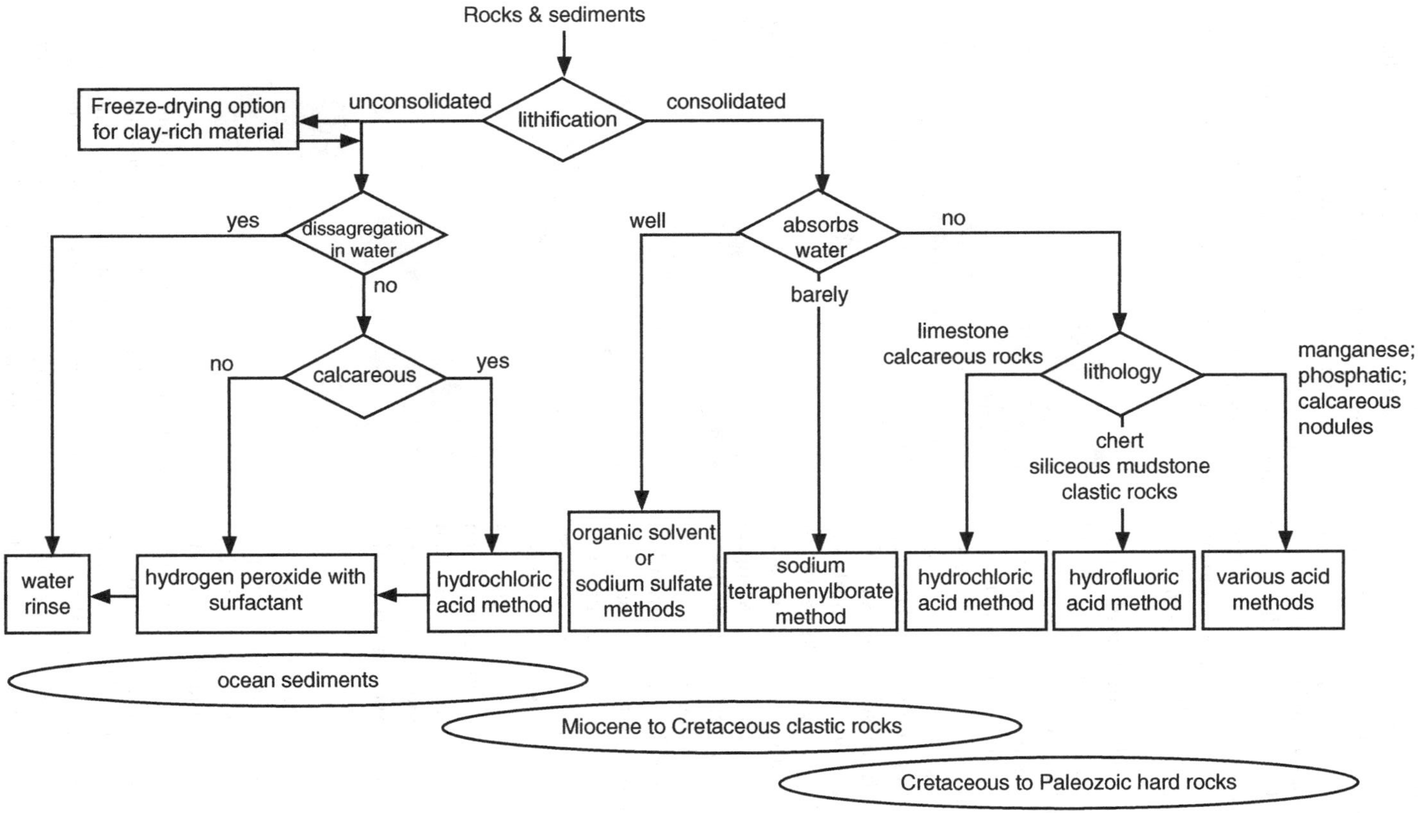

Figure 6.17 Flow chart of possible sample processing procedures in lab to extract radiolarians embedded in other materials. Ovals at bottom indicate the most likely types of procedures for different sources of material.

Radiolarian skeletons collected from plankton need to be mostly free of organic material before study. Boltovskoy et al. (1983) give details for various methods. Commonly material is treated with hydrogen peroxide (H_2O_2) which optionally can be catalyzed, e.g. with potassium dichromate (Welling and Pisias, 1995). Other cleaning solutions can be used: nitric, sulfuric, or hydrochloric acid and potassium permanganate (Boltovskoy, 1998; Matsuoka, 2009).

Radiolarians in sediment samples normally have little organic matter associated with them and have many other sediment components. H_2O_2 is used both to remove any organics and, via the bubbling reaction, to break up sediment. Acids – though a weak acid such as 10% HCl is sufficient – are also used to remove carbonate.

Removing clay from sediment samples can be done using a variety of methods. Dispersants such as Quaternium-O or sodium metaphosphate, as found in sedimentology labs, are often used for samples with only moderate to small amounts of clays. Roelofs and Pisias (1986) provide details and alternative dispersant formulas. Adding organic solvents to dried samples (kerosene, grill starter fluid, or "lamp fuel" used for decorative home lighting are best, as lower flashpoint solvents, e.g. auto gasoline, may work better but are dangerous in the laboratory), soaking, draining, then adding water is effective for more clay-rich samples. Abelmann (1988) reported that freeze-drying samples first improves the clay removal efficiency of chemical treatments.

Ultrasonic baths are very useful for improving sediment dispersal and in particular to clear sediment from the interiors of shells. However, depending on model and settings, ultrasound also can break shells apart, so must be used with care and only after testing to see the effects of different lengths of treatment. Often only a few seconds is needed. Particular caution is needed with older material where the radiolarian shell has been replaced by opal CT or quartz, as these shells, being made out of numerous loosely bound microcrystals, are far less robust than the massive, glassy opaline original shell structure.

All of these treatments are normally done using fine-meshed sieves (stainless steel or plastic mesh) to separate the residues from the reacted material (Figure 6.18)

Figure 6.18 Sieving sediment to recover radiolarians and other particles after disaggregating a deep-sea sediment sample. Residue retained on sieve (above) is washed into a small beaker (below) using a squirt bottle with water for further processing. Source: Courtesy of S. Salzmnann.

and, until the chemical solutions have been removed, under a fume hood for safety. 63 μm sieves were used in earlier studies but to capture smaller specimens (more common in the Neogene–Recent) 45 μm or even smaller sizes should be used. Smaller sieves however will also retain more (usually unwanted) finer particles such as diatoms, to the point where prepared slides will be so diluted that sufficient numbers of radiolarians are hard to find. A variety of methods have been employed to better separate diatoms and other material from radiolarians. Although the separation efficiency cannot be easily quantified, simply spreading dry residue on a clean glass slide then tapping the looser material off again can cause many of the (typically very flat) diatoms to adhere to the slide by surface adhesion while the more three-dimensional radiolarians come free. More sophisticated methods involve repeated settling of residues through a column of water to separate faster-sinking radiolarians from slower-settling diatoms, although small radiolarians may sink more slowly than larger diatom specimens, so separation by settling is only moderately useful. Alternatively, one can make slides from different size fractions of the same sample to capture both small species and, using a coarse sieve e.g. 100 μm, concentrate larger but rarer forms.

More indurated rocks often require stronger chemical solutions and/or more processing time, although the chemicals are mostly the same as for non-indurated sediments. More general information on chemical and mechanical processing of rocks and the risks for microfossils is given by Hodgkinson (1991). An exception is the extraction method for siliceous rocks – where the rock matrix, not just the radiolarians, is partly or entirely composed of silica. The most important such rock is chert. Chert and similar rocks are prepared by a process of repeated etching of the surface by weak solutions of hydrofluoric acid (HF, ca. 5–10%) and collecting the radiolarian fossils as they are freed from the rock matrix, but before they too are dissolved by the corrosive HF (Pessagno and Newport, 1972; De Wever

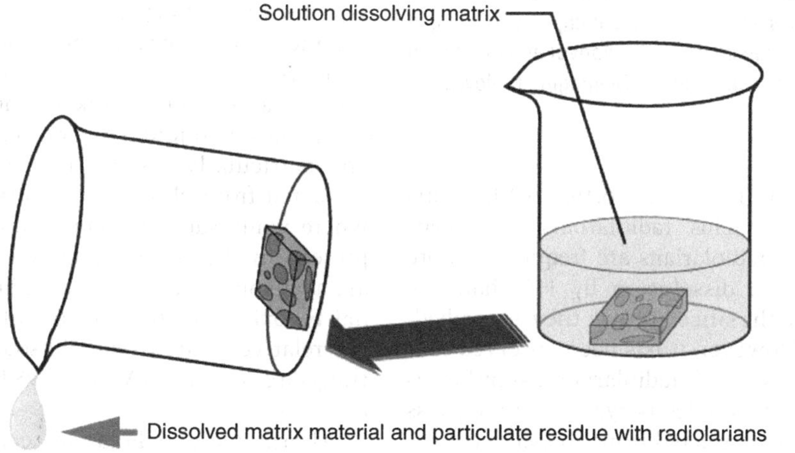

Figure 6.19 Rock samples, typically broken into small chips to increase surface area, are immersed in a solution that dissolves the matrix. At regular intervals the solution, including the small radiolarians as particulate matter, is decanted and filtered as for sediment processing to recover the radiolarians. The cycle is repeated until enough specimens have been recovered.

Figure 6.20 Radiolarians as they appear in indurated rocks in thin section imaged with simple brightfield transmitted light (above) and on etched surfaces of rocks using SEM imaging (below).

et al., 2001; figures 6.19, 6.20). This procedure yields radiolarians as, apparently, the radiolarians are frequently more resistant to dissolution by HF than the matrix of the siliceous rock they are embedded in. However, it has been observed that the diversity of radiolarian assemblages from chert samples is typically much less than those extracted from co-eval, ecologically equivalent material that has been preserved in other rock facies, particularly concretions. It is assumed that in reality not all radiolarian specimens are actually more resistant than the matrix, and probably the degree of resistance depends also on the species (Blome and Reed, 1993). Despite this limitation, the method has been extremely important in the study of older radiolarian faunas, as due to diagenesis, dissolution, and re-deposition of primary opal A, it is often the case that radiolarian-bearing rocks have at least a partially siliceous matrix. Without the HF method, many such faunas would not have been recoverable for research.

Separation of radiolarians from other chemically resistant similar-sized components of residue

As noted above, radiolarians are often not the only particles in the residue extracted by chemicals and sieving. When the unwanted non-radiolarian particles are very abundant it can be difficult to work with the material due to their dilution and masking effects. When the unwanted particles are not made from the same material as the radiolarians (various phases of opal, or quartz, vs. other mineral grains) the differences in particle density can be used to separate them from the opaline material. This is done by suspending the residue in a liquid with a density in between that of the opaline radiolarians and the minerals to be removed. The radiolarians float to the top, the unwanted grains sink to the bottom. Such "heavy liquid" separation methods are particularly useful for radiolarians extracted from clastic sediments or rocks where many sand-size grains may also be present in the residues. Different liquids are available for use, including bromoform (fairly toxic), zinc bromide or chloride, and the relatively safe to handle sodium polytungstate solution (Ashby, 1983; Madella et al., 1998).

Alternatively, radiolarians can be separated by hand, once cleaned and freed from matrix, using fine brushes and a reflecting light microscope to individually pick up specimens (which hold to the brush tip by surface adhesion, sometimes enhanced by wetting the brush). This method is mostly used for older radiolarian materials, which tend to be more robustly constructed and less easily broken than Oligocene to Recent radiolarians, which have much thinner, more easily broken shells (Lazarus et al., 2009). For picking any finely tipped object can be used – a very fine brush or needle, or a single human hair attached to a handle. The latter, particularly e.g. eyelashes, are naturally tapered and thus well suited to small, delicate specimens. Dry residues can be spread on a tray, or for more fragile specimens, residue can be immersed in a shallow tray of distilled water (De Wever et al., 2001).

Mounting Radiolarians

Methods for preparing extracted radiolarian specimens for imaging and analysis depend mostly on the imaging method to be used: transmitted or reflected light microscopy (LM), transmission electron microscope (TEM), or scanning electron microscope (SEM). Reflected light mounting is perhaps the simplest – specimens can either be observed directly by strewing dried residue on a tray, or, after picking, permanently mounted on paper/cardboard microscope slides marked with grids, using gum-tragacanth or other glues to fix the specimens to the backing (Harris and Sweet, 1989).

In early studies of radiolarians, indurated rock samples were not processed as described above and the radiolarians were sometimes studied by the use of rock thin sections (e.g. Hojnos, 1916), but the difficulty of visualizing shell morphology, and the time required to prepare the thin section, means this method has largely been abandoned.

Preparing transmitted light radiolarian slides from extracted shells is complicated, and a distinction is made between strewn slides and random-settled slides. Strewn slides are made by spreading wet, suspended residue over a glass slide surface using a pipette or eyedropper, letting it dry, covering with a transmitted light microscopy optical mounting medium, and covering this with a glass cover slip. Radiolarian slides made this way have, however, a non-uniform distribution of radiolarians on the slide surface. The hydrodynamic properties of large vs. small, flat vs. globular, etc., specimens cause different types to preferentially settle out of the suspension in different areas on the slide. If only a small area of the slide is then examined (for example, to count 300 specimens for quantitative analyses, see e.g. Chapter 7: Paleoceanography), unless a special randomized process is used to select areas for counting, the counts obtained will not usually be representative of the average composition of the assemblage, adding substantial random biases to the data collected. Methods are available that randomly settle radiolarians onto the slides (or the cover slips), usually by pouring the residue into the top of a column of water, with the slides or coverslips sitting at the bottom (Moore, 1973, Figure 6.21). The water is briefly stirred then left standing for some hours. The random movements of particles as they settle through the water column ensure a random distribution of specimens on the slides that they settle on (Figure 6.22). There are many technical details and variants of the method available, all meant to ensure e.g. that a coverslip stays on the bottom when stirring (Figure 6.23), and that water can be removed from the slide without disturbing the distribution, ensuring that, once settled, the radiolarians do not move when mounting media is added, etc. (Roelofs and Pisias, 1986; Lazarus, 1994). Similar randomization problems are encountered in other areas of micropaleontology and analogous methods have been developed. One, originally developed for sedimentology and widely used for foraminiferal studies, is a splitter – a device that equally divides (dry or wet) bulk granular material into smaller samples, a process that is repeated until the required random sample size needed has

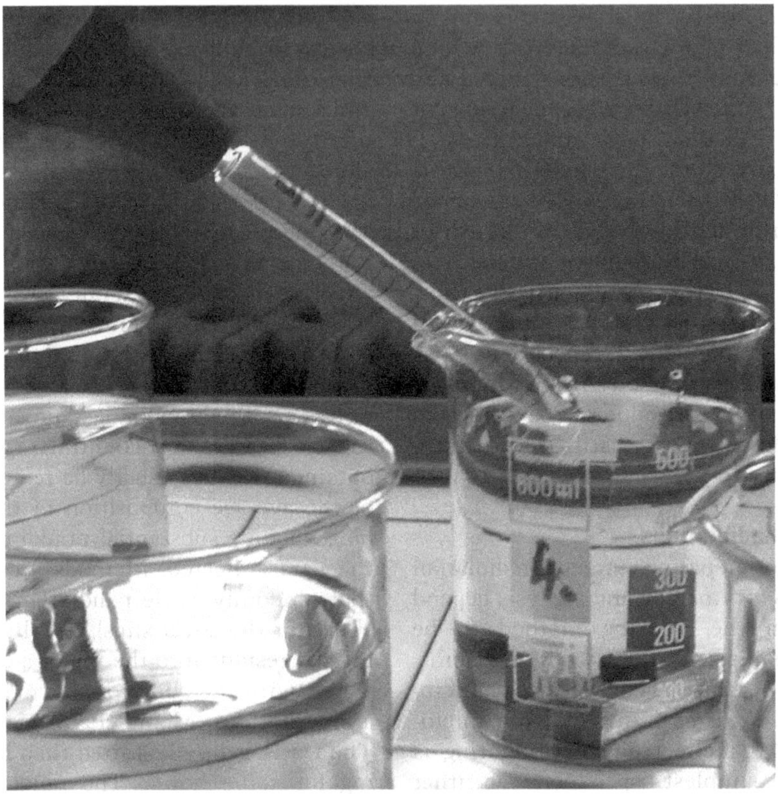

Figure 6.21 Measured volume of suspended sieved sediment containing radiolarians being added via pipette to a 600 ml beaker filled with di-ionized water. Cover slip holder (Figure 6.23) visible at bottom of beaker. After adding residue the water column is gently stirred with a plunger (a kitchen sink drain strainer attached to a metal rod) and radiolarians are left for several hours to settle onto the cover slips and bottom of beaker. Source: Courtesy of S. Salzmann.

been obtained. Splitters are sometimes used in radiolarian work as well (e.g. Takahashi, 1997).

Slides for transmitted light examination, except for temporary use, where plain water can be used, need to be permanently fixed using optical mounting media. Canada Balsam (conifer tree sap, typically from Canadian trees) is the traditional media. It is usually dissolved to a syrup consistency in a solvent such as xylene. Canada Balsam has a refractive index of 1.55 which works well for thick shelled (Paleogene or older) radiolarians, and more recent radiolarians from polar regions which, for unknown reasons, have a different refractive index than those from low latitudes. Neogene–Recent radiolarians from low latitudes have a refractive index that is nearly identical to Canada Balsam and thus can be very difficult to see if it is used to prepare this material. Synthetic alternatives such as Naphrax and Pleurax which have a higher refractive index (1.6–1.7) can be used to produce preparations with good contrast. Lastly, there are many specialized glues used in the optics industry (optical cements) which come in a variety

Figure 6.22 After water removed by siphoning, then drying away the last ca. 1 cm of water above the slip surfaces under heat lamps, the now-dry cover slips are coated with a uniform distribution of radiolarians and other sediment particles. Holders are removed from the beaker using vertical handles. Cover slips are mounted on glass slides using Canada Balsam, Naphrax, or optical cement, depending on refractive index of radiolarians and storage needs. Source: Courtesy of S. Salzmann.

of formulations, viscosity, etc., some with a refractive index similar to Canada Balsam, others with higher values similar to e.g. Naphrax. These are typically hardened using intense ultraviolet light, and thus are ready to use almost immediately (Canada Balsam slides typically need several days to dry to the point where they can be safely handled). The long-term stability of most of these mounting media is not known, and at least some commercial synthetic media introduced in the 1980s were withdrawn from the market, in part because they proved to be unstable, becoming cloudy after a few years of storage. At least some optical cements have remained stable for 30 years (Lazarus, unpub. obs.). Canada Balsam is stable indefinitely. Microscopic preparations made with it 150 years ago are still optically transparent (Lazarus, 1998), and, as amber is simply geologically aged versions of the same material, such slides should last until the glass itself disintegrates (e.g. thousands of years).

It should also be noted that many of the chemicals used in radiolarian preparation are dangerous – toxic, carginogenic, or capable of causing lethal chemical burns (most particularly HF). Preparing radiolarians is thus best left to those properly trained and equipped (laboratories with fume hoods, protective clothing, etc.) to do the work safely.

Live Preparations

Living radiolarians can be examined with the same microscopes used to extract them from plankton samples. However, better-quality images can be obtained by using higher power magnification transmitted

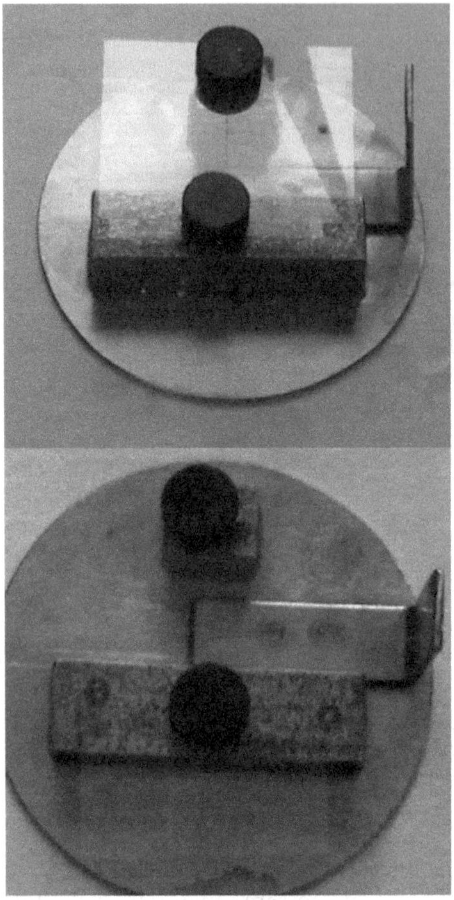

Figure 6.23 Two cover slip holders used to make uniformly distributed ("random settled") radiolarian slides (Moore, 1973). Many variations of holders exist, both for cover slips and entire slides. The ones shown here (Lazarus, 1994) can hold three cover slips and thus make three slides from each sample, reducing the loss of material that otherwise falls to the bottom of the beaker (see Figure 6.22). The slips are held onto the stainless steel metal support blocks with small ceramic magnets. Source: Courtesy of S. Salzmann.

light microscopes, generally equipped with water immersion objective lenses.

Plankton material may be stained with various compounds to highlight cytoplasmic features of interest. Rose Bengal is often used to distinguish living from dead specimens (Boltovskoy, 1998), while fluorescent markers such as DAPI can identify symbionts or other cytoplasmic components (Zhang et al., 2017). For physiologic studies, living specimens may be dosed with radiochemical tracers (Anderson, 1983), or fluorescent markers like PBMPO and HCK-123 for siliceous metabolism (Ogane et al., 2010; Hori et al., 2017).

Examination using the TEM requires a complex set of preparation steps, usually described at least briefly in most papers, e.g. Anderson (1976). Most of these are similar to those used for other protist organisms. The presence of an opal shell, however, is a complication as it is difficult to create clean thin slices of tissue with a microtome when substantial amounts of skeletal material must be cut at the same time. For this reason, many of the TEM studies of polycystines have been done on taxa with no, or only a weakly developed, shell structure (Anderson, 1983). Ultrastructural details of the cytoplasm of more heavily silicified species are therefore relatively poorly known.

SEM imaging can be done from material that is simply strewn over the surface of an SEM stub, or from specimens that are handpicked and placed on the stub. Normal coating (e.g. sputter coating with gold) is typical (Anderson, 1976; De Wever et al., 2001)).

Dissection and Serial Sectioning

The morphology of radiolarian shells is normally much more complex than simply their external surface. Internal structures are frequently important for species identification, and, despite uncertainty in how to evaluate them, also clearly very important in higher-level classification (see Chapter 5: Taxonomy). In (mostly Cenozoic) specimens where the primary opal has not been altered, the outer shell is often transparent enough to permit transmitted light microscopic images of interior structures. If the opal has been even slightly altered, the transparency and ability to view the interior is lost. Even when primary opal is present, more complex shells with multiple interior layers can effectively prevent viewing structures near the center of the shell. Views of these structures are then limited to the, normally far too rare, chance specimens that are broken in such a way as to expose the characters of interest in the interior. Breaking specimens deliberately can sometimes expose the desired morphology but usually many specimens must be destroyed before a suitably broken shell is obtained. Although newer methods (confocal laser microscopy and micro-CT; see below) are gradually supplanting them, dissection methods have provided over many years an effective, inexpensive way to routinely examine the interior structures of radiolarian specimens. Their main disadvantages are the time and skill required to use the methods, and the unavoidable destruction of the shell being examined. Two methods are known: dissection using a fine needle of specimens embedded in resin, and serial grinding of specimens embedded in ice. The resin method, developed by Dumitrica in the 1980s, is described in detail in De Wever et al. (2001), and many of the interior structure images in their volume were made by this method. The ice method (Stanek and Kiessling, 2002) is a relatively recent development and has not been used so far in routine work. The radiolarian specimen is placed on a slide in a drop of distilled water, and the water is shock frozen into glassy ice extremely rapidly by using a simple, inexpensive Peltier chip to rapidly cool the slide. The ice surface is ground away using a fine polishing brush (the authors suggest a trimmed electric toothbrush), the Peltier chip is turned off to melt the ice, and the specimen, now with a surface partially removed, is available for imaging using normal transmitted light microscopy or SEM. The procedure can be repeated as necessary to successively reveal the interior structure of the radiolarian.

Imaging radiolarians

Digital images of radiolarians and all that can be derived from them using computer technology is a very broad subject, and one that is evolving extremely rapidly. It is also very important, as digital images are the foundation of most subsequent research, and changes here consequently affect most other aspects of research on radiolarians.

Radiolarians, due to their small size, and frequently also fragility, are difficult to physically pick, sort, or manipulate. Much of the work with them thus is either done on the "live" image of the radiolarian as it is initially encountered in a microscopic preparation, or on an archived image. This contrasts to work on many other types of biologically collected objects, e.g. specimens of macroscopic organisms in museums, where the actual specimens are both directly visible and the primary focus of study, and images of them are secondary – mostly for documentation. Radiolarian image archives were previously physical volumes of photographs made with film cameras, today they are digital image libraries made with digital microscope cameras. Imaging living radiolarians can be challenging due to movements both of the radiolarian and the medium they are floating in. Fossils, while conveniently stationary, are still challenging to image well. Images of radiolarians (with or without protoplasm) that are taken with optical microscopes are characterized by either very low resolution (<100X magnification) or limited depth of field (200–400X magnification, which is most typical). Except for those few radiolarians which are very flat in shape (and even for these, only those specimens oriented horizontally on the slide), only ca. 10% of the vertical dimension is in sharp focus at any one time.

Experienced workers thus sweep the focus vertically through the specimen, building a mental image of the three-dimensional structure. As microphotographs are usually only of a single focus it is usually necessary to take at least three, and sometimes up to five, images to adequately capture all morphologic features of interest.

Radiolarians are prepared for SEM imaging using standard methods. Picked specimens are usually used. They are mounted on the SEM specimen holder ("stub") using adhesive or double-sided tape, coated with gold, and mounted in the SEM specimen chamber. Details are given in De Wever et al. (2001).

Visualization (enhanced imagery)

The simplest form of image processing is image stacking, occasionally used with transmitted light images. As noted above, these normally have a much shallower depth of field (vertical region of sharp focus) than the size of the radiolarian shell. By extracting only the sharply focused part of each of several images taken in a series of focal planes through the specimen and compositing them together using an image editing program, a single image can be produced that shows the entire specimen in sharp focus. A more sophisticated version of this is an automatic image-stacking program which composites together a stack of images, ideally taken at a constant vertical step size, which can be most easily obtained with a motorized microscope with an automated vertical stage adjuster (Tetard et al., 2020). Images created this way have not however found much favor with radiolarian specialists, and not just because of the time involved, or the relatively high cost of automated microscopes. The stacking algorithms tend to pick out the sharpest focused set of local pixels from the stack and sometimes have a hard time correctly compositing radiolarian shells that consist of multiple layers. Even if this problem is avoided, the resultant image is dominated by the most external layers, and appears to be very flat – it is difficult to distinguish a spherical from a flat, discoidal morphology. A carefully chosen, limited set of single focal plane images generally gives a better representation of the radiolarian structure. If compositing is desired, it is still best to do this from these selected single focal plane images. In this book the composite images shown in Chapter 5 (Taxonomy) were mostly generated by hand compositing, athough image-stacking software could also have been used.

A variety of three-dimensional (3D) imaging methods have been developed in biological microscopy, and several have been used with radiolarians. Confocal microscopy (O'Connor, 1996); Colin et al., 2017, 2018 and 3D Micro Computed Tomography (Micro-CT; Matsuoka et al., 2012) can in principle overcome the limitations of SEM, individual, or stacked transmitted light images by offering a 3D computer representation of the specimen which can be virtually manipulated or dissected interactively and non-destructively. Confocal laser microscopy makes use of a modified light microscrope, together with computer imaging and compositing software. The method is something of a hybrid in concept between conventional LM and SEM. The light source is a carefully focused laser that only illuminates a small point on the specimen. The optics are designed to *minimize* depth of field, so that at any moment only a very small (ca. .5 μm) volume of the specimen is in view. The entire specimen is scanned in a single depth plane, and the process repeated until the entire specimen has been imaged. The image is stored as a 3D data volume in the computer, where the image can be seen and manipulated. A conceptually related method, which also delivers a 3D volume image in a computer as its output, makes use of a different type of imaging technology – computed tomography. This images three-dimensional objects by reconstructing the visual density of each small subvolume of the specimen based on how effectively beams of radiation pass through the object at different angles. Computed tomography is widely employed in the medical sciences, and many readers probably have themselves been the subject of such an imaging process. Micro-CT scanners are simply miniaturized versions that can image details down to ca. .5 μm or less. Micro-CT can even be used to obtain 3D images of specimens that are still embedded in rock matrix and which would otherwise be entirely invisible, or only visible in cross-section at exposed surfaces. The main limitation of all these methods are the very high relative cost (in particular to routine transmitted light microscopy) of acquiring images, in equipment, time, and specialized skills. They have thus until now largely been confined to exploratory studies, or to analyses of key morphology where the character data could not be obtained in any other way (e.g. Kachovich et al., 2019).

Morphometrics

Measurement of characters to provide quantitative metrics of shell morphology has always been a part of radiolarian observations. The simplest form of linear measurements of overall shell dimensions was practiced by the earliest radiolarian researchers, e.g. Ehrenberg, and similar simple size or dimensional ratio values are given in most taxonomic papers. Beginning in the 1980s and the spread of inexpensive personal computers, more sophisticated types of morphologic measurements were employed. One used for radiolaria, albeit for a single study (Lazarus, 1986), is the extension of linear measurements into 3D by use of a digitizer of the vertical position of the microscope stage. This was done with a custom design (Lazarus, 1983), although similar measurements can be achieved with various subsequently available commercial microscope systems. During the same time period, image analysis systems became common in both academia and industry. These systems worked with digital images (x-y pixel arrays where each pixel possessed at least greyscale, or full color information), were, and are still, used to take simple linear measurements (e.g. MacLeod, 1988; Hatakeda et al., 2007; Kamikuri, 2012). From these images automatic analyses are however also possible. In early studies generally only the outline of the objects were programmatically extracted and analyzed to obtain a variety of linear or 2D generic morphologic descriptors (size, size ratio, diameter of enclosing circle, perimeter, etc.), or processed by more complex algorithims such as fourier descriptors (Lestrel, 1997), eigenshape analysis (Lohmann, 1983; MacLeod, 1999), or geometric methods (Zelditch et al., 2014); Danelian and MacLeod, 2019, the latter supporting also (manually) selected internal points (landmarks) from the image. Radiolarian outlines, or full sets of landmark locations, are often difficult to obtain from transmitted light images for many species due to depth of field and specimen orientation problems, and outline image analysis in particular has been used in only a few radiolarian studies (e.g. Granlund, 1986).

Automatic Identification

Computer classification systems for biologic objects, including microfossils, have developed gradually over many decades, but the complexity of the computing problem meant that until fairly recently, expensive special-purpose hardware, mini, or even mainframe computers were required to achieve usable results. These systems worked with a limited range of morphologically rather distinct classes, either species (e.g. Brechner, 2000) or higher taxa, such as higher taxa of living plankton (Davis et al., 1996; Lindsay et al., 2014; Nakamura et al., 2017). Image analysis is based on feature (character) extraction and multivariate data analyses (e.g. discriminant analysis).

Although the accuracy is good enough for routine use, the effort of programming such systems for each taxon, and the difficulty of scaling them to handle larger numbers of taxa, has severely limited their use for individual species identification.

In recent years, advances in program algorithms which make use of more affordable computing power have created new opportunities. Neural network or other "machine learning" systems are based on a fundamentally different approach to solving the computing problem of automatic identification (and many other computing tasks as well). Instead of defining a solution based on externally defined sets of morphologic features, which must be then individually hand-coded, neural network systems have no pre-defined understanding of the problem, or its solution. Instead they are programs that learn by training. Large sets of training images, tagged with the taxa names, are shown to the system and from these, the program itself learns to identify features in the images that result in a correct classification of the images. Once trained, the network can be used to classify new images in routine use. Further, so long as the "type" of images remain fairly similar, the same program, with no, or only minor modification, can be trained with a different set of tagged images and used to automatically classify images of a different set of taxa. The most ambitious study of this sort so far published is on Recent planktonic foraminifera, where a team of 23 specialists collected over 34,000 carefully isolated, multiply identified images of 34 species of planktonic foraminifera and used this to achieve > 87% accuracy in specimen identification (Hsiang et al., 2019).

At the time of writing this book, only a few pilot studies using this new technology to identify radiolarians have been published 2013. Early studies (Fimbres-Castro et al., 2013; Apostol et al., 2016; Keçeli et al., 2017) used only a limited number of morphologically distinctive taxa, taken from "ideal" image sources – published catalogs where isolated, pre-oriented specimens are imaged against a neutral background. Under these conditions the classifiers perform well (95% or better accuracy), but very few studies have yet been done on how well they work with the much more diverse taxa and overlapping individuals as seen in "raw" images through the microscope by working paleontologists. Itaki et al. (2020) have successfully used a proprietary machine-learning system to count the relative abundance of a single morphologically distinctive radiolarian species (*Cycladophora davisiana*) in several normal microscope slides made from Pleistocene Southern Ocean sediments: >90% confidently (p>.95) identified and very high correlation (r=.98) with counts done by a human expert. Renaudie et al. (in press) were able to achieve high rates of accuracy – also close to 90% for confidently identified images – in two sets of closely related species pairs using raw images from transmitted light microscopy, also suggesting that such systems have the potential to automatically identify specimens in routine work. Lastly Tetard et al. (2020) were able to create an automated workflow to identify entire radiolarian species assemblages, albeit with variable degrees of accuracy, using stacked images of automatically identified individual specimens. The system automatically scans entire slides, locates the objects on them, creates stacked focal plane images, and identifies each object. Similar systems, albeit without the radiolarian specific software, are increasingly becoming available as off-the-shelf commercial products, primarily for bio-medical uses. Accurate, fully automated systems would open the way to collecting vastly greater amounts of occurrence data for species than at present and also collect vast numbers of digital images suited for morphometric analyses. At least in Cenozoic samples, there is virtually no limitation on the number of specimens that can be viewed, only on the numbers that can be identified and recorded by the human observer. General limitations of machine learning systems include the vast numbers of images of each species needed to properly train the software. The example of Hsiang et al. (2019): nearly two dozen experts and 34,000 tagged images for 34 species, when scaled to the thousands of species of just Cenozoic radiolaria, illustrates the problem: there are currently not enough trained radiolarian specialists to train the systems that would ease the bottleneck in availability of trained specialists. Another problem is that it is still difficult to understand, in terms understandable to humans, what explicitly the system used to classify the images. "Maps" showing the general areas of the image most important to the software can be obtained, but how the morphology in these areas is evaluated and combined is still obscure. These methods are thus not yet useful for creating quantitative metrics of classic character based morphology, which are needed for many types of analyses (e.g. growth models, comparison to environmental parameters).

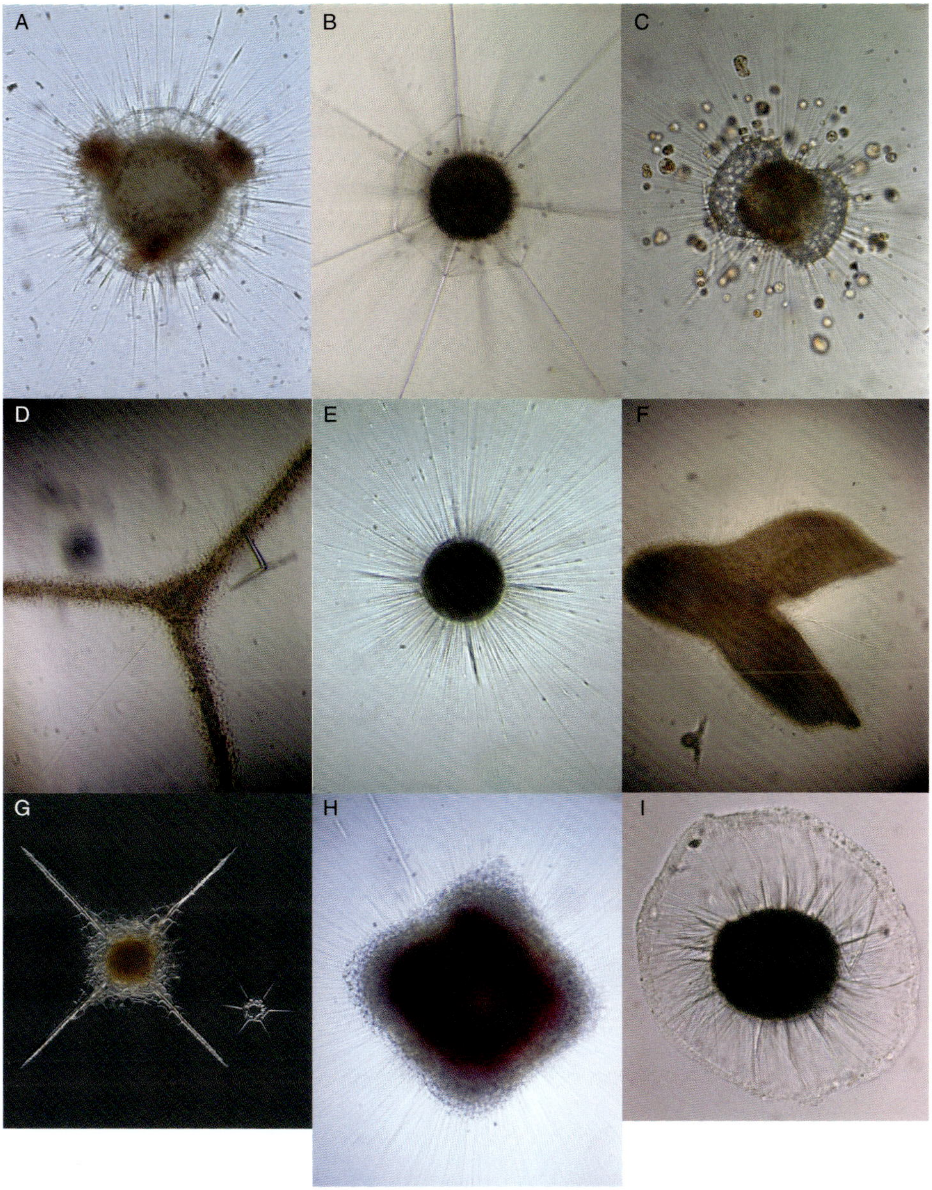

Figure 2.1 Living spumellarians. A) *Arachnosphaera hexasphaera* Popofsky. B) *Astrosphaera hexagonalis* Haeckel. C) *Didymocyrtis tetrathalamus tetrathalamus* (Haeckel). D) *Myelastrum lobum* (Swanberg, Anderson, and Bennett). E) Hexastylidae gen. et sp. indet. F) *Myelastrum* aff. *trinibrachium* Takahashi. G) *Cleveiplegma boreale* (Cleve). H) *Spongaster tetras tetras* Ehrenberg. I) Spongodiscid with gelatinous envelope. Most images show external radiating pseudopods and actinopods; pseudoflagellum clearly visible in D, F and H; symbionts in Source: C. A–F, H courtesy Dr. A. Matsuoka; G by John Dolan/Inigo Montoya; I by Y. Ishitani.

Paleobiology of the Polycystine Radiolaria, First Edition. David Lazarus, Noritoshi Suzuki, Yoshiyuki Ishitani and Kozo Takahashi.

Figure 9.21 (*Continued*)w

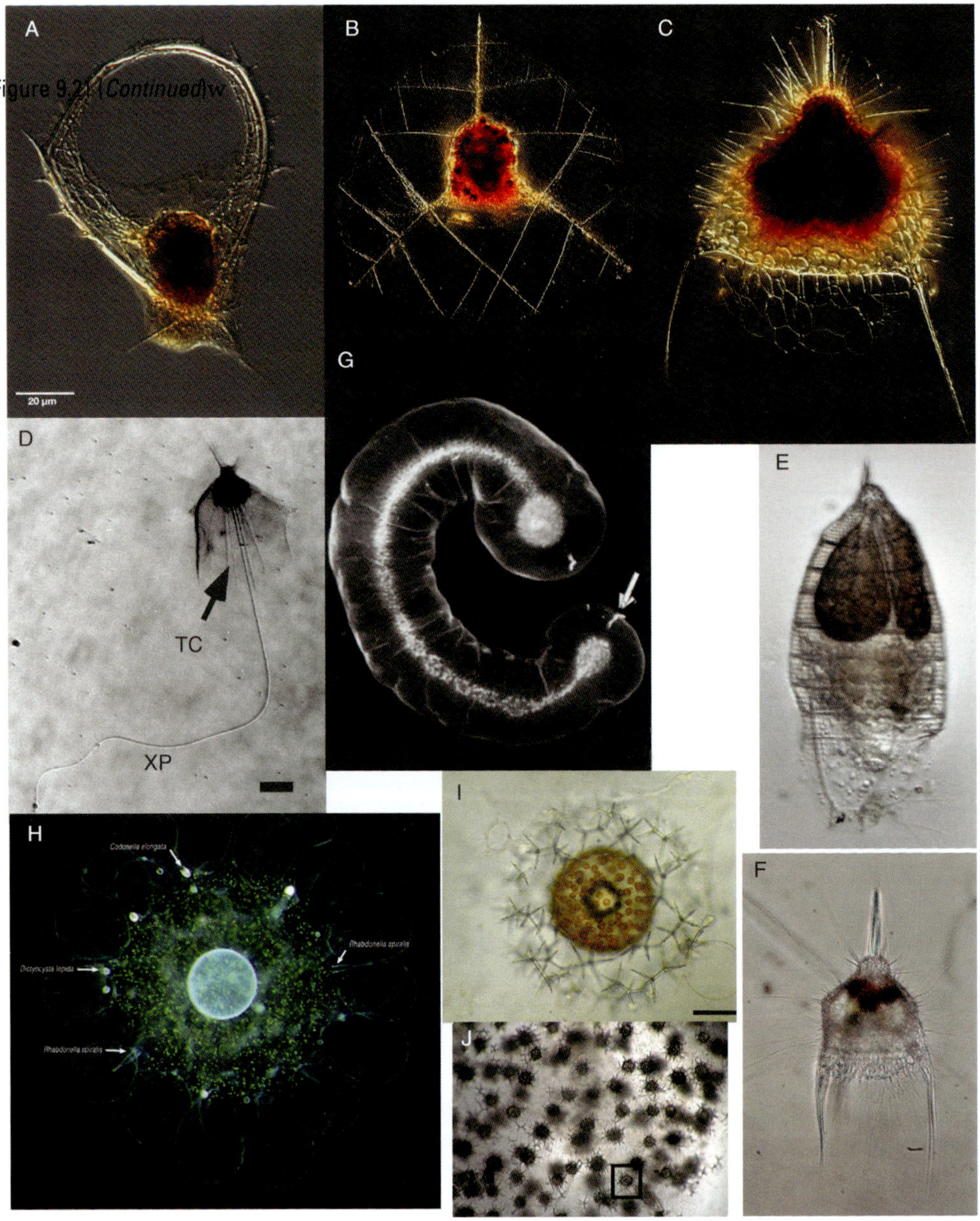

Figure 2.2 Living nassellarians. A) *Zygocircus* sp. B) *Callimitra* sp. C) *Pterocanium trilobum* with numerous fine external spines supporting axopods. A–C are Rugol-stained specimens that show dense colored central capsules in central part of shell, much thinner cytoplasm with granules extending over distal skeletal elements. D) *Dictyocodon prometheus*, showing terminal cone of axopods ("TC") and long pseudoflagellum "XP." E) *Eucyrtidium* sp. showing multilobed nucleus in apex of shell and less dense cytoplasm extending to base. F) *Pterocanium praetextum praetextum*, with short terminal cone of axopods. G–J: Collodaria. G) Colony of *Collozoum caudatum*, with cells (white spots), thick gelatinous sheath, and excreted waste material (white threads at either end [arrow]). H) Solitary form *Thallassicolla pelagica* with outer layer of alveoli, symbiont cells (green dots), central capsule (blue sphere), and captured prey (labeled). I) single-cell of collodarian from a colony (J: partial view of a similar colony with single cell marked by box), showing outer envelope with spicules, yellow central capsule, and orange symbiont cells. Source: A–C, H from the online image archive of the Observatoire Océanologique de Villefranche Sur Mer (John Dolan/Inigo Montoya) (use under Creative Commons licence); D from Sugiyama et al. (2008); E by Y. Ishitani; F by N. Suzuki; G from Swanberg and Anderson (1981); I from Yuasa and Takahashi (2014).

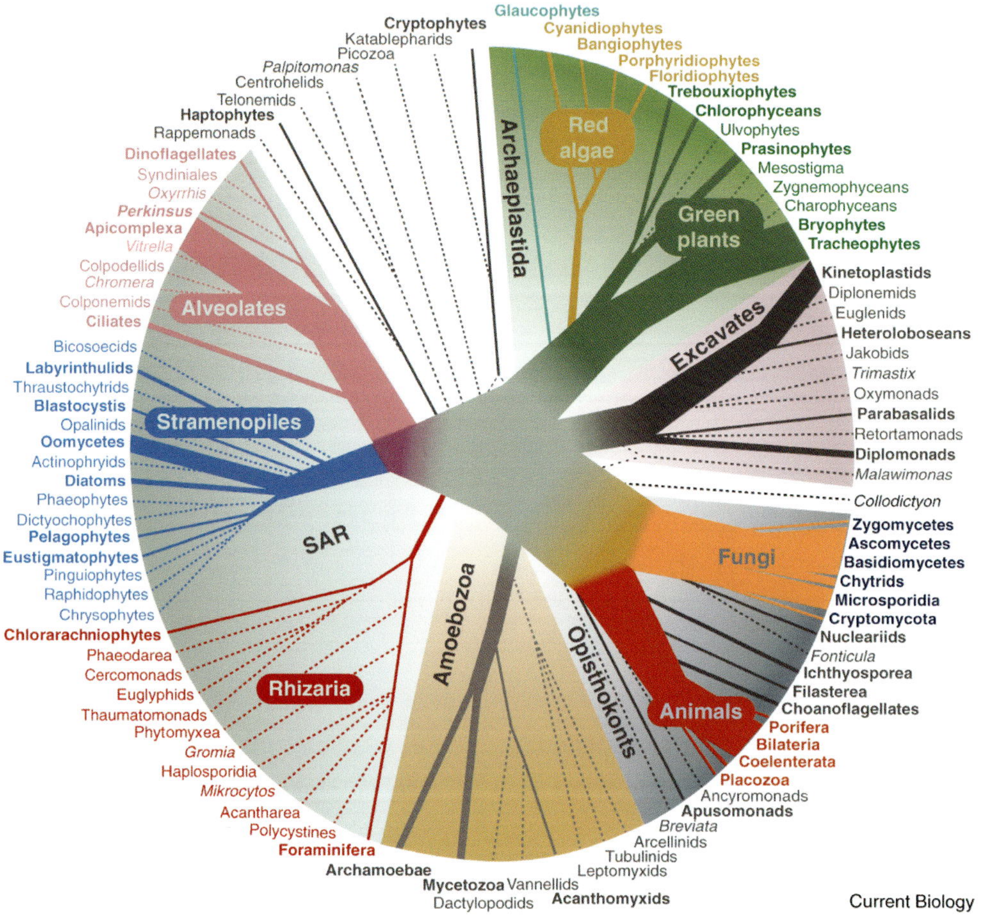

Figure 2.3 A phylogeny of eukaryotic life (e.g. without bacterial groups), showing the placement of polycystine radiolarians within the broader clade of Rhizaria. Polycystines are closely related to acantharians and foraminifera, but not to phaeodarians. The term "radiolarians" is still widely used for acantharians, polycystines, and phaeodarians, but in this book will mean only polycystines. The width of the branching lines in this figure is proportional to the amount of molecular information available for each group, which makes clear that relatively little is known yet about rhizarians in general. Source: Burki, Current Biology, 2014. Reproduced with permission of Elsevier.

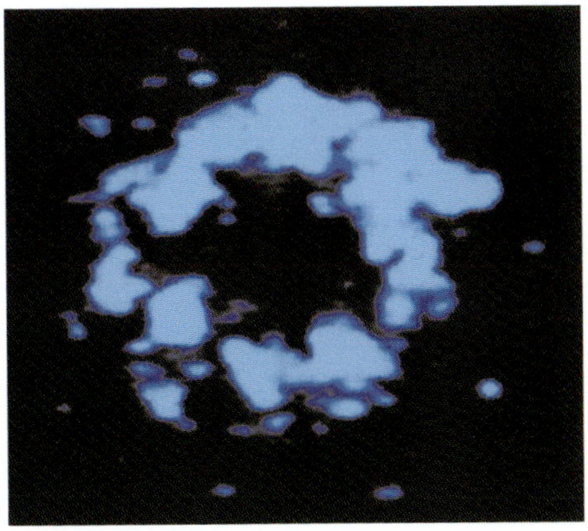

Figure 2.21 Bioluminescence – Image of blue light emission from the extracapsular layer of a large (ca. 2 mm total diameter) solitary collodarian (*Thalassicola nucleata* – Figure 2.2H). Light production is invoked by mechanical stimulation. Source: Widder (2010), after Latz, Journal of Plankton Research, 1991. Reproduced with permission of Oxford.

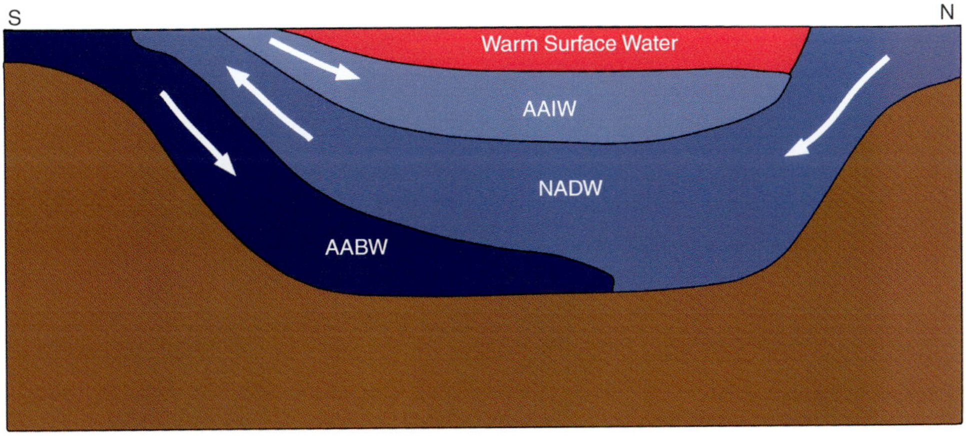

Figure 3.1 Atlantic Ocean deep-water mass circulation in latitudinal cross-section. Cold Antarctic Bottom Water (AABW) forms near the Antarctic continent in winter and spreads northward along the ocean bottom, displacing the slightly less dense North Atlantic Deep Water (NADW) that forms in the Norwegian-Greenland Sea and flows south. NADW flows upward and mixes into the surface in the Southern Ocean. Near the Antarctic Polar Front cold, but less dense surface water submerges and flows as Antarctic Intermediate Water (AAIW) north between the top of NADW and the warm surface water "lid" of low- to mid-latitudes. A similar, if less complete pattern (due to less than full pole-to-pole extent) is seen in other ocean basins. Note diagram is extremely vertically exaggerated (>1000X); in reality the main water masses at the scale of this printed diagram would be thinner than the paper of the page they are printed on.

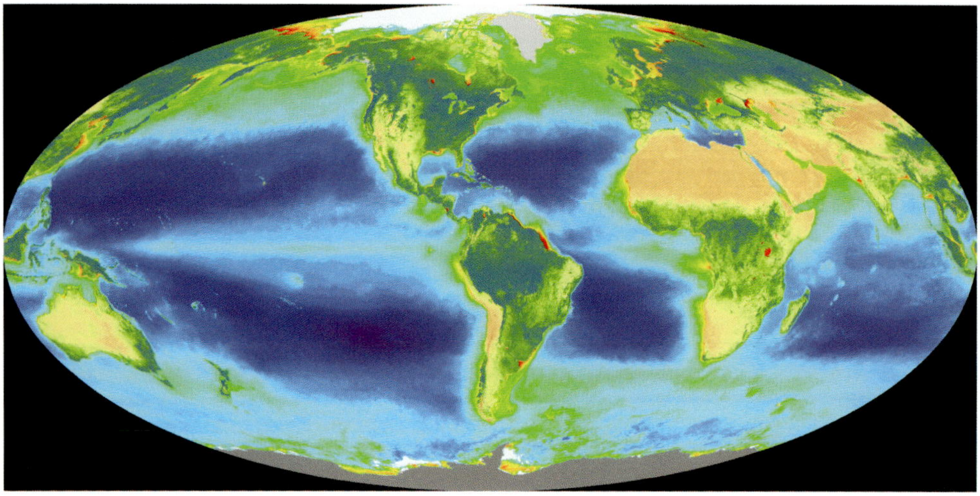

Figure 3.5 Map of global chlorophyll concentration in the oceans (and on land), averaged over several years of satellite imaging. Chlorophyll concentration is a reasonable proxy for primary ocean productivity (the total amount of organic carbon fixed in the near surface waters), although not necessarily for net (aka new) productivity, which measures how much organic material is exported into deeper water layers vs. being consumed and recycled in the surface water. False color image – highest concentrations in red, grading through orange, yellow light blue, dark blue and (lowest concentration) purple. Concentrations vary substantially seasonally and yearly. Source: NASA/SeaWifs (public domain).

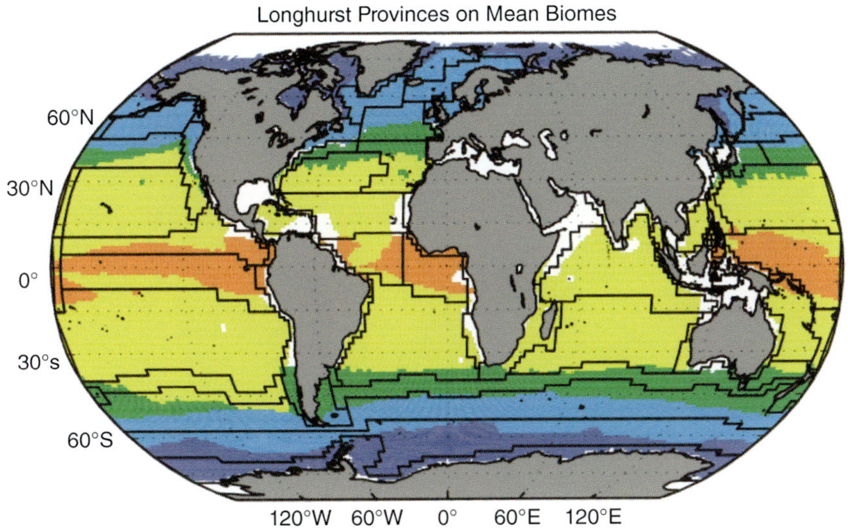

Figure 3.6 Global map of major pelagic biogeographic regions, showing two different classification schemes, both based on physical and chemical ocean characteristics. Black outlined areas are ocean provinces of Longhurst (1998), colored regions ocean biomes from Fay and McKinley (2014). Orange: equatorial; yellow: subtropical permanently stratified; green: subtropical seasonally stratified; light blue: subpolar seasonally stratified; dark blue: (sea) ice. White areas are unclassified and mostly correspond to upwelling regions, coastal areas and marginal seas. Longhurst provinces are subdivisions of his coarser scale biomes, which more closely match those of Fay and McKinley. Note similarity in structure, despite minor differences in placement of boundaries. Source: Earth System Science Data. Licensed under CC BY 4.0.

Euchitonia elegans-furcata

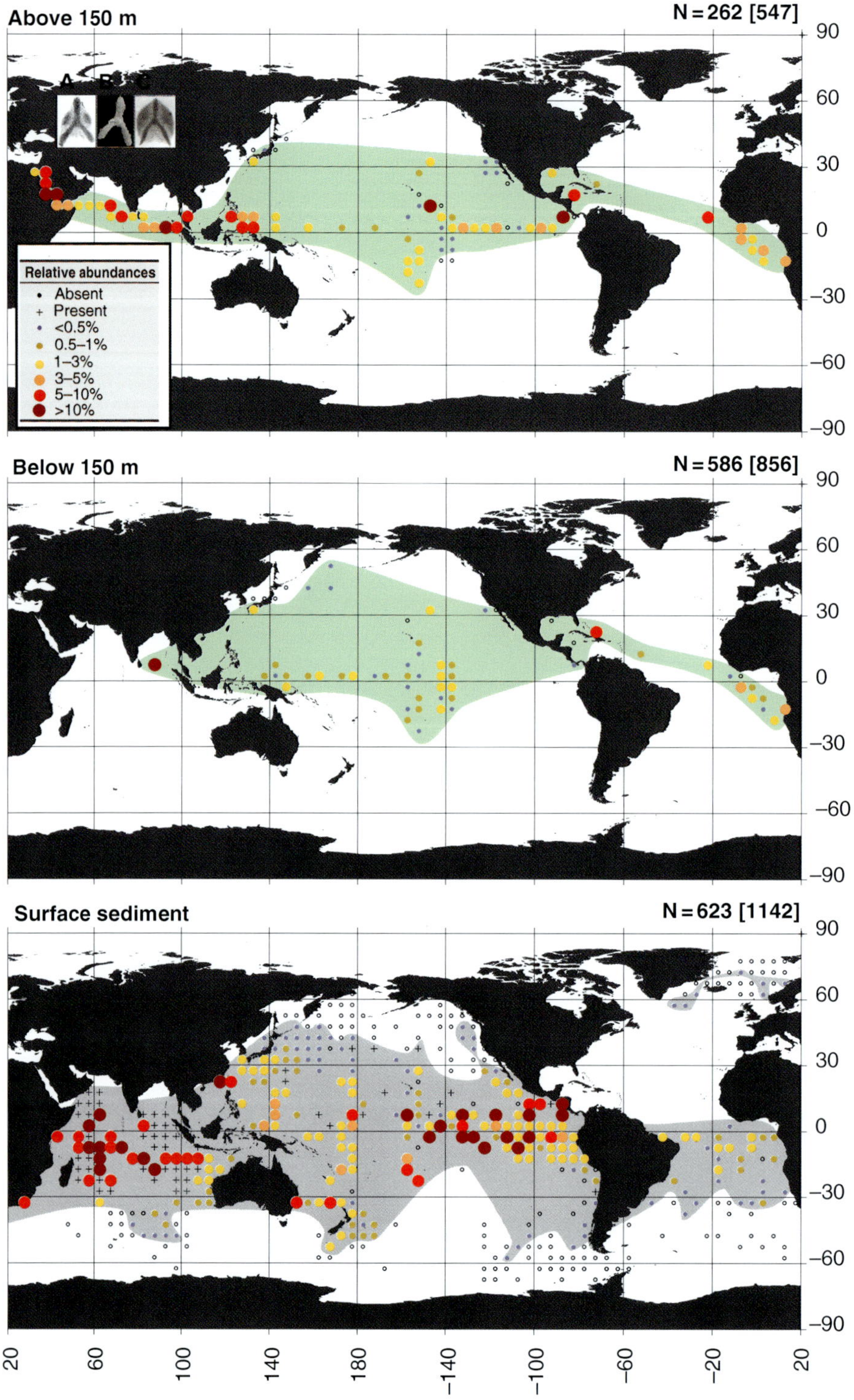

Figure 3.7 Distribution of a tropical species of radiolarian (*Euchitonia elegans*), based on a synthesis of virtually all available published data (Boltovskoy et al., 2010). Shown are relative abundances in color and size coded dots (key in upper panel lower left corner) representing percent categories for surface water (above 150 m), deep waters (below 150 m), and in surface sediments. Simple presence/absence shown by crosses and open circles, respectively. Absence only shown for locations where based on source publication taxa was looked for but not found. Data for distribution in the water column is much less complete than from surface sediments. *E. elegans* is most common within ca. 20° of the equator. It is also present in lower abundances throughout the subtropics (ca. 40°N/S), but only rarely and sporadically in cold-water regions near the poles. Source: Boltovskoy et al. (2010). Reproduced with permission of The Palaeontological Association.

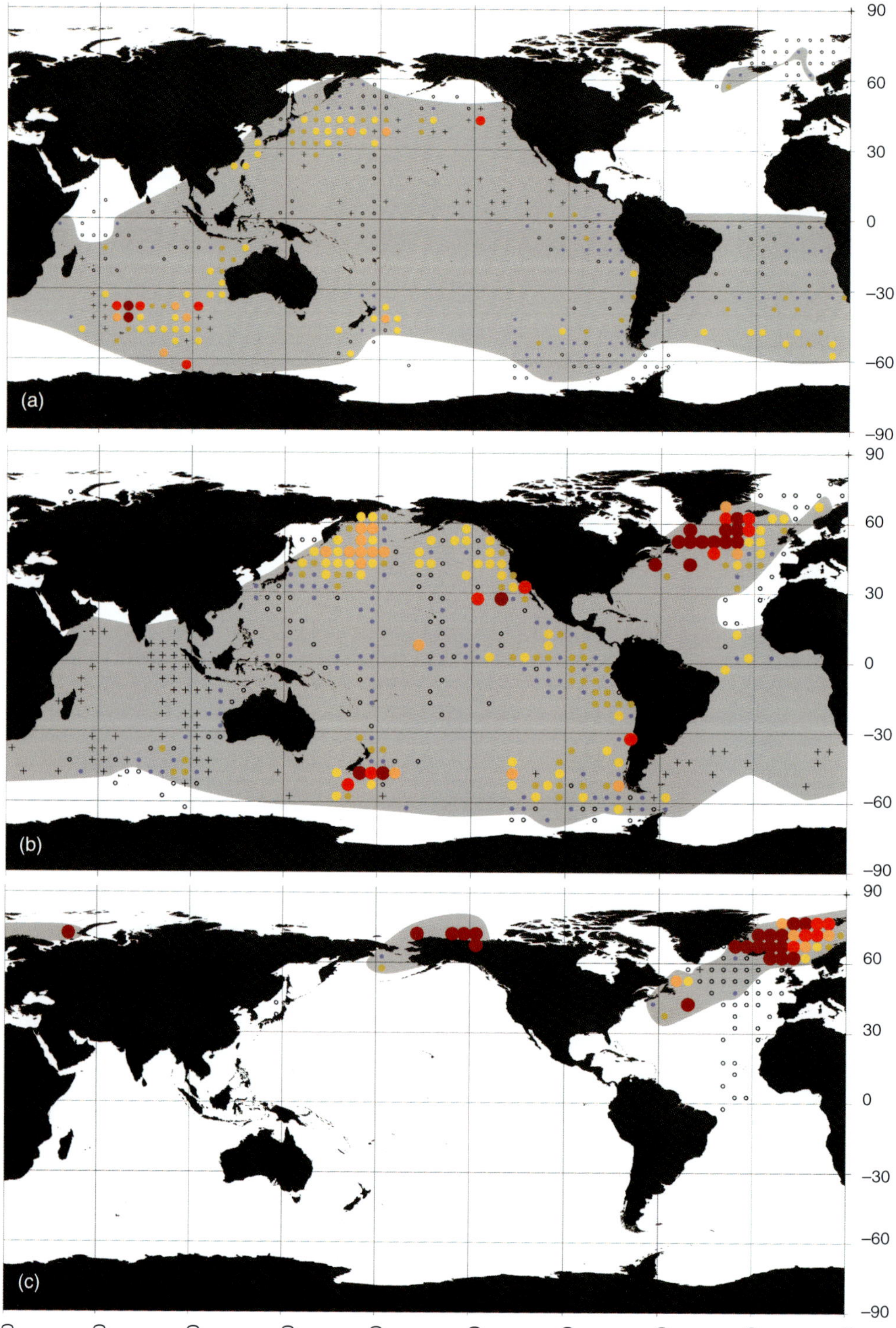

90
60
30
0
−30
−60
−90
(a)
(b)
(c)
20
60
100
140
180
−140
−100
−60
−20
20

Figure 3.8 Plotting conventions as in Figure 3.7. All data shown is from surface sediments only. A – Distribution of the actinommid spumellarian *Actinomma medianum*. Although present over a broad latitudinal range, this typical subtropical cosmopolitan species is most abundant in mid latitudes, ca. 20–40° N/S, and is much less common both in the tropics or in more polar regions. B – Distribution of *Botryostrobus aquilonaris*, an artostrobid nassellarian. This species is primarily found in subpolar regions between 40–60° N/S but also to some extent in low- to mid-latitude areas of upwelling such as the Peru–Chile current region and the eastern Equatorial Pacific. C – An example of a polar endemic species, the cannobotryid nassellarian *Amphimelissa setosa*. This species is a northern polar endemic, mostly in the Arctic Ocean but also found where currents such as the East Greenland Current transport cold water southwards along the eastern margin of ocean basins. All figures from Boltovskoy et al. (2010). Source: Boltovskoy et al. (2010). Reproduced with permission of The Palaeontological Association.

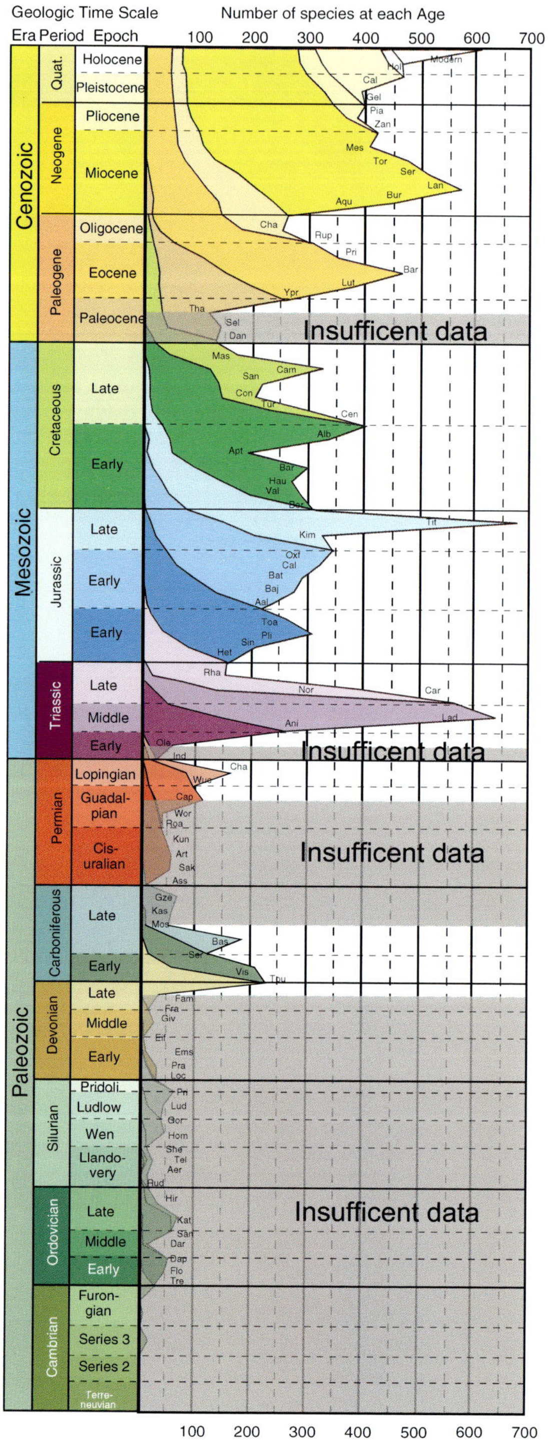
Geologic Time Scale
Era Period Epoch
Number of species at each Age
100 200 300 400 500 600 700
Cenozoic
Quat.
Holocene
Pleistocene
Neogene
Pliocene
Miocene
Paleogene
Oligocene
Eocene
Paleocene
Mesozoic
Cretaceous
Late
Early
Jurassic
Late
Early
Early
Triassic
Late
Middle
Early
Paleozoic
Permian
Lopingian
Guadal-pian
Cis-uralian
Carboniferous
Late
Early
Devonian
Late
Middle
Early
Silurian
Pridoli
Ludlow
Wen
Llando-very
Ordovician
Late
Middle
Early
Cambrian
Furon-gian
Series 3
Series 2
Terre-neuvian
Insufficent data
Insufficent data
Insufficent data
Insufficent data
Modern
Hol
Cal
Gel
Pia
Zan
Mes
Tor
Ser
Lan
Bur
Aqu
Cha
Rup
Pri
Bar
Lut
Ypr
Tha
Sel
Dan
Mas
Cam
San
Con
Tur
Cen
Alb
Apt
Bar
Hau
Val
Ber
Tit
Kim
Oxf
Cal
Bat
Baj
Aal
Toa
Pli
Sin
Het
Rha
Nor
Car
Lad
Ani
Ole
Ind
Cha
Wuc
Cap
Wor
Roa
Kun
Art
Sak
Ass
Gze
Kas
Mos
Bas
Ser
Vis
Tou
Fam
Fra
Giv
Eif
Ems
Pra
Loc
Pri
Lud
Gor
Hom
She
Tel
Aer
Rud
Hir
Kat
San
Dar
Dap
Flo
Tre
100 200 300 400 500 600 700

Figure 5.3 Total number of available species names in the Suzuki (2016) database (9,870), with the diversity history of each geologic epoch, as determined by the recorded total range of each species, shown as a colored cohort. Species only reported as living are marked separately as "Modern." Note that available names are simply those formally published according to the Rules of Zoologic Nomenclature. Many of these are known to be synonyms, and there are many other biases such as unequal preservation or recording effort by time interval, so this figure is not a direct estimate of biologic diversity. Source: Suzuki (2016). Reproduced with permission of Paleontological Society of Japan.

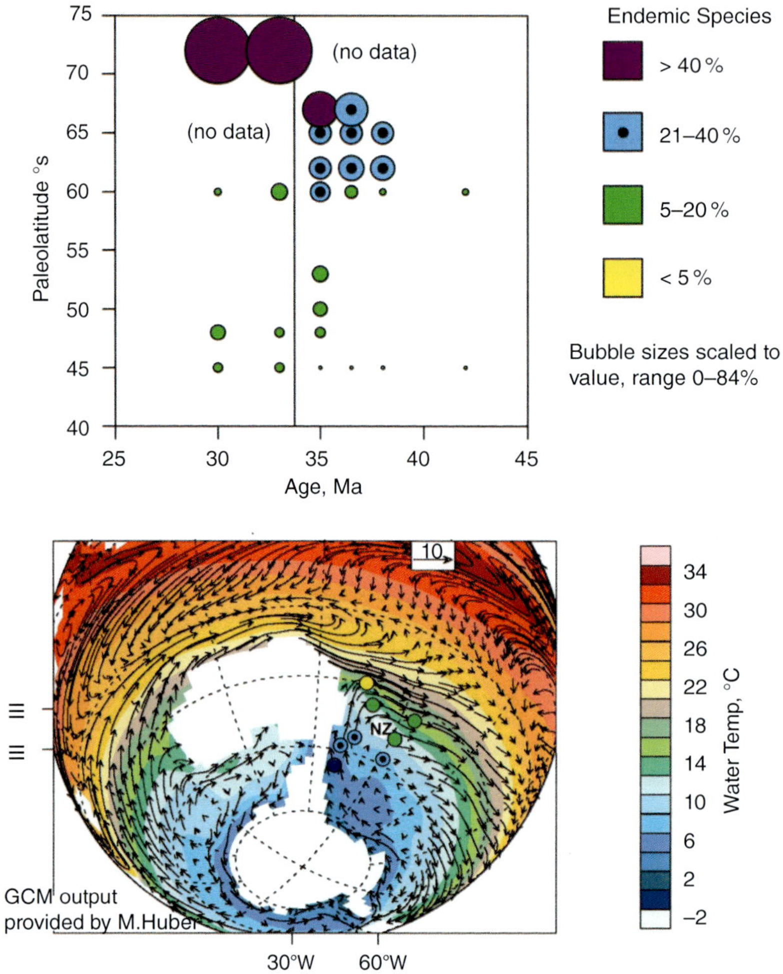

Figure 5.39 Gradual development of endemic austral radiolarian faunas in the late Eocene to early Oligocene in the southwest Pacific sector of the Southern Ocean and their correlation to the physical environment. On left, the degree of endemism is strongly related to latitude and (although geographic coverage in the SW sector is incomplete) increases over time. Increasing endemism in the Southern Ocean across the Eocene–Oligocene boundary is documented in other ocean sectors as well (Lazarus and Caulet, 1994). On right, data (colored dots off eastern Australia–New Zealand region) from a single time interval (ca. 35 Ma) are overlain on an ocean temperature and circulation reconstruction to show close correspondence of endemism and oceanographic conditions. Source: Lazarus et al. (2008), ocean circulation/temperature reconstruction by Huber et al. (2004). Reproduced with permission of Micropress.org.

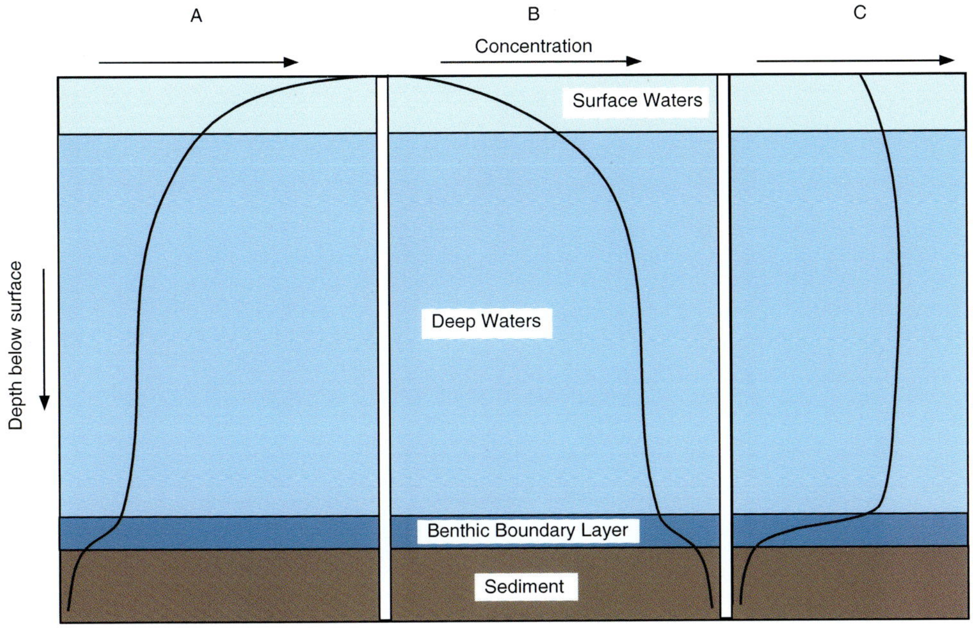

Figure 6.2 Concentrations of key biologically active elements vs. water depth in the ocean. A) Biologic source materials produced in the surface water layer decline in abundance with increasing water depth as they are degraded by biologic activity, chemical dissolution, and other processes. The pattern is non-linear, with much higher rates of decline in upper waters than deep. Relatively intense decline also occurs in the thin benthic boundary layer (nepheloid layer and mixed layer of surface sediment), with only much slower rates of change in the deeper layers of sediment below. Examples of such biologic particles include particulate organic carbon and poorly silicified shells of phytoplankton. B) Dissolved products of degradation of biologic materials in "A" show the inverse pattern of increase. Dissolved silica is an example. C) Refractory components such as the shells of many species of radiolarians show little decay with depth in the water column and may even show slight increase if some production occurs only in deep waters. Most of the decline in these components occurs within the benthic boundary layer, e.g. prolonged exposure of radiolarian shells to bottom waters undersaturated in silica.

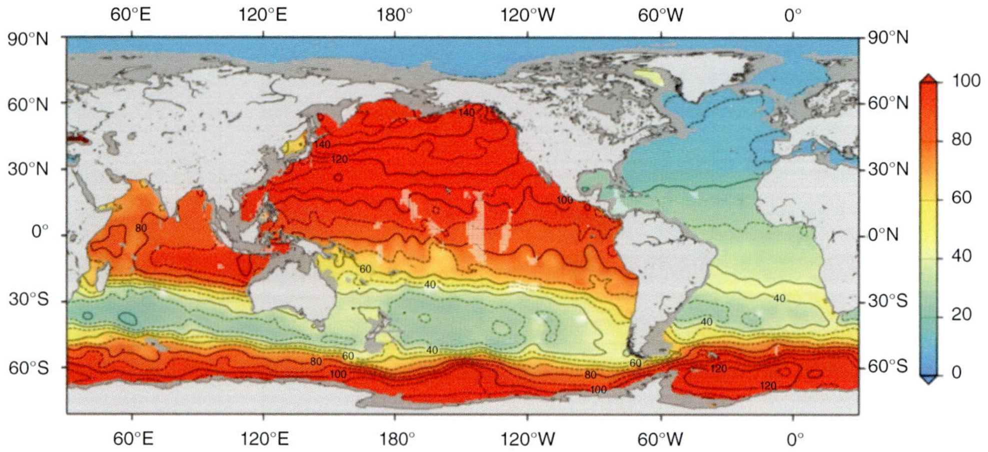

Figure 6.3 Dissolved silica concentration in ocean water at 1 km water depth, showing gradient from North Atlantic to North Pacific due to accumulation along path of deep water flow, and influence of export of biogenic silica into deep water from areas of high export productivity such as the Southern Ocean and equatorial regions. Source: World Ocean Atlas 2018 map.

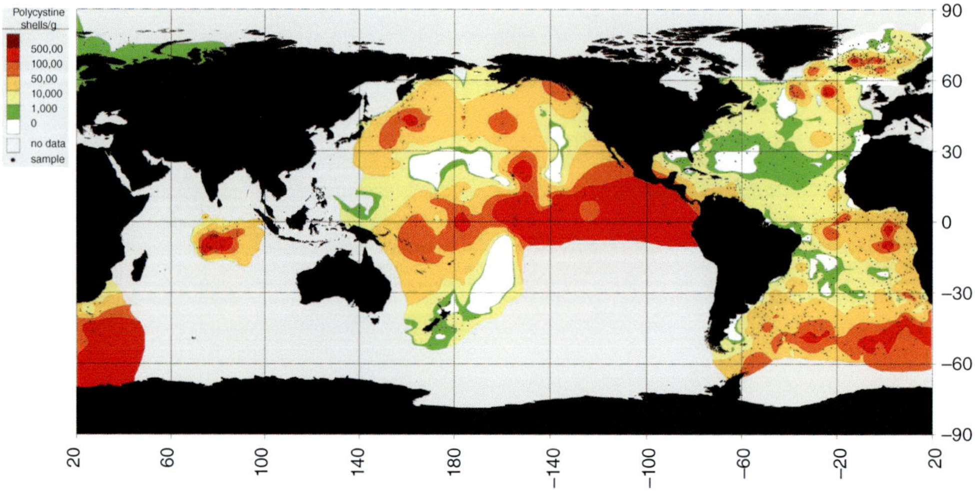

Figure 6.4 Radiolarian concentration in ocean surface sediments on a carbonate free basis. Pattern reflects input flux from surface waters, dissolution in water column and surface layer, and dilution by diatoms, and in some coastal regions, clastics. Source: Boltovskoy et al. (2010). Reproduced with permission of The Palaeontological Association.

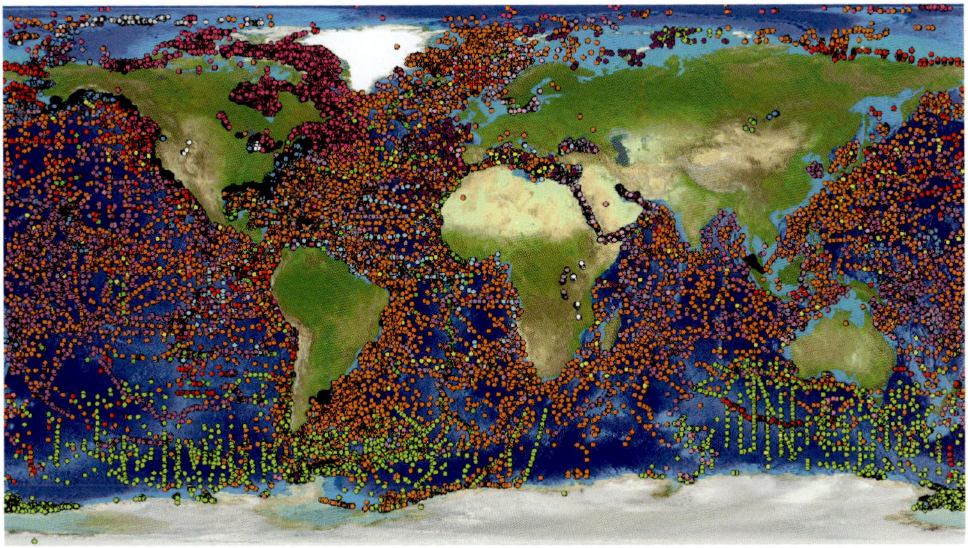

Figure 6.11 Map of deep-sea sediment samples, mostly piston cores. Different color dots in original refer to curation institution. Map shows only cores taken by North American, German, France and UK ships. Similar numbers of samples have been taken by other countries. Source: National Geophysical Data Center, NOAA.

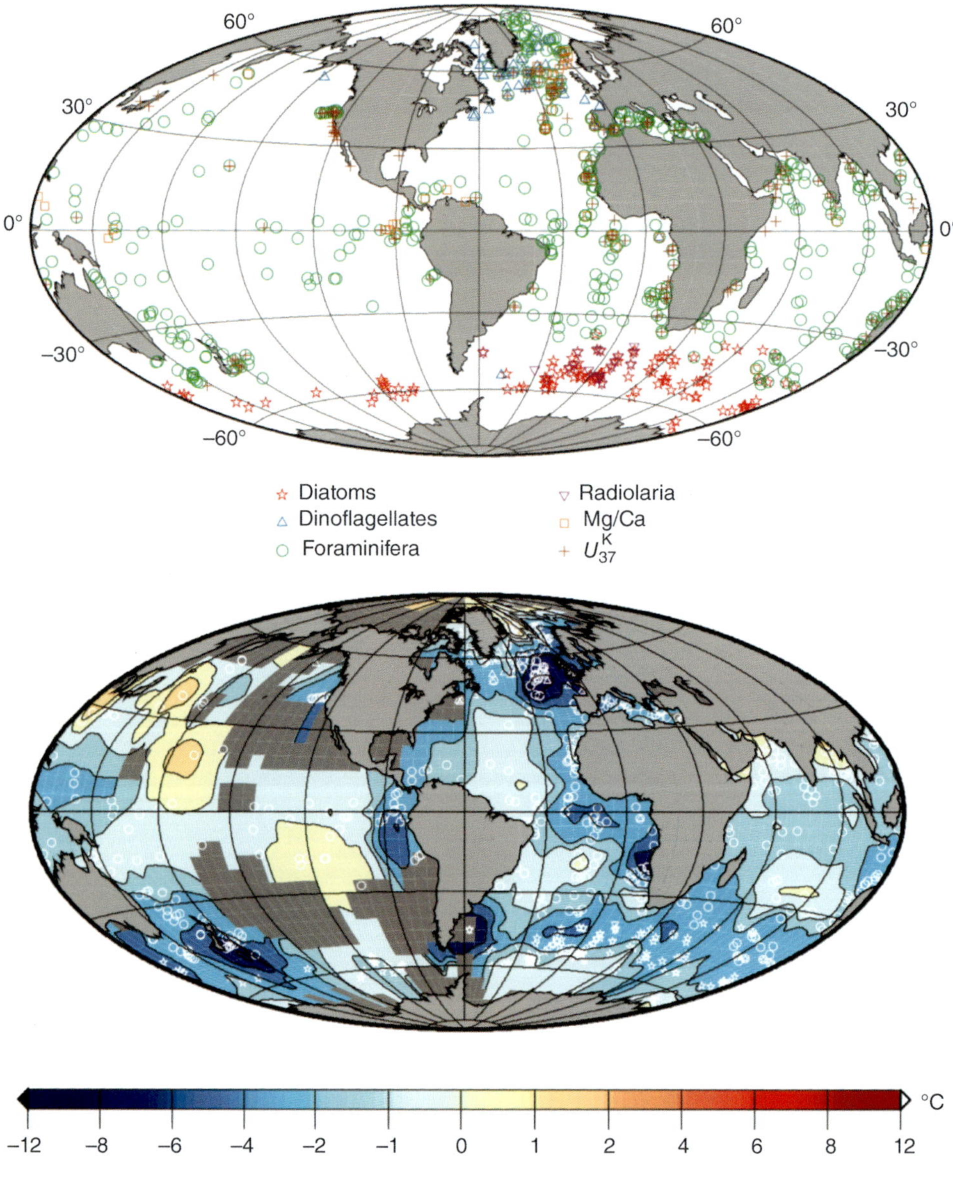

Figure 7.1 Example of modern quantitative paleoceanographic analysis: data and reconstructed temperature changes in the global oceans between the last glacial maximum and modern conditions. Upper panel: locations and types of data used. Radiolarians are used in the southern Atlantic, together with more extensive data from several other microfossil groups and two types of geochemical temperature estimates. Lower panel: difference in temperature from the modern (colder than modern values are negative). Note concentration of temperature change in subpolar and mid-latitude upwelling regions. Source: Waelbroeck et al. (2009). Reproduced with permission of Springer Nature.

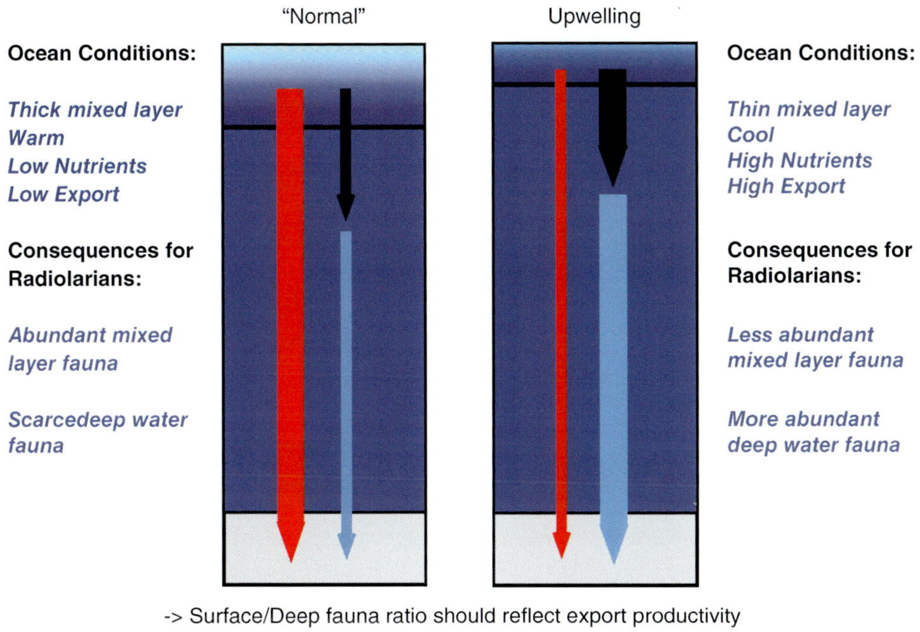

Figure 7.5 Concept underlying using water depth preferences of radiolarians to estimate export productivity. Under "normal," well-stratified ocean conditions there is a large flux of microfossils from the well mixed surface-water layer (thick left-hand arrow), while low export of organic carbon (thin black right-hand arrow) supports similarly low amounts of deep water taxa (lower light arrow on right). In areas of enhanced export productivity such as in an upwelling zone (right panel) these relations are reversed, resulting in a proportionally larger flux of deep water species into the sediments. Source: Lazarus, Paläontologische Zeitschrift, 2005. Reproduced with permission of Springer.

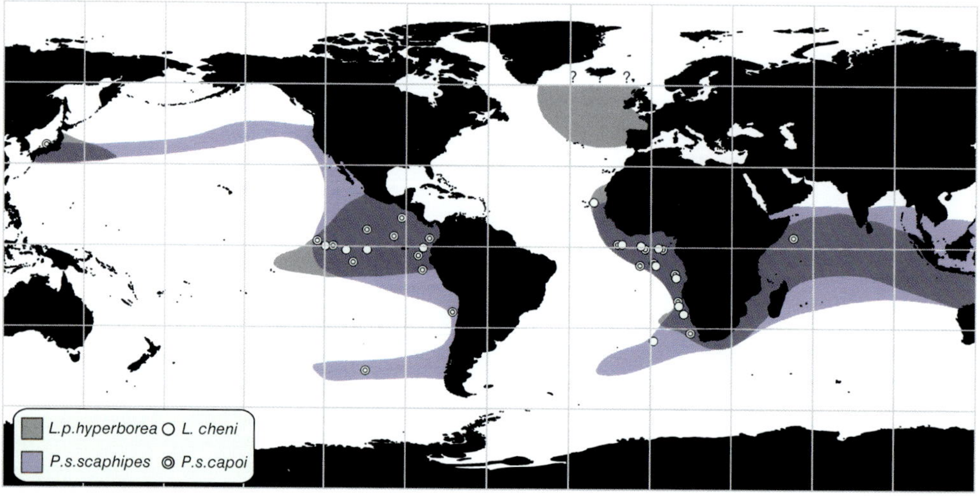

Figure 9.9 Geographic distribution in global surface sediments of four putative hybrid species of trissocyclid radiolarians (ancestral stocks have cosmopolitan distributions and are not shown). Putative hybrid taxa are found mostly in regions where water masses are mixed due to (seasonal) upwelling (eastern Equatorial Pacific, Benguela Upwelling System, equatorial Indian Ocean).

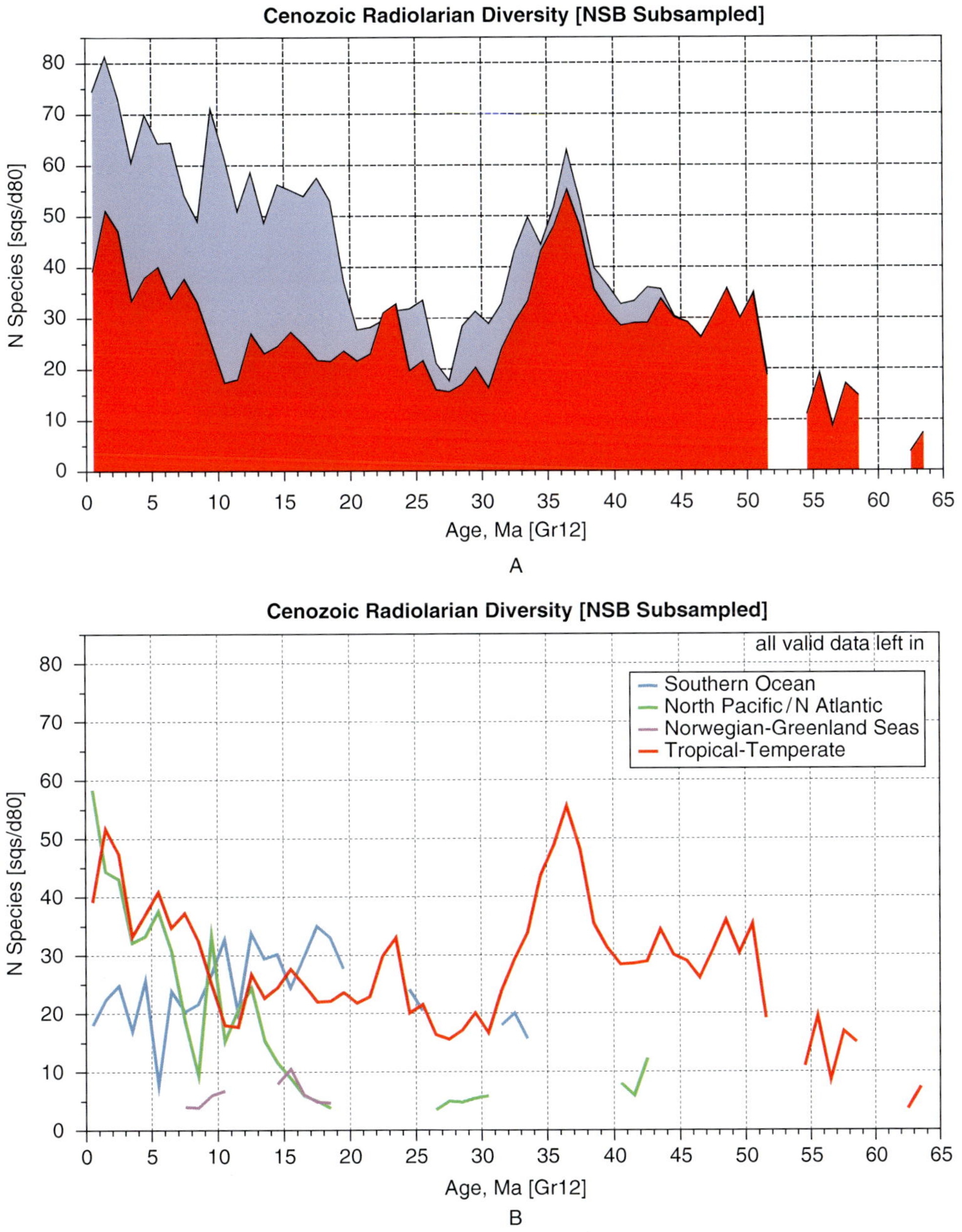

Figure 9.21 Cenozoic radiolarian diversity and turnover. A) The upper blue shaded curve is the global diversity, the red curve all tropical + temperate diversity – here defined as 45S–40N (all analyses by paleolatitude). B) diversity history of each region: tropical–temperate; Southern Ocean – here >55S; North Pacific and North Atlantic (40N–65N); and Norwegian-Greenland Seas – >65N. Because some taxa are cosmopolitan and found in more than one region the diversities of the regions usually add up to more than the global estimate for any time bin. C) – origination and extinction rates (global) for Cenozoic radiolaria. Time series have been smoothed with a three-point moving average function

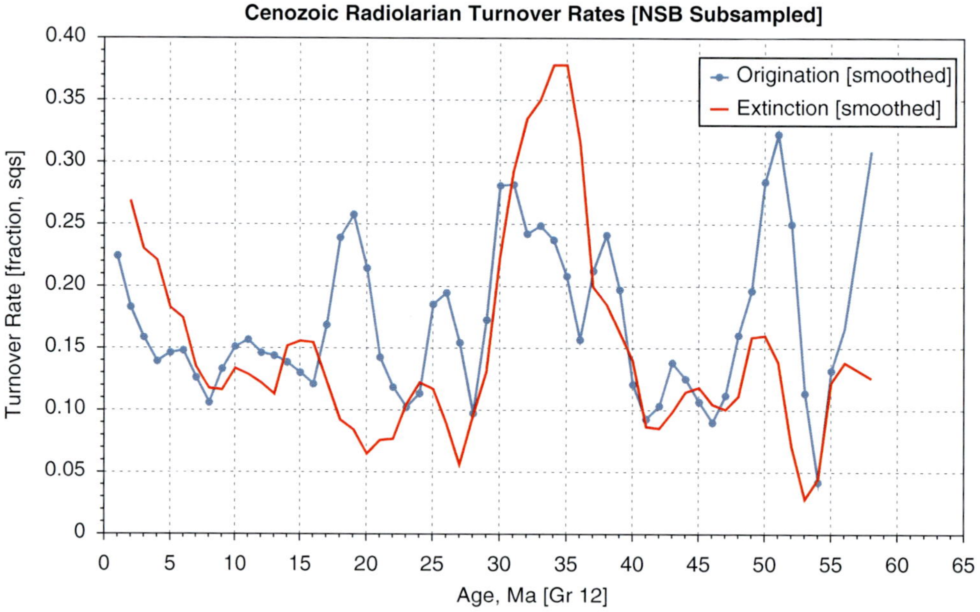

Cenozoic Radiolarian Turnover Rates [NSB Subsampled]
Origination [smoothed]
Extinction [smoothed]
Turnover Rate [fraction, sqs]
0.40
0.35
0.30
0.25
0.20
0.15
0.10
0.05
0
0
5
10
15
20
25
30
35
40
45
50
55
60
65
Age, Ma [Gr 12]

C

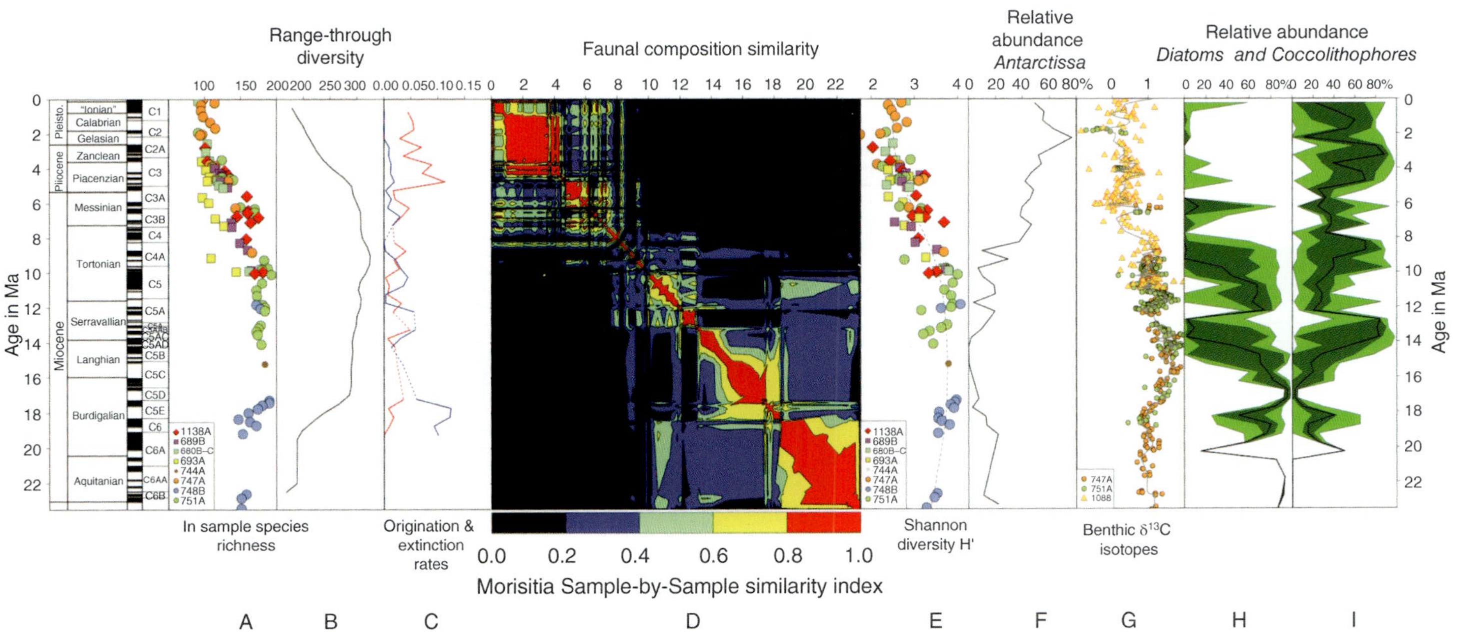

Figure 9.27 Diversity and evolutionary turnover in the Neogene Southern Ocean radiolarian fauna, based on exhaustive counts of diversity in ca. 100 samples in the Atlantic and Indian Ocean sectors of the Antarctic (total of >500k individual specimens). Age scale on left. A) in-sample diversity (symbols show sections), B) total range-through diversity, C) rates of origination (blue) and extinction (red), D) cross-plot of faunal similarity, E) Shannon-Wiener diversity index H , F) relative abundance of species of the genus Antarctissa, and compilations of environmental parameters [G) benthic 13C, composition of phytoplankton in bulk sediment: H) percent coccolithophores, G) percent diatoms]. Similarity cross-plot between samples (similar sample pairings: bright colors; dissimilar pairs, dark color) shows major break at ca. 9 Ma, corresponding in time to change in Antarctissa abundance and change in environmental parameters. Diversity rises by ca. 50% in the early Miocene and, accompanied by high extinction rates, declines again in the Plio-Pleistocene. Source: Composite of figures from Renaudie and Lazarus (2013).

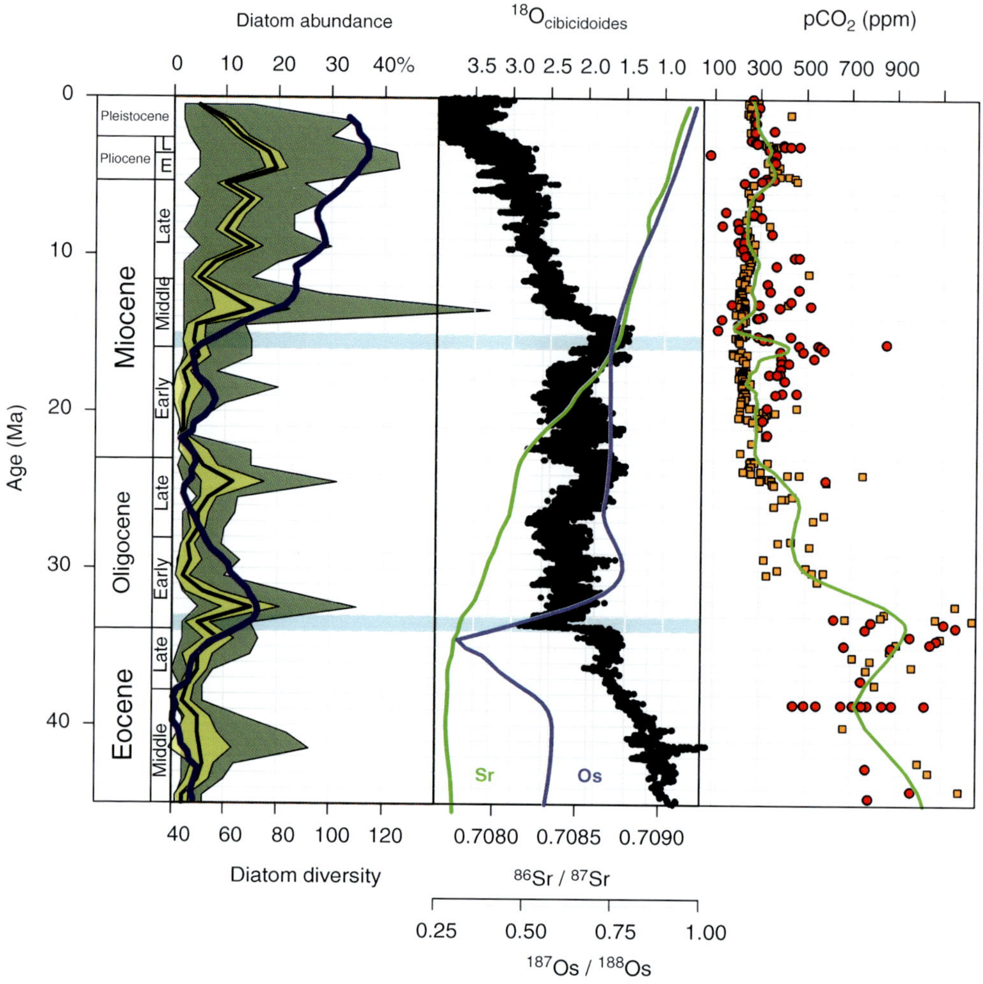

Figure 9.29 Cenozoic ocean silica export (diatom abundance in global deep-sea sediments, diversity); global weathering (Sr and Os), and climate (∂18O), and atmospheric carbon dioxide concentrations (pCO2). Left: abundance median – black line, light shading 95% median confidence interval, dark shading interquartile range, blue line – diversity; Middle: black dots – oxygen isotopes, blue line – Os isotopes, green – Sr isotopes; Right: pCO2 estimates and smoothed curve in green. Horizontal blue shaded intervals show correlated peaks and/or changes in rates of change between parameters. Diatom abundance and diversity, both measures of silica availability in upper ocean waters, increased near the Eocene–Oligocene boundary, and in the middle Miocene. Radiolarian response to the inferred change in silica availability shown in Figure 9.28. Source: Figure and synthesis of diatom abundances from Renaudie (2016), other data sources as given in that publication. Renaudie (2016). Licensed under CC BY 3.0.

7 Paleoceanography

Introduction

Paleoceanography is the study of past oceans: their extent, patterns of circulation, water column structure, and physical/chemical properties, including distribution of nutrients and patterns of biologic productivity. Paleoceanography only developed as a distinct subject in the second half of the twentieth century. A convergence of many factors were responsible for this, including improved technologies for recovering ocean-bottom sediments (described in Chapter 6), industrial (mostly the oil industry) and academic support for micropaleontology studies, the development (using marine geologic data) of the theory of plate tectonics, and others. Last but not least, the field grew due to much higher levels of interest and funding for oceanographic research in general by national military and civilian agencies. Despite its youth, paleoceanography has produced many important scientific insights. The history of the oceans was largely unknown until paleoceanographic research determined its major outlines in the last few decades. Through paleoceanographic studies, we now have been able to reconstruct the major features of ocean evolution over the last ca. 100 my – the interval of geologic time which has been primarily examined by the field, as this is the time interval for which large amounts of deep-sea sediment are available for examination. Paleoceanographic studies have

Paleobiology of the Polycystine Radiolaria, First Edition. David Lazarus, Noritoshi Suzuki, Yoshiyuki Ishitani and Kozo Takahashi.

been a key element in developing the field of paleoclimatology, and in particular, our understanding of the ice ages. Lastly, paleoceanography is one of the disciplines that has contributed to the still-emerging paradigm of earth system science. More detail on the history of the field and how this relates both to general developments in science, and specifically to the history of radiolarian research, is given in Chapter 1.

One of the characteristics of paleoceanography as a discipline which distinguishes from many other disciplines is that the primary subjects of study – past oceans – are not preserved in the geologic record. This is in contrast to, for example, radiolarian research, where at least the shells of the radiolarians are available for examination. Paleoceanography therefore makes use of secondary traces of past ocean conditions, or "proxies"; for the most part things that were formed in past ocean waters and are now preserved in the deep-sea ocean sediments that formed in these oceans. Examples of paleoceanographic proxies include overall sediment composition, microfossil assemblages, stable isotope ratios (e.g. oxygen, carbon) and other element concentrations and ratios of elements, organic carbon abundance and the characteristics of various biogeochemicals contained in the organic fraction, sediment structures on various scales, and many others. New proxies are developed frequently – and new complexities in interpreting existing proxies are discovered frequently as well. Most paleoceanographic studies make use of multiple proxies to obtain a better picture of the ocean for the target study interval, and the research typically involves teams of specialists in different areas of earth science research: e.g. geochemists, sedimentologists, micropaleontologists, and geophysicists.

Paleoceanographic studies usually combine proxy data from a large number of individual samples to reconstruct patterns of variation in past ocean conditions. This can either be a series of reconstructed values over a period of time – a time series – or a set of values from a geographically distributed set of samples, all from a relatively narrow time interval – a time slice (Figure 7.1). Both of these require age estimates for each sample, and to evaluate errors, rates of change, or other properties these need to be expressed numerically. Errors in age estimation in a time slice, for example, limit the accuracy of geographic pattern reconstruction.

For example, mixing glacial and interglacial conditions together would largely compromise the usefulness of a time slice for studying late Pleistocene paleoceanography. Similarly, one of the goals of studying a time series is to determine the rate at which processes take place, so that they can be compared to quantitative models, based on modern oceanographic knowledge of how these processes work. Age errors here can result in erroneous rate estimates and lead to the wrong conclusions in a study. Paleoceanography is thus built in part on the foundation of numerical age estimation of geologic samples, a subject covered in Chapter 8.

In addition to studying rates of change in ocean conditions, time series data (radiolarian or other paleoceanographic proxy data) has often been used to extract frequencies from the history of ocean variability to test theories about orbital influence on paleoceanographic change over time. Some paleoceanographic insight has been gained from the study of orbital frequencies, particularly when such time series are compared to time equivalent series of other types of paleoceanographic proxy data. This can show how different components of the system tend to lead or lag each other in responding to climate change over time (e.g. Cortese et al., 2007). It has been difficult, though, to relate such orbital signals to specific, geographically localized processes that can be directly evaluated with the principles of physical or chemical oceanography. One important spin-off of such studies is the use of orbital frequencies to "tune" geochronologic age models for sections (see Chapter 8).

Paleontologists sometimes make use of specific features of fossil morphology to reconstruct past environments. Examples include teeth and limb structure in mammals, which give information on plant communities (e.g. forest vs. grassland), or plant morphology characteristics, such as leaf structures that are used to quantitatively estimate temperature and precipitation in Cenozoic floras. Micropaleontologists in general, and radiolarian researchers in particular, have not yet been able to employ such methods, as we know far too little about the functional morphology of microfossil shells. Microfossils are employed in paleoceanographic research in two distinct ways. In the first, the chemical composition of their shells can be analyzed, a procedure that is the starting point for most of the geochemical proxies used in paleoceanography. In most cases, multiple specimens belonging to just a single species are used from each sample to extract the geochemical signal, as the signals of water conditions such as temperature vary according to the ecologic habitat of each species in the water column while it lived. The large majority of such geochemical measurements are however not made with radiolarians but with foraminifera, as the mineral calcite, which makes up foraminiferal shells, has been found to be a much better geochemical archive than the opaline silica that makes up shells of radiolarians (and diatoms). There have been some attempts to measure geochemical proxies from opaline silica (e.g. $\partial^{18}O$: Shemesh et al., 1995; $\partial^{30}Si$: Fontorbe et al., 2017), but these have been beset by numerous difficulties in the preservation of the geochemical signal, and as well by the very small size (relative to foraminifera) of radiolarian and diatom shells, which make measurements on a collection of picked specimens of single species very difficult. Radiolarians' use in paleoceanography is thus almost entirely via the second approach. This relies on a taxonomic/ecologic interpretation via the analysis of the relative abundances of different species in individual samples. This method is based on the observation in the modern ocean that the composition of plankton assemblages in the water column is consistently related to ocean-water conditions of interest to paleoceanographers, e.g. temperature, nutrient conditions, and others (Chapter 3). It does not require understanding of the adaptations of each species, only that the occurrences of species are consistently related to environmental conditions. How this observation is translated into paleoceanographic research is detailed below. This chapter, however, will not attempt to summarize the scientific findings of paleoceanography. These are far too vast, complex, and are based on many different types of data and analysis, of which radiolarian studies are only a small part. We will instead examine the specific methods and results from representative publications that use radiolarians as a paleoceanographic tool.

The use of radiolarians in paleoceanography developed in parallel to the use of other microfossils for paleoceanographic research. Early studies in the 1950s of foraminifera from short sediment cores obtained using piston-core methods made it clear that microfossil assemblages could be used to distinguish different past ocean conditions between glacial and interglacial periods of the latest Pleistocene (Thomas, 2008; see also Chapter 1). Studies of radiolarian changes over geologic time from such cores in the late 1950s and early 1960s (e.g. Riedel and Funnell, 1964; Hays, 1965) had difficulty distinguishing biostratigraphic (evolutionary) changes from environmentally driven changes in assemblages of the sort used for paleoenvironmental reconstruction. Separation of evolutionary from ecologic change is essential in paleoceanographic research. Although environmental change may be the cause of evolutionary change, evolutionary responses are one-off events, and will not, for any given species, re-occur even if the same environmental event re-occurs. As an analogy, a seismograph that breaks part-way through a nearby earthquake will record the earthquake signal only up to the point of breakage. Continued shaking will not produce any further response on the graph of the recorder. Extinctions or originations of species present a similar problem to paleoceanographic environmental reconstruction. Thus, only as evolutionary changes and stratigraphic zonations became better defined in the early 1970s (Riedel and Sanfilippo, 1970, 1971; see also Chapters 1 and 6) was it possible to better separate evolutionary from environmental signals. The 1960s had also witnessed the development of biogeographic studies of living and surface sediment distributions of microfossils (see Chapter 3: Ecology), and with these, a much better understanding of how individual species and whole assemblages were related to environmental conditions. These modern biogeographic studies also explored quantitative methods of characterizing assemblages, tools which were then extended to the study of past environmental conditions as a part of developing paleoceanographic research.

Radiolarians are primarily employed in paleoceanographic studies in regions where they are better preserved than other microfossils, particularly in polar regions and areas of upwelling, or where organic carbon concentrations in sediment, and thus dissolution of carbonate, are high. As these regions are also those which respond most strongly to climate change (Ruddiman, 2000; Figure 7.1) they form an important component in paleoceanographic research. Modern paleoceanographic research data from radiolarians is, however, generally combined with other types of data in joint syntheses (Figure 7.1).

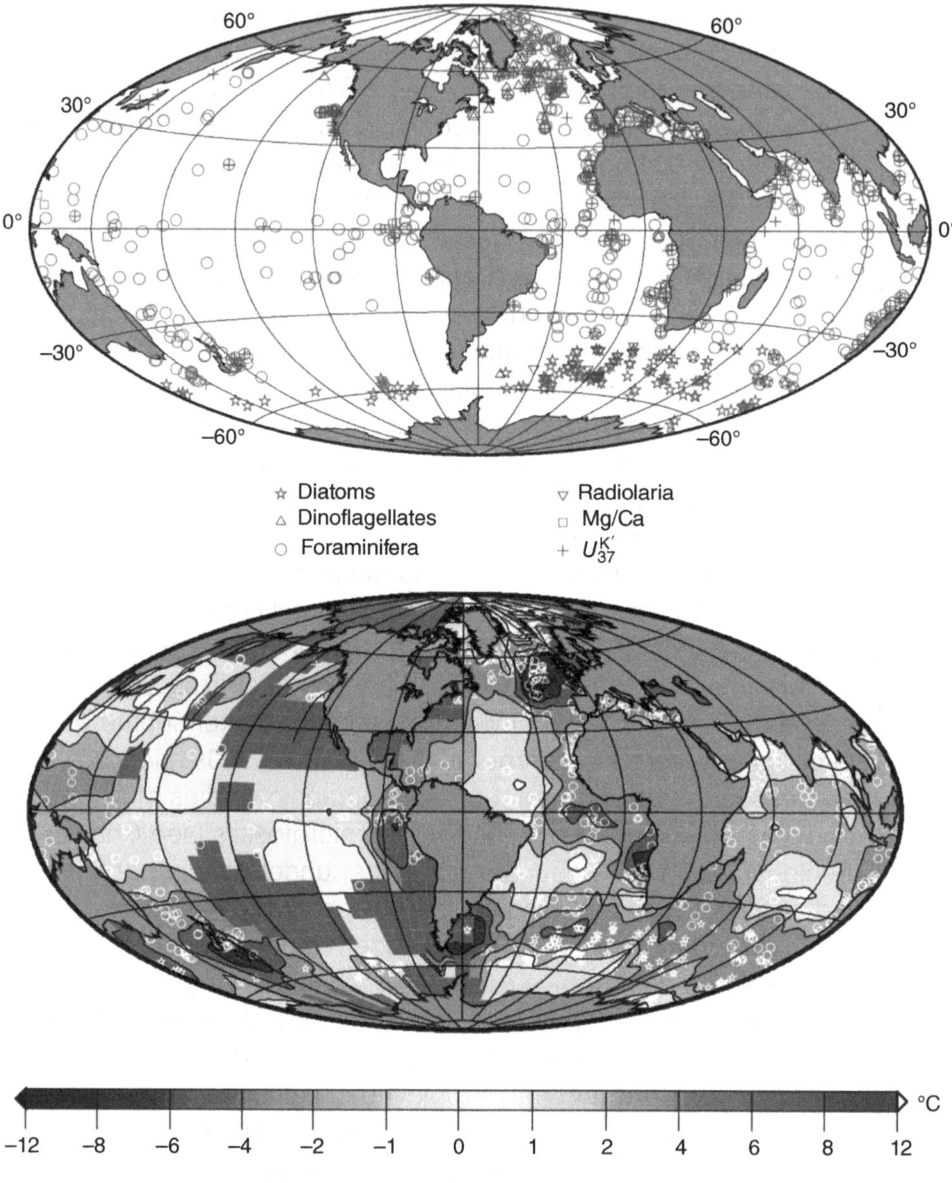

Figure 7.1 Example of modern quantitative paleoceanographic analysis: data and reconstructed temperature changes in the global oceans between the last glacial maximum and modern conditions. Upper panel: locations and types of data used. Radiolarians are used in the southern Atlantic, together with more extensive data from several other microfossil groups and two types of geochemical temperature estimates. Lower panel: difference in temperature from the modern (colder than modern values are negative). Note concentration of temperature change in subpolar and mid-latitude upwelling regions. Source: Waelbroeck et al. (2009). Reproduced with permission of Springer Nature. (*See color plate section for the color representation of this figure.*)

Radiolarians as Tracers of Water Masses

Radiolarians in both modern (Chapter 3) and in many ancient oceans (Chapters 5 and 9) have displayed distinct, partially, or largely endemic biotas associated with major ocean water masses and circulation patterns. In addition to their importance for understanding evolution, the presence of such biotas in sediments provide a simple and robust way to qualitatively trace the movements of water masses and circulation in past times. This is a particularly useful technique in that it makes very few assumptions about individual species biology, and so can be applied in a variety of environments and past time intervals, where the species in the assemblages are extinct and their detailed ecologic preferences are not known. The method is widely employed in paleontology, but radiolarians are particularly useful in paleoceanographic studies because they are one of the few microfossil groups that are present over long geologic time periods and which tend to form distinct biotas in polar regions during colder climate intervals. They are thus very well suited to studying polar water mass development over various time scales. Baumgartner (1992), for example, used this method to interpret Cretaceous water masses and currents off the west coast of Australia. The URI (Upwelling Radiolarian Index) method for estimating upwelling in lower latitudes (described later in this chapter) can be seen as an extension of this approach as well.

Assemblage-Based Methods of Paleoceanographic Analysis

Most paleoceanographic studies using assemblages of radiolarians or other microfossils that go beyond a purely qualitative approach have used a method first developed by Imbrie and Kipp (1971) for planktonic foraminifera. This method (the "Imbrie-Kipp" or "Factor Analytic" method) is based on several distinct steps of data analysis, and is based on the consistent correlation of species abundances in the modern ocean to significant ocean characteristics. Although in principle microfossil assemblages using this method can be used in paleoceanographic reconstruction of any environmental characteristic, the overwhelming majority of research has been in reconstruction of past ocean-surface water temperatures. There is in fact no real theoretical justification for preferring temperature over other possible water column characteristics – as reviewed in Chapter 3, the occurrences of radiolarian species in the water column are regulated by many factors, not just temperature. However, so long as a strong statistical correlation between temperature and radiolarian assemblages can be established using modern correlations, and (critically) the biologic and oceanographic mechanisms that create these correlations do not change over time, a purely statistical approach can be used to estimate past temperature using fossil assemblages. The steps used (highly simplified) are as follows (Figure 7.2):

A, B. Collect quantitative data on the abundances of different species in a set of samples from different locations in the modern ocean. These samples will be used to establish a reference set of how assemblages vary under modern conditions. Usually, only the more common species in the assemblages are counted, with a typical cut-off of at least 5% abundance in at least one sample in the sample set. Depending on how diverse the assemblages are, anything from a dozen to >100 species can be used.

C. Collect information about the environmental characteristic to be reconstructed from the same locations used in the first step. Typically, these are taken

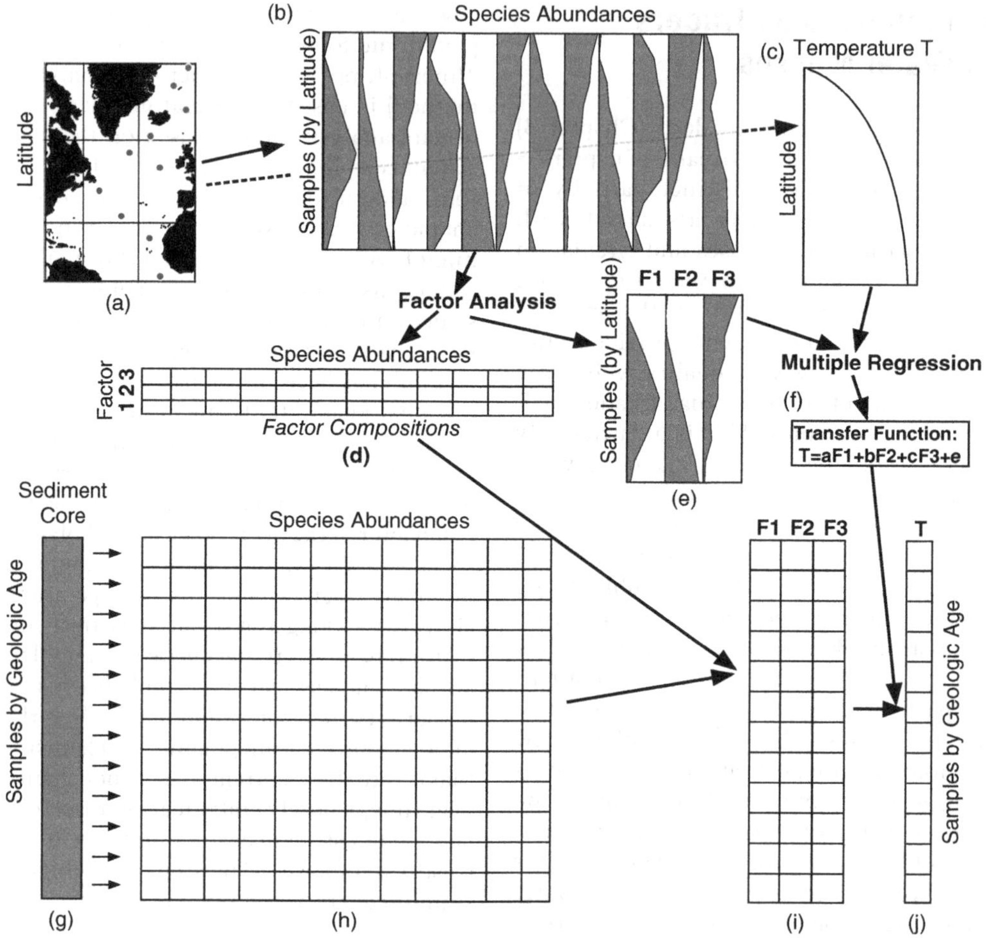

Figure 7.2 Details of the Imbrie-Kipp method of reconstructing past ocean conditions using modern assemblages of microfossils. A) A set of reference samples from surface sediments yields B) a dataset of individual species abundances in each sample (here plotted by sample latitude) and C) values of the environmental parameter (here surface-water temperature T) at each sample location. Factor analysis of the microfossil dataset yields D) a definition, in terms of individual species abundances, for a set of factors (a factor "scores" matrix), and E) the relative abundance of each factor in each of the surface sediment samples (a factor "loadings" matrix). These factor abundances in each sample are compared using multiple regression to the temperature values associated with each sample to create F) a transfer function, which calculates estimated temperatures based on the relative abundances of factors in a sample. A new set of samples is taken from G) a sediment section, and the abundances of the species are counted to create H) a down-core sample by species data matrix. The factor scores matrix from D is used to convert this new data matrix into I) a down-core factor "loadings" matrix. Lastly, the transfer functions from F are used to calculate J) the estimated surface water temperature for each sample in the sediment section.

from atlases or online global compilations of modern ocean data (e.g. http://www.nodc.noaa.gov/OC5/WOD/pr_wod.html).

D. Reduce the complexity of the resultant data matrix (species × samples) to a simpler factor × samples data matrix using a data reduction method called factor analysis. Factor analysis, originally developed by social scientists in the early twentieth century, makes use of the characteristic in most datasets that many variables covary – in this case, that many species tend to have much the same patterns of being relatively rare or common between the different samples analyzed. A small number of new variables (factors) are determined that can display at least most of the variation in the dataset, by replacing several similar (covarying) original variables in each case with a single new one. The methods for doing this generally require a computer and many different solutions ("best" fit number and orientation of new variables) are possible. Factor analysis, and related methods, are very widely used in data analysis, and which methods are most appropriate and other details can be found in any one of hundreds of books that have been published on the subject. In most cases in paleoceanography, only a few factors (almost always fewer than 10) are retained in the analysis, as each factor should ideally correspond to, and be interpretable as, a feature of the modern environment, such as a water mass, seasonal condition, etc. This last point must be understood in a fairly liberal way. Factors (defined in a factor "scores" matrix) are simply a list of weightings to apply to actual abundance values of the species in samples, but, being generated by a statistical method, the weightings are not always easy to understand in a biologic sense. One aspect of many computer program's factor outputs is the existence of negative values for some species in the list of weightings. This means that the factor is defined, at least for the species with a negative weight, by the absence of the species, not by its presence. While not difficult to understand as a statistical definition, clearly this not the same as defining an ecologic assemblage of species for a given water mass or region, where the ecologic interactions, or shared common ecologic traits, are not based on species not even present.

E. Compute the abundance of each factor in each sample in the modern dataset (this table, the factor "loadings" matrix, is generally provided as an output from most factor analysis programs).

F. Use multiple regression to calculate an equation that best estimates the environmental parameter at each sample location based on the relative abundances of each factor in each sample. Multiple regression is also a very widely used statistical method, and details can be found in any textbook. The result of this calculation is what is known as a "transfer function" – it can be used to calculate the environmental characteristic from species abundance data, so long as the species data is first converted to factor representation (again, using a matrix of values that is usually provided by the factor analysis output). The transfer function is generally a linear equation with an error term ("e" in this figure) to account for any mismatch between the calculated and actual value. The magnitude of these error terms can be used to check how well the method is working.

G, H. Collect a new set of species abundance data, using the same species list as before, but for a new set of samples from which one wants to estimate the environmental parameter. These samples could be either a series of samples taken from a single deep-sea sediment core – a time series, or a set of samples from different locations in the oceans, all representing the same short interval of time in the past – a time slice.

I, J. Use the factor compositions matrix developed in "D" above and the transfer function in step "F" to calculate the environmental characteristic for each sample in the new dataset. One can also calculate a statistic ("communality," not shown in this figure) that gives the fraction (between 0 and 1) of the total sample abundance variation "explained" by the set of factors developed earlier. Communality values for each sample in the new data should be within the range of the communalities for the samples used in the modern data, or nearly so. If they are much lower, the environmental estimates are likely to be not very robust. An example of an actual study with radiolarians using the Imbrie-Kipp method is given as Figure 7.3a–d.

The above short summary of the factor analysis method is little more than an outline, and there is a large body of literature discussing details and debating different strategies. Ignoring all the potential statistical and computational complications, a few more general points should nonetheless be mentioned, most of which are equally applicable to the other quantitative methods described in this chapter. First, although it was noted in Chapter 6 that the relative abundance of species in sediments are typically a noticeably distorted picture of the abundances of species in the plankton, this does not necessarily affect the effectiveness of the method. So long as the biases between plankton and sediment are constant, statistical correlations between sediment abundances and water properties will still exist which can be exploited. Second, quantitative methods can be no better than the quality of data used with them, and seemingly minor differences in data generation methods can significantly alter the results obtained. In radiolarian studies, sieve sizes and taxonomic concepts are both critical. Many small species (particularly, in the modern ocean, plagiacanthids and artostrobids) have mean shell dimensions smaller than the 63 μm size used in many studies, and are partially lost when sieving. Thus, counts made of the same species list but with different sieve sizes (e.g. 45 μm, which captures almost all complete radiolarian specimens) will give significantly different results. A transfer function created from a reference modern sample set counted on one sieve size cannot be used for another data set counted on samples prepared with a different sieve size. Taxonomy is obviously also critically important. In stratigraphic

Figure 7.3 Example of radiolarian paleoceanographic temperature reconstruction using the Imbrie-Kipp method, from a Southern Ocean study by Abelmann et al. (1999). Compare to Figure B. A) Location of 53 surface sediment samples, from which the relative abundance of 23 radiolarian species was counted (ca. 400 specimens per sample). B) Factor abundances (loadings) vs. temperature for each of four factors identified by the factor analysis (smoothed curves drawn through original data plots). C) Comparison of actual temperature values at sample locations to temperatures estimated from the surface sediment samples using the transfer function calculated in the study. The standard error in estimation is 1.2°C. D) Paleoceanographic temperature estimates from a piston core near the Polar Front in the South Atlantic, showing both estimates made using downcore radiolarian faunas and the results above, and estimates made using similar methods but with diatom floral assemblages. The core extends from the Holocene at the top to the last glacial maximum (isotope stage 2). Source: A–C from Lazarus (2005). Reproduced with permission of John Wiley and Sons.

studies, "questionable" or incompletely preserved specimens can usually simply be ignored, and only specimens that can be unambiguously identified recorded. In the faunal census counts used in paleoceanographic studies, it is not possible to use such a simple rule of thumb. Although one does not need to unambiguously identify every specimen encountered in the sample, very clear reproducible rules are needed not

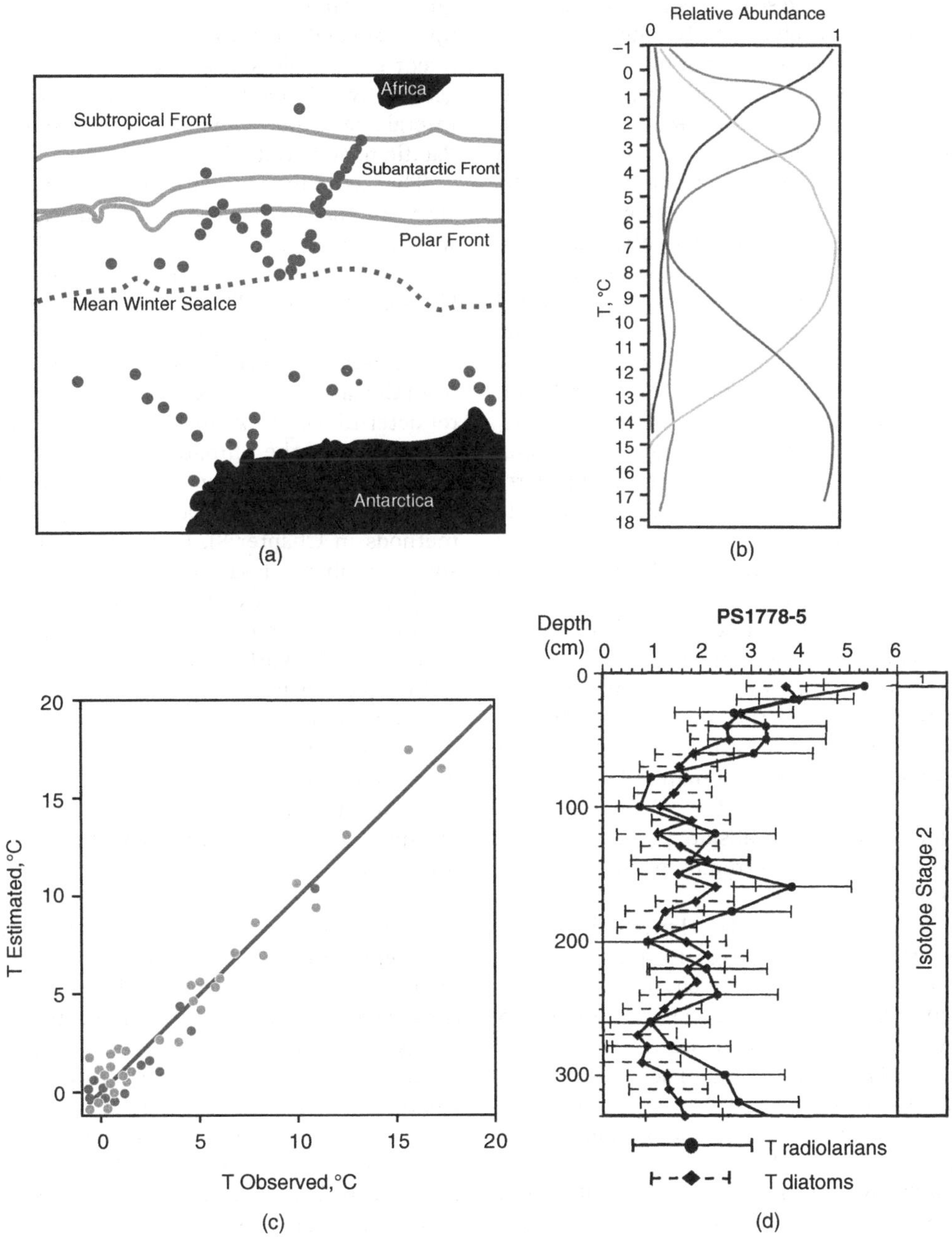

only for the identification of each species, but also for which specimens are to be assigned to "unknown," or "broken." In order to ensure data quality, it is advisable to occasionally do "blind" recounts of samples to confirm that the counts are reproducible. Many researchers even try to have all data for a given study counted by the same individual, so that personal differences in identification are at least consistent throughout the studied dataset. Another question is the trade-off between numbers of individuals counted and the statistical accuracy of the counts simply due to sample size. Commonly, a target of 300 individual specimens is used. This, with a 5% cutoff for species as described earlier, means that, in at least one sample, 15 specimens of the species will have been counted, a number which gives, in relation to count uncertainties due to other factors, an adequate degree of accuracy in the data. A related issue is ensuring that the material counted is representative of the sample from which it was extracted, or, if any biases exist in the sample preparation methods, that these are at least consistent and do not vary between samples. Sample preparation methods are discussed in detail in Chapter 6. For geologically younger materials where the shells are still preserved as opaline silica, random settled slides from wet sieved material is the normal method used. For older materials where the radiolarians are picked from dry sample residues, sample splitters are the preferred method.

The Imbrie-Kipp (1971) method of using microfossil assemblages to quantitatively investigate past ocean environments (i.e. temperatures) was immediately taken up by several groups of researchers, particularly a large consortium of scientists in the newly launched CLIMAP project (see Chapter 1). The first radiolarian papers using the method were published by Sachs (1973a, b) and Moore (1973), all submitted in January of the same year to the then-new journal *Quaternary Research*. The basics of methodology: count numbers (300–500 specimens), sieve sizes (63 μm), etc., and procedures for choosing parameter values in the statistical analysis were established that have largely been followed by subsequent researchers. Temperature changes of up to several degrees (Sachs, 1973b) and major geographic shifts in patterns of biogeography (Moore 1973) over past glacial-interglacial cycles were seen in the North Pacific region studied.

The past climatic history of the climatically important Southern Ocean region was first analyzed using the Imbrie-Kipp method by Lozano and Hays (1976) and Hays et al. (1976). This study used a relatively large set of both reference cores from surface sediments (145) as well as samples from the last glacial maximum (34), the latter determined by examining time series of abundance in the climate-sensitive radiolarian species *Cycladophora davisiana* in dozens of cores (see also stratigraphic methods in Chapter 8). In addition to the sheer numbers of samples and data, this study provides an excellent example of the importance, and methods, of data cleaning. Although 145 surface samples were initially used, an iterative series of analyses and consistency checks were used to identify several large clusters of samples which were anomalous. These samples, mostly in the Crozet and Argentine Basins, were interpreted as being influenced by erosion, reworking, and lateral syn-depositional northward transport of radiolarians by bottom currents. By such analyses, the surface sediment data was filtered down to just 72 high-quality assemblages, which yielded three factors that could be related to the overlying oceanography, and creation of ocean temperature transfer functions. Analyses of the 34 last glacial maximum samples showed northward shifts of several degrees in the position of the Polar Front, while the position of the Subtropical Front did not change significantly.

Molina-Cruz (1977a, b) used factor analyses and transfer functions to examine paleoclimates in the eastern Equatorial Pacific but, finding problems correlating assemblages to temperature, considered other factors, particularly wind stress on the ocean surface, and estimated wind intensity changes over the last 75 kyr. In a later study (Molina-Cruz, 1984), he more directly linked radiolarian assemblages to upwelling and coastal current systems. The Indian Ocean sector of the Southern Ocean and subantarctic region northwards was studied by Dow (1978), who used the Imbrie-Kipp method to relate three surface-water radiolarian factors to temperature, and to estimate temperature changes along the major high-latitude frontal systems during the last glacial. Dow also identified, as had Sachs (1973a) a factor more related to bottom water processes (large-shelled species geographically located near areas of inferred high-bottom water flow). All of these studies, together with data from many other microfossils, isotopes, and geochemical studies, were used in the major syntheses of the CLIMAP project (e.g. CLIMAP Project Members, 1976; Moore et al., 1980 and several subsequent publications). Since then, several other Imbrie-Kipp type paleoceanographic studies using radiolarians have been published, mostly in polar regions where other, geochemical paleoceanographic methods for temperature estimation are difficult to use due to the lack of preserved microfossil carbonate, e.g. Abelmann et al. (1999); Pisias et al. (1997, 2001); Dolven et al. (2002); Cortese et al. (2007) and Lüer et al. (2009).

Although the Imbrie-Kipp method is the most widely used, there are many other methods that have been employed to quantitatively analyse faunal assemblages of radiolarians, and several have been employed in paleoceanographic studies. The first method, recurrent group analysis, was used briefly in the 1960s. Coming from ecology, this method looks for recurring patterns of co-occurrence in data (and thus shares some similarity to the Unitary Associations method in stratigraphy: see Chapter 8). It was successfully used in studies of species in surface sediments to demonstrate coherent groupings of similar species composition that could be related to major surface-water masses in both the North Pacific and Indian Oceans (Nigrini, 1967; 1970; Johnson and Nigrini, 1980; 1982). While having the advantage of working with simple presence-absence information, the output of such analyses is simply a set of grouped taxa names, without quantitative relative contribution information, and with no quantiative procedure to evaluate the recurrent group composition of individual samples. Without these abilities no method was developed that would provide quantitative information on ocean properties, and recurrent group analysis was soon abandoned in favor of the Imbrie-Kipp procedure, which did.

Numerous studies of paleoceanographic change using the Imbrie-Kipp method with various microfossil groups however established that errors in temperature estimates were typically ± 2°C, and that in some ocean regions results were systematically different than what many expected them to be. Estimates by the CLIMAP project of nearly no change in tropical ocean temperatures were particularly controversial, as many climate models suggested changes of up to several degrees. Concerns that the method might be unreliable, and the desire for greater precision, led to the development of several alternate methods of fossil assemblage-based quantitative paleoenvironmental reconstruction. These include the modern analog method "MAT" (Hutson, 1977) and extensions of this (the "SIMMAX" method: Pflaumann et al., 1996). Modern analog methods are conceptually simple (Ortiz and Mix, 1997). A metric that measures overall assemblage similarity is computed between each "new" sample that should become an environmental estimate, and each of the assemblages in a reference dataset, where each reference assemblage is (as in the Imbrie-Kipp method) associated with a value for the environmental parameter. A small number of the samples in the reference dataset that are compositionally most similar to the new sample are selected, and an average of the environmental values associated with these assemblages are used to calculate the environmental value for the new sample. Different variations of this method exist, based on which metric to use to calculate similarity, or how to average the environmental values in the selected samples. Morley (1989) used this method to calculate the latitudinally most similar modern fauna in a reference set of surface sediment samples to downcore sample series in three south Indian Ocean Pleistocene sediment cores from near the modern position of the Antarctic Polar Front. He was able to show that the water mass positions, or environmental gradients, have shifted in this region by several degrees of latitude north or south in response to Pleistocene ice ages. More recently, artificial neural networks (ANNs) have been employed to relate assemblages to environmental parameters (Cortese et al., 2005). These are computer programs that internally simulate, to some degree, biologic neural structures, e.g. several layers of nodes connected with differently weighted interaction links. Such a network can be trained to accurately perform many types of tasks (including classifying radiolarian images: see Chapter 6), by using a training data set. This consists of two parts: first, a reference set of data, which is then compared repeatedly to a second "target" data set. After each comparison, the errors between the predicted and actual values for the target data are used to adjust the weightings between nodes using an algorithm. With the correct algorithm, each new training run tends to produce lower overall errors between the reference and target data. When the system has reached a stable "optimum" state with minimum prediction errors, it can be used to estimate values for new data. The details of how the adjustment algorithms function, or how the network makes its internal adjustments as it develops its capability are generally opaque to the user, so that such systems, while capable of rather accurately estimating e.g. temperature from assemblage composition, provide very little insight as to how assemblages are related to temperature. These newer methods seem to provide somewhat better precision and reliability, but have not yet been as widely used in radiolarian studies as the Imbrie-Kipp method. An example of MAT and ANN results is given in Figure 7.4. Rogers and De Deckker (2007, 2011) compare several of these methods (though not the Imbrie-Kipp method) using data from the Indian and Southern Oceans. Software supporting the Imbrie-Kipp method changes frequently. The original FORTRAN programs are rarely used. At the time of writing, the "rioja" package by Juggins (2015), supporting Imbrie-Kipp, MAT, and some other methods, is available in the open-source, cross-platform computing environment R, which also offers an extremely rich library for other types of data analysis. The free PAST package (Hammer and Harper, 2001, 2008) offers a more limited set of analyses, but with more detailed explanations.

In older sediments, the Imbrie-Kipp method, or any other method based on calibrations to modern ocean conditions, cannot be reliably used. This is because, as noted above, the methods depend on the fossil assemblages having the same relation to oceanographic parameters, e.g. temperature (expressed by the transfer functions), which can only strictly be true if the fossil assemblage consists of the same set of species as the reference set from surface sediments, if none of the species used has

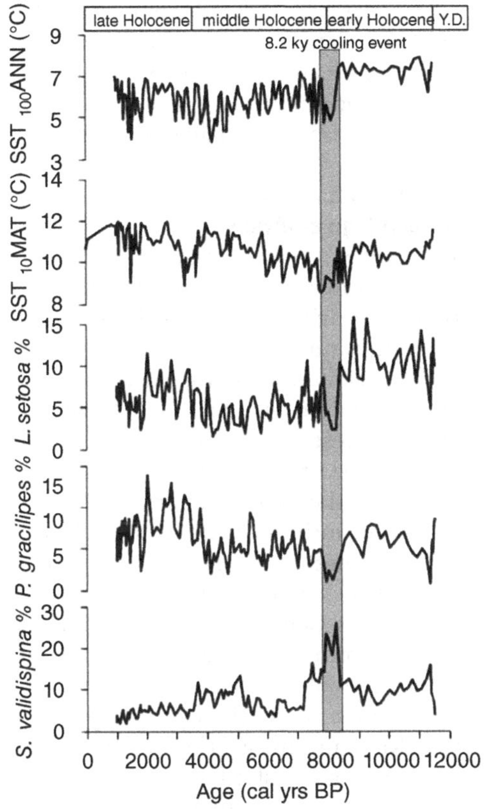

Figure 7.4 Water temperature estimates over the last ca. 12,000 years from a sediment core in the Norwegian-Greenland Sea, along with abundance fluctuations in three radiolarian species used to make one of the estimates. Top panel: estimated water temperature at 100 m water depth, using radiolarian assemblages and an artificial neural network (ANN) program. Second panel: water temperature estimated for 10 m water depth from the same core, but using planktonic foraminiferal assemblages and the modern analog technique (MAT). Lower three panels: abundances of three of the radiolarian species used to calculate the radiolarian ANN temperature estimate. Source: Cortese et al., 2005. Reproduced with permission of Elsevier.

evolved in some fashion that would alter its ecologic response behavior (Imbrie and Kipp, 1971), and that the physical-chemical ocean system has also largely not changed in its fundamental behavior. For example, if the correlation between species abundances and the target paleoenvironmental parameter (e.g. temperature) are in fact due to both being correlated to another intermediate parameter (such as a correlation that both temperature and species abundance might have to nutrient concentrations), then it must be assumed that this pattern of covariation remains constant in the past, even if the specific values do change. The large majority of studies, regardless of fossil group or precise method used, have been made for the late Pleistocene (ca. <500 kyr), where it seems safe to assume that evolution has made at most only minimal alteration to the ecologic behavior of the assemblages being used for analysis, and one can hope that the relationship between direct and indirect parameters has largely remained stable. These assumptions can be at least indirectly checked in the Imbrie-Kipp method by examining the extent to which the factors, defined from modern assemblages, successfully explain (in the statistical sense) the composition of fossil assemblages. This parameter, the "commonality," should remain nearly constant in older samples, and close to the value it has for the modern assemblages. Should it decline significantly, it is likely that the assemblages have a substantial aspect to their ecologic behavior that is not captured by the factors defined for modern environments. Temperature or other paleoenvironmental estimates made using these fossil assemblages are not very trustworthy. A few authors have attempted to extend the Imbrie-Kipp method into the Neogene (e.g. Moore and Lombari, 1981) by restricting the modern analysis and transfer function creation to only those species which are also present in the target older time interval. This however excludes any older taxa with ecologic adaptations that were so different from the modern conditions that they no longer exist. By analogy, it is like using only brackish water fish species – common to both saltwater and freshwater environments – to estimate paleosalinity. A major shift, e.g. from saltwater to freshwater, would be overlooked. A comparison to independent data on paleotemperature, such as from stable isotopes or alkenones that would be needed to test the results of such extended usage has also not been made.

A less problematic approach to quantitative study of paleoceanographic changes in older time intervals is to forgo a direct calibration to absolute values of environmental conditions using modern species occurrences, and instead attempt to only recover patterns of relative change, using only the patterns of distribution in time and space of factors or other statistical summaries of quantitative species' abundance data. The paleoenvironmental interpretation of each factor is then done by indirect reference to other available proxy data, the geographic distribution of taxa, or other means. This method is very general and can be employed in principle with any fossil data. In radiolarian studies it has been used frequently. For example, in Keany (1978), radiolarian occurrences in several early Pliocene Southern Ocean cores were summarized using factor and principal components analyses (the latter being a special case of the former), and, using several sources of information (correlation to stable isotopes, geographic preferences for a few, still living species, and others), temporal changes in principal component abundances were interpreted as showing a significant cooling during the early Pliocene of this region from prior warmer conditions. Cooling after an early Pliocene warm interval has subsequently been confirmed globally, and the warm interval is now the subject of much study as part of global climate change research (e.g. Federov et al., 2013). There are numerous other quantitative methods that can be used to examine how different types of microfossil data are related to each other, or to environmental parameters (e.g. canonical correlation analysis, Pisias et al., 2001), many drawn from ecologic research.

Non-temperature uses of assemblage analyses

Until now we have reviewed the use of radiolarians in reconstructing, for the most part, past ocean temperatures. Temperature is the best understood characteristic of modern oceans, and is of great importance in understanding past oceans. It is particularly of interest in using past ocean conditions to test the computer models of modern global climate that are used to predict future anthropogenic global climate change. Nonetheless, radiolarian assemblages reflect many other aspects of the ocean environment, and can in principle be used to examine past ocean conditions for these aspects of the environment as well. Given the purely statistical nature of the Imbrie-Kipp and most other methods, one can generate a transfer function for almost any desired ocean parameter, though if the parameter chosen is only very indirectly linked to the actual proximal controls on species abundances, the reliability of any function derived is rather questionable. Ideally, any statistical method correlating assemblage composition to environmental conditions should attempt to identify the most proximal controls on this composition, and develop correlations only to these proximal controls. In some cases, preliminary tests of the strength of correlation between assemblages and different potential proximal controls suggest that indeed, temperature is most important, but in other cases other oceanic characteristics appear to more important, or that subsurface temperature rather than surface temperature is (Cortese et al., 2005). However, other than in the pioneering studies by Molina-Cruz (1977a, b; 1984), radiolarian workers have not tried to develop statistical correlation methods for ocean parameters other than temperature, even though it has been done using other microfossil groups (e.g., using the Imbrie-Kipp method with planktonic foraminifera to reconstruct thermocline depth in the tropical Pacific: Andreasen and Ravelo, 1997). Instead, methods have been developed to reconstruct ocean conditions based on a priori knowledge of the species' ecology, with three methods in particular of interest: the Upwelling Radiolarian Index (URI), the Thermocline-Surface Radiolarian Index (TSRI), and the Water Depth Ecology (WADE) method.

The URI method was developed as an offshoot of earlier observations on the biogeography of species in surface sediment studies of low to mid latitude upwelling regions. Several studies showed that there were distinct species which, in low to mid-latitudes, shared a distribution pattern restricted to several disjunct regions, all in areas of strong offshore coastal upwelling (work summarized in Nigrini and Caulet, 1992). Some of these species appeared to be endemic to such regions, others had broader distributions in subpolar regions, but were otherwise absent from low to mid-latitude locations. Caulet et al. (1992) proposed using the relative abundance of these species in radiolarian assemblages as an index of upwelling activity. In this study, abundances of URI species in a sediment core in the Somalian upwelling region were shown to vary over time between 1% and 6%, and to largely covary with carbonate concentration in the sediment. The latter in this location was interpreted, based on comparison to a dissolution index of foraminifera, to be largely a reflection of surface water productivity, with higher carbonate reflecting greater fluxes from surface waters into the sediment. This correspondence, as well as other indicators, led to the conclusion that the URI index could be used as a proxy for upwelling strength. Several subsequent papers expanded this idea, all from studies of upwelling regions in the western Indian Ocean (Vénec-Péyre and Caulet, 2000 and earlier papers cited therein). The URI method has been used in only a few other studies. Jacot Des Combes and Abelmann (2007) for example calculated the URI for a time series of radiolarian assemblages over the last 350 kyr from a core in the well-studied Benguela Upwelling System (BUS) off of western Africa, but found the pattern complex, and not closely related either to total organic carbon (TOC) in the sediments (a commonly used geochemical indicator of paleoproductivity), or to the glacial-interglacial cycle. Rogers and De Deckker (2007) calculated the URI for a large latitudinal transect of surface sediment samples in the eastern and southern Indian Ocean but found no correlations to surface-water conditions, possibly due in part to missing rare URI species from counting only 300 specimens/sample.

The URI method is essentially a biogeographic method, using an assemblage associated with a specific water mass to trace changes in water mass influence over time. It is thus a specific example of a more general approach which can be applied to most ocean regions and intervals of geologic time. Limitations of the URI method include the counting aspect just alluded to: because most URI species are rather rare, at least several hundred, and usually several thousand specimens need to be counted to yield meaningful percentage values for the index. This is a much larger number than is needed for most other quantitative paleoceanographic methods. Because the URI consists of a mixture of endemic low-latitude upwelling region species, and species that are widespread in polar regions, it is not easily applied in high latitudes, or even as an upwelling indicator in low latitudes outside of major upwelling zones. Lastly, the URI, like other similar assemblage-to-water mass associations, is based only on a biogeographic observation, and the physical oceanographic parameters that are responsible for the presence and/or increased abundance of these species are unknown. It is thus difficult to interpret the URI in terms of physical or biologic oceanographic processes, or to convert the index into a quantitative proxy of such processes.

A different approach to understanding past ocean productivity was introduced by Jacot Des Combes (Jacot Des Combes et al., 1999a, b; 2005), based on an idea first suggested by Casey (1977). Casey had noted that in mid- to low latitudes, many radiolarian species were present at thermocline or subthermocline depths, where they fed on the flux of organic material from surface water production. He speculated that, in regions of elevated export productivity, one would expect to find also relatively higher abundances of deep-water radiolarian species, at least in comparison to other low to mid-latitude regions with strongly stratified, oligotrophic surface waters, where most production and consumption took place above the thermocline (Figure 7.5). This idea is similar to using the rate of benthic foraminifera accumulation to estimate paleoproductivity (Berger et al., 1994), although benthic foraminifera at the sea floor receive only a much smaller fraction of exported carbon than do organisms located at the thermocline, or midwater depths. Jacot Des Combes created a simple index of deep- to surface-water radiolarians based on the relative abundances of six (deeper water) "Thermocline" species:

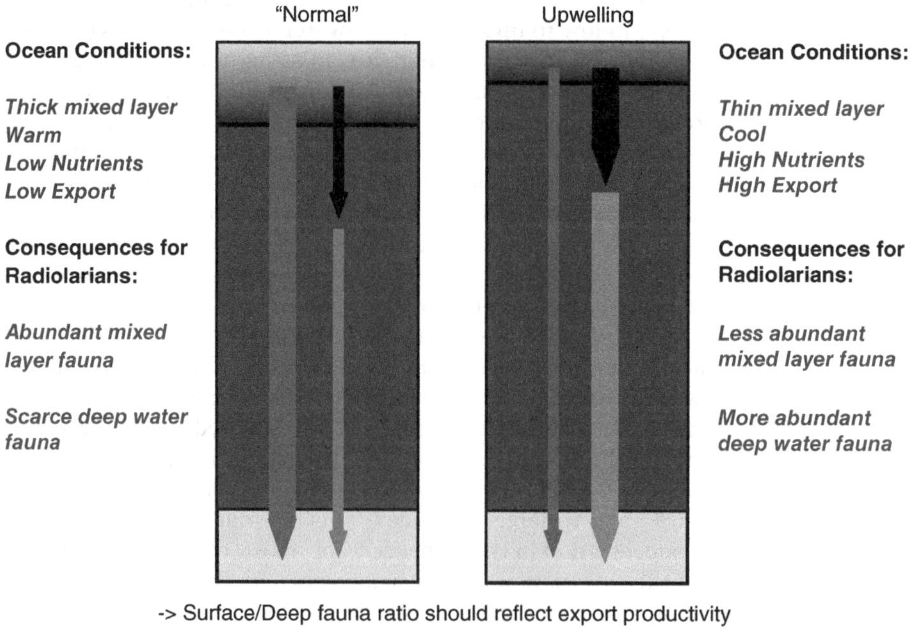

Figure 7.5 Concept underlying using water depth preferences of radiolarians to estimate export productivity. Under "normal," well-stratified ocean conditions there is a large flux of microfossils from the well mixed surface-water layer (thick left-hand arrow), while low export of organic carbon (thin black right-hand arrow) supports similarly low amounts of deep water taxa (lower light arrow on right). In areas of enhanced export productivity such as in an upwelling zone (right panel) these relations are reversed, resulting in a proportionally larger flux of deep water species into the sediments. Source: Lazarus, Paläontologische Zeitschrift, 2005. Reproduced with permission of Springer. (*See color plate section for the color representation of this figure.*)

Cycladophora davisiana, C. cosma (as "*C. davisiana cornutoides*"), *Cyrtolagena laguncula, Eucyrtidium calvertense, Larcopyle buetschlii*, and *Spongodiscus osculosus* (as "*Spongopyle oculosa*") and three surface-water species: *Heliodiscus asteriscus, Lithopera bacca*, and *Siphonosphaera polysiphonia*. This "Thermocline to Surface water Radiolarian Index," or TSRI, would increase in value as the relative abundance of deeper water radiolarians increased, and should, in Casey's model, reflect higher export productivity. In the above papers, time series of TSRI in several piston cores from the equatorial Indian Ocean were compared to other indicators of past productivity, including bulk (carbonate dominated) accumulation rate, organic carbon (TOC), and various bulk sediment geochemical parameters. The correspondence of the TSRI to other indicators, or to the glacial-interglacial cycles, varied according to the section studied; matching one or the other in one section (Figure 7.6), but not in another section. It could not be determined if this reflected the regional complexities of productivity history, problems with the other proxies, or if the TSRI itself was not reflecting export productivity reliably.

One clear limitation of the TSRI is the reliance on only a very small number of species. Abundance variations in even one species due to factors other than export productivity (particularly surface-water species) could significantly modify the TSRI values. For this reason, a broader assemblage-based index would be likely to be more reliable. A first step in this direction was taken by Weinheimer (2001) who used extensive data on depth preferences for radiolarian species from numerous prior studies from both Atlantic and Pacific to calculate the changing ratio of "intermediate" water species vs. "warm" or "transitional" species in an ODP section from the Benguela Upwelling System (Figure 7.7). She briefly suggested that the pattern might be used as a proxy for upwelling

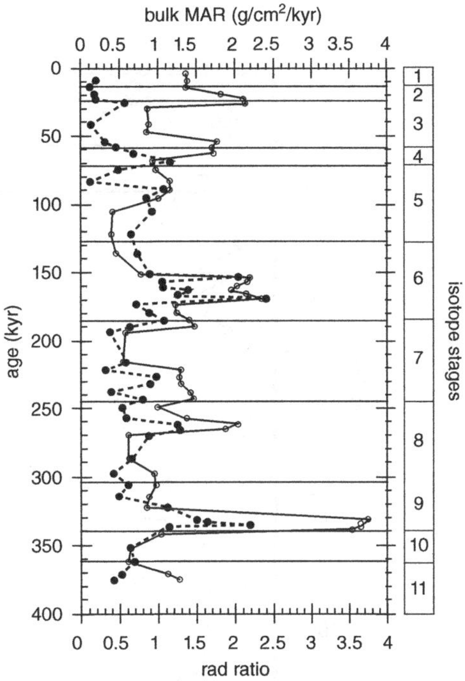

Figure 7.6 Jacot Des Combes' TSRI productivity proxy (solid line with open circles, ratio scale on lower x-axis) and another proxy for productivity (dotted line and solid circles, bulk accumulation of sediment "MAR") in a piston core from the equatorial Indian Ocean. Geologic age in kyr, with marine isotope stages (glacial-interglacial) on right. Source: Jacot des Combes, Marine Geol., 1999. Reproduced with permission of Springer.

intensity. This idea was formalized under the name the "Water Depth Ecology" method (WADE) by Lazarus (2005). He defined the index as the ratio of warm to intermediate water species (the scale thus being the inverse of the TSRI). This was applied in two more detailed studies of the BUS (Lazarus et al., 2006, 2008), covering the latest Miocene–Recent. The WADE index, based on but extended from the set of species defined by Weinheimer, was found to correlate well with other proxies of export productivity such as TOC, and generally to show higher export productivity

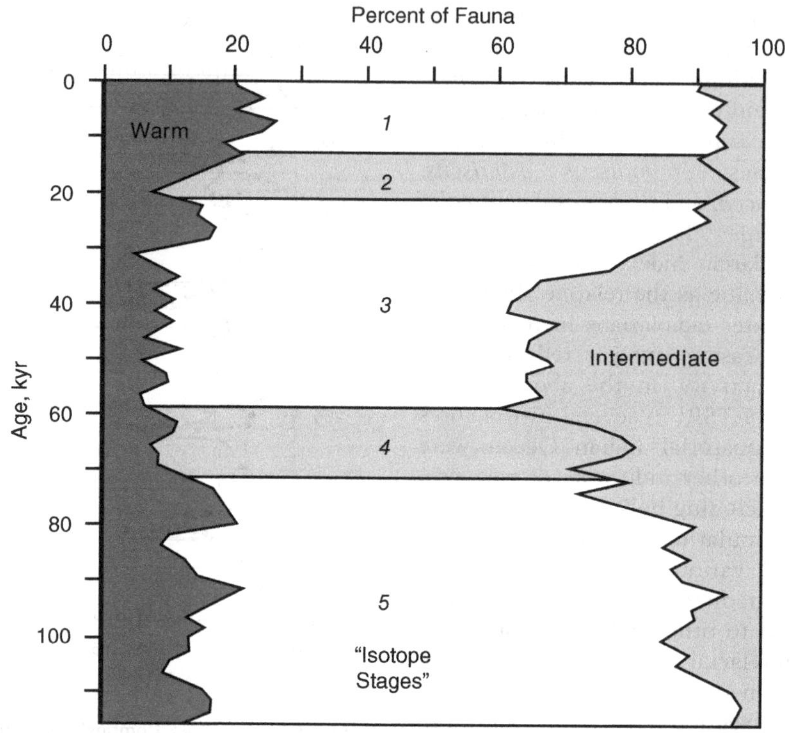

Figure 7.7 Increasing relative abundance of intermediate (deeper) water species in radiolarian assemblages from the last glacial maximum (isotope stages 3 and 4) from an ODP hole drilled into sediments underlying the Benguela Upwelling System, coastal Atlantic, southern Africa. Higher relative abundances of deeper water species are interpreted as indicating stronger upwelling and increased export productivity from surface to deep waters. Source: Lazarus, Paläontologische Zeitschrift, 2005 after data given in Weinheimer (2001). Reproduced with permission of Springer.

during glacials (Figures 7.8, 7.9). Further study of this method is needed, however, to understand the reasons for the correlation, and to quantify the relationship to productivity if any. Periods of upwelling in the BUS system for example are associated with cooler surface waters, and the ratio could simply reflect a reduction in warm-water taxa, rather than an increase in deeper water taxa due to increased organic matter export.

Lastly, individual species can be used for paleoceanographic purposes when the ecology of the species is well enough known. This is so far largely restricted to just a few Recent species whose distribution in the water column is well studied. Hays and Morley (2004) and Hays et al. (2013) used the high abundances of *Cycladophora davisiana* and selected other species in high-latitude polar oceans to argue that these oceans had, in glacial times, a water column structure similar to the modern Sea of Okhotsk (a cold, low-salinity stable water layer from 0–200 m, with warmer, more saline waters below). This inference is based on the ecology of these species, which are largely restricted to intermediate depths in the Sea of Okhotsk today, where precisely such unusual (for the modern

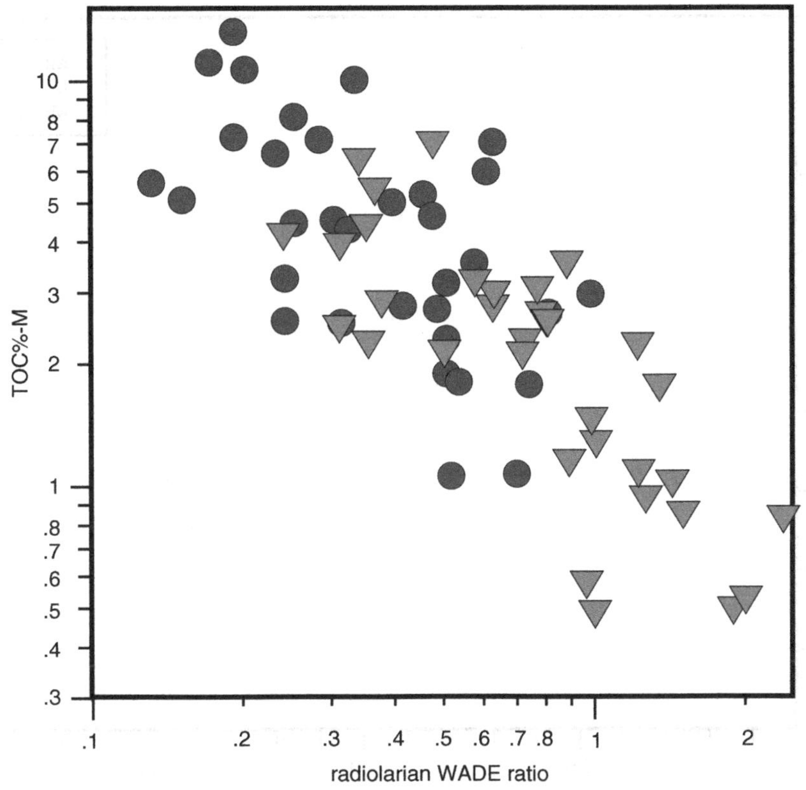

Figure 7.8 Correlation of radiolarian Water Depth Ecology ["WADE"] index (lower values indicate higher relative abundances of deep water species) and Total Organic Carbon concentrations ["TOC," a common paleoproductivity proxy] in the same samples. Samples range in age from latest Miocene to Recent and come from two nearby deep-sea drilling sites (circles and triangles) in the Benguela Upwelling System, coastal Atlantic, southern Africa. Source: Lazarus et al., 2006. Reproduced with permission of Elsevier.

ocean) water column conditions are found. Under these conditions, radiolarian fluxes into the sediments are dominated by production of these intermediate water forms.

Radiolarians in Bulk: Summary Indices and Non-Taxonomic Uses of Radiolarians in Paleoceanography

Until now, all described uses of radiolarians in paleoceanographic research have been based on how radiolarians behave in modern oceans, with many of the methods explicitly linked to precise calibrations between living radiolarian assemblages and modern ocean conditions. However, many uses of radiolarians as paleoceanographic indicators are carried out with much older assemblages. For the most part, the primary method is simply to use the presence of radiolarians, in bulk, without regard to the assemblage composition, as an indicator of high productivity. There are likely hundreds of papers which have interpreted radiolarian occurrences in older sediments

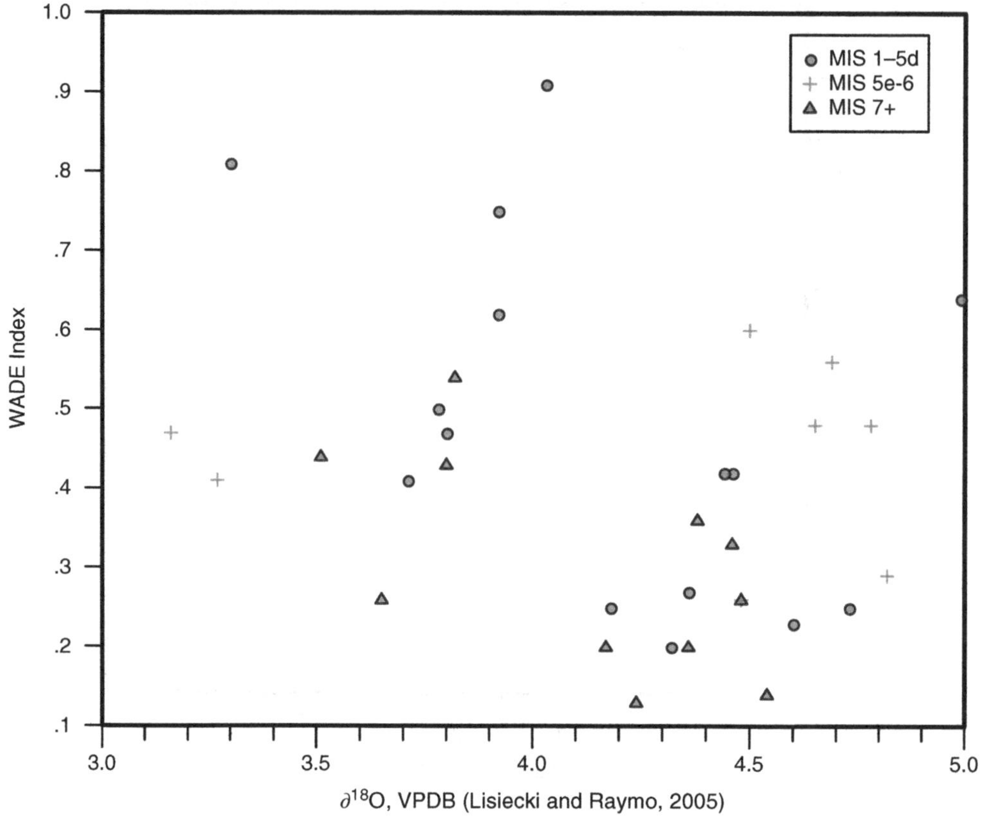

Figure 7.9 Comparison of radiolarian WADE index values to approximate glacial-interglacial state (using inferred $\partial^{18}O$ values) for samples covering the last few glacial-interglacial cycles in an ODP Hole drilled in the Benguela Upwelling System, coastal Atlantic, southern Africa. Lower ("lighter") oxygen isotope values are characteristic of warm interglacial states. These samples generally have higher values of the WADE index than for glacial samples (higher oxygen isotope values), suggesting lower relative abundances of deep water species and thus lower export productivity. Source: Lazarus et al., 2008. Reproduced with permission of Micropress.org.

this way. The justification for this is only occasionally stated, but almost invariably claimed to be based on the fact that occurrences of radiolarians in modern ocean sediments are primarily in regions of relatively high productivity. Independent evidence for high export productivity is only rarely available and generally rather indirect, although in these cases, the pattern of radiolarian occurrence does indeed seem to suggest that they are related to above-average export productivity, although often nearly equally due to the absence of carbonate, or existence of dysoxic depositional conditions (Bak, 2007; Aguado et al., 2008; Figures 7.10, 7.11). Sedimentation rates for Mesozoic radiolarian-rich sediments are also generally much lower than those seen for modern areas of upwelling (Bak, 2007), suggesting that bottom-water controls, or sedimentary controls (winnowing or downslope concentration of skeletons, Chiari et al., 2007) may at least sometimes be the primary reason for radiolarian preservation

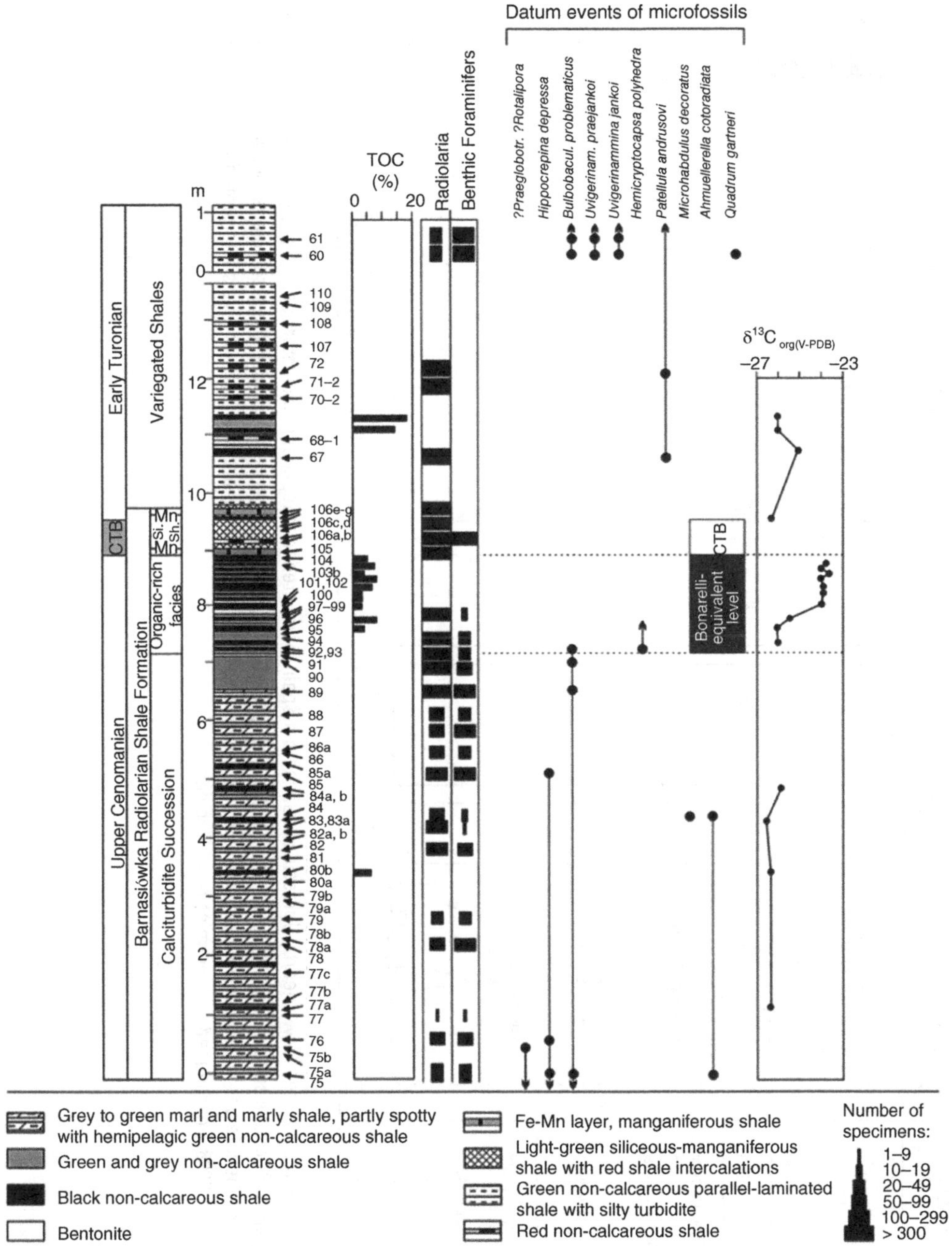

Figure 7.10 Radiolarian abundances in a late Cretaceous section from Poland crossing the black shale level (OAE event) at the Cenomanian–Turonian boundary, compared to bulk lithology and indicators of productivity or bottom water dysoxia (Total Organic Carbon: TOC). Radiolarians are most abundant across the interval with elevated TOC, but also show a strong correlation to sediments without carbonate, which dilutes biogenic opal and hinders preservation in modern sedimentary environments. Source: Bak, 2007. Reproduced with permission of Elsevier.

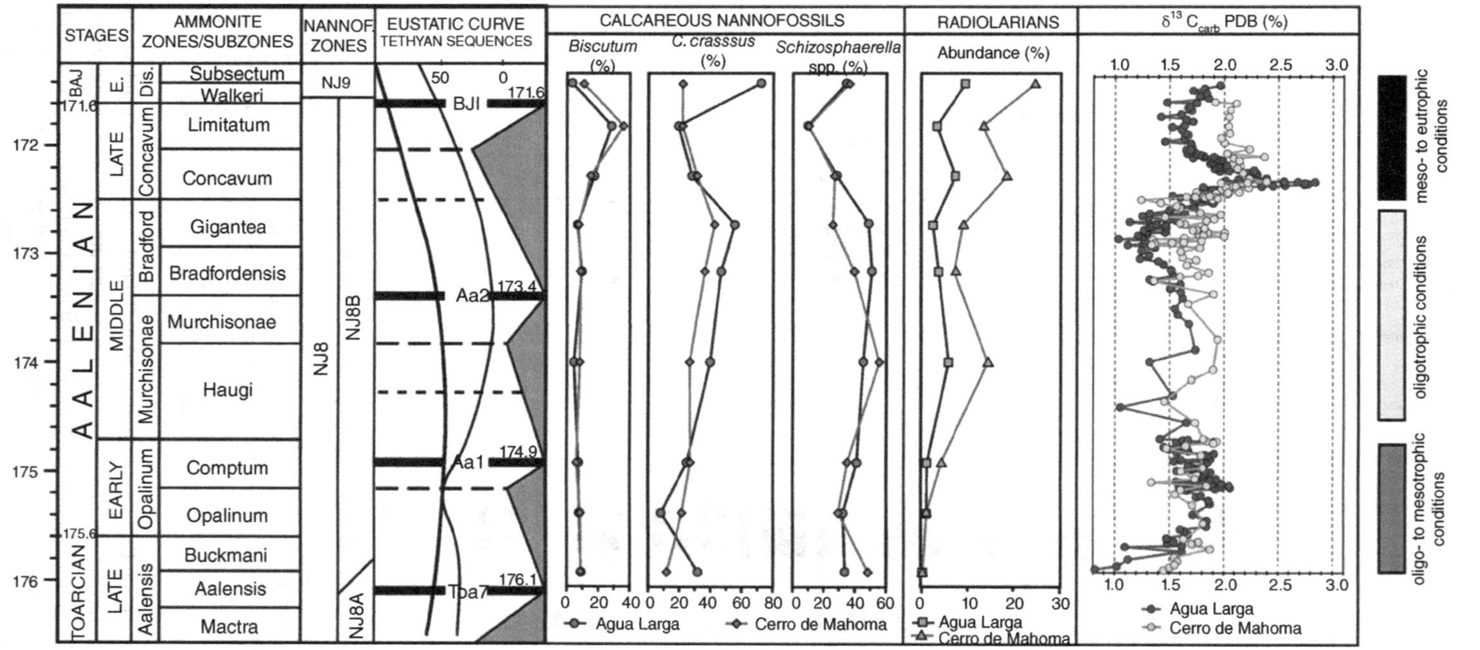

Figure 7.11 Radiolarian abundances in two nearby Jurassic sections from Spain, compared to independent indicators of high productivity (the calcareous nannofossil genus *Biscutum* and carbon isotope shift). Summary interpretation of productivity given on right. Highest radiolarian values are seen in the most productive time interval (meso-eutrophic), but moderately high radiolarian abundances are also seen at levels interpreted as oligotrophic. Source: Aguado et al., 2008. Reproduced with permission of Elsevier.

rather than high export productivity. A further complicating factor is that high export productivity can, by the consumption of organic matter in deep waters and surface sediments, itself create dysoxic or anoxic conditions that dissolve carbonate in sediments while enhancing preservation of opaline silica, thus indirectly linking productivity to the preservation of radiolarians in sediments.

In older geologic intervals, the basic relationships between radiolarians and their occurrences in the geologic record may however not be the same as in modern ocean environments. This is because in the modern oceans, the occurrences of radiolarians in sediments is largely controlled by the distribution of the far more abundant biogenic opal from marine diatoms, a group which first became globally prominent in the Eocene. In modern oceans it is the high flux of diatoms into the sediments that provides sufficient opal to prevent complete dissolution of the opaline skeletons of the radiolarians that also are present. High diatom fluxes in modern oceans directly reflect high export productivity in surface waters, as diatoms tend to out-compete other phytoplankton in highly productive waters, and more readily form mats and aggregates that sink rapidly, efficiently exporting organic materials into the deep ocean and sea floor (Longhurst, 1998; see also Chapters 2 and 6). What were the conditions of opal preservation in older oceans? This is not very well known, but the basic principles of preservation (Chapter 6), and knowledge of how ocean sediments and chemistry have evolved over the Phanerozoic, allow at least a rough characterization to be made (Maliva et al. 1989; Racki and Cordey, 2000).

In the geologic time interval between the early Paleocene and the late Triassic, diatoms were either absent, or not sufficiently abundant in most open ocean regions to create diatom rich sediments, let alone diatom oozes. With the absence of substantial numbers of diatoms, less opal was exported from the surface ocean, and the concentration of dissolved silica in ocean waters was likely to be substantially higher than in modern oceans. Importantly there is no evidence during this time period for significant amounts direct precipitation of silica from ocean waters. Thus, silica was being removed, and held at concentrations below the saturated state that would permit direct precipitation by other factors. The only known alternative is removal by organisms, and in the absence of diatoms, this would have been the radiolarians, and to a lesser degree siliceous sponges. Thus assuming a steady state, radiolarian export into sediments would have fulfilled the same geochemical function as diatoms do today, and at least some radiolarians would need to be preserved in sediments, to ensure that a balance was maintained to newly weathered dissolved silicon continuously being input into the ocean. Where in the oceans would this be most likely to occur? Coccolithophores (and, from the Jurassic, also planktonic foraminifera) were sufficiently abundant in these oceans to create a significant flux of carbonate into the deep ocean, and, in regions above the carbonate compensation depth where all carbonate dissolves, to create carbonate-rich deep ocean sediments. Deep-ocean waters were also about 10°C warmer than those of today. All of these factors must have significantly altered the processes which regulated the preservation of radiolarians in sediments. Presumably, as is true in modern oceans, radiolarian shells, often aggregated with other organic materials, sank fairly quickly to the ocean bottom (see also Chapter 6: Preservation). Here, in the surficial sediment layer, most would dissolve due to the corrosive effects of bottom water. However, higher dissolved silicon concentrations reduced the rate of dissolution of opal shells, even though this was to some extent counterbalanced by the higher dissolution rate of opal due to warmer water (about 2X for the 10°C temperature difference). As the partially dissolved opal became buried by further material, the final fate of the radiolarian shells would depend on the sediment pore waters becoming saturated with respect to dissolved silicon before the last shells were dissolved. This would be more likely to happen in sediments consisting mostly of radiolarians or other opaline material (e.g. sponge spicules), and less likely to occur in sediments with a large carbonate component. Radiolarian preservation would be most likely in regions where the absolute flux of radiolarians was above the ocean average, and where the preservation of carbonate (or influx of other diluting material) into the sediments was fairly low. If radiolarian absolute fluxes were in these oceans still proportional to export productivity, this suggests that radiolarian occurrences in sediments in these older oceans indeed reflected above-average export productivity, although it is not clear to what degree this reflected the highly focused, localized patterns of export due to upwelling that we see today, vs. a much more diffuse pattern of surface-water productivity that may have been present in older oceans (Muttoni and Kent, 2007; Lazarus et al., 2009). It is also important to recognize that unlike diatoms, which are the export productivity, there is no similar automatic link between high ocean productivity and high productivity of radiolarians. Despite high export productivity, other (mostly unfossilizable) zooplankton groups could locally have largely displaced radiolarians in the water column, so that there would be no radiolarian productivity signal in the sediments. In this model, radiolarian occurrences were also relatively strongly influenced by dilution by other sedimentary components, since the primary flux of primary producer coccolith carbonate would normally be much larger than the opal flux from zooplanktonic radiolarians, higher in the food chain. Radiolarian occurrences in sediments would be influenced by secondary control, e.g. water depth dissolution of carbonate. Many regions of above-average surface-water productivity may have been present that exported high fluxes of both radiolarians and carbonate to the sea floor, but only in regions where the carbonate was dissolved would radiolarians be likely to be preserved in the sediments, which is in accord with the common pattern of higher radiolarian abundances in carbonate poor sediments as reviewed above.

In even older oceans (early Triassic to the first widespread occurrences of radiolarians in the Ordovician), the controls on radiolarian deposition are even less clear. In these oceans most carbonate was still deposited in shallow waters on continental shelves, as open-ocean carbonate-secreting plankton had not yet evolved (Ridgwell and Zeebe, 2005). Radiolarians were probably not as abundant as in the later Mesozoic oceans, and relatively more opal export occurred via shallow (and probably deep) water siliceous sponges. Dissolved silicon concentrations in seawater were probably even higher than in the later Mesozoic oceans. Under these conditions, radiolarian

Figure 7.12 Using an integrated multiproxy approach to interpret paleoceanographic signals in radiolarians. In this study of Permian radiolarians from epipieric sea sediments, radiolarian bulk abundance ("richness" in upper panel), and ratio of spumellarians to other radiolarian taxa ("sphaerellarians %") are compared to changes in sediments (left-hand column) and various geochemical parameters. Thin black lines in the data plot columns are 7-point moving averages. Together, the sedimentologic, radiolarian, and geochemical data yield a model for fluctuating changes in ocean conditions (lower panel) in this ancient seaway. Source: Jin et al., 2012. Reproduced with permission of Elsevier.

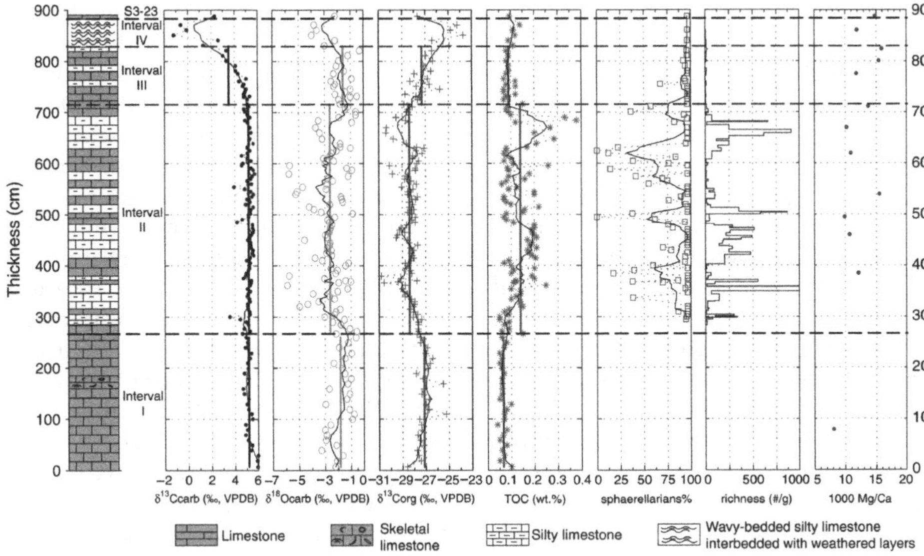
S3-23
Interval IV
Interval III
Interval II
Interval I
Thickness (cm)
δ13Ccarb (‰, VPDB)
δ18Ocarb (‰, VPDB)
δ13Corg (‰, VPDB)
TOC (wt.%)
sphaerellarians%
richness (#/g)
1000 Mg/Ca
Limestone
Skeletal limestone
Silty limestone
Wavy-bedded silty limestone interbedded with weathered layers

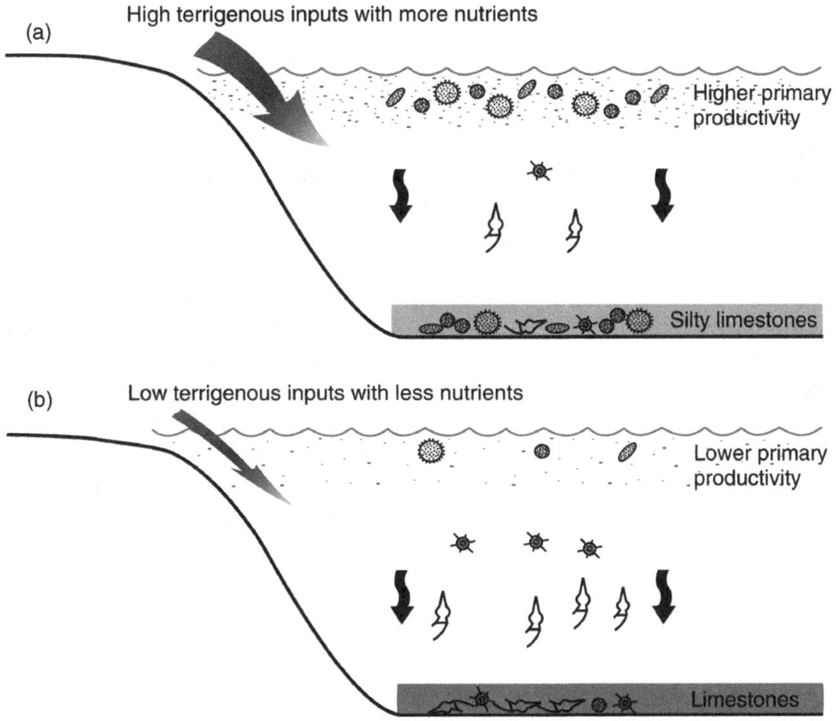
(a)
High terrigenous inputs with more nutrients
Higher primary productivity
Silty limestones
(b)
Low terrigenous inputs with less nutrients
Lower primary productivity
Limestones

preservation in sediments may have had as much to do with the presence of sponge spicules and the absence of local dilution by other sedimentary components as it had to do with primary productivity.

Lastly, there are a few methods which employ bulk properties of radiolarian assemblages, and thus lie between using (as above) the abundance of "radiolaria" as a single object, and studies of individual species composition. The most popular of these has been using the ratio between Spumellaria and Nassellaria as a rough estimate of water depth (in post-Paleozoic sections, where these two groups are the dominant component of the preserved assemblages). This is based on the early observation by Haeckel (1887) that in shallow waters, spumellarians were more common than nassellarians, whereas in deep-water sediments the reverse was true. This observation was given some biologic support by studies of radiolarian occurrences in the water column (Chapter 3; Casey 1993), where modern ocean subsurface waters below the thermocline (which are absent in shallow-water areas, e.g. inner shelf regions) have a relatively large number of small nassellarian species, in comparison to surface water layers. Based on these observations, many authors have calculated and interpreted the spumellarian/nassellarian ratio as a type of paleoceanographic water depth (e.g. Vishnevskaya and De Wever, 1998) or productivity indicator (e.g. Bartolini et al., 1999). In older geologic time intervals where radiolarians are embedded in rock, not sediment, diagenesis may have in many cases however differentially preserved radiolarian skeletons, causing higher or lower ratios due simply to the different degree of the nassellarian or spumellarian shell robustness, rather than due to any primary ecologic signal (O'Dogherty et al., 2003). Nor is the current relative abundance of nassellarians in deep-water necessarily a fundamental feature of radiolarian biology that can be assumed to be valid in the deep past, but may simply represent relatively recent diversification of nassellarian-dominated deep-water faunas, particularly Plagiacanthidae, in response to late Cenozoic deep-water cooling. The generality of this method is thus not yet known. Other non-specific characteristics of radiolarian assemblages have occasionally been used as qualitative proxies for paleoceanography, including assemblage diversity or the relative abundance structure of species in assemblages (e.g. evenness or dominance: Chapter 3), although links to oceanographic conditions are complex and thus difficult to interpret. Despite these limitations, when used in combination with other paleoceanographic proxies, such "bulk" properties of radiolarians can be interpreted in a local context and are often very useful in interpreting paleoceanographic changes in individual sections. Kamikuri et al. (2005) and Jin et al. (2012) for example, use bulk radiolarian abundance, diversity, and in the latter, spumellarian/nassellarian ratios, together with geochemical and sedimentary paleoceanographic proxies, to interpret, respectively, the paleoceanographic development of the Oligocene–Miocene boundary interval in the equatorial Pacific Ocean, and fluctuating oceanographic conditions in a Permian epipieric sea (Figure 7.12).

8

Radiolarian Biostratigraphy

Introduction

Biostratigraphy (reviews by McGowran, 2005; Aubry, 2014) uses the distribution of fossils in time to determine the relative ages of rocks and sediments. It is closely related to the science of physical stratigraphy (often called just stratigraphy), which determines the distribution of sedimentary rocks in time and space, and to geochronology, which determines the absolute ages of geologic materials. Traditionally, biostratigraphy has been done by specialists working on a single group of fossil organisms, although the results from several such studies are often combined, together with other stratigraphic and geochronologic information, in determining geologic ages. Biostratigraphy is a major research theme in radiolarian studies, although its importance varies somewhat according to geologic time interval. Generally, Paleozoic and Mesozoic radiolarian studies are dominantly for biostratigraphic purposes, while Cenozoic radiolarian research, although frequently biostratigraphic in nature, tends to include relatively more work on other themes (e.g. paleoceanography or evolution).

This chapter provides an overview of biostratigraphic methods, and how they relate to, and are integrated with, the sciences of physical stratigraphy and geochronology. A detailed listing

Paleobiology of the Polycystine Radiolaria, First Edition. David Lazarus, Noritoshi Suzuki, Yoshiyuki Ishitani and Kozo Takahashi.

and discussion of all of the individual biostratigraphic datums, however, is not attempted. Instead these are summarized in the form of figures for each geologic time interval or region, with comments on the state of the art for these intervals given in the accompanying text. Someone wishing to actually carry out biostratigraphic work should consult the primary literature cited in this chapter, where detailed information is given for each of the many defined datums, including the essential descriptions and illustrations of the radiolarian taxa that form the base of biostratigraphic studies.

Biostratigraphy is based on the evolution and extinction of organisms, and the irreversibility of evolution, so that species occurrences over geologic time create a pattern which is unique for each geologic time interval. The first biostratigraphic studies date back to the early nineteenth century, when William Smith noted the consistency of fossil occurrences in a sequence of successive rock formations in England and realized that the fossils could be used to estimate equivalent relative ages for the formations. Determining the sequential distribution of (mostly invertebrate) fossils and using this information to estimate relative geologic age soon became one of the central research themes in paleontology, since no other method has yet been developed that can so routinely provide geologic age estimates for sedimentary rocks. Biostratigraphy consists three main components: determining the general relative sequence of fossil occurrences over time; correlating individual geologic sections to each other based on this knowledge; and lastly, by cross-correlations to geologic samples whose absolute age is known, to provide absolute age estimates for these sections.

A complex terminology and set of concepts are associated with biostratigraphy. The total interval over which an individual taxon (usually species) occurs is called the stratigraphic range of the species. The stratigraphic ranges of one or more species in traditional biostratigraphy are used to define a biostratigraphic zone (aka biozone): an interval in the fossil record with a defined composition of fossils, which is assumed to represent a specific interval of geologic time. Although many types of zones have been defined, only a few are widely used. An interval zone, for example, is defined as the interval of overlap in the stratigraphic ranges of two different taxa. The ends of a species stratigraphic range, created by its origination and extinction, are called biostratigraphic first and last occurrence or appearance "events" (typically abbreviated as FAD and LAD), and are also used, independently or in conjunction with zonal concepts, in some types of biostratigraphic work. Changes in absolute or relative abundance within the range of a species, if prominent and consistent from location to location, can also be used in stratigraphy. Such "acme" events however are mostly usable only over relatively restricted geographic regions.

Because of their close relationships, some understanding of stratigraphy and geochronology is needed in order to understand how biostratigraphy is done. Equally important is that the basic methods used for radiolarian biostratigraphy vary quite a bit according to the type of physical stratigraphy and geochronologic information available. Typically, older geologic rocks have distinctly different physical characteristics and types of geochronologic information available than younger rocks or unconsolidated sediments, and thus require quite different methods than those used in studying younger rocks. This difference in methodologies, together with differences in the higher-level taxonomic composition of the radiolarian faunas (Chapter 5) has led to a certain degree of fragmentation among radiolarian workers, with most being specialized either in the study of Cenozoic or living materials, or in the study of pre-Cenozoic materials.

Biostratigraphy in Shallow Marine Rocks: General Aspects

As described in more detail in Chapter 6, the geologic record of shallow marine environments is far from uniform or continuous (Holland, 1995; Miall, 2015). Shelf sediments deposited in shallow waters vary in their lithology (composition) over time according to how close they are to the shoreline, and the record is frequently interrupted by intervals of non-deposition at various time scales, such as gaps of hundreds of thousands to millions of years during sea-level lowstands. As sediments undergo diagenesis and lithify (are altered to rocks), they are physically compacted by the pressure of overlying sediments, and chemically altered as well, causing the loss of many fossils, with the extent of loss partially determined by the lithology of the rocks and the robustness of individual species' morphologies. Sedimentary rocks once formed along the edges of continents are frequently at least partially eroded away as they are periodically exposed above sea level either by changing global sea level, or by more local uplift of the region to higher elevations by tectonic processes. Tectonic processes can further alter the original rock record by folding and faulting of sediments, breaking up or even inverting the normal sequence of older rocks being lower in the geologic sections.

The fossil record contained in shallow marine rocks is also only a partial record of the original biota. Fossilization for most organisms is relatively infrequent and fossils only make up an exceedingly small fraction of primary sediments. Even when preserved, the fossil can be so rare that it is not recovered in the necessarily limited volume of geologic samples used for a biostratigraphic study. The original ecologic history of organisms also complicates the use of their fossils in biostratigraphy. In the above definitions of species ranges, or first and last occurrence events, it is implied (rock and fossil preservation permitting) that a species' stratigraphic range should be correctly recorded in local sedimentary sections. However, many species appear to have changed their geographic range over geologic time, presumably in response to changing environmental conditions. Thus, the total stratigraphic range of a species is the sum of a series of overlapping but partially different local stratigraphic ranges, and the first and/or last occcurence of a species is said to be diachronous – e.g. variable in time (Figure 8.1).

The consequence of all of the above for biostratigraphy in shallow marine rocks is that the available data on the occurrences of fossils is extremely incomplete. The presence of a fossil of a species provides the information that the rocks enclosing it were deposited at some point in time during the total temporal range of the species. By contrast, the absence of a particular species fossil in any given rock or sample tells us rather little – its absence could mean the rock comes from a time interval outside of the stratigraphic range of the species, or it could just be absent for any of the numerous reasons outlined above. Traditional biostratigraphic methods are optimized to deal with this asymmetric type of information. Zones in particular encapsulate this by stating a geologic age interval for any given sample

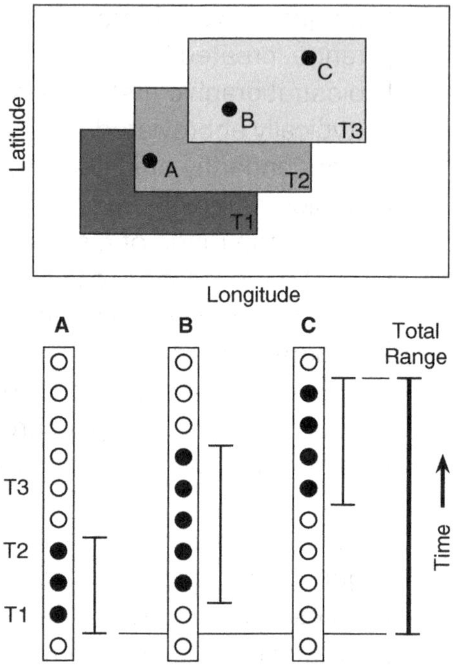

Figure 8.1 Relationship between the stratigraphic range of a species in three geologic sections (A–C) over time (three moments in time T1–T3 shown). Upper panel: species geographic range shift over time (shifting locations of shaded range area) causes species to no longer be present in geographic location A by time period T3. Lower panel: (assuming perfect preservation) occurrences of the species in geologic sections from the three localities A–C. The local range of the species (black = species present, white = species absent) in samples taken at various times (T) will be different in each of the three sections. The Total Range of the species is the total time between its origination and extinction and is equal to, or greater than, the temporal extent of the species composited across sections from different geographic locations.

that contains the fossils defining it, without making any statement about the geologic age of any other rock samples, even others from the same geologic section (Figure 8.2, upper panel). The precision of an age call for a sample is thus that imposed by the age range of the zone. Although these can in theory be very precise, in practice they are usually not much narrower than ca. 0.5 my, and are often well over 1 my in duration. This limits the precision of the age model that can be constructed for the section. The major advantage is that the age estimates are independent of the sample's location in a section, and work equally well for isolated samples that are not found in a sedimentary section.

Biostratigraphy in Deep-Sea Sediment Sections

The deep-sea fossil record, as explained in Chapter 6, is rather different from the shallow marine rock record. First, deep-sea sediments are much more continous records of geologic time than those from shallow marine waters. Changing sea level has no significant effect on the depth of the water in the open ocean, or on sedimentation in deep-sea sections, and ocean-bottom sediments are not exposed periodically to erosion, unlike those on land. Although deep-sea currents can erode sediments, and local sediment slumping can create local hiatuses and re-deposition, these processes are more the exception than the rule. Deep-sea sediments can be considered to be mostly continuous, at all time scales above the annual cycle of falling particulate matter from the surface waters that creates the deep-sea sediment record, up to the total duration of the geologic section being studied. Although it is a bit extreme, one can think of shallow marine rocks as a series of isolated snapshots of past time at any given locality, and deep-sea sediments as a continuous recorder, albeit one that runs at different speeds and occasionally stops completely, leaving a gap in the record. The sections are themselves often quite different than those of shallow marine rocks, now exposed on land. Unlike the latter, which usually represent only a fairly short interval of geologic time (e.g. a few million years), and are bounded by larger hiatuses above and below the section due to the discontinuous nature of deposition, deep-sea sections, particularly those recovered by deep-sea ocean drilling (Chapter 6) may contain an relatively uninterrupted record of many millions, or tens of millions, of years of earth history. Second, the occurrences of species in individual samples from such sediments are much more reliable indicators of the true pattern of the species' original occurrences over time. Unlike shallow marine sediments, most deep-sea sediments are vastly richer in fossils: indeed, deep-sea oozes are a common sediment that are primarily composed of microfossils, often including radiolarians.

The combination of the above factors means that, unlike most fossil occurrences in shallow marine sections, deep-sea microfossil occurrences are not fortunate exceptions to a more general pattern of absence, but a normal expectation, within the broader constraints of deep-sea sedimentation detailed in Chapter 6. Thus, not only the presence of a species, but also the absence of a species in a given sample, can be taken as reliable information on the stratigraphic range of a species in a section. The ability to use absences as useful information as well as occurrences is the basis for using only the first and last occurrence of a species in a section as the basis for biostratigraphic study – although the species occurs in numerous samples within the section between these points, the information is redundant, and the presence of species in such intermediate samples gives only imprecise range of species age information. The first and last occurrences

(FADs and LADs) in a section, so long as they are not biased in some way (e.g. reworking, taxonomic error, or diachroniety), however, represent a unique moment in geologic time. If the age of this geologic moment has been, via other means, previously precisely determined, each such biostratigraphic event provides a precise, point-like geologic age determination for a geologic section (Figure 8.2, bottom). If a sufficient number of such events can be determined for a section, a highly precise history of how geologic age changes over depth in the section can be constructed – an age model for the section. This is particularly practical in deep-sea sections that have recovered long intervals of geologic time, and thus have the potential to yield

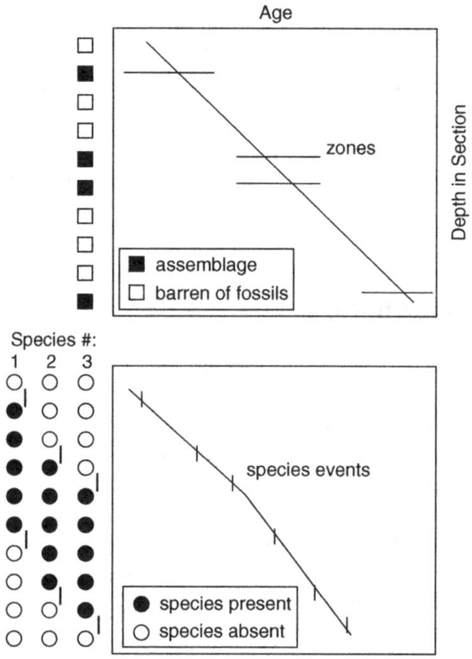

Figure 8.2 Geologic age dating of sections using zones (top panel) or events (bottom panel). Top: In zonal concept, one or more species' presence together in an assemblage of fossils found in a sample are used to estimate a range of geologic age for the sample (broad horizontal lines). Samples without fossils are considered to have no geologic age information. If one can assume on other grounds that sedimentation is reasonably continuous (which is frequently not true, see text) an age model can be constructed for the section (straight line), although this is limited by the typically broad age ranges of the zonal age assignments. Bottom: In sediments where fossil preservation is assumed to be continuous, the absence of a species in samples can be used along with the presence of the species in other samples to define the datum (position) in the section where the species first appears or last appears (FAD or LAD). The precision in this placement, assuming the "event" was effectively instantaneous in geologic time, is determined by the spacing between samples in the section, and is thus amenable to improvement by examination of additional samples. The high precision of age estimates at several points in the section allow a more precise age model for the section to be constructed (solid line of differing slopes, with each slope representing a different rate of sedimentation).

longer sequences of biostratigraphic events. This is in fact the normal way in which deep-sea biostratigraphy is done. Age models are important in that they provide an unbiased, continuously varying, absolute age estimate for *all* intervals in the section, not just at those where fossil information is available, or where biostratigraphic events occur. This type of *continuous* age information, which allows computing detailed histories of rates of change, is essential in many areas of research, e.g. paleoceanography (Chapter 7) and in studies of microevolution (Chapter 9).

The high degree of completeness of the microfossil record in deep-sea sediments causes complications for stratigraphy as well. As noted in Chapter 6, reworking of fossils into younger sediment layers is relatively common, since fossils are so abundant. This can be dealt with by various means. The most common is to collect quantitative abundance data on species occurrences, which distinguishes between reliable occurrences based on many specimens and occurrences based on rare, possibly reworked individual shells. As noted in Chapter 5 (Taxonomy and Fossil Record), continuous records of species lineages over time have revealed numerous examples of substantial gradual evolutionary change, and many of these lineages have been subdivided into numerous morphospecies. High completeness of faunal preservation means that sister species, often similar in morphology, are also preserved. Many of these lineage and sister group morphospecies are used in Cenozoic deep-sea sediment biostratigraphy. The gradual nature of transitions between lineage morphospecies, or similarities to sister species, however makes it often difficult to precisely define species boundaries – e.g. first and last occurrence events – for use in biostratigraphic zonation, and thus places a limit to the degree of precision in their use in geologic age estimation. In older rocks, less continuous, less faunally complete records only rarely show such gradual transitions, so that the impact of fine-scale species morphologic variation on biostratigraphic precision is less problematic.

Other Types of Geochronologic Information

Radiometric Dating and Absolute Age

Early in the history of geology, estimating absolute (not relative) geologic age was attempted by many methods, such as extrapolating determinable short-term rates of sedimentation to the estimated cumulative thickness of all sedimentary rocks. Nowadays, essentially all absolute geologic age estimates come from radiometric age measurements, and are based on the known rates of decay of certain element isotopes into their daughter products. Radiometric dating is a large subject in itself and beyond the scope of this book. However, certain key characteristics should be mentioned here. First, rocks suitable for radiometric dating are quite rare, and the vast majority of geologic samples, including most of those with radiolarians, are not directly datable. Most radiometric dates come not from sedimentary rocks, but from igneous or volcanic rocks, in particular those that were occasionally extruded at times in the past onto or into recently deposited sediments, and where the sediments so affected contain fossils that can be placed into a biostratigraphic scale. Most such interleaved rock/sediment sequences are found on land, and usually the fossils in these rocks are not radiolarians. Only a very few radiometric dates are available in direct association with radiolarian occurrences, e.g. from volcanic ash layers found in deep-sea sediments. Thus converting a relative age scale based on radiolarian biostratigraphic zones or events into an absolute age scale requires a correlation, or often complex series of correlations, between the original radiolarian scale, via one or more other relative age sequences (typically biostratigraphic scales made with other types of fossils), ending in the biostratigraphic sequence that is actually present in the sedimentary rocks where the radiometric date is obtained. In the latter part of this chapter we indicate which parts of the radiolarian zonations have absolute age estimates attached to them and the nature of the calibration, but do not try to explain the complexities of how these "tie-points" were achieved. In this book we primarily use Gradstein et al. (2004), a relatively recent widely accepted geochronologic age scale as the default value for reporting absolute geologic age, although in a few cases, when ages have changed significantly, we use the newer Gradstein et al. (2012) scale. The Gradstein et al. volumes should be consulted for a comprehensive review of how geochronologic scales are made, and details of many of the methods summarized briefly in the introductory text below.

In addition to absolute age, there are several other methods, not based on fossils, that are used to estimate relative ages of sediments. These methods include paleomagnetic stratigraphy, patterns of stable isotope change over time, and counting numbers of cycles in sedimentary rock components that are known to vary somehow at a regular frequency over geologic time.

Paleomagnetic Stratigraphy

The earth's magnetic field approximates that of a simple dipole magnet, with a magnetic north pointing to the earth's geographic North Pole, and is present at all points on the earth's surface. As sediments initially accumulate into loose, watery aggregates on the bottom of water bodies, sedimentary particles are still largely free to move about in the watery matrix, and very small magnetic grains that are sometimes present tend to orient themselves like tiny compasses in the direction of the earth's magnetic field. As further sediment accumulates, the sediment layer is compacted, and the grains' orientation becomes firmly fixed as other grains press against them and prevent movement. The sediment layer formed thus has a slight asymmetry in its magnetic properties, due to the cumulative effect of the tiny oriented magnetic particles contained in it, and the asymmetry is aligned with the magnetic field that produced it. This effectively provides a "fossilized" record of the orientation of the earth's magnetic field. In the 1950s it was discovered that the polarity of the earth's magnetic field has changed possibly thousands of times, but at irregular, largely random, intervals in the earth's geologic past. For periods of a few thousand to a few hundred thousand years at a time, the magnetic north pole has pointed first in one geographic direction (e.g. North), then rapidly (in just a few thousand years) flipped so that the magnetic north pole has pointed to the earth's opposite geographic Pole (e.g. South). Sediments that accumulate during each of these paleomagnetic polarity intervals thus have a matching pattern of asymmetry in magnetic properties, with the sediment magnetization pointing respectively either to the geographic North ("normal polarity") or South ("reversed polarity"). Through much effort, a detailed history of the polarity of the earth's magnetic field over most of geologic time has been determined. Although the numerous individual polarity intervals are not unique in character (they are all one of only two types: normal or reversed), the pattern over time is distinctive, due to the largely random nature of polarity change. Thus, the geologic age of a sedimentary section can in principle be estimated if one has a long enough, accurate history of the paleomagnetic polarity history held in the sediment magnetization which can then be matched to the global reference pattern of the earth's magnetic polarity changes (Figure 8.3).

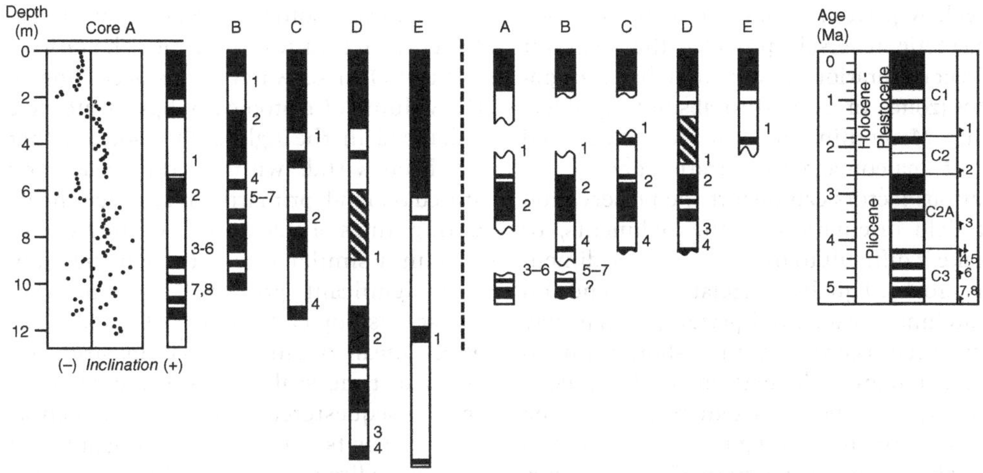

Figure 8.3 Use of paleomagnetic polarity in sedimentary sections in conjunction with biostratigraphy to estimate geologic age of sediments. Left: Measured paleopolarity of sediment samples in piston core A shows intervals of consistent paleopolarity; here only the inclination component is shown. Intervals of negative inclination at the location where the core was taken correspond to the modern earth's magnetic field orientation; positive inclinations to reversed polarity. These are shown as, respectively, black and white intervals vs. depth in core. Similarly, polarity patterns are seen in other cores B–E, all from the equatorial Indian Ocean. Biostratigraphic events (first and last occurrences of radiolarian species: numbers 1–8) are also seen in the cores. Right: Assuming biostratigraphic events are approximately isochronous each core's sedimentation history is reinterpreted and the magnetic polarity intervals are matched to absolute time via the known global paleomagnetic reversal history of the earth, and prior calibrations of the geologic age of each biostratigraphic event (rightmost plot of age, geomagnetic polarity, and event ages). Wavy intervals with gaps are inferred hiatuses in the individual sections. Diagonally striped interval in Core D was lost at sea during core recovery. Cores A–E are respectively: V19–169, V19–170, V19–171, V29–40, RC14–22; radiolarian biostratigraphic events are respectively: top *Pterocanium prismatium*, top *Stichocorys peregrina*, top *Spongaster pentas*, top *Didymocyrtis penultima*, base *Spongaster tetras*, top *Spongaster berminghami*, top *Solenosphaera omnitubus*. Source: Data and interpretations from Opdyke and Glass (1969), Burkle and Opdyke (1977), and Lazarus et al. (1985); radiolarian event calibrations from Theyer and Hammond (1974), Theyer et al. (1978), and Johnson and Nigrini (1985); geochronologic scale after Gradstein et al. (2004). Redrafted in part from Lazarus et al. (1985).

Paleomagnetic stratigraphy (see also Ogg and Smith, 2004) has some unique advantages. Because the earth's field is global so is the polarity change signal: in principle the same signal is recorded anywhere on Earth, in both marine and non-marine sedimentary rocks, as well as other rocks that can preserve magnetism (e.g. volcanic rocks as they cool and solidify). Further, because of the rapidity of individual polarity changes, each reversal can be treated as essentially a globally isochronous event in time. Lastly, because the paleomagnetic reversal history of the earth has been correlated to absolute geologic time, if a magnetic reversal in a section containing radiolarians can be uniquely identified, an absolute age can be directly assigned to the radiolarians at that level. A pattern of several such identified magnetic reversals provides the basis for direct calibration of radiolarian biostratigraphy (zones or events) to absolute geologic time. Many things however complicate and limit paleomagnetic stratigraphy's use. Various factors can distort the preservation of field orientation in the sediments, or make it difficult to measure in the sedimentary rocks. Equally, correlation depends on matching a sequential pattern, which may have been recorded over too short a time to yield a matchable pattern, or have been recorded at a non-constant rate, distorting the pattern and making it hard to match to the global reference pattern. Changing rates of sedimentation are thus challenges to the method, particularly the loss of section due to hiatuses or erosion, which splice together different, non-adjacent parts of the sequence without any internal indication (e.g. in the paleomagnetic data itself) of a discrepancy. In practice, one needs a fairly good independent geologic age framework (e.g. accurate to better than 1 my, on average) in order to unambiguously interpret most paleomagnetic patterns. Lastly, some intervals of Earth history have no known paleomagnetic reversals (e.g., a 52 my long normal interval in the Cretaceous), and thus provide no opportunity to use reversal patterns.

Stable Isotope Stratigraphy

Changes in the proportion of stable isotopes within certain elements (mostly C, O and Sr) can provide an important set of constraints on geologic age (as well as providing information about past ocean environments, as explained in Chapter 7) (Hayes et al., 1999; McArthur and Howarth, 2004; Bassinot, 2007). These isotopic values are usually measured from carbonate in the sediments, and ideally from single species of organisms, e.g. specific planktonic or benthic species of foraminifera. The carbon and oxygen isotope composition of marine carbonate reflects a variety of local processes but also the global mean isotopic composition of seawater, which varies over time. The ratio of ^{18}O to ^{16}O in seawater varies according to the volume of isotopically light water (e.g enhanced in the lighter ^{16}O isotope) that has been withdrawn from the oceans and stored on land, primarily in large ice sheets during times of glaciation. ^{13}C and ^{12}C also vary in a similar way, although the globally significant primary reservoirs that extract isotopically light ^{12}C and thus affect mean ocean-water composition on geologic time scales are organic material that is sequestered in sediments, such as coal deposits on land or organic-rich marine sediments deposited on the ocean margins. The changes in these storage reservoirs can be large enough and rapid enough that the mean oxygen and carbon isotope composition of seawater can change sufficiently to be detected and used for correlation purposes over even a few thousand years. Lastly, the 87 and 86 isotopes of Sr vary gradually over longer geologic time periods (hundreds of thousands to millions of years) as the relative rates of continental erosion and leaching of ocean basement rocks changes. Since the oceans are normally uniformly mixed by ocean circulation on scales greater than a few hundred years, all of these changes in mean ocean water are effectively isochronous and present throughout the global ocean. If a series of isotope measurements are available from a geologic section, the local effects on isotope composition can be factored out, and the resulting pattern can be accurately matched to the known global history for this elemental isotope system, a highly precise age estimate can be made. However the same problem that affects paleomagnetic stratigraphy is present here – the isotope values are not chronologically unique, and to accurately match any given fluctuation in the values of isotopes within a geologic section, one must have a sufficiently accurate idea already of the geologic time interval.

Cyclostratigraphy

Cyclostratigraphy is the last of the major supporting methods that is widely used in conjunction with radiolarian biostratigraphy. It is based on finding some characteristic of the sediments or rocks that varies along depth in section in a regular, periodic fashion; identifying the frequency of the periodicity, and assigning numeric ages (relative to some absolute age point in the section determined by other means) by counting the number of cycles vs. depth in section. The cyclicity normally used is created by changes in the earth's climate over tens of thousands of years due to periodic changes in the earth's orbit and axis of rotation, and the expression of these periodic climate changes in climate sensitive components of the sediments. These orbital changes are due in part to regular, long-term periodic changes in gravitational interactions between the earth and other planets, and method is often referred to as orbital tuning. The sedimentary components that record orbital signals can be quite diverse. Examples of what has been used include stable isotopes of oxygen; regular changes in sediment color, caused by changing proportions of biogenic and terrigenous materials due to changes in marine productivity and weathering input into the sediment from land; changes in the strength of magnetization of sediment due to climate-driven alterations in the amount of magnetic minerals deposited as trace components of the sediment, and of course, fluctuations in the abundance of climatically sensitive species. *Cycladophora davisiana*, a living intermediate water species, is today a relatively uncommon member of the preserved sedimentary assemblage of radiolarians, but over the latter part of the Pleistocene has shown a regularly repeating pattern of being substantially more common during glacial periods. This pattern – a type of recurring "acme" in abundance, has been used in improving resolution of age models for Pleistocene sections (Hays et al., 1976b; Hays et al., 1976a; Morley and Hays, 1979).

Orbital tuning of sedimentary sections (Figure 8.4) was first developed as an offshoot of late Pleistocene paleoclimate studies of deep-sea sediments in the 1970s. This early work demonstrated that the frequency and duration of Pleistocene ice ages were paced by specific frequencies in the earth's orbital variations: the precession cycle at 21 kyr; the obliquity cycle at 41 kyr, and the eccentricity cycle at ca. 100 kyr (Hays et al., 1976). The importance of these orbital frequencies as a regulator of Pleistocene climate was proposed by scientists such as J. Croll in the late nineteenth century but first fully analyzed by M. Milankovitch (1941, and earlier studies) and they are thus often called the Milankovitch frequencies. Later work (e.g. Lourens et al., 2004) calculated the expected frequencies for these, and other orbital variations, much further back in time (currently, to the top of the Paleozoic [Huang et al., 2011; Laskar et al., 2011]), and showed that the frequencies are very stable and only alter slightly with increasing geologic age. Many studies have shown that these orbital frequencies are often well expressed in changing sediment composition in these older rocks, and thus the variations have been used to calculate more precise ages for the sections containing them. Because multiple simultaneous frequencies create more complex patterns, special software using the Fourier transform is used to decompose the composite pattern of change in orbitally sensitive rock characteristics into a set of primary orbital frequencies. Much of this type of data analysis is derived from other areas of science and engineering where spectral analysis is well developed. Orbital tuning is currently the most precise method available for age determination below the late Pleistocene, where radiocarbon and similar methods allow extremely high precision.

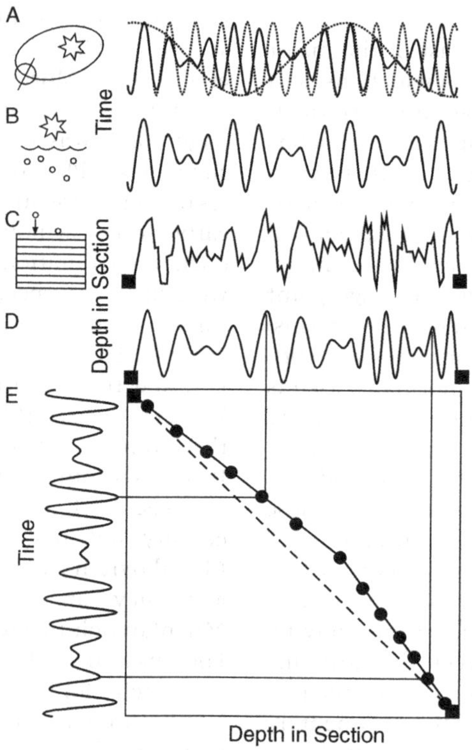

Figure 8.4 Cyclostratigraphy and orbital tuning, here shown in very simplified diagrammatic form with only two orbital components and no discussion of geographic or long-term temporal variation. (A) Long-cycle eccentricity (ca. 100 kyr, dashed line) and short-term cyclicity (here, precession, 21 kyr, thin line) interact to create a composite modulation of solar insolation (solid line) that affects (B) plankton communities in the ocean, and thus their export signal of productivity and community composition. These variations affect sediment composition and thus sediment proxies such as color (C), although the signal is distorted by various processes and the frequency vs. depth in section is affected by overall sedimentation rates. Through a series of data processing steps, the cyclic signal in the measured proxy is extracted (D) and compared to the expected signal over time computed directly from the orbital equations (E). Known ages (solid squares), typically from bio- or magnetostratigraphy mark ends of interval and provide essential scale to estimate approximate frequencies. Cycles are matched to create age and depth point pairs (dots), which in turn can be used to create a detailed age model for the section (solid line) which is more refined vs. the age model based only on prior known ages (dashed line).

Orbital tuning can be accurate to just a few thousand years, so long as all assumptions of the methods are met. One area of uncertainty is how the sediment characteristic being used for tuning is related to the forcing signal of global climate. Different sediment signals can differ by up to ca. 200,000 years from each other in the same section, due to major leads and lags in how different parts of the ocean respond to climate (Beddow et al., 2018).

Quantitative Biostratigraphy

As the above review makes clear, there are many types of data that are used in radiolarian biostratigraphy, including different quality data on radiolarian occurrences, paleomagnetic, isotope, cyclostratigraphic, and occasionally even radiometric data. Evaluating and integrating such large, complex data sets is assisted by several methods that together are called quantitative biostratigraphy. The oldest and most widely used of these has already been introduced above – a graphical plot by depth and age of the available data for a section, and a line of correlation drawn through the data that represents the section's age model. Age depth plots, first promoted by Shaw (1964), are widely used in Cenozoic stratigraphy to integrate a variety of different types of age information for a single section (Figure 8.5),

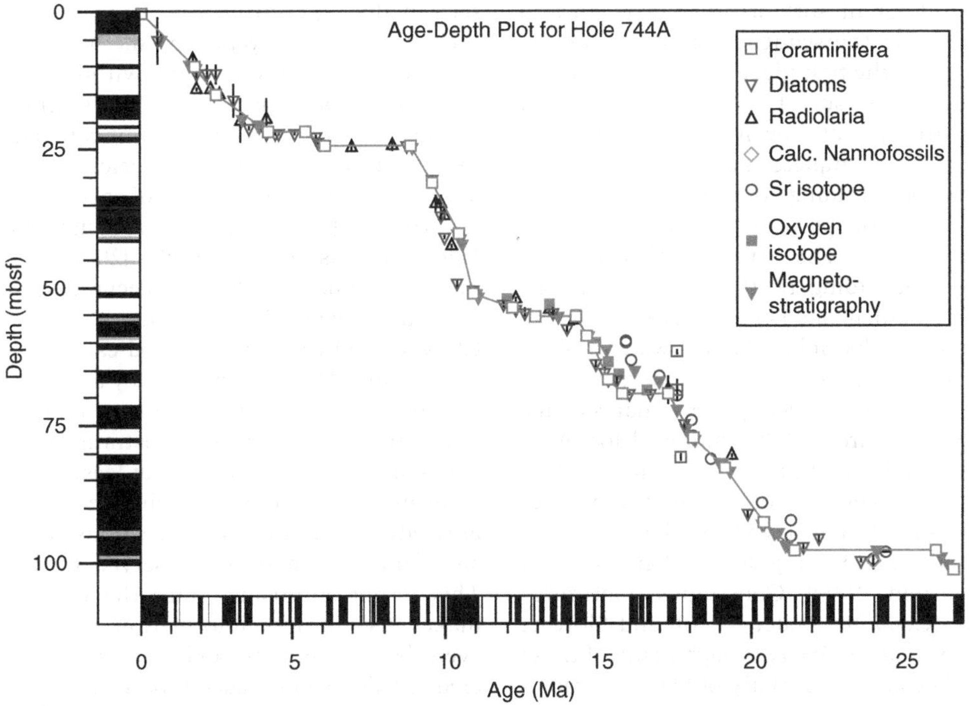

Figure 8.5 An age-depth plot for a deep-sea drilling section (ODP Site 744, Kerguelen Plateau, Southern Ocean) showing how biostratigraphic event data from multiple microfossil groups including radiolaria (upright black triangle symbols) are integrated with geochemical event data (oxygen isotope glacial events), strontium stable isotope events, and paleomagnetic stratigraphy to create an age model (green line with white squares marking inflection points between linear sections) that provides a continuous geologic age vs. depth relationship. Mean age uncertainty of ca. 0.5 my in individual events is reduced by use of multiple event data and resulting age model is sufficiently precise to provide framework for orbital tuning (as shown in Figure 8.4), although not all depth intervals are equally well determined. Paleomagnetic polarity as measured by shipboard studies is shown on the left vertical axis (black: normal, white: reversed, grey: unclear), the reference polarity age scale (Berggren et al., 1985) is on lower x axis. Data is inconsistent between ca. 60 and 70 mbsf. Hiatuses are in this interpretation present at around 25, 55, 70, and 100 mbsf. Source: Data sources and original interpretation given in Keating and Sakai (1991) and Lazarus et al. (1995). Note newer age model for the lower part of this site, based on newer data and calibrations, is also available (Florindo et al., 2013).

and are based on the assumption that all plotted data has already been associated with a numeric absolute age or age range. Age depth plots do allow, however, estimating the age of as-yet uncalibrated fossil zones or events in a section, so long as other data is available and provides the basis for drawing a line of correlation, by projecting the depths of these unknown events onto the age axis via the line of correlation.

A different sort of problem is how to determine the sequence of species occurrences in the first place. As noted above, in almost all actual sections, individual occurrences of species are to varying degrees incomplete or sometimes even erroneous, leading to partially conflicting sequences of species occurrences in different sections. The extent of this problem varies according to the type of sections being studied, and is generally more severe in older rocks and in shallow water or shelfal marine sections. Several different methods have been developed to address this, and many are widely employed for other types of biostratigraphic data (e.g. occurrences of benthic foraminifera and other fossils in oil industry wells). These include Ranking and Scaling (RASC: Agterberg and Gradstein, 1999), Constrained Optimization (CONOP: Sadler, 2010), and Slotting (Gary et al., 2005). Although many of these methods are potentially of value for radiolarian biostratigraphic work, until now they have not been exploited. Instead, radiolarian workers studying deep-sea sediments have mostly made use of age-depth plots and graphical analysis as these are sufficient for most of their relatively high-quality Cenozoic radiolarian occurrence data, while workers on older material have made use of a method that was developed in association with radiolarian research: Unitary Associations (Guex, 1991).

In older, less well-preserved sections, methods that stress the occurrence, rather than the absence, of fossils are preferred. The Unitary Association method adheres to this idea – the co-occurrences of species in samples are the basis for analysis, not the absence of fossils in any given sample. Unitary Association, however, also takes into account the often discontinuous nature of fossil sample occurrences in such sections, including (in the extreme) isolated single samples, or samples whose original increasing relative age vs depth in section relationship has been inverted in a local section due to faulting and movement of the original rock layers. The method uses only co-occurrences of fossils without any information on relative order in a section, and an underlying assumption of coherent individual species ranges, to calculate an optimum ordering of a series of species co-occurrences (e.g. the unitary associations: Figure 8.6). As noted by Sadler (2010), this method can be considered a special case of the more general algorithms used in CONOP, and the latter method can be set to emulate a UA style analysis by appropriate choice of parameters. Unitary associations in practice yield biostratigraphic zones of fairly low resolution, at least when compared to the resolution obtained using biostratigraphic events in sections with abundant, continuously preserved fossils. This is not surprising, since the method makes use of less information than do event-based methods, and can be considered as a deliberate trade-off between precision and usability that is optimal for less well-preserved materials. Lastly, unitary association zones do not necessarily have a continuous transition relationship to each other and assigning precise boundary dates to them can be difficult as in reality no specific bio-events (first or last appearances of species) are defined by the method.

One characteristic of many of these quantitative methods is that they use a combination of random permutations and accuracy of fit metrics to gradually find an optimum solution from an extremely large potential number of possible solutions. Such methods often have many parameters

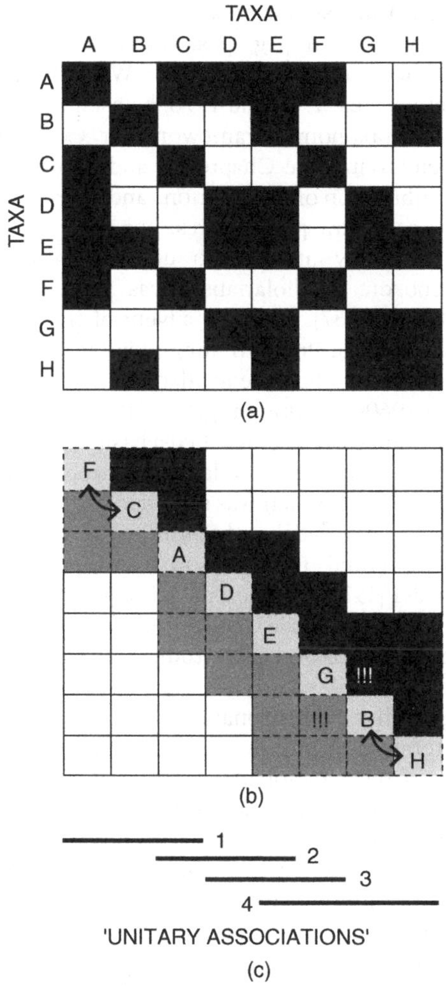

Figure 8.6 Unitary associations. (A) Co-occurrences of species compiled from several sections (black – co-occur in at least one sample). (B) Matrix rearranged to maximize co-occurrences along the diagonal. Exclamation points mark co-occurrence implied by other data but not seen in original matrix, exchange arrows show unresolved relative positions along diagonal. Relative positions along diagonal can be used, or C) blocks of co-occurrence (dashed lines) can be used to define assemblage zones (e.g. "unitary associations" of Guex, 1991). Source: Sadler (2004). Reproduced with permission of John Wiley and Sons.

that need to be correctly set, sometimes yield wrong results due to subtle errors or other problems in the input data, can occasionally get "trapped" in a less than optimal solution, and also tend to be rather computationally intensive (run times of several hours on 2012 vintage desktop computers were typical for larger datasets). The details on how to use the programs effectively, and how to avoid obtaining misleading outputs, are beyond the scope of this book. As is often true of scientific computing, they are best learned by practice, and consulting other more experienced users.

In the remainder of this chapter we discuss the radiolarian zonations of the Phanerozoic, beginning with the Cenozoic. Cenozoic radiolarian stratigraphy serves as a reference template for older time intervals, and is also the interval that provides the majority of data for evolutionary studies. Succeedingly older intervals are then covered, together with the additional challenges due to preservation, tectonics, and other factors that they present to biostratigraphic practice.

Cenozoic Radiolarian Stratigraphy

The oceans of the Cenozoic were very similar in basic biogeography and circulation to modern oceans (Chapters 3 and 7). A tropical province existed along the equator, with a broader warm- water subtropical province extending to ca. 40° N and S. Polar biotas existed over much of the Cenozoic further north or south, respectively. Until the closure of Panama in the early Pliocene, the equatorial provinces were interconnected. Upwelling regions existed to various degrees along the eastern low to mid-latitude margins of the ocean basins. These primary oceanographic patterns largely determine the structure of Cenozoic radiolarian stratigraphy, with separate faunas and thus zonations for four main provinces: low latitudes, the Southern Ocean, the North Pacific, and the North Atlantic (including the Norwegian-Greenland seas).

Ocean circulation, and changes in circulation, affected the distribution of silica and thus preservation patterns of radiolarians over the Cenozoic (Chapter 7). The subtropical regions were generally oligotrophic throughout the Cenozoic and silica preservation poor. Tropical silica preservation was generally good in the older Atlantic Cenozoic but declined in the mid Miocene, while preservation of silica improved at this time in the North Pacific. Sea levels in the Cenozoic were generally much lower than in much of the Mesozoic or Paleozoic (Miller et al., 2005), and thus deep-water pelagic marine sediments with biogenic silica are relatively rare on land. This latter point, together with generally low preservation in the mid-latitudes, means that Cenozoic radiolarian exposures on land are few and mostly found either in the tropics or mid- to high latitudes. In practice, relatively poor knowledge of geology in tropical countries means that with few exceptions (e.g. Eocene radiolarians from Barbados) most early studies of pre-Holocene Cenozoic radiolarians were from isolated outcrops of mid-latitude or polar faunas from North America, Europe, Japan and the Soviet Union. More extensive knowledge of low-latitude Cenozoic faunas, in particular in longer sections suitable for stratigraphic analysis, came only from deep-sea sediment cores after World War II. Stratigraphic analysis was also aided by the rapid development between ca. 1940 and 1960 of global marine Cenozoic stratigraphy based on other microfossils, particularly planktonic foraminifera, which were extensively employed by the oil industry (Bolli et al., 1985), and which were frequently available in the same sections as radiolarians. This provided a framework for developing and calibrating a Cenozoic radiolarian zonation.

History of Development

Early modern (e.g. post-WWII) studies of radiolarians were led by W.R. Riedel. Between ca. 1950 and 1970, he established a basic taxonomic framework for Cenozoic radiolarians (see Chapters 1 and 5), using a combination of land sections and short deep-sea sediment piston cores, but only a first, preliminary attempt to define zones for Cenozoic radiolarians was attempted (Riedel, 1957). With the advent of the Deep Sea Drilling Program, long radiolarian-bearing sections became available at the end of the 1960s. Working together, Riedel and A. Sanfilippo soon established a basic Cenozoic zonation for the low latitudes (tropics and subtropics), which was summarized and stabilized by Riedel and Sanfilippo (1978) and documented in detail in Sanfilippo and Riedel (1985). Numerous subsequent analyses of low latitude Cenozoic sediments – primarily DSDP and ODP studies from deep-sea sediments – mostly served to validate the generality of this zonation, with only minor alterations being introduced. Modern studies of polar radiolarian faunas began in the late 1950s (Nakaseko, 1954; Riedel, 1958). The first zonations were published soon after (Nakaseko 1960, Nakaseko et al. 1965) for sections in Japan, and Lipman (1960, 1965) for Cenozoic land sections of the Soviet Union. These authors proposed only a small number of very general assemblage zones, e.g. "the Eocene assemblage," and thus were of somewhat limited use. Southern Ocean late Neogene zones by Hays (1965) were by contrast event based and much more precise (see section on Southern Ocean in this chapter), similar to the low-latitude zonation then under development by Riedel.

The early DSDP sediment sections that were becoming available at the end of the 1960s used relatively primitive rotary drilling methods that disturbed sediments (see Chapter 6) and destroyed paleomagnetic properties. Piston cores recover undisturbed sediment suitable for paleomagnetic measurements. The first calibrations of radiolarian zones to the paleomagnetic time scale (GPTS) were thus done by Hays in collaboration with paleomagnetic specialist N. Opdyke and other co-workers at the Lamont Observatory in the late 1960s to 1970s (Hays and Opdyke, 1967; Hays et al., 1969) using late Neogene piston cores, with additional work on the Neogene using piston cores by Theyer and Hammond 1974; Theyer et al., 1978). Longer sections with paleomagnetic records, particularly in carbonate rich low-latitude sediments in the early 1980s, became available only with improved DSDP drilling techniques (Chapter 6), and more sensitive paleomagnetometers capable of detecting the relatively weak paleomagnetic signal in carbonate sediments (e.g. Tauxe et al., 1984). These newer sections have been used to improve correlation of zones to the GPTS and thus geologic time, as detailed in the individual sections below. Basic information on the calibrated age and reported occurences of Cenozoic radiolarian (and other marine microfossil) biostratigraphic events are also available from the NSB database.

Tropical Cenozoic Radiolarian Stratigraphy

The current version of the tropical Cenozoic radiolarian zonation is given in Figure 8.7. It is based on the zonal framework of Riedel and Sanfilippo (1978) and Sanfilippo and Riedel (1985), with some minor alterations to zone definitions, and calibrations to the GPTS as given by Moore et al. (1995) and Nigrini et al. (2006) in, respectively, ODP Legs 138 and 199. Both of these studies are based on deep-sea sediment sections from the tropical Pacific. This region has yielded the most complete sections for low-latitude siliceous microfossils, and these sections, together with earlier sections from the same region (e.g. DSDP Leg 85: Nigrini, 1985) can be considered the standard for low-latitude radiolarian zonation. Most other Cenozoic microfossil zonations (e.g. planktonic foraminifera, calcareous nannofossils) have long made use of numeric codes for zones, as these are easier for non-paleontologists to use, and the relative age is clear by the code's number. Codes for the tropical Cenozoic radiolarian zonation were first assigned by Sanfilippo and Nigrini (1998). Figure 8.7 gives both the traditional taxonomic zone names as well as the number code names, as the latter was introduced only recently and most literature thus still uses the older taxon-based zone names. As the geochronologic age scale used in ODP Legs 138 and 199 is that of Berggren et al. (1995), the ages in Figure 8.7 have been updated to the Gradstein et al. (2004) scale used in this book. A partial update for the early Eocene to early Miocene time interval, including calibrations for many secondary events (226 events in all) is given in Kamikuri et al. (2012).

The tropical Cenozoic zonation is applicable throughout the tropics and subtropics. In addition to 39 formal zones (17 Neogene and 22 Paleogene) defined by 37 primary biostratigraphic events, there are around 300 additional events which have been identified and which can be used as secondary markers (see summaries in Moore et al. (1995) and Nigrini et al. (2006)). Zonal resolution is thus about 1.7 my, although with secondary markers the mean resolution is approximately 200 kyr. The resolution is not uniform over the Cenozoic, being higher in the mid- to late Eocene, and in the Neogene and lower in the Oligocene. This reflects primary underlying differences in the diversity and rates of evolutionary turnover in the assemblages (Chapter 9, Evolution). The majority of the species used in this zonation were first defined by modern authors, although many species originally described by early authors such as Haeckel and Ehrenberg are also used. Most importantly, many of the species are members of lineages – series of morphologically defined taxa within what appears to be a single gradually evolving species (see also Chapters 5 and 9). This includes many species within, among others, the genera *Podocyrtis, Thyrsocyrtis, Calocyclas, Lithocyclia, Dorcadospyris, Calocycletta,*

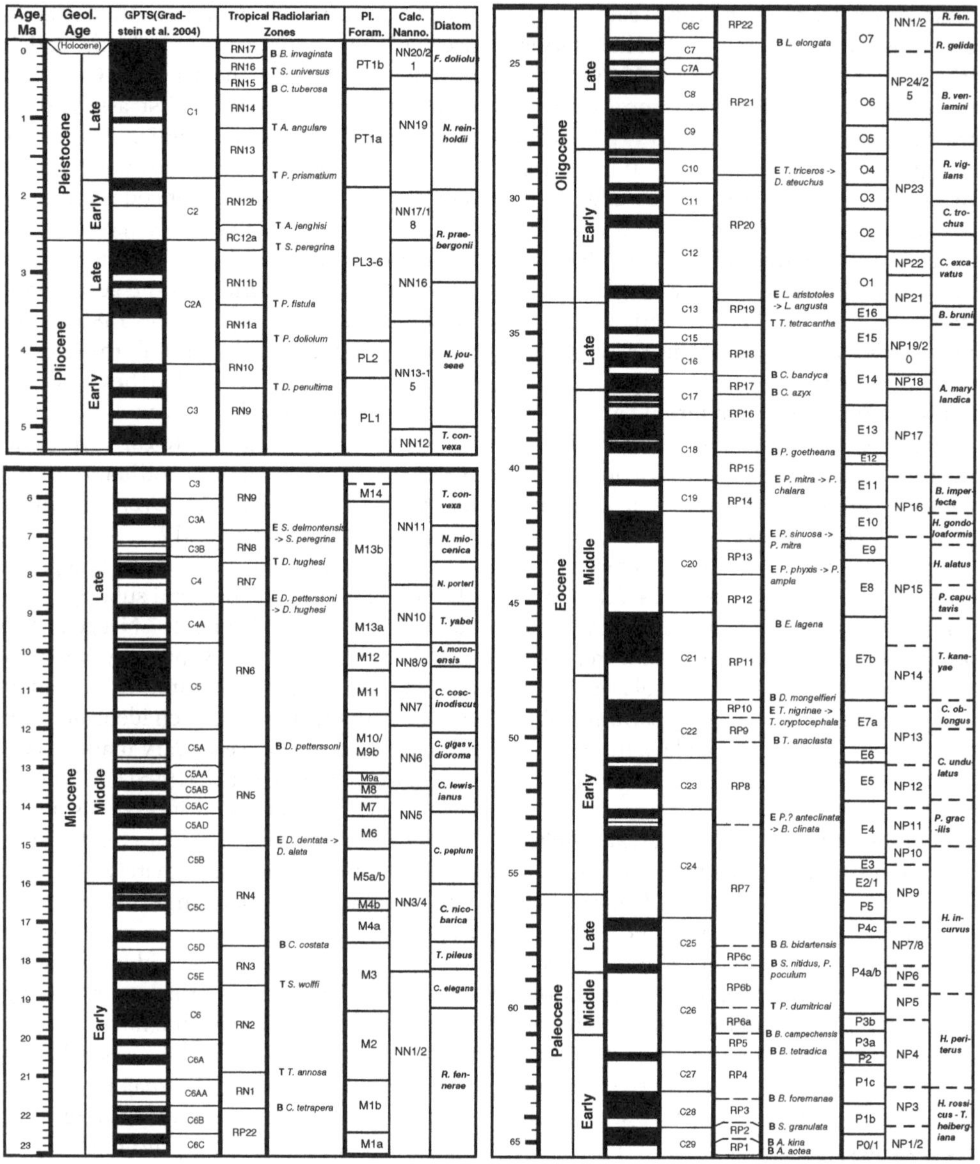

Figure 8.7 Cenozoic tropical rad zonation. Timescale and GPTS are after Gradstein et al. (2004). Zonal names and numeric code equivalents after Sanfilippo and Nigrini (1998). Ages for radiolarian events younger than T (top e.g. LAD) *D. alata* are from Moore (1995), older events from Nigrini et al. (2006), updated to the Gradstein et al. (2004) time scale. Ages of planktonic foraminifera zones from Wade et al. (2011), calcareous nannofossil and diatom zones younger than ca. 40 Ma from Lyle et al. (2009), older nannofossil zone ages from Gradstein et al. (2004), older diatom zones from Barron and Baldauf (1995). Notes: due to different age-averaging methods used by Moore et al. (1995), ages for events here are not always the same as given in Shackleton et al. (1995) for Leg 138; age of evolutionary transition *Diartus petterssoni* > *D. hughesi* as given in Nigrini et al. (2006) appears to be in error. Minor differences in GPTS scale (ca. +/- 200 kyr) exist between Lyle et al. (2009) and Gradstein et al. (2004), particularly for Paleogene – differences averaged for this figure. Calibration of most zones to GPTS below ca. 40 Ma are poor and shown as dashed lines.

Cyrtocapsella, Stichocorys, Diartus, and *Didymocyrtis.* While traditional first and last appearances of morphospecies within these lineages are used for some stratigraphic events, in a substantial number of cases the stratigraphic event is defined as an evolutionary transition between the ancestor and descendant form, with the evolutionary event being the point where the populations shift from dominantly ancestral to dominantly descendent morphology. It is not clear whether in practice this leads to more accurate or reproducible events, since it requires enough individuals to judge relative abundances, and classification of closely related transitional forms into different categories. The difference in event definition is however stratigraphically important: in most cases in the Cenozoic there is approximately a 0.5 my difference in time between an event defined as an evolutionary transition and the same event defined as the first or last appearance of either of the two taxon morphospecies (compare events tabulated in e.g. Nigrini et al., 2006).

Cenozoic radiolarian biostratigraphy is marked by an abundance of fossils and sections, and the frequent availability of independent age information in the form of other microfossil groups, paleomagnetic stratigraphy, and other methods. Low-latitude Cenozoic deep-sea sedimentation has also been relatively continuous, with fewer hiatuses than in the polar regions. These conditions are ideal for studying the degree to which individual biostratigraphic events are isochronous or diachronous. Early work by Hays and Shackleton (1976) on the last appearance of *Stylatractus universus* showed that this species disappeared isochronously throughout the global ocean in the late Pleistocene, at least to the resolution of a few thousand years (Figure 8.8). This was shown by comparing the last appearance level in several cores from different ocean basins where precise ages had been determined using high-resolution cyclostratigraphy based on glacial-interglacial stable oxygen isotope data. Similar studies have shown that the first or last appearances of other microfossil species are also isochronous (Thierstein et al., 1977). Most microfossil species' first and last occurrences have not been examined at this level of resolution, but many have been evaluated against a geologic time framework based on paleomagnetically dated sections, or sections with multiple, calibrated biostratigraphy, which provide age information precise enough to determine isochrony or diachrony at the scale of a few 100 kyrs or more (Hills and Thierstein, 1989). Studies of radiolarian events using these methods by Johnson and Nigrini (1985), Spencer-Cervato et al. (1994), and Moore et al. (1995) all show that while many events appear to be isochronous at scales of a few 100 kyr, or less many events are clearly not (Figure 8.9). Some of these show clear patterns of diachrony vs. latitude or longitude, while others show a more chaotic pattern vs. geography (Spencer-Cervato et al., 1994). Geographically correlated patterns of diachrony for species' first or last occurrences may be related to ecologic or evolutionary processes, as discussed in Chapter 9. The more chaotic patterns may reflect taxonomic or other problems in primary event data, although additional studies are still needed to confirm this.

Subtropical North Atlantic to Arctic

Plankton biogeography of the modern ocean (see also Chapter 3) includes several distinct latitudinal provinces in the North Atlantic, with tropical faunas ranging up to ca. 30° N, subtropical/temperate faunas dominated by more cosmopolitan forms up to ca. 45° N, and a separate polar assemblage, which because of permanent ice cover in the Arctic Ocean is found mostly in the Norwegian-Greenland Seas (NGS) between Greenland and Scandinavia. The Mediterranean, as a largely enclosed mid-latitude marginal sea, has a plankton composition that is derived from, but differs somewhat from, the

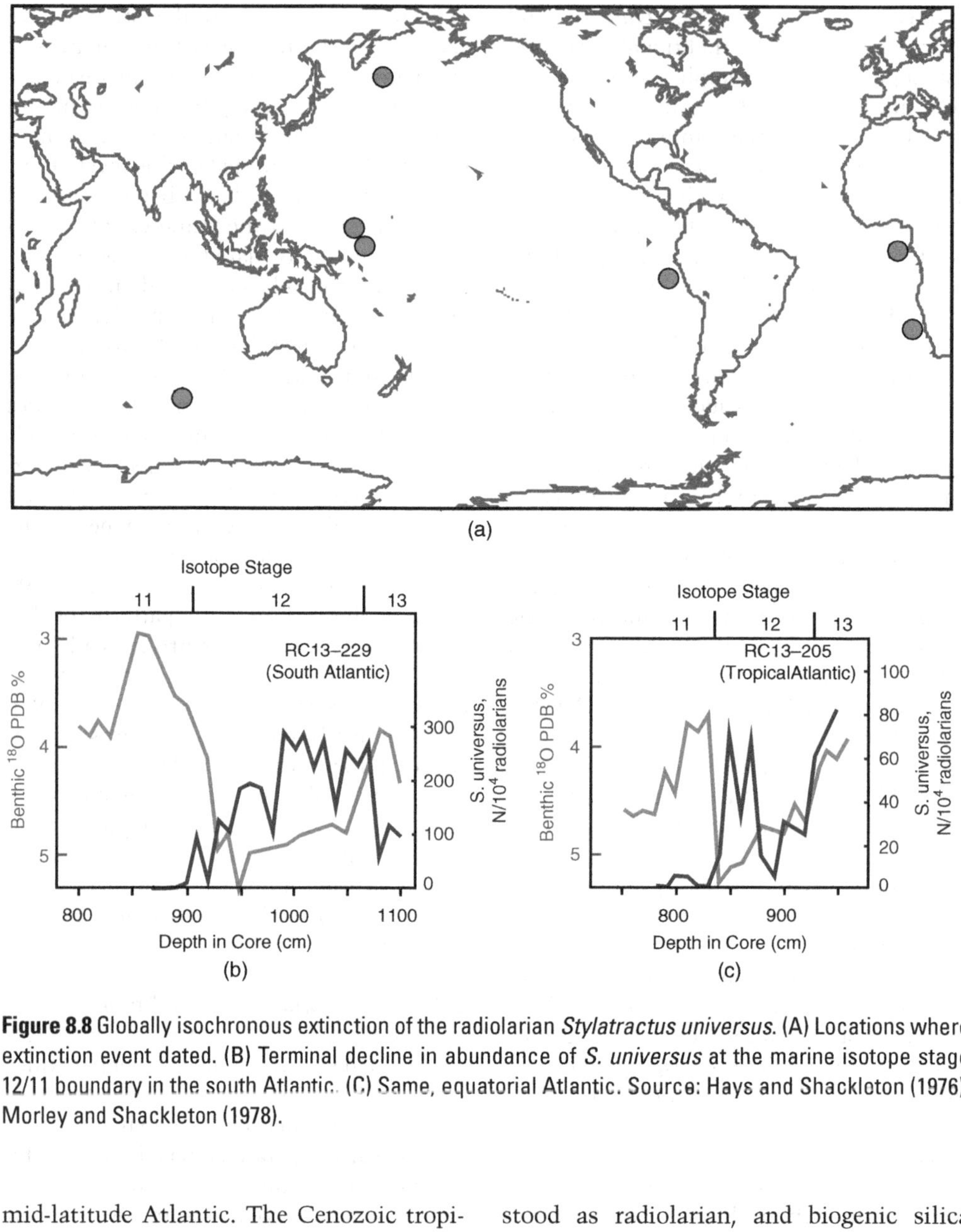

Figure 8.8 Globally isochronous extinction of the radiolarian *Stylatractus universus*. (A) Locations where extinction event dated. (B) Terminal decline in abundance of *S. universus* at the marine isotope stage 12/11 boundary in the south Atlantic. (C) Same, equatorial Atlantic. Source: Hays and Shackleton (1976), Morley and Shackleton (1978).

mid-latitude Atlantic. The Cenozoic tropical zonation can generally be used with little problem between 0° and from 20–30° N, but some zonal markers are either absent or diachronous in the temperate oceans beyond. The utility of radiolarian zonations however in both the subtropical and temperate North Atlantic is not that well understood as radiolarian, and biogenic silica preservation in general, is rather patchy and variable throughout the Cenozoic. This fact, plus the availability of well-preserved calcareous microfossils in most sections from this region, has meant that no specific regional radiolarian zonation has been developed. Papers reporting Cenozoic

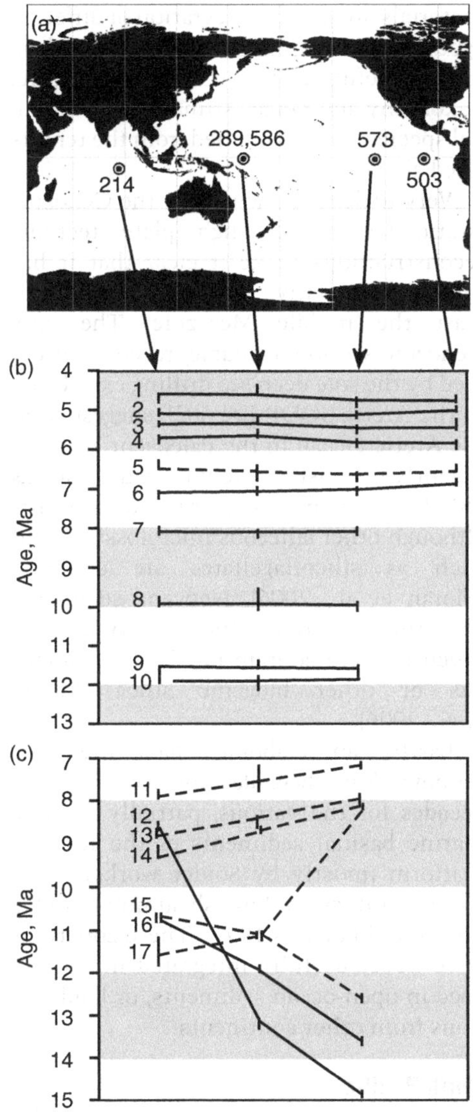

Figure 8.9 Isochronous (B) and diachronous (C) selected radiolarian biostratigraphic events (1–17) in the Neogene tropical oceans. From DSDP Sites shown in A. Vertical lines in B and C are uncertainties in event age in each section, mostly due to sample spacing. Last occurrences – solid, first occurrences – dashed lines. Source: Replotted from, and with age scale of, Johnson and Nigrini, Mar. Micro., (1985). Reproduced with permission of Elsevier.

radiolarian stratigraphy from this region include Benson (1972), Johnson (1978 – Site 369), Weaver and Dinkelman (1978), Sanfilippo and Riedel (1979), Ling (1979), Westberg et al. (1980), Labracherie (1984), Palmer (1986, 1987, 1988), Westberg-Smith and Riedel (1985), Westberg-Smith et al. (1986), Nishimura (1987, 1992, 2001), Lazarus and Pallant (1989), and Sanfilippo and Blome (2001) for the subtropical to temperate North Atlantic, and Sanfilippo et al. (1973, 1978), Sanfilippo (1988), and Khokhlova et al. (1994) for the Mediterranean. Abundance variations of *Cycladophora davisiana* have also been used for late Pleistocene correlation in the North Atlantic (Morley and Hays, 1979).

The Cenozoic microfossil assemblages of the NGS are, like those of their austral biogeographic counterpart the Southern Ocean, highly endemic and largely lack calcareous fossils. Unlike the Southern Ocean however, preservation of biogenic silica in Cenozoic NGS sediments is highly variable in both time and space, making construction of a useful radiolarian zonation difficult. Sedimentation in the narrow NGS basin, being adjacent to major landmasses and ice-sheets, has been dominated by terrigenous material over most of the Cenozoic, and these sediments often have well-preserved dinoflagellates and other palynomorphs. These appear to be the most generally useful group for zonation of NGS sediments and can be related to palynologic studies done for oil industry drilling in the North Atlantic region, although this work is still in its early stages (see Mudie et al., 1990; Eldrett et al., 2004 for summaries).

NGS sediments have very high average sedimentation rates in the Pleistocene and regular piston cores do not penetrate into older sediment layers. Most recovered Cenozoic NGS radiolarian material comes from just four deep-sea sediment drilling legs: 38, 104, 151, and 162. Two parallel, often discrepant reports were published for DSDP Leg 38: Bjørklund (1976) and

Dzinoridze et al. (1978), with most subsequent citations being only to Bjørklund (1976). Poor core recovery during Leg 38 precluded detailed analysis of radiolarian stratigraphic distributions, although Bjørklund (1976) provided a provisional zonation. Much better recovery of sections by Leg 104 allowed also good recovery of paleomagnetic stratigraphy (Bleil, 1989), but the different microfossil study results, with differing interpretations of sediment ages, were not successfully integrated into a generally supported synthesis report – the biostratigraphic synthesis chapter is notable for having only a single author (Goll, 1989). Goll and Bjørklund's own radiolarian zonation for this leg (1989) includes many inferred hiatuses which are difficult to reconcile with the paleomagnetic stratigraphy (Bleil, 1989). ODP Leg 151 with at least one excellent Neogene section (Site 907, with nearly continuously preserved biogenic silica from the mid-Miocene to Recent, and an equally good paleomagnetic record (Myhre et al., 1995; Channell et al., 1999), has no Neogene radiolarian report, but instead a detailed study of the unique Paleogene assemblages, fortunately present in the relatively short intervals of core recovery (Hull, 1996). Hull did note however (Hull et al., 1996) that the radiolarian zonation proposed by Goll and Bjørklund (1989) was at best only partially usable for Leg 151 sections. Leg 162 recovered several good Pliocene-Recent sections, some with biogenic silica, but no radiolarian studies were carried out.

Given the above problems of variable preservation, incomplete recovery, gaps in primary analysis of the material, and the resulting instability in zonation efforts, we can only echo Hull's (1996) comment: "Efforts to use radiolarians in resolving biostratigraphic and paleoceanographic questions of the Norwegian-Greenland Sea have been challenging at the least" – and forego presenting a zonation here. Hopefully, additional study of existing materials and better integration of different types of stratigraphic information will lead to a useful zonation in the future. Taxonomy and range estimates for individual species can be obtained from the reports cited above.

Very little is known about the Cenozoic Arctic Ocean, although plate tectonic reconstructions make it clear that it has existed as a relative isolated basin since at least the middle Mesozoic. The only Cenozoic section available, recently recovered by the sole deep-sea drilling expedition to the Arctic (IODP Leg 302), suggests that the Arctic Ocean in the Paleogene was sufficiently brackish to exclude groups such as radiolarians that need near-normal salinity, although other siliceous microfossil groups such as silicoflagellates are common (Moran et al., 2006). Neogene sediments from this Leg are too fragmentary to be of much use and contain nearly no radiolarians or other biogenic silica (Moran et al., 2006).

Lastly, an elaborate biostratigraphic framework has been developed over several decades for the various, partially isolated marine basinal sediments of the Russian Platform (mostly by Soviet workers publishing in Russian; English summary in De Wever and Popova, 1997). These zonations were never usefully integrated into those used in open-ocean sediments, or land sections from other continents.

North Pacific

This region of the Pacific ocean includes warm temperate waters of the oligotrophic province north of the tropical zone; cool temperate, more productive waters of the North Pacific Gyre; and north of ca. 40° N, marginal seas such as the Sea of Okhotsk, Japan Sea, and Bering Sea; plus upwelling influenced near to offshore environments since the late Neogene off of North America. Given the great oceanographic and environmental complexity of this region it is not surprising that no single stratigraphic zonation has been shown to work across the entire region. In addition, onshore Cenozoic sections are widely distributed in Japan and the biostratigraphy of these has been developed to some extent in parallel, with publications primarily in Japanese, to the study of open ocean sections where studies have been published in English. In recent years, the different sources have become better integrated, although languages continue to pose a barrier to understanding.

Although biosiliceous sediments of all Cenozoic ages are known from the North Pacific region, Paleogene sections are mostly found on land (e.g. Clark and Campell, 1942; Blueford, 1988; Suzuki, 1988) and are relatively poorly known, with no agreed-upon specific regional zonation other than two Paleocene zones proposed by Hollis and Kimura (2001). Neogene sections by contrast are much better preserved, particularly from the middle Miocene–Recent, and are more intensively studied. The most recent summary of Neogene zonations is by Motoyama (2019, in Japanese; Figure 8.10).

The history of stratigraphic study of these materials began with Hays (1970), who used paleomagnetically dated deep-sea sediment piston cores to define four zones covering the mid-Pliocene–Recent interval, and Nakaseko (e..g. Nakaseko et al., 1972) who, having studied their taxonomy since the 1950s, began to propose zones for Neogene sections onshore in Japan. Numerous taxonomic/applied stratigraphic studies of Japanese sections by a wide range of authors followed, and have continued to the present day (more than 100 papers known in literature databases, brief summary in Motoyama, 2019, and earlier reviews by Motoyama cited therein). Studies of sections recovered by early DSDP coring initially tried to graft the late Neogene zones of Hays (1970) to the standard tropical zonations of Riedel and Sanfilippo (e.g. Reynolds et al., 1980) but this proved unsatisfactory, as it soon became clear that many tropical forms were in higher latitudes either absent or, as subsequently documented by Spencer-Cervato et al. (1993), strongly diachronous. A general problem in biostratigraphy of polar regions is the absence or diachroniety of low-latitude stratigraphic markers, not only for radiolarians, but also for other microfossil groups. Calibration of events has thus been very difficult and has relied on recovery of particularly well-defined, complete paleomagnetically dated sections, with rare other events occasionally providing additional age information. Motoyama (1993, 1996) provided the first calibrated events for the late Neogene using the somewhat better dated (via somewhat questionable paleomagnetic data) regional diatom stratigraphy, while Morley and Nigrini (1995) using better recovered sections from DSDP Leg 86, albeit without paleomagnetic stratigraphy, systematically documented regional ranges and estimated ages for many, partially endemic North Pacific Neogene species. Kamikuri and co-authors, in a series of papers starting in the 2000s (e.g. Kamikuri et al., 2004; Kamikuri, 2017, 2019) further revised and integrated the regional events from both numerous Japanese land studies and studies of deep-sea sediments into calibrated zonations, including the Japan Sea (summarized in Motoyama, 2019).

Important taxa for these zonations include *Cycladophora*, *Larcopyle*, *Lychnocanium*/*Lychnocanoma*, and *Eucyrtidium*, which are partially the same genera as used in the Southern Ocean (see below), although mostly using different species. Several of the species used in the two regions are similar enough that there has also been some confusion as to the correct name(s). In some cases, the forms are indeed indistinguishable and apparently conspecific, with similar, but non-identical stratigraphic ranges. In other cases, although very similar, the morphologies are distinctive enough on close examination that they

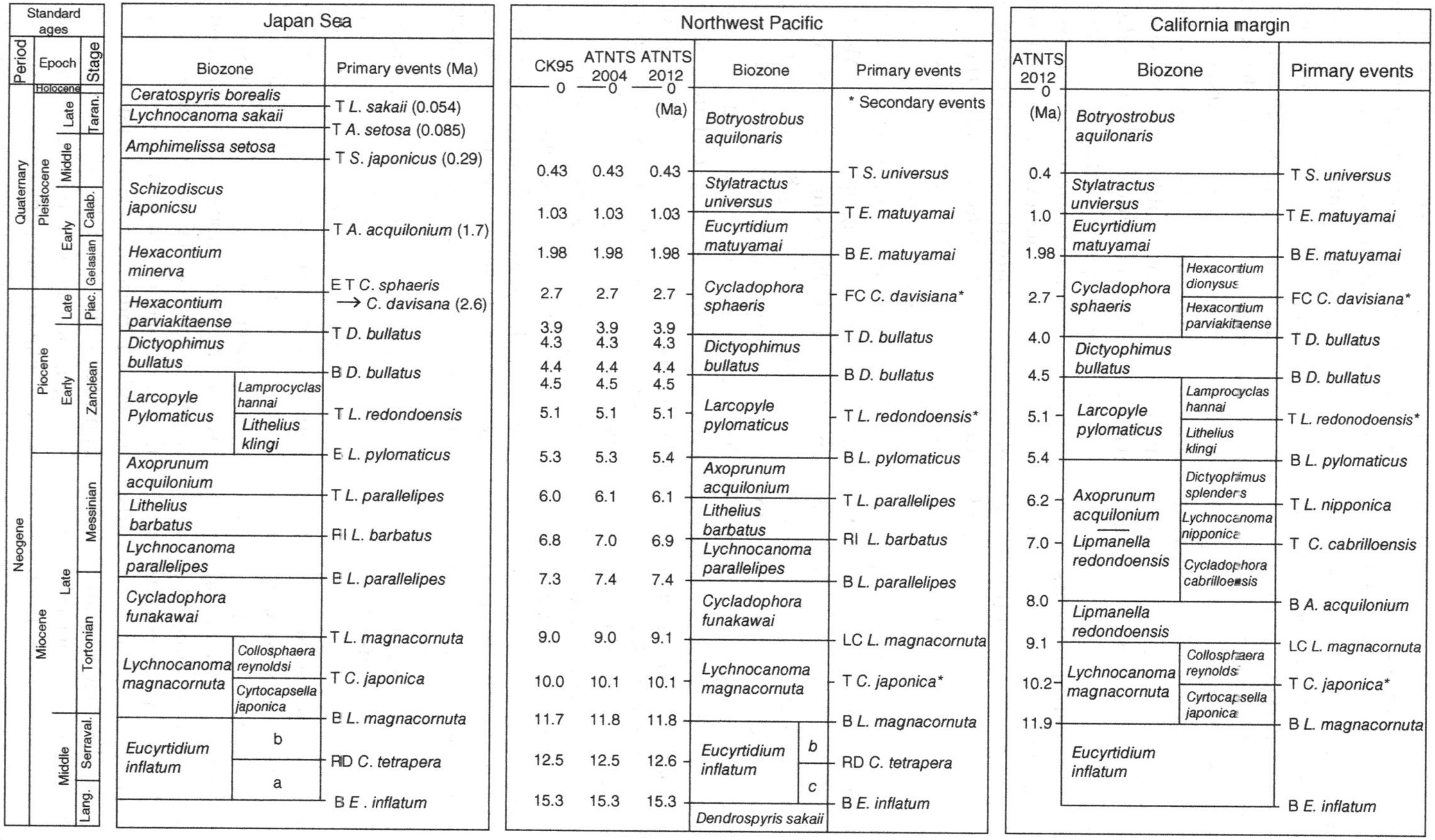

Figure 8.10 North Pacific Neogene radiolarian zonations. CK, ATNTS ("astronomically tuned Neogene time scale") 2004, 2012 refer to the Cande and Kent (1995) and Gradstein et al. (2004, 2012) geochronologic time scales, respectively. Source: Motoyama (2019). Licensed under CC BY 4.0.

should be seen as closely related species, possibly being part of species pair relationships formed by allopatric speciation (see Chapter 9: Evolution).

Southern Ocean

The Southern Ocean region, with its highly endemic fauna and flora, has existed, albeit in different form, since at least the late Eocene (Lazarus and Caulet, 1994; Lazarus et al., 2008; Bijl et al., 2011). Biogenic opal is generally well preserved in Southern Ocean deep-sea marine sediments since the late Eocene, with older sediments being barren of radiolarians except in short time intervals. Neogene Southern Ocean sediments have only low-diversity carbonate microfossil assemblages, or are purely biosiliceous, so that Southern Ocean Neogene geochronology relies heavily on siliceous microfossils, primarily diatoms and radiolarians. Radiolarian biostratigraphy of the Southern Ocean region currently is based on zones defined by first and last appearances, and covers the late Eocene to Recent (Figure 8.11). Although some Cretaceous faunas have been recovered (Ling and Lazarus; 1990) no biostratigraphy has been derived from them. A late Pleistocene-Recent cyclostratigraphy based on abundance variations of *Cycladophora davisiana* has also been used (Morley and Hays, 1979).

History

Radiolarians from the Southern Ocean have been studied since the beginning of radiolarian research (e.g. Ehrenberg, 1844), but primarily from surface sediments and the water column (see Chapter 1). With the collection of numerous piston cores by post-WWII oceanographic institutions, significant numbers of pre-Quaternary sections became available, and were used by Hays (1965; Hays and Opdyke, 1967) to create a first zonation for the Southern Ocean region. Hays' zones extended to the basal Pliocene, and were given Greek letter names (Alpha to Tau). The zones were dated using paleomagnetic polarity data, and thus were also pioneering cooperative studies integrating bio- and magnetostratigraphic events.

Early DSDP Sites provided much older and more continuous material, and Hays' student P.H. Chen extended the zonation to the base of the Miocene using material from DSDP Leg 28, although his zones were given more standard taxonomic names. (Chen, 1975) Chen also described Paleogene faunas, but did not develop a Paleogene zonation. A parallel study by Petrushevskaya (1975) from Leg 29, actually published some months before that of Chen, gave extensive data on species distributions in time, but did not precisely link these to sample depths or well defined stratigraphic zones. Subsequent studies have built on, and extended the zonations first established by Hays and Chen. Although Weaver (1976) and Keany (1978) used piston cores to refine the Neogene zonation, later work has all been based on deep-sea drilling materials. These include DSDP Legs studied by Weaver – Legs 35 (1976), and 71 (1983), and Caulet – Leg 90 (1986). Better sections became available, and thus more detailed zonations, from ODP Sites. ODP Legs 113 (Weddell Sea) and 119 and 120 (both Kerguelen Plateau) provided the first sections with radiolarian zones cross-calibrated to substantial Miocene paleomagnetic data (Caulet, 1991; Lazarus, 1992; Abelmann, 1992). The first formal zonations for Paleogene material (late Eocene to Oligocene) were also established from these Legs (Takemura, 1992; Takemura and Ling 1997). More recent studies of Neogene zonation include those of Lazarus (2001), Leg 178 and Vigour and Lazarus (2002), Leg 183. Apel et al. (2002) described Paleogene faunas from Leg 183 but were unable to establish a zonation, while Funakawa and Nishi (2005) re-examined the Paleogene stratigraphy from Leg 113 Site 689 (and followed this Funakawa

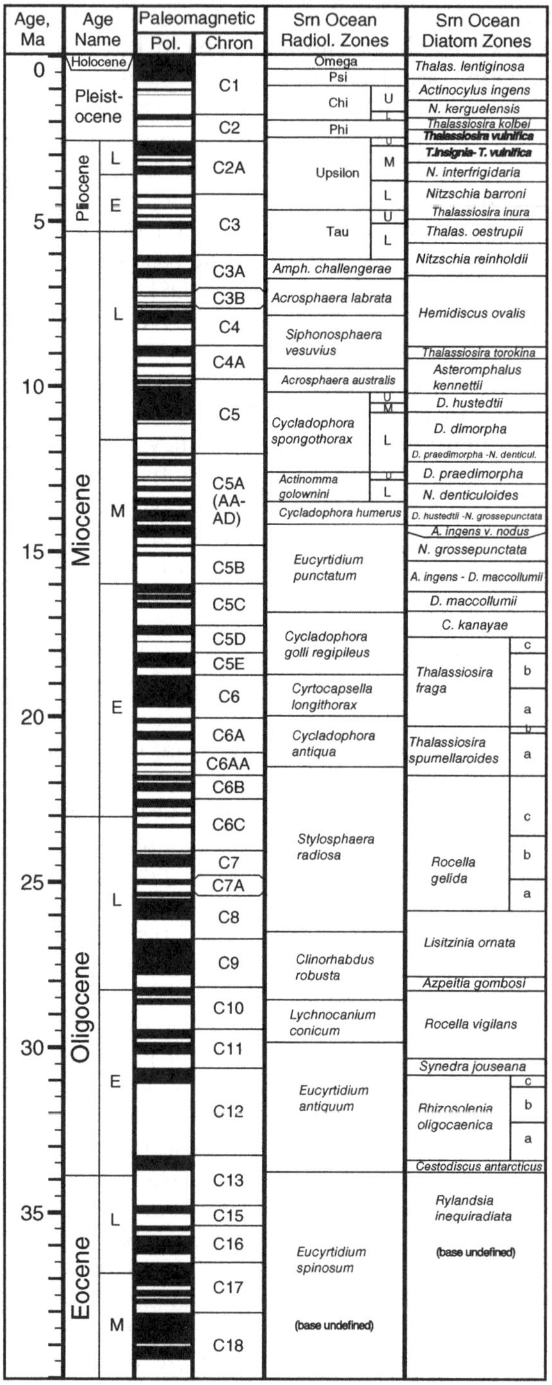

Figure 8.11 Southern Ocean Cenozoic radiolarian zonation, based on Lazarus (1992) for the Neogene and Funakawa and Nishi (2005) for the Paleogene. Diatom zonation, updated to the Gradstein et al. (2004) timescale, is from Barron and Baldauf (1995).

and Nishi 2008] with a paper on paleoceanography using the same dataset). Renaudie (2012) restudied the Neogene from several sites and noted many potential new biostratigraphic events.

It should also be noted that, despite the presence of well-preserved radiolarians, no radiolarian biostratigraphic reports were ever published for DSDP Leg 36, or ODP Legs 114, 177, or 188, although limited event data by Whalen is given in tabular form in an extended abstract by Florindo et al. (2003).

Characteristics

There are several important features of Southern Ocean biostratigraphy. There are relatively few sections available for study and none from land; and as noted, not all available sections have been examined. Southern Ocean radiolarian faunas have been highly endemic since the early Oligocene (see also Chapters 3 and 9). As noted above the sections lack, at least in much of the Neogene, lower-latitude microfossils with well-established event ages that can be used for calibration. Lastly, sections usually include several hiatuses, some of regional extent. All these features act together to make correlation, and in particular, calibration of Southern Ocean zonations difficult. Although paleomagnetic polarity data is generally of high quality, gaps in recovery, hiatuses, and, in lower Neogene and older sediments, loss of primary magnetization make calibration solely from polarity patterns difficult and error-prone. Paleogene sediments are more easily calibrated with low-latitude carbonate microfossils but too few sections have been studied to create a proven radiolarian zonation scheme, and thus the available zonation must be considered quite provisional. Currently the Neogene is divided into 17 zones with 12 subzones, and the mid-late Paleogene into 5 zones, providing an average resolution of ca. 1 and 3 my, respectively. This compares reasonably well, at least in the Neogene, to the resolution of diatom zones for the Southern Ocean, although these are not as well standardized as for radiolarians, with nearly a dozen different Neogene zonations having been proposed. In Figure 8.11 the Southern Ocean radiolarian zones are compared to the diatom zonal synthesis of Barron and Baldauf (1995) as these are derived from pelagic Southern Ocean sections, unlike several other diatom zonations based either on near-shore sections or sections at, or even north of the biogeographic boundary of the Southern Ocean. Cody et al. (2008) and Crampton et al. (2016) provide useful syntheses of individual diatom species ranges which is easier to interpret than the often inconsistent diatom zonal names.

Important Sections

Although they are the original sections used to establish the zonal framework for the Southern Ocean Neogene, the piston cores and DSDP sections used by Hays and Chen are not optimal. The former are short, low sedimentation rate cores and the latter are incompletely recovered, rotary drilled sections. Leg 113 Sites 689 and 690 are probably the best overall sections available, although sedimentation rates are low and the sections have several hiatuses in the Neogene. More generally it is best said that there are no single sections that are reasonably complete and well preserved throughout either for the Neogene or Paleogene, and only a composite section (e.g. Renaude and Lazarus, 2013) can be used as a reference frame.

Important Species

Most Southern Ocean biostratigraphic marker taxa are not only different species, but are from different genera than those used in lower latitudes. This is a consequence of the highly endemic faunas throughout much of the Cenozoic.

Several species from the genus *Antarctissa* are both stratigraphically important and common (*A. deflandrei, A.*

robusta – mid- to late Miocene; *A. strelkovi, A. denticulata* – late Miocene–Recent; *A. cylindrica* – latest Miocene–Pleistocene). The systematics of this group is unclear and some stratigraphically important, apparently closely related species such as *Helotholus vema* (Pliocene) exist as well. The validity of a few antarctissid species names such as *A. longa* and *H. praevema* are uncertain due to inadequate differential diagnoses to other taxa, and they are sometimes lumped with other species. Gradual evolutionary transitions from Miocene to Pliocene forms are seen, making it difficult to define precise marker horizons.

Cycladophora is another important genus for Southern Ocean biostratigraphy, including *C. antiqua* – early Miocene; *C. humerus* – mid-Miocene; *C.* Miocene; and *C. pliocenica* – Pliocene. Evolutionary transitions between these forms have not been documented and marker events (tops or bottoms) are fairly distinctive. Other genera or families appear to have high biostratigraphic potential but have not been studied in enough detail to make full use in zonation. The genus *Larcopyle* for example includes several stratigraphically restricted species, including *L. polyacantha titan* – late Miocene–Pliocene and *L. hayesi* – mid–late Miocene. Collosphaerids also yield several useful biostratigraphic markers, including *Siphonosphaera vesuvius, Acrosphaera labrata,* and *A. australis.* Collosphaerids, usually defined by distinctive pore structures, are particularly useful as most species are identifiable even as small fragments, which indeed is their most common mode of preservation.

Mesozoic Radiolarian Stratigraphy

Cretaceous

The Cretaceous was a very warm period in Earth history, with wide areas of shallow water epicontinental seas around the world, and several periods of organic-rich marine sedimentation between the middle Valanginian and Turonian. In the Early Cretaceous, West Gondwana broke up to form a seaway, the progenitor to today's South Atlantic Ocean, between Africa and South America. At the same time today's India began to move northward by separating from East Gondwana (Antarctica plus Australia). In the Late Cretaceous, Australia started to separate from the Antarctic part of East Gondwana to form a deep gulf extending to Tasmania. These seas were delimited in the Austral region by the South America–Antarctic and Australia–Antarctic (Tasmanian) land bridges. In the later Cretaceous, the South and North Atlantic basins were connected by further spreading of the Atlantic Ocean basin. Cretaceous ocean circulation models (Poulsen et al., 2001; Otto-Bliesner et al., 2002) suggest that a set of surface circulation gyre systems similar to those of the modern oceans existed. However, no circumpolar Antarctic circulation existed, and, although not well understood, some degree of circumtropical circulation was possible via the open ocean gateways between the Pacific, Carribean, and Tethys. The early Cretaceous isolated North and South Atlantic basins had relatively extreme water conditions of high temperature and salinity, which became more typical of general ocean conditions as a meridional (along longitude) deep circulation between the basins developed in the later Cretaceous. These modelled ocean circulation results are constrained by biogeographic patterns that have been recognized in several groups of planktonic marine organisms including calcareous nannofossils (Street and Bown, 2000), planktonic foraminifers (Malmgren, 1991; Nishi et al., 2003), and macrofossils such as ammonoids and inoceramids (Zakharov et al., 2002). Biogeographic differences in the Cretaceous are also recognized in radiolarians (Empson-

Morin, 1984; Okamoto et al., 1994; Bak, 1999; Vishnevskaya and Murchey, 2002). These biogeographic constraints on species distributions affect, as in the Cenozoic, the scope and content of biostratigraphic zone definitions.

Radiolarian biostratigraphic schemes for the Cretaceous (Figure 8.12) have been proposed from several regional datasets – the western Tethys (Jud, 1994: O'Dogherty, 1994; Baumgartner et al., 1995; Bak, 1999; Bragina, 2004), Japan (Taketani, 1982, Tumanda, 1989), Sakhalin (Zonova et al., 1993), Kamchatka and Koryak (Vishnevskaya, 2001), the east European and west Siberian platforms (Amon, 2000; Vishnevskaya, 2001; Vishnevskaya and Kazintsova, 1990), and the California Coast Ranges (Pessagno, 1976, 1977). Several variations of each have been published, with a total of over 30 proposed biostratigraphic zonations. This number, much higher than that for the Cenozoic clearly is not only due to faunal endemism but also human factors.

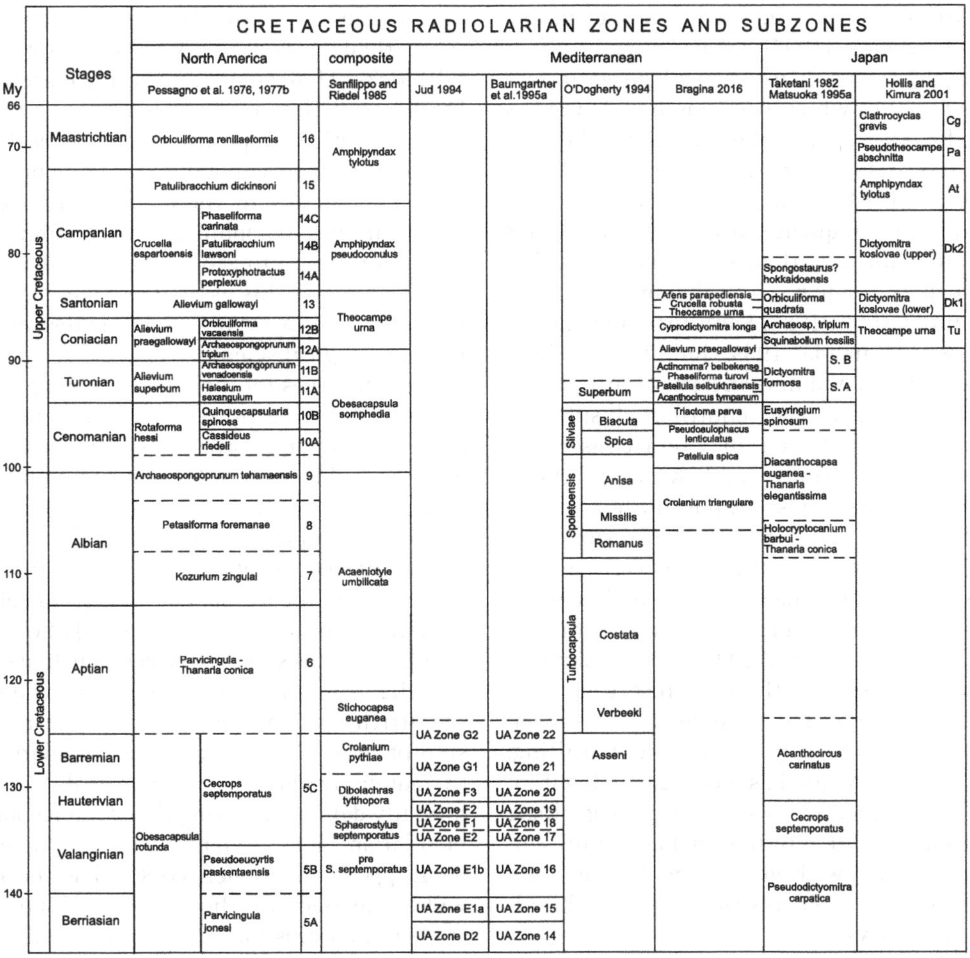

Figure 8.12 Cretaceous radiolarian zonations.
Source: Goričan, Rev. Micro, 2018. Reproduced with permission of Elsevier.

One is the use of different types of species concepts (Hollis and Kimura, 2001); in particular, the differential use of narrow vs. broad concepts (aka splitters vs. lumpers). Pessagno (1976, 1977) for example introduced many relatively narrowly defined Cretaceous genera and species. O'Dogherty (1994) combined most of these genera into just three: *Archaeodictyomitra, Dictyomitra,* and *Stichomitra*. He also combined many of the narrowly defined species into a significantly smaller number of species, but with broader definitions and more intraspecies variation. In an extreme example, O'Dogherty (1994) synonymized into a single species (with recognized variation) ten species which Pessagno had defined as valid. Other types of nomenclatural differences also exist. For example, Schaaf (1981, 1984), while not changing species concepts, frequently used different species names for taxa than those used by other workers (e.g. Taketani, 1982, Baumgartner, 1984, Sanfilippo and Riedel, 1985, Tumanda, 1989, Jud, 1994). These conceptual and nomenclatural differences in taxonomy have contributed substantially to confusion in Cretaceous zonal definitions (Hollis and Kimura, 2001). Possibly the most important reason for the relatively fragmented state of Cretaceous zonation, however, is the nature of the materials studied. Unlike the Cenozoic, Cretaceous, and older intervals are largely studied from rocks exposed on land, as too little well preserved material of Cretaceous or older age has been recovered by deep-sea sediment drilling. Thus, the work has been carried out in close connection with the field geologic studies of different national and continental regions, with all the incoherency and correlation problems that land-based geology displays.

A further complication comes from how tectonics has affected geologic sections on land. Section locations today can be significantly displaced by plate tectonics from their original locations (sometimes by several thousand kilometers: Stampfli and Borel, 2002), via the transport of entire geologic "terranes" by plate movements across ocean basins. Thus multiple, sometimes oceanographically unrelated faunas can be found in close proximity in land section studies. This problem is much larger with radiolarians than with many other groups of fossils, as many radiolarian-bearing rocks are found in accreted terranes along the edges of continents which were originally deposited in tectonically mobile open-ocean environments on oceanic crust, rather than on geologically stable continental blocks. Understanding the meaning of local faunas thus requires unravelling the biogeographic juxtapositions caused by the complex geologic context. Once done however, radiolarians in these rocks provide, via their biogeographic affinities, important information to geologists on past tectonic activity. This is in fact one major factor for the support of radiolarian research in the geological sciences.

Absolute ages for most Cretaceous radiolarian zones are primarily derived from cross-correlation to other biostratigraphies, e.g. those for ammonoids, calpionellids, inoceramids, calcareous nannofossils, and planktonic foraminifera, although some direct magnetostratigraphic calibrations also exist. As the formal definition (via definition of type sections, i.e. GSSPs) of only four Cretaceous stages have been ratified so far (the base of Cenomanian, Turonian, Maastrichtian, and Danian), the relationship of most Cretaceous stage boundaries to radiolarian biostratigraphy is also not formally defined. Although the Cretaceous radiolarian biostratigraphy synthesized in Sanfilippo and Riedel (1985) has been widely applied for the DSDP and ODP studies, this biostratigraphic scheme does not incorporate subsequent progress on Cretaceous geochronology and taxonomic revisions. Most importantly this zonation does not incorporate the more recent recognition of strong endemism in Cretaceous radiolarian biogeography, reflecting presumably oceanic circulation. It is thus necessary, as in the Cenozoic, to consider each biogeographic region separately. However, as noted above, currently a full synthesis at the level of ocean biogeography is not yet possible and instead we must review a mixture of oceanographic areas and in part, also the various regional zonations from land-based studies.

Europe and Southwest North America

As they were historically the earliest regions on land to be extensively studied, many of our ideas on Cretaceous stratigraphy have evolved from work on North American and European sections. Europe and North America in the Cretaceous were bounded by the Tethys/Caribbean, southern extensions of Arctic Ocean, and the east margin of Pacific (Ogg and Hinnov, 2012). Tethyan radiolarian-bearing sedimentary rocks are found in western Europe, the Caucasus Mountains (mainly in Azerbaijan and Georgia), Inner and Outer Carpathians (from Austria, through southern Poland and southwest Ukraine, to Romania); sections bordering the Arctic Ocean are found in Denmark, Germany, and Russia; and in the Great Valley Sequence of the California Coast Ranges. Various Cretaceous radiolarian biostratigraphic schemes have repeatedly been proposed from these sections (summarized in Goričan et al., 2018). Efforts to integrate these biostratigraphic schemes include O'Dogherty (1994), Baumgartner et al. (1995), Vishnevskaya (2001), and Goričan et al. (2018). Reconstructed maps of these locations in the Cretaceous (Ogg and Hinnov, 2012) suggest that the biostratigraphic schemes established in Europe and southwest North America can be classified into four regions, namely the low-latitude western part of Mesotethys (the north part of Neotethys), the mid-latitude northern part of Mesotethys, epicontinental seas of the Russian Platform, and the mid-latitude East Pacific.

Low-Latitude Western Part of Mesotethys

Radiolarian-bearing sedimentary rocks found in the *Préalpes médianes*, Southern Alps, Umbria Marche Apennines, Northern Apennines, and Betic Cordillera (Jud, 1994; O'Dogherty, 1994) were deposited at 10–20°N paleolatitudes (Jud, 1994). The Berriasian–lower Barremian radiolarian biostratigraphy from these sections is divided into nine Unitary Association Zones, and is well dated via calcareous nannofossils, calpionellids, and the magnetostratigraphy in the same sections (Jud, 1994). The upper Barremian–lower Turonian radiolarian biostratigraphy is divided into five zones and seven subzones using the Unitary Association method (O'Dogherty, 1994). As this geologic time interval is just within the Cretaceous Normal Super-Chron ("Cretaceous magnetostratigraphic Quiet Zone"), the age of the radiolarian biostratigraphy in this interval is based entirely on associated macrofossils. The Berriasian–lower Barremian radiolarian biostratigraphy established by Jud (1994) in Tethyan sections is applicable in the tropical Pacific region due to presence of common species, while the upper Barremian–lower Turonian zonation in O'Dogherty (1994) cannot be easily applied in the Pacific because few of the marker species are found in the Pacific. Several important studies on Lower Cretaceous radiolarians were done as well in the Greater and Lesser Caucasus in the 1960s (Aliev, 1965, 1967), which at the time of deposition was also located in the low-latitude western Mesotethys.

Mid-Ltitude Northern Part of Mesotethys

Radiolarian sedimentary rocks found in the Pieniny Klippen Belt were deposited in deep longitudinal troughs, separated by shallow submarine or emergent ridges near the northern margin of the western Mesotethys in the Cretaceous (Bak, 2006), at ca. 30–35°N paleolatitude. In this region, three zones and five subzones were proposed in the mid-Albian to lower Turonian intervals using the Unitary Association method (Bak, 1999). The species composition in the Pieniny Klippe Belt is similar to the low-latitude Pacific and the low-latitude western part of Mesotethys, but the stratigraphic ranges of several species are reported to be diachronous (Bak, 1999).

Russian Epicontinental Seas

Shallow (generally less than a few hundred meters water depth) epicontinental seas were widely developed over vast areas of Denmark, Germany, the East European platform, and the west Siberian platform, and were connected both to the Atlantic and Arctic Oceans (Zakharov et al., 2002). Owing possibly to strong endemism in radiolarian faunas, more than 15 different radiolarian biostratigraphic schemes have been proposed for these seas, including zonations from sections near Moscow and Voronezh in the Moscow Basin, Ul'yanovsk in the Volgo-Ural Basin, Pechora in the Timan-Pechora Basin, and west Siberia (Vishnevskaya, 2001). In this book, the radiolarian biostratigraphy for these regions follows the recent summary by Goričan et al. (2018).

East Margin of the Mid-Latitude Pacific

The Cretaceous radiolarian biostratigraphy of the east margin of the mid-latitude Pacific was first established in the Great Valley Sequence and Franciscan Complex of the California Coast Ranges (Pessagno, 1976, 1977). The depositional site of the Great Valley Sequence is part of a terrane that was at approximately 35°N paleolatitude near the western margin of Laurentia since the late Jurassic (Hagstrum and Murchey, 1996). This radiolarian zonation was largely dated with the then-current planktonic foraminiferal biostratigraphy. As planktonic foraminiferal zonation calibrations have changed (Ogg and Hinnov, 2012) there is a need to update the radiolarian zonation calibrations as well. For instance, the planktonic foraminiferal *Globotruncanella havanesis* Zone correlated to the Maastrichtian in Pessagno (1976) is now considered to be correlated to the upper Campanian in Ogg and Hinnov (2012), and thus the radiolarian zone correlated to this foraminiferal zone is automatically changed to the upper Campanian. The correlation between the geologic time scale and Pessagnos's biostratigraphy in the intervals of the Barremian–Turonian was also revised by O'Dogherty (1994).

Northwest Pacific

More than ten biostratigraphic schemes for Cretaceous radiolarians have been proposed in Japan and Russia (see summaries by Hollis and Kimura, 2001, and Vishnevskaya 2001). One reason for this complexity is the geology. Cretaceous geology of the eastern margin of the Eurasia is complicated due to the presence of fore-arc basin deposits (e.g. Yezo Group: Taketani, 1982; Zonova et al., 1993), accreted sediments on the Kula and Pacific plates (e.g. Chichibu and Shimanto belts: Okamura, 1992; Takahashi and Ishii, 1995), and an inland sedimentary basin over partially emergent accretionary complexes (e.g. Izumi and Soto-izumi groups: Hollis and Kimura, 2001). All of these disparate geologic environments contain Cretaceous radiolarian-bearing formations. Tectonic reconstructions for the Northwest Pacific (Natal'in, 1993; Levashova, 2000), suggest that Cretaceous radiolarian biostratigraphy in the Northwest Pacific can be separated into tropical regions (Cretaceous accretionary complexes in Japan and Russia), mid-latitude regions (inland sea sediments in Japan), transitional regions (fore-arc sediments in Japan), and high latitudes (central to west Sakhalin). The biostratigraphy in the tropical regions consists of deep to shallow water radiolarians. The biostratigraphy established in the inland sea sections may contain shallower water radiolarians than used in the tropical regions. The sediments in Hokkaido, Japan, used to define fore-arc radiolarian biostratigraphy (e.g. Taketani, 1982) are well studied and cross-correlated to transitional planktonic foraminiferal faunas (Nishi et al., 2003) and paleomagnetic data (Tamaki and Itoh, 2008). Although the inoceramid and ammonoid faunas in the forearc sediments in west and central Sakhalin are similar to those in Hokkaido (e.g. Yazykova et al., 2004), a different radiolarian biostratigraphy was defined, because important marker species for the biostratigraphy in Hokkaido such as *Diacanthocapsa euganea, Thanarla elegantissima,* and *Theocampe urna* have not been found in west and central Sakhalin (Zonova et al., 1993).

Other Regions

For most remaining regions in reconstructed Cretaceous paleomaps such as the Cenotethys (the south part of Neotethys), the Atlantic seaway, and Austral high-latitude region (e.g. Metcalfe, 1996), radiolarian biostratigraphic schemes have either not been proposed or are still very preliminary.

The Upper Cretaceous sequence in Oman, which was deposited in the southern Cenotethys, contains an endemic radiolarian species, *Anikaella omanensis*, but a suitable biostratigraphy has not been proposed. In the Atlantic seaway, Thurow (1988) revised the biostratigraphic scheme of Schaaf (1984), although as noted above, this scheme needs further work both due to age calibrations and different taxonomic nomenclature.

Radiolarian faunas from Austral high paleolatitude regions are reported in the Berriasian–Barremian interval from DSDP Leg 27 (Renz, 1974) and ODP Leg 123 (Baumgartner, 1993; Ishii et al., 2009), Aptian-Albian and Maastrichtian of Leg 113 from the Weddell Sea (Ling and Lazarus, 1990), and from the Lower Cretaceous in Indonesia (Tan, 1927). Although Renz (1976) established three assemblage zones, this has not yet been updated or tested in other locations.

The Jurassic–Cretaceous Boundary (Tithonian–Berriasian Boundary)

The GSSP of the Tithonian–Berriasian boundary has not been formally defined, although a working definition equates the base of magnetic Chron M18r as equal to the base of Berriasian (Ogg and Hinnov, 2012). This is approximately equal to biostratigraphic markers in several fossil groups, e.g. calcareous nannofossils, calpionellids, and ammonoids.

The correlation of this magnetic event to radiolarian biostratigraphy has been studied in Europe, (Ondrejickova et al., 1993, Jud, 1994, Mekik et al., 1999), Japan and the western Pacific (Matsuoka, 1995); Antarctica (Kiessling, 1999), and North America and the Caribbean (Pessagno et al., 2009). In Europe the Jurassic-Cretaceous boundary is within the lower part of Zone D2 of Jud (1994) and species associated with the underlying Zone D1 (*Aacanthocircus sublongus, Ristola altissima altissima, Triactoma blakei, Triactoma jonesi,* and *Podobursa spinosa*) are assignable to the Jurassic. Species that are first found in Zone D2 such as *Sethocapsa kitoi, Wrangellium depressum, Obesacapsula rusconensis rusconensis, Obesacapsula morroensis,* and *Syringocapsa limatum* are mostly of Cretaceous age. Pessagno et al. (2009) placed the Jurassic-Cretaceous boundary in the boundary horizon between their Subzone 4β2 and Subzone 4α1, Kiessling (1999) in the boundary between Zone 4 and Zone 5, and Matsuoka (1995) in the lower part of the *Pseudodictyomitra carpatica* Zone. The exact correlation among these radiolarian zones is still unclear.

Jurassic

The Jurassic (see also Chapter 5: Taxonomy and Fossil Record) was a period in Earth history of warm humid climates and moderately high sea level, particularly in the later Jurassic. Although a narrow North Atlantic ocean existed by the late Jurassic, the early Jurassic continents were fully assembled into a single supercontinent – Pangea – although the Tethys made a deep embayment in the east. Jurassic physical geography is thus somewhat simpler than in the Cretaceous, and, particularly in the early Jurassic, biogeographic provinces for groups such as ammonoids, while clearly present, do not show the same degree of endemism as they do in the later Jurassic or Cretaceous. Biostratigraphic correlation between different latitudinal realms however is still challenging.

Many Jurassic radiolarian biostratigraphies have been proposed (Figure 8.13): from European sections (Baumgartner et al., 1984), western North America (e.g. Carter et al., 1988, 1998, Pessagno et al., 1987, 2009), and Japan and the western Pacific (e.g. Yao, 1982, Aita, 1987, Hori, 1990, Matsuoka, 1995a, Nishizono et al., 1997). An integrated biostratigraphy between different regions was also proposed for the Early Jurassic (Carter et al., 2010) and the Middle to Late Jurassic (Baumgartner et al., 1995), respectively. The biozones of Baumgartner et al. (1995), Matsuoka (1995a), Yang and Matsuoka (1997), Pessagno et al. (2009), and Carter et al. (1998, 2010) are among those most commonly used.

Middle and Late Jurassic

The Late Jurassic is divided, from top to bottom, into the Tithonian, Kimmeridgian, and Oxfordian while the Middle Jurassic into the Callovian, Bathonian, Bajocian, and Aalenian (Figure 8.13, Gradstein et al., 2012). The bases of these geologic stages are formally defined by first occurrences of particular ammonoids, but these ammonoid events have never been correlated to Middle to Late Jurassic-aged radiolarian-bearing formations. Users of Jurassic radiolarian biostratigraphy need to be aware of substantial differences in published age calibrations (mainly at the stage level) for many zones. For example, in the Aalenian and Tithonian there is a strong discrepancy in correlation between the geologic time scale and radiolarian biostratigraphy between the Japan-western Pacific regions (Matsuoka, 1995, Yang and Matsuoka, 1997), the Tethys (Baumgartner et al., 1995), and western North America (Pessagno and Hull, 1996; Pessagno et al., 1999, 2009). A specific example is the *Striatojaponocapsa plicarum* and *Striatojaponocapsa conexa* zones of Matsuoka (1995). Matsuoka (1995) correlated these two biozones to the Bajocian to the middle Callovian; Baumgartner et al. (1995) correlated them to the Bajocian to the lower Bathonian, and Pessagno et al. (1993, 2009) correlated them to the Oxfordian: an age range of at least 10 my (Gradstein et al., 2012). Resolving these differences will most likely only be possible once the radiolarians can be more directly correlated to ammonites or other groups that are tied to GSSPs. Until then users need to keep in mind that stage name ages like the Oxfordian are, if determined by radiolarian zones, not really the same age, or unable to be correlated across the Pacific-western Pacific, the Tethys, and western North America.

An additional problem is found in the use in the Jurassic (and in other time intervals below the Cenozoic) of very different methodologies for analyzing radiolarian occurrence data and creating zonations. Zonations have been made using first and last occurrence events of species, unitary associations, and also with what are effectively assemblage zones based on the joint first and last appearances of several different species. As the fundamental units that define the zones are different, correlating between them is difficult. Examples include: the event-based zonation of Japan and the

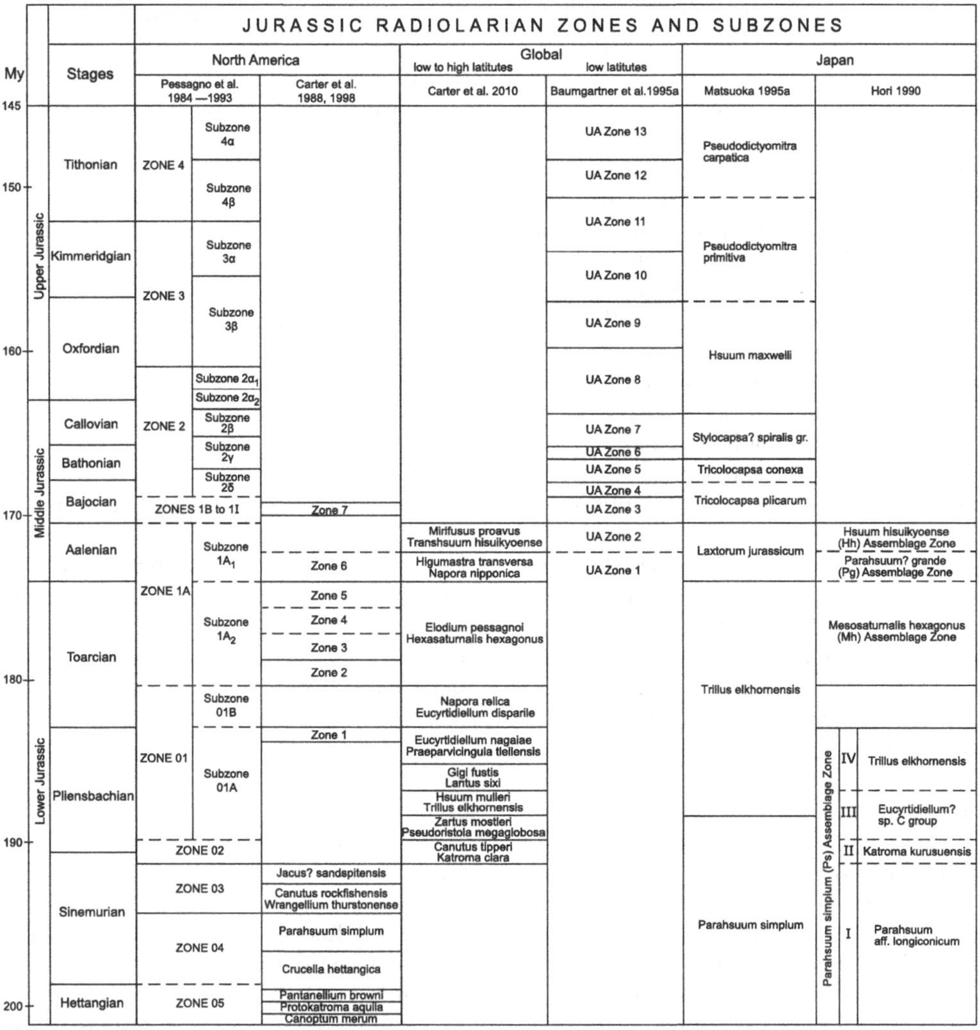

Figure 8.13 Jurassic radiolarian zonations.

Source: Goričan, Rev. Micro, 2018. Reproduced with permission of Elsevier.

western Pacific (Matsuoka, 1995; Nishizono et al., 1997) which has been verified in many local sections, but as the sections lack other fossils, is not well cross-calibrated to the geologic scale; the Tethyan zonation based on unitary associations (Baumgartner 1984; Baumgartner et al., 1995; Carter et al., 1998, 2010); and lastly, the "assemblage" type zones defined for North America (Pessagno et al., 1987, 1993, 2009). Assemblage type zones are particularly problematic as, although the multiple bio-events that define them may co-occur in the original sections, they may well show diachrony in other sections. This has in fact been documented for Jurassic radiolarian assemblage zones by Yang and Matsuoka (1997). These methodologic problems would be less acute if the primary data were published and not only the interpretations (e.g.

in the above-cited papers by Pessagno) as were the full data available, it would be possible to reanalyze it and identify how each dataset is correlated to the others.

Lower Jurassic

The Early Jurassic is subdivided into the Toarcian, Pliensbachian, Sinemurian, and Hettangian. All the calibration reference sections (established GSSPs and candidates) are biostratigraphically defined by ammonoids. The Lower Jurassic radiolarian biostratigraphy established in Queen Charlotte, Canada, is well correlated to associated ammonoid fossils (Carter et al., 1998, 2010). Although these ammonoid assemblages are cosmopolitan or Tethyan faunas, they unfortunately do not include exactly the same taxa which define the GSSPs, except for the Triassic–Jurassic boundary. Thus, the correlation between the radiolarian biostratigraphy and the geologic time scale at the stage level is indirect, via ammonite secondary marker taxa. According to Carter et al. (1998, 2010), the Hettangian–Sinemurian, Sinemurian–Pliensbachian, and Pliensbachian–Toarcian boundaries are placed within the *Crucella hettangica* Interval Zone, within *Canutus tipperi–Katroma clara* Concurrent Zone, and the base of *Napora relica–Eucyrtidiellum disparile* Concurrent Zone, respectively. The first occurrence of *Parahsuum simplum* was previously used to mark the base of the Jurassic (e.g. Yao, 1990), but this datum is now placed in the early Sinemurian by Sugiyama (1997) and Carter et al. (2010). The Toarcian black shale – a widely correlatable event in the Jurassic – is correlated to the *Napora relica–Eucyrtidiellum disparile* Concurrent Zone (Carter et al., 2010).

Triassic–Jurassic Boundary

The Triassic-Jurassic boundary is defined by the first appearance datum (FAD) of the first species of *Psiloceras* ammnonites (e.g. *P. spelae*), significantly above a sharp negative carbon-isotope excursion in the Kuhjoch section, Tyrol, Austria. The presence in the North American sections of this Tethyan ammonoid genus makes it possible to correlate the boundary (in Austria) to the radiolarian biostratigraphy defined in North America (Carter et al., 1998). The approximate position of the boundary in most radiolarian faunas is very easy to see, since 20 genera and over 130 species of the Rhaetian in age disappeared at the end of the Triassic and several new genera such as *Charlottea, Udalia,* and *Parahsuum* (sensu lato) first appeared in the lowest Hettangian (Carter and Hori, 2005; see Chapter 5: Taxonomy and Fossil Record). For more precise recognition of the Triassic-Jurassic boundary by radiolarians, the final occurrences of several taxa, e.g. the genera *Betraccium* and *Risella,* and the species *Globolaxtorum tozeri, Livarella valida,* and *Pseudohagiastrum giganteum* are useful, according to Carter and Hori (2005).

Triassic

Several contrasting Triassic radiolarian biostratigraphic schemes (Figure 8.14) were proposed in Japan in the 1980s (See the summaries by Yao, 1990, and O'Dogherty et al., 2010). Recent papers on Triassic radiolarians often use the biozones proposed by Sugiyama (1997), Carter and Hori (2005), and Kozur (2003). Radiolarian faunas are so different among the low-latitudinal Panthalassa (Sugiyama, 1997), high-latitudinal Panthalassa (Carter and Hori, 2005), and Tethys regions (Kozur and Mostler, 1994, 1996) that different biozones are used in each region. No practical biostratigraphy for the high latitudes has been proposed yet.

Triassic radiolarian biostratigraphy is relatively well correlated with the geologic time scale, largely via conodont and ammonoid biostratigraphy (Carter 1993, Kozur and Mostler 1994, 1996, Sugiyama, 1997). However, its age precision is different

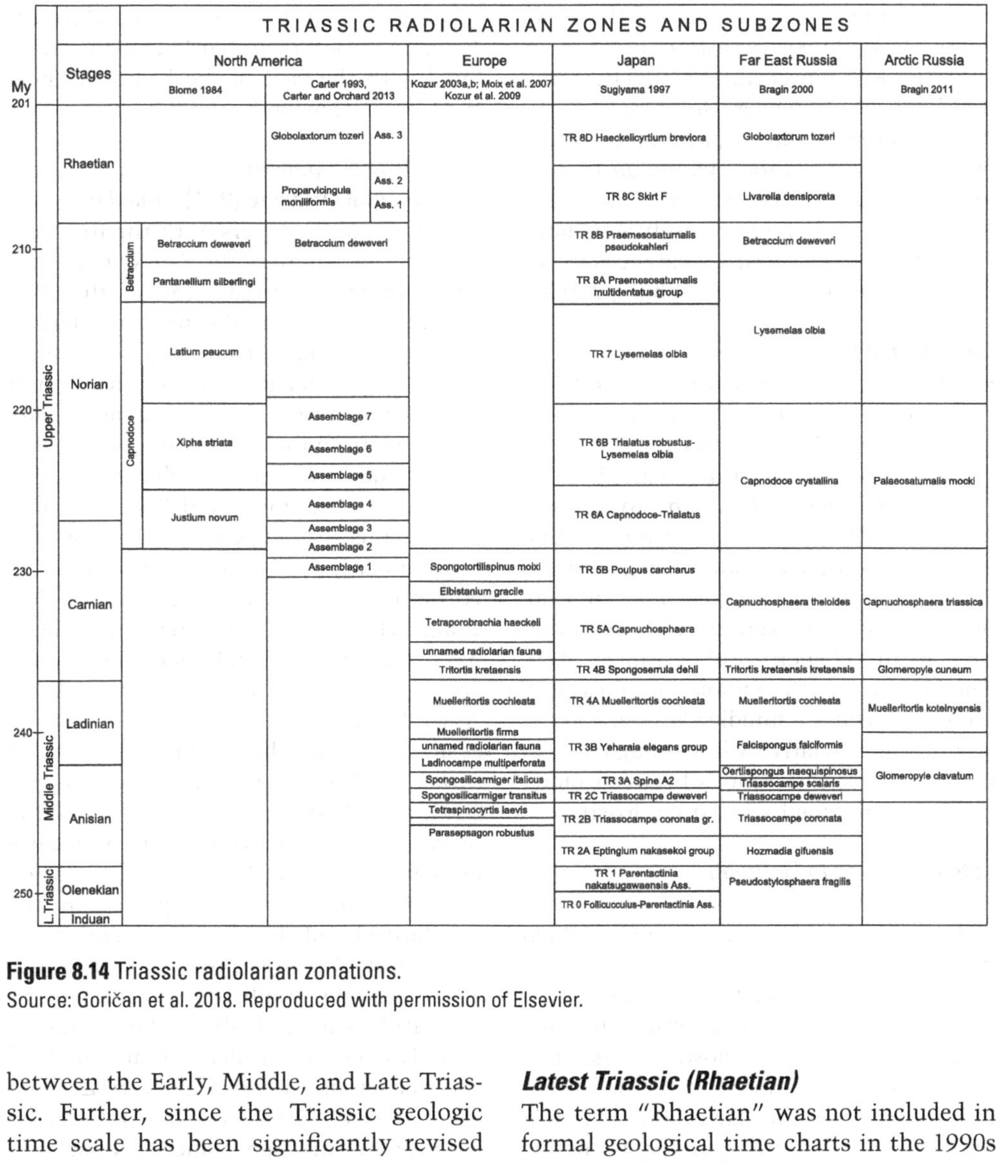

Figure 8.14 Triassic radiolarian zonations.
Source: Goričan et al. 2018. Reproduced with permission of Elsevier.

between the Early, Middle, and Late Triassic. Further, since the Triassic geologic time scale has been significantly revised after defining or proposing the GSSP sections, the correlations between the geologic time scale and numerical age have also been significantly changed. In particular, the total time duration from the start of the Carnian to the end of Rhaetian has is now twice as long (Mundil et al., 2010; Ogg, 2012) than in earlier calibrations e.g. Gradstein et al. (2004). This book follows the more recent Triassic age calibrations.

Latest Triassic (Rhaetian)

The term "Rhaetian" was not included in formal geological time charts in the 1990s because the underlying interval – the Norian – was considered to substantially overlap it (e.g. Tozer, 1990). Early radiolarian biostratigraphic studies in the western North America (Blome, 1984; Pessagno et al., 1979) thus used the term "Norian" in the sense of the formal geologic time chart, including Rhaetian. Although the GSSP of the base of the Rhaetian Stage has not been ratified yet, it has been proposed to be placed at the first appearance of the conodonts *Misikella posthernsteini* or *M. hernsteini* by Krystyn et al. (2007). The genus *Misikella* is however so rare in the Americas that the first appearance horizon of the radiolarian *Proparvicingula moniliformis* is used to recognize the base of the Rhaetian, as this event is closely associated with the *Misikella* first appearances in the Tethys region (Carter and Orchard, 2007).

Carnian and Norian

In the Triassic, the distinctive twisted radial spines of *Capnuchosphaera* (middle Carnian–upper Norian) and hollow radial spines with internal three partitions of *Capnodoce* (upper Carnian to middle Norian) can be used to date even rocks that have undergone substantial diagenesis as these features are still preserved and are unique characters appearing at no other time in the Phanerozic (Sugiyama, 1997). The precision of the radiolarian biostratigraphy between the middle Carnian and the late Norian is relatively low though, due to too few direct correlations between radiolarian biostratigraphy and other fossils.

Late Olenekian to Ladinian

Biostratigraphy of radiolarians, conodonts, and ammonoids from the upper Olenekian (Smithian) to the Ladinian are well correlated to each other. The correlation between radiolarian biostratigraphy and geologic time scale is however disputed, due to different opinions on the definition of the base of the Ladinian. The ICS (International Commission on Stratigraphy) ratified the base of the Ladinian as the lowest occurrence horizon of the ammonoid *Eoprotrachyceras curionii* (Brack et al., 2005). Kozur (2003) however, in one of the major papers on Triassic biostratigraphy, defined the base of the Ladinian at the base of the ammonoid *Reitziites reitzi* Zone, older than the *E. curionii* Zone. Thus, in the sense of Kozur (2003), the base of the Ladinian correlates to the *Oertlispongus primitivus* Zone, while the *Ladinocampe vicentinensis* Zone is correlated with the base of the Ladinian under the formal decision by the ICS.

Basal Triassic (Induan) and Permian–Triassic (P–T) Boundary

Major faunal turnovers with the first substantial appearances of Nassellaria and Spumellaria occurred around the P–T boundary. However, the precise relationship of these faunal changes to geologic events are unclear due to the "chert gap" – radiolarian fossils are practically unknown from rocks between the end-Permian and the early Olenekian (see Chapter 5: Taxonomy and Fossil Record). Continuous lithologic sections across the P–T boundary are found in the Oruatemanu Formation in New Zealand (Hori et al., 2011) but no biostratigraphic scheme has been proposed yet. Induan radiolarians are reported as the *Entactinia nikorni* Assemblage in Thailand (Sashida et al., 1998) and the *Stigmosphaerostylus turkensis* Assemblage in Turkey (Kozur, 2003). The first appearance of Nassellaria in the Triassic is recorded in the early Induan (Griesbachian), when *Tripedocorbis? blackae* appeared in the mid-latitudes of the southern Hemisphere (Hori et al., 2011). Permian-type radiolarians, consisting of albaillellids and latentifistularids, survived till the latest Induan (late Dienerian) in Peninsular Thailand (Sashida et al., 1998) and New Zealand (Kamata et al., 2007). For biostratigraphic purposes, the first appearance of Nassellaria needs to be used to avoid incorrectly assigning basal Triassic rocks with "Permian-type" radiolarians to the latest Permian. These Permian-type radiolarians are definitely survivors after the end-Permian mass extinction event because several taxa like *Albaillella aotearoa* and a new latentifistulid genus, *Oruatemanua*, newly appeared in the Induan (Takemura et al., 2007). Occurrences of other species, e.g. *Cauletella manica* (Latentifistularia), *Hegleria mammila* (Spumellaria), and *Thaisphaera minuta* (Spumellaria) are not indicative of the earliest Triassic, as their biostratigraphic ranges go back to the late Permian at least.

Paleozoic Radiolarian Stratigraphy

Paleozoic radiolarian biostratigraphy has only been studied for a comparatively short interval of time and is the least well-developed of the Phanerozoic, with the first stratigraphic papers appearing in the 1970s, e.g. Nazarov (1973). Paleozoic stratigraphic work is confronted by even more extreme material limitations than in the Mesozoic. Current zonations have thus been generally established from a limited number of localities, few stratigraphic levels, and with only sporadic age-control data. As is also true for much of the Mesozoic, our knowledge of individual assemblages is biased by the loss of specimens during diagenesis and possible additional loss of specimens while extracting radiolarians from the rock matrix. Another problem is the relatively low numbers of identifiable specimens of most species recovered in individual samples, which adds to uncertainties on species ranges and patterns of co-occurrence. Despite these challenges, Paleozoic stratigraphic knowledge has advanced rapidly in the last 30 years, from an exotic subject with only a few radiolarian workers at the beginning of the 1980s to one of the major areas of radiolarian research, as indicated by the numbers of results presented at international radiolarian meetings. Advances in preparation methods, the use of advanced imaging systems (micro-CT, see Chapter 6) to decipher the morphology of specimens embedded in matrix, and the sheer accumulation of data over the years have all helped to advance the field.

Despite these advances, our knowledge of radiolarian stratigraphic ranges in the Paleozoic are much less detailed than for the Mesozoic or Cenozoic, and the methodologies used thus also differ. While event-based stratigraphy is standard in Cenozoic studies, and unitary associations with well-defined zones are typical for Mesozoic work, Paleozoic studies tend to refer to assemblages, and in older parts of the Paleozoic, may define these using genera rather than only species. The most recent summary of Paleozoic radiolarian stratigraphy is given by Aitchison et al. (2017).

Permian biostratigraphy is mainly constructed with species belonging to the Order Albaillellaria. Carboniferous biostratigraphy has also been established with Albaillellaria, but radiolarian zones have not been formally proposed in the mid-Carboniferous Serpukhovian and Bashkirian intervals and the basal Carboniferous (lower Tournaisian). Devonian and Silurian biostratigraphy has been defined in detail in limited intervals like Famennian (uppermost Devonian) and the Middle Upper Silurian, but the correlation of these proposed biozones to the geologic time is still under discussion. Ordovician radiolarian biostratigraphy is fragmentally proposed in several papers. Cambrian radiolarian biostratigraphy has only very recently started to be established: Maletz (2011) recently summarized the biozones published so far for the Cambrian.

Permian

Permian radiolarian biostratigraphy (Figure 8.15) has been well established by the examination of stratigraphically continuous chert sequences in Japan (Ishiga, 1982, and subsequent papers), from shallow-water (shelf) limestone and chert sequences in South China (Yu, 1996, Wang et al., 1994, Wang and Yang, 2007), and from shelf limestone and chert in the Southern Urals of central Asia and in Nevada (Nazarov and Ormiston, 1985,

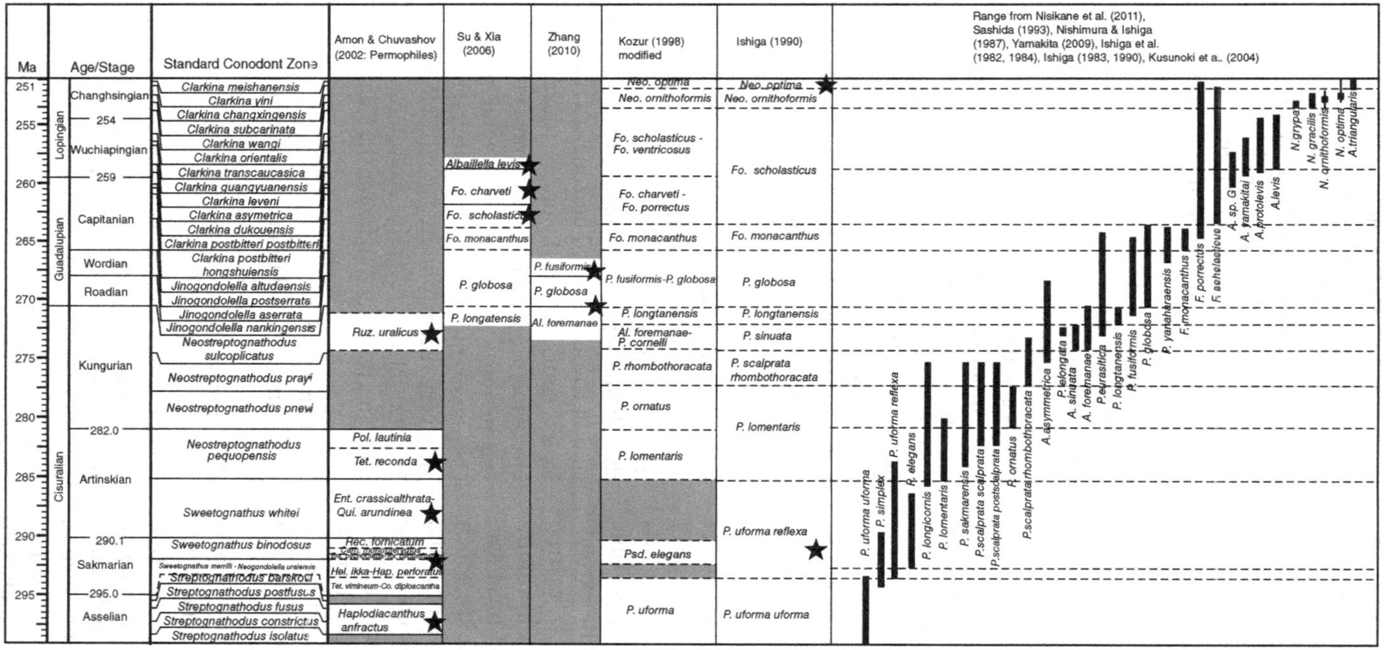

Figure 8.15 Permian radiolarian zonations and ranges of key marker species. Stars mark locations of age calibration data. Numerical age of the Asselian–Sakmarian and Wuchianpingian–Changhsingian intervals are from Schmitz and Davydov (2012) and Shen et al. (2010).

1993). A continuous [?] radiolarian biostratigraphy covering almost the entire Permian from Asselian to Changhsingian has been established in the Tethys region and Japan (Ishiga, 1990). Geochronologic calibration of Permian radiolarian biostratigraphy is however still sporadic and is based on a mixture of conodonts, smaller foraminifers, and fusulinids (Isakova and Nazarov, 1986, Maldonado and Noble, 2010, Nestell and Nestell, 2010, Sun and Xia, 2006, Zhang et al., 2010).

As examples of the current state of understanding, we mention three problems that have been recognized in Permian radiolarian biozones: (1) the relative age of the latest Permian zones (*Neoalbaillella optima* and *Neo. ornithoformis* zones) were reversed due to studies by Kuwahara et al. (1998); (2) *Albaillella yamakitai* (=*Albaillella* sp. F of Kuwahara et al., 1998) was considered to have occurred from the late Permian middle to late Wuchiapingian (Kuwahara et al., 2003), but was later shown to have occurred only over a short interval just above the basal Wuchiapingian in Japan, Texas and China (e.g. Maldonado and Noble, 2010); (3) in the early Permian (i.e. Cisuralian), De Wever et al. (2001) considered the *Pseudoalbaillella lomentaria* Zone as being within the late Asselian, whereas earlier authors assigned it to a several million-year younger age in the late Artiskian. Radiolarian faunas in the Permian are similar among Japan, south China, Thailand, Oregon, Sikhote-Alin, and Texas, although different biozones have been proposed in these regions. Permian faunas by contrast are so different in the South Urals that different biozones have been proposed by Nazarov and Ormiston (1993) and Amon and Chuvashov (2002).

Carboniferous

Radiolarian biostratigraphy for the Carboniferous (Figure 8.16) is still patchy. Zones have been established in the intervals between Moscovian and Gzhelian (upper Carboniferous: Isakova and Nazarov, 1986, Ishiga, 1982) and for the lower-middle Carboniferous between the middle Tournaisian and the lowermost Serpukhovian (Braun and Schmidt-Effing, 1993, Won 1991, 1998, Wonganan et al., 2007), respectively. In contrast, little is known of the entire upper-mid Carboniferous (Serpukhovian and Bashkirian) or basal Carboniferous intervals due to rare occurrences of radiolarians. The age of the Carboniferous radiolarian biostratigraphy is well calibrated with conodont biostratigraphy, but different species concepts affect the identification of *Albaillella* species, and thus also the recognition of the radiolarian zones. For instance, the *Albaillella paradoxa* Zone is defined in a different manner between different authors (See Figure 8.16). Problems in the Carboniferous biostratigraphy are commented on in detail by Nazarov and Ormiston (1993) and Won (1998). In particular, Won (1998) summarized the many conceptual differences for *Albaillella* species in previous papers.

Devonian and Silurian

Devonian and Silurian radiolarian biostratigraphy (Figure 8.17) has been established from sections in Australia (Aitchison et al., 1999), China (Wang et al., 2000, Zhang et al., 2008), Japan (Umeda, 2002, Kurihara, 2004), and the USA (Noble, 1994, Schwartzapfel and Holdsworth, 1996). Although well-known Silurian radiolarians have been reported from the Cape Phillips Formation of the Canadian Arctic Archipelago (e.g. Goodbody, 1986), formal biozones have not been established. All the Devonian to Silurian radiolarian zonations, with a few exceptions (Noble, 1994, Schwartzapfel and Holdsworth, 1996) were established as assemblage zones due to sporadic occurrences of radiolarians. Although the geologic age of these biozones have been carefully estimated with conodonts and

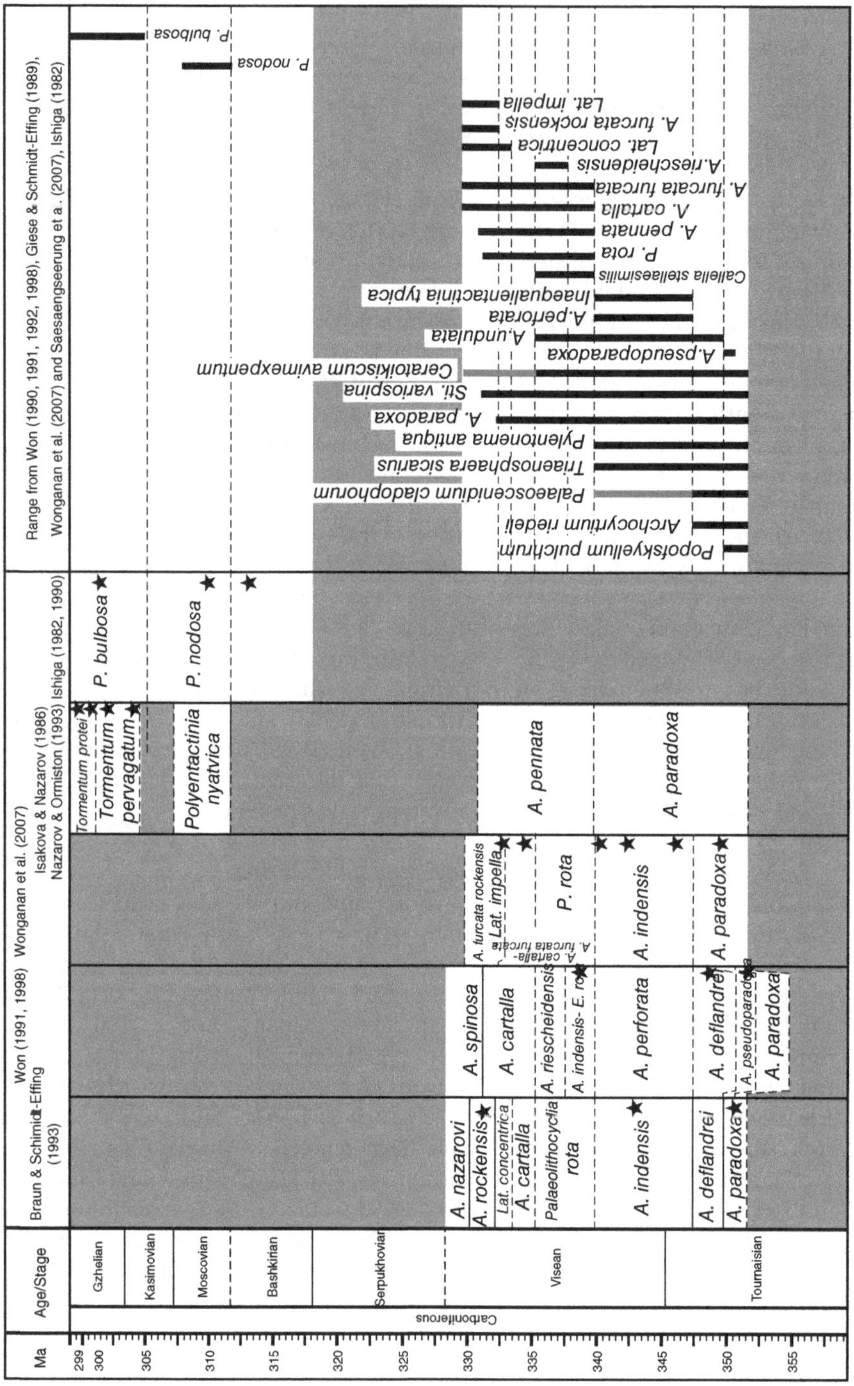

Figure 8.16 Carboniferous radiolarian zonations and ranges of key marker species. Stars mark locations of age calibration data.

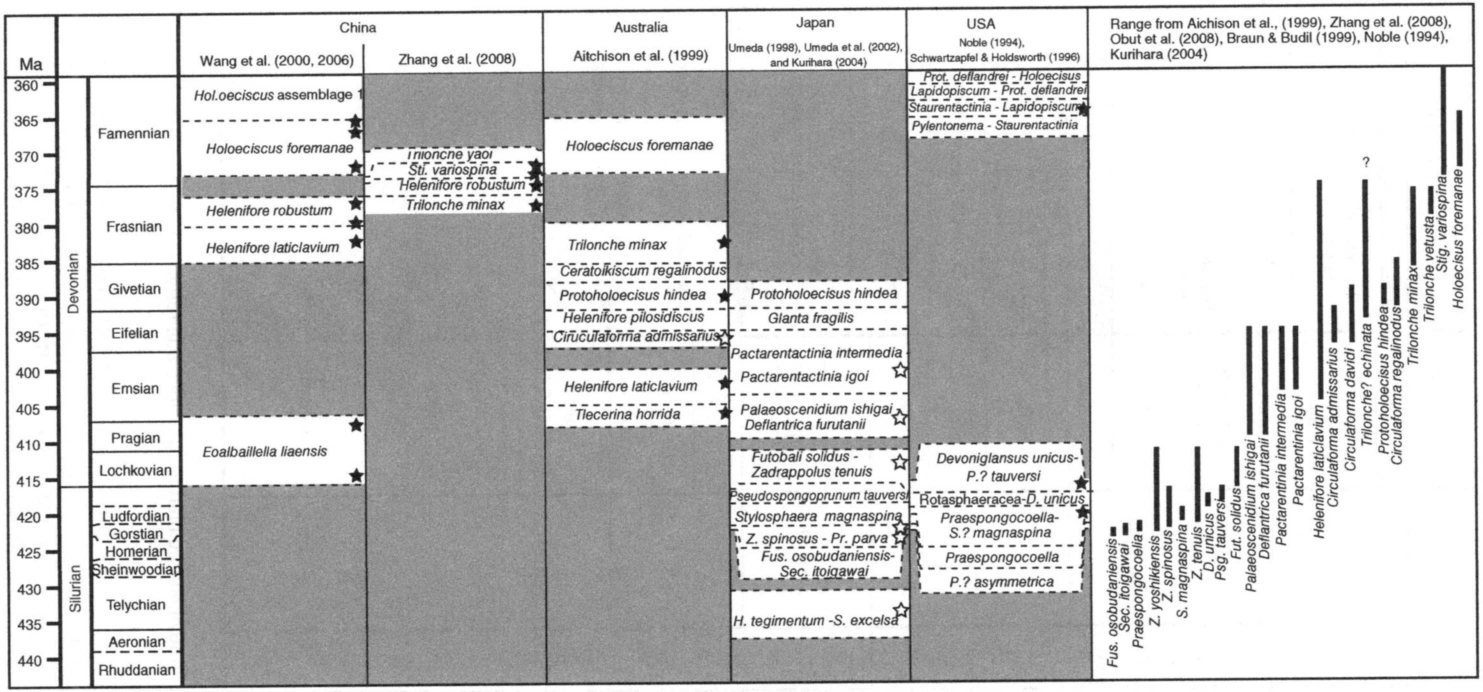

Figure 8.17 Silurian–Devonian radiolarian zonations and ranges of key marker species. Stars mark locations of age calibration data: filled for calibrations based on direct fossil evidence, open for indirect calibrations based on stratigraphic position and general faunal similarity.

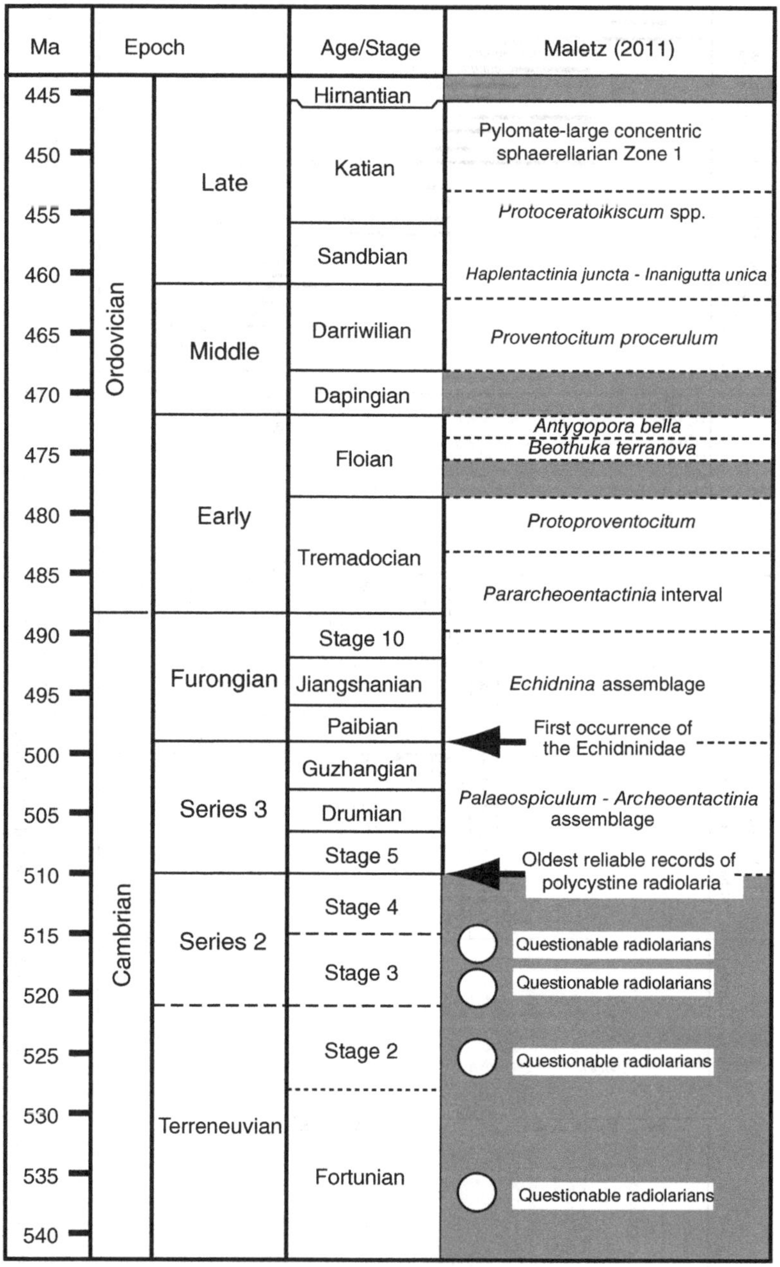

Figure 8.18 Cambrian–Ordovician radiolarian zonations and ranges of key marker species.

graptolites, if available, the estimated age has often been substantially revised by subsequent studies. For instance, the *Devoniglansus unicus* assemblage, a biostratigraphically important biozone, was originally dated as Early Devonian, albeit without associated fossil evidence, by Wakamatsu et al. (1990). Later, Noble (1994) established the *Devoniglansus unicus–Pseudospongoprunum (?) tauversi* Interval Zone in Caballos Range in Texas, and revised the age of this zone as late Silurian (Ludlovian to Pridolian) based on associated fossils. Similar refinement of assigned ages to Devonian–Silurian radiolarian biozones has been repeatedly proposed as new evidence is found. Currently, the geologic age of the Devonian–Silurian radiolarian biostratigraphy is provisional, and should not be accepted without first carefully examining the methods used in the original references to estimate ages.

Ordovician and Cambrian

Cambrian radiolarians, including doubtful identifications, have been reported in Australia, Newfoundland, China, and west Siberia (Obut and Iwata, 2000), whereas Ordovician radiolarians are found in China, Estonia, Kazakhstan, Nevada, New York, and Newfoundland. The most useful biostratigraphy (Figure 8.18) is that proposed by Maletz (2011) who summarized current knowledge of radiolarian biostratigraphy and evolutionary history between the Cambrian and Ordovician.

9 Evolution

Introduction and General Principles

Evolution is one of the great unifying concepts in biology, and, because of the philosophic implications, one that is of great interest to the general public. Traditionally, studies of evolution (history summarized in Chapter 1) have involved both living materials and, to the extent available, the fossil record. For most groups of organisms, living material (unless the group is extinct) is more readily available and easier to study than the very incomplete, or sometimes completely absent fossil record. For radiolarians however, our current understanding and ability to observe living material is still limited (see Chapters 2–4), and most studies of radiolarian evolution therefore have come from micropaleontologic research. This chapter thus largely reviews the existing knowledge of radiolarian evolution from fossil studies.

Paleontology has played a significant role in the development of evolutionary theory. In the nineteenth century, perhaps its most important contribution was to demonstrate that evolution had occurred, by the presence in older geologic layers of species no longer present on earth, and the restriction of living species to only younger geologic strata. Paleontology has also in more recent years played a role in documenting the importance of mass extinctions in the evolution of life, and in principle should be a major source of information on how evolution works, as it is

Paleobiology of the Polycystine Radiolaria, First Edition. David Lazarus, Noritoshi Suzuki, Yoshiyuki Ishitani and Kozo Takahashi.

a process that proceeds so slowly that it cannot be easily studied on human time scales. Paleontologic studies of evolution are limited by the historical nature of the data; it is not an experimental science. The geologic record of evolution can nonetheless be considered as a series of (uncontrolled) natural experiments. If enough information on how these experiments developed can be extracted from the fossils and rocks, they can provide insight into how evolution has functioned as a process over longer time periods. Analysis is generally done by comparing observed patterns – of morphology or taxa abundances and their distribution over space and time – to the patterns expected on theoretical grounds. The patterns can be documented qualitatively, and many such are summarized in Chapter 5, but such observations are subjective and not numeric, so hypotheses cannot be tested statistically. Most studies thus attempt to document patterns quantitatively, using measurements of morphology ("morphometrics") or counts of taxa abundances, either in individual samples, or the number of occurrences per interval of time. There are several limitations to these methods, as noted below, but in many cases a useful quantitative pattern can be obtained for analysis and correlation to inferred causal factors. Correlations are taken to support the hypothesis, although of course, secondary correlations due to unrecognized common causes are always a potential pitfall in such studies. Specific hypotheses about radiolarian evolution are usually derived from more general biologic theories of evolution, ecology, and ecosystems, and from our understanding of the pelagic marine environment and its evolution over geologic time.

The other main limitation of paleontologic evolution studies is the incomplete fossil record. Complete organisms, with soft tissues and genes, are not preserved, only hard parts. For radiolarians and other microfossils this means, among other things, that only shell morphology-based definitions of taxa are possible. As discussed in Chapters 2–4, most of our knowledge of even living radiolarian taxa are based on shell morphology, so this is not a particular limitation of paleontologic data. However, there is substantial evidence from other marine protist plankton groups for the widespread existence of morphologically indistinguishable closely related species ("cryptic species"), and thus there is the real possibility that many fossil radiolarian morphospecies may in fact be clusters of closely related but indistinguishable cryptic species. This obviously would have an effect on our ability to apply general biologic theories about the evolution of individual species to fossil morphospecies data. Single morphospecies might mostly be sets of independently evolving cryptic species, although in this case, it is hard to understand why most fossil microfossil morphospecies maintain a single characteristic morphology over time, and not gradually diverge into a broad range of more or less similar forms. One possible explanation is a phenomenon that is increasingly known to be a feature of many single-celled organisms, as well as many more complex organisms, particularly those such as plants, that can reproduce asexually: the transfer of genes between species, either through sexual reproduction (hybridization), or via other mechanisms, such as viral transfer. This type of between species gene transfer can have various effects, including the creation of new species, but may also act to maintain a degree of coherence over time in the evolution of a cluster of species which are exchanging genetic material with each other. The concept of a linked, co-evolving species cluster has been labeled a "super-species" (de Vargas et al., 2004). Although the nature of radiolarian morphospecies remain a major unresolved issue, for the moment it is best to accept the fossil record of morphospecies as equivalent to the similarly defined biologic units used in most plankton research, keeping in mind that the details of their structure and modes of evolution may well differ from those of the better known organisms that evolutionary theory is largely derived from (mostly terrestrial vertebrates and invertebrates).

Another aspect of protist plankton biology that needs to be kept in mind is the role of population size, and the structure of gene flow within species' populations. Radiolarians and other marine planktonic protists generally have huge population sizes. Even rare species, if sexual reproduction is part of the life cycle, must maintain a minimum population density, and thus large absolute population size if they are not to become so diluted in the ocean water as to make zygote fusion improbable. Evolutionary theories that draw on population bottlenecks, such as Mayr's classic speciation in small population isolates (Mayr, 1963; Futuyma, 1998), are unlikely to be applicable to radiolaria. Gene flow in pelagic radiolarians is also different than for terrestrial environments, due to the enormous size of the regions occupied by species. Complete mixing of surface-water masses on the scale of decades, and at least some transfer of waters between remote surface locations on the scale of years (Chapter 2), means that true isolation by simple physical barriers does not exist in the ocean. However, despite physical mixing, pelagic environments show a distinctive, stable large-scale structure of provinces and subprovinces, each with distinctive environmental properties, and numerous gradients in properties within them (Chapters 2 and 3). These patterns of environmental difference act as substantial barriers to gene flow, particularly in that, given typical life spans of weeks for individual cells, even transport times of a few years represent multiple generations, each of which needs to not only survive, but reproduce in the series of changing environments current flow exposes them to. Thus, although plankton species are much more connected on a global scale than species on land, they are far from being uniformly mixed.

Features of the Deep-Sea Microfossil Record Relevant to the Study of Evolution

In addition to biologic features, the fossil record of radiolarian evolution has several distinctive characteristics. Some are common to other marine microfossil groups, and distinguish them from most other types of fossil evolutionary information (Prothero and Lazarus, 1980; Lazarus, 2011; see also Chapter 6), while other aspects are unique to radiolarians themselves.

Radiolarians, at least since the late Paleozoic, were part of the open ocean plankton (Chapter 5), and thus have, in comparison to shallow marine benthos or terrestrial organisms, an extremely simple biogeographic structure. Individual plankton species in the modern ocean are distributed over very large areas within a very small number of biogeographic provinces. These provinces are created by the large-scale surface ocean circulation which in turn is driven by simple physical laws. The significance for evolution is that, in contrast to the complex, and over time highly dynamic biogeography of other organisms, only a few locations scattered over the ocean's area are sufficient to sample the full diversity of radiolarians at any point in time, and this simple structure, with minor modifications due to changing climate, can be safely assumed to have existed also in the earth's geologic past.

Although the large majority of individual radiolarian shells are dissolved in the ocean (Chapter 6), even the small fraction that are preserved results in large numbers of specimens in individual samples, and widespread pelagic sediment distributions (or rock exposures for older materials now exposed on land) mean that the number of specimens that can be obtained and examined for study is limited in many cases only by investigator effort. For many other types of fossils, the numbers of specimens

available are in practice much more limited. This statement though must be qualified by limits in the availability of sediment or rock to sample. Deep-sea sediments, particularly from more deeply buried layers, are extremely expensive and difficult to sample and thus the number of sections is relatively few. Once sediments are tectonically transferred to continental locations they are exposed to the same geologic forces of erosion and loss of record as for other types of fossiliferous rock, and thus the quality of the record deteriorates increasingly with longer geologic exposure times. Radiolarians are also susceptible to post-burial diagenesis (Chapter 6), which means that older time intervals can have relatively few well-preserved assemblages of species for study.

Our current knowledge of radiolarian biology and taxonomy is still very limited, and this also affects studies of radiolarian evolution. Very little is known about most radiolarian species other than the morphology of the shell, and we have as yet virtually no understanding of the shell's biologic meaning: directly functional; indirectly associated with functional cell structures; or some other reason for existence (Chapter 2). Thus, we cannot generally create hypotheses about adaptation or selection for specific morphologies, as is possible with better understood organisms (giraffe's necks, horse's teeth, and the like). Lastly, because we understand the higher-level systematics of radiolarians only very poorly, it is difficult to use higher-level taxa in evolutionary studies. This is particularly true for Cenozoic genera, which still consist of a mixture of carefully defined modern concepts and numerous older names, most of which appear to have little biologic meaning (Chapter 5). Our poor understanding of both the functional biology of radiolarian shells and their systematic taxonomy mean that radiolarian evolutionary studies have so far contributed little to understanding the evolution of biologic novelty, traditionally seen as a major subject for paleontologic studies of evolution (e.g. Levinton, 1988).

With these limits in mind we can now look at the available studies. These are organized into studies of individual species evolution, including origination, change within species, and extinction ("microevolution"); and larger-scale phenomena such as the evolution of total diversity over time ("macroevolution"), the latter being in effect an aggregate of the changes due to processes acting on multiple individual species at the level of microevolution.

Microevolution

Darwin (1859) established individual species evolution as central to evolutionary theory with the title of his famous book *On the Origin of Species*, and studies of species evolution remain the central focus of modern biologic research on evolution. Although small evolutionary changes within species can be observed over periods of years to decades, most biologic evolution research uses the method also used in paleontologic evolution study: comparing observed patterns – of genes, morphology or taxa abundances, and their distribution within the organisms or over space (though not time) – to the patterns expected on theoretical grounds. The cumulative literature is vast and can best be approached using general textbooks (e.g. Futuyma, 1998), although there are no biologic studies of species evolution specific to radiolaria. Several fossil studies of radiolarian species evolution exist, and cover the range of phenomena that together define individual species evolution: origination of new species ("cladogenesis"), evolutionary change within a single species ("anagenesis"), and the extinction of species.

Pattern and Processes

Cladogenesis: Historically the central focus of evolutionary biology, biologists have developed an extensive set of theories, or models, for speciation (cladogenesis), in which an ancestral species splits into two (or more) sister species. These are derived from ideas about how fundamental biologic processes such as gene flow, selection, and adaptation work over evolutionary time. For paleontologists, who cannot observe these processes, it is only essential that the models make different predictions about observable patterns of how populations' geographic distributions change over time, in concert with the development of a new species. The combination of changing species characteristics, geography, and time, if observable in the fossil record, can thus identify which model best fits the data, and thus contribute to our understanding of evolutionary processes. The most important speciation models include classical allopatric, peripheral isolate allopatric, parapatric, and sympatric. In allopatric speciation, a species becomes divided into two geographically isolated populations which subsequently, and fairly gradually, diverge into distinct species. The peripheral isolate model proposes that new species arise fairly rapidly, but only from small populations that are isolated from the main species, usually near the periphery of the species' geographic range. In parapatric speciation, divergence into separate species takes place over a spatial gradient in characteristics in a species (a "cline"), despite the existence of population mixing at all times between the diverging populations. In sympatric speciation, separation and divergent evolution occur within the geographic range of the species as different populations begin to live preferentially in different habitats. Each of these models can in principle be tested with deep-sea microfossil data (Lazarus 1983).

Anagenesis: Evolutionary change over time within a single species creates a *lineage*, a temporal series of linked populations with at least partially different characteristics. The process can occur gradually or rapidly, more or less continuously or only intermittently. Change can be unidirectional, oscillating, or seemingly without direction (a "random walk"). If the amount of evolutionary change is so great that it produces differences between early and later members of the lineage of the same magnitude that typically exist between cladogenetically created sister species, the different members of the lineage may be given different names and are called chronospecies. Relatively little theoretical work has been done on anagenetic change other than a claim, derived from the peripheral isolate model of speciation, that it largely should not happen. In this theory of species evolution (Mayr, 1963), established species are internally regulated by gene flow and other processes so that they are not capable of substantial change; only in small isolated populations do these regulatory processes break down and permit change sufficient to create new species. Eldredge (1971) noted that the prediction of this speciation model would be extreme rarity of species transitions and a great preponderance of morphological stasis for species in the fossil record. This proposed pattern was called "punctuated equilibrium," and many studies of fossil species were made to determine if stasis truly dominated, or if gradually changing lineages were common. Interest in this subject was driven mostly by the claim of Gould and Eldredge (Eldredge and Gould, 1972; Gould and Eldredge, 1977, and many later papers by these and other authors) that a predominance of punctuated equilibrium implied a hierarchy to evolutionary processes via emergent properties such as species selection. These claims were seen as a major challenge to evolutionary theory and thus attracted a great deal of attention, as well as controversy. Although claims for hierarchy of process were not successfully developed, the discussion started a renewed cooperation between paleontology and evolutionary biology that continues today.

Extinction: The complete disappearance of a species, extinction can only be recognized by information on the occurrences of a species over time, and by definition, no such species is still present today. Many species have become extinct during human history, but only a very few extinction events have been scientifically studied, and the reasons for extinction have usually been local human disturbance of habitats, or even deliberate extermination, rather than naturally occurring processes. Our understanding of extinction as a normal part of evolution is thus still very limited. Moreover, the large majority of information about extinction is for terrestrial organisms, while the information available for marine organisms is mostly about species decline rather than extinction, and is for either shallow shelf benthos, or larger nekton, such as fish. Virtually nothing is known about extinction in plankton from biologic research. However, since the marine plankton play an important role in global geochemistry, and extinction can potentially alter this behavior, understanding extinction is of some importance. The record of extinction in radiolarians and other microfossils is extensively documented via biostratigraphic research, but detailed studies of extinction events as a biologic process are few. One must also distinguish between local extinction ("extirpation") – the disappearance of a species from one region, with persistence elsewhere – and true disappearance of the species globally. A further distinction is needed for fossil data in that many named species are chronospecies – arbitrarily defined segments of evolving lineages (see below) – and the "disappearance" of such a form in the fossil record merely means it has been replaced by the next named morphology in the lineage – no biologic extinction has actually taken place.

Hybridization: A last process to be considered defies easy categorization into the simple origin–change–end history of individual species, but may be an important feature of species evolution in radiolaria. Hybridization, or more broadly, any other mechanism that allows exchange of genetic material between species, can have rapid, substantial effects on species evolution. Hybridization occurs when gametes from two distinct sexually reproducing species fuse and produce viable progeny. If these then reproduce with either of the parent species some of the genetic material from one parent species may be transferred into the other parent ("introgression"). The hybrid progeny may be able to reproduce on its own, and, with a mixture of characteristics from the two ancestral species become a new hybrid species. Hybridization is increasingly documented for marine organisms although it is not common (Arnold 2006). It is quite common in vascular plants, possibly because hybrids can often also reproduce by asexual means (Arnold, 2006). Almost no studies of hybridization exist for protists, although at least one marine diatom example has been reported (Casteleyn et al., 2009). More generally, single-celled organisms, including protists, can exchange genetic material without using sexual reproduction, although the frequency and significance of this for typical species of protists is not known, and is mostly documented for relatively ancient prokaryote–eukaryote transfers (Arnold, 2006).

Examples of Microevolution

Cladogenesis

This has been documented for many species of radiolaria. The majority of these are purely qualitative and are part of taxonomic and/or stratigraphic research. Many have been summarized by Sanfilippo et al. (1985), such as the middle Miocene origin of *Diartus petterssoni* from *Didymocyrtis laticonus*. Given the stratigraphic and/or taxonomic nature of these studies, evolutionary hypotheses are normally neither proposed or tested, but the documentation of these events is still important in demonstrating that speciation is commonly preserved in the marine microfossil record of radiolarian evolution, a situation which, if the record is as complete as claimed above, indeed should be seen. Of these numerous observed instances of cladogenesis, only a very few have been explicitly studied as evolutionary phenomena, two of which are reviewed here.

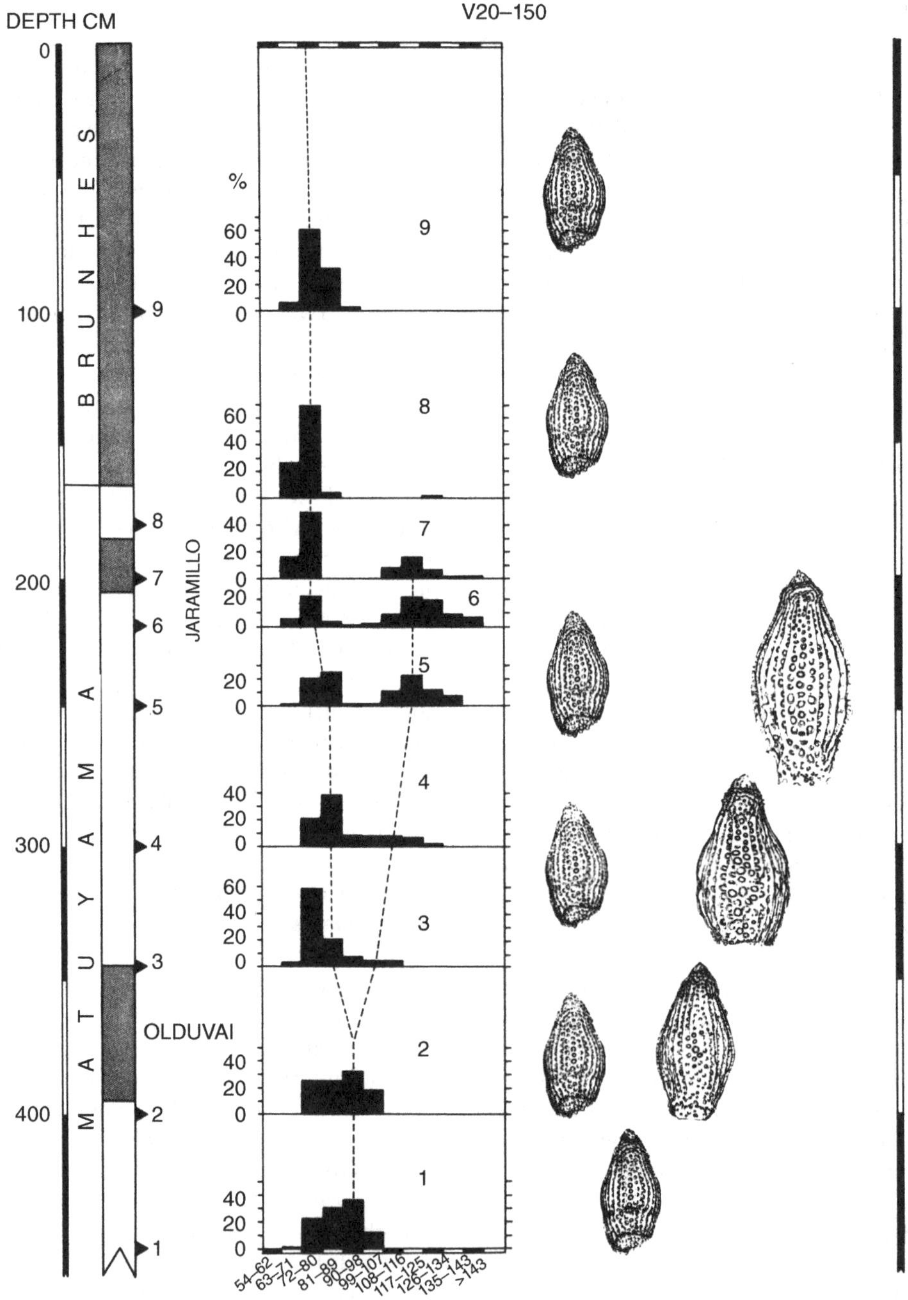

Hays (1970) carried out the first study, examining the origin at the base of the Pleistocene of *Eucyrtidium matuyamai* from its ancestor *E. calvertense*. *E. calvertense* is a fairly cosmopolitan species. Although its geographic distribution has not been explicitly mapped in modern sediments, it is found in both low-latitude sediments (tropical and subtropical biomes) and polar (Southern Ocean) sediments up to the base of the Pleistocene, after which it became restricted to low- to mid-latitude regions (north of the Antarctic Front and south of the southern edge of the North Pacific province). *E. calvertense* first appeared in the mid Miocene (Lazarus et al. 2012, supplemental table 2). One of many species assigned to the important Cenozoic genus *Eucyrtidium* (Nassellaria: Theoperidae, Figure 5.26), it has a spindle-shaped multisegmented biconical shell with a reduced cephalis, and no significant external spines or differentiated basal structures. *E. matuyamai* is found only in deep-sea sediments from the North Pacific. It became extinct ca. 1 my ago and has virtually the same general morphology as *E. calvertense*, but is noticeably larger than its ancestor. Hays documented the gradual divergence of *E. matuyamai* from *E. calvertense* by measuring the maximum width of numerous specimens in series of samples from several deep-sea sediment cores. A shift from unimodal to bimodal, and eventually completely non-overlapping size-differentiated populations was shown from a core (V20-105) at the southern edge of the North Pacific biogeographic province, in which both ancestral and descendant species can be found (Figure 9.1). Kellogg and Hays (1975) provided additional size data for older Pliocene populations of *E. calvertense* in the same core. Hays also documented that the average size of *E. matuyamai*, although always significantly larger than *E. calvertense*, had, by the end of its evolution, developed into a clear size gradient with latitude: the more northerly populations in the North Pacific at 1 ma were ca. 10% larger than the more southerly populations (Figure 9.2). Hays interpreted this pattern as a case of allopatric speciation, in which a population of a species becomes detached from the main population along the periphery of its habitat, and then evolves into a new species (this is the typical description of how speciation occurs that is found in most standard textbooks on evolution). The presence of both ancestor and descendant in the same sediment core was interpreted as re-invasion of the Subtropical Province by descendant species from the North Pacific province once it had established its identity as a separate species. Kellogg and Hays (1975) documented a small shift towards smaller size and lower coefficient of variation in size in the ancestral species *E. calvertense* in core V20-105 after the emergence of a separately measured *E. matuyamai* population, and argued that this represented an example of character displacement in *E. calvertense* caused by competition between the two species. Although this interpretation is a possibility, there are many unresolved questions, including the possibility that the shift in mean size in *E. calvertense* only represents

Figure 9.1 Cladogenetic speciation of *Eucyrtidium matuyamai* from its ancestor *E. calvertense* as seen in a series of nine latest Pliocene and Pleistocene samples from piston core V20-105 located in the transitional region between the Subtropical and North Pacific provinces. Depth in core section in cm and paleomagnetic polarity stratigraphy on left, histograms of maximum shell width showing gradual separation of the two forms in middle, and drawings of representative specimens on right. Source: Hays (1970). The Geological Society of America.

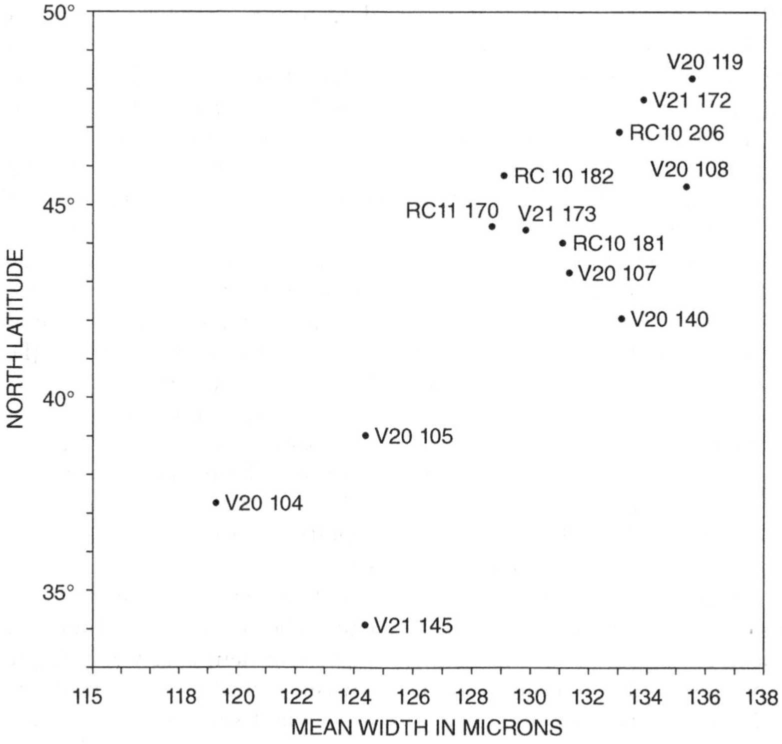

Figure 9.2 Mean width of populations of *Eucyrtidium matuyami* from 1.0 my old samples of 12 cores at different latitudes in the North Pacific. Source: From Hays (1970). The Geological Society of America.

the ability at this point to partition larger specimens into a separately measured *E. matuyamai* population, and thus is no more than a methodologic artifact. The generality of the shift in morphology of *E. calvertense* is also not known (only data from one core is shown), nor is the statistical significance of the shifts toward smaller size and lower coefficient of variation in size tested, versus longer-term trends toward lower values in both characteristics in this species. Lastly, Lazarus (1983) argued that the interpretation of the speciation event as being due to allopatry was unlikely, and instead a parapatric model was more realistic. This was based on oceanographic principles, e.g. that continuous mixing along the frontal zone of North Pacific and Subtropical waters must have existed throughout the time during which the speciation event took place, and the existence of clinal variation in *E. matuyamai* (i.e. *E. matuyamai*'s tendency to be larger with increasing latitude), similar to the characteristic that distinguishes the two species (e.g. size).

The origin of the tropical Pliocene species *Pterocanium prismatium* (Nassellaria: Theoperidae) from its more cosmopolitan ancestor *P. charybdeum* is the second example of cladogenesis reviewed. Species of *Pterocanium* consist primarily of a well-developed thorax and three prominant, divergent basal feet. As in most theoperids, the cephalis is small and globular. *P. prismatium* differs from its ancestor in several ways, including having a generally flatter, almost prismatic shape to the thorax and straight rather than (as in *P. charybdeum*) curved feet. Kellogg and Hays (1975) first studied evolutionary change in *P. prismatium*, documenting a gradual increase in thorax length between ca. 4.5 and 3.5 Ma, followed by ca. 1.5 my of little further net change. Lazarus et al. (1985) surveyed the species-level taxonomy, distribution, and phylogeny of Pliocene–Recent *Pterocanium* and first proposed the cladogenetic origin of *P. prismatium*. This identified the interval of change reported by Kellogg and Hays (1975) as being associated with cladogenesis, and thus not an example of anagenesis. Lazarus et al. (1985) documented for several species the overall pattern of evolution with a few simple measurements in two sediment sample series from the tropical Pacific and tropical Indian oceans, demonstrating, at the ca. 0.5 my spacing of the measurements, that the evolutionary pattern was similar. Details of the cladogenetic event were examined in Lazarus (1986). In this study only samples from the tropical Pacific were examined, but at a much higher temporal resolution (ca. 50 ky) and with a much larger number of morphologic measurements, including thorax and foot shape, the most important characters distinguishing the ancestor and descendent species. This study verified the gradual change in *P. prismatium* between its origin at ca. 4.3 Ma and 3.5 Ma, and the lack of net change thereafter. The analysis however showed an almost equal coeval shift, in the opposite direction, in the morphology of the ancestral *P. charybdeum* lineage, also with a cessation of net change after ca. 3.5 Ma. Lazarus (1986) interpreted this pattern, due to the presence of both ancestor and descendant in the same geographic region, to be another example of parapatric speciation, as first proposed for this event in Lazarus et al. (1985), and specifically speciation along a water depth gradient. This was elaborated on in Lazarus (2001; Figure 9.3). Although not discussed in the original study, the ecologic context of this speciation event can at least be partially determined, due to extensive recent paleoceanographic research in the tropical Pacific. This shows a relatively rapid, major shift at ca. 4.2 Ma towards a shallower thermocline in the tropical Pacific and an increasing east–west gradient in nutrients/productivity (Cannariato and Ravelo, 1997), which would be compatible with a deep-to-surface speciation event tracking the resultant changes in vertical gradients in food supply. Without any data on *P. prismatium*'s ecology, this however remains speculation. One other feature of this speciation event not brought out in the original study is apparent longitudinal diachrony. *P. prismatium*'s divergence in the equatorial Pacific is preceded by its earlier appearance in even older sediments to the west. According to Johnson and Nigrini (1985), *P. prismatium* first appears in the equatorial Indian ocean at ca. 5.7–5.8 Ma, in the western Equatorial Pacific at 4.9–5.0 Ma, between 4.7–4.8 Ma in the central Equatorial Pacific (at DSDP Site 573, the same as used in Lazarus, 1986), and 4.2–4.3 in the eastern Equatorial Pacific. Both studies use the same time scale, and only sections with paleomagnetic stratigraphy, so the reported difference in pattern must lie in some combination of differential taxonomy or true diachrony of the speciation event. Some of the difference may be the use of a morphotypic species concept by Johnson and Nigrini (1985), as is typical for biostratigraphic studies (see Chapters 5 and 8), vs the population concept used in Lazarus (1986). If there however is true diachrony in the timing of speciation, then the speciation of *P. prismatium* must be the temporal expression of the evolution of a ring species. These are species pairs that are fully distinct from each other at the extreme ends of their distribution but are connected via a cline of intermediate populations (Irwin, 2002). In *P. prismatium*, the distinct end-members would be the differentiated (over water depth) populations to

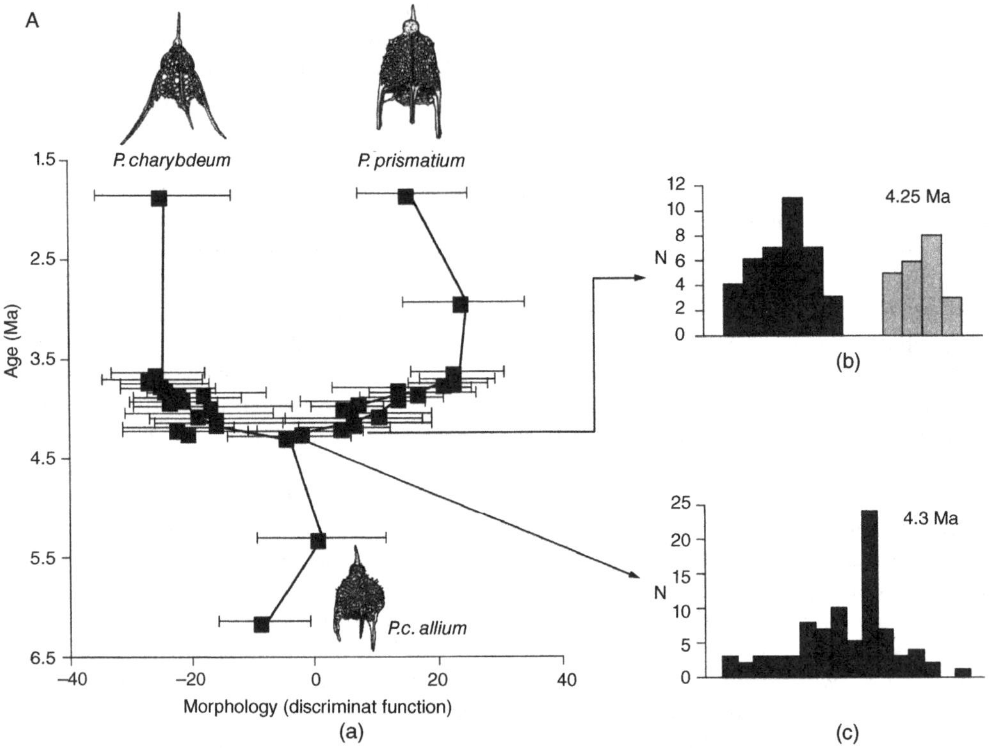

Figure 9.3 A) Cladogenetic speciation of *Pterocanium prismatium* from its ancestor *P. charybdeum* as measured in samples from tropical Pacific DSDP Site 573. The ancestral species *P. charybdeum* is highly variable in morphology and is subdivided into several subspecies, while *P. prismatium* is morphologically more distinct and uniform in character. Geologic age (Berggren et al. 1985 scale) on left, morphologic change, with drawings of each species in middle, histograms of morphologic measurements for many individuals in a population in two representative samples on right. Morphologic measurements of a wide range of characters on each specimen were combined into a single composite value (a "discriminant score") for each specimen, based on the ability of the combined measurements to differentiate between two defined populations (here the two species, from the time interval at 3.7 Ma when they are easily distinguished). B) (a) Model of environmental context of speciation in *Pterocanium*. Schematic 3D structure of global ocean on left, showing nested half-cylindrical slabs of water, with warm, nutrient-poor water in tropical surface water (light shading, centered on equator), global deep water ocean (outcropping at surface in high latitudes), and intermediate water bands (see also Chapter 3). Based on latitudinal distribution of ancestor (*P. charybdeum*) and descendant (*P. prismatium*), shown by double-headed arrows, ancestor is thought to have occupied intermediate waters, and descendant shallower tropical water. Evolution of the new species thus occurred along a depth gradient. In the sedimentary record in a tropical location (black dot), specimens settle through water column (downward pointing arrow) from various water depths, compositing a gradient of different populations into a single, highly variable sediment assemblage (b) upper diagram of morphologic distributions. (c) Over time, break-up of the depth-clinally varying populations leads to a sediment record over time of low variance, then high variance single distributions eventually separating into two separate species, as in fact seen in *Pterocanium* data. Source: Lazarus (2001), based on original study of Lazarus (1986). Reproduced with permission of John Wiley and Sons.

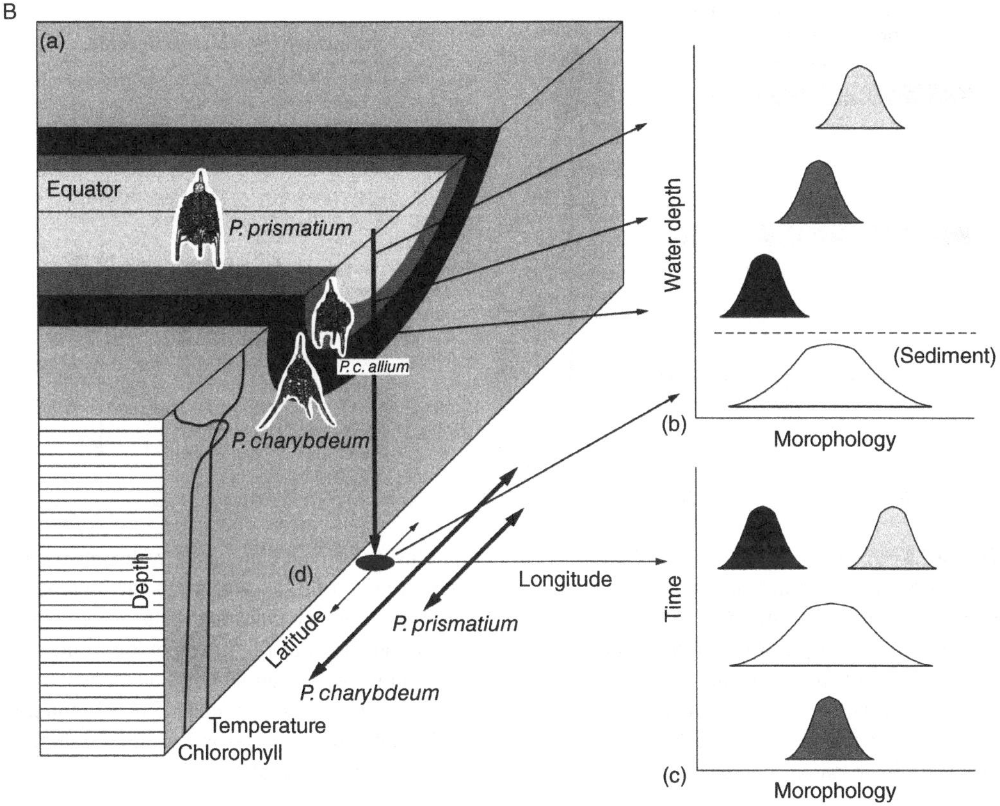

Figure 9.3 (*Continued*)

the west, which are still connected by intermediate populations to the east (Figure 9.4). This possibility has not however been investigated further.

A last example of cladogenetic study using radiolarians comes from the molecular analysis of *Spongotrochus glacialis* by Ishitani et al. (2014), mentioned previously in Chapter 4. These sibling species of *S. glacialis* are, in low- to mid-latitudes, vertically segregated, as measured by chlorophyll-a concentration, suggesting that these are sister species separated by ecological partitioning (Figure 9.5). The molecular clock analysis demonstrated that the timing of the divergence of the sibling species is (roughly) coincident with a major turnover in general in radiolarian assemblages in the Equatorial Pacific (Kamikuri et al., 2005). The speciation may thus have occurred along an, for radiolarians, important change in ecologic, possibly depth-related gradients. This model is similar to that proposed earlier for *P. prismatium* (above), although as *S. glacialis* is mostly common in polar waters the geographic aspects of speciation in this case are not clear.

Anagenesis

Anagenesis occurs in many – possibly most – species of marine microfossils. Changes in morphology within a lineage over time are frequently substantial enough that the series can be divided into a sequence of named morphologic chronospecies, even if the boundaries between different named

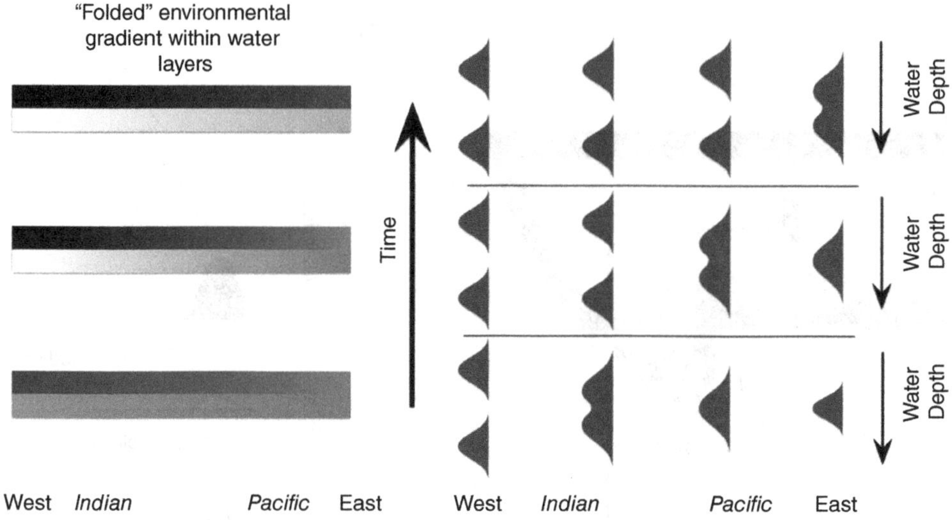

Figure 9.4 Hypothetical model of diachronous depth parapatric speciation compatible with known data for *Pterocanium prismatium*. Evolutionary pattern on right is driven by (on left) gradually increasing contrast in vertical environmental gradients between water layers, which individually also show a geographic (west–east) gradient in environmental properties. Populations of *Pterocanium* (right) diverge into two distinct species along the vertical gradient, but at succeedingly later times as one moves from west to east. The divergence was documented in Figure 9.3 at a location in the Pacific. The first appearance of distinct populations of the descendant species *P. prismatium* (the first appearance datum or "FAD") is diachronous, becoming progressively younger in the eastward locations.

steps in the evolutionary sequence are necessarily somewhat arbitrary. Numerous lineages in radiolarians (and other marine microfossils) have been used in this way to define series of morphospecies that characterize specific geologic time intervals, and make up a large fraction of the species that are used to define the Cenozoic low-latitude radiolarian stratigraphic zones (see Chapters 5 and 8 for discussion, and Sanfilippo et al., 1985 for extensive stratigraphic and descriptive documentation). Many of these series show consistent trends in morphologic change. In *Didymocyrtis* for example, a trend persists over >10 my and a half dozen named species towards increased chamber development and reduction of spongy material in the polar caps and the ends of the shells. Quantitative documentation of anagenesis in radiolarians is much less common, and some of the published studies are ambiguous. One problem is that to establish that change is occurring within a single species, the taxonomy of that species and the immediate phylogeny to any sister species must first be established, else any pattern observed may turn out to be part of a cladogenetic event, as was shown for morphologic evolution in *Pterocanium prismatium*. Another is that, should a species exhibit clinal variation, shifts in the geographic extent of a cline over time (possibly due to changing climate conditions) can cause the morphology seen at a single location to change as the cline migrates over the site. Clinal variation does exist in radiolaria, including many of the lineages described above, as the diachronous stratigraphic ranges of many lineage morphospecies members with latitude implies (Spencer-Cervato et al, 1993). Clinal variation has been analyzed however only in a

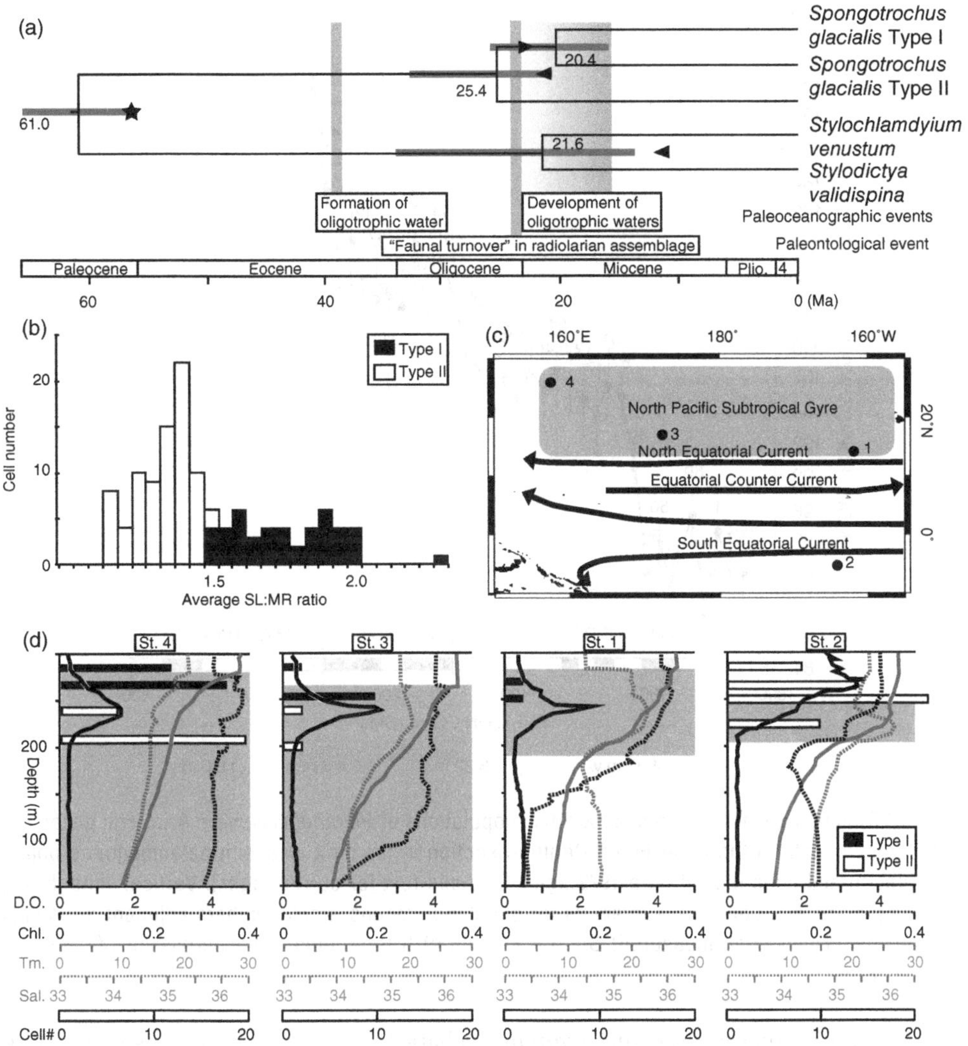

Figure 9.5 (a) Bayesian phylogeny and divergence time estimation of *Spongotrochus glacialis* based on SSU + LSU + ITS rRNA (9 taxa, 5156 sites). Numbers on the nodes indicate estimated divergence time. Upper and lower constraints are indicated by the right and left arrows, respectively. Dark-shaded (horizontal) bars around nodes are indicated 95% credible intervals. Dark-shaded (vertical) bars in the geological time scale indicate inferred timing of paleoceanographic and paleontological events (Ishitani et al., 2014). (b) Histogram of the mean ratios of spine length (SL) to mantle radius (MR) for each cell of *S. glacialis* type I (black) and II (white). (c) Location of the four sampling stations (black circles) in the Central Pacific Ocean. The black arrows indicate the circulation pattern. The dark-shaded area represents the North Pacific Subtropical Gyre. (d) Vertical profiles of environmental parameters (dissolved oxygen concentration, chlorophyll density, temperature, and salinity) shown as black, black-dashed, gray, and gray-dashed lines, respectively. The dark-shaded area indicates the pycnocline. The number of individuals collected of *S. glacialis* types I (black bars) and II (white bars) are shown at each sampling depth. Source: Ishitani, Molecular Phyl. Evol., 2014. Reproduced with permission of Elsevier.

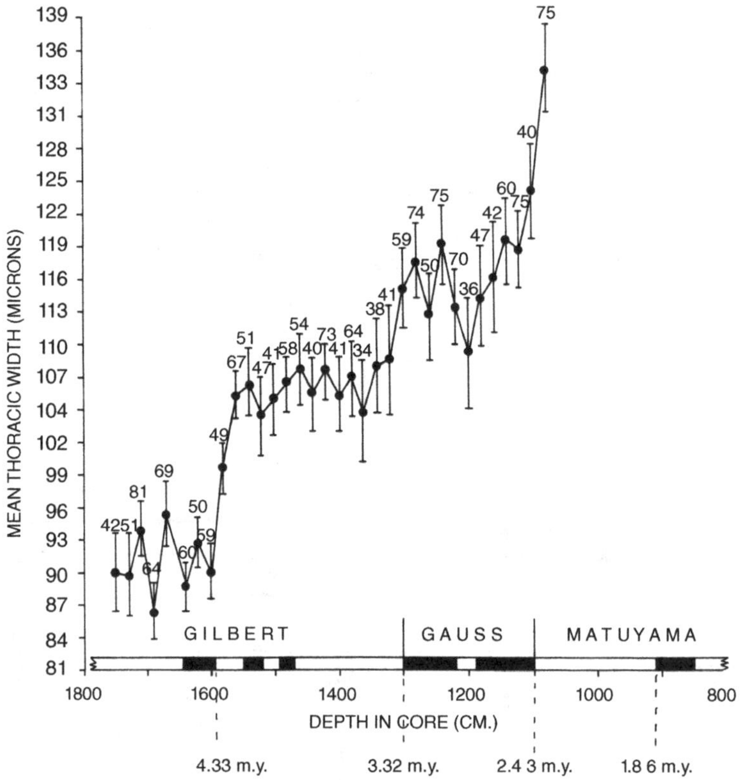

Figure 9.6 Evolutionary increase in shell width in populations of *Helotholus vema* in Antarctic piston core E14-8, located south of the Polar Front. Depth in section in cm on x axis with paleomagnetic polarity interpretation and ages (based on a 1972 age scale: see text for current ages). Vertical lines through each sample are 95% confidence intervals for population means, small numbers adjacent are sample sizes (number of individuals measured). Source: Kellogg 1975. Reproduced with permission of Cambridge University Press.

few species (e.g. *Eucyrtidium matuyamai*, *Actinomma haysi*: Hays, 1970; Bjørklund, 1970), and the only example of shifting clines affecting local patterns of change over time in radiolaria documented so far is for the *Theocorythium vetulum-trachelium* lineage (Baker, 1983).

The most widely cited example of quantitatively documented anagenesis in radiolarians is gradual, if also partly static and episodic, change in shell width in the Pliocene radiolarian *Helotholus vema* (=*Pseudocubus vema*; Kellogg, 1975; Figure 9.6). This extinct nassellarian species was endemic to the Southern Ocean and is characterized by having a small caplike cephalis at least halfway embedded in a much broader short cylindrical thorax, and possessing a set of internal radial bars connected to a central ring (Lazarus 1990). Kellogg's data show initial stasis in the earliest Pliocene (magnetochron C3n3n and older, ca. 5.0 to 4.6 Ma, a quite rapid jump to larger widths near the base of magnetochron C3n2n (Nunivak) magnetochron at 4.6 Ma, again stasis for nearly 1 my, followed by irregular but, particularly towards the top of the species range in the late Pliocene, substantial increase in thorax width. However, in standard radiolarian zonations of the

Southern Ocean (Lazarus 1992; see Chapter 8) the first appearance of *H. vema* marks the base of the Upsilon Zone and is placed at the base of the C3n2n. The large jump in thorax width values seen in Kellogg's data at this level thus seems to correspond to a shift in which species is being measured: *H. vema*, with large values above, and some other species below. Several other species similar in morphology to *H. vema* exist in the early Pliocene of the Southern Ocean (Lazarus, 1992) and presumably one or more of these was measured instead. It has not been established what the ancestor of *H. vema* is, so it is not clear if the older measurements are actually part of an evolutionary series. However, several of these species continue into the younger Pliocene, so that the rapid shift at ca. 4.6 Ma may reflect change associated with a rapid cladogenetic event, and thus a similar shift in interpretation between earlier and later study of the evolution of *Pterocanium prismatium* (above). The increase in size reported by Kellogg in the latter part of the *H. vema* lineage by contrast appears to reflect anagenetic change.

A second example comes from study of much older material: the important Jurassic species in the genus *Striatojaponocapsa* (Hatakeda et al., 2007, Figure 9.7). In this study a combination of linear measurements and meristic characters such as the number of plicae on the abdominal segment were used in two sections to show that the transition from the older *S. plicatum* via a stratigraphically short-ranging from *S. synconexa* to *S. convexa* occurred via anagenesis. The appearance of *S. riri* somewhat later was interpreted as an example of cladogenesis, as *S. convexa* continued to exist as a distinct species coeval with *S. riri*. Although sample sizes were limited, counts of the discrete character (number of irregular pore frames) which is the primary difference between the species document not only a distinct,

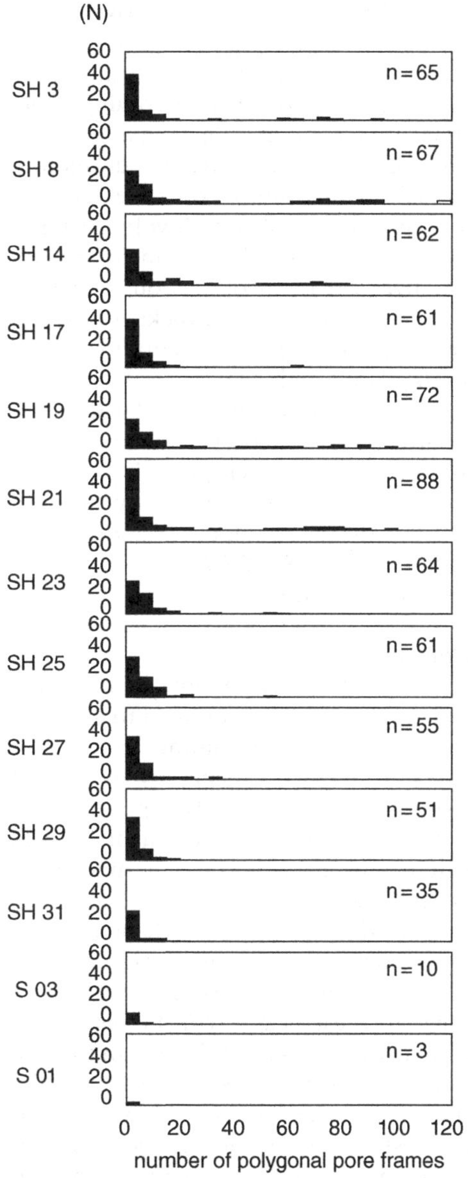

Figure 9.7 Gradual change of a taxonomic character (number of irregular polygonal pore frames on a half hemisphere) in the divergence of the new species *Striatojaponocapsa riri* from *S. conexa* in a series of samples (SH n) from a middle Jurassic geologic section in Japan. N gives number of specimens. Source: Hatakeda, 2007, Reproduced with permission of Elsevier.

non-overlapping difference between the two coeval species, but also hint at a gradual cladogenetic event, with the ancestral species (*S. convexa*) becoming more variable in pore frames, leading to a (due to small sample sizes, weakly developed) bimodal distribution over a series of samples. The amount of time involved cannot be precisely estimated, the transition takes place over ca. 10 m of section, which in these chert-rich Mesozoic rocks is likely to be around 1 my (e.g. a sedimentation rate of 1 cm/kyr) or even longer.

There are a few other, similar quantiative studies of anagenetic change in radiolarians (*Calocycletta caepa* in Kellogg and Hays, 1975; *Didymocyrtis* species in Kellogg, 1980; *Cycladophora davisiana* in Motoyama, 1997; *Stichocorys peregrina* in Kamikuri, 2012).

Extinction

Detailed studies of extinction for individual species are essentially confined to determinations of extinction timing in different geographic locations, generally as part of biostratigraphic research. For species living in surface waters, the environment consists of numerous gradients – in temperature, nutrients, depth to thermocline, seasonality, and other factors (see Chapter 3). As environments change in response to changing ocean conditions, specific values of change in these factors will be reached at different times along gradients, and to the extent that such factors are critical to the survival of species, one would expect critical threshold values causing local extirpation to be reached at different times in different locations. Extinctions should thus, for surface dwelling species, normally be to at least some degree diachronous, and in fact for many species this is so (Johnson and Nigrini, 1985; Spencer-Cervato et al., 1993, 1994; Moore et al., 1995; Figure 8.9). These same studies also however show that some species are, within the resolution of the analyses (ca. 200 kyr), isochronous. Hays and Shackleton (1976) showed that the extinction of *Stylatractus universus* was globally synchronous to within ca. 10 ky, during a specific glacial-interglacial climate transition (figure 8.8). *S. universus* appears to have lived in intermediate waters, and thus may have experienced a more uniform environmental change than a surface dwelling species. However, it is hard to believe that all other species showing an isochronous pattern (op. cit.) were subsurface-dwellers, and the explanation for this behavior is still unknown. Liow et al. (2007; 2010) examined the mean trend in frequency of occurrence over the history of individual species in radiolarians and other microfossil species in the NSB (Neptune) database and concluded that there was a tendency, in all groups, towards lower frequency of occurrences well prior to extinction, which suggests that on average, species were in some sense in ecologic decline long before they became extinct. However, the mean values for groups hide a great deal of individual variation in behavior (Lazarus et al., 2012) and the effects of data outliers in these analyses have not yet been evaluated.

Hybridization

The existence of hybridization in radiolaria has been promoted in several studies by Goll (1976, 1979, 1985), primarily on the distribution of character states between different species of Trissocyclidae, and in his 1976 study, by the geographic distribution of morphologically intermediate forms. Goll examined several species of trissocyclids in surface sediments of the world ocean and interpreted the morphology of four taxa (*Phormospyris* [=*Dendrospyris*] *stabilis scaphipes, P. s. capoi, Lophospyris pentagona hyperborea,* and *L. cheni*) as being intermediate to two proposed ancestral species, both of these being quite cosmopolitan in distribution and both showing a latitudinal gradient between two subspecies (*Phormospyris*

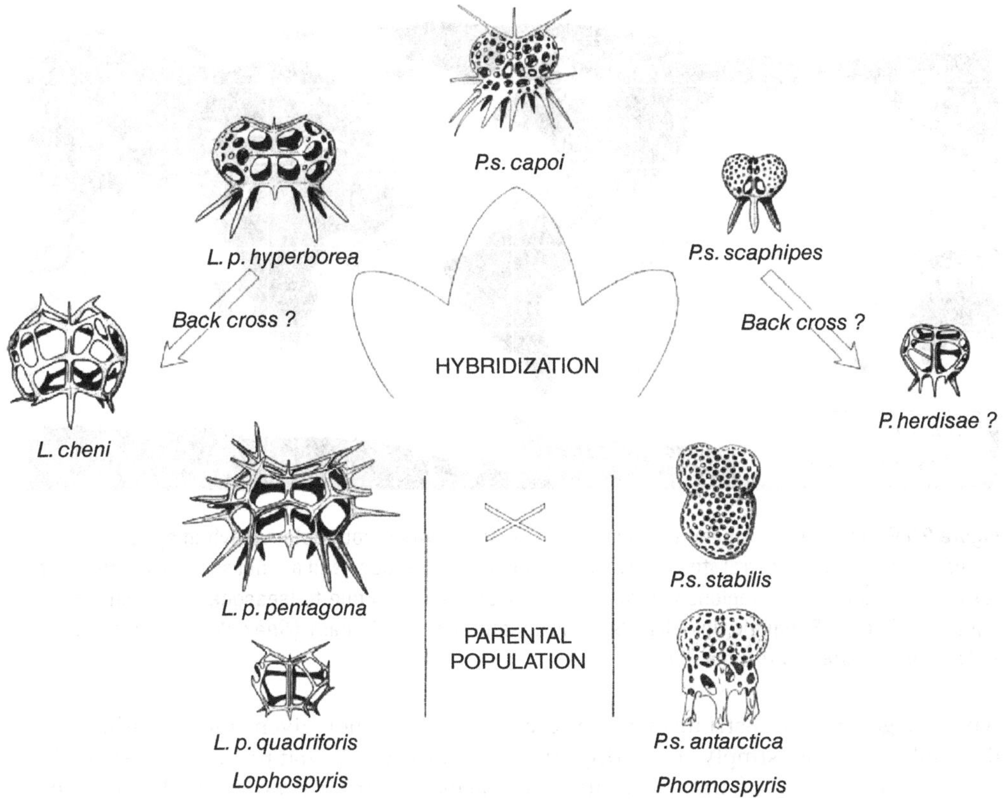

Figure 9.8 Line drawings and proposed hybrid relationships of nine species of trissocyclid radiolarians found in global ocean surface sediments. Source: Goll (1976). Reproduced with permisison of micropress.org.

stabilis stabilis and *P. s. antarctica*; and *Lophospyris pentagona pentagona* and *L. p. quadriforis*, the latter forms in each case being the polar morph) (Figure 9.8). Goll assumed that the two ancestral species occupy different environments, e.g. water layers in the ocean. The four intermediate forms occur in regions where upwelling causes mixing of water layers (Figure 9.9), which is indeed the geographic pattern expected by hybridizing species. Further, the two forms (*P. s. capoi* and *L. cheni*) that Goll, on morphologic grounds, interpreted as intermediate between primary hybrids (*P. s. scaphipes* and *L. p. hyperborea*) and ancestors, and thus potential "backcrosses," occur primarily in regions where the "primary hybrids" overlap. Goll (1976) was thus able to show that the geographic patterns in these taxa do indeed suggest the possibility of hybridization. Difficulties however in understanding the complex morphology and characters of these taxa, missing information on the actual habitats of the taxa in the water column, the absence of objective, quantitative morphologic analyses, and the lack of genetic data have limited the acceptance of this idea. Without such data, it is equally parsimonious to hypothesize that the four "intermediate" forms evolved via non-hybrid mechanisms from one or the other ancestor,

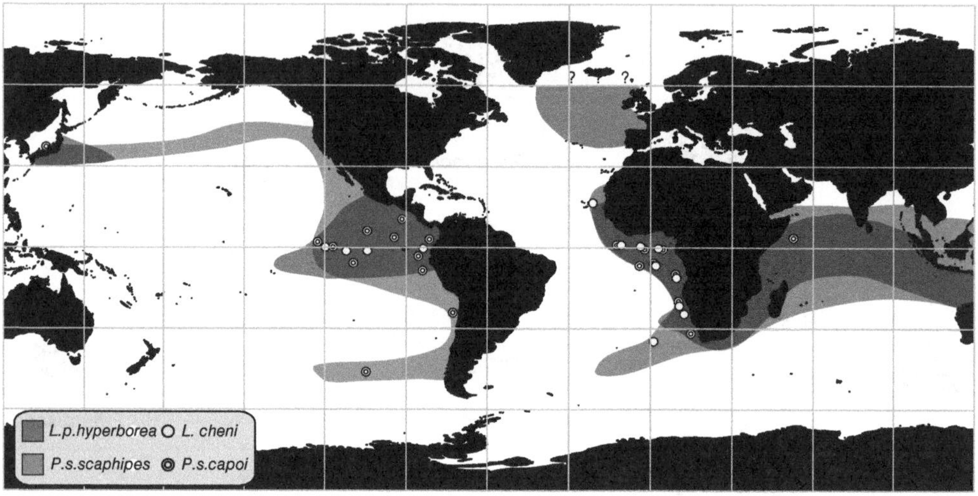

Figure 9.9 Geographic distribution in global surface sediments of four putative hybrid species of trissocyclid radiolarians (ancestral stocks have cosmopolitan distributions and are not shown). Putative hybrid taxa are found mostly in regions where water masses are mixed due to (seasonal) upwelling (eastern Equatorial Pacific, Benguela Upwelling System, equatorial Indian Ocean). (*See color plate section for the color representation of this figure.*)

with the geographic distributions of these descendant forms simply reflecting the intermediate nature of the environments that the taxa evolved in, together with an environmental correlation for the characters examined. Ideally, the existence of hybridization should be readily apparent by detailed, quantitative study of the time history of development of character distributions in the species involved; or by (so far lacking) genetic studies. Detailed quantitative morphologic analysis was not done in detail by Goll, and only possible examples exist from study of other materials. For example, the genus *Theocorythium* appears to have evolved near the base of the Pliocene (Baker, 1983; Lazarus et al., 2012), from members of the genus *Lamprocyclas*, according to Baker (1983). Baker describes and illustrates specimens which, in his opinion, represent morphologically intermediate forms between these genera (specifically, between *T. trachelium dianae* and *L. m. maritalis*) in the late Pliocene and Pleistocene. A quantitative analysis of the pattern is not given. Other authors have also suggested that previously distinct lineages at certain time intervals show intermediate morphologies (Lazarus et al., 1985; Caulet and Nigrini, 1988) but detailed analyses are not available in these studies either.

Macroevolution

Definitions and Theory

Evolutionary theory has classically concentrated on evolution of populations and the processes of speciation. All other aspects of evolution have been effectively assumed to be completely explainable by extrapolation from the cumulative effects of these fundamental processes. Some scientists, particularly paleontologists and developmental biologists, have questioned this "reductionist" approach to understanding evolution, and have proposed instead that a full theory of evolution will need to incorporate additional processes

that affect evolution, such as "species selection," or macromutational origin of novelties via non-microevolutionary mechanisms, e.g. developmental genetics ("evo-devo") (Stanley, 1979; Gould, 1980; Carroll, 2002; Carroll, 2008; Pigliucci and Müller, 2010). Other scientists do not claim that additional fundamental processes exist, but feel that extrapolation of low-level processes to larger-scale phenomena is impractical without the use of concepts of higher-level behavior which effectively abstract the former, and which can more easily be used to describe emergent properties of complexly organized systems (Mayr, 1982). This is part of a long tradition of the meaning of reduction and hierarchy in scientific thought. A much-used analogy (MacCorquodale and Meehl, 1948) is the relationship in physics between the ideal gas law, that abstracts and better describes the emergent properties of pressure and temperature, than the more fundamental processes of individual molecule motion that collectively produce the behavior of gases. In the remainder of this chapter we discuss macroevolution in this latter, purely practical sense.

It should be noted however that, even if no additional processes have yet been identified that underly a distinct set of mechanisms for macroevolution, this does not at all mean that they do not exist. Although the chromosome-driven "macromutations" of an early proponent of macroevolutionary mechanisms (Goldschmidt, 1940) are no longer thought to be realistic, other developmental and molecular mechanisms that can create "macro"-scale evolutionary novelties, such as introns (Mattick and Gagen, 2001) or horizontal gene transfer (Arnold, 2006), have been shown to exist throughout the biologic world, including protists (Haugen et al, 2005; Carroll, 2002; Carroll, 2008; Keeling and Palmer, 2008). Further, one of the main objections to macromutations – the inability of the new form to survive by sexual reproduction with an individual not possessing the mutation (Stanley, 1979; Mayr, 1982) – is largely removed in organisms (most likely including radiolarians) that can reproduce asexually (Chapter 2).

This approach to studying evolution would emphasize the origin of novel characteristics internal to organisms themselves (for example, adaptation and origination of functional novelty and the distribution of such characters in the group's phylogenetic tree). This type of evolutionary analysis is indeed often what the general public thinks of by "evolution" – e.g. adaptations that define higher taxa and the novel adaptations they possess: the conquest of terrestrial environments by different groups of plants and animals in the early Paleozoic; the origin of vertebrate flight; the evolution of humans and intelligence. Unfortunately, although the distinctive, diverse intrinsic functional adaptations of radiolaria are in principle very important (and the variety of morphologic, presumably functional, features immense), we currently know far too little about radiolarian functional biology, their higher-level systematics, or their origins via molecular biology to discuss this aspect of their evolution in much detail.

Thus, to organize our discussion of macroevolution, we are required to choose available concepts that effectively describe properties of radiolarian evolution beyond the microevolutionary level described in the first part of this chapter, and for which there is enough data to make a discussion meaningful. Although some examples of intrinsic characters are cited, we mostly use various aspects of environmental control of larger-scale evolutionary patterns, where the evolutionary pattern is largely measured by taxonomic diversity at various levels (species, genera, or families). This is essentially an ecologic view of evolutionary change, where the radiolarians at any taxonomic level (e.g. species) are treated simply as equivalent units of biodiversity.

Theories of Diversity and Evolution

Unlike microevolutionary theory, which has concentrated on adaptational changes within species and speciation, and has long had an highly inter-related, if not entirely uniform, theoretical framework (the modern, or neo-Darwinian synthesis: Mayr, 1963, 1982; Futuyma, 1998), ecologic theory has developed in a less uniform way, with a variety of patterns and processes in nature inspiring different theoretical ideas. Food web models and behavioral ecology are examples of theoretical ecologic research that are not specifically focused on biodiversity. Bodies of ecologic theory have also developed that focus on particular levels of biologic organization, such as population ecology, or community ecology. As these various phenomena are linked in nature, there are normally overlaps between theoretical approaches as well, which have caused both inspiration and confusion among biologists. Although many of these theories have been used on occasion in paleontology, theories about two questions on the processes that regulate biodiversity have played particularly prominent roles in guiding paleontologic research.

The first question is whether biodiversity is, for any given environment fundamentally limited or not. In the former, a particular environment should have a maximum "carrying capacity" and diversity should eventually, given enough evolutionary time reach an equilibrium level (the environmental diversity is "saturated"). The opposing view is that biotic interactions can continue indefinitely to divide or create new niches for species to occupy, so that diversity should continue to grow over time without any upper limit. The idea of diversity having a maximum for any given environment comes from studies of biodiversity on islands, and the theory created to explain the patterns seen of different diversities on different islands (MacArthur and Wilson, 1967; Ricklefs and Schluter, 1993). The opposing view that diversity is in practice unbounded is based in part on negative arguments about the robustnesss of data cited in support of limits, interpreting the Phanerozoic diversity history without full regard to sampling effects (e.g. Benton, 1995), and in part from various ecologic models and biologic data that suggest that many living biotas are not currently at an equilibrium (Harmon and Harrison, 2015). In practice, using the fossil record to determine if diversity approaches equilbrium in a given environment is difficult without knowing how stable the environment was. Assuming a constant carrying capacity without independent ecologic justification, or post-hoc explanations (after the diversity is shown not to be constant over time) as to why it wasn't, undermine tests of these theories. Marshall and Quental (2016) recently developed models for the more realistic situation of changing carrying capacity over time, although how such an abstract concept can be determined in the geologic record is still unclear.

The other major theme is the role of biotic interactions vs. the influence of the physical environment as drivers of evolutionary change, and thus, at the level of speciation and extinction, as primary regulators of diversity. Although the significance of both types of processes were recognized as early as Darwin, more modern formulations try to emphasize the dominance of one or the other mechanism in regulating the evolution of diversity. The most famous modern version of the biotic control hypothesis is the Red Queen hypothesis proposed by Van Valen (1973), and has the distinction of being the only major theoretical formulation in macroevolution/ecology to have been proposed by a paleontologist. Van Valen compiled taxa survivorship data from the paleontologic literature and noted that the chance of going extinct for a very wide variety of taxa is constant over time – from species groups to families; protists, invertebrate, vertebrate; marine and terrestrial. He argued that taxa do not evolve to become better adapted over time but instead evolve simply not to become more poorly adapted in an ever-changing evolutionary landscape. This he believed to be most easily explained as being made mostly of other, similarly evolving taxa that interacted with each other. He called this the Red Queen, after the character in Lewis Carroll's *Through the Looking Glass* story that had to keep running constantly simply to stay in the same place. Van Valen, while accepting that physical changes could also affect evolution, felt that the near constancy of extinction rate seen in most of his data suggested that this mechanism was "relatively unimportant on the evolutionary time scale" (Van Valen, 1973, p. 19). The contrasting view of physical (abiotic) controls was dubbed "The Court Jester" by Barnosky (2001), alluding to the arbitrary changes (from the viewpoint of the organisms) that physical change causes in the "playing field" within which taxa interact with each other. As noted by Barnosky (2001) in practice it is very difficult to distinguish between biotic or abiotic controls without very good information on the environmental history, together with paleontologic data collected at the appropriate (for the environmental factors) temporal and spatial scale. Van Valen's (1973) assumption was that physical events would be infrequent and large enough to create clear disjunctions in survivorship patterns, and thus be distinguishable from virtually constant rates of biotic change. Neither assumption about abiotic or biotic rates however is really justified, given the very complex history of the earth's environmental evolution, and the possibility of relatively rare, but highly influential introductions of adaptational novelties in other groups of interacting organisms.

Macroevolutionary Patterns in Radiolaria

Our discussion of macroevolution in radiolaria reviews important patterns and some of the various proposed mechanisms, both intrinsic and ecologic, that may have caused them. Our review focuses on the following themes: origin of higher-level clades where some understanding of the functional adaptations exist – radiolarians themselves, and clades with unique ecologic characteristics such as the colonial Collodaria; the diversity history of radiolarians and its control by changing biogeography and ocean circulation (organized by geologic time interval, from oldest to youngest); mass extinctions and recoveries; and silica availability and co-evolution with other silica-using groups of organisms.

Origin of Radiolarians

The first, and in some ways most important, macroevolutionary event concerns the origin and earliest evolution of radiolarians as a new clade of eukaryotic organisms. Although very little is known, major advances in earth history research and molecular biology in the last couple of decades now provide at least a framework within which plausible speculations may be discussed.

The earliest preserved radiolarians (see also Chapter 5) are known from lower or mid-Cambrian sediments (Maletz, 2011; Pouille et al., 2011). Radiolarian fossils from the Cambrian are however very scarce and the state of preservation of the fossils found is also very poor. It is thus entirely possible that radiolarians could have existed significantly earlier in geologic time than the oldest fossils yet recovered. Knowing the timing of initial radiolarian evolution is essential if we want to compare this biologic event to the environmental conditions within which it took place. There are three major constraints on this: the strong minimum age constraint of the oldest (ca. late early Cambrian) fossils; divergence times estimated from molecular biologic studies, which are currently too imprecise to be of much use (e.g. DeCelle et al., 2012, see Chapter 4); and Earth history, which limits the range of time in which suitable environments existed in which radiolarians could have lived.

The early history of the earth belongs to the geologically long Precambrian time interval (ca. 4,500–542 Ma), e.g. below the base of the Cambrian (542 Ma). The Cambrian is the lowermost Period of the Phanerozoic Eon, with the oldest record of abundant fossilized life, including the first occurrence of many mineralized forms. Free oxygen was essentially absent in the oceans and atmosphere of the earlier part of the Precambrian, the Archean (ca. 4,500–2,500 Ma), and early life consisted mostly of anaerobic prokaryotes. Around 2,400 Ma, the Archean Eon ended with the initial development of oxygen in the atmosphere and gave way to the Proterozoic Eon. Eukaryotes, which contain many essentially aerobic organelles (peroxisomes, mitochondria, and plastids), are thought to have evolved originally as aerobic forms, and their initial diversification, and probably origin, must thus have occurred in the Proterozoic. Molecular phylogenies are broadly in agreement with this but show a wide scatter of ages. The concentration of oxygen in the Proterozoic atmosphere was still very low ($\leq$ 10% of modern values), and the oceans were still partially anoxic, with sulfidic, anoxic, or euxinic deeper waters underlying a weakly oxic surface layer (Johnston et al. 2009). Productivity in the Proterozoic ocean was low, with limited cycling of nutrients. Much of the Proterozoic was also marked by glaciations, the more extreme of which resulted in a nearly completely ice-covered ("snowball") earth. These conditions still favored prokaryotes over eukaryotes; eukaryotic diversity is thought to have been low, and the fossil record of eukaryotic plankton is restricted to small phytoplankton (Acritarchs – cells or cysts of various affinities, at least partially dinoflagellate). Only with the end of the last major glaciations near the end of the Proterozoic (end of the Cryogenian at ca. 620 Ma) did ocean chemistry begin to change towards more oxic conditions, with higher marine productivity. These changes set the stage for a rapid diversification of eukaryotic life in the Ediacaran, last Period of the Precambrian (Johnston et al., 2009; Butterfield, 2011). Molecular phylogenies also now place the origins of most modern eukaryotic groups, including the radiolarians, near the end of the Proterozoic, although there is still considerable scatter in the age estimates (see Chapter 4). Radiolarians thus probably first evolved within a relatively narrow time window between the early Ediacaran and earliest fossil radiolarians from the early Cambrian, e.g. ca. 620–520 Ma. Ediacaran and basal Cambrian sediments have been documented from a variety of geologic facies, and suggest that life at this time was evolving rapidly in both benthic and planktonic environments. Well-oxygenated waters, with higher primary productivity, were still more restricted to shallow water, coastal settings. In particular, ecosytems in these enviroments, both benthic and pelagic, were, supported by the higher productivity, also becoming more complex, with longer food chains and higher-level predators (Butterfield 1997, 2009; Porter 2011). An early "Ediacaran" biota of largely benthic, sessile, non-mineralized metazoans, protists, and bacteria was replaced in the early Cambrian by a wide variety of metazoan forms belonging to still-existing phyla, including many with mineralized body parts, higher mobility, and clear adaptations for predation. True planktonic metazoans were scarce in the early Cambrian, although several groups are thought to have been meroplanktic (e.g. planktotrophic larvae of benthic groups), and these, together with pelagic nekton such as chaetognaths (Vannier et al. 2007) played an increasing role in grazing of other, larger planktonic organisms (Rigby and Milsom 2000; Butterfield 2009; Figure 9.10).

Radiolarians are relatively large for protists, and were significantly larger than most other bacteria or protists reported from, or thought to have existed in, Ediacaran and lower Cambrian time. Their large size, purely planktonic mode of life, and primary adaptions for predatory feeding (see Chapter 2) suggest that they evolved as higher trophic level planktonic predators as part of the general radiation of protists and ecosystems at this time. They are the oldest group of larger planktonic heterotrophic (predatory) protists known in the fossil record. Radiolarians feed on a range of prey including many smaller (e.g. <10 μm) protists and in turn serve as potential food sources for larger, metazoan plankton, which normally do not capture smaller protists efficiently (e.g. copepods and most other metazoan suspension feeders feed mostly on particles >5 to 10 μm; see Chapter 3). They thus may have played a key role as the dominant microzooplankton in Ediacaran/basal Cambrian pelagic ecosystems, in the absence of other larger microzooplankton and mesozooplankton groups. The absolute and relative origin times of non-fossilizing eukaryotic heterotrophic protist clades are still too poorly constrained by molecular methods to determine if they were also present at the time when radiolarians first evolved.

Rapid evolution of taxa and ecosystems during the later Cambrian and subsequent periods of the lower Paleozoic soon resulted in the appearance of a full range of other groups of pelagic predators on, and competitors to heterotrophic protists, including, at some point, other zooplanktic protists, copepod-like organisms in the mid to late Cambrian (Harvey et al., 2012), as well as many other metazoan groups

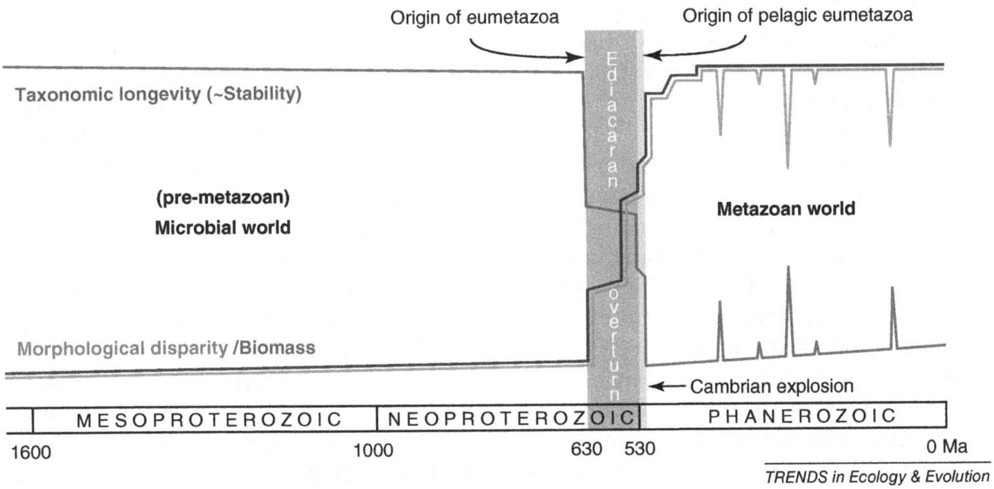

Figure 9.10 Schematic representation of major biologic events at the Precambrian–Cambrian transition, at which time radiolarians are thought to have first evolved as early pelagic protistan predators. Biomass, ecologic complexity, and rates of evolution all increased extremely rapidly at the end of the Neoproterozoic in both benthic and pelagic environments. Source: Butterfield, 2011, Reproduced with permission of CELL.

(Rigby and Milsom, 1996) and tiering of food chains to larger prey sizes (e.g. large nektonic predators like ammonites: Klug et al., 2010). To understand the effect that this increased ecosystem complexity and potential competition may have had on later Paleozoic radiolarians however we need to first determine the diversity history of radiolarians, as this provides the primary means by which to judge the impact of external events on radiolarian evolution. Unfortunately, as explained below, our knowledge is still too incomplete to carry out such comparisons.

Origin of Collodaria and Colonial Radiolaria

Ishitani et al. (2012) suggested an evolutionary history for the Collodaria based on the calculated divergence times of their collodarian phylogeny (Figure 9.11; see also Chapter 4). They proposed that Collodaria originated from Nassellaria in the middle Eocene and, due to increasing competition for silica in the pelagic environment from the then-rapidly expanding pelagic diatoms, suppressed their ability to deposit silica shells. Colonial forms are proposed to have originated at the Eocene–Oligocene boundary in response to the development of more oligotrophic surface waters in low latitudes (Lazarus et al., 2009).

This study is very useful, as it illustrates both the strengths and weaknesses of a molecular approach to understanding macroevolutionary processes in radiolarians. Limitations include the very high uncertainty in dating divergence times, and dependence of these on the specific tree topology, which may well change with additional taxonomic sampling. Estimating the last common ancestor of a clade to an outgroup is also not the same as estimating the time at which the clade actually acquired the functional adaptations that are central to evolutionary hypotheses of change linked to the history of the environment. These temporal uncertainties make it nearly impossible to test the proposed model against the geologic record using only the molecular date data, since changes of even a few my in the appearance of key adaptations might well associate the evolutionary events with environmental changes essentially opposite in nature (in this example, the origin of colonial forms with the mesotrophic Eocene vs. the more oligotrophic Oligocene). In this particular taxonomic group, however, the early fossil record is also very poor and the origination times are also uncertain by at least several my.

These limitations should not, however, be allowed to obscure the real contributions of such studies. Molecular methods provide crucial information on the phylogenetic links between higher taxa which, due to the extreme differences in shell topology in radiolarians, are difficult to obtain with only morphologic methods. Further molecular methods can, however roughly, provide estimates for evolutionary events in groups, such as the Collodaria, where the fossil record itself is even less useful, due to lack of preservation, in providing age estimates for evolutionary events. For most groups of polycystines the fossil record is more robust and can provide strong tests of the hypotheses proposed by molecular analysis. Because molecular studies are done with living materials, they are automatically linked to the biology and adaptations (so far as known) of the organisms under study, thus providing an adaptational framework for understanding the phylogeny of the group. They have, in addition to predictions about the fossil record, their own internal inferences that can serve as tests. For example, the idea that the earliest collodarians suppressed shell formation in response to declining silica availability in the Eocene begs the question: why did not the ancestral Nassellarians also do this? These questions can be examined, and used to test the idea, by further study of the biology and silica use in the two different groups. Thus,

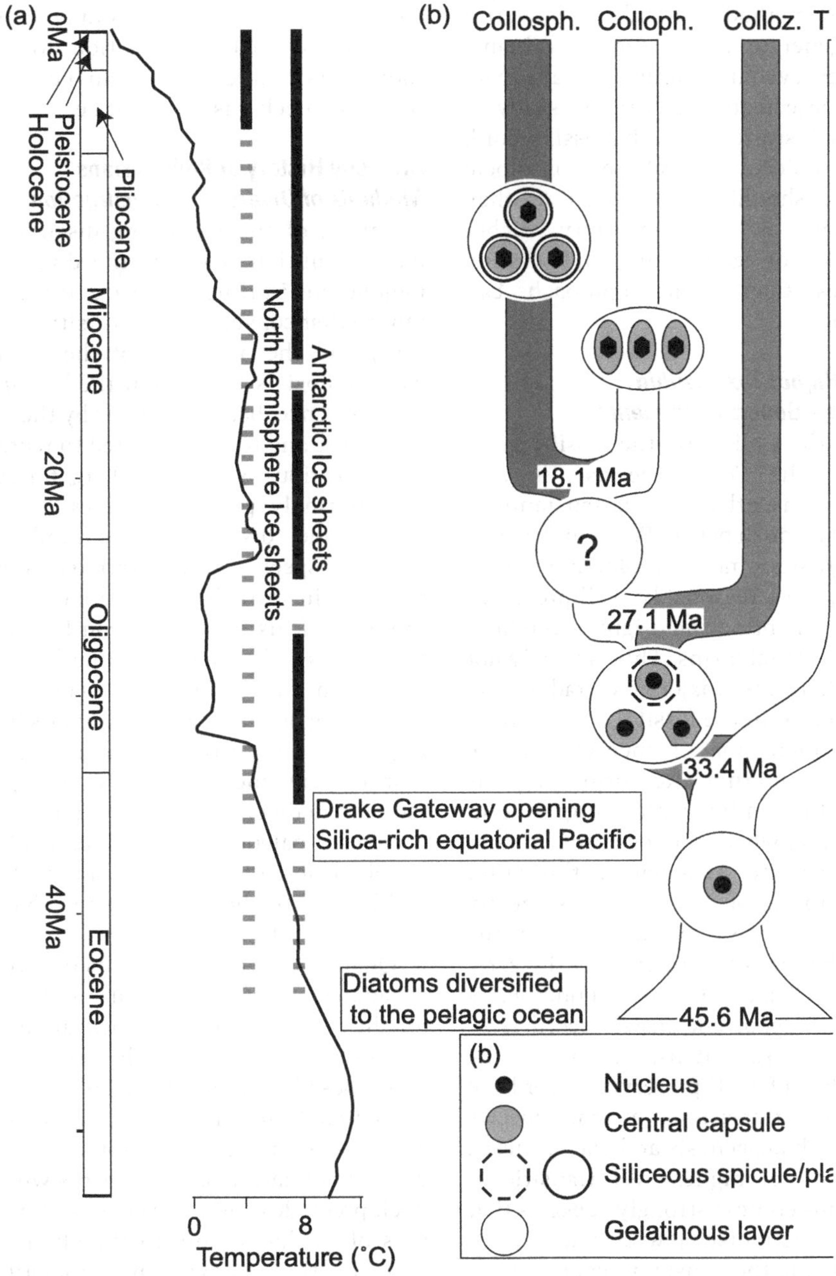

Figure 9.11 Evolutionary scenario for the evolution of collodarian radiolaria. Phylogenetic tree from molecular analysis on right with mean estimates of divergence times in Ma. General environmental history of oceans and climate change on left. Abbreviations: Colloz, Collozoidae; Colloph, Collophidae; Collosph, Collosphaeridae; Thalla, Thalassicollidae. Source: Ishitani et al., 2012. Licensed under CC BY 4.0.

molecular methods provide a powerful way to generate evolutionarily valuable hypotheses, even if in some cases adequate tests of these require additional study of the clades' histories using the fossil record. Ideally, molecular and paleontologic approaches should be combined, as they can provide a complementary, richer framework for generating and testing hypotheses than either approach can yield alone.

Origin of Higher Taxa within Radiolaria – General Comments

While for the Collodaria the fossil record, except for the Collosphaeridae, is very poor, for many other radiolarian families the record is much better. We would expect therefore that for most radiolarian families we would have reasonably well-preserved fossils documenting the origin of the family from ancestral forms. This is largely not the case. The vast majority of radiolarian families appear in the fossil record without any clear indication of ancestry via the presence of intermediate forms. This is true even for families which originated in the Cenozoic, where the fossil record is in general excellent. Although in this latter case the timing of family origins mostly are within a preservational gap in the record – the low silica interval of the early Paleocene (Chapter 5) – it is nonetheless striking how poor the fossil record is of higher-level taxa origins, despite having any number of well-preserved records of evolutionary transitions at the species level for both anagenesis and cladogenesis. The absence of higher-level transitional forms in this context strongly suggests that a purely reductionist view of macroevolution – as being the constant accumulation of normal population and speciation type change – is inadequate. Higher taxa must have evolved much more rapidly, and in such ways as to make their transitional forms hard to recover even in a remarkably complete fossil record. Without at least one example of such a transition to study, though, not much more can be said on what the mechanisms might be.

Diversity History of Radiolarians

Methods of Diversity Reconstruction

As much of the remaining discussion of macroevolution refers to the diversity of radiolarians in time and space, it is important to first consider how diversity is measured, and the adequacy of the available data. Biologists have, from the beginnings of the science, been fascinated by the diversity of life and have attempted to represent this in various ways. Ehrenberg, for example, routinely presented tables of occurrences and discussed the numbers of different species in his various samples, inspired in part by the strong ecologic interests of his mentor A. von Humboldt. Ecology as a distinct discipline has struggled with the problem of quantitatively estimating diversity almost from its founding at the end of the nineteenth century. Our current methods for studying past radiolarian diversity are derived from more general paleontologic methodologies largely adapted from ecology in the 1970s, and later with the growth of paleobiologic research (Smith, 2001).

Quantitative measurements of diversity all address the fundamental problem that for the most part a direct measurement of true diversity, at least at the species level, is not possible, and thus diversity must be estimated from incomplete observations. This is due in part to the sometimes high diversity of natural environments which in itself precludes complete census. However, it is often due simply to the observation that, in the overwhelming majority of cases, there is a dramatic difference in the relative abundances of species in natural environments: a small number of species are fairly common, but most species are rare. As a result, it is very difficult to observe enough specimens, most of which will be the few common species, to also observe at least one specimen of each of all of the rare species. Thus, in biodiversity studies any normal counting of species will be incomplete, and also have a sample-size bias: the observed number of species increases with the number of observations made. With very small sample sizes, increasing sample size results in a relatively rapid increase in the number of species seen, but, as an ever-larger fraction of total diversity is observed, it becomes more likely that an additional observation will be another specimen of a taxon already seen; finding the remaining rare species takes proportionally ever more effort. The relationship thus is non-linear, with the sampling-diversity curve having a rapid rise and a slow flattening of the curve which approaches full diversity as the asymptope only at very large sample sizes (Figure 9.12A). The diversity one sees and "knows" is not the diversity that exists. Ecologists and paleobiologists have dealt with this problem in various ways (summarized below; see review by Gotelli and Colwell, 2011).

One method is to include the relative abundance information in the estimation of diversity in the form of a diversity index. There are many such indices, and though many still vary with sample size, most are weighted toward the more common (and thus easily sampled) species, and thereby reduce the effect of sample size on the diversity estimate. Such indices however have two problems. In practice, one does not always have quantitative relative abundance data available for species, only presence/absence information, which limits their use. More importantly, such indices conflate two very different biologic phenomena: the relative abundance structure of biotas (termed "evenness"), and the total pool of species (termed "species richness"). The former is the result of ecologic processes while the latter is ultimately due to evolutionary processes, and these two fundamentally different characteristics of biotas should be separately measured and analyzed.

A second approach tries to estimate species richness by extrapolation from the incomplete values seen in the data collected. A mathematical function (that has an asymptote at large sample sizes) relating diversity to numbers of observations is fit to the data and the diversity at the asymptote is used as the estimate of total true diversity (species richness). There are many functions in use, but none seem to be a good fit to more than a fraction of the datasets studied. Alternatively, the relative frequencies of rare species in the data can be used to estimate the total true diversity (based on a principle discovered by Alan Turing in breaking the Nazi Enigma code: Chao et al., 2020): these latter, nonparametric methods are now the most often used by ecologists to estimate total diversity. Both methods (also the "nonparametric") assume some sort of underlying model of how diversity is related to numbers of specimens, and extrapolations are generally rather imprecise unless the sample sizes are large enough so that the observations vs. total observed diversity curve is already beginning to flatten out.

A third approach attempts to deal with the inadequate size of typical samples of species by counting instead higher taxonomic units, which are normally much fewer in number than species. A sample which recovers only a small fraction of all the species present may well be large enough to recover all of the phyla or families present. The main problems with this approach are twofold. Higher taxa may be a poor indicator of significant differences in lower level (e.g. species) diversity (Wiese et al., 2016); and higher taxa are can often be, to a significant degree, arbitrary taxonomic categories, being neither stable in

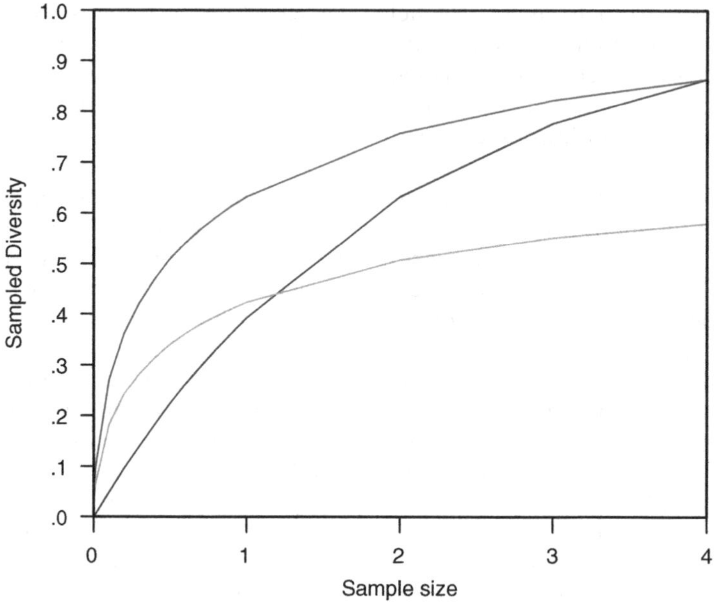
1.0
.9
.8
.7
.6
.5
.4
.3
.2
.1
.0
Sampled Diversity
0
1
2
3
4
Sample size

(a)

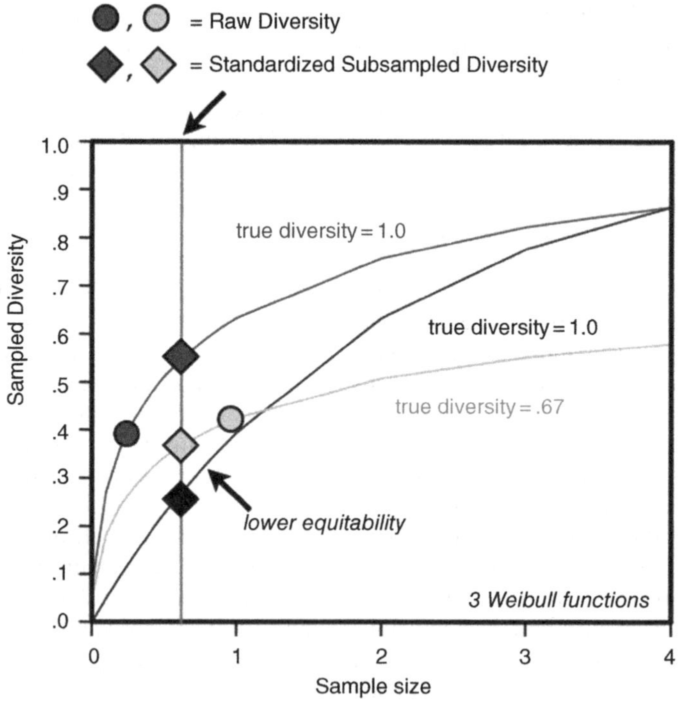
= Raw Diversity
= Standardized Subsampled Diversity
true diversity = 1.0
true diversity = 1.0
true diversity = .67
lower equitability
3 Weibull functions
Sampled Diversity
Sample size
1.0
.9
.8
.7
.6
.5
.4
.3
.2
.1
.0
0
1
2
3
4

(b)

their definition between authors, or consistently applied between different groups of organisms (Hendricks et al., 2014). Diversity computed from such abstract, heterogenous units is thus often hard to interpret biologically.

The last main approach derived from ecology is standardized sampling or "subsampling," originally developed by ecologists in the 1960s and introduced into paleontology in the 1970s (e.g. Raup, 1975). Subsampling only however became a standard method in paleontology with the development of better algorithms and data in the late 1990s; new refinements are still being introduced (see Alroy et al., 2001; Alroy, 2010). Subsampling attempts to accurately measure relative diversity between different populations. In standardized sampling, identical-sized samples are taken from each of several populations that are to be compared, and the diversity of each sample is counted. These values are considered to be comparable estimates of diversity between the populations, at the sample size used, and thus suited to measuring relative diversity differences between populations (Figure 9.12B). There are several potential problems with this method, although it is one of the most used in paleontology. First, although the relative diversity estimates in any one study are, in principle, not affected by sample size, direct comparisons of diversity values between studies are not possible, unless both studies have used the same sample sizes. Standardization by subsampling also discards much data and forces a tradeoff. Unless one sets the standard sample size to the size of the smallest sample in the raw data, intervals with small data amounts will be excluded from the analysis. But by choosing a small sample size, all intervals with more data will be degraded to that of the weakest data in the series, a problem that can only be partially addressed by repeated runs of subsampling and averaging. Lastly, the ability to compare relative diversity assumes that both the evenness structures (equitability), and the degree of geographic clustering of diversity (e.g. endemism) of the populations being compared are the same. If these characteristics are not constant, then the method can give seriously erroneous results (Figure 9.12B). Unfortunately, in most real-world data, evenness does differ between populations, and there is often change in the degree of geographic clustering. Methods to compensate for the effect these have on subsampling methods are only now being developed.

In addition to these ecologic approaches, there are some that are unique to paleontology. The most important of these is the range-through method, which uses the stratigraphically oldest and youngest occurrences of species in the fossil record,

Figure 9.12 A) Plots of how total observed diversity (y axis, arbitrary scale) increases in three different samples as more observations are made (x axis, arbitrary scale). The two curves uppermost on the right side of the plot eventually plateau (at very large sample sizes, not shown) at true diversity=1.0, the other curve has a true diversity 2/3 this. The relative diversity ratio between samples can either be constant, or may vary with changing sample size. B) Same as "A," but showing sources of curve differences and effects of sample size and subsampling. Two samples with the same evenness structure but different true relative diversities (the upper two curves – true scaled diversity of 1 or .67), if sampled with different sample sizes (circles) may show incorrect relative diversity values. Taking a standard-sized subsample of these two samples returns the correct relative diversity values (upper two diamonds). If one of the samples however has a different evenness structure (curve labeled "lower equitability," which also has a true diversity at large sample sizes of 1.0), the subsampled diversity value (lowermost black diamond), and thus the ratio of diversities between can be substantially different than the true values.

and calculates diversity for a given time level as all species whose stratigraphic ranges cross this level, even if the species has not been reported from this level. This type of calculation can also be sample-size biased, if there are systematic differences in the numbers of geologic samples, or larger numbers of observations per sample in different time intervals. Sample-size bias effects though are more subtle in that the diversity estimate for any one time level is derived from data drawn from many different sample levels, and only occurrences of species that extend the stratigraphic range have any effect on the diversity calculation – a large sample with more species has no effect if the additional species in it are within their known total ranges.

Other Problems of Diversity Reconstruction

In addition to the statistical problems above, diversity data can also be biased systematically. In the fossil record, this is typically from the changing average quality of preservation over geologic time. This is a particular problem for radiolarians, especially when comparing data from different geologic Eras, where preservation quality differs dramatically (Chapter 6). Systematic biases in observation also exist; data collected for Cenozoic studies typically report primarily biostratigraphic zonal marker species, rather than the full assemblages, as is done in Mesozoic and Paleozoic research. Data compilations of Cenozoic data thus tend to underrepresent the diversity of this time interval (Lazarus 2011).

Data for Diversity Reconstruction

Two basic types of data can be used to calculate past diversity. Catalogs that summarize information about species often contain estimates of species first and last stratigraphic occurrence, and these values can be used to calculate range-through diversity. Catalog data usually has the advantage that the taxonomy used in the catalog has been standardized in some way, at least for the individual species; and that many other types of potential data problems, such as inaccurately dated samples, have been identified and excluded from the geologic age range estimates for the taxa. Occurrence data by contrast is a compilation (typically in a database system) of the actual occurrences of species in a large number of geologic samples from different locations and age intervals. These data can be analyzed using subsampling methods, as described above. In such studies the data is binned by time interval or geographic region, and a standard-sized subsample of the individual occurrence records of species in specific geologic samples is taken from each bin. This is analogous, although not identical, to taking subsamples of a standard number of individual specimens from an environmental collection of species from a locality in ecologic diversity studies. Occurrence data compilations typically have the problem of containing often incompletely evaluated composites of data from many different sources, with various types of errors: taxonomic, stratigraphic, and sedimentologic (reworked specimens). Microfossil data can also be biased by inclusion of varying numbers of biostratigraphic sample records, which may report only a small number of species, even from diverse samples. As a result, the data is somewhat noisy and has a significant number of outliers, and may hide significant biases (Lazarus, 2011). Diversity analyses thus need to use methods to remove or minimize data errors (e.g. Lazarus et al., 2012), or use methods that are not sensitive to them.

Global Phanerozoic Diversity

Over the last ca. 20 years there have been several attempts to summarize the overall diversity history of radiolarians from their origins to the Recent. These have been replotted, so far as possible, to a common age scale in Figure 9.13, together with one study covering only the Paleozoic (Umeda 2002), as this interval is not well covered in some of the other studies. Also included is a diversity estimate made for this book using the Paleobiology Database (PBDB). This database is the paleontologic community's main Phanerozoic occurrence style database and has enough radiolarian data in it to provide at least some estimate of diversity over at least some of the Phanerozoic time interval. Although the majority of the studies examined species or genus diversity, a family-level diversity is provided by De Wever et al. (2001, 2003). If the results of these analyses all reflect the actual diversity history of radiolarians we would expect each to show very similar patterns of diversity change over time, at least for studies done at the same taxonomic level, and, for the broadest features over the Phanerozoic, also between taxonomic levels, e.g. species vs. genera or families. This is clearly not the case. The only feature common to most curves is a substantial increase in diversity in the Mesozoic compared to the Paleozoic. All but the Vishnevskaya (2009) curve also show an increase between the earlier and later Paleozoic in the Devonian, and three out of five curves with data show a slight diversity decline between the Mesozoic and Cenozoic. These inconsistencies are so major that even the one feature they do largely share (Paleozoic–Mesozoic diversity increase) must be questionable.

To understand why these diversity estimates vary so much, we can evaluate them based on the prior review of the methods and pitfalls of diversity reconstruction. All of these curves are based on a simple compilation of species diversity per geologic interval, which as noted above, can be substantially biased by changes in the amount of data available per time interval. The Sepkoski (2002) and De Wever curves are based only on catalog-style listings of first and last appearances, so there is no way to check the possible effect of changing data availability. The Sepkoski data however is accessible online and can be used to illustrate the effects on bin size on results. Using "substage" resolution instead of stage resolution produces rather different curves (Figure 9.13). The overall shapes are very similar, but the magnitudes are not; and there are some significant differences in detail (e.g. timing of increase in the Devonian, between 400 and 350 Ma). The main difference – absolute magnitude – reflects the much larger numbers of genera that could not be assigned to the finer substage resolution and were thus excluded from the analysis (422 unresolved in the substage analysis vs. 146 – out of a total of 709 – in the stage-level analysis). Kiessling (2002), using a third "hybrid" version of the Sepkoski genus curve (Figure 9.13), argued that 80 Tithonian genera known to him were not given in the Sepkoski database and if these were added, Tithonian genus diversity would have been higher than the Recent. This conclusion however is not valid using the non-hybrid, stage-only version of the curve, which has fewer missing data. The situation is further confused by using genera (as a proxy for diversity of species). As noted in Chapter 5, Cenozoic genera names are more numerous relative to species than for older time intervals, but their use is fairly arbitrary, and still reflects the unrevised legacy of Haeckel's artificial approach to higher level taxa. Thus, a species-level compilation of unvetted taxonomic names might well show lower relative Cenozoic diversity than the genus curves of Figure 9.13. There are, as already noted, basic doubts as to how well genera can be used as proxies for species in biodiversity studies (Hendricks et al., 2014), and specifically for marine microfossils (Wiese et al., 2016).

The estimates by Umeda and Vishnevskaya were made by unspecified methods and no further evaluation is possible. The Afanasieva curve comes from a database which (apparently; details are very

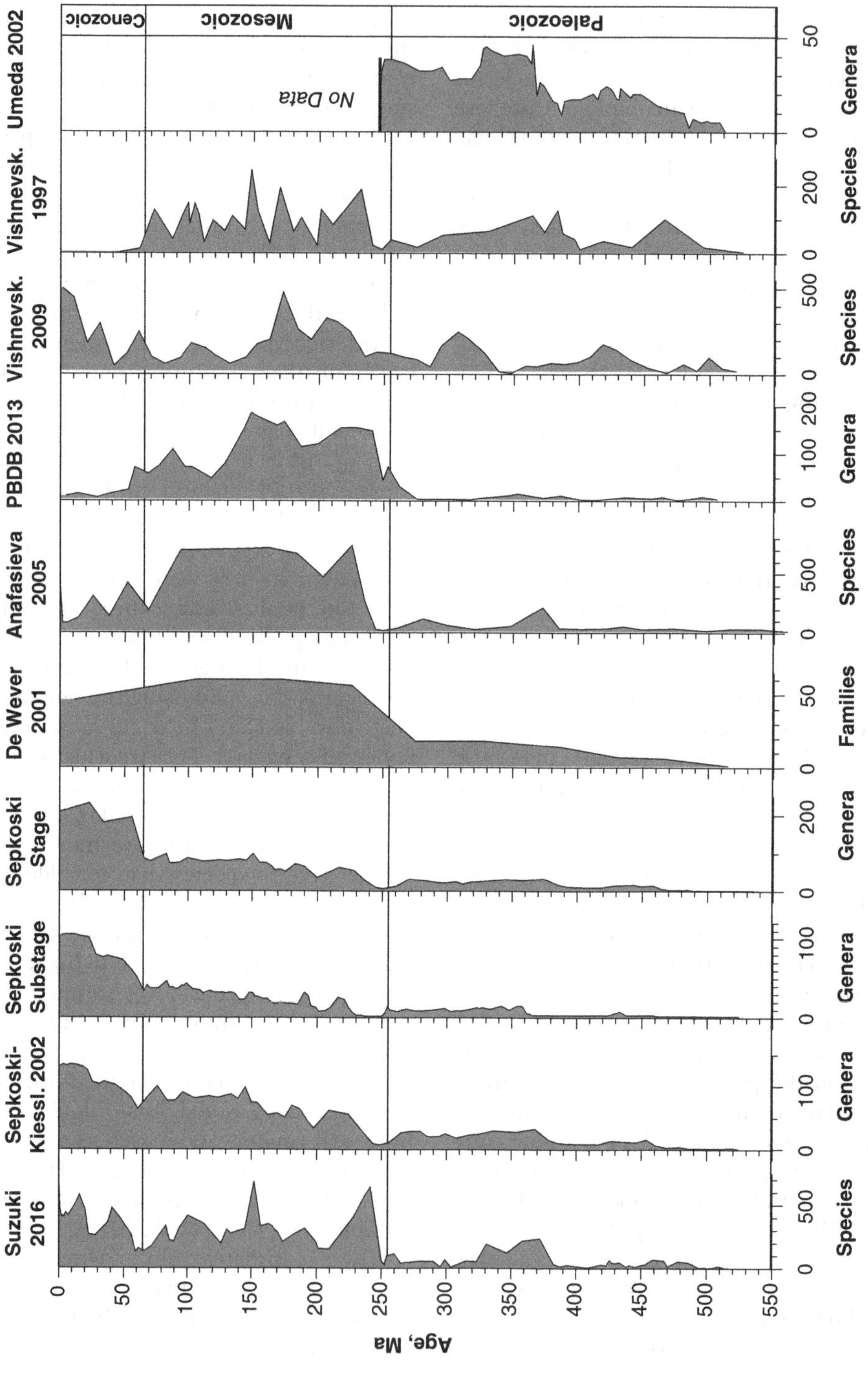
Suzuki 2016
Sepkoski-Kiessl. 2002
Sepkoski Substage
Sepkoski Stage
De Wever 2001
Anafasieva 2005
PBDB 2013
Vishnevsk. 2009
Vishnevsk. 1997
Umeda 2002
Cenozoic
Mesozoic
Paleozoic
No Data
Species
Genera
Genera
Genera
Families
Species
Genera
Species
Species
Genera
Age, Ma
0
50
100
150
200
250
300
350
400
450
500
550

Figure 9.13 Different published estimates of Phanerozoic radiolarian diversity (sources given as plot headers) and age assignments, where possible, to the Gradstein et al. (2004) scale. Suzuki (2016): simplified from same figure in Chapter 5, age scale is Gradstein et al. (2012); Sepkoski db: three different calculations of genus diversity (see text discussion) using online version at http://strata.geology.wisc.edu/jack/ and predefined search function class "radiolaria," total diversity in bin; De Wever et al. (2001): raw family diversity in-bin per period, based on families given in De Wever et al. (2001), numeric diversity values given per geologic period in De Wever et al. (2003) and reassigned ages; Afanasieva (2005): species diversity digitized from their figure 52, with ages reassigned based on geologic named intervals given in their similar plot in figure 51; PBDB: total genus diversity per bin, raw data downloaded 12.3.2013, and separate calculation of unique resolved genus names per standard PBDB (ca. 10 my) age bins; Vishnevskaya (1997 – species diversity): digitized from figures and reassigned ages based on given geologic interval names; Vishnevskaya (2009): species diversity, digitized from figure, no detailed age scale information given and thus not altered, but Era boundary ages do not match the Gradstein et al. (2004) scale; Umeda (2002) genus diversity digitized from figure.

sketchy) has individual occurrence information, but this was not used to evaluate the effect of data size variation. Only the Paleobiology Database (PBDB) Phanerozoic diversity provides information on both diversity and the amount of data used to compile it, and thus permits testing if the diversity values are correlated to the amount of data available. As shown in Figure 9.14, there is in fact a very strong correlation between data availability and the diversity found. Note also that the simple fitted relationship, though it appears linear on the plot, is actually a power function (the plot axes are logs): diversity increases as the 0.6 power of amount of data used to calculate it. Albeit only a simple non-asymptotic fitted function, this is a non-linear relationship analogous to those seen in ecologic data relating sample size and observed diversity (Figure 9.12). Most of the diversity variation seen in the PBDB results can be thus explained as a simple function of the amount of data used: there is no good evidence for changing diversity over the Phanerozoic in this simple compilation of diversity data. Similar biases are likely to exist in the other curves as well, even if they cannot be tested. Figure 9.14 also shows why subsampling the PBDB database is not yet (plot shown is from 2013 but as of Dec. 2018 content virtually unchanged) practical for computing radiolarian diversity for the Phanerozoic – with but few exceptions, only Mesozoic data has even a minimal (>100) number of occurrences per ca. 10 my time bin, and subsampling would need to be reduced to trivially small values – just a couple dozen occurrences/bin (also in the Mesozoic) – to compute a Phanerozoic curve.

What then can be said about the evolution of radiolarian diversity over the Phanerozoic? Unfortunately, from such global compilation curves as yet not much – most of the available reconstructions are inconsistent and most likely are too biased by internal errors to be used. Thus detailed comparisons and postulated scenarios of how Phanerozoic diversity has evolved in response to controlling factors (e.g. environment), which are very common in the literature (Racki, 1999; O'Dogherty and Guex, 2002; Afanasieva et al., 2005, to cite just a few), are to some extent premature. A possible exception to this problem is the family level diversity curve of De Wever et al. (2001, 2003). At this high taxonomic level, it is unlikely that many families have not been sampled yet, so the reported diversity should not be significantly biased by data availability.

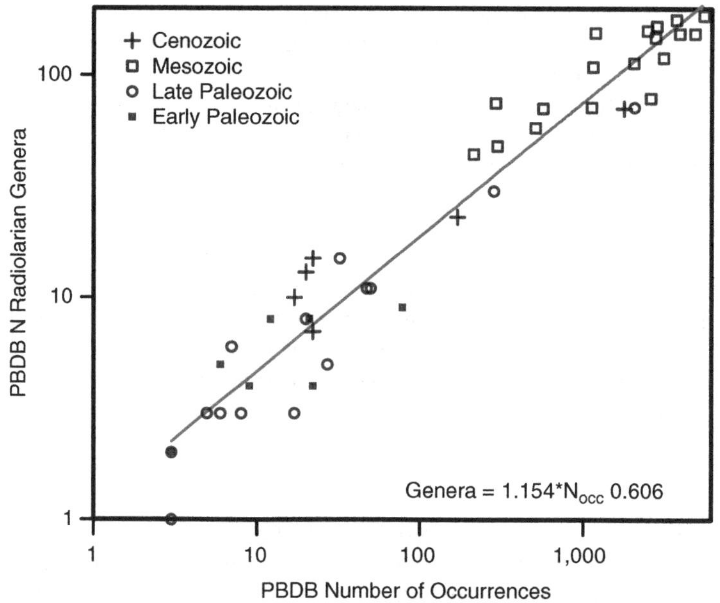

Figure 9.14 PBDB radiolarian generic diversity (raw, sampled in bin) for standard PBDB Phanerozoic bins (variable duration but ca. 10 my long) as a function of number of radiolarian occurrences per time bin. Same data (downloaded 12.3.2013) as in Figure 9.13. Note log scales. Equation of fitted line on lower right.

However, there are possibly significant, systematic differences in the underlying taxonomy used to construct this curve. Many of the families defined for Mesozoic material, particularly those defined on inferred central ontogenetic structures, would, in our opinion, be grouped into single Cenozoic-level families (or the Cenozoic families would need to be split to match the concepts used in the Mesozoic): this reflects long-standing differences in approach to higher-level systematics between Cenozoic and Mesozoic/Paleozoic workers (see Chapter 5). The apparent slightly higher Mesozoic vs. Cenozoic family diversity may thus be an artefact of inconsistent higher-level taxonomy. Even within the Mesozoic the family-level systematics present numerous problems, as many show discontinuous ranges within the Mesozoic, suggesting either incorrect family-level systematics, or larger than previously recognized problems with the quality of the Mesozoic stratigraphic record (O'Dogherty et al. 2011). Despite these problems, the major increase in family-level diversity from the Paleozoic to the Mesozoic is probably not an artefact: even in the less well-sampled late Paleozoic, it is likely that most families have been sampled. Given the amount of Phanerozoic species-level data now available, and largely also compiled, with adequate methods of analysis, a usable estimate of Phanerozoic species diversity history should be possible. This is shown by considering smaller-scale studies, where the problems inherent in diversity reconstruction are already being addressed. These results are both very scattered and provisional but point the way forward for radiolarian diversity research.

Diversity and possible correlations to environmental change are often reported in regional or short time interval studies of radiolarians. An overview of the record has already been given in Chapter 5. Here we summarize the patterns so far reported. Although in many cases full data is not available, and thus subsampling is not possible, within single studies sample sizes are often, if not constant, at least similar, so relative trends in diversity are in principle interpretable.

Paleozoic

The Paleozoic radiolarian record has been summarized in Chapter 5, where it was noted that diversity increased over the era, and was marked by multiple changes in taxic composition and the origin of new forms. However, as all but a few rare blocks of Paleozoic open-ocean sea floor have been lost to plate tectonic subduction of ocean crust, most of the Paleozoic radiolarian record comes from shallower water sediments originally located on the edges of continents or from epicontinental settings, and little is known in detail of their Paleozoic open-ocean history. The numerous preservational biases for individual radiolarian assemblages (described in Chapter 6), and the more general problems of diversity reconstruction, suggest that, with the exception of the late Permian and its extinction event, it is not yet possible to do a useful analysis of Paleozoic radiolarian diversity or possible causes, such as its correlation to environmental change. Useful syntheses of available data are however being published for individual time intervals, e.g. Tetard et al. (2017).

The major extinction event of the Permo-Triassic boundary that affected all forms of life caused a complex pattern of extinction, origination, and evolutionary change at high taxonomic levels in radiolarians (summary in Chapter 5). Extinctions were extended over a substantial time interval in the late Permian, eliminating at first only species, then genera. Despite the poor fossil record caused by the "chert gap" of the earliest Triassic it is clear that most Permian diversity, including many families, had been lost during the extended time period of adverse environments that mark the Permo-Triassic boundary. These clearly primarily are of an external physical nature (anoxia and euxinia, ocean acidification, warming). They began in the late Permian but reached a maximum at the boundary, continued intermittently for some million years beyond it, and are at this point robustly linked to massive Siberian Trap volcanism (Payne and Clapham, 2012; Schootbrugge and Wignall, 2015). However, the long period of recovery in the early Triassic of several million years, beyond the main period of volcanic disturbances, points to a long-term ecologic effect as well. This may be a combination of a less functional carbon cycle due to ecologic damage to the oceanic carbon pump, and a variety of more direct biotic interactions among survivor species, which are interpreted to tended to have been less specialized (i.e. "r" vs. "k" mode taxa) and having existed in relatively "primitive" ecosystem structures (Tong et al., 2007; though see Foster and Twitchett, 2014, for data supporting a less extreme view). The existence of a very low-diversity radiolarian fauna of mostly small, morphologically simple taxa in the basal Triassic (see Chapter 5) is significant, as it indicates that radiolarians responded similarly to, and thus were similarly affected by, early Triassic ecosystem changes as in better understood taxonomic groups. These include many groups of shallow-water invertebrates which are known to have undergone significant dwarfing in size, often accompanied by a reduction in morphologic complexity in response to the preceding boundary extinction event (Twitchett, 2007). This pattern is interpreted as in part a direct response to low oxygen, and in part to low food supply levels (Twitchett, 2007), both of which would be caused by a substantial long-term decrease in organic carbon creation by reduced ocean primary productivity. Lastly, although some elements of the dominant Mesozoic families were present in the late

Paleozoic, most typical Mesozoic forms evolved gradually throughout the lower part of the Triassic, in combination with gradual recovery of ecosystems, at least in part after the main phase of physical impacts to the environment (Chapter 5). Taken together, radiolarians appear to have been affected both by physical as well as biotic changes during the Permo-Triassic extinction and recovery, although the available data on radiolarians is much too sparse to provide useful details. The pattern, despite the evidence for a massive crisis in the physical environment, does not clearly support either a purely Court Jester or Red Queen model of macroevolutionary change but instead suggests elements of both.

Mesozoic

The Triassic ocean hosted the dramatic radiation of radiolarians described in Chapter 5 and documented at the family level in the De Wever family curve (Chapter 5; Figure 9.13), including the first unquestionable Nassellarians, but no evolutionary studies of this radiation exist. Other than the Phanerozoic compilations discussed earlier, no quantitative studies of early Mesozoic (Triassic–early Jurassic) radiolarian diversity are available, except those of Kiessling and Danelian (2011) and Kocsis et al. (2014). These studies were focused only on turnover rates, and show the average "background" rate of extinction in radiolarians declined significantly from the Triassic into the Jurassic (sample-standardized; Figure 9.15). They note that this pattern is also seen in coeval macrobenthos. One can only speculate but possibly the pattern may reflect global changes in marine environmental stability, or gradual stabilisation of the global ecosystem after recovery from the Permo-Triassic mass extinction. The Triassic ended with a major mass extinction event affecting the entire earth's biota, including radiolarians (see Chapter 5), although discrepancies between local sections and global syntheses have not been resolved so the picture is unclear.

In a study of mid-Mesozoic (Aalenian–Aptian: mid-Jurassic–early Cretaceous) radiolaria from the Tethys (tropical proto-Mediterranean), Danelian and Johnson (2001) compiled occurrence data for >400 species and analyzed diversity and rates of evolutionary change, with a bin resolution of 3 my (Figure 9.16). Although this study appeared before subsampling became common, the authors checked their data for sample-size biases (which are suggested by visual correlation of diversity and sample numbers in Figure 9.16) by computing diversity vs. cumulative sample size for each of the time intervals (Figure 9.17). These show that the relative ranking of diversity in their diversity vs. time curve is generally matched by their ranking at equivalent sample size: the Callovian is relatively diverse, the Aalenian–Bajocian interval among the least, and the early Cretaceous diversity is intermediate between the two (low Tithonian diversity though may be a sample-size artefact). Although a broad correlation between their diversity history and Jurassic–early Cretaceous paleoceanographic history can be discerned, most of the variation in Jurassic paleoceanographic data is short-term (Dera et al., 2011) – less than the 3 my scale of the diversity data – and thus meaningful comparisons are difficult. Interestingly, diversity and rates of change patterns for different families were computed and no significant differences were found, suggesting that family-level characteristics of radiolarian biology were not differentially affected by environmental change in the Tethys over this ca. 60 my interval of geologic time.

Jurassic radiolarians are well enough known that biogeographic provinces can be readily identified, if definitions not always agreed on (Chapter 5). Kiessling (1999) defined five separate regions: a broad Tethyan province, a poorly known Gyral

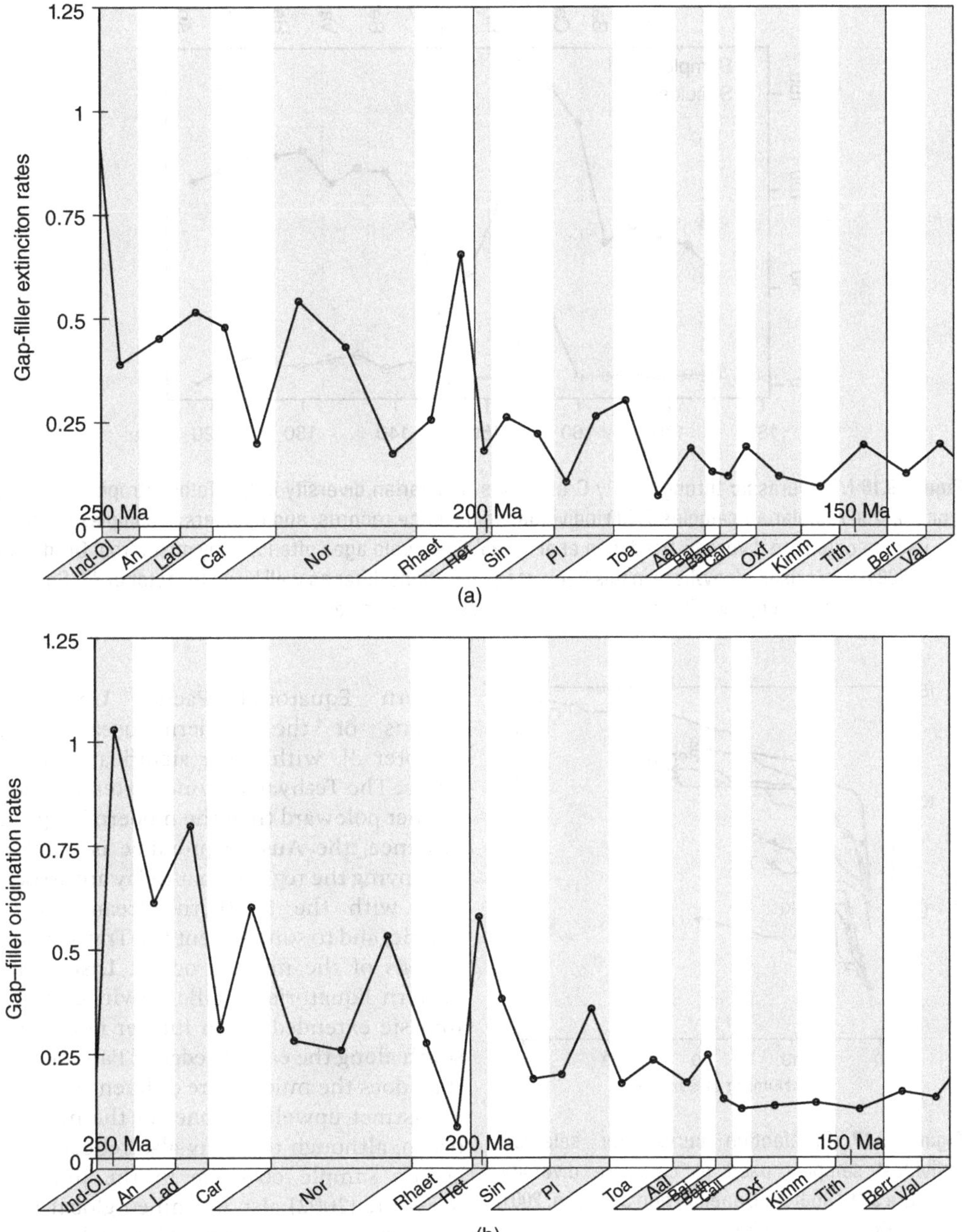

Figure 9.15 Decreasing rates of genus extinction [A] and origination [B] in early vs. mid-Mesozoic radiolaria, from a peak value near 1 (=100%) across the Permo-Triassic boundary. Shaded bands mark geologic epochs. Age scale is Gradstein et al. (2012); geographic scope of data global though primarily tropics and mid-northern latitudes [modern positions], Tethys and Panthalassa in original environments. Source: Kocsis et al. (2014). Reproduced with permission of Cambridge University Press.

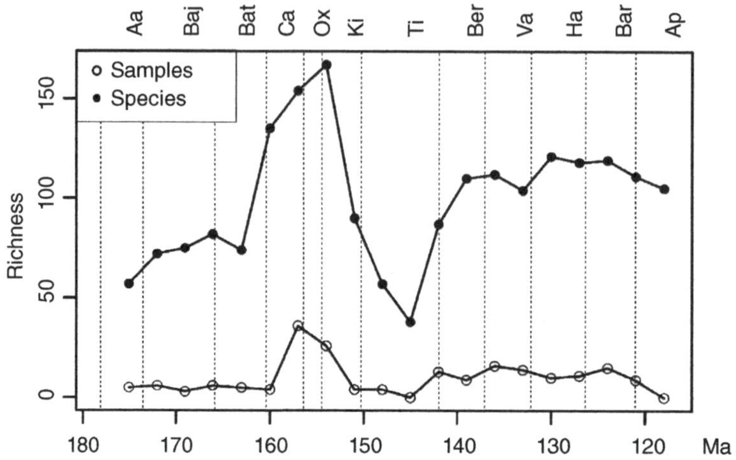

Figure 9.16 Mid-Jurassic through Early Cretaceous radiolarian diversity in the Tethys (tropical ocean), from ca. 200 radiolarian samples/7,500 individual occurrence records, and numbers of samples per time bin used to calculate diversity. Gradstein et al. (1995) time scale ages often 2–4 my offset from Gradstein et al. (2004) scale used elsewhere in this book; stage durations differ as well by up to factor of 2X. Source: Danelian and Johnson, 2001 Reproduced with permission of Elsevier.

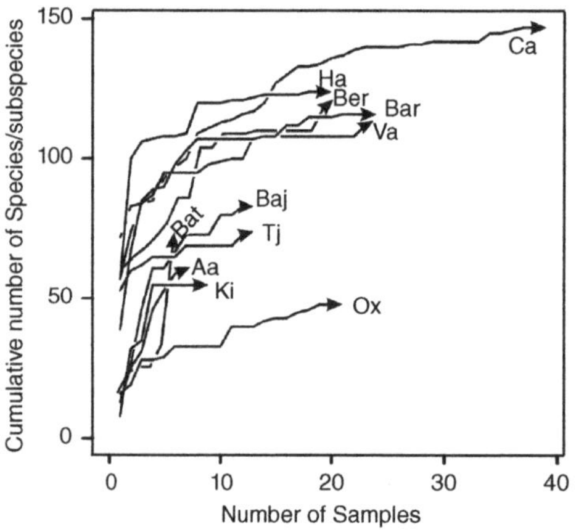

Figure 9.17 Rarefaction curves for selected individual samples used to calculate diversity in Figure 9.16. Source: Danelian and Johnson, 2001 pattern Reproduced with permission of Elsevier.

province, two polar provinces (Boreal and Austral), and an East Equatorial (Figure 5.37). These provinces are somewhat similar to the Tropical, Subtropical, Norwegian-Greenland Sea, Southern Ocean, and Eastern Equatorial Pacific Upwelling regions of the modern oceans (see Chapter 3), with some significant differences. The Tethyan province extends much further poleward than the modern Tropical province; the Austral province is bipolar, occupying the regions that today are associated with the Southern Ocean, North Pacific, and to some extent the Transitional regions of the modern ocean. Lastly, the Eastern Equatorial Pacific province of the Jurassic extended much further north and south along the eastern edge of Panthalassa than does the much more differentiated set of distinct upwelling zones of the modern ocean, although this may also reflect inadequate sample coverage in the Jurassic. Kiessling (2002) also examined global biogeographic patterns in radiolarian diversity for a short time interval near the Jurassic–Cretaceous boundary, and found that – corrected for sample size and preservation biases – polar faunas were less diverse than tropical or subtropical ones. Polar faunas were however not nearly as endemic as those of modern oceans, suggesting that low- to high-latitude faunal exchange was common and did not frequently trigger allopatric speciation. Nonetheless, many species at this time were restricted to one or the other province, and illustrates the extremely important role that ocean provinciality played, even in the Mesozoic, in regulating radiolarian diversity.

Cretaceous radiolarian species diversity for the Western Tethys was summarized by O'Dogherty and Guex (2002) using a form of range-through method derived from a unitary associations analysis. A total of 431 species covering the latest Jurassic (Tithonian) to early late Cretaceous (Turonian) showed diversity near 100 species in the early Cretaceous, a brief but sharp drop to <70 species in the late Barremian to Aptian, and rebound to >130 species in the Albian-Cenomanian (Figure 9.18). Neither the absolute values nor the trends through time in this analysis match those reported for global diversity in other studies (e.g. Figure 9.13; see also comment by Kiessling (2002)), although given the different geographic scope and numerous problems with the global compilations this in itself is not particularly problematic. Range-through methods are however very sensitive to data outliers and changes in data density, and it is impossible to judge if the diversity history reported by these authors reflects

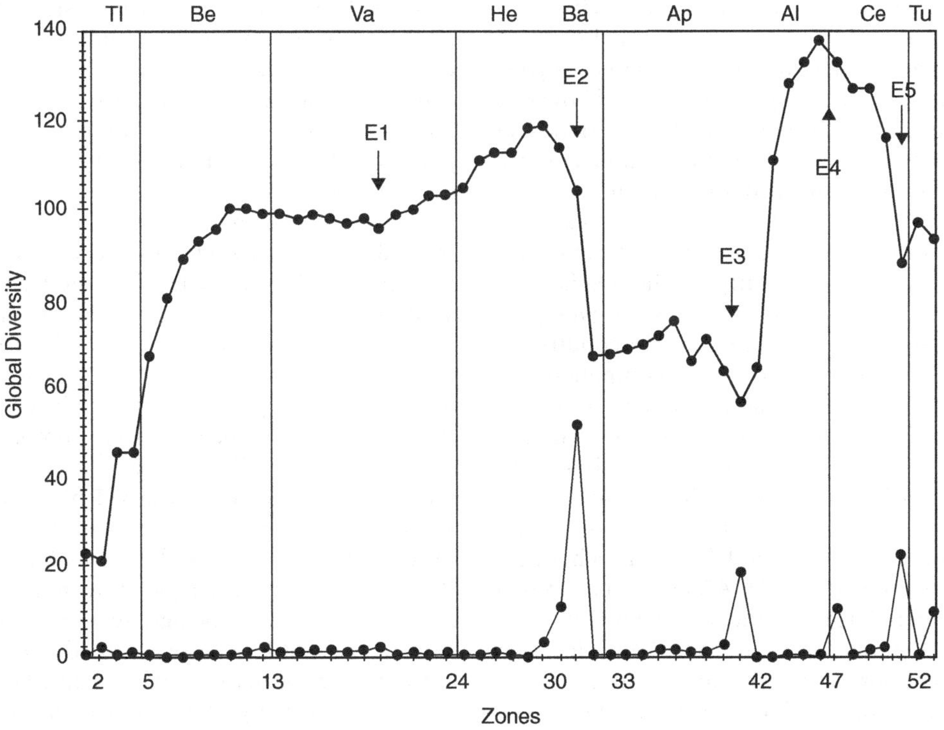

Figure 9.18 Radiolarian diversity (approximately range-through of raw data, though from assemblages ordinated into zones using unitary associations) in the western Tethys region of the tropical ocean, latest Jurassic to early Late Cretaceous. Note horizontal (x) axis is categorical (zone numbers), not numeric. Source: O'Dogherty and Guex (2002). Reproduced with permission of micropress.org.

biologic history or only artefacts of the data compilation. The authors give an extensive interpretation of inferred changes over time in diversity and environment, with two major diversity shifts at the ends of the Aptian. Diversity drops during the Barremian development of the "Faranoni" OAE and recovers by the late Albian, an interval of time without OAE development. Erbacher et al. (1996) had earlier summarized radiolarian diversity dynamics (extinctions and radiations, not absolute diversity) in relation to Cretaceous OAEs and suggested a similar relationship, with OAE-linked extinction and subsequent recovery of diversity afterwards. Erbacher et al. (1996) also proposed an environmental model for the cause of the relationship: mid-water anoxia during periods of high marine productivity leading to loss of water-column habitat and extinction (Figure 9.19A–C). This is a good example, next to provinciality, of how one of the most important characteristics of the pelagic environment – water column structure – can regulate radiolarian evolution. How well this model explains Cretaceous low-latitude radiolarian evolution depends critically on the precise correlation between extinctions, recovery, the development of anoxic water conditions, and the response of other components of the plankton. These relationships are not adequately resolved as yet (see Leckie et al., 2002, for a summary), but with increasing geochronologic resolution, due in part to better sections, use of orbital tuning methods, and better geochemical characterization of OAE intervals, a much clearer picture should soon be possible.

Diatoms have in the Cenozoic been important factors in radiolarian evolution (see below). Today they form one of the main groups of marine plankton, and dominate the marine silica cycle. Diatoms may have evolved as early as the Jurassic, based on molecular phylogenies with substantial age uncertainties. They are first known as fossils in the early Cretaceous but appear to have been rare, although their relatively thin siliceous shells preserve poorly in older sediments and their record is very patchy. There are no substantial records of open-ocean pelagic forms prior to the Cenozoic: Cretaceous fossils appear to have been primarily benthic or coastal plankton (Harwood and Nikolaev, 1995).

Cretaceous–Tertiary Boundary

The Cretaceous period ended with a mass extinction event (the 'K/T') – the much-popularized asteroid impact – which exterminated almost all calcareous plankton species and also brought an end to the non-avian dinosaurs (see Chapter 5: Taxonomy and Fossil Record). This, the most recent mass-extinction event, is also the best studied. The oceans were affected strongly, and most of the calcareous plankton went extinct. Dinoflagellates and diatoms showed little extinction, although the record is sparse. One problem for siliceous plankton studies, including studies of radiolarians, is the near absence of preserved siliceous rocks from the earliest Paleocene, immediately following the impact event. This pattern is similar to the "chert gap" seen at the Permo-Triassic boundary, and for the K/T event is thought to be due to a major decline, and multi-phased recovery of marine export productivity and the functioning of the open-ocean carbon pump.

Reasonably well-preserved records of radiolarian faunas across the K/T boundary are known only from a few locations, the most important of these being, as for the Permo-Triassic boundary, shallow to intermediate water depth pelagic sections, now exposed on land in New Zealand. These sections have been extensively studied in a series of papers by Hollis and colleagues (e.g. Hollis, 1993, 1997, 2003). In contrast to most other regions of the ocean, the early Paleocene in these sections records localized high diatom productivity, resulting in mixed calc-silica pelagic sediments

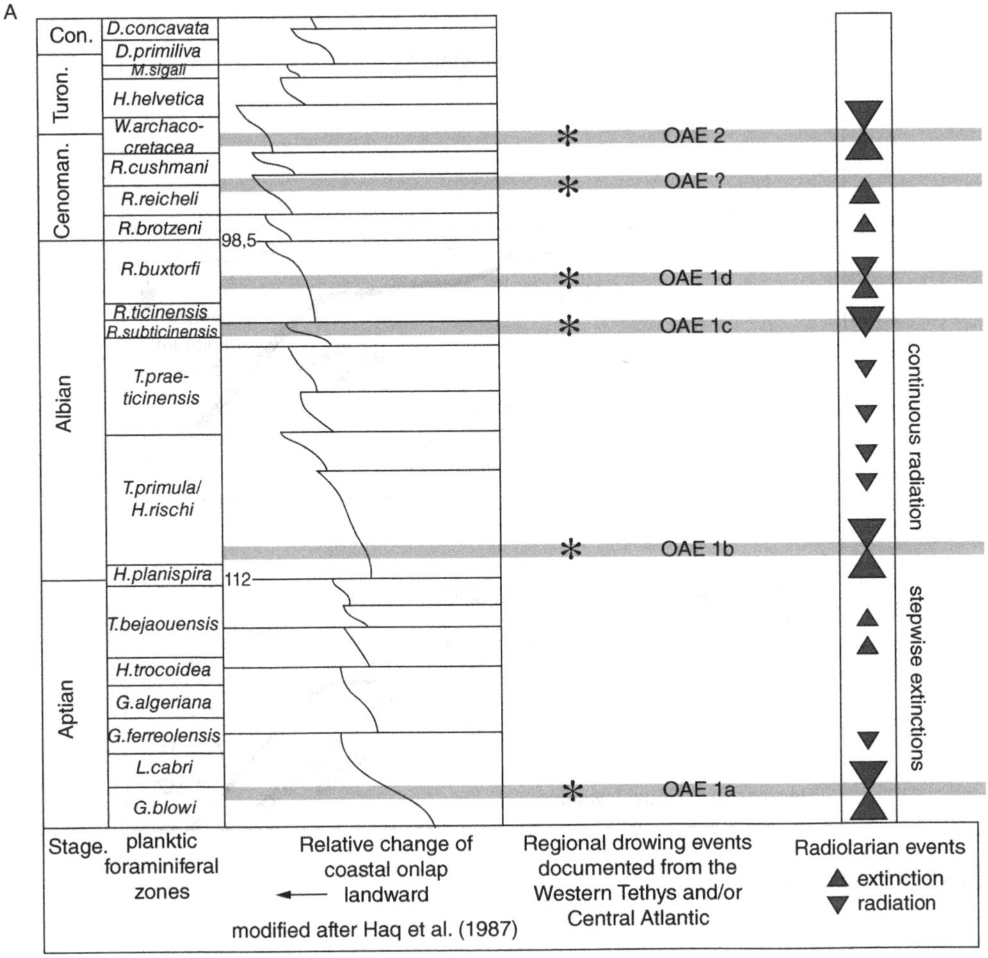

Figure 9.19 a) Correlation of radiolarian evolutionary turnover (originations and extinctions, diagrammatic only, right side of figure) with oceanic anoxic events (OAEs, grey zones, middle) and sea-level change (left) in mid-Cretaceous tropical ocean regions (Tethys and Central Atlantic). Based on thesis data of Erbacher (1994). b) Model relating two different types of nutrient flux due to sea-level change (A and B) to development of ocean anoxia and radiolarian turnover. A: high sea levels remobilize shelf nutrients, and export of resultant high ocean-surface productivity fixed organic carbon creates broad anoxic zone in ocean mid-waters, with extinction of many deeper-water radiolarian taxa. B: high erosion rates of former shallow water, organic carbon-rich sediments during sea-level low stands creates organic-rich sediments from transported detrital matter on margins, but ocean productivity remains moderate and oxygen levels in the water column are not affected. Radiolarian diversity is high. Source: Erbacher et al. (1996). The Geological Society of America.

B

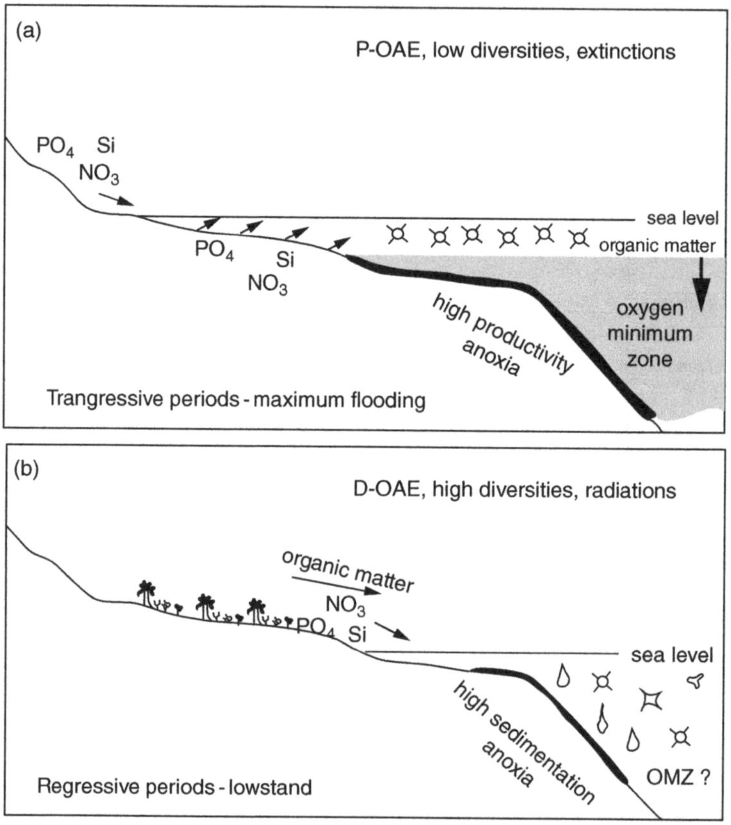

Figure 9.19 (*Continued*)

(now siliceous chalks and beds of chert). Radiolarians, although they show substantial changes in relative species abundance, and soon after the boundary, the origin of many new species, show little evidence for extinction at the K/T boundary (Figure 5.38). Radiolarian faunas from these relatively shallow sections document major extinction of pre-Paleocene forms and the development of faunas typical for the later Paleocene only ca. 2–3 my later, near the end of the early Paleocene. This pattern is similar to the less well-documented pattern reported for dinoflagellates (e.g. Brinkhuis and Zachariasse, 1988; Bowman et al., 2012). In particular, the radiolarian pattern of change only well into the Paleocene is strikingly similar to that of the diatoms. Early Paleocene (Danian) diatoms are almost entirely Cretaceous forms, and are replaced, originally gradually, but at an increasing pace in the mid-Paleocene, by newly evolved Cenozoic forms (Renaudie et al., 2018). Interestingly, the radiolarian, and diatom, turnovers are close in time to when the open-ocean carbon pump finally recovered to pre-K/T levels of functioning (D'Hondt et al., 1998; Figure 9.20), suggesting that the extinction of "Mesozoic" radiolarians and diatoms was due to changing ecologic conditions in the recovering early Cenozoic ocean,

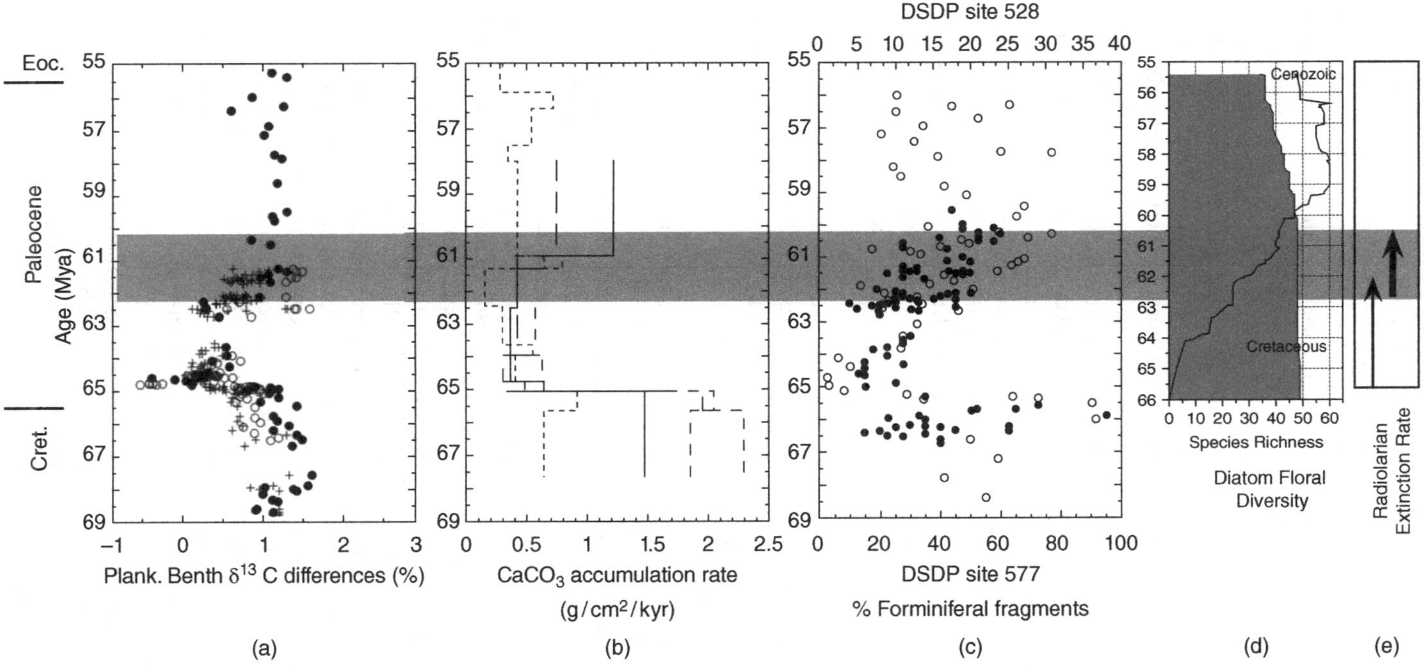

Figure 9.20 Radiolarian macroevolutionary turnover and Paleocene ecosystem recovery. Shaded grey interval between ca. 62–60 Ma marks interval of maximum turnover in Paleocene radiolaria (see Ch. 5 and panel E). A: Surface-deep carbon isotope gradient, central Pacific Site 577 as measured by black –phytoplankon/benthos difference; white – surface/benthic foraminifera; crosses – subsurface/benthic foraminifera. Low values indicate little organic carbon flux to ocean bottom; B: accumulation rate of carbonate in ocean bottom sediments from several deep-sea sections. Low values, together with indicator of low dissolution [panel C] indicate low production/export from surface waters; C: foraminiferal fragments (as % total foraminiferal material, Site 528 with black dots). Low values indicate low dissolution and low organic carbon in bottom waters; D: Diatom species richness of Cretaceous survivors and newly evolving Cenozoic flora from compilation of published data by Renaudie et al. (2018); E: Radiolarian extinction rates (diagrammatic) in Paleocene Woodbourne Creek section, NZ. Gradual turnover in earlier Paleocene (interval with thin arrow) accelerates in mid-Paleocene (heavy arrow). Source: A–C from D'Hondt (2005), D replotted from data provided by J. Renaudie, E from results of Hollis (1997), taken from Chapter 5, Figure 5.38. D'Hondt 2005, Renaudie et al. 2018.

rather than to the direct effects of the impact event itself.

Cenozoic

The Cenozoic radiolarian record is by far the best preserved time interval of the entire Phanerozoic. Here it should be possible to more robustly reconstruct the diversity of the group and how it has changed over time. While this is in principle entirely correct, the reality of what has been done so far is rather different. This has to do with how differences in preservation, sample availability, and, in particular, differences in research priorities, largely set by funding agencies, have affected the type of data collected (see also Lazarus, 2011). In the Mesozoic, and particularly Paleozoic, radiolarian-bearing samples require considerable work to find and prepare, and only a fraction of these will have well-preserved material. These specimens, often only a few thousands to a few tens of thousands in number in an individual study, are then typically examined thoroughly. The large majority of taxa found are recorded, and in some form also reported, in the resultant publications. Research on Paleozoic and Mesozoic radiolarians is largely in service of biostratigraphic dating, and in these time intervals use a whole-assemblage approach to biostratigraphic data analysis, with sample relative age ordination methods such as RASC or Unitary Associations (see Chapter 8). Cenozoic studies by contrast are mostly based on the virtually unlimited numbers of samples obtainable from deep-sea sediments, and biostratigraphic work, while still a main driver of research, is based on events (e.g. first and last appearances) from a small number of selected species that are believed to be reliable age markers (Chapter 8). This focus is set by the need for Cenozoic workers to collaborate with large international research programs such as the deep-sea drilling projects, which have fostered large, multispecialty team type research (Chapter 1: History). Most Cenozoic studies make use of many more radiolarian-rich samples than is typical of Mesozoic or Paleozoic work, and have access in these samples to a vastly greater number of specimens – typically several hundred thousand, or several million, specimens – available for examination on the prepared project microslides. Given the focus on recording the range of only stratigraphic forms, and the vast amount of other material, most species seen on slides are simply ignored, or recorded only sporadically, and with little effort to resolve their taxonomy or stratigraphic range. The large majority of published Cenozoic radiolarian occurrences literature thus provides accurate data just for stratigraphic markers, with only sketchy, incomplete data on the occurrences of other forms. As a consequence, the published biodiversity data for the Cenozoic is, paradoxically, for the most part worse than for the older, less poorly preserved faunas of the Mesozoic and Paleozoic.

For these reasons it is only recently that Cenozoic, global-scale synthesis of radiolarian diversity history has been attempted (Renaudie et al., in prep.; Figure 9.21). These study use the NSB database (Lazarus, 1994; Spencer-Cervato, 1999; Renaudie et al., 2020), which provides detailed global occurrence data for (mostly) Cenozoic microfossils, including radiolarians, from deep-sea drilling sections, linked to numeric age models and taxa synonymies. This type of data is suited to subsampling and outlier removal methods as discussed above. A typical global Cenozoic diversity curve from these analyses is shown in Figure 9.21A, and additionally diversity only in warm-water regions (tropical and temperate). A more detailed breakdown of diversity by geographic region is given in Figure 9.21B, while the turnover rates (origination and extinction) are shown in Figure 9.21C.

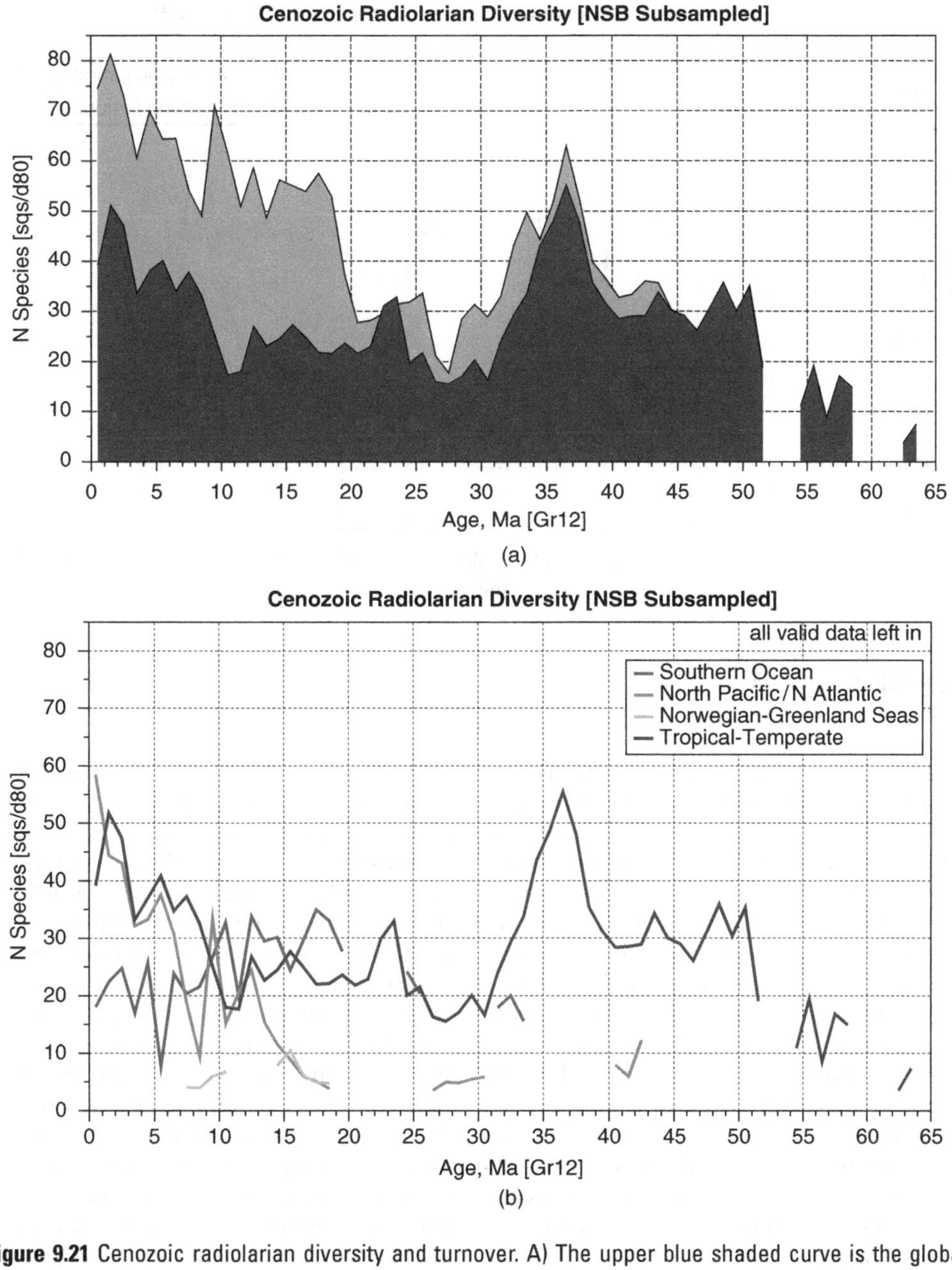

Figure 9.21 Cenozoic radiolarian diversity and turnover. A) The upper blue shaded curve is the global diversity, the red curve all tropical + temperate diversity – here defined as 45S–40N (all analyses by paleolatitude). B) diversity history of each region: tropical–temperate; Southern Ocean – here >55S; North Pacific and North Atlantic (40N–65N); and Norwegian-Greenland Seas – >65N. Because some taxa are cosmopolitan and found in more than one region the diversities of the regions usually add up to more than the global estimate for any time bin. C) – origination and extinction rates (global) for Cenozoic radiolaria. Time series have been smoothed with a three-point moving average function. (*See color plate section for the color representation of this figure.*)

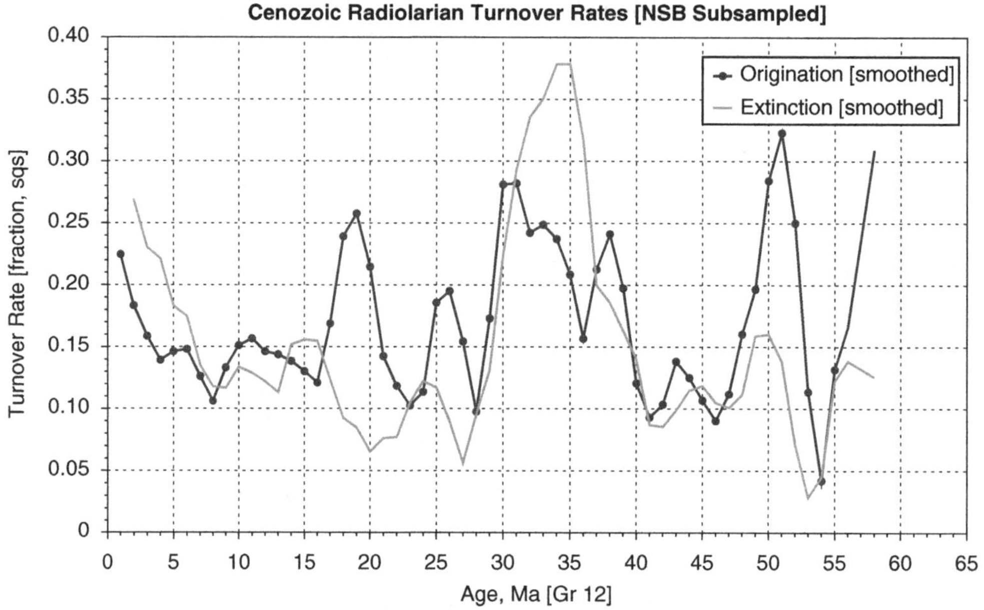

(c)

Figure 9.21 (*Continued*)

The main conclusions from these analyses are that radiolarian diversity has increased substantially over the Cenozoic; Paleocene–early Eocene diversity was lower than subsequently in the same tropical–temperate regions; and diversity growth since then globally has mostly been due to diversification of polar assemblages, mostly in the Southern Ocean and North Pacific. There is also an Oligocene "dip" in diversity which separates higher later Eocene and Neogene diversity levels. This parallels the drop in numbers of stratigraphic events and thus temporal resolution seen in Eocene vs. Oligocene radiolarian stratigraphy (see Chapter 8 and below).

These conclusions appear to be robust, as they are based on patterns that appear in numerous alternate analyses of NSB data, using gradually improving data over a period of years, different sampling procedures, and data selection rules. However it is important to note that these estimates are nonetheless only approximations to what a comprehensive diversity survey would show, as to some degree they must be assumed to be affected by the problems of subsampling large data compilations, as mentioned previously. The only way to be sure that the patterns recovered are reliable is to compare them, at least for some time intervals and/or regions, with full diversity data (all forms seen counted, large sample sizes, and all size fractions). Given the high diversity of the actual assemblages, this is challenging and so far has only been done for a couple time intervals/regions – the Neogene Southern Ocean (Renaudie and Lazarus, 2013), and a new study of the late Neogene tropics that is currently (2020) nearing completion (Trubovitz et al., 2019, in prep.). These authors note that although the general patterns seen in subsampling NSB data are correct, the details are not always identical, e.g. in the precise timing of diversity change (see below). More detailed studies are needed to better understand the patterns of change in Cenozoic radiolarians, particularly for the earlier Paleogene where the compiled data in NSB is still very patchy. A few such studies are reviewed below.

Several studies of purely stratigraphic or taxonomic nature are available for Early Paleocene to mid-Eocene radiolarians, but accessible data suited to diversity analysis is relatively rare, and mostly not yet synthesized in the NSB database. The most detailed, well dated, complete, accessible occurrence data from a section covering almost the entire time interval is from the mid-latitude North Atlantic (Sanfilippo and Blome 2001). Although the study emphasized stratigraphically useful species and is only from one region, and thus is only a subset of both local and global diversity, most taxa which could be clearly identified and which showed a stratigraphic signal were tallied, allowing at least a relative diversity change estimate (Figure 9.22) and shows a substantial increase in diversity from the early to late Paleocene, with a substantial turnover (peaks in both extinction and origination rates) in the mid to

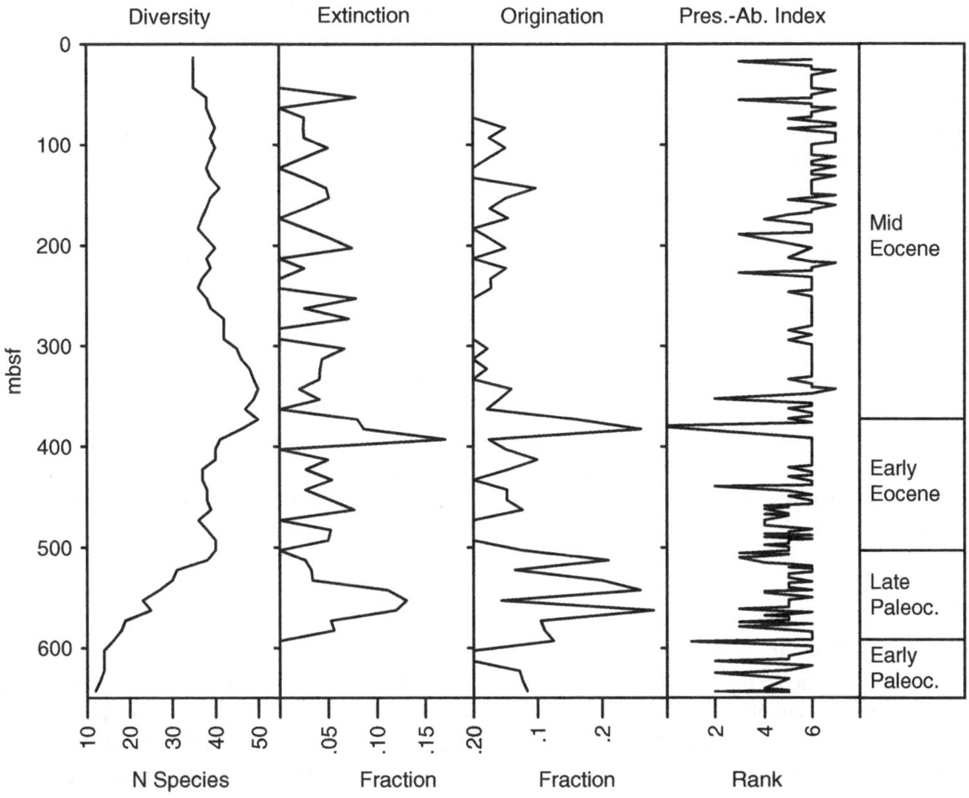

Figure 9.22 Diversity (range-through), extinction and origination rates, and semi-quantitative sample quality index (combination of categorical values for radiolarian abundance and preservation - values below 5 are relatively poor) vs. depth in section (meters below sea floor, mbsf) for Paleocene–mid Eocene radiolarians from the mid-North Atlantic. Spikes in indices near 400 mbsf (early–mid Eocene boundary) are due to a short hiatus in the section.

late Paleocene. Diversity and rates of change are then nearly constant until the late mid Eocene – the spike in extinction and origination rates at the mid/early Eocene boundary is due to a hiatus. The diversity/rates record at this location is not probably all that much influenced by preservation, as indicated by the preservation-abundance index. This only is consistently poor (and thus capable of affecting range through diversity estimates) in the very lowest part of the section. Early to mid-Eocene radiolarian diversity and turnover has also been studied in tropical Pacific material Kamikuri et al. (2013). The data (from Kamikuri et al., 2012) is of particular importance in that an attempt was made to record all species encountered (though only in the >63 μm fraction), and sample sizes (individuals per sample) varied only to a moderate degree (mean ca. 1,600, s.d. ca. 350). Kamikuri et al. (2013) noted several minor episodes of taxa turnover (a few species each) which they related to possible oceanographic changes. Diversity appeared to be relatively constant (mean 81 species, s.d. 9 species) throughout the study interval, with most variation being either random or due to sample size variation. This constancy in diversity is in accord with the results of Sanfilippo and Blome (2001). It is worth noting that the mean diversity per sample in this study is approximately twice that of other studies, or for time-equivalent data in the NSB database. This illustrates the degree to which much current data, at least for the Paleogene, is still incomplete.

In the NSB analyses (Figure 9.21) the late Eocene represents a peak in radiolarian diversity, and parallels the late Eocene diversity maximum seen also in planktonic marine diatoms (Lazarus et al., 2014). Major increases in the latest Eocene and earliest Oligocene extinction rates (Figure 9.21C) cause a major drop in diversity.

The dominant climatic and paleoceanographic event of the Cenozoic – Late Eocene cooling and its culmination in the formation of the Southern Ocean and glaciation of Antarctica at the Eocene–Oligocene boundary – has attracted much research, including studies of radiolarian faunal change. A detailed study by Funakawa et al. (2006) from tropical Pacific samples (Figure 9.23), although based on only partial diversity data, showed major turnover in faunas, mostly concentrated at two levels: ca. 35 Ma (late Late Eocene), and ca. 33 Ma (immediately above the Eocene–Oligocene boundary). The authors record ca. 60 extinction or origination events during this ca. 3 my interval – a majority of the recorded standing diversity of 100 species. Diversity however changed little, as overall rates of extinction and origination were similar. Funakawa et al. (2006) also noted that the faunas became noticeably less equable, and that many of the new Oligocene species were of a more

Figure 9.23 Evolutionary change across the Eocene–Oligocene transition in the equatorial Pacific ODP Site 199-2018A. Data plotted vs depth in section, (mbsf, left), along with core recovery, lithology, geologic time intervals, and sample locations. Qualitative estimate of sample preservation shows modest decline in the early Early Oligocene. Two of the three brief drops in otherwise nearly constant diversity ("species richness" – Sp1 and Sp2; ca. 35 and 33 Ma, respectively) are associated with substantial evolutionary turnover (Tm: "top morphotypic," e.g. extinction; Bm: "base morphotypic," e.g. origination events) – here labeled TM1, TM3, and BM1. Evenness ("equitability") declines from the late Eocene to early Oligocene, as shown by example histograms of relative abundances for most common taxa in samples (right column), which influences the Shannon-Wiener "diversity" index H′. Data for two other, similar sections not shown. Source: Funakawa et al., 2006 Reproduced with permission of Elsevier.

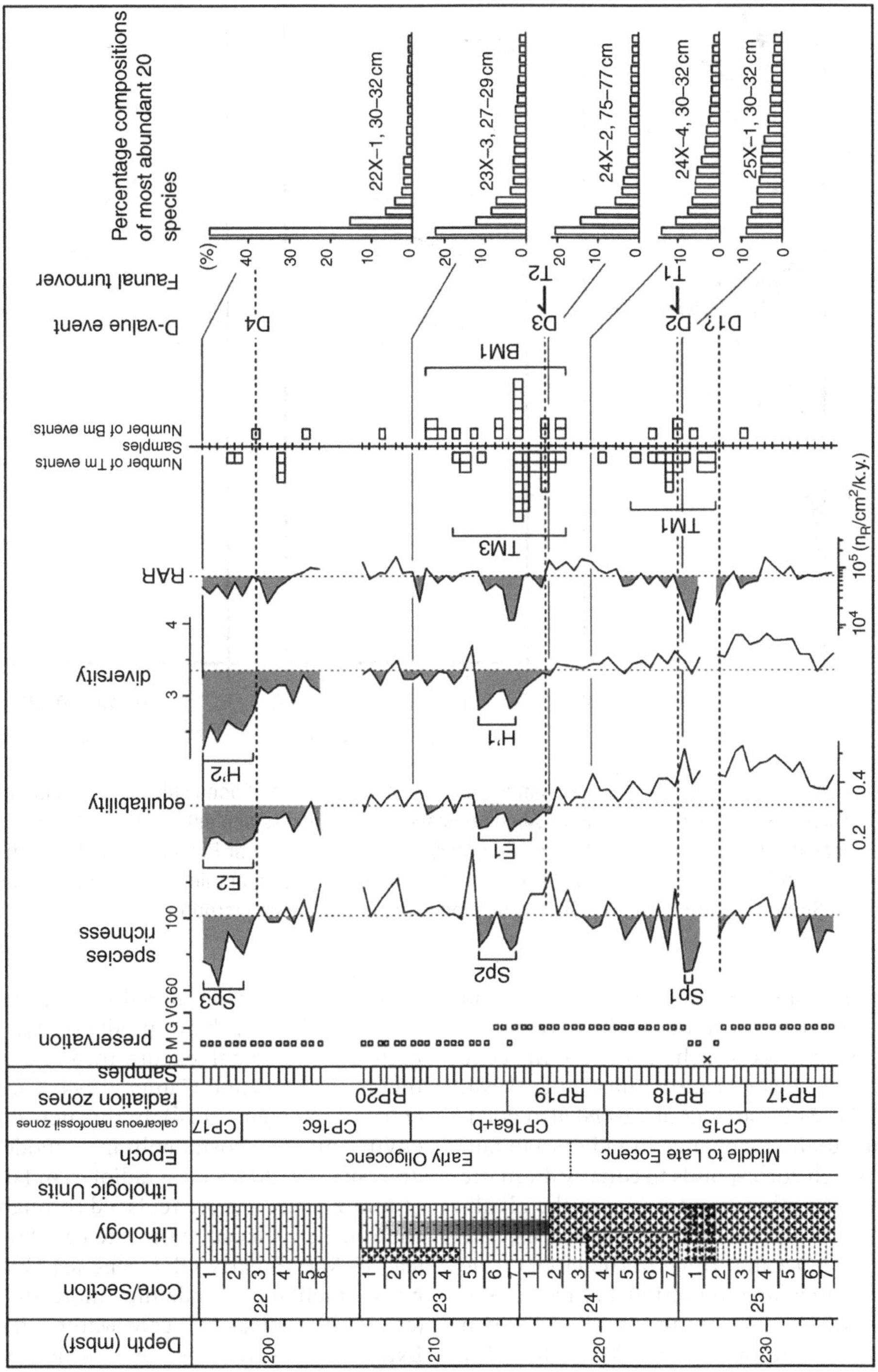
Depth (mbsf)
Core/Section
Lithology
Lithologic Units
Epoch
calcareous nanofossil zones
radiation zones
Samples
preservation
species richness
equitability
diversity
RAR
Number of Tm events
Samples
Number of Bm events
D-value event
Faunal turnover
Percentage compositions of most abundant 20 species
Early Oligocene
Middle to Late Eocene
CP17
CP16c
CP16a+b
CP15
RP20
RP19
RP18
RP17
Sp3
Sp2
Sp1
E2
E1
H'2
H'1
TM3
TM1
BM1
D4
D3
D2
D1?
T2
T1
22X–1, 30–32 cm
23X–3, 27–29 cm
24X–2, 75–77 cm
24X–4, 30–32 cm
25X–1, 30–32 cm
(%)
B M G VG
60
100
0.2
0.4
3
4
10^4
10^5 (n_R/cm^2/k.y.)
200
210
220
230
22
23
24
25

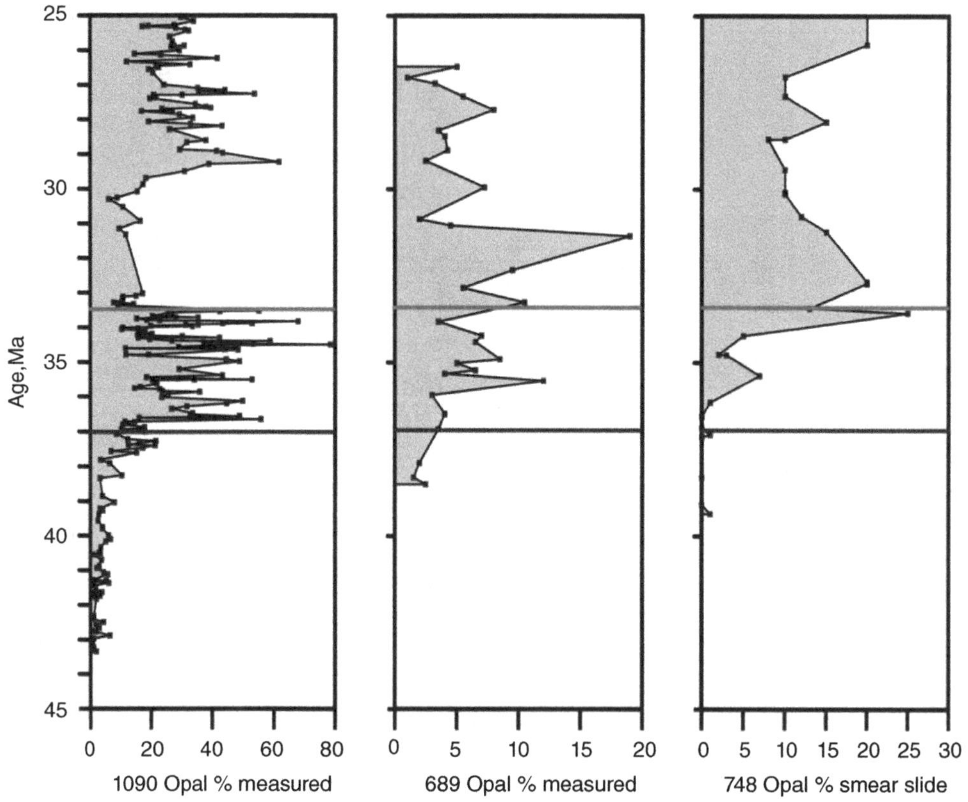

Figure 9.24 Increase in biogenic opal accumulation beginning in the Late Eocene (dominantly diatoms but also radiolarians and minor amounts of silicoflagellates and sponge spicules) at sites in the Southern Ocean opal belt: 689 – Maud Rise, Weddell Sea; 1090 – South Atlantic near Polar Front; 748 – Kerguelen Plateau, Indian Sector. Lower line marks Mid/Late Eocene boundary, upper line the Eocene–Oligocene boundary. Source: Lazarus et al. (2008). Reproduced with permission of micropress.org.

cosmopolitan, cooler water nature than the late Eocene faunas they replaced. Integrated studies of the Eocene–Oligocene Southern Ocean (Lazarus and Caulet, 1994; Apel, 2005; Lazarus et al., 2008) also document a dramatic change in radiolarian faunas, which corresponds to equally dramatic changes in other components of the plankton. In the southern high latitudes, the Southern Ocean formed in the late Eocene as circumpolar circulation became possible with the opening of the Tasman Gateway around 35 Ma (Stickley et al., 2004), creating an isolated circulation system that promoted new, endemic plankton biotas, high productivity, and high biogenic opal export into the sediments (Lazarus and Caulet, 1994; Lazarus et al., 2008; Figure 9.24). In this new ocean environment, cosmopolitan, moderate diversity (ca. 80 species) radiolarian faunas of mid-Eocene age are replaced throughout the Southern Ocean in the late Late Eocene (ca. 35 Ma) to early Oligocene (ca. 30 Ma) by a much more endemic, more diverse (ca. 130 species) cold-water fauna. Extinctions were more or less uniformly distributed throughout this ca. 5 Myr-long

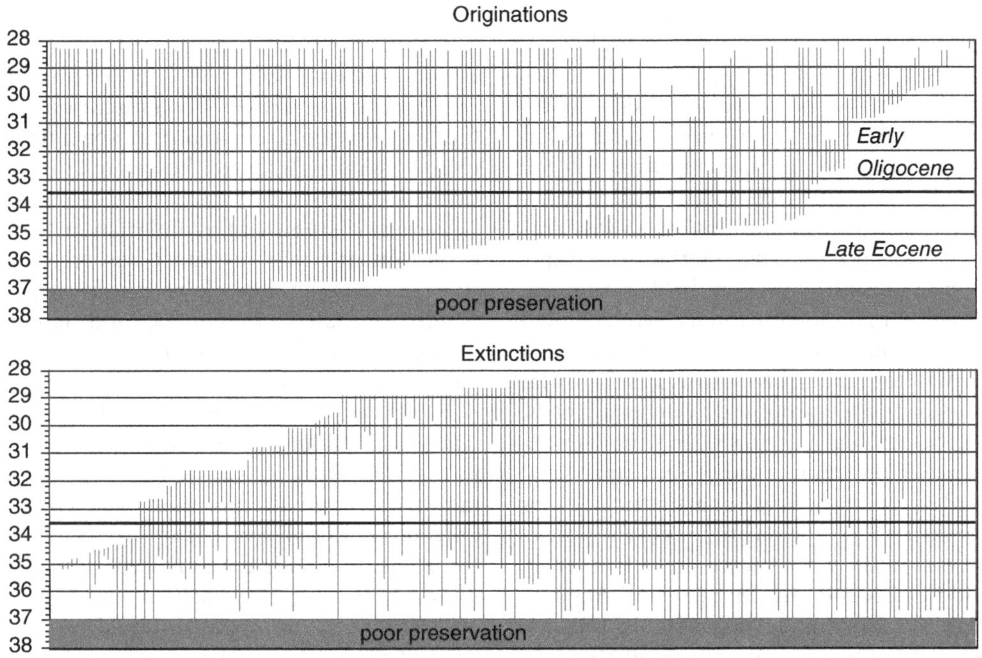

Figure 9.25 Ranges of late Eocene and early Oligocene radiolarian species in the Southern Ocean, sorted by origination time (top) and extinction time (bottom). Ranges are based on composite of data from several sections in different sectors of the Southern Ocean (Apel, 2005), ages are referenced to the Berggren et al. (1995) time scale. Outside of edge effects and poor radiolarian preservation at the base of the section, extinctions appear to have occurred at a fairly uniform rate throughout, while originations show a strong pulse at ca. 35 Ma. Source: Lazarus et al. (2008). Reproduced with permission of micropress.org.

interval, but originations were strongly concentrated near 35 Ma (Figure 9.25). Overall diversity increased moderately from the late Eocene to the early Oligocene and declined slightly into the late Oligocene. Taken together, these results confirm the high rates of late Eocene–early Oligocene turnover seen in the global NSB analysis, and show that the restructuring of faunas was global in scope, affecting both low and high latitudes. The overall decrease in global diversity across the Eocene–Oligocene boundary, despite the development of the largely endemic polar fauna, reflects the dramatic decrease in tropical endemic diversity. This is in accord with the observations by Funakawa et al. (2006) of more Oligocene cosmopolitan vs. Eocene endemic species in the tropics. The response of faunas in both low and high latitudes to environmental changes is complex – the origination pulse in the new Southern Ocean for example precedes a much more drawn-out extinction phase by several million years, a pattern opposite to that usually seen in mass extinction events. The mechanisms controlling macroevolutionary turnover of radiolarians in this time interval are not yet clear, but they are difficult to explain by either a purely biotic or purely abiotic model such as the Red Queen or Court Jester. The remainder of the Oligocene is, in the global data, an interval of relatively static diversity, and no detailed studies of radiolarian diversity dynamics are available.

The effect of changing oceanic biogeography on evolutionary mechanisms within single clades of related species has been examined in the evolution of the largely Paleogene radiolarian group lophocyrtids (Sanfilippo and Caulet, 1998). This study thus provides a (so far) rare example of how fossil data can bridge the gap between macroevolutionary whole fauna diversity analyses and microevolutionary studies of individual lineages (similar ideas have been proposed by other authors, e.g. Casey et al., 1983, but not documented). In their study, Sanfilippo and Caulet defined four anagenetically (or sympatrically speciating) evolving lineages, three of which were present in early Middle Eocene, in the tropical regions (Figure 9.26). In the late Middle Eocene, most of the lineage ranges spread also into the high southern latitudes. Near the Middle–Late Eocene boundary, as the Southern Ocean formed, most of the high-latitude populations of the cosmopolitan species were extirpated, and a new, Southern Ocean endemic lineage appeared. In addition, one of the tropical lineages became restricted to the Southern Ocean and evolved anagenetically into new species. Diversity declined over the late Oligocene, with loss of the remaining tropical forms, and the last Southern Ocean member of the lineage became extinct in the early Oligocene. The group thus shifted from a purely tropical clade of moderate diversity in the mid-Eocene, to a more diverse clade with both low- and high-latitude lineages in the late Eocene and early Oligocene, as geographic differentiation drove diversification of faunas. Anagenetic/sympatric speciation change within lineages however continued throughout.

Neogene studies of radiolarian macroevolution are rare, as most research on these assemblages is done for biostratigraphy or paleoceanography, and faunal change is largely evaluated as a form of paleoceanographic indicator. A few studies, however, explicitly examine evolutionary change, including diversity and turnover rates. Kamikuri et al. (2005) examined radiolarian faunal evolution across the Oligocene–Miocene boundary in the equatorial Pacific. They documented a modest increase in diversity (origin of 7 species – ca. 20%) during an interval broadly coinciding with a major glacial event (Mi1, at the Oligocene–Miocene boundary) which marks the end of a long interval of relative climate stability during the late Oligocene. Evidence for rapid (relative to the Oligocene) shifting loci of global marine opal deposition in the early Miocene was also presented, suggesting rapidly varying global patterns in abundance vs. biogeography in siliceous plankton, which presumably drove evolutionary change in the faunas. Generally, the global pattern in the early Miocene is one of major diversification in radiolaria, with originations dramatically outpacing extinctions in this interval, driven by polar regions (Figure 9.21A).

Figure 9.26 Diagrammatic summary of geographic and stratigraphic distribution of individual species in the four main subgeneric lineages of the radiolarian genus *Lophocyrtis*. Named species within each of the four lineages are morphotypic and in many cases probably do not represent distinct biologic species when two morphotypes overlap in time. Between lineage differences are assumed to represent distinct biologic species. For each named species the range of the morphotype in both low (dashed line) and high (solid line) latitude regions of the ocean are given. Question marks indicate uncertain range ends. Source: Sanfilippo and Caulet (1998). Reproduced with permission of micropress.org.

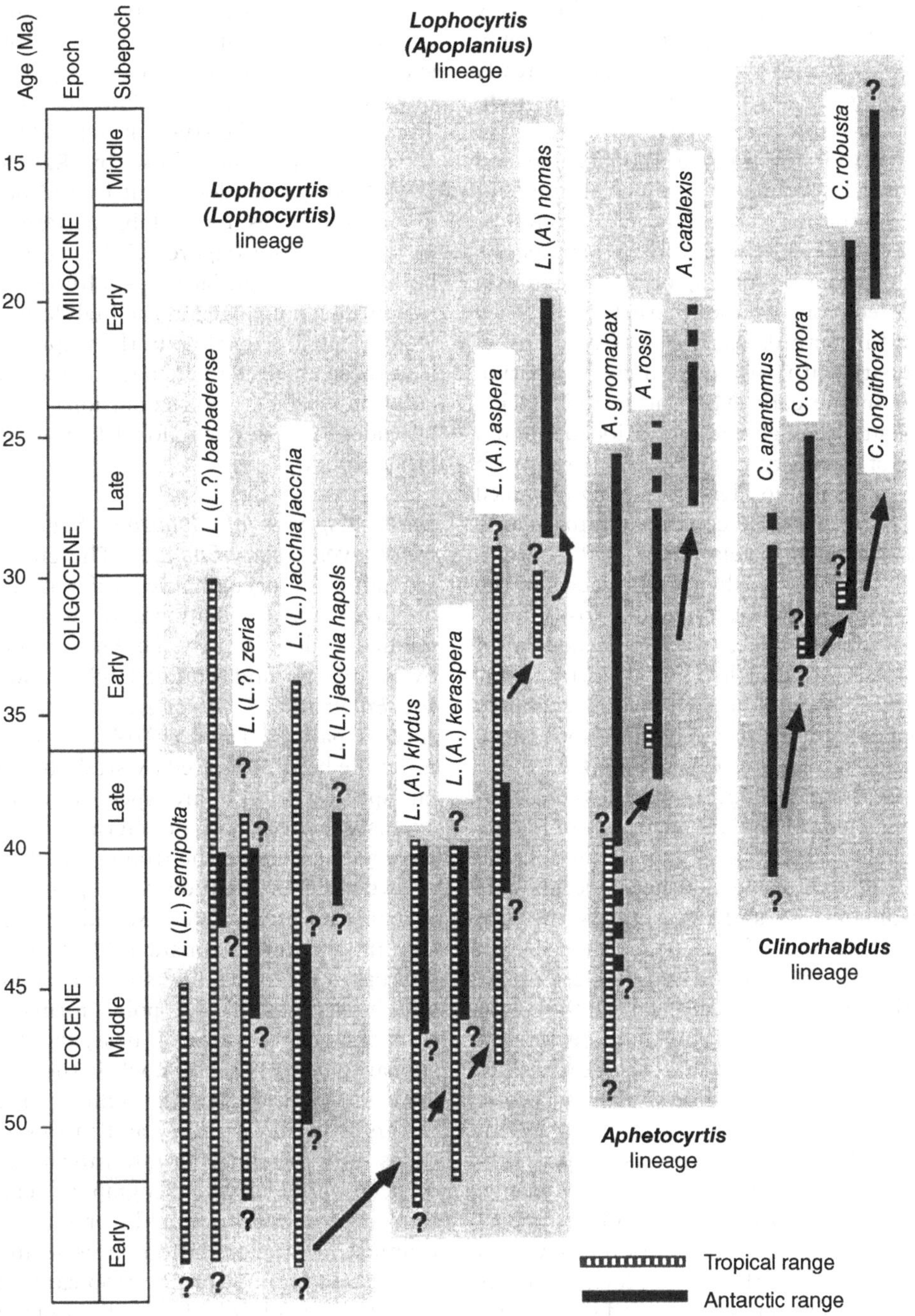

Age (Ma)
Epoch
Subepoch
15
20
25
30
35
40
45
50
MIIOCENE
OLIGOCENE
EOCENE
Middle
Early
Late
Early
Late
Middle
Early
Lophocyrtis (Lophocyrtis) lineage
Lophocyrtis (Apoplanius) lineage
Aphetocyrtis lineage
Clinorhabdus lineage
L. (L.) semipolta
L. (L.?) barbadense
L. (L.?) zeria
L. (L.) jacchia jacchia
L. (L.) jacchia hapsis
L. (A.) klydus
L. (A.) keraspera
L. (A.) aspera
L. (A.) nomas
A. gnomabax
A. rossi
A. catalexis
C. anantomus
C. ocymora
C. robusta
C. longithorax
Tropical range
Antarctic range

The evolution of the Neogene faunas of the Southern Ocean were studied by Renaudie and Lazarus (2013). This study not only used subsampling of the NSB database, but also collected taxonomically complete data (as determined from individual sample specimens vs. cumulative diversity curves) from ca. 100 large (several thousand specimens) samples throughout the Southern Ocean. The more complete data of this study found many rare species not previously known, or not previously reported from the Southern Ocean. They found in both datasets a near doubling of diversity with sustained high origination rates in the early Miocene, largely stable diversity through the late Miocene, and a loss of diversity and elevated extinction rates in the Plio-Pleistocene that reduced diversity in the Recent to approximately early Miocene levels (Figure 9.27). Maximum diversity reached >300 species in the early late Miocene. Diversity loss in the last few million years was preceded by an early late Miocene (ca. 9 Ma) shift in overall assemblage composition, as evenness dropped with the rise in relative abundance of a group of species in a single genus: *Antarctissa*. The shift in evenness was tentatively correlated to changing nutrients and phytoplankton composition in the early late Miocene Southern Ocean, due presumably to shifting sources of deep water and loci of opal deposition which occurred at this time (Billups, 2002; Cortese et al., 2004). Kamikuri et al. (2009) documented a co-eval, although more modest radiolarian turnover in the early late Miocene of the equatorial Pacific, while Kamikuri et al. (2007) documented a gradual, modest (ca. 10%) decrease in diversity in the North Pacific over the late Miocene, with little change in the Plio-Pleistocene. The cause of the major Southern Ocean extinction of the Plio-Pleistocene is unknown, but might be related to the Plio-Pleistocene spread of sea ice and loss of primary productivity south of the Polar Front. Renaudie and Lazarus (2012) speculated that earlier increasing dominance of *Antarctissa* in the late Miocene reduced the relative population size of many other species, making them more susceptible to later extinction in the Plio-Pleistocene. Taken together, the study of Renaudie and Lazarus (2013), and those by Kamikuri et al. (2005, 2009) document that Miocene diversity changes in the global pattern seen in the NSB data (Figure 9.21) occurred in both low and high latitudes, although the magnitude of changes in the high latitudes were much greater than in the tropics, and link these changes to changes in ocean circulation, surface water conditions in high latitudes, and shifting global loci of silica productivity.

Other Aspects of Cenozoic Radiolarian Macroevolutionary Change The effect of latitudinal biogeographic differentiation on diversity, as documented above, and studied in detail for the *Lophocyrtis* clade by Sanfilippo and Caulet (1998), and for the *Eucyrtidium calvertense-matuyamai* species pair by Hays (1970) can also in principle be illuminated by studying bipolar species, which are species reported exclusively from both northern and southern regions outside the tropics (temperate or even higher latitude polar provinces). Stepanjants et al. (2006; see also discussion in Chapter 3), consider 30–40 species (out of a total living diversity of ca. 600, of which ca 100-200 are polar forms) are apparently bipolar, and of these, about half appear to be truly bipolar, while the others are cosmopolitan, being present in surface waters in high latitudes, and in low latitudes only as relatively rare intermediate or deep water forms. Geographic separation of species into weakly or non-communicating subpopulations thus seems to be not uncommon for non-tropical forms. The effect this has had on radiolarian evolution is likely to have been substantial but however is largely unknown, as there

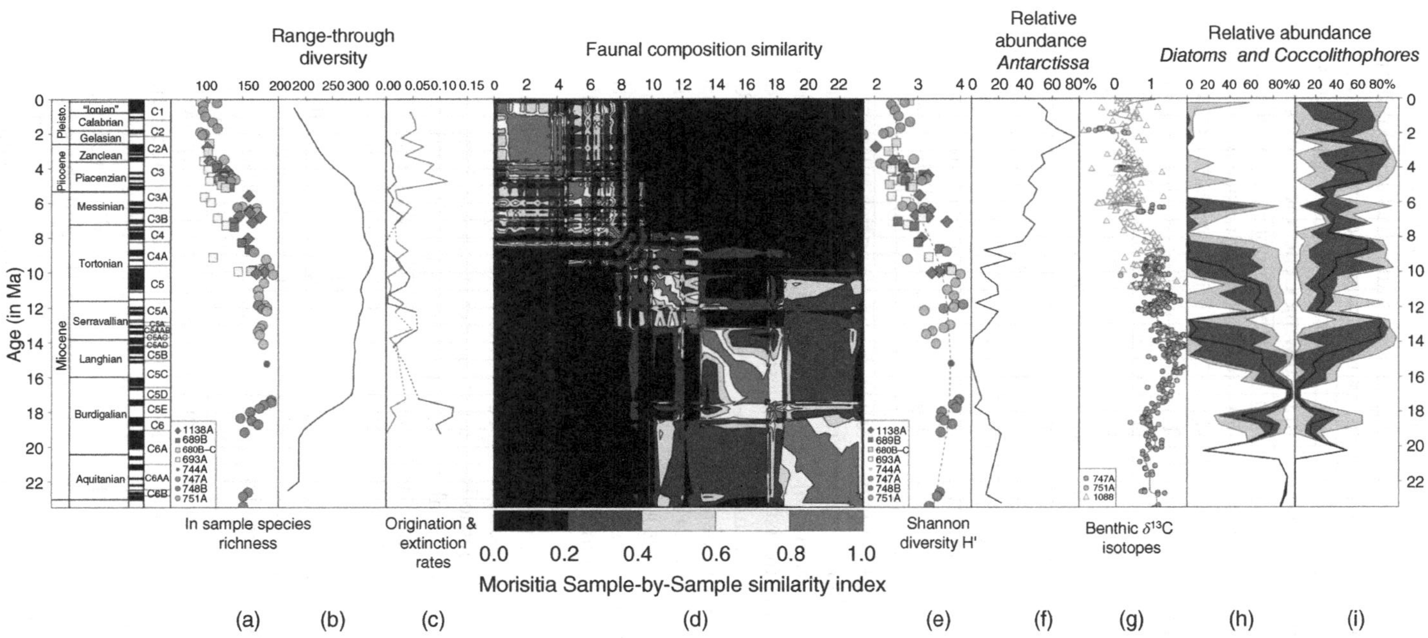

Figure 9.27 Diversity and evolutionary turnover in the Neogene Southern Ocean radiolarian fauna, based on exhaustive counts of diversity in ca. 100 samples in the Atlantic and Indian Ocean sectors of the Antarctic (total of >500k individual specimens). Age scale on left. A) in-sample diversity (symbols show sections), B) total range-through diversity, C) rates of origination (blue) and extinction (red), D) cross-plot of faunal similarity, E) Shannon-Wiener diversity index H , F) relative abundance of species of the genus Antarctissa, and compilations of environmental parameters [G) benthic 13C, composition of phytoplankton in bulk sediment: H) percent coccolithophores, G) percent diatoms]. Similarity cross-plot between samples (similar sample pairings: bright colors; dissimilar pairs, dark color) shows major break at ca. 9 Ma, corresponding in time to change in Antarctissa abundance and change in environmental parameters. Diversity rises by ca. 50% in the early Miocene and, accompanied by high extinction rates, declines again in the Plio-Pleistocene. Source: Composite of figures from Renaudie and Lazarus (2013). (*See color plate section for the color representation of this figure.*

are no detailed studies of how bipolarity within single radiolarian species translates into speciation and diversification between species. Studies of morphologically cryptic species in a closely related group, planktonic foraminifera, have shown both that true, genetically bipolar species exist (Darling et al. 2000), and that bipolar distributions can lead to between-pole differentiation and evolution (Darling et al. 2004).

It is also worth noting the very limited effect of emergence of geographic barriers on diversity or the distribution of individual species in the Cenozoic between oceans in the tropics, particularly the closure of the Tethys in the mid Cenozoic and the Pacific–Atlantic connection in the late Neogene (Kennett, 1982). Despite the existence of these barriers, tropical faunas remain very similar to each other in each ocean (Boltovskoy et al., 2010; Chapter 5). The minimal impact of geographic isolation on tropical faunas is also known from other plankton groups such as the planktonic foraminifera, where species pairs are known but generally rare (Norris, 2000). Although there is no suitable data for radiolarians at the finest level of (cryptic) molecular species, in better studied groups such as the foraminifera evidence even in these taxa for restriction to single-ocean basins is limited and dependent on the methods used to calculate similarity (André et al., 2014). The reasons for this are unknown and could include the relatively minimal amount of change in environments in these tropical oceans since the emergence of the barriers. In addition, there may be, however rare, largely zonal (parallel to latitude) transport of viable organisms across land barriers by winds. Long-distance transport of microbes, including particles tens of micrometers in size, is known to occur over distances of thousands of km both from continental regions to the oceans (Gorbushina et al., 2007), and from ocean waters to land (Sassen et al., 2003). Such events could, if also capable of radiolarian transport, re-seed individual ocean basins with species and thus act to maintain similarity in taxonomic composition.

One theme that has appeared repeatedly in reviewing Cenozoic studies of macroevolution is the importance of changing nutrients, due to changing ocean circulation, as a significant driver of radiolarian evolution. The history of ocean fertility is in general poorly known, even in the Cenozoic, but at least in this time period the major patterns of change have been determined by decades of ocean drilling. This work has also revealed that the Cenozoic oceans underwent a major change in ocean silica chemistry, due to the rise of planktonic diatoms to a major component of the ocean plankton (Spencer-Cervato 1999; Renaudie, 2016). Silica is an essential nutrient needed for radiolarian shell growth, and in an important early paper, Harper and Knoll (1975) suggested that radiolarians and diatoms had co-evolved over the Cenozoic. Radiolarians, they hypothesized, evolved to use less silica as diatoms diversified and increasingly reduced average ocean concentrations of dissolved nutrient silica – via their ability to capture dissolved silica rapidly into their shells, and sequester the silica in the deep sea and in biogenic silica-rich diatomaceous sediments. This hypothesis was corroborated in later studies by Lazarus et al. (2009) and Renaudie (2016), which used more detailed data on radiolarian silica use and diatom diversification. In Lazarus et al. (2009), radiolarian silica use – per unit volume – was shown to have decreased substantially over the Cenozoic, but only in those ocean regions (low- to mid-latitude surface waters) where dissolved silica concentrations in the modern ocean are very low (Figure 9.28). Low silica in these regions is due to the combined action of diatoms that convert the dissolved surface water silica into shells, and the sequestration of silica via dissolution of sinking diatom shells in deep waters, which in low lati-

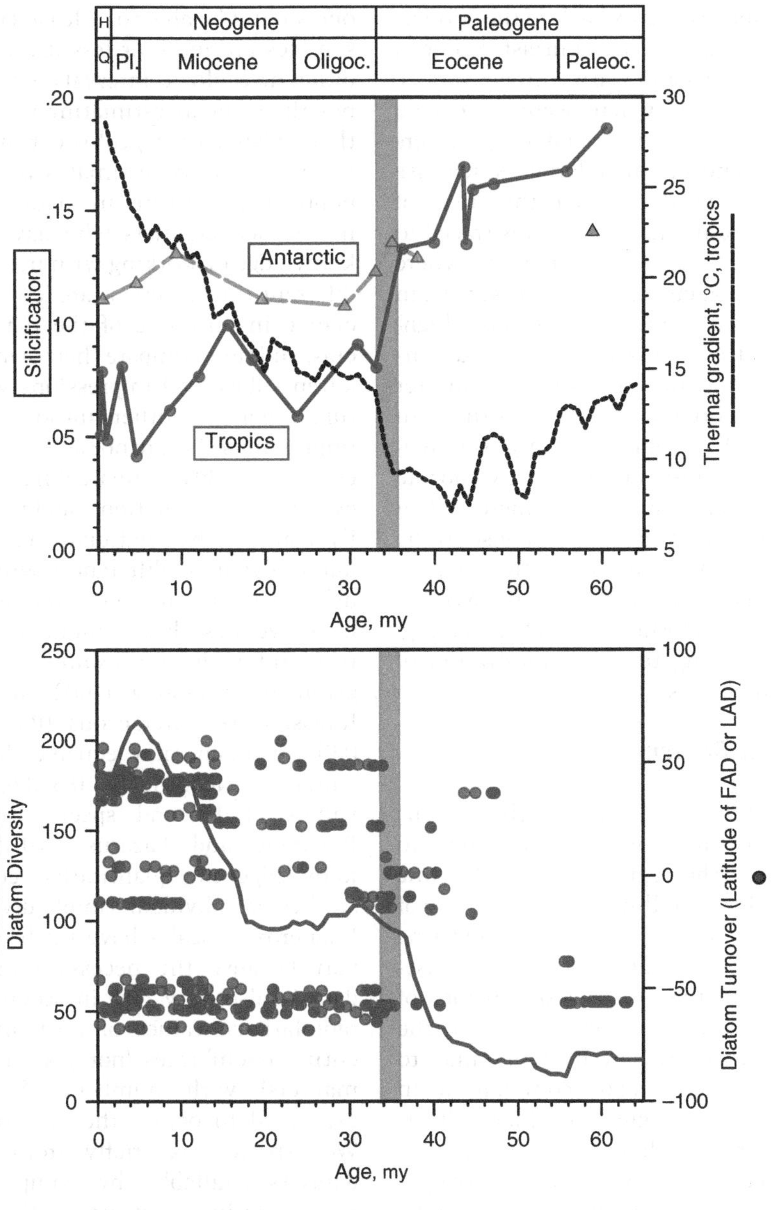

Figure 9.28 a) Change in radiolarian silica use over the Cenozoic in tropical and polar (Antarctic) regions, measured as "silification": fraction of opal shell volume vs. total volume of outermost shell. Based on morphometric measurements of several thousand shells from ca. 30 samples. Silica use declines substantially in the tropics (solid line and circles) at the Eocene–Oligocene boundary (grey vertical band), at the same time that low-latitude water column stratification increases (vertical temperature gradient, black dashed line). Silica use does not change significantly in the polar regions (long-dashed line and diamonds). b) Cenozoic diatom diversity (solid line, from Spencer-Cervato 1999); see Figure 9.29 for updated curve and time/latitude of individual diatom origination/extinction events (dots, from Cervato and Burckle, 2003). Berggren et al. (1995) age scale. Source: Lazarus et al. (2009). Licensed under CC BY 4.0.

tudes cannot easily mix back to the surface, given the high-density contrast between cold deep waters and warm surface waters. In high latitudes, where waters are cold throughout and surface and deep waters mix, no Cenozoic radiolarian silica use decrease was noted. Further, the origin of low silica use in radiolarians was traced to the Eocene–Oligocene boundary, where polar glaciation created, for the first time in the Cenozoic, cold deep waters, high-density contrasts to warm low-latitude surface waters, and thus the ability to sequester nutrients exported to deep waters there. In Renaudie (2016, Figure 9.29) a global synthesis of opal preservation in the Cenozoic oceans showed that diatoms increased in abundance and diversity in response to the changing ocean conditions at the Eocene–Oligocene boundary, as well as the M1 glaciation and mid-Miocene global cooling, thus contributing to the sequestration of silica from ocean surface waters.

Phanerozoic Diversity – A More Modest View

Before leaving this section, with its focus on diversity, it is worth considering to what degree the regional, or single time period studies can illuminate the history of Phanerozoic diversification, as existing Phanerozoic compilations are not yet useful for all but the broadest suggestions of diversity change over time. Some of the regional studies reviewed are similar to those used in Phanerozoic compilations in providing only a single, unevaluable diversity number, but others give values for the number of occurrences per time interval, or the number of individuals per sample. Diversity in individual samples for example can provide an alternate way to compare diversity, and such comparisons are often made, e.g. to infer that Mesozoic radiolarian diversity was substantially higher than living (e.g. Kiessling 2002). Assuming that a similar number of individuals are compared, and that the evenness is similar, one should be able to at least say if typical samples are more or less diverse between different geologic intervals, even if this is not the same as estimating total diversity that existed in a geologic time interval. Such comparative estimates as made in the published literature are frequently asymmetric, as scientists who have first-hand knowledge of working with materials from different time intervals are few – most specialize in only one of the main geologic eras, and thus compare their own first-hand (often subjective) impressions with literature reports of other authors. Subjective impressions also are not able, in our experience, to reliably estimate differences in the evenness of populations, and thus can easily confuse observed diversity differences due to evenness differences with diversity differences due to true total species richness. Evenness does seem to differ between time intervals: for example, the values given by Kiessling (2002) for the end Jurassic/early Cretaceous (Pielou's J, ca. 0.9) are much more equable than for the Antarctic Neogene (J= 0.5–0.6, based on values of H′ and species richness in Renaudie and Lazarus, 2013). To our knowledge no quantitative comparative studies of individual sample diversity over Phanerozoic scales have yet been done, in part because the necessary quantitative data is also very rare in studies of fossil radiolarians: values of taxa diversity for entire assemblages (not just stratigraphic markers) with numbers of specimens examined to obtain the diversity values. We explore this briefly, from those few datasets available, by comparing three representative datasets: the Jurassic (Kiessling, 2002, using his "lumped" taxonomy, which he suggests is best comparable to Cenozoic taxonomy), the early Cenozoic (Hollis 2006) and late Cenozoic (Kamikuri et al., 2009; Figure 9.30). Within-sample diversity appears to be significantly lower in the early Cenozoic compared to the late Cenozoic, in accord with both the

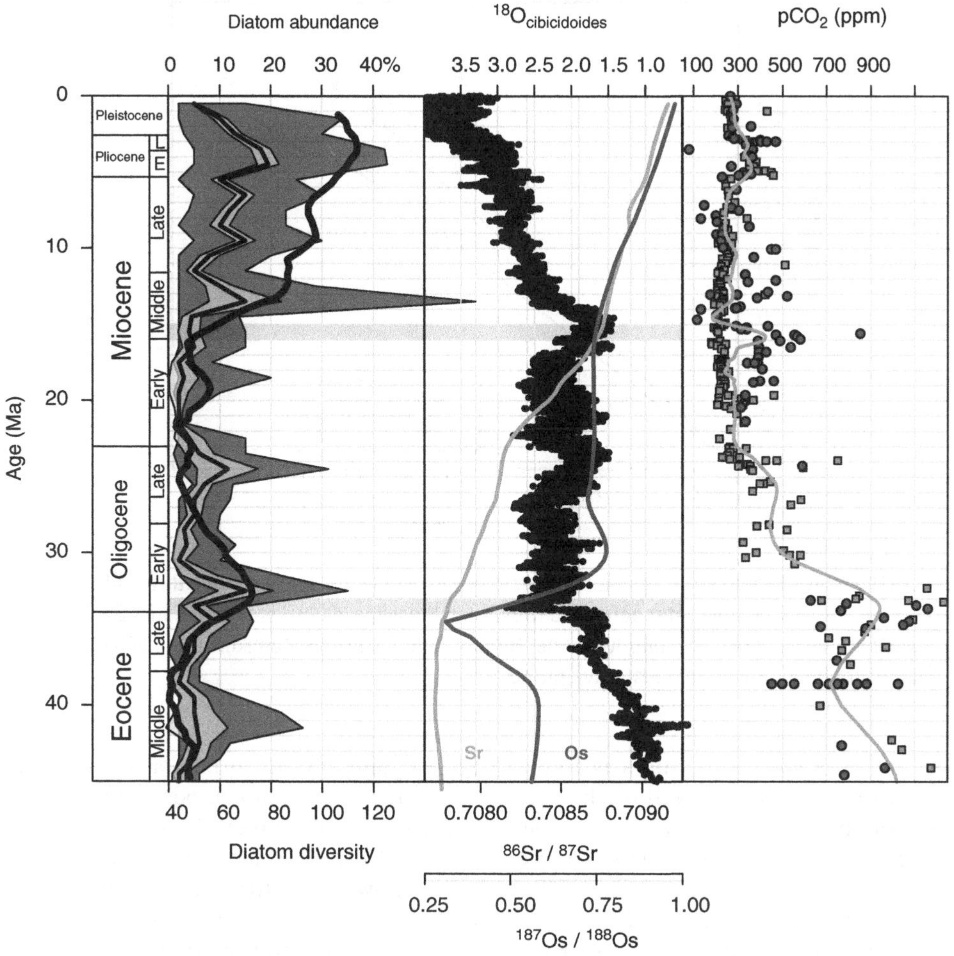

Figure 9.29 Cenozoic ocean silica export (diatom abundance in global deep-sea sediments, diversity); global weathering (Sr and Os), and climate ($\partial^{18}O$), and atmospheric carbon dioxide concentrations (pCO_2). Left: abundance median – black line, light shading 95% median confidence interval, dark shading interquartile range, blue line – diversity; Middle: black dots – oxygen isotopes, blue line – Os isotopes, green – Sr isotopes; Right: pCO_2 estimates and smoothed curve in green. Horizontal blue shaded intervals show correlated peaks and/or changes in rates of change between parameters. Diatom abundance and diversity, both measures of silica availability in upper ocean waters, increased near the Eocene–Oligocene boundary, and in the middle Miocene. Radiolarian response to the inferred change in silica availability shown in Figure 9.28. Source: Figure and synthesis of diatom abundances from Renaudie (2016), other data sources as given in that publication. Licensed under CC BY 3.0.) (*See color plate section for the color representation of this figure.*

global and regional analyses reviewed above. Interestingly, the Jurassic does not, in this example, appear to be any more diverse than the late Cenozoic, despite frequent claims to the contrary. Given the higher evenness of the Jurassic data (see

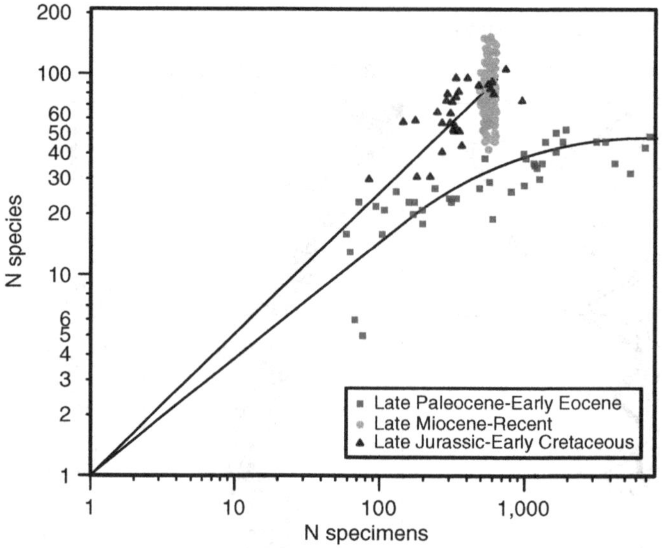

Figure 9.30 Diversity of samples vs. number of specimens in sample for sections from three different time intervals: Late Jurassic–Early Cretaceous (mid Mesozoic, Kiessling 2002); Late Paleocene–Early Eocene (early Cenozoic, Hollis 2006); and late Miocene–Recent (late Cenozoic, Kamikuri et al. 2009). Note log axes. Curves are hand drawn, for illustrative purposes.

above), the true species richness is likely an overestimate, so it may well be that the Jurassic samples are actually less diverse than the late Cenozoic ones. This simple example serves to underline the need for better data to determine the history of species diversity in pre-Cenozoic time intervals.

Summary Discussion

Studies of microevolution using radiolarians, and other microfossils have been quite successful and have been able to test various models of how e.g. species arise. They have established that for these organisms microevolutionary change is largely created by gradual anagenetic change in species, and parapatric cladogenesis along environmental gradients. The ability to achieve these insights has depended on having on the one hand, an excellent fossil record, and on the other, a set of well-defined evolutionary models that make explicit, testable predictions for the fossil record. Despite these achievements, we still know very little about either the details or the underlying mechanisms of selection or adaptation that have created the documented patterns. This is most likely due to our still extremely primitive knowledge of the functional biology and ecology of radiolarians, either as a whole, or for individual species. Nor do we know enough yet about how patterns of evolution at the level of morphospecies are correlated to subscale processes of evolution between (possible) cryptic genetic species.

Generally, studies to date of radiolarian macroevolution, while broadly addressing various factors that theoretical ecology has identified as controls of macroevolution, e.g. environmental change, competition and co-evolution – do not explicitly frame their analyses as tests of formal ecologic theories. It is nonetheless clear that both abiotic and biotic factors must play important roles in regulating global radiolarian diversity over geologic time. This can be inferred from combining observations of the importance of gradients in the physical environment as drivers of speciation, the increase in geographic and water depth physical environmental gradients in Cenozoic oceans due to global cooling, and the complex diversity response signal of radiolarians. Overall, the expectation of higher diversity in a world with stronger environmental gradients is seen – the global diversity of radiolarians increased dramatically over the Cenozoic. However, this was largely due to the development of new polar endemic faunas. In the low latitudes the initial response of radiolarians to stronger vertical and longitudinal environmental gradients beginning in the late Eocene to early Oligocene was in fact the opposite of what a purely abiotic model would have expected. Instead of an increase in diversity, what is seen is a dramatic drop in diversity. Possibly the reduction in the availability of nutrients from water stratification and, via oligotrophy, a reduction in the production of food which occurred at this time, not only affected radiolarian shell morphology, but also negatively affected tropical radiolarian diversity, implicating co-evolutionary interactions in low latitudes with diatoms and other primary producers as the cause of radiolarian diversity decrease. This hypothesis has not, however, been confirmed by any studies.

Much work is still needed to better link ecologic theories of diversity to the types of data actually available from the fossil record of radiolarians (and other groups). Some work of this sort has already been attempted for other groups of microfossils (Ezard et al., 2012, 2013; Ezard and Purvis, 2016) and needs to be extended to radiolarians. A challenge will be obtaining robust data on the biologic characteristics of radiolarians, particularly extinct forms, that are needed to develop and test such models. Similarly, better understanding of radiolarian biology, behavior, and molecular biology are essential if macroevolutionary study of intrinsic factors is to progress.

Mass extinctions suggest that radiolaria are relatively robust to proximal extinction forces, but that such events nonetheless have a major impact. This is likely due to subsequent change in biologic environment and in turn suggests that biotic interactions also play a major role shaping radiolarian evolution. Lastly, the studies to date suggest that a strict separation between biotic and abiotic models is at least sometimes artificial, as biotic interactions can be mediated by changes in the physical environment, as shown by integrated radiolarian-diatom-dissolved ocean water silica studies.

Most useful data for evolutionary studies so far are from the Cenozoic, where data is more extensive, records better preserved, detailed, temporally correlated paleoenvironmental data available, and at least some biologic data can be inferred from living representatives. Pre-Cenozoic studies so far mostly have only confirmed mechanisms known from studies of the Cenozoic, although occasionally evolutionary factors and behaviors not seen in the Cenozoic are uniquely documented in older time intervals. Among these are water column anoxia as an evolutionary factor which has come from Mesozoic research; and the origin of radiolarians in latest pre-Cambrian/Cambrian oceans which were both physically and biologically not analogous to Cenozoic or Recent oceans.

The integration of microevolutionary insights with macroevolutionary patterns and processes leaves much to be desired, although this problem is common to evolutionary biology as a whole. Some general observations are however possible. Much evolutionary change occurs in-situ within biogeographic provinces as radiolarians adapt to physical environmental changes in the water column. This is seen both from detailed studies of lineage evolution and relation to ocean conditions, and the simple fact that anagenetic evolution within a single biogeographic province is very common in radiolarians. As is true for macroevolutionary patterns, it is largely unknown whether the role of physical environmental change is direct or is instead usually indirect, being mediated by biologic interactions. Clearly biogeographic change is also very important. This is seen both at the level of individual speciation of related forms (low–high latitude differentiation of clades, complementary bipolar species pairs), and in the dominant influence of high-latitude biogeographic development in the regulation of Cenozoic diversity. It should be noted, however, that geographic separation is, by itself, apparently not a major force for evolutionary change. This is inferred from the fact that, despite major physical barriers to faunal exchange between ocean basins (particularly the low-latitude Atlantic and Pacific today), few distinct species pairs are known, either in the modern plankton or in the fossil record. Only when geographic differences also include environmental differences (such as occur along latitudinal or longitudinal gradients) does evolution act to create new forms and modify diversity.

Lastly, and beyond comments specific to studies of evolution, we would like to briefly suggest, based on the current state of knowledge as documented in this book, how the field of radiolarian research is likely to develop in the next decade(s). As already stated in the introduction, radiolarian research stands at the beginning of a major transformation. Traditional methods of individual specimen and assemblages study, and traditional 'service' oriented themes within the earth sciences such as biostratigraphy are rapidly being enriched by the opportunities provided by the spectacular growth of new technologies: molecular biology; AI driven automatic identification; and large, networked databases of biologic and earth science data. Although these developments will to some degree affect most themes in radiolarian research, their impact is likely to be much greater in more biologically oriented subjects, and thus these areas are the ones likely to see the greatest growth in the next decades. Molecular biology is for the first time providing an objective, data-based framework on which to build a higher level systematics for radiolarians, and as well, the ability to robustly and broadly survey the distribution and abundance of radiolarians in ocean waters. This new knowledge will create many new opportunities for biologic and paleobiologic studies of radiolarians as has been the case for other, better studied groups of fossilizing plankton where such data is already available, such as planktonic foraminifera. Molecular data has also now revealed the importance of radiolarians in plankton ecosystems, which we expect will lead to more intensive work on the living forms, and in particular their role in ocean ecology and element cycling. The effect of automatic species identification systems promises to be equally important. Despite the almost unimaginable richness of the radiolarian fossil record, particularly in deep sea sediments, the actual amount of observational data gathered on fossil radiolarians has been little different than those for many other fossil groups that are far more depauperate in fossilized material (e.g. in vertebrate and invertebrate paleontology). The main limiting factor of course has been manpower - the requirement for each specimen, be it a radiolarian or a mollusc, to be identified by a human expert. A single fully automated AI identification system should, in the not too far future, be able to generate millions of new observations of species occurrences, together with rich, quantitative morphologic data on each of them, in a single year. It will thus soon be possible to study both individual radiolarian species and (assuming the taxonomy has been done) entire assemblages at unprecedented temporal resolution and geographic breadth, and over time scales ranging from hundreds, to many millions of years. These data should yield major new insights into how ecologic and evolutionary processes interact and scale, from decades, to millennia, and into to the vast intervals of geologic time. Studies of other microfossil groups will also benefit of course, but radiolarian studies will be major beneficiaries, due to radiolarians' diversity and broad range of ecologic behaviors. Finally, the flood of new data from both molecular and fossil studies will feed the further development of database systems and 'big data' analyses, often in conjunction with data from oceanography and paleoceanography, supporting the integration of radiolarian biologic data into larger scale, earth system science studies. The future of radiolarian research thus looks to us to be very bright: the next 20 years may be the most exciting of all nearly two hundred years over which science has been studying these fascinating forms.

References

Abelmann, A., 1992. Early to mid-Miocene radiolarian stratigraphy of the Kerguelen Plateau (ODP Leg 120), in: Wise, S.W., Schlich, R. (Eds.), *Proceedings of the Ocean Drilling Program, Scientific Results*. Ocean Drilling Program, College Station, pp. 757–784.

Abelmann, A., 1988. Freeze-drying simplifies the preparation of microfossils. *Micropaleontology* 34, 361.

Abelmann, A., Brathauer, U., Gersonde, R., et al., 1999. Radiolarian-based transfer function for the estimation of sea surface temperatures in the Southern Ocean (Atlantic sector). *Paleoceanography* 14, 410–421.

Abelmann, A., Gersonde, R., 1991. Biosiliceous particle flux in the Southern Ocean. *Marine Chemistry* 35, 503–538.

Aberdeen, E., 1940. Radiolarian fauna of the Caballos formation, Marathon Basin, Texas. *Journal of Paleontology* 14, 127–139.

Adl, S.M., Simpson, A.G.B., Lane, C.E., et al., 2012. The revised classification of eukaryotes. *Journal of Eukaryotic Microbiology* 59, 429–514.

Afanasieva, M.S., Amon, E.O., Agarkov, Y.V., Boltovskoy, D., 2005. Radiolarians in the Geological Record. *Paleontological Journal* 39, 135–392.

Agarkov, Y.V., Agarkov, A.Y., Gorbunova, T.L., 2019. Analysis of species richness dynamic of Mesozoic and Cenozoic radiolarians. IOP Conf. Series: Earth and Environmental Science 272, 1–7. https://doi.org/10.1088/1755-1315/272/2/022243

Agnini, C., Monechi, S., Raffi, I., 2017. Calcareous nannofossil biostratigraphy: historical background and application in Cenozoic chronostratigraphy. *Lethaia* 50, 447–463. https://doi.org/10.1111/let.12218

Agterberg, F., Gradstein, F., 1999. The RASC method for ranking and scaling of biostratigraphic events. *Earth-Science Reviews* 46, 1–25.

Agterberg, F.P., Gradstein, F.M., 1988. Recent developments in quantitative stratigraphy. *Earth-Science Reviews* 25, 1–73. https://doi.org/10.1016/0012-8252(88)90099-2

Aguado, R., O'Dogherty, L., Sandoval, J., 2008. Fertility changes in surface waters during the Aalenian (mid-Jurassic) of the Western Tethys as revealed by calcareous nannofossils and carbon-cycle perturbations. *Marine Micropalaeontology* 68, 268–285.

Aita, Y., 1987. Middle Jurassic to Lower Cretaceous radiolarian biostratigraphy of Shikoku with reference to selected

Paleobiology of the Polycystine Radiolaria, First Edition. David Lazarus, Noritoshi Suzuki, Yoshiyuki Ishitani and Kozo Takahashi.

sections in Lombardy Basin and Sicily. *Science Reports of the Tohoku University*, Series 2: Geology 58, 1–91.

Aita, Y., Bragin, N.Y., 1999. Non-Tethyan Triassic radiolaria from New Zealand and northeastern Siberia. *Geodiversitas* 21, 503–526.

Aitchison, J.C., 1998. A Lower Ordovician (Arenig) Radiolarian Fauna from the Ballantrae Complex, Scotland. *Scottish Journal of Geology* 34, 73–81.

Aitchison, J.C., 1993. Albaillellaria from New England Orogen, Eastern NSW. *Marine Micropalaeontology* 21, 353–367.

Aitchison, J.C., Davis, A.M., Stratford, J.M.C., Spiller, F.C.P., 1999. Lower and Middle Devonian radiolarian biozonation of the Gamilaroi terrane New England Orogen, eastern Australia. *Micropaleontology* 45, 138–162.

Aitchison, J.C., Suzuki, N., Caridroit, M., et al., 2017a. Paleozoic radiolarian biostratigraphy. *Geodiversitas* 39, 503–531.

Aitchison, J.C., Suzuki, N., O'Dogherty, L., 2017b. Inventory of Paleozoic radiolarian species. *Geodiversitas* 39, 533–637.

Aliev, Kh.Sh., 1967. Melovye Otlozheniya Vostochnogo Kavkaza i Prilegayushchikh Oblastei New radiolarian species of the Valanginian and Albian stages of northeastern Azerbaidzhan, in: Aliev, M.M. (Ed.),. *Nauka*, pp. 23–30.

Aliev, Kh.Sh., 1965. Radiolarians of the Lower Cretaceous deposits of northeastern Azerbaidzhan and their stratigraphic significance. Izdatel'stvo Akademii Nauk, Azerbaidzhanskoi SSR, Baku.

Alroy, J., 2010. Fair sampling of taxonomic richness and unbiased estimation of origination and extinction rates, in: Alroy, J., Hunt, G. (Eds.), *Quantitative Methods in Paleobiology*. The Paleontological Society, pp. 55–80.

Alroy, J., Marshall, C.R., Bambach, R.K., et al., 2001. Effects of sampling standardization on estimates of Phanerozoic marine diversification. *Proceedings of the National Academy of Sciences (USA)* 98, 6261–6266.

Amaral-Zettler, L., Sogin, M.L., Caron, D.A., 1997. Phylogenetic relationships between the Acantharea and the Polycystinea: A molecular perspective on Haeckel's Radiolaria. *Proceedings of the National Academy of Sciences USA* 94, 11411–11416.

Amaral-Zettler, L.A., Anderson, O.R., Caron, D.A., 1999. Towards a molecular phylogeny of colonial spumellarian radiolaria. *Marine Micropaleontology* 36, 67–79.

Amaral-Zettler, L.A., Anderson, O.R., Caron, D.A., 1998. Insights on the diversity within a "species" of Thalassicolla (Spumellarida) based on 16S-like ribosomal RNA gene sequencing. *Journal of Eukaryotic Microbiology* 45, 488–496.

Amon, E.O., 2000. Upper Cretaceous radiolarians of Urals region. Rossiyskaya Akademiya Nauk, Ural'skoye Otdeleniye, Institut Geologii i Geokhimii. Yekaterinburg, Russian Federation.

Amon, E.O., Chuvashov, B.I., 2002. Radiolarian biostratigraphy of the Sakmarian Stage (Lower Permian) in southern Urals. *Permophile* 41, 16–26.

Anderson, O.R., 1996. The physiological ecology of planktonic sarcodines with applications to paleoecology: patterns in space and time. *Journal of Eukaryotic Microbiolology* 43, 261–274.

Anderson, O.R., 1994. Cytoplasmic origin and surface deposition of siliceous structures in Sarcodina. *Protoplasma* 181, 61–77.

Anderson, O.R., 1993. The trophic role of planktonic foraminifera and Radiolaria. *Marine Microbial Food Webs* 7, 31–51.

Anderson, O.R., 1983. *Radiolaria*. Springer, New York.

Anderson, O.R., 1981. Radiolarian fine structure and silica deposition, in: Simpson, T.L., Volcani, B.E. (Eds.), *Silicon and Siliceous Structures in Biological Systems*. Springer-Verlag, pp. 347–379.

Anderson, O.R., 1978. Light and electron microscopic observations of feeding behavior, nutrition, and reproduction in laboratory cultures of Thalassicolla nucleata. *Tissue & Cell* 10, 401–412.

Anderson, O.R., 1976. Ultrastructure of a colonial radiolarian Collozoum inerme and a cytochemical determination of the role of its Zooanthellae. *Tissue & Cell* 8, 195–208.

Anderson, O.R., Bennett, P., 1985. A conceptual and quantitative analysis of skeletal morphogenesis in living species of solitary Radiolaria: Euchitonia elegans and Spongaster tetras. *Marine Micropaleontology* 9, 441–454.

Anderson, O.R., Gastrch, M.D., Amaral-Zettler, L.A., 1999. Fine structure of the colonial radiolarian Collozoum serpentinum (Polycystinea: Spumellaria) with a reconsideration of its taxonomic status and re-establishment of the genus Collophidium (Haeckel). *Marine Micropaleontology* 36, 81–89.

Anderson, O.R., Gupta, S.M., 1998. Evidence of binary division in mature central capsules of a collosphaerid colonial radiolarian: implications for shell ontogenetic patterns in modern and fossil species. *Palaeontologia Electronica* 1, 1–13.

Anderson, O.R., Swanberg, N.R., Bennett, P., 1985. Laboratory studies of the ecological significance of host-algal nutritional associations in solitary and colonial Radiolaria. *Journal of the Marine Biological Association of the United Kingdom* 65, 263–272.

Ando, H., Kunitomo, Y., Sarashina, I., et al., 2009. Intraspecific variations in the ITS region of Recent radiolarians. *Earth Evolution Sciences* 3, 37–44.

André, A., Quillévéré, F., Morard, R., et al., 2014. SSU rDNA divergence in planktonic foraminifera: molecular taxonomy and biogeographic implications. PLOS One 9, 1–19 (web).

Andreasen, D.J., Ravelo, A.C., 1997. Tropical Pacific Ocean thermocline depth reconstructions for the Last Glacial Maximum. *Paleoceanography* 12, 395–413.

Anonymous, 2006. Dedication: Catherine Nigrini (1939–2005), in: *Proceedings of the Ocean Drilling Program, Scientific Results, Leg 199*. Ocean Drilling Program, College Station, TX.

Anonymous, 1932. "Bug-Men" lead hunt for black gold: searching for oil with a microscope. *Popular Mechanics* 370–374.

Apel, M., 2005. Taxonomie, Evolution und Diversitätsparameter von obereozänen bis unteroligozänen Radiolarien im Südozean: Einfluss von Paläoozeanographie und Paläoökologie (PhD Thesis). Humboldt University, Berlin.

Apel, M., Kiessling, W., Böhm, F., Lazarus, D., 2002. Radiolarian faunal characteristics in Oligocene sediments of the Kerguelen Plateau, Leg 183, Site 1138, in: Frey, F.A., Coffin, M.F., Wallace, P.J., Quilty, P.G. (Eds.), *Proceedings of the Ocean Drilling Program, Scientific Results*. pp. 1–48.

Appeltans, W. et al. (151 authors), 2012. The Magnitude of Global Marine Species Diversity. Current Biology in press.

Araujo, R., Moreno, D., Ramos, M.A., 1993. The Asiatic clam Corbicula fluminea (Müller, 1774) (Bivalvia: Corbiculidae) in Europe. *American Malacological Bulletin* 10, 39–49.

Arche, A., López-Gómez, J., 2014. The Carnian Pluvial Event in Western Europe: New data from Iberia and correlation with the Western Neotethys and Eastern North America–NW Africa regions. *Earth-Science Reviews* 128, 196–231.

Armentrout, J., 2018. AAPG Top 100 Papers [WWW Document]. URL

http://100years.aapg.org/top-100-papers (accessed 6.26.18).
Arnold, M.L., 2006. *Evolution through Genetic Exchange*. Oxford University Press, New York.
Ashby, J.N., 1983. Improvements in the technique for the extraction and statistical faunal analysis of radiolarians from Neogene mudstones. *Micropaleontology* 29, 146–149.
Aubry, M.-P., 2014. Biostratigraphy, in: Rink, W.J., Thompson, J. (Eds.), *Encyclopedia of Scientific Dating Methods*. Springer Netherlands, Dordrecht, pp. 1–35. https://doi.org/10.1007/978-94-007-6326-5_192-1
Aze, T., Ezard, T.H.G., Purvis, A., et al., 2011. A phylogeny of Cenozoic macroperforate planktonic foraminifera from fossil data. *Biological Reviews* 86, 900–927.
Bachmann, G.H., 2014. Heinz W. Kozur (1942–2013). Hallesches Jahrbuch für Geowissenschaften 141–142.
Bailey, E., McCallien, J.W., 1950. The Ankara Mëlange and the Anatolian Thrust. *Nature* 166, 938–940.
Bąk, K., 2011. Tethyan radiolarians at the Cenomanian-Turonian Anoxic Event from the Apennines (Umbria-Marche) and the Outer Carpathians: palaeoecological and palaeoenvironmental implications. *Studia Geologica Polonica* 134, 5–279.
Bąk, K., 2007. Organic-rich and manganese sedimentation during the Cenomanian-Turonian boundary event in the Outher Carpathian basins; a rew record from the Skole Nappe, Poland. *Palaeogeography, Palaeoclimatology, Palaeoecology* 256, 21–46.
Bąk, K., 2006. Sedimentological, geochemical and microfaunal responses to environmental changes around the Cenomanian-Turonian boundary in the Outer Carpathian Basin; a record from the Subsilesian Nappe, Poland. *Palaeogeography, Palaeoclimatology, Palaeoecology* 237, 335–358.
Bąk, M., 2004. Radiolarian biostratigraphy of the Upper Cenomanian-Lower Turonian deposits in the Subsilesian Nappe (Outer Western Carpathians). *Geologica Carpathica* 55, 239–250.
Bąk, M., 2000. Radiolaria from the Upper Cenomanian-Lower Turonian Deposits of the Silesian Unit (Polish Flysch Carpathians). *Geologica Carpathica* 51, 309–324.
Bąk, M., 1999. Cretaceous radiolarian zonation in the Polish part of the Pieniny Klippen belt (Western Carpathians). *Geologica Carpathica* 50, 21–31.
Baker, C., 1983. Evolution and hybridization in the radiolarian genera Theocorythium and Lamprocyclas. *Paleobiology* 9, 341–354.
Baker, C.L., Bowman, W.F., 1917. Geologic exploration of the southeastern Front Range of Trans-Pecos, Texas. *University of Texas Bulletin* 1753, 93–101.
Balini, M., Ferretti, A., Finney, S., Monechi, S., 2017. The contribution of fossils to chronostratigraphy, 150 years after Albert Oppel. *Lethaia* 50, 323–335. https://doi.org/10.1111/let.12224
Balzano, S., Corre, E., Decelle, J., et al., 2015. Transcriptomic analyses to investigate symbiotic relationship between marine protists. *Frontiers in Microbiology* 6, 1–14.
Bandini, A.N., Baumgartner, P.O., Caron, M., 2006. Turonian Radiolarians from Karnezeika, Argolis Peninsula, Peloponnesus (Greece). *Eclogae geologicae Helvetiae* 99, S1–S20.
Bandini, A.N., Flores, K., Baumgartner, P.O., et al., 2008. Late Cretaceous and Paleogene Radiolaria from the Nicoya Peninsula, Costa Rica: a tectonostratigraphic application. *Stratigraphy* 5, 3–21.
Barnosky, A.D., 2001. Distinguishing the effects of the Red Queen and Court Jester

on Miocene mammal evolution in the northern Rocky Mountains. *Journal of Vertebrate Paleontology* 21, 172–185.

Barron, J.A., Baldauf, J.G., 1995. Cenozoic marine diatom biostratigraphy and applications to paleoclimatology and paleoceanography, in: Blome, C.D., Whalen, P.M., Reed, K.M. (Eds.), *Siliceous Microfossils*. The Paleontological Society, pp. 107–118.

Bartolini, A., 2003. Cretaceous radiolarian biochronology and carbon isotope stratigraphy of ODP Site 1149 (northwestern Pacific, Nadezhda Basin). *Proceedings Ocean Drilling Program, Scientific Results* 185, web.

Bartolini, A., Baumgartner, P.O., Guex, J., 1999. Middle and Late Jurassic radiolarian palaeoecology versus carbon-isotope stratigraphy. *Palaeogeography, Palaeoclimatology, Palaeoecology* 145, 43–60.

Bascom, W., 1961. *A Hole in the Bottom of the Sea*. Weidenfeld & Nicolson, London.

Bass, D., Moreira, D., Lopez-Garcia, P., et al., 2005. Polyubiquitin insertions and phylogeny of Cercozoa and Rhizaria. *Protist* 156, 149–161.

Bassen, D.M., Hou, Y., Bowser, S.S., Banavali, N.K., 2016. Maintenance of electrostatic stabilization in altered tubulin lateral contacts may faciliate formation of helical filaments in foraminifera. *Scientific Reports* 6, 31723.

Bassinot, F.C., 2007. Oxygen Isotope Stratigraphy of the Oceans, in: Elias, S.A. (Ed.), *Encyclopedia of Quaternary Science*. Elsevier, Oxford, pp. 1740–1748.

Baumgartner, P.O., 2013. Mesozoic radiolarites – accumulation as a function of sea surface fertility on Tethyan margins and in ocean basins. *Sedimentology* 60, 292–318.

Baumgartner, P.O., 1993. Early Cretaceous radiolarians of the Northeast Indian Ocean (Leg 123: Sites 765, 766 and DSDP Sites 261): The Antarctic-Tethys connection. *Marine Micropaleontology* 21, 329–352.

Baumgartner, P.O., 1992. Lower Cretaceous radiolarian biostratigraphy and biogeography off northwestern Australian (ODP sites 765 and 766 and DSDP site 261), Argo abyssal plain and lower Exmouth Plateau, in: Gradstein, F.M., Ludden, J.N., et al., (Eds.), *Proceedings of the Ocean Drilling Program, Scientific Results*. College Station, TX (Ocean Drilling Program), pp. 299–342.

Baumgartner, P.O., 1984. A Middle Jurassic - Early Cretaceous low latitude radiolarian zonation based on unitary associations and age of Tethyan radiolarites. *Eclogae geologicae Helvetiae* 77, 729–841.

Baumgartner, P.O., 1980. Late Jurassic Hagiastridae and Patulibracchiidae (Radiolaria) from the Argolis Peninsula (Peleponnesus, Greece). *Micropaleontology* 26, 274–322.

Baumgartner, P.O., O'Dogherty, L., Goričan, S., et al., 1995. Middle Jurassic to Lower Cretaceous Radiolaria of Tethys: Occurrences, Systematics, Biochronology. *Memoires de Geologie (Lausanne)*, Lausanne, Switzerland.

Beauchamp, B., Baud, A., 2002. Growth and demise of Permian biogenic chert along northwest Pangea: Evidence for end-Permian collapse of thermohaline circulation. *Palaeogeography, Palaeoclimatology, Palaeoecology* 184, 37–63.

Beddow, H.M., Liebrand, D., Wilson, D.S., et al., 2018. Astronomical tunings of the Oligocene–Miocene transition from Pacific Ocean Site U1334 and implications for the carbon cycle. *Clim. Past* 14, 255–270.

Beerling, D.J., Brentnall, S.M., 2007. Numerical evaluation of mechanisms driving Early Jurassic changes in global carbon cycling. *Geology* 35, 247–250.

Behrensmeyer, A.K., Kidwell, S.M., Gastaldo, R.A., 2000. Taphonomy and paleobiology. *Paleobiology 26 Supplement*, 103–147.

Ben Fadhel, M., Layeb, M., Ben Youssef, M., 2010. Upper Albian planktic foraminifera and radiolarian biostratigraphy (Nebeur - northern Tunisia). *Comptes Rendus Palevol* 9, 73–81.

Ben Fadhel, M., Zouaghi, T., Amri, A., Ben Youssef, M., 2014. Radiolarian and planktic foraminifera biostratigraphy of the Early Albian organic rich beds of Fahdene Formation, Northern Tunisia. *Journal of Earth Science* 25, 45–63.

Benson, R.N., 1972. Radiolaria, Leg 12, Deep Sea Drilling Project, in: Laughton, A.S., Berggren, W.A., et al., (Eds.), *Initial Reports of the Deep Sea Drilling Project*. U. S. Government Printing Office, Washington, D.C., pp. 1085–1113.

Benton, M.J., 1995. Diversification and extinction in the history of life. *Science* 268, 52–58.

Benton, M.J., Harper, D.A.T., 2009. *Introduction to Paleobiology and the Fossil Record*. Wiley-Blackwell, Chicester, UK.

Benyamovskiy, V.N., Alekseev, A.S., Ovechkina, M.N., et al., 2012. Upper Campanian-lower Maastrichtian sections of the northwestern Rostov region. Article 1. Description, paleontological assemblages, and lithobiostratigraphy. *Stratigr. Geol. Correl.* 20, 346–379. https://doi.org/10.1134/S0869593812040028

Berger, W.H., Herguera, J.C., Lange, C.B., Schneider, R., 1994. Paleoproductivity: flux proxies versus nutrient proxies and other problems concerning the Quaternary productivity record, in: Zahn, R., et al., (Eds.), *Carbon Cycling in the Glacial Ocean: Constraints on the Ocean's Role in Global Changes*. NATO ASI Series 117. Springer, Berlin, pp. 385–412.

Berggren, W.A., Kent, D.V., Flynn, J.J., Van Couvering, J.A., 1985. Cenozoic geochronology. *Geological Society of America Bulletin* 96, 1407–1418.

Berggren, W.A., Kent, D.V., Swisher, C.C., III, Aubry, M.-P., 1995. A revised Cenozoic geochronology and chronostratigraphy, in: Berggren, W.A., Kent, D.V., Aubry, M.P., Hardenbol, J. (Eds.), *Geochronology, Time Scales and Stratigraphic Correlation: Framework for an Historical Geology*. Society of Economic Paleontologists and Mineralogists, Tulsa, pp. 129–212.

Bernoulli, D, Jenkyns, H.C., 2009. Ancient oceans and continental margins of the Alpine- Mediterranean Tethys: deciphering clues from Mesozoic pelagic sediments and ophiolites. *Sedimentology* 56, 149–190.

Bernoulli, D., Jenkyns, H.C., 2009. Ophiolites in ocean–continent transitions: From the Steinmann Trinity to sea-floor spreading. *C. R. Geoscience* 341, 363–381.

Bernstein, T., 1932. Ueber einige arktische Radiolarien (On some Arctic radiolarians). *Archiv fur Protistenkunde* 76, 217–227.

Biard, T., Bigeard, E., Audic, S., et al., 2017. Biogeography and diversity of Collodaria (Radiolaria) in the global ocean. *ISMEJour.* 1–17.

Biard, T., Pillet, L., Decelle, J., et al., 2015. Towards an integrative morpho-molecular classification of the Collodaria (Polycystinea, Radiolaria). *Protist* 166, 374–388.

Bickert, T., Henrich, R., 2011. Climate records of deep-sea sediments: towards the Cenozoic ice house, in: Hüneke, H., Mulder, T. (Eds.), *Developments in Sedimentology*, Vol. 63. pp. 793–823.

Bijl, P.K., Pross, J., Warnaar, J., et al., 2011. Environmental forcings of Paleogene Southern Ocean dinoflagellate biogeography. *Paleoceanography* 26, 1–12 (web).

Bjørklund, K., Goll, R.M., 1979. Internal Skeletal Structures of Collosphaera

and Trisolenia: a case of repertitative evolution in the Collosphaeridae (Radiolaria). *Journal of Paleontology* 53, 1293–1326.
Bjørklund, K., Kruglikova, S.B., 2003. Polycystine radiolarians in surface sediments in the Arctic Ocean basins and marginal seas. *Marine Micropaleontology* 49, 231–273.
Bjørklund, K.R., 1977. Actinomma haysi, n.sp., its Holocene distribution and size variation in Atlantic Ocean sediments. *Micropaleontology* 23, 114–126.
Bjørklund, K.R., 1976. Radiolaria from the Norwegian Sea, Leg 38 of the Deep Sea Drilling Project, in: *Initial Reports of the Deep Sea Drilling Project*. U. S. Government Printing Office, Washington, D. C., pp. 1101–1168.
Bjørklund, K.R., Swanberg, N.R., 1987. The distribution of two morphotypes of the radiolaria Amphimelissa setosa Cleve (Nassellarida): a result of environmental variability? *Sarsia* 72, 245–254.
Bleil, U., 1989. Magnetostratigraphy of Neogene and Quaternary sediment series from the Norwegian Sea: Ocean Drilling Program, Leg 104, in: Eldholm, O., Thiede, J., Taylor, E., et al., (Eds.), *Proceedings of the Ocean Drilling Program, Scientific Results*. Ocean Drilling Program, College Station, Texas, pp. 829–902.
Blome, C., 1985. Middle Jurassic (Callovian) radiolarians from carbonate concretions, Alaska and Oregon. *Micropaleontology* 30, 343–389.
Blome, C., 1984. Upper Triassic Radiolaria and radiolarian zonation from western North America. *Bulletins of American Paleontology* 85, 88.
Blome, C.D., Hull, D.M., Pessagno, E.A., Reed, K.M., 1995. Mesozoic Radiolaria, in: Blome, C.D., Whalen, P.M., Reed, K.M. (Eds.), *Siliceous Microfossils: Short Courses in Paleontology* No. 8. pp. 31–60.
Blome, C.D., Reed, K.M., 1993. Acid processing of pre-Tertiary radiolarian cherts and its impact on faunal content and biozonal correlation. *Geology* 21, 177–180.
Blueford, J., 1988. Radiolarian biostratigraphy of siliceous Eocene deposits in central California. *Micropaleontology* 34, 236–258.
Blueford, J.R., 1982. Miocene actinommid Radiolaria from the Equatorial Pacific. *Micropaleontology* 28, 189–213.
Bock, W., Hay, W.W., Lee, J., 1985. Order Foraminiferida D'Orbigny, 1826, in: Lee, J.J., Hutner, S.H., Bovee, E.C. (Eds.), *Illustrated Guide to the Protozoa*. Society of Protozoologists, Lawrence, KS, pp. 252–273.
Boeuf, D., Edwards, B.E., Eppley, J.M., et al., 2019. Biological composition and microbial dynamics of sinking particulate organic matter at abyssal depths in the oligotrophic open ocean. *Proceedings of the Natural Academy of Sciences of the United States of America* 116, 11824–11832.
Bohrmann, G., Spiess, V., Hinze, H., Kuhn, G., 1992. Reflector "Pc" a prominent feature in the Maud Rise sediment sequence (eastern Weddell Sea): Occurrence, regional distribution and implications to silica diagenesis. *Marine Geology* 106, 69–87.
Bolli, H.M., Saunders, J.B., Perch-Nielsen, K., 1985. *Plankton Stratigraphy*. Cambridge University Press, Cambridge, UK.
Boltovskoy, D., 2017. Vertical distribution patterns of Radiolaria Polycystina (Protista) in the World Ocean: living ranges, isothermal submersion and settling shells. Journal of Plankton Research 39, 330–349. https://doi.org/10.1093/plankt/fbx003
Boltovskoy, D., 1998. Classification and distribution of south Atlantic recent polycystine radiolaria. *Palaeontologia Electronica* 2, web.

Boltovskoy, D., 1994. The sedimentary record of pelagic biogeography. *Progress in Oceanography* 34, 135–160.

Boltovskoy, D., 1988. Equatorial sedimental shadows of near-surface oceanographic patterns. *Speculations in Science and Technology* 11, 219–232.

Boltovskoy, D., Bender Kotzian, S., Pedrozo, F.L., 1983. Some new techniques for the preparation of Polycystina (Radiolaria). *Micropaleontology* 29, 382–390.

Boltovskoy, D., Correa, N., 2016. Planktonic equatorial diversity troughs: fact or artifact? Latitudinal diversity gradients in Radiolaria. *Ecology.*

Boltovskoy, D., Jankilevich, S.S., 1985. Radiolarian distribution in East equatorial Pacific plankton. *Oceanologica Acta* 8, 101–123.

Boltovskoy, D., Kling, S.A., Takahashi, K., Bjørklund, K., 2010. World atlas of distribution of living radiolaria. *Palaeontologia Electronica* 13, web only.

Boltovskoy, D., Kogan, M., Alder, V.A., Mianzan, H., 2003. First record of a brackish radiolarian (Polycystina): Lophophaena rioplatensis n. sp. in the Rio de la Plata estuary. *Journal of Plankton Research* 25, 1551–1559.

Boltovskoy, D., Oberhänsli, H., Wefer, G., 1996. Radiolarian assemblages in the eastern tropical Atlantic: patterns in the plankton and in sediment trap samples. *Journal of Marine Systems* 8, 31–51.

Boltovskoy, D., Riedel, W.R., 1980. Polycystine radiolaria from the southwestern Atlantic Ocean plankton. *Revista España Micropaleontologia* 12, 99–146.

Booth, J.D., Burckle, L.H., 1976. Displaced Antarctic diatoms in the southwestern and central Pacific. *Pacific Geology* 11, 99–108.

Bowden, A.J., Gregory, F.J., Henderson, A.S., 2013. *Landmarks in Foraminiferal Micropalaeontology: History and Development*. Geological Society of London.

Bowman, V.C., Francis, J.E., Riding, J.B., et al., 2012. A latest Cretaceous to earliest Paleogene dinoflagellate cyst zonation from Antarctica, and implications for phytoprovincialism in the high southern latitudes. *Review of Palaeobotany and Palynology* 171, 40–56.

Bown, P.R., Lees, J.A., Young, J.R., 2004. Calcareous nannoplankton evolution and diversity through time, in: Thierstein, H.R., Young, J.R. (Eds.), *Coccolithophores: From Molecular Processes to Global Impact*. Springer, Berlin, pp. 481–508.

Brack, P., Rieber, H., Nicora, A., Mundil, R., 2005. The Global boundary Stratotype Section and Point (GSSP) of the Ladinian Stage (Middle Triassic) at Bagolino (Southern Alps, Northern Italy) and its implications for the Triassic time scale. *Episodes* 28, 233–244.

Bragin, N.Y., 2011. Triassic radiolarians of Kotel'nyi Island (New Siberian Islands, Arctic). *Paleontological Journal* 45, 711–778.

Bragin, N.Y., 2009. Echinocampidae fam. nov., a new family of Late Jurassic-Early Cretaceous radiolarians of Arctic Siberia. *Paleontological Journal* 43, 356–369.

Bragin, N.Y., 2000. Triassic radiolarian zonation in the Far East of Russia. *Stratigraphy and Geological Correlation* 8, 579–592.

Bragin, N.Y., Konstantinov, A.G., Sobolev, E.S., 2012. Upper Triassic Stratigraphy and Paleobiogeography of Kotel'nyi Island (New Siberian Islands). *Stratigraphy and Geological Correlation* 20, 541–566.

Bragina, L.G., 2016. Radiolarian-based zonal scheme of the Cretaceous (Albian–Santonian) of the Tethyan regions of Eurasia. *Stratigraphy and Geological Correlation* 24, 141–166.

Bragina, L.G., 2012. Radiolarian biostratigraphy of the Perapedhi Formation (Cyprus): implications for the geological evolution of the Troodos Ophiolite. *Bulletin de la Société Géologique de France* 183, 343–348. https://doi.org/10.2113/gssgfbull.183.4.343

Bragina, L.G., 2004. Cenomanian-Turonian radiolarians of northern Turkey and the Crimean Mountains. *Paleontological Journal* 38, 325–456.

Bragina, L.G., Agarkov, Yu.V., Bragin, N.Yu., 2007. Radiolarians of the Upper Cenomanian and Lower Turonian from Deposits of the Ananuri Formation, the Western Caucasus (Lazarevskoe Area). *Stratigraphy and Geological Correlation* 15, 310–320.

Bralower, T.J., Premoli Silva, I., Malone, M.J., 2006. Leg 198 synthesis: A remarkable 120-m.y. record of climate and oceanography from Shatsky Rise, northwest Pacific Ocean. *Proceedings Ocean Drilling Program, Scientific Results* 198, web.

Bramlette, M.N., Riedel, W.R., 1954. Stratigraphic value of discoasters and some other microfossils related to recent coccolithophores. *Journal of Paleontology* 28, 385–403.

Braun, A., Budil, P., 1999. A Middle Devonian radiolarian fauna from the Chotec Limestone (Eifelian) of the Prague Basin (Barrandian, Czech Republic). *Geodiversitas* 21, 581–592.

Braun, A., Chen, J., Waloszek, D., Maas, A., 2007. First Early Cambrian Radiolaria. *Geological Society* 286, 143–149.

Braun, A., Schmidt-Effing, R., 1993. Biozonation, diagenesis and evolution of radiolarians in the Lower Carboniferous of Germany. *Marine Micropaleontology* 21, 369–383.

Brechner, S.A., 2000. Automatic coccolith classification and extraction of morphological features in SEM images (PhD Thesis). ETH, Zürich.

Brinkhuis, H., Zachariasse, W.J., 1988. Dinoflagellate cysts, sea level changes and planktonic foraminifera across the Cretaceous/Tertiary boundary at El Haria, northwest Tunisia. *Marine Micropaleontology* 13, 153–190.

Brown, J., Colling, A., Park, D., et al., 1989. *Ocean Circulation.* Pergamon Press, The open University, Oxford, Walton Hall.

Bryan, S.E., Ferrari, L., 2013. Large igneous provinces and silicic large igneous provinces: Progress in our understanding over the last 25 years. *Geological Society of America Bulletin* 125, 1053–1078.

Burckle, L., 1978. Marine Diatoms, in: Haq, B.U., Boersma, A. (Eds.), *An Introduction to Marine Micropaleontology*. Elsevier, New York, pp. 245–266.

Burckle, L.H., Opdyke, N.D., 1977. Late Neogene diatom correlations in the Circum-Pacific, in: *Proceedings of the First International Congress on Pacific Neogene Stratigraphy*. Kaiyo Shuppan Co. Ltd., Tokyo.

Burki, F., Keeling, P.J., 2014. Rhizaria. *Current Biology* 24, 103–107.

Burki, F., Kudravtsev, A., Matx, M.W., et al., 2010. Evolution of Rhizaria: new insights drom phylogenomic analysis of uncultivated protists. *BMC Evolutionary Biology* 10, 377.

Bütschli, O., 1882. Beitrage zur Kenntnis der Radiolarienskelette, insbesondere der der Cyrtida. *Zeitschrift fur Wissenschaftliche Zoologie* 36, 485–540.

Butterfield, N.J., 2011. Animals and the invention of the Phanerozoic Earth system. *Trends in Ecology and Evolution* 26, 81–87.

Butterfield, N.J., 2009. Macroevolutionary turnover through the Ediacaran transition: ecological and biogeochemical implications. *Geological Society, London, Special Publications* 326, 55–66.

Butterfield, N.J., 1997. Plankton ecology and the Proterozoic-Phanerozoic transition. *Paleobiology* 23, 247–262.
Cachon, J., 1964. Contribution – l'etude des Peridiniens parasites. Cytologie, cycles evolutifs. Annales des Sciences Naturelles. Serie 12, *Zoologie et Biologie Animale* 6, 1–158.
Cachon, J., Cachon, M., 1978. Sticholonche zanclea Hertwig: A reinterpretation of its phylogenetic position based upon new observations on its ultrastructure. *Archiv für Protistenkunde* 120, 148–168.
Cachon, J., Cachon, M., 1972. Le systeme axopodial des Radiolaries Nassellaires. *Archiv für Protistenkunde* 113, 80–97.
Cachon, J., Cachon, M., 1971. Ultrastructures du genre Oodinium Chatton. Differentiations cellulaires en apport avec la vie parasitaire. *Archiv für Protistenkunde* 7, 153–169.
Campbell, A.K., Herring, P.J., 1990. Imidazolopyrazine bioluminescence in copepods and other marine organisms. *Marine Biology* 104, 219–225.
Campbell, A.S., 1954. Radiolaria, in: Moore, R.C. (Ed.), *Treatise on Invertebrate Paleontology*. Geological Society of America and University of Kansas Press, Lawrence, Kansas, USA, pp. 11–195.
Campbell, A.S., Clark, B.L., 1944a. Miocene radiolarian faunas from southern California. *Geological Society of America Special Paper* 51, 1–76.
Campbell, A.S., Clark, B.L., 1944b. Radiolaria from Upper Cretaceous of Middle California. *Geological Society of America Special Paper* 57, 1–61.
Cande, S.C., Kent, D.V., 1995. Revised calibration of the geomagnetic polarity timescale for the Late Cretaceous and Cenozoic. *Journal of Geophysical Research: Solid Earth* 100, 6093–6095.
Cannariato, K.G., Ravelo, A.C., 1997. Pliocene-Pleistocene evolution of eastern tropical Pacific surface water circulation and thermocline depth. *Paleoceanography* 12, 805–820.
Cao, W., Feng, Q., Feng, F., Ling, W., 2014. Radiolarian Kalimnasphaera from the Cambrian Shuijingtuo Formation in South China. *Marine Micropalaeontology* 110, 3–7.
Caridroit, M., Danelian, T., O'Dogherty, L., et al., 2017. An illustrated catalogue and revised classifcation of Paleozoic radiolarian genera. *Geodiversitas* 39, 363–417.
Caridroit, M., De Wever, P., Dumitrica, P., 1999. Un nouvel ordre, une nouvelle famille et un nouveau genre de Radiolaires du Paléozoïque : Latentifistularia, Cauletellidae et Cauletella. *Comptes Rendus de l'Académie des Sciences de Paris, Série IIA - Sciences de la Terre et des Planètes* 329, 603–608.
Caron, D.A., Swanberg, N.R., 1990. The ecology of planktonic sarcodines. *Reviews in Aquatic Sciences* 3, 147–180.
Carroll, R.L., 2002. Evolution of the capacity to evolve. *Journal of Evolutionary Biology* 15, 911–921.
Carroll, S.B., 2008. Evo-Devo and an expanding evolutionary synthesis: a genetic theory of morphological evolution. *Cell* 134, 25–36.
Carter, E., Orchard, M.J., 2007. Radiolarian-conodont-ammonoid intercalibration around the Norian-Rhaetian boundary and implication for trans-Panthalassan correlation. *Albertiana* 36, 149–163.
Carter, E.S., 1993. Biochronology and paleontology of uppermost triassic (Rhaetian) radiolarians, Queen Charlotte Islands, British Columbia, Canada. *Mémoires de Géologie (Lausanne)* 11, 1–175.
Carter, E.S., Cameron, B.E.B., Smith, P.L., 1988. Lower and Middle Jurassic radiolarian biostratigraphy and systematic paleontology, Queen Charlotte Islands, British Columbia.

Geological Survey of Canada, Bulletin 386, 1–109.
Carter, E.S., Goričan, S., Guex, J., et al., 2010. Global radiolarian zonation for the Pliensbachian, Toarcian and Aalenian. *Palaeogeography, Palaeoclimatology, Palaeoecology* 297, 401–419.
Carter, E.S., Hori, R.S., 2005. Global correlation of the radiolarian faunal change across the Triassic-Jurassic boundary. *Canadian Journal of Earth Sciences* 42, 777–790.
Carter, E.S., Orchard, M.J., 2013. Intercalibration of conodont and radiolarian faunas from the Carnian-Norian boundary interval in Haida Gwaii, British Columbia, Canada, in: Tanner, L.H., Spielmann, J.A., Lucas, S.G. (Eds.), *The Triassic System: New Developments in Stratigraphy and Paleontology: Bulletin 61*. New Mexico Museum of Natural History and Science, pp. 67–92.
Carter, E.S., Whalen, P.A., Guex, J., 1998. Biochronology and paleontology of Lower Jurassic (Hettangian and Sinemurian) radiolarians, Queen Charlotte Islands, British Columbia. *Geological Survey of Canada, Bulletin* 496, 1–162.
Casey, R.E., 1993. Radiolaria, in: Lipps, J.E. (Ed.), *Fossil Procaryotes and Protists*. Blackwell, Boston, pp. 249–284.
Casey, R.E., 1987. Radiolaria, in: Broadhead, T.W. (Ed.), *Fossil Prokaryotes and Protists; Notes for a Short Course*. Department of Geological Sciences, University of Tennessee, Knoxville, Tennessee, USA, pp. 213–247.
Casey, R.E., 1977. The ecology and distribution of Recent Radiolaria, in: Ramsay, A.T.S. (Ed.), *Oceanic Micropalaeontology*. Academic Press, London, UK, pp. 809–845.
Casey, R.E., 1971. Distribution of polycystine Radiolaria in the oceans in relation to physical and chemical conditions, in: Funnell, B.M., Riedel, W.R. (Eds.), *The Micropalaeontology of Oceans*. Cambridge University Press, Cambridge, UK, pp. 151–159.
Casey, R. E., Gust, L., Leavesley, A., et al., 1979. Ecological niches of radiolarians, planktonic foraminiferans and pteropods inferred from studies on living forms in the Gulf of Mexico and adjacent waters. *Transactions of the Gulf Coast Association of Geological Societies* 29, 216–223.
Casey, R.E., Partridge, T.M., Sloan, J.R., 1971. Radiolarian life spans, mortality rates and seasonality gained from sediment and plankton samples, in: Farinacci, A. (Ed.), *Proceedings of the II Planktonic Conference, Roma 1970*. Edizioni Tecnoscienza, Roma, Italy, pp. 159–165.
Casey, R.E., Spaw, J.M., Kunze, F.R., 1982. Polycystine radiolarian distributions and enhancements related to oceanographic conditions in a hypothetical Ocean. *Transactions-Gulf Coast Association of Geological Societies* 32, 319–332.
Casey, R.E., Spaw, J.M., Kunze, F.R., et al., 1979. Radiolarian ecology and the development of the radiolarian component in Holocene sediments, Gulf of Mexico and adjacent seas with potential paleontological applications. *Transactions of Gulf Coast Association of Geological Societies* 29, 228–237.
Casey, R.E., Wigley, C.R., Perez-Guzman, A.M., 1983. Biogeographic and ecologic perspective on polycystine radiolarian evolution. *Paleobiology* 9, 363–376.
Casteleyn, G., Adams, N.G., Vanormelingen, P., et al., 2009. Natural hybrids in the marine diatom Pseudo-nitzchia pungens (Bacillariophyceae): genetic and morphological evidence. *Protist* 160, 343–354.
Caulet, J.P., 1991. Radiolarians from the Kerguelen Plateau, ODP Leg 119, in: Barron, J., Larsen, B. (Eds.), *Proceedings of the Ocean Drilling Program, Scientific Results*. Ocean Drilling Program, College Station, pp. 513–546.

Caulet, J.P., 1986. Radiolarians from the southwest Pacific, in: Kennett, J.P., von der Borch, C.C. (Eds.), *Initial Reports of the Deep Sea Drilling Project*. U. S. Government Printing Office, Washington, D. C., pp. 835–861.

Caulet, J.P., Nigrini, C., 1988. The genus Pterocorys (Radiolaria) from the tropical late Neogene of the Indian and Pacific Oceans. *Micropaleontology* 34, 217–235.

Caulet, J.-P., O'Dogherty, L., Dumitrica, P., Suzuki, N., 2017. An up-to-date catalogue of Cenozoic radiolarian genera and families: a review with illustrations of type species. Presented at the Interrad XV, Niigata, Japan.

Caulet, J.P., Venec-Peyre, M.-T., Vergnaud-Grazzini, C., Nigrini, C., 1992. Variation of South Somalian upwelling during the last 160 ka: radiolarian and foraminifera records in core MD 85674, in: Summerhayes, C.P., Prell, W.L., Emeis, K.C. (Eds.), *Upwelling Systems: Evolution Since the Early Miocene*. Geological Society of London, London, pp. 379–389.

Cavalier-Smith, T., 2003. Protist phylogeny and the high-level classification of Protozoa. *European Journal of Protistology* 39, 338–348.

Cavalier-Smith, T., 1993. Kingdom Protozoa and its 18 phyla. *Microbiological reviews* 57, 953–994.

Cervato, C., Burckle, L., 2003. Patten of first and last appearance in diatoms: Oceanic circulation and the position of polar fronts during the Cenozoic. *Paleoceanography* 18, web.

Channell, J.E.T., Amigo, A.E., Fronval, T., et al., 1999. Magnetic stratigraphy at Sites 907 and 985 in the Norwegian-Greenland Sea and a revision of the Site 907 composite section, in: Raymo, M.E., Jansen, E., Blum, P., Herbert, T.D. (Eds.), *Proceedings of the Ocean Drilling Program, Scientific Results, Volume 162*. Ocean Drilling Program, College Station, Texas, pp. 131–148.

Chao, A., Kubota, Y., Zelený, D., et al., 2020. Quantifying sample completeness and comparing diversities among assemblages. *Ecological Research* 35, 292–314. https://doi.org/10.1111/1440-1703.12102

Chen, P.H., 1975. Antarctic radiolaria, in: *Initial Reports of the Deep Sea Drilling Project*. U. S. Government Printing Office, Washington, D. C., pp. 437–513.

Cheng, Y.-n, 1986. Taxonomic studies on Upper Paleozoic Radiolaria. *National Museum of Natural Science*, Special Publication 1–213.

Chiari, M., Cobianchi, M., Picotti, V., 2007. Integrated stratigraphy (radiolarians and calcareous nannofossils) of the Middle to Upper Jurassic Alpine radiolarites (Lombardian Basin, Italy): Constraints to their genetic interpretation. *Palaeogeography, Palaeoclimatology, Palaeoecology* 249, 233–270.

Cienkowski, L., 1871. Uber Schwarmerbildung bei Radiolarien (Zoospore formation in Radiolaria). *Archiv fur Mikroskopische Anatomie* 7, 372–381.

Clapham, M.E., Shen, S.H., Bottjer, D.J., 2009. The double mass extinction revisited: reassessing the severity, selectivity, and causes of the end-Guadalupian biotic crisis (Late Permian). *Paleobiology* 35, 32–50.

Clark, B.L., Campbell, A.S., 1942. Eocene radiolarian faunas from the Monte Diablo area, California. *Geological Society of America Special Paper* 39, 1–112.

Cleve, P.T., 1899. Plankton collected by the Swedish expedition to Spitzbergen in 1898. *Kongliga Svenska Vetenskapsakademiens handlingar* 32, 1–51.

CLIMAP project members, 1976. The surface of the ice-age earth. *Science* 191, 1131–1137.

Cody, R.D., Levy, R.H., Harwood, D.M., Sadler, P.M., 2008. Thinking

outside the zone: High-resolution quantitative diatom biochronology for the Antarctic Neogene. *Palaeogeography, Palaeoclimatology, Palaeoecology* 260, 92–121.
Coe, A.L., 2010. *Geological Field Techniques*. Wiley-Blackwell.
Colin, S., 2018. Simple 3D imaging of biogenic silica structures by fluorescent microscopy. https://doi.org/10.17504/protocols.io.vd8e29w
Colin, S., Coelho, L.P., Sunagawa, S., Bowler, C., Karsenti, E., Bork, P., Pepperkok, R., Vargas, C. de, 2017. Quantitative 3D-imaging for cell biology and ecology of environmental microbial eukaryotes [WWW Document]. *eLife*. https://doi.org/10.7554/eLife.26066
Corliss, J.O., 1978. A salute to fifty-four great microscopists of the past: a pictorial footnote to the history of protozoology. Part 1. *Transactions of the American Microscopical Society* 97, 419–548.
Corliss, J.O., 1975. Three Centuries of Protozoology: A Brief Tribute to its Founding Father, A. van Leeuwenhoek of Delft. *The Journal of Protozoology* 22, 3–7. https://doi.org/10.1111/j.1550-7408.1975.tb00934.x
Cortese, G., 2014a. Editorial: Marine micropaleontology special issue dedicated to Paulian Dumitrică. *Marine Micropaleontology* 110, 1–2.
Cortese, G., 2014b. Radiolarian researchers based in Italy during the late nineteenth and early twentieth centuries. *Journal of Micropalaeontology* 33, 95–103.
Cortese, G., Abelmann, A., Gersonde, R., 2007. The last five glacial-interglacial transitions; a high-resolution 450,000-year record from the subantarctic Atlantic. *Paleoceanography* 22, PA4203.
Cortese, G., Bjørklund, K.R., 1998. Morphometry and taxonomy of Hexacontium species from western Norwegian fjords. *Micropaleontology* 44, 161–172.
Cortese, G., Dolven, J.K., Bjørklund, K.R., Malmgren, B.A., 2005. Late Pleistocene-Holocene radiolarian paleotemperatures in the Norwegian Sea based on artificial neural networks. *Palaeogeography, Palaeoclimatology, Palaeoecology* 224, 311–332.
Cortese, G., Gersonde, R., Hillenbrand, C.-D., Kuhn, G., 2004. Opal sedimentation shifts in the World Ocean over the last 15 Myr. *Earth and Planetary Science Letters* 224, 509–527.
Crame, J.A., 1996. Antarctica and the evolution of taxonomic diversity gradients in the marine realm. *Terra Antarctica* 3, 121–134.
Crampton, J.S., Cody, R.D., R., L., Harwood, D.M., McKay, R., Naish, T.R., 2016. Southern Ocean phytoplankton turnover in response to stepwise Antarctic cooling over the past 15 million years. *Proceedings of the National Academy of Sciences of the United States of America*.
Croft, W.J., 2006. *Under the Microscope: A Brief History of Microscopy*. World Scientific Publishing, Singapore.
Cui, Z., Hua, H., Song, Q., 2000. The Late Ordovician radiolarian assemblage of the North Qinling backarc basin. China. *Acta Geologica Sinica* 74, 254–258.
Cushman, J.A., 1928. *Foraminifera, Their Classification and Economic Use*. Special Publications. Cushman Foundation for Foraminiferal Research.
Czarniecki, S., 1993. Grzybowski and his school: the beginnings of applied mimpalaeontology in Poland at the turn of the 19th and 20th centuries, in: Kaminski, M.A., Geroch, S., Kaminski, D.G. (Eds.), *The Origins of Applied Micropalaeontology: The School of Jozef Grzybowski*, The Grzybowski Foundation Special Publication. The Micropalaeontological Society, London.
Danelian, T., Asatryan, G., Sosson, M., et al., 2008. Discovery of Middle Jurassic (Bajocian) Radiolaria from the

sedimentary cover of the Vedi ophiolite (Lesser Caucasus, Armenia). *Comptes Rendus Palevol* 7, 327–334.
Danelian, T., Baudin, F., Gardin, S., et al., 2007. The record of mid Cretaceous oceanic anoxic events from the Ionian zone of southern Albania. *Revue de Micropaléontologie* 50, 225–237.
Danelian, T., Caridroit, M., Noble, P., Aitchison, J.C., 2017. Catalogue of Paleozoic radiolarian genera: *Geodiversitas Special Issue*, pp. 345–637.
Danelian, T., Clarkson, E.N.K., 1998. Ordovician radiolaria from bedded cherts of the Southern Uplands. *Scottish Journal of Geology* 34, 133–137.
Danelian, T., Baudin, F., Gardin, S., et al., 2002. Early Aptian productivity increase as recorded in the Fourcade level of the Ionian zone of Greece. *Comptes Rendus Geoscience* 334, 1087–1093.
Danelian, T., Johnson, K.G., 2001. Patterns of biotic change in Middle Jurassic to Early Cretaceous Tethyan radiolaria. *Marine Micropaleontology* 43, 239–260.
Danelian, T., Macleod, N., 2019. Morphometric Analysis of Two Eocene Related Radiolarian Species of the Podocyrtis (Lampterium) Lineage. *Paleontological Research* 23, 314–330. https://doi.org/10.2517/2019PR007
Danelian, T., Popov, L., 2003. Ordovician radiolarian biodiversity: insights based on new and revised data from Kazakhstan. *Bulletin de la Société Géologique de France* 174, 325–335.
Danelian, T., Tsikos, H., Gardin, S., et al., 2004. Global and regional palaeoceanographic changes as recorded in the mid-Cretaceous (Aptian–Albian) sequence of the Ionian zone (NW Greece). *Journal of the Geological Society of London* 161, 703–709.
Darling, K., Kucera, M., Pudsey, C.J., Wade, C.M., 2004. Molecular evidence links cryptic diversification in polar planktonic protists to Quaternary climate dynamics. *Proceedings of the National Academy of Sciences, USA* 101, 7657–7662.
Darling, K.F., Wade, C.M., Stewart, I.A., et al., 2000. Molecular evidence for genetic mixing of Arctic and Antarctic subpolar populations of planktonic foraminifers. *Nature* 405, 43–47.
Darwin, C., 1859. *The Origin of Species.*
Davis, C.S., Gallager, S.M., Marra, M., Stewart, W.K., 1996. Rapid visualization of plankton abundance and taxonomic composition using the Video Plankton Recorder. *Journal of Plankton Research* 24, 797–805.
Dayan, T., Wool, D., Simberloff, D., 2002. Variation and covariation of skulls and teeth: modern carnivores and the interpretation of fossil mammals. *Paleobiology* 284, 508–526.
de Vargas, C., Audic, S., Henry, N., et al., 2015. Eukaryotic plankton diversity in the sunlit ocean. *Science* 348.
de Vargas, C., Saez, A.G., Medlin, L.K., Thierstein, H.R., 2004. Super-Species in the calcareous plankton, in: Thierstein, H.R., Young, J.R. (Eds.), *Coccolithophores: From Molecular Processes to Global Impact.* Springer, Berlin, pp. 271–298.
De Wever, P., 1982. Nassellaria (Radiolaires Polycystines) du Lias de Turquie. *Revue de Micropaléontologie* 24, 189–232.
De Wever, P., 1981. Une nouvelle sous-famille, les Poulpinae, et quatre nouvelles espèces de Saitoum radiolaires mésozoïques tethysiens. *Geobios* 14, 5–15.
De Wever, P., Caridroit, M., 1984. Description de quelques nouveaux Latentifistulidae (radiolaires polycystines) Paléozoïques du Japon. *Revue de Micropaléontologie* 27, 98–106.
De Wever, P., Cordey, F., 1986. Datation par les Radiolaires de la Formation des Radiolarites s.s. de la Serie du Pinde-Olonos (Grece): Bajocien (?)-Tithonique. *Marine Micropaleontology* 11, 113–127.

De Wever, P., Dumitrica, P., Caulet, J.P., et al., 2001. *Radiolarians in the Sedimentary Record*. Gordon and Breach, Amsterdam.

De Wever, P., Baudin, F., Azéma, J., Fourcade, E., 1995. Radiolarians and Tethyan radiolarites from primary production to their paleogeography, in: Narin (Ed.), *The Ocean Basins and Margins, Volume 8: The Tethys Ocean*. Plenum Press, New York, pp. 267–318.

De Wever, P., O'Dogherty, L., Caridroit, M., et al., 2003. Diversity of radiolarian families through time. *Bulletin de la Societe Geologique de France* 174, 453–469.

De Wever, P., O'Dogherty, L., Goričan, S., 2007. The plankton turnover at the Permo-Triassic boundary, emphasis on radiolarians. *Eclogae Geologicae Helvetiae* 99, 49–62.

De Wever, P., Popova, I.M., 1997. Cenozoic radiolaria from European Platform: a review. *Geodiversitas* 19, 383–469.

De Wever, P., Ricou, L., Fourcade, E., 1986. La fin brutale de l'optimum radiolaritique au Jurassique terminal; l'effet de la circulation oceanique (The end of the radiolaritic optimum at the uppermost Jurassic; the effect of oceanic circulation). *Comptes rendus des seances de l'Academie des Sciences, serie II: Mecanique-Physique, Chimie, Sciences de l'Univers, Sciences de la Terre* 302, 665–670.

De Wever, P., Sanfilippo, A., Riedel, W.R., Gruber, B., 1979. Triassic radiolarians from Greece, Sicily and Turkey. *Micropaleontology* 25, 75–110.

De Wever, P., Thiébault, F., 1981. Les radiolaires d'âge Jurassique supérieur à Crétacé supérieur dans les radiolarites du Pinde-Olonos (prequ'île de Koroni; Péloponnèse méridional, Grèce). *Géobios* 14, 577–609.

Decelle, J., Colin, S., Foster, R.A., 2015. Photosymbiosis in marine planktonic protists, in: Ohtsuka, S., Suzaki, T., Horiguchi, T., Suzuki, N., Not, F. (Eds.), *Marine Protists: Diversity and Dynamics*. pp. 465–500.

Decelle, J., Probert, I., Bittner, L., et al., 2012a. An original mode of symbiosis in open ocean plankton. *Proceedings of the National Academy of Sciences of the United States of America* 109, 18000–18005.

Decelle, J., Suzuki, N., Mahe, F., et al., 2012b. Molecular phylogeny and morphological evolution of the Acantharia (Radiolaria). *Protist* 163, 435–450.

DeConto, R.M., Pollard, D., 2003. Rapid Cenozoic glaciation of Antarctica induced by declining atmospheric pCO2. *Nature* 421, 245–249.

Deflandre, G., 1963. Pylentonema, nouveau genre de Radiolaire du Viséen. *Comptes Rendus hebdomadaires des Séances de l'Académie des Sciences (Paris), Série D : Sciences naturelles* 257, 3981–3984.

Deflandre, G., 1960. A propos du developpment des recherches sur les Radiolaires fossiles (Remarks about the expansion of radiolarian research). *Revue de micropaleontologie* 2, 212–218.

Deflandre, G., 1953. Radiolaires fossiles, in: Grassé, P.P. (Ed.), *Traité de Zoologie*. Masson, Paris, pp. 389–436.

DeMaster, D.J., 2003. The diagenesis of biogenic silica: chemical transformations occurring in the water column, seabed, and crust, in: Turekian, K.K., Holland, H.D., Mackenzie, F.T. (Eds.), *Treatise on Geochemistry*. Elsevier, pp. 87–98.

Dera, G., Brigaud, B., Monna, F., et al., 2011. Climatic ups and downs in a disturbed Jurassic world. *Geology* 39, 215–218.

D'Hondt, S., 2005. Consequences of the Cretaceous/Paleogene mass extinction for marine ecosystems. *Annual Review of Ecology Evolution and Systematics* 36, 295–317.

D'Hondt, S., Donaghay, P., Zachos, J., et al., 1998. Organic carbon fluxes and ecological recovery from the Cretaceous-Tertiary mass extinction. *Science* 282, 276–278.

Dietz, L.F., Sarjeant, W.A.S., Mitchell, T.A., 1999. The dreamer and the pragmatist: a joint biography of Walter Wetzel and Otto Wetzel, with a survey of their contributions to geology and micropaleontology. *Earth Sciences History* 18, 4–50.

Díez, B., Pedrós-Slió, C., Massana, R., 2001. Study of genetic diversity of eukaryotic picoplankton in different oceanic regions by small subunit rRNA gene cloning and sequencing. *Applied Environmental Microbiology* 67, 2932–2941.

Dolven, J.K., 2001. www.radiolaria.org. Radiolaria: Newsletter of the International Association of Radiolarian Paleontologists 19, 7.

Dolven, J.K., Cortese, G., Bjørklund, K.R., 2002. A high-resolution radiolarian derived paleotemperature record for the Late Pleistocene-Holocene in the Norwegian Sea. *Paleoceanography* 17, 1072 (13 pp.).

Dong, X.P., Knoll, A.H., Lipps, J.H., 1997. Late Cambrian Radiolaria from Hunan, China. *Journal of Paleontology* 71, 753–758.

Donofrio, D., Mostler, H., 1978. Zur Verbreitung der Saturnalidae (Radiolaria) im Mesozoikum der Nördlichen Kalkalpan und Südalpen. *Geologisch Paläontologische Mitteilungen Innsbruck* 7, 1–55.

Donovan, S.K., 2005. The geology of Barbados: a field guide. *Caribbean Journal of Earth Science* 38, 21–33.

Dow, R.L., 1978. Radiolarian distribution and the Late Pleistocene history of the Southeastern Indian Ocean. *Marine Micropaleontology* 3, 203–227.

Dreyer, F., 1889. Morphologischen Radiolarienstudien. 1. Die Pylombildungen in vergleichend-anatomischer und entwicklungs geschichtlicher Beziehung bei Radiolarien und bei Protisten uberhaupt nebst System und Beschreibung neuer und der bis jetzt bekannten pylomatischen Spumellarien. Jenaische Zeitschrift fur Naturwissenschaft 23, 1–138.

Dujardin, F., 1835. Recherches sur les organisms inférieurs. *Annales des Sciences Naturelle, Zoologie*. Series 2 4, 343–377.

Dumitrică, P., 2020. Some new or newly interpreted Cenozoic larnacillid radiolarian taxa. *Revue de Micropaléontologie* 66, 100405. https://doi.org/10.1016/j.revmic.2019.100405

Dumitrica, P., 2017. On the status of the Triassic Nassellarian radiolarian family Tetraspinocyrtiidae Kozur and Mostler and description of some related taxa. *Revue de Micropaléontologie* 60, 33–85.

Dumitrica, P., 2013. Early Tithonian entactinarian Radiolaria from the Solnhofen area (southern Germany). Part I. *Revue de Micropaléontologie* 56, 75–95.

Dumitrica, P., 2004. New Mesozoic and early Cenozoic spicular Nassellaria and Nassellaria-like Radiolaria. *Revue de Micropaléontologie* 47, 193–224.

Dumitrica, P., 1995. Systematic framework of Jurassic and Cretaceous Radiolaria, in: Baumgartner, P.O., O'Dogherty, L., Goričan, S., Urquhart, E., Pillevuit, A., De Wever, P. (Eds.), *Middle Jurassic to Lower Cretaceous Radiolaria of Tethys: Occurrences, Systematics, Biochronology*. pp. 19–35.

Dumitrica, P., 1991. Middle Triassic Tripedurnulidae, n. fam. (Radiolaria) from the eastern Carpathians (Romania) and Vicentinian Alps (Italy). *Revue de Micropaléontologie* 34, 261–278.

Dumitrica, P., 1989. Internal skeletal structures of the superfamily Pyloniacea (Radiolaria), a basis of a new systematics.

Revista española de Micropaleontología 21, 207–264.
Dumitrica, P., 1988. New families and subfamilies of Pyloniacea (Radiolaria). *Revue de Micropaléontologie* 31, 178–195.
Dumitrica, P., 1982a. Foremanellinidae, a new family of Triassic Radiolaria. *Dari de Seama ale Sedintelor, Institutul de Geologie si Geofizica, Bucaresti* 67, 75–82.
Dumitrica, P., 1982b. Middle Triassic spicular Radiolaria. *Revista española de Micropaleontología* 14, 401–428.
Dumitrica, P., 1978a. Badenian Radiolaria from central Paratethys, in: Brestenska, E. (Ed.), *Chronostratigraphie Und Neostratotypen; Miozaen Der Zentralen Paratethys*. VEDA, Verlag der Slowakischen Akademie der Wissenschaften, Bratislava, Czechoslovakia, pp. 231–261.
Dumitrica, P., 1978b. Family Eptingiidae n. fam., extinct Nassellaria (Radiolaria) with sagittal ring. *Dari de Seama ale Sedintelor, Institutul de Geologie si Geofizica, Bucaresti* 64, 27–38.
Dumitrica, P., 1978c. Triassic Palaeoscenidiidae and Entactiniidae from the Vicentinian Alps (Italy) and eastern Carpathians (Romania). *Dari de Seama ale Sedintelor, Institutul de Geologie si Geofizica, Bucaresti* 64, 39–54.
Dumitrica, P., 1970. Cryptocephalic and cryptothoracic Nassellaria in some Mesozoic deposits of Romania. *Revue Roumaine de Géologie, Géophysique et Géographie (série Géologie)* 14, 45–124.
Dumitrica, P., Caridroit, M., De Wever, P., 2000. Archaeospicularia, ordre nouveau de radiolaires: une nouvelle étape pour la classification des radiolaires du Paléozoïque inférieur. *Comptes Rendus de l'Académie des Sciences Paris, Sciences de la Terre et des planètes* 330, 563–569.
Dumitrica, P., Carter, E.S., 1999. Family Kungalariidae, n. fam., new Mesozoic entactinarian Radiolaria with a nassellarian-type initial spicule. *Micropaleontology* 45, 418–428.
Dumitrica, P., Immenhauser, A., Dumitrica-Jud, R., 1997. Mesozoic radiolarian biostratigraphy from Masirah ophiolite, Sultanate of Oman. Part 1: Middle Triassic, uppermost Jurassic and Lower Cretaceous Spumellarians and multisegmented Nassellarians. *Bulletin of the National Museum of Natural Science, Taiwan* 9, 1–105.
Dumitrica, P., Kozur, H., Mostler, H., 1980. Contribution to the radiolarian fauna of the Middle Triassic of the Southern Alps. *Geologisch Paläontologische Mitteilungen Innsbruck* 10, 1–46.
Dumitrica, P., Tekin, U.K., 2014. Trimeridianellidae nov. fam., a Triassic spumellarian radiolarian family with simple initial skeleton and pyloniacean mode of growth. *Revue de Micropaléontologie* 57, 45–56.
Dumitrica, P., Zügel, P., 2003. Lower Tithonian mono- and dicyrtid Nassellaria (Radiolaria) from the Solnhofen area (southern Germany). *Geodiversitas* 25, 5–72.
Dumitrica, P., Zügel, P., 2002. Mendacastrum n. gen. and Domuzdagia n. gen., two Jurassic spherical Spumellaria (Radiolaria) with hagiastrid medullary shell. *Micropaleontology* 48, 23–34.
Dumitrica, P., Zügel, P., 1998. Hexapylocapsa anachoreta n. gen., n. sp., type of a new monocyrtid nassellarian family (Radiolaria) with double-shelled cephalis. *Paläontologische Zeitschrift* 72, 249–256.
Dunham, J.B., Murphy, M.A., 1976. An Occurrence of Well Preserved Radiolaria from the Upper Ordovician (Caradocian), Eureka County, Nevada. *Journal of Paleontology* 50, 882–887.
Dzinoridze, R.N., Jousé, A.P., Koroleva-Golikova, G.S., et al., 1978. Diatom

and radiolarian Cenozoic stratigraphy, Norwegian Basin, DSDP Leg 38, in: Talwani, M., Udintsev, G., et al., (Eds.), *Initial Reports of the Deep Sea Drilling Project, Supplement Volume.* U. S. Government Printing Office, Washington, D. C., pp. 289–428.

Ehrenberg, C.G., 1876. Fortsetzung der mikrogeologischen Studien als Gesammt-Uebersichtder mikroskopischen Palaontologie gleichartig analysirter Gebirgsarten der Erde, mit specieller Rucksicht auf den Polycystinen-Mergel von Barbados. Konigliche Akademie der Wissenschaften zu Berlin, Abhandlungen, Jahre 1875. 1–225.

Ehrenberg, C.G., 1847. Uber die mikroskopischen kieselschaligen Polycystinen als machtige Gebirgsmasse von Barbados und uber das Verhaltniss deraus mehr als 300 neuen Arten bestehenden ganz eigenthumlichen Formengruppe jener Felsmasse zu den jetzt lebenden Thieren und zur Kreidebildung. Eine neue Anregung zur Erforschung des Erdlebens. Konigliche Preussische Akademie der Wissenschaften zu Berlin, Bericht, Jahre 1847 40–60.

Ehrenberg, C.G., 1846. Uber eine halibiolithische, von Herrn R. Schomburgk entdeckte, vorherrschend aus mikroskopischen Polycystinen gebildete, Gebirgsmasse von Barbados. Konigliche Preussische Akademie der Wissenschaften zu Berlin, Bericht, Jahre 1846 382–385.

Ehrenberg, C.G., 1844. Einige vorlaufige Resultate seiner Untersuchungen der ihm von der Sudpolreise des Capitain Ross, so wie von den Herren Schanyer und Darwin zugekommenen Materialien uber das Verhalten des kleinsten Lebens in den Oceanen und den Grossten bisher zuganglich. Bericht uber die zu Bekanntmachung geeigneten Verhandlungen der Koniglichen Preufs Akademie der Wissenschaften zu Berlin 1844, 182–207.

Ehrenberg, C.G., 1839. Uber die Bildung der Kreidefelsen und des Kreidemergels durch unsichtbare Organismen. Konigliche Akademie der Wissenschaften zu Berlin, Abhandlungen, Jahre 1838 59–147.

Eisenack, A., 1971. Die Mikrofauna der Ostseekalke (Ordovizium). 3. Graptolithen, Melanoskleriten, Spongien, Radiolarien, Problematika nebst 2 Nachtragen uber Foraminiferen und Phytoplankton. *Neues Jahrbuch für Geologie und Paläeontologie, Monatschaft* 137, 337–357.

Eldredge, N., 1971. The allopatric model and phylogeny in Paleozoic invertebrates. *Evolution* 25, 156–167.

Eldredge, N., Gould, S.J., 1972. Punctuated equilibria: an alternative to phyletic gradualism, in: Schopf, T.J.M. (Ed.), *Models in Paleobiology*. Freeman, San Francisco, pp. 82–115.

Eldrett, J.S., Harding, I., Firth, J., Roberts, A.P., 2004. Magnetostratigraphic calibration of Eocene-Oligocene dinoflagellate biostratigraphy from the Norwegian-Greenland Sea. *Marine Geology* 204, 91–127.

Emiliani, C., 1955. Pleistocene Temperatures. *The Journal of Geology* 63, 538–578.

Empson-Morin, K., 1984. Depth and latitude distribution of Radiolaria in Campanian (Late Cretaceous) tropical and subtropical oceans. *Micropaleontology* 30, 87–115.

Empson-Morin, K., 1981. Campanian Radiolaria from DSDP Site 313, Mid-Pacific Mountains. *Micropaleontology* 27, 249–292.

Enerson, O.D., 2018. Rudolf Ludwig Karl Virchow [WWW Document]. whonamedit.com. URL http://www.

whonamedit.com/doctor.cfm/912.html (accessed 6.21.18).
Enriques, P., 1919. Ricerche sui Radiolari Coloniali. R. Comitato Talassografico Italiano, Mem. 71, pt. 1 1–177.
Entzeroth, A., Hausmann, K., 1993. 175 Years of Protozoology in Germany. Presented at the IX. International Congress of Protozoology, Freie Universitat Berlin, 25th July–l st August 1993 Berlin.
Erba, E., 2004. Calcareous nannofossils and Mesozoic oceanic anoxic events. *Marine Micropaleontology* 52, 85–106.
Erba, E., Bartolini, A., Larson, R.L., 1999. Valanginian Weissert oceanic anoxic event. *Geology* 32, 149–152.
Erbacher, J., 1994. Entwicklung und Palbozeanographie mittelkretazischer Radiolarien der westlichen Tethys (Italien) und des Nordatlantiks. *Tübinger Mikropalaontol. Mitt.* 12, 1–120.
Erbacher, J., Thurow, J., 1997. Influence of Oceanic Anoxic Events on the evolution of mid-Cretaceous Radiolaria in the North Atlantic and Western Tethys. *Marine Micropaleontology* 30, 139–158.
Erbacher, J., Thurow, J., Littke, R., 1996. Evolution patterns of radiolaria and organic matter variations: A new approach to identify sea-level changes in mid-Cretaceous pelagic environments. *Geology* 24, 499–502.
Ezard, T.H.G., Pearson, P.N., Aze, T., Purvis, A., 2012. The meaning of birth and death (in macroevolutionary birth-death models). *Biology Letters* 8, 139–142.
Ezard, T.H.G., Purvis, A., 2016. Environmental changes define ecological limits to species richness and reveal the mode of macroevolutionary competition. *Ecology Letters* 1–8 (web).
Ezard, T.H.G., Thomas, G.H., Purvis, A., 2013. Inclusion of a near-complete fossil record reveals speciation-related molecular evolution. *Methods in Ecology and Evolution* 4, 745–753.
Falkowski, P., Schofield, O., Katz, M.E., Van de Schootbrugge, B., Knoll, A., 2004. Why is the land green and the ocean red?, in: Thierstein, H., Young, J.R. (Eds.), *Coccolithophores: From Molecular Processes to Global Impact*. Springer, Berlin, pp. 429–453.
Fay, A.R., McKinley, G.A., 2014. Global open-ocean biomes: mean and temporal variability. *Earth System Science Data* 6, 273–284.
Fenchel, T., 1987. *Ecology of the Protozoa*. Science Tech Pub./Springer, Madison WI/Berlin.
Feng, Q., He, W., Gu, S., et al., 2007. Radiolarian evolution during the latest Permian in South China. *Global and Planetary Change* 55, 177–192.
Fenner, J., 1985. Cretaceous to Oligocene planktic diatoms, in: Bolli, H.M., Saunders, J.B., Perch-Nielsen, K. (Eds.), *Plankton Stratigraphy*. Cambridge University, Cambridge, pp. 713–762.
Fensome, R.A., MacRae, R.A., Moldowan, J.M., et al., 1996. The early Mesozoic radiation of dinoflagellates. *Paleobiology* 22, 329–338.
Fimbres-Castro, C., Álvarez-Borrego, J., Vázquez-Martínez, I., et al., 2013. Nonlinear correlation by using invariant identity vectors signatures to identify plankton. *Gayana* 77, 105–124.
Finlay, B.J., Esteban, G.F., 2001. Ubiquitous microbes and ecosystem function. *Limnetica* 20, 31–43.
Florindo, F., Bohaty, S.M., Erwin, P.S., et al., 2003. Magnetobiostratigraphic chronology and palaeoenvironmental history of Cenozoic sequences from ODP sites 1165 and 1166, Prydz Bay, Antarctica. *Palaeogeography, Palaeoclimatology, Palaeoecology* 198, 69–100.
Florindo, F., Farmer, R.K., Harwood, D.M., et al., 2013. Paleomagnetism

and biostratigraphy of sediments from Southern Ocean ODP Site 744 (southern Kerguelen Plateau): Implications for early-to-middle Miocene climate in Antarctica. *Global and Planetary Change, Magnetic iron minerals in sediments and their relation to geologic processes, climate, and the geomagnetic field* 110, 434–454. https://doi.org/10.1016/j.gloplacha.2013.05.004

Fontorbe, G., Frings, P.J., De La Rocha, C.L., et al., 2017. Enrichment of dissolved silica in the deep Equatorial Pacific during the Eocene- Oligocene. *Paleoceanography* 32, 848–863.

Foreman, H.P., 1975. Radiolaria from the North Pacific, Deep Sea Drilling Project, Leg 32, in: Larson, R.L., Moberly, R., et al. (Eds.), *Initial Reports of the Deep Sea Drilling Project Radiolaria from the North Pacific, Deep Sea Drilling Project, Leg 32*. U.S. Government Printing Office, pp. 579–676.

Foreman, H.P., 1973. Radiolaria from DSDP Leg 20, in: Heezen, B.C., MacGregor, J.D., et al., (Eds.), *Initial Reports of the Deep Sea Drilling Project Radiolaria from DSDP Leg 20*. U.S. Government Printing Office, pp. 249–305.

Foreman, H.P., 1968. Upper Maestrichtian Radiolaria of California. *Special Papers in Palaeontology* 3, 1–82.

Foreman, H.P., Riedel, W.R., 1972. *Catalogue of Polycystine Radiolaria, Series 1 (1834–1900)*. American Museum of Natural History, Special Publication.

Foster, W.J., Twitchett, R.J., 2014. Functional diversity of marine ecosystems after the Late Permian mass extinction event. *Nature Geoscience* 7, 233–238. https://doi.org/10.1038/ngeo2079

Fraass, A.J., Kelly, D.C., Peters, S.E., 2015. Macroevolutionary history of the planktic Foraminifera. *Annual Review of Earth and Planetary Sciences* 43, 139–166.

Fraguas, A., Comas-Rengifo, M.J., Gómez, J.J., Goy, A., 2012. The calcareous nannofossil crisis in Northern Spain (Asturias province) linked to the Early Toarcian warming-driven mass extinction. *Marine Micropaleontology* 94–95, 58–71.

Friedrich, O., Erbacher, J., Moriya, K., Wilson, P.A., 2008. Warm saline intermediate waters in the Cretaceous tropical Atlantic Ocean. *Nature Geoscience* 1, 453–457.

Fujimoto, H., 1938. Radiolarian remains discovered in a crystalline schist of the Sambagawa system. *Proceedings of the Imperial Academy of Japan* 14, 252–254.

Funakawa, S., Nishi, H., 2008. Radiolarian faunal changes during the Eocene-Oligocene transition in the Southern Ocean (Maud Rise, ODP Leg 113, Site 689) and its significance in paleoceanographic change. *Micropaleontology* 54, 15–26.

Funakawa, S., Nishi, H., 2005. Late middle Eocene to late Oligocene radiolarian biostratigraphy in the Southern Ocean (Maud Rise, ODP Leg 113, Site 689). *Marine Micropaleontology* 54, 213–247.

Funakawa, S., Nishi, H., Moore, T.C., Nigrini, C.A., 2006. Radiolarian faunal turnover end paleoceanographic change around Eocene/Oligocene boundary in the cental equatorial Pacific, ODP Leg 199, Holes 1218A, 1219A, and 1220A. *Palaeogeography, Palaeoclimatology, Palaeoecology* 230, 183–203.

Funnell, B.M., Riedel, W.R. (Eds.), 1971. *The Micropaleontology of Oceans*. Cambridge University Press, London.

Furbish, D.J., Arnold, A.J., 1997. Hydrodynamic strategies in the morphological evolution of spinose planktonic foraminifera. *Geological Society of America Bulletin* 109, 1055–1072.

Futuyma, D.J., 1998. *Evolutionary Biology*. Sinauer, Sunderland, MA.

Galli, M.T., Jadoul, F., Bernasconi, S.M., Weissert, H., 2005. Anomalies in global carbon cycling and extinction at the Triassic/Jurassic boundary: evidence from a marine C-isotope record. *Palaeogeography, Palaeoclimatology, Palaeoecology* 216, 203–214.

Gary, A.C., Johnson, G.W., Ekart, D.D., et al., 2005. A method for two-well correlation using multivariate biostratigraphical data, in: Powell, A.J., Riding, J.B. (Eds.), *Recent Developments in Applied Biostratigraphy*. Geological Society, London, pp. 205–217.

Gasman, D., 2004. *The Scientific Origins of National Socialism*, new edition. Transaction, New Brunswick.

Gibbs, M.T., Bluth, G.J.S., Fawcett, P.J., Kump, L.R., 1999. Global chemical erosion over the last 250 my: Variations due to changes in paleogeography, paleoclimate, and paleogeology. *American Journal of Science* 299, 611–651.

Gibbs, S.J., Brown, P.R., Sessa, J.A., Bralower, T.J., 2006. Nannoplankton Extinction and Origination Across the Paleocene- Eocene Thermal Maximum. *Science* 314, 22.

Giese, M., Schmidt-Effing, R., 1989. Eine Radiolarienfauna aus dem Unter-Karbon von Amonau bei Wetter (Rheinisches Schiefergebirge/Hessen). *Geologica et Palaeontologica* 23, 71–81.

Gilg, I.C., Amaral-Zettler, L.A., Counway, P.D., et al., 2010. Phylogenetic affiliations of mesopelagic Acantharia and Acantharian-like environmental 18S rRNA genes off the Southern California Coast. *Protist* 161, 197–211.

Gohau, G., 1990. *A History of Geology*. Rutgers University Press, New Brunswick.

Goldfuss, G.A., 1820. *Handbuch der Zoologie*. J. L. Schrag, Nuremberg.

Goldschmidt, R., 1940. *The Material Basis of Evolution*. Yale University Press, New Haven, CT, USA.

Goll, R.M., 1989. A synthesis of Norwegian Sea biostratigraphies: ODP Leg 104 on the Vøring Plateau, in: *Proceedings of the Ocean Drilling Program, Scientific Results*. Ocean Drilling Program, College Station, TX, pp. 777–828.

Goll, R.M., 1985. Exotic outcross between late Miocene polycystine radiolaria from the central equatorial Pacific Ocean. *Geology* 13.

Goll, R.M., 1979. The Neogene evolution of Zygocircus, Neosemantis and Callimitra: their bearing on nassellarina classification A revision of the Plagiacanthoidea. *Micropaleontology* 25, 365–396.

Goll, R.M., 1976. Morphological intergradation between modern populations of Lophospyris and Phormospyris (Trissocyclidae, Radiolaria). *Micropaleontology* 22, 379–418.

Goll, R.M., 1968. Classification and phylogeny of Cenozoic Trissocyclidae (Radiolaria) in the Pacific and Caribbean Basins. Part I. *Journal of Paleontology* 42, 1409–1432.

Goll, R.M., Bjørklund, K.R., 1989. A new radiolarian biostratigraphy of Norwegian Sea sediments: ODP Leg 104, in: Eldholm, O., Thiede, J., Taylor, E. (Eds.), *Proceedings of the Ocean Drilling Program, Scientific Results*. Ocean Drilling Program, College Station, pp. 697–738.

Goll, R.M., Bjørklund, K.R., 1974. Radiolaria in surface sediments of the South Atlantic. *Micropaleontology* 20, 38–75.

Goll, R.M., Bjørklund, K.R., 1971. Radiolaria in surface sediments of the North Atlantic Ocean. *Micropaleontology* 17, 434–454.

Goll, R.M., Caulet, J.P., 1985. Sur la présence d'un Collosphaeridae dans les sédiments Eocene supérieur du Plateau de Kerguelen-Heard. *Comptes rendus de*

l'Académie des Sciences (Paris), série II 301, 1375–1380.
Goodbody, Q.H., 1986. Wenlock Palaeoscenidiidae and Entactiniidae (Radiolaria) from the Cape Phillips Formation of the Canadian Arctic Archipelago. *Micropaleontology* 32, 129–157.
Gorbushina, A.A., Kort, R., Schulte, A., et al., 2007. Life in Darwin's dust: intercontinental transport and survival of microbes in the nineteenth century. *Environmental Microbiology* 9, 2911–2922.
Goričan, S., 1994. Jurassic and Cretaceous radiolarian biostratigraphy and sedimentary evolution of the Budva Zone (Dinarides, Montenegro). *Memoires de Geologie (Lausanne)* 18, 1–120.
Goričan, S., Carter, E.S., Guex, J., et al., 2013. Evolutionary patterns and palaeobiogeography of Pliensbachian and Toarcian (Early Jurassic) Radiolaria. *Palaeogeography, Palaeoclimatology, Palaeoecology* 386, 620–636.
Goričan, S., O'Dogherty, L., Baumgartner, P.O., et al., 2018. Mesozoic radiolarian biochronology – current status and future directions. *Revue de Micropaléontologie* 61, 165–189.
Gotelli, N.J., Colwell, R.K., 2011. Estimating species richness, in: Magurran, A.E., McGill, B.J. (Eds.), *Frontiers in Measuring Biodiversity.* Oxford University Press, New York, pp. 39–54.
Gould, S.J., 1980. Is a new and general theory of evolution emerging? *Paleobiology* 6, 119–130.
Gould, S.J., 1977. *Ontogeny and Phylogeny*. Harvard University Press, Cambridge, MA.
Gould, S.J., Eldredge, N., 1977. Punctuated Equilibria: the tempo and mode of evolution reconsiderd. *Paleobiology* 3, 115–151.
Gowing, M.A., Bentham, W.N., 1994. Feeding ecology of phaeodarian radiolarians at the VERTEX North Pacific time series site. *Journal of Plankton Research* 16, 707–719.
Gradstein, F., Agterberg, F., Ogg, G.J., et al., 1995. A Triassic, Jurassic, and Cretaceous time scale, in: Berggren, W.A., Kent, D.V., Aubry, M.-P., Hardenbol, J. (Eds.), *Geochronology, Time Scales and Global Stratigraphic Correlations.* SEPM, Tulsa, OK, pp. 95–128.
Gradstein, F.M., Ogg, J.G., Schmitz, M.D., Ogg, G.M., 2012. *The Geologic Time Scale.* Elsevier, Amsterdam.
Gradstein, F.M., Ogg, J.G., Smith, A.G., 2004. *A Geologic Time Scale 2004.* Cambridge University Press, Cambridge.
Granlund, A., 1986. Size and shape patterns in the recent radiolarian genus Antarctissa from a south Indian Ocean transect. *Marine Micropaleontology* 11, 243–250.
Gries, R.R., 2017. Three women, one breakthrough: unsung heroines who transformed oil exploration. *AAPG Explorer* 38, 20–21, 27.
Groom, S., Sathyendranath, S., Ban, Y., et al., 2019. Satellite Ocean Colour: Current Status and Future Perspective. *Frontiers in Marine Science* 6. https://doi.org/10.3389/fmars.2019.00485
Guex, J., 2011. Some recent "refinements" of the Unitary Association Method: a short discussion. *Lethaia* 44, 247–249.
Guex, J., 1991. *Biochronological Correlations*. Springer Verlag.
Guex, J., O'Dogherty, L., Carter, E.S., et al., 2012. Geometrical transformations of selected Mesozoic radiolarians. *Geobios* 45, 541–554.
Guidi, L., Chaffron, S., Bittner, L., et al., 2016. Plankton networks driving carbon export in the oligotrophic ocean. *Nature* 532, 465–470. https://doi.org/10.1038/nature16942
Gutierrez-Rodriguez, A., Stukel, M.R., dos Santos, A.L., et al., 2018. High contribution of Rhizaria (Radiolaria) to vertical export in the California

Current Ecosystem revealed by DNA metabarcoding. *ISMEJournal* 13, 964–976.
Habura, A., Wegener, L., Travis, J.L., Bowser, S.S., 2005. Structural and Functional Implications of an Unusual Foraminiferal β-Tubulin. *Molecular Biology and Evolution* 22, 2000–2009. https://doi.org/10.1093/molbev/msi190
Haddock, S.H.D., Moline, M.A., Case, J.F., 2010. Bioluminescence in the sea. *Annual Review of Marine Science* 2, 443–493.
Haeckel, E., 1904. *Kunstformen der Natur.* Bibliographisches Institut, Leipzig.
Haeckel, E., 1887. Report on the Radiolaria collected by H.M.S. Challenger during the years 1873–1876. Report on the Scientific Results of the Voyage of the H.M.S. Challenger, *Zoology* 18, clxxxviii + 1803.
Haeckel, E., 1881. Entwurf eines Radiolarien-Systems auf Grund von Studien der Challenger-Radiolarien (Basis for a radiolarian classification from the study of Radiolaria of the Challenger collection). *Jenaische Zeitschrift fur Naturwissenschaft* 15, 418–472.
Haeckel, E., 1866. *Generelle Morphologie der Organismen: allgemeine Grundzüge der organischen Formen-Wissenschaft, mechanisch begründet durch die von Charles Darwin reformirte Descendenz-Theorie.* George Reimer, Berlin.
Haeckel, E., 1862. *Die Radiolarien (Rhizopoda Radiaria). Eine monographie.* Reimer, Berlin, Germany.
Haecker, V., 1906. Über einige grosse Tiefsee-Radiolarien. *Zoologischer Anzeiger* 30, 878–895.
Hagstrum, J.T., Murchey, B.L., 1996. Paleomagnetism of Jurassic radiolarian chert above the Coast Range ophiolite at Stanley Mountain, California, and implications for its paleogeographic origins. *Geological Society of America, Bulletin* 108, 643–652.
Hallam, A., 1994. Jurassic climates as inferred from the sedimentary and fossil record, in: Allen, J.R.L., Hoskins, B.J., Sellwood, B.W., Spicer, R.A., Valdes, P.J. (Eds.), *Palaeoclimates and Their Modelling: With Special Reference to the Mesozoic Era.* Springer, Dordrecht, pp. 79–88.
Hamblin, J.D., 2005. *Oceanographers and the Cold War: Disciples of Marine Science.* University of Washington Press, Seattle.
Hammer, Ø., Harper, D.A.T., 2008. *Paleontological Data Analysis.* John Wiley & Sons.
Hammer, O., Harper, D.A.T., 2001. PAST: paleontological statistics software package for education and data analysis. *Palaeontologia Electronica* 4, web.
Hancock, J.M., 1977. The historical development of concepts of biostratigraphic correlation, in: Kauffman, E., Hazel, J. (Eds.), *Concepts and Methods of Biostratigraphy.* Hutchinson Ross Publishing Company, Stroudsburg, Pannsylvania, pp. 3–22.
Harmon, L.J., Harrison, S., 2015. Species diversity is dynamic and unbounded at local and continental scales. *The American Naturalist* 185, 584–593. https://doi.org/10.1086/680859
Harper, H.E., Knoll, A.H., 1975. Silica, diatoms and Cenozoic radiolarian evolution. *Geology* 3, 175–177.
Harris, A., Sweet, W.C., 1989. Mechanical and chemical techniques for separating microfossils from rock, sediment and residue matrix. *Paleontological Society Special Publications* 4, 70–86.
Harvey, T.H.P., Vélez, M.I., Butterfield, N.J., 2012. Exceptionally preserved crustaceans from western Canada reveal a cryptic Cambrian radiation. *Proceedings of the National Academy of Sciences of the United States of America* 109, 1589–1594. https://doi.org/10.1073/pnas.1115244109

Harwood, D.M., Nikolaev, V.A., 1995. Cretaceous diatoms: morphology, taxonomy, biostratigraphy, in: Blome, C.D., Whalen, P.M., Reed, K.M. (Eds.), *Siliceous Microfossils*. Paleontological Society, Knoxville, pp. 81–106.

Hatakeda, K., Suzuki, N., Matsuoka, A., 2007. Quantitative morphological analyses and evolutionary history of the Middle Jurassic polycystine radiolarian genus Striatojaponocapsa Kozur. *Marine Micropalaeontology* 63, 39–56.

Haugen, P., Simon, D.M., Bhattacharya, D., 2005. The natural history of group I introns. *Trends in Genetics* 21, 111–119. https://doi.org/10.1016/j.tig.2004.12.007

Hay, W.W., 2008. Evolving ideas about the Cretaceous climate and ocean circulation. *Cretaceous Research* 29, 725–753.

Hay, W.W., Floegel, S., 2012. New thoughts about the Cretaceous climate and oceans. *Earth-Science Reviews* 115, 262–272.

Hay, W.W., Southam, J.R., 1978. Quantifying Biostratigraphic Correlation. *Annual Review of Earth and Planetary Sciences* 6, 353–375. https://doi.org/10.1146/annurev.ea.06.050178.002033

Hayes, J.M., Strauss, H., Kaufman, A.J., 1999. The abundance of 13C in marine organic matter and isotopic fractionation in the global biogeochemical cycle of carbon during the past 800 Ma. *Chemical Geology* 161, 103–125.

Hays, J.D., 1970. Stratigraphy and evolutionary trends of radiolaria in North Pacific deep-sea sediments, in: Hays, J.D. (Ed.), *Geological Investigations of the North Pacific, Memoirs*. Geological Society of America, pp. 185–218.

Hays, J.D., 1965. Radiolaria and late Tertiary and Quaternary history of Antarctic Seas, in: Llano, G.A. (Ed.), *Biology of Antarctic Seas II*. American Geophysical Union, Antarctic Research Series, pp. 125–184.

Hays, J.D., Cook, H.E.I., Jenkins, G., et al., 1972. *Initial Reports of the Deep Sea Drilling Project, Volume IX*. U. S. Government Printing Office, Washington, D. C.

Hays, J.D., Imbrie, J., Shackleton, N.J., 1976a. Variations in the earth's orbit: pacemaker of the ice ages. *Science* 194, 1121–1132.

Hays, J.D., Lozano, J.A., Shackleton, N., Irving, G., 1976b. Reconstruction of the Atlantic and western Indian Ocean sectors of the 18,000 B.P. Antarctic Ocean, in: Cline, R.M., Hays, J.D. (Eds.), *Investigation of Late Quaternary Paleoceanography and Paleoclimatology*. Geological Society of America Memoir, Boulder, CO, USA, pp. 337–372.

Hays, J.D., Martinson, D.G., Morley, J.J., 2013. Biological and climatic consequences of a cold, stratified, high latitude ocean. *Quaternary Science Reviews* 82, 78–92.

Hays, J.D., Morley, J.J., 2004. The Sea of Okhotsk: A window on the Ice Age Ocean. *Deep Sea Research Part I: Oceanographic Research Papers* 51, 593–618.

Hays, J.D., Opdyke, N.D., 1967. Antarctic Radiolaria, magnetic reversals and climatic change. *Science* 158, 1001–1011.

Hays, J.D., Saito, T., Opdyke, N.D., Burckle, L.H., 1969. Plio-Pleistocene sediments of the equatorial Pacific: their paleomagnetic, biostratigraphic, and climatic record. *Geological Society of America, Bulletin* 80, 1481–1514.

Hays, J.D., Shackleton, N.J., 1976. Globally synchronous extinction of the radiolarian Stylatractus universus. *Geology* 4, 649–652.

Hein, J.R., Yeh, H.W., 1983. Oxygen-isotope composition of secondary silica phases, Costa Rica Rift, Deep Sea Drilling Project Leg 69, in: Cann, J.R., Langseth, M.G., Honnorez, J., von Herzen, R.P., White, S.M. (Eds.), *Initial Reports of the Deep Sea Drilling Project*.

U. S. Government Printing Office, Washington, D. C., pp. 395–422.
Hemleben, C., Spindler, M., Anderson, O.R., 1989. *Modern Planktonic Foraminifera*. Springer Verlag, New York.
Hendricks, J.R., Saupe, E.E., Myers, C.E., et al., 2014. The generification of the fossil record. *Paleobiology* 40, 511–528.
Henjes, J., Assmy, P., Klaas, C., Smetacek, V., 2007. Response of the larger protozooplankton to an iron-induced phytoplankton bloom in the Polar Frontal Zone of the Southern Ocean (EisenEx). *Deep-Sea Research* 54, 774–791.
Hertwig, R., 1932. Uber den Bau der Perpyleen (Sphaeroideen). Abhandlungen der Bayerischen Akademie der Wissenschaften, *Mathematisch-naturwissenschaftliche Abteilung*, new ser. 1–40.
Hertwig, R., 1879. *Der Organismus der Radiolarien* iv + 149.
Hertwig, R., 1876. *Zur Histologie der Radiolarien*. Untersuchungen über den Bau und die Entwicklung der Sphaerozoiden und Thalassicolliden. W. Engelmann, Leipzig.
Hills, S.J., Thierstein, H.R., 1989. Plio-Pleistocene calcareous plankton biochronology. *Marine Micropaleontology* 14, 67–96.
Hinde, G.J., 1890. Notes on Radiolaria from the Lower Palaeozoic rocks (Llandeilo-Caradoc) of the south of Scotland. *Annals and Magazine of Natural History* 6, 40–59.
Hinnov, 2004. Earth's orbital parameters and cycle stratigraphy, in: Gradstein, F.M., Ogg, J.G., Smith, A.G. (Eds.), *A Geologic Time Scale 2004*. Cambridge University, Cambridge, pp. 55–62.
Ho, W.-C., Ohya, Y., Zhang, J., 2017. Testing the neutral hypothesis of phenotypic evolution. *Proceedings of the National Academy of Sciences U S A* 114, 12219–12224.
Hodgkinson, R.L., 1991. Microfossil processing: a damage report. *Micropaleontology* 37, 320–326.
Hojnos, R., 1916. Beiträge zur Kenntnis der ungarischen fossilen Radiolarien. *Földtany Közlöny* 46, 340–364.
Holdsworth, B.K., 1977. Paleozoic Radiolaria: Stratigraphic distribution in Atlantic Borderlands, in: Swain, F.M. (Ed.), *Stratigraphic Micropaleontology of Atlantic Basin and Borderlands*. Elsevier, Amsterdam, pp. 167–184.
Holdsworth, B.K., 1969. Namurian Radiolaria of the genus Ceratoikiscum from Staffordshire and Derbyshire, England. *Micropaleontology* 15, 221–229.
Holdsworth, B.K., 1966. Radiolaria from the Namurian of Derbyshire. *Palaeontology* 9, 319–329.
Holdsworth, B.K., Jones, D.L., 1980. Preliminary radiolarian zonation for late Devonian through Permian time. *Geology* 8, 281–285.
Holland, S.M., 1995. The Stratigraphic Distribution of Fossils. *Paleobiology* 21, 92–109.
Hollande, A., 1953. Complements sur la cytologie des Acanthaires et des Radiolaires, in: Grasse, P.P. (Ed.), *Traitë de Zoologie*. Masson et Cie, Paris, pp. 1089–1100.
Hollande, A., Enjumet, M., 1960. Cytologie, évolution et systématique des Sphaeroides (Radiolaires). *Archives du Museum national d'histoire naturelle, Paris* 7, 134.
Hollis, C., 1997. Cretaceous-Paleocene Radiolaria from Eastern Marlborough, New Zealand, *Institute of Geological and Nuclear Sciences. Monograph*, GNS Science, Lower Hutt, New Zealand.
Hollis, C., Rogers, K.A., Parker, R.J., 1995. Siliceous plankton bloom in the earliest Tertiary of Marlborough, New Zealand. *Geology* 23, 835–838.
Hollis, C.J., 2006. Radiolarian faunal turnover through the Paleocene-Eocene transition, Mead Stream, New Zealand.

Eclogae geologicae Helvetiae 99, S79–S99.
Hollis, C.J., 2003. The Cretaceous/Tertiary boundary event in New Zealand: profiling mass extinction. *New Zealand Journal of Geology & Geophysics* 46, 307–321.
Hollis, C.J., 1993a. Latest Cretaceous to Late Paleocene radiolarian biostratigraphy: A new zonation from the New Zealand region. *Marine Micropaleontology* 21, 295–328.
Hollis, C.J., 1993b. Radiolarian faunal turnover through the Paleocene-Eocene transition, Mead Stream, New Zealand. *Eclogae geologicae Helvetiae* 99, S79–S99.
Hollis, C.J., Kimura, K., 2001. A unified radiolarian zonation for the Late Cretaceous and Paleocene of Japan. *Micropaleontology* 47, 235–255.
Honjo, S., Connell, J.F., Sachs, P.L., 1980. Deep-ocean sediment trap; design and function of PARFLUX Mark II. *Deep Sea Research Part A* 27, 745–753.
Honjo, S., Doherty, K.W., 1988. Large aperture time-series sediment traps; design objectives, construction and application. *Deep Sea Research Part A* 35, 133–149.
Hori, R., 1990. Lower Jurassic Radiolarian Zones of SW Japan. *Transactions and Proceedings of the Palaeontological Society of Japan, new Series* 159, 562–586.
Hori, R.S., Cho, C.-f, Umeda, H., 1993. Origin of cyclicity in Triassic – Jurassic radiolarian bedded cherts of the Mino accretionary complex from Japan. *Island Arc* 3, 170–180.
Hori, R.S., Higuchi, Y., Fujiki, T., et al., 2007. Geochemistry of the Oruatemanu Formation, Arrow Rocks, Northland, New Zealand. *GNS Science Monograph* 24, 123–156.
Hori, R.S., Shinki, T., Iwakiri, A., et al., 2017. A growth model and strategy of forming siliceous skeletons of living Spumellaria (Radiolaria). Presented at the Interrad XV, Niigata, Japan.
Hori, R.S., Yamakita, S., Ikehara, M., et al., 2011. Early Triassic (Induan) Radiolaria and carbon-isotope ratios of a deep-sea sequence from Waiheke Island. North Island, New Zealand. *Palaeoworld* 20, 166–178.
Hou, Y., Sierra, R., Bassen, D., et al., 2013. Molecular evidence for beta-tubulin neofunctionalization in Retaria (Foraminifera and radiolarians). *Molecular Biology and Evolution* 30, 2487–2493.
Hovasse, R., Brown, E.M., 1953. Contribution a la connaissance des Radiolaires et de leurs parasites syndiniens. *Annales des Sciences Naturelles, Zoologie er Biologie Animale*, ser. 11 15, 405–437.
Hsiang, A.Y., Brombacher, A., Rillo, M.C., et al., 2019. Endless Forams: >34,000 Modern Planktonic Foraminiferal Images for Taxonomic Training and Automated Species Recognition Using Convolutional Neural Networks. *Paleoceanography and Paleoclimatology* 34, 1157–1177. https://doi.org/10.1029/2019PA003612
Huang, C., Tong, J., Hinnov, L., Zhong, Q.C., 2011. Did the great dying of life take 700 k.y.? Evidence from global astronomical correlation of the Permian-Triassic boundary interval. *Geology* 39, 779–782.
Huber, B.T., Buzas, M.A., Drew, A., Jett, J.A., 2013. History of foraminiferal collections at the Smithsonian Institution, in: Bowden, A.J., Gregory, F.J., Henderson, A.S. (Eds.), *Landmarks in Foraminiferal Micropaleontology*, The Micropalaeontological Society, Special Publications. The Geological Society, London, pp. 71–83.
Huber, M., Brinkhuis, H., Stickley, C.E., et al., 2004. Eocene circulation of the Southern Ocean: Was Antarctica kept warm by subtropical waters? *Paleoceanography* 19, 1–12 (web).

Huisman, J., Weissing, F.J., 1999. Biodiversity of plankton by species oscillations and chaos. *Nature* 402, 407–410.

Hull, D.M., 1997. Upper Jurassic Tethyan and southern boreal radiolarians from western North America. *Micropaleontology* 43, 1–202.

Hull, D.M., 1996. Paleoceanography and biostratigraphy of Paleogene radiolarians from the Norwegian-Greenland Sea, in: Thiede, J., Myhre, A.M., Firth, J.V., Johnson, G.L., Ruddiman, W. (Eds.), *Proceedings of the Ocean Drilling Program, Scientific Results, Volume 151*. Ocean Drilling Program, College Station, Texas, pp. 125–152.

Hull, D.M., Osterman, L.E., Thiede, J., 1996. Biostratigraphic synthesis of Leg 151, North Atlantic-Arctic Gateways, in: Thiede, J., Myhre, A.M., Firth, J.V., Johnson, G.L., Ruddiman, W. (Eds.), *Proceedings of the Ocean Drilling Program, Scientific Results, Volume 151*. Ocean Drilling Program, College Station, Texas, pp. 627–644.

Hulsemann, K., 1963. Radiolaria in plankton from the Arctic drifting station T-3,including the description of three new species, in: Arctic Institute of North America Technical Paper No. 13. p. 52.

Hüneke, H., Henrich, R., 2011. Pelagic sedimentation in modern and ancient oceans, in: Hüneke, H., Mulder, T. (Eds.), *Deep-Sea Sediments*. pp. 215–351.

Hurd, D.C., 1973. Interactions of biogenic opal, sediment and seawater in the Central Equatorial Pacific. *Geochimica et Cosmochimica Acta* 37, 2257–2282.

Hurd, D.C., 1972. Factors affecting solution rate of biogenic opal in seawater. *Earth and Planetary Science Letters* 15, 411–417.

Hutchinson, G.E., 1961. The paradox of the plankton. *The American Naturalist* 95, 137–145.

Hutson, W., 1977. Transfer functions under no-analog conditions: experiments with Indian Ocean planktonic foraminifera. *Quaternary Research* 8, 355–367.

Huxley, T.H., 1851. Zoological notes and observations made on board H.M.S. Rattlesnake. *Annals and Magazine of Natural History*, series 2 viii, 28; 434–443, 1 plate.

Ichikawa, K., 1982. History of paleontology on Mesozoic and Paleozoic radiolarians in Japan. *News of Osaka Micropaleontologists*, special volume 5, 1–9.

Ichikawa, K., 1946. A study on the Radiolaria as index-fossils. *Journal of the Geological Society of Japan (Chishitsugaku Zasshi)* 52, 25–28.

Ihoriya, N., Hori, R.S., Ikehara, M., 2009. Constraints upon anoxic water mass and variations in carbon isotopes during OAE1a (early Cretaceous) from deep-sea sedimentary rocks of the Pacific Ocean. *News of Osaka Micropaleontologists* 14, 297–315.

Ikeda, M., Hori, R.S., 2014. Effects of Karoo–Ferrar volcanism and astronomical cycles on the Toarcian Oceanic Anoxic Events (Early Jurassic). *Palaeogeography, Palaeoclimatology, Palaeoecology* 410, 134–142.

Ikenoue, T., Bjørklund, K., Kruglikova, S.B., et al., 2015. Flux variations and vertical distributions of siliceous Rhizaria (Radiolaria and Phaeodaria) in the western Arctic Ocean: indices of environmental changes. *Biogeosciences* 12, 2019–2046.

Ikenoue, T., Takahashi, K., Tanaka, S., 2012. Fifteen year time-series of radiolarian fluxes and environmental conditions in the Bering Sea and the central subarctic Pacific, 1990–2005. *Deep Sea Research Part II: Topical Studies in Oceanography* 61–64, 17–49.

Imbrie, J., Imbrie, K.P., 1979. *Ice Ages*. Enslow, Short Hills, New Jerey.

Imbrie, J., Kipp, N.G., 1971. A new micropaleontological method for quantitative paleoclimatology: application to a late Pleistocene Carribean core, in: Turekian, K.K. (Ed.), *Late Cenozoic Glacial Ages*. Yale University Press, New Haven, pp. 71–181.

Irwin, D.E., Irwin, J.H., Price, T.D., 2002. Ring species as bridges between microevolution and speciation, in: Hendry, A.P., Kinnison, M.T. (Eds.), *Microevolution Rate, Pattern, Process*. Springer, Netherlands, pp. 223–243.

Isakova, T.N., Nazarov, B.B., 1986. Stratigrafiya i mikrofauna pozdnego karbona-ranney permi Yuzhnogo Urala. *Akademiiya Nauk SSSR, Ordena Trudovogo Krasnogo Znameni Geologicheskiy Institut, Nauka, Moskow* 402, 1–183.

Ishida, K., Hashimoto, H., Yamasaki, T., et al., 2010. Significance of the direct correlation of ammonite and radiolarian zones in the Izumi Group for integrated biostratigraphy of Late Cretaceous NW paleo-Pacific region. *Natural Science Research University Tokushima* 24, 27–31.

Ishida, N., 2009. Jurassic to Early Cretaceous accretionary complexes and Upper Jurassic trench-slope basin deposits of the Southern Chichibu Terrane in the Itsuki-Gokanosho area, western Kyushu. *News of Osaka Micropaleontologists (NOM), Special Volume* 14, 375–403.

Ishiga, H., 1990. Paleozoic radiolarians, in: Ichikawa, K. (Ed.), Pre-Cretaceous Terranes of Japan, IGCP Project 224: Pre-Jurassic Evolution of Eastern Asia. *IGCP Project 224*, Osaka, Japan, pp. 285–295.

Ishiga, H., 1983. Morphological change in the Permian Radiolaria, Pseudoalbaillella scalprata in Japan. *Transactions and Proceedings of the Palaeontological Society of Japan*, new series 1–8.

Ishiga, H., 1982. Late Carboniferous and Early Permian radiolarians from the TambaBelt, Southwest Japan. Earth Science, *Journal of the Association for the Geological Collaboration in Japan* 36, 333–339.

Ishiga, H., Imoto, N., Yoshida, M., Tanabe, T., 1984. Early Permian radiolarians from the Tamba Belt, Southwest Japan. *Earth Science (Chikyu Kagaku)* 38, 44–54.

Ishiga, H., Kito, T., Imoto, N., 1982. Permian radiolarian biostratigraphy. *News of Osaka Micropaleontologists, special volume* 5, 17–26.

Ishii, Y., Suzuki, N., Kano, H., 2009. Berriasian–Barremian (Early Cretaceous) radiolarians from paleo-Pacific regions (DSDP and ODP Holes 463, 800A, 801B, 765C, 1213B, and the Goshikigahama bedded red shale of the Northern Shimanto Belt. *News of Osaka Micropaleontologists* 14, 317–373.

Ishitani, Y., Ishikawa, S., Inagaki, Y., Tsuchiya, M., Takahashi, K., Takishita, K., 2011. Multigene phylogenetic analyses including diverse radiolarian species support the "Retaria" hypothesis – The sister relationship of Radiolaria and Foraminifera. *Marine Micropaleontology* 81, 32–42.

Ishitani, Y., Takahashi, K., 2007. The vertical distribution of Radiolaria in the waters surrounding Japan. *Marine Micropalaeontology* 65, 113–136.

Ishitani, Y., Takahashi, K., Okazaki, Y., Tanaka, S., 2008. Vertical and geographic distribution of selected radiolarian species in the North Pacific. *Micropaleontology* 54, 27–40.

Ishitani, Y., Takishita, K., 2015. Molecular evidence for wide vertical distribution of the marine planktonic protist Larcopyle buetschlii (Radiolaria) in a semi-enclosed marginal sea. *Journal of Plankton Research* 37, 851–856.

Ishitani, Y., Ujié, Y., De Vardas, C., et al., 2012a. Phylogenetic relationships and evolutionary patterns of the order Collodaria (Radiolaria). PLoS ONE 7, e35775.

Ishitani, Y., Ujiié, Y., De Vardas, C., et al., 2012b. Two distinct lineages in the radiolarian Order Spumellaria having different ecological niches. *Deep Sea Research II* 61–64, 172–178.

Ishitani, Y., Ujiié, Y., K., T., 2014. Uncovering sibling species in Radiolaria: Evidence for ecological partitioning in a marine planktonic protist. *Molecular Phylogenetics and Evolution* 78, 215–222.

Isozaki, Y., 2009. Illawarra Reversal: the fingerprint of a superplume that triggered Pangena breakup and the end-Guadalupian (Permian) mass extinction. *Gondwana Research* 15, 421–432.

Isozaki, Y., Kawahata, H., Minoshima, K., 2012. Capitanian (Middle Permian) mass extinction and recovery in western Tethys: a fossil, facies and ∂13C study from Hungary and Hydra Island (Greece). *Palaios* 27, 78–89.

Isozaki, Y., Kawahata, H., Minoshima, K., 2007. The Capitanian (Permian) Kamura cooling event: the beginning of the Paleozoic-Mesozoic transition. *Palaeoworld* 16, 16–30.

Itaki, T., Taira, Y., Kuwamori, N., et al., 2020. Automated collection of single species of microfossils using a deep learning–micromanipulator system. *Progress in Earth and Planetary Science* 7, 19. https://doi.org/10.1186/s40645-020-00332-4

Itaki, T., Taira, Y., Kuwamori, N., et al., 2018. A new tool for microfossil analysis in the Southern Ocean – automatic image collector equipped with deep learning. *American Geophysical Union Annual Fall Meeting Abstracts.*

Itsweire, E.C., Osborn, T.R., 1988. Microstructure and vertical velocity shear distribution in Monterey Bay, in: Nihoul, J.C.J., Jamart, B.M. (Eds.), *Small-Scale Turbulence and Mixing in the Ocean*. Elsevier, Amsterdam, pp. 213–228.

Jablonski, D., 1994. Extinctions in the fossil record. *Philosophical Transactions of the Royal Society B* 344, 11–17.

Jackson, G.A., 1993. Flux feeding as a mechanism for zooplankton grazing and its implications for vertical particulate flux 1. *Limnology and Oceanography* 38, 1328–1331. https://doi.org/10.4319/lo.1993.38.6.1328

Jacot des Combes, H., Abelmann, A., 2007. A 350-ky radiolarian record off Lueeritz, Namibia; evidence for changes in the upwelling regime. *Marine Micropaleontology* 62, 194–210.

Jacot Des Combes, H., Caulet, J.P., Tribovillard, N.P., 2005. Monitoring the variations of the Socotra upwelling system during the last 250 kyr: A biogenic and geochemical approach. *Palaeogeography, Palaeoclimatology, Palaeoecology* 223, 243–259.

Jacot Des Combes, H., Caulet, J.-P., Tribovillard, N.P., 1999. Pelagic productivity changes in the equatorial area of the northwest Indian Ocean during the last 400,000 years. *Marine Geology* 158, 27–55.

Jacot des Combes, H., Tribovillard, N.P., Caulet, J.-P., 1999. Paleoproductivity and paleoceanographic changes in the Amirante Passage area (Equatorial Indian Ocean): a 200 kyr interval of lower pelagic productivity. *Bulletin de la Societé Geologique de France* 170, 899–914.

Jahn, I., 1990. *Grundzüge der Biologiegeschichte*. G. Fischer, Jena.

Jahn, R., 1998. C. G. Ehrenberg: The man and his contribution to botanical science, in: Williams, D.M., Huxley, R. (Eds.), *Christian Gottfried Ehrenburg [Sic]: The Man and His Legacy*. Academic, pp. 14–29.

Jahn, R., Landsberg, H., 2000. Christian Gottfried Ehrenberg – Entdeckung der Biodiversität im Mikro-Kosmos, in: Bredekamp, H., Brüning, J., Weber, C. (Eds.), *Theatrum Naturae et Artis – Theater Der Natur Und Kunst. Wunderkammern Des Wissens. Ausstellung Vom 10.12. 2000 – 4.3. 2001 Im Martin-Gropius-Bau Berlin. Essays.* pp. 219–225.

Jenkyns, H.C., 2010. Geochemistry of oceanic anoxic events. *Geochemistry Geophysics Geosystems* 11, n/a.

Jenkyns, H.C., Clayton, C.J., 1997. Lower Jurassic epicontinental carbonates and mudstones from England and Wales: Chemostratigraphic signals and the early Toarcian anoxic event. *Sedimentology* 44, 687–706.

Jiang, H., Lai, X., Luo, G., et al., 2007. Restudy of conodont zonation and evolution across the P/T boundary at Meishan section. *Global and Planetary Changes* 55, 39–55.

Jin, Y., Noble, P.J., Poulson, S.R., 2012. Paleoenvironmental and paleoecological implications of Permian (Guadalupian) radiolarian and geochemical variations in the Lamar Limestone, Delaware Basin, West Texas (USA). *Palaeogeography, Palaeoclimatology, Palaeoecology* 346–347, 37–53.

Jinha, A.E., 2010. Article 50 million: an estimate of the number of scholarly articles in existence. *Learned Publishing* 23, 258–263. https://doi.org/10.1087/20100308

Johnson, D.A., 1978. Cenozoic Radiolaria from the eastern tropical Atlantic, DSDP Leg 41, in: Lancelot, Y., Seibold, E., et al., (Eds.), *Initial Reports of the Deep Sea Drilling Project*. U. S. Government Printing Office, Washington, D.C., pp. 763–789.

Johnson, D.A., Nigrini, C., 1982. Radiolarian biogeography in surface sediments of the eastern Indian Ocean. *Marine Micropaleontology* 7, 237–281.

Johnson, D.A., Nigrini, C., 1980. Radiolarian biogeography in surface sediments of the western Indian Ocean. *Marine Micropaleontology* 5, 111–152.

Johnson, D.A., Nigrini, C.A., 1985. Synchronous and time-transgressive Neogene radiolarian datum levels in the equatorial Indian and Pacific Oceans. *Marine Micropaleontology* 9, 489–523.

Johnson, M.W., Brinton, E., 1963. Biological species, water masses and currents, in: Hill, M.N. (Ed.), *The Sea*, Volume 2. Wiley Interscience, New York, pp. 381–414.

Johnston, D.T., Wolfe-Simon, F., Pearson, A., Knoll, A., 2009. Anoxygenic photosynthesis modulated Proterozoic oxygen and sustained Earth's middle age. *Proceedings National Academy of Sciences* 106, 16925–16929.

Jonsson, B.F., Watson, J.R., 2016. The timescales of global surface-ocean connectivity. *Nature Communications* 7, 11239.

Jorgensen, E., 1905. The protist plankton and the diatoms in bottom samples. *Bergens museums skrifter,* Bergen, 1905 49–151.

Jud, R., 1994. Biochronology and systematics of Early Cretaceous Radiolaria of the Western Tethys. *Memoires de Geologie (Lausanne)* 19, 1–147.

Juggins, S., 2015. rioja: *Analysis of Quaternary Science Data.* https://eprints.ncl.ac.uk.

Jürgens, K., Massana, R., 2008. Protist grazing on marine bacterioplankton, in: Kirchman, D.L. (Ed.), *Microbial Ecology of the Oceans*. Wiley, pp. 383–441.

Kachovich, S., Sheng, J., Aitchison, J.C., 2019. Adding a new dimension to investigations of early radiolarian evolution. *Scientific Reports* 9, 6450. https://doi.org/10.1038/s41598-019-42771-0

Kamata, Y., 2007. Systematic description of primitive Triassic nassellarians. *GNS Science Monograph* 207–211.

Kamata, Y., Matsuo, A., Takemura, A., et al., 2007. The oceanic Permian/Triassic boundary sequence at Arrow

Rocks (Oruatemanu), Northland, New Zealand Late Induan (Dienerian) primitive nassellarians from Arrow Rocks, Northland, New Zealand, in: Spörli, K.B., Takemura, A., Hori, R. S. (Eds.), 24. *GNS Science*, pp. 109–116.

Kamikuri, S., 2019. Middle to late Miocene radiolarians from ODP Site 1021 in the eastern North Pacific. *Bulletin of the Geological Survey of Japan* 70, 163–194.

Kamikuri, S., 2017. Late Neogene radiolarian biostratigraphy of the eastern North Pacific ODP Sites 1020/1021. *Paleontological Research* 21, 230–253.

Kamikuri, S., 2012. Evolutionary changes in the biometry of the fossil radiolarian *Stichocorys peregrina* lineage in the eastern equatorial and eastern North Pacific. *Marine Micropaleontology* 90–91, 13–28.

Kamikuri, S., Moore Jr., T.C., Lyle, M., et al., 2013. Early and Middle Eocene radiolarian assemblages in the eastern equatorial Pacific Ocean (IODP Leg 320 Site U1331): Faunal changes and implications for paleoceanography. *Marine Micropaleontology* 98, 1–13.

Kamikuri, S., Moore, T.C., Ogane, K., et al., 2012. Data report: Early to Middle Eocene radiolarian biostratigraphy, IODP Expedition 320 Site U1331, eastern equatorial Pacific, in: Pälike, H., Lyle, M., Nishi, H., Raffi, I., Gamage, K., Klaus, A., Scientists, E. 320/321 (Eds.), *Proceedings IODP 320/321*. Integrated Ocean Drilling Program Management International, Inc., Tokyo.

Kamikuri, Shin-ichi, Moore, T.C., Ogane, K., Suzuki, N., Palike, H., Nishi, H., 2012. Early Eocene to early Miocene radiolarian biostratigraphy for the low-latitude Pacific Ocean. *Stratigraphy* 9, 77–108.

Kamikuri, S., Motoyama, I., Nishi, H., Iwai, M., 2009. Neogene radiolarian biostratigraphy and faunal evolution rates in the eastern equatorial Pacific ODP Sites 845 and 1241. *Acta Palaeontologica Polonica* 54, 713–742.

Kamikuri, S., Motoyama, I., Nishimura, A., 2008. Radiolarian assemblages in surface sediments along longitude 175° E in the Pacific Ocean. *Marine Micropaleontology* 69, 151–172.

Kamikuri, S., Nishi, H., Moore Jr., T.C., et al., 2005. Radiolarian faunal turnover across the Oligocene/Miocene boundary in the equatorial Pacific Ocean. *Marine Micropalaeontology* 57, 74–96.

Kamikuri, S., Nishi, H., Motoyama, I., 2007. Effects of late Neogene climatic cooling on North Pacific radiolarian assemblages and oceanographic conditions. *Palaeogeography, Palaeoclimatology, Palaeoecology* 249, 370–392.

Kamikuri, S., Nishi, H., Motoyama, I., Saito, S., 2004. Middle Miocene to Pleistocene radiolarian biostratigraphy in the Northwest Pacific Ocean, ODP Leg 186. *The Island Arc* 13, 191–226.

Kasai, K., Sakai, T., Aita, Y., Amano, K., 2000. Chert-clastic sequences in the central Yamizo Mountains, northeast Japan. *Journal of Geological Society of Japan* 106, 1–13.

Katz, M.E., Finkel, Z.V., Grzebyk, D., et al., 2004. Evolutionary trajectories and biogeochemical impacts of marine eukaryotic phytoplankton. *Annual Review of Ecology, Evolution, and Systematics* 35, 523–556.

Keany, J., 1978. Paleoclimatic trends in early and middle Pliocene deep-sea sediments of the Antarctic. *Marine Micropalaeontology* 3, 35–49.

Keating, B., Sakai, H., 1991. Magnetostratigraphic studies of sediments from Site 744, southern Kerguelen Plateau, in: Barron, J., Larsen, B. (Eds.), *Proceedings of the Ocean Drilling Program, Scientific Results*, 119. Ocean Drilling Program, College Station, TX, pp. 771–794.

Keeling, P.J., Palmer, J.D., 2008. Horizontal gene transfer in eukaryotic evolution. *Nature Reviews Genetics* 9, 605–618.

Kellogg, D.E., 1980. Character displacement and phyletic change in the evolution of the radiolarian subfamily Artiscinae. *Micropaleontology* 26, 196–210.

Kellogg, D.E., 1975. The role of phyletic change in the evolution of Pseudocubus vema (Radiolaria). *Paleobiology* 1, 359–370.

Kellogg, D.E., Hays, J.D., 1975. Microevolutionary patterns in Late Cenozoic Radiolaria. *Paleobiology* 1, 150–160.

Kelly, D.C., Bralower, T.J., Zachos, J.C., et al., 1996. Rapid diversification of planktonic foraminifera in the tropical Pacific (ODP Site 865) during the late Paleocene thermal maximum. *Geology* 24, 423–426.

Kemp, A.E.S., Pike, J., Pearce, R.B., Lange, C.B., 2000. The "Fall dump" — a new perspective on the role of a "shade flora" in the annual cycle of diatom production and export flux. *Deep Sea Research II* 47, 2129–2154.

Kennett, J.P., 1982. *Marine Geology*. Prentice-Hall, Englewood Cliffs, N.J.

Khokhlova, I.E., Bragina, L.G., Krasheninnikov, V.A., 1994. Zonal stratigraphy of the Upper Cretaceous and Paleogene deposits of the key Perapedhi section (Southern Cyprus) by means of radiolarians and correlation with the foraminiferal zones, in: Krasheninnikov, V.A., Hall, J.K. (Eds.), *Geological Structure of Northeast Mediterranean (Cruise 5 of the Research Vessel 'Akademik Nikolai Strakhov.'* Historical Productions-Hall Ltd., pp. 219–250.

Khudyaev, I.E., 1931. On the Radiolaria in phosporites of the Sysoljsk area. *Trudy Glavnogo Gelogo-Razvedochnogo Upravleniya VSNKh, SSSR* 46, 1–48.

Kidder, D.L., Erwin, D.H., 2001. Secular distribution of biogenic silica through the Phanerozoic: comparison of silica-replaced fossils and bedded cherts at the series level. *Journal of Geology* 109, 509–522.

Kiessling, W., 2002. Radiolarian diversity patterns in the latest Jurassic-earliest Cretaceous. *Palaeogeography, Palaeoclimatology, Palaeoecology* 187, 179–206.

Kiessling, W., 1999. Late Jurassic radiolarians from the Antarctic Peninsula. *Micropaleontology* 45, supplement 1, 1–96.

Kiessling, W., Danelian, T., 2011. Trajectories of Late Permian-Jurassic radiolarian extinction rates: no evidence for an end-Triassic mass extinction. *Fossil Record* 14, 95–101.

Kimoto, K., Yuasa, T., Takahashi, O., 2011. Molecular identification of reproductive cells released Cypassis irregularis Nigrini (Radiolaria). *Environmental Microbiology Reports* 3, 86–90.

Kiørboe, T., 2008. *A Mechanistic Approach to Plankton Ecology*. Princeton University Press, Princeton, NJ.

Kling, S.A., 1979. Vertical distribution of polycystine radiolarians in the central North Pacific. *Marine Micropaleontology* 4, 295–318.

Kling, S.A., Boltovskoy, D., 1995. Radiolarian vertical distribution patterns across the southern California Current. *Deep Sea Research Part I: Oceanographic Research Papers* 42, 191–231.

Klug, C., Kröger, B., Kiessling, W., Mullins, G.L., Servais, T., Fryda, J., Korn, D., Turner, S., 2010. The Devonian nekton revolution. *Lethaia* 43, 465–477.

Kochhann, K.G.D., Baecker-Fauth, S., Fauth, G., 2013. Late Maastrichtian Radiolaria from ODP Leg 165 — Site 999B, Colombian Basin. *Geodiversitas* 35, 5–20. https://doi.org/10.5252/g2013n1a1

Kockerbeck, C., 1986. *Ernst Haeckels "Kunstformen der Natur" und ihr Einfluß auf die deutsche bildende*

Kunst der Jahrhundertwende : Studie zum Verhältnis von Kunst und Naturwissenschaften im Wilhelminischen Zeitalter. Lang, Frankfurt am Main.

Kocsis, A.T., Kiessling, W., Pálfy, J., 2014. Radiolarian biodiversity dynamics through the Triassic and Jurassic: implications for proximate causes of the end-Triassic mass extinction. *Paleobiology* 40, 625–639.

Korovnikov, I., Sennikov, N., Danelian, T., et al., 2013. The biostratigraphic and palaeoenvironmental significance of Lower Cambrian (Botomian) trilobites from the Ak-Kaya section of the Altai Mountains (southern Siberia, Russia). *Annales de Paléontologie 99, 79–89.*

Kozur, H., 1988. Muelleritortiidae n. fam., eine charakteristische longobardische (oberladinische) Radiolarienfamilie, Teil I. Freiberger Forschungshefte. C, geowissenschaften *Geologie* 419, 51–61.

Kozur, H., 1984. New radiolarian taxa from the Triassic and Jurassic. *Geologisch Paläontologische Mitteilungen Innsbruck* 13, 49–88.

Kozur, H., 1981. Albaillellidae (Radiolaria) aus dem Unterperm des Vorurals. *Geologisch Paläontologische Mitteilungen Innsbruck* 10, 263–274.

Kozur, H., Mostler, H., 1994. Anisian to middle Carnian radiolarian zonation and description of some stratigraphically important radiolarians. *Geologisch Paläontologische Mitteilungen Innsbruck, Sonderband* 3, 39–255.

Kozur, H., Mostler, H., 1983. The polyphyletic origin and the classification of the Mesozoic saturnalids (Radiolaria). *Geologisch Paläontologische Mitteilungen Innsbruck* 13, 1–47.

Kozur, H., Mostler, H., 1981. Beiträge zur Erforschung der mesozoischen Radiolarien. Teil IV: Thalassosphaeracea HAECKEL, 1862, Hexastylacea HAECKEL, 1862 emend. PETRUSHEVSKAJA, 1979, Sponguracea HAECKEL, 1862 emend. und weitere triassische Lithocycliacea, Trematodiscacea, Actinommacea und Nassellaria. *Geologisch Paläontologische Mitteilungen Innsbruck, Sonderband* 1, 1–208.

Kozur, H., Mostler, H., 1979. Beiträge zur Erforschung der mesozoischen Radiolarien. Teil III: Die Oberfamilien Actinommacea HAECKEL 1862 emend., Artiscacea HAECKEL1882, Multiarcusellacea nov. der Spumellaria und triassische Nassellaria. *Geologisch Paläontologische Mitteilungen Innsbruck* 9, 1–132.

Kozur, H., Mostler, H., 1978. Beiträge zur Erforschung der mesozoischen Radiolarien Teil II: Oberfamilie Trematodiscacea HAECKEL 1862 emend. und Beschreibung ihrer triassischen Vertreter. *Geologisch Paläontologische Mitteilungen Innsbruck* 8, 123–182.

Kozur, H., Mostler, H., 1972. Beiträge zur Erforschung der mesozoischen Radiolarien. Teil I: Revision der Oberfamilie Coccodiscacea HAECKEL 1862 emend. und Beschreibung ihrer triassischen Vertreter. *Geologisch Paläontologische Mitteilungen Innsbruck* 2, 1–60.

Kozur, H.W., 2003a. Integrated ammonoid-, conodont and radiolarian zonation of the Triassic. *Hallesches Jahrbuch für Geowissenschaften B* 25, 49–79.

Kozur, H.W., 2003b. Integrated ammonoid, conodont and radiolarian zonation of the Triassic and some remarks to stage/substage subdivision and the numeric age of the Triassic stages. *Albertiana* 28, 57–83.

Kozur, H.W., 1998. Some aspects of the Permian-Triassic boundary (PTB) and the possible causes for the biotic crisis around this boundary. The Permian-Triassic boundary and global Triassic correlations. *Palaeogeography,*

Palaeoclimatology, Palaeoecology 143, 227–272.
Kozur, H.W., 1980. Ruzhencevispongidae, eine neue Spumellaria Familie aus dem oberen Kungurian (Leonardian) und Sakmarian des Vorurals. *Geologisch Paläontologische Mitteilungen Innsbruck* 10, 235–242.
Kozur, H.W., Bachmann, G.H., 2010. The middle Carnian wet intermezzo of the Stuttgart Formation (Schilfsandstein), Germanic Basin. *Palaeogeography, Palaeoclimatology, Palaeoecology* 290, 107–119.
Kozur, Heinz W., Krainer, K., Mostler, H., 1996. Radiolarians and Facies of the Middle Triassic Loibl Formation, South Alpine Karawanken Mountains (Carinthia, Austria). *Geologisch Paläontologische Mitteilungen Innsbruck, Sonderband* 4, 195–269.
Kozur, H.W., Moix, P., Ozsvárt, P., 2009. New Spumellaria (Radiolaria) from the Early Tuvalian Spongotortilispinus moixi Zone of Southeastern Turkey, with some Remarks on the Age of this Fauna. *Jahrbuch der geologischen Bundesanstalt. Wien* 149, 25–59.
Kozur, H.W., Moix, P., Ozsvárt, P., 2007. Characteristic Nassellaria of the lower Tuvalian (Upper Triassic) Spongotortilispinus moixi Zone of the Hu\uglu Unit in the Mersin Mélange, southeastern Turkey. *Bulletin de la Société vaudoise des Sciences naturelles* 90, 151–173.
Kozur, H.W., Mostler, H., 2006. Radiolarien aus dem Longobard der Dinariden. *Hallesches Jahrbuch für Geowissenschaften* 28, 23–91.
Kozur, H.W., Mostler, H., 1989. Radiolarien und schwammskleren aus dem Unterperm des Vorurals. *Geologisch Paläontologische Mitteilungen Innsbruck, Sonderband* 2, 147–275.
Kozur, H.W., Mostler, H., 1982. Entactinaria subordo nov., a new radiolarian suborder. *Geologisch Paläontologische Mitteilungen Innsbruck* 11, 399–414.
Kozur, H. W., Mostler, H., Repetski, J.E., 1996. Well-preserved Tremadocian primitive Radiolaria from the Windfall Formation of the Antelope Range, Eureka County, Nevada, USA. *Geologisch-Paläontologische Mitteilungen Innsbruck* 21, 245–271.
Krabberød, A.K., Bråte, J., Dolven, J.K., et al., 2011. Radiolaria divided into Polycystina and Spasmaria in combined 18S and 28S rDNA phylogeny. PLoS ONE 6, e23526.
Krabberød, A.K., Orr, R.J.S., Bråte, J., et al., 2017. Single Cell Transcriptomics, Mega-Phylogeny, and the Genetic Basis of Morphological Innovations in Rhizaria. *Molecular Biology and Evolution* 34, 1557–1573.
Krystyn, L., Boquerel, H., Kuerschner, W., et al., 2007. Proposal for a candidate GSSP for the base of the Rhaetian Stage. *New Mexico Museum of Natural History and Science Bulletin* 41, 189–199.
Kullenberg, B., 1947. The piston core sampler. *Swensca Hydrografisc-Biologiska Kommissionens Skrifter* 1, 2–46.
Kunitomo, Y., Sarashina, I., Iijima, M., et al., 2006. Molecular phylogeny of acantharian and polycystine radiolarians based on ribosomal DNA sequences, and some comparisons with data from the fossil record. *European Journal of Protistology* 42, 143–153.
Kurihara, T., 2004. Silurian and Devonian radiolarian biostratigraphy of the Hida Gaien belt, central Japan. *The Journal of the Geological Society of Japan* 110, 620–639.
Kurilov, D.V., Vishnevskaya, V.S., 2011. Early Cretaceous radiolarian assemblages from the east Sakhalin Mountains. *Stratigraphy and Geological Correlation* 19, 44–62.
Kusunoki, T., Ohara, M., Musashino, M., 2004. Carboniferous-Permian

microbiostratigraphy in chert sequence from the southeastern part of the Tamba Belt, Shizukawa District, Uji City (online note). *Earth Science, Journal of the Association for the Geological Collaboration in Japan* 58, 37–54.

Kuwahara, K., Yao, A., Ezaki, Y., et al., 2003. Occurrence of Late Permian radiolarians from the Chituao section, Laibin, Guangxi, China. *Journal of Geosciences, Osaka City University* 46, 13–23.

Kuwahara, K., Yao, A., Yamakita, S., 1998. Reexamination of Upper Permian radiolarian biostratigraphy. *Earth Science, Journal of the Association for the Geological Collaboration in Japan* 52, 51–64.

Labracherie, M., 1984. Neogene and Pleistocene radiolarians from Deep Sea Drilling Project Leg 79 in the region of the Mazagan Plateau, off central Morocco, in: Hinz, K., Winterer, E.L., et al., (Eds.), *Initial Reports of the Deep Sea Drilling Project, Volume 79*. United States Government Printing Office, Washington, D. C., pp. 557–562.

Lahm, B., 1984. Spumellarienfaunen (Radiolaria) aus den mitteltriassischen Buchensteiner-Schichten von Recoaro (Norditalien) und den obertriassischen Reiflingerkalken von Grosreifling (Österreich). Systematik, Stratigraphie. Münchner geowissenschaftliche Abhandlungen. *Reihe A, Geologie und Paläontologie* 1, 1–161.

Lahr, D.J.G., Parfrey, L.W., Mitchell, E.A.D., et al., 2011. The chastity of amoebae: re-evaluating evidence for sex in amoeboid organisms. *Proceedings of the Royal Society London, B* 278, 2081–2090.

Lalli, M.C., Parsons, T.R., 1993. *Biological Oceanography: An Introduction* (First Edition). Elsevier, Amsterdam.

Lampitt, R.S., Salter, I., Johns, D., 2009. Radiolaria: Major exporters of organic carbon to the deep ocean. *Global Biogeochemical Cycles* 23.

Larsen, P.O., von Ins, M., 2010. The rate of growth in scientific publication and the decline in coverage provided by Science Citation Index. *Scientometrics* 84, 575–603. https://doi.org/10.1007/s11192-010-0202-z

Laskar, J., Fienga, A., Gastineau, M., Manche, H., 2011. La2010: a new orbital solution for the long-term motion of the Earth. A&A 532, A89. https://doi.org/10.1051/0004-6361/201116836

Latz, M.I., Bowlby, M.R., Case, J.F., 1991. Bioluminescence of the solitary spumellarian radiolarian, Thalassicolla nucleata (Huxley). *Journal of Plankton Research* 13, 1187–1201.

Lazarus, D., 2014. The legacy of early radiolarian taxonomists, with a focus on the species published by early German workers. *Journal of Micropalaentology* 33, 3–19.

Lazarus, D., 2011. The deep-sea microfossil record of macroevolutionary change in plankton and its study, in: Smith, A., McGowan, A. (Eds.), *Comparing the Geological and Fossil Records: Implications for Biodiversity Studies*. The Geological Society, London, pp. 141–166.

Lazarus, D., 2005. A brief review of radiolarian research. *Paläontologische Zeitschrift* 79/1, 183–200.

Lazarus, D., 1994. An improved cover slip holder for preparing microslides of randomly distributed particles. *Journal of Sedimentary Petrology* A64, 686.

Lazarus, D., 1986. Tempo and mode of morphologic evolution near the origin of the radiolarian lineage Pterocanium prismatium. *Paleobiology* 12/2, 175–189.

Lazarus, D., 1983. Speciation in pelagic protista and its study in the planktonic microfossil record: a review. *Paleobiology* 9, 327–340.

Lazarus, D., Barron, J., Renaudie, J., Diver, P., Türke, A., 2014. Cenozoic diatom diversity and correlation to climate change. *PLOS One* 9, 1–18.

Lazarus, D., Bittniok, B., Diester-Haass, L., Meyers, P., Billups, K., 2006. Comparison of radiolarian and sedimentologic paleoproductivity proxies in the latest Miocene-Recent Benguela Upwelling System. *Marine Micropaleontology* 60, 269–294.

Lazarus, D., Caulet, J.P., 1994. Cenozoic Southern Ocean reconstructions from sedimentologic, radiolarian, and other microfossil data, in: Kennett, J.P., Warnke, D.A. (Eds.), *The Antarctic Paleonvironment: A Perspective on Global Change*. Part 2. Antarctic Research Series. pp. 145–174.

Lazarus, D., Kotrc, B., Wulf, G., Schmidt, D.N., 2009. Radiolarians decreased silicification as an evolutionary response to reduced Cenozoic ocean silica availability. *Proceedings of the National Academy of Sciences USA* 106, 9333–9338.

Lazarus, D., Pallant, A., 1989. Oligocene and Neogene radiolarians from the Labrador Sea, ODP Leg 105, in: Srivastava, S.P., Arthur, M., Clement, B., et al., (Eds.), *Proceedings of the Ocean Drilling Program, Scientific Results*. College Station, TX (Ocean Drilling Program), pp. 349–380.

Lazarus, D., Suzuki, N., Caulet, J.-P., et al., 2015. An evaluated list of Cenozic-Recent radiolarian species names (Polycystinea), based on those used in the DSDP, ODP and IODP deep-sea drilling programs. *Zootaxa* 3999, 301–333.

Lazarus, D., Weinkauf, M., Diver, P., 2012. Pacman profiling: a simple procedure to identify stratigraphic outliers in high density deep-sea microfossil data. *Paleobiology* 38, 144–161.

Lazarus, D.B., 2001a. Late Miocene to early Pliocene radiolarians from glaciomarine drift deposits, ODP Leg 178, Hole 1095B (Bellinghausen Basin, Antarctic Ocean). *Proceedings of the Ocean Drilling Program, Scientific Results* 178, 1–22.

Lazarus, D.B., 2001b. Speciation, in: Briggs, D., Crowther, P. (Eds.), *Paleobiology II*. Blackwell, London, pp. 133–137.

Lazarus, D.B., 1998. The Ehrenberg Collection and its curation, in: Williams, D.M., Huxley, R. (Eds.), *Christian Gottfried Ehrenburg [Sic] (1795-1876): The Man and His Legacy*. The Linnean Society, pp. 31–48.

Lazarus, D.B., 1994. The Neptune Project – a marine micropaleontology database. *Mathematical Geology* 26, 817–832.

Lazarus, D.B., 1992. Antarctic Neogene radiolarians from the Kerguelen Plateau, ODP Legs 119 and 120, in: Wise, S.W., Schlich, R. (Eds.), *Proceedings of the Ocean Drilling Program, Scientific Results*. Ocean Drilling Program, College Station, pp. 785–810.

Lazarus, D.B., 1990. Middle Miocene to Recent radiolarians from the Weddell Sea, Antarctica, ODP Leg 113, in: Barker, P.F., Kennett, J.P. (Eds.), *Proceedings of the Ocean Drilling Program, Scientific Results*. Ocean Drilling Program, College Station, TX, pp. 709–728.

Lazarus, D.B., 1983. Speciation and Phyletic Evolution in Pterocanium (Radiolaria) (PhD Thesis). Columbia, New York.

Lazarus, D.B., Hollis, C., Apel, M., 2008. Patterns of opal and radiolarian change in the Antarctic mid-Paleogene: clues to the origin of the Southern Ocean. *Micropaleontology* 54, 41–48.

Lazarus, D.B., Jahn, R., 1998. Using the Ehrenberg Collection. *Diatom Research* 13, 273–291.

Lazarus, D.B., Scherer, R.P., Prothero, D.R., 1985. Evolution of the radiolarian species-complex Pterocanium: A preliminary survey. *Journal of Paleontology* 59, 183–220.

Lazarus, D.B., Spencer-Cervato, C., Pianka-Biolzi, M., et al., 1995. Revised chronology of Neogene DSDP holes from the world ocean. *Ocean Drilling Program*, College Station.

Lear, C.H., Rosenthal, Y., Wright, J.D., 2003. The closing of a seaway: ocean water masses and global climate change. *Earth and Planetary Science Letters* 210, 425–436.

Leckie, M., Bralower, T.J., Cashman, R., 2002. Oceanic anoxic events and plankton evolution: Biotic response to tectonic forcing during the mid-Cretaceous. *Paleoceanography* 17.

Lenz, A.C., Noble, P.J., Masiak, M., et al., 2006. The lundgreni Extinction Event: Integration of paleontological and geochemical data from Arctic Canada. *GFF* 128, 153–158.

Lestrel, P.E., 1997. *Fourier Descriptors and their Applications in Biology*. Cambridge University Press, Cambridge.

Levashova, N.M., 2000. Paleomagnetism and geochronology of the Late Cretaceous-Paleogene island arc complex of the Kronotsky Peninsula, Kamchatka, Russia: Kinematic implications. *Tectonics* 19, 834–851.

Levinton, J.S., 1988. *Genetics, Paleontology, and Macroevolution*. Cambridge University Press, Cambridge, UK.

Li, R.Q., Sashida, K., 2012. Additional note on earliest Cretaceous entactinarians (radiolaria) from the Mariana Trench. *Paleontological Research* 16, 26–36.

Lindsay, D.J., Yamaguchi, A., Grossmann, M.M., et al., 2014. Vertical profiles of marine particulates: a step towards global scale comparisons using an Autonomous Visual Plankton Recorder. *Bulletin of the Plankton Society of Japan* 61, 1–8.

Ling, H.-Y., 1979. Radiolarians from the west flank of Reykjanes Ridge, Leg 49 of the Deep Sea Drilling Project, in: Luyendyk, B.P., Cann, J.R., et al., (Eds.), *Initial Reports of the Deep Sea Drilling Project*. U. S. Government Printing Office, Washington, D.C., pp. 583–588.

Ling, J., Lazarus, D.B., 1990. Cretaceous radiolaria from the Weddell Sea: Leg 113 of the Ocean Drilling Program, in: Barker, P., Kennett, J.P., et al., (Eds.), *Proceedings of the Ocean Drilling Program, Scientific Results*. Ocean Drilling Program, College Station, TX, pp. 353–364.

Liow, L.H., Skaug, H.J., Ergon, T., Schweder, T., 2010. Global occurrence trajectories of microfossils: environmental volatility and the rise and fall of individual species. *Paleobiology* 36, 224–252.

Liow, L.H., Stenseth, N.C., 2007. The rise and fall of species: implications for macroevolutionary and macroecological studies. *Proceedings of the Royal Society B* 274, 2745–2752.

Lipman, R.Kh., 1969. A new genus and new species of Eocene radiolarians in the USSR. *Trudy Vsesoyuznogo Nauchno-Issledovatelskogo Geologicheskogo Instituta* 130, 180–200.

Lipman, R.Kh., 1965. Kompleksy radiolyarii v paleogenovykh otlozheniyakh Turgaiskogo progiba i severnogo Priaralya (Radiolarian complexes in the Paleogene deposits of the Turgaisk Trough and the Northern Pre-Aral). *Trudy vsesoyvznogo Nauchno-Issledovatelskogo Geologicheskogo Instituta (VSEGEI) (Transactions of the All Union Scientific Research Institute of Geology (VSEGEI)*, new ser. 115, 78–89.

Lipman, R.Kh., 1960a. Radiolaria, in: Stratigrafiya i Fauna Paleogenovykh Otlozhenii Zapadno-Sibirskoinizhmennosti (Stratigraphy and Fauna of the Paleogene Deposits of the Western Siberian Depression). *Trudy VSEGEI*, new series, pp. 69–98, 161.

Lipman, R.Kh., 1960b. Radiolaria, in: Stratigrafiya i Fauna Melovykh Otlozhenii Zapadno-Sibirskoi

Nizhmennosti (Stratigraphy and Fauna of Cretaceous Deposits of the Western Siberian Depression). *Trudy VSEGEI*, new series, pp. 124–134.

Lipman, R.Kh., 1959. Zachenie radiolyarii dlya stratigraficheskogo rascheleniya osadochnykh porod (The significance of radiolarians for the stratigraphical division of sedimentary rocks). *Byulleten Moskovskogo Obshchestva Ispytatelei Prirody, Otd. Ottisk*. (Bulletin of the Moscow Society of Naturalists, Department of Geology) 34, 67–88.

Lipman, R.Kh., 1948. O stratigraficheskom znachenii radiolyarii (On the stratigraphical importance of Radiolaria). *Materialy Vsesoyuznogo Nauchno-Issledovatelskogo GeologicheskogoInstituta, Obshchaya Seriya* 89–92.

Little, C.T.S., Benton, M.J., 1995. Early Jurassic mass extinction: A global long-term event. *Geology* 23, 495–498.

Littler, K., Robinson, S.A., Bown, P.R., et al., 2011. High sea-surface temperatures during the Early Cretaceous Epoch. *Nature Geoscience* 4, 169–172.

Llopis-Monferrer, N., Boltovskoy, D., Tréguer, P., et al., 2020. Estimating Biogenic Silica Production of Rhizaria in the Global Ocean. *Global Biogeochemical Cycles* 34, e2019GB006286. https://doi.org/10.1029/2019GB006286

Lohmann, G.P., 1983. Eigenshape analysis of microfossils: a general morphometric procedure for describing changes in shape. Mathematical Geology 15, 659–672.

Lohmann, G.P., Malmgren, B.A., 1983. Equatorward migration of Globorotalia truncatulinoides ecophenotypes through the Late Pleistocene: gradual evolution or ocean change? *Paleobiology* 9, 414–421.

Lombard, F., Boss, E., Waite, A.M., et al., 2019. Globally Consistent Quantitative Observations of Planktonic Ecosystems. *Frontiers in Marine Science* 6. https://doi.org/10.3389/fmars.2019.00196

Lombari, G., Boden, G., 1985. *Modern radiolarian global distributions*. Cushman Foundation for Foraminiferal Research. Special Publication.

Lombari, G., Lazarus, D., 1988. Neogene cycladophorid radiolarians from North Atlantic, Antarctic, and Noth Pacific deep-sea sediments. *Micropaleontology* 34, 97–135.

Longhurst, A.R., 1998. *Ecological Geography of the Sea*. Academic Press, San Diego.

Longridge, L.M., Carter, E.S., Smith, P.L., Tipper, H.W., 2007. Early Hettangian ammonites and radiolarians from the Queen Charlotte Islands, British Columbia and their bearing on the definition of the Triassic–Jurassic boundary. *Palaeogeography, Palaeoclimatology, Palaeoecology* 244, 142–169.

López-Garciá, P., Rodriguez-Valera, F., Moreira, D., 2002. Toward the monophyly of Haeckel's Radiolaria: 18S rRNA environmental data support the sisterhood of Polycystinea and Acantharea. *Molecular Biology and Evolution* 19, 118–121.

Löser, H., 2004. PaleoTax – a database program for paleontological data. *Computers and Geosciences* 30, 513–521.

Lourens, L., Hilgen, F., Shackleton, N.J., et al., 2004. Neogene Period, in: Gradstein, F.M., Ogg, J.G., Smith, A.G. (Eds.), *A Geologic Time Scale 2004*. Cambridge University, Cambridge, pp. 409–440.

Lozano, J.A., Hays, J.D., 1976. Relationship of radiolarian assemblages to sediment types and physical oceanography in the Atlantic and Western Indian Ocean sectors of the Antarctic Ocean, in: Cline, R.M., Hays, J.D. (Eds.), *Investigation of Late Quaternary Paleoceanography and Paleoclimatology*. Geological Society

of America Memoir, Boulder, CO, USA, pp. 303–336.
Lucas, S.G., Tanner, L.H., Donohoo-Hurley, L.L., et al., 2011. Position of the Triassic-Jurassic boundary and timing of the end-Triassic extinctions on land. *Data from the Moenave Formation on the southern Colorado Plateau, USA* 302, 194–205.
Lüer, V., Cortese, G., Neil, H.L., et al., 2009. Radiolarian-based sea surface temperatures and paleoceanographic changes during the late Pleistocene-Holocene in the subantarctic Southwest Pacific. *Marine Micropaleontology* 70, 151–165.
Lyle, M., Drury, A.J., Tian, J., et al., 2019. Late Miocene to Recent high resolution Eastern Equatorial Pacific carbonate records: stratigraphy linked by dissolution and paleoproductivity. *Climates of the Past* 15, 1715–1739.
Lyle, M., Moore Jr., T.C., 2002. Dedication: Catherine Nigrini 1939-2005, in: *Proceedings of the Ocean Drilling Program*. Ocean Drilling Program, College Station, TX.
Lyle, M., Raffi, I., Pälike, H., et al., Scientists, Expedition 320/321, 2009. Pacific Equatorial Transect. *IODP Preliminary Report* 321, web.
MacArthur, R.H., Wilson, E.O., 1967. *The Theory of Island Biogeography*. Princeton Univ. Press, Princeton, NJ.
MacCorquodale, K., Meehl, P.E., 1948. On a distinction between hypothetical constructs and intervening variables. *Psychological Review* 55, 95–107.
MacLeod, N., 1999. Generalizing and extending the eigenshape method of shape space visualization and analysis. *Paleobiology* 25, 107–138.
Macleod, N., 1988. Lower and Middle Jurassic Perisphyridium (Radiolaria) from the Snowshoe Formation, east-central Orgeon. *Micropaleontology* 34, 289–315.
Macleod, N., Rawson, P.F., Forey, P.L., et al., 1997. The Cretaceous–Tertiary biotic transition. *Journal of the Geological Society of London* 154, 265–292.
Madella, M., Powers-Jones, A.H., Jones, M.K., 1998. A simple method of extraction of opal phytoliths from sediments using a non-toxic heavy liquid. *Journal of Archaelogical Science* 25, 801–803.
Maldonado, A.L., Noble, P.J., 2010. Radiolarians from the upper Guadalupian (Middle Permian) Reef Trail Member of the Bell Canyon Formation, West Texas and their biostratigraphic implications. *Micropaleontology* 56, 69–115.
Maletz, J., 2011. Radiolarian skeletal structures and biostratigraphy in the early Palaeozoic (Cambrian-Ordovician). *Palaeoworld* 20, 116–133.
Maletz, J., 2007. Overlooked and ignored: Radiolarians and their biostratigraphic potential in the Ordovician. *Acta Palaeontologica Sinica* 46, 291–297.
Maletz, J., Bruton, D.L., 2007. Lower Ordovician (Chewtonian to Castemainian) radiolarians of Spitsbergen. *Journal of Systematic Palaeontology* 5, 245–288.
Maletz, J., Bruton, D.L., 2005. The Beothuka terranova (Radiolaria) assemblage and its importance for the understanding of early Ordovician radiolarian evolution. *Geological Magazine* 142, 395–411.
Maliva, R.G., Knoll, A.H., Siever, R., 1989. Secular change in chert distribution: a reflection of evolving biological participtation in the silica cycle. *Palaios* 4, 519–532.
Malmgren, B.A., 1991. Biogeographic patterns in terminal Cretaceous planktonic foraminifera from Tethyan and warm transitional waters. *Marine Micropaleontology* 18, 73–99.
Marcon, F., 2015. *The Knowledge of Nature and the Nature of Knowledge in Early Modern Japan*. University Chicago Press.

Marshall, C.R., Quental, T.B., 2016. The uncertain role of diversity dependence in species diversification and the need to incorporate time-varying carrying capacities. *Philosophical Transactions of the Royal Society B*. 371, 1–13 (web).

Martin, R.E., 2013. The evolution of Gulf Coast micropalaeontology: from biostratigraphy to chronostratigraphy, in: Bowden, A.J., Gregory, F.J., Henderson, A.S. (Eds.), *Landmarks in Foraminiferal Micropaleontology, The Micropalaeontological Society*, Special Publications. The Geological Society, London, pp. 103–124.

Marzoli, A., Renne, P.E., Piccirillo, E.M., et al., 1999. Extensive 200-million-year-old continental flood basalts of the Central Atlantic Magmatic Province. *Science* 284, 616–618.

Mast, H., 1910. Die Astrophaeriden der Deutschen Tiefsee-Expedition. Wissenschaftliche Ergebnisse der Deutschen Expedition auf dem Dampfer "Valdivia" 1898-1899. Inaugural-Dissertation zur Erlangung der Doktorwurde, Universitat Tubingen 19, 1–68.

Matsuoka, A., 2009. Late autumn living radiolarian fauna from subtropical surface waters in the East China Sea off Sesoko Island, Okinawa, southwest Japan. *News of Osaka Micropaleontologists, special volume* 14, 11–29.

Matsuoka, A., 2007. Living radiolarian feeding mechanisms: new light on past marine ecosystems. *Swiss Journal of Geosciences* 100, 273–279.

Matsuoka, A., 1995. Jurassic and Lower Cretaceous radiolarian zonation in Japan and in the western Pacific. *The Island Arc* 4, 140–153.

Matsuoka, A., 1993. Observations of living radiolarians of the surface water in the Caribbean Sea. *News of Osaka Micropaleontologists* 9, 349– 363.

Matsuoka, A., 1992. Skeletal growth of a spongiose radiolarian Dictyocorne truncatum in laboratory culture. *Marine Micropaleontology* 19, 287–297.

Matsuoka, A., Anderson, O.R., 1992. Experimental and observational studies of radiolarian physiological ecology: 5. Temperature and salinity tolerance of Dictyocoryne truncatum. *Marine Micropaleontology* 19, 299–313.

Matsuoka, A., Suzuki, N., Ito, T., et al., 2017. Excursion guide to the radiolarians of the East China Sea near Sesoko Island, Okinawa, Japan: An important research station for living radiolarian studies. *Science Reports of Niigata University (Geology)* 32.

Matsuoka, A., Yoshino, T., Kishimoto, N., et al., 2012. Exact number of pore frames and their configuration in the Mesozoic radiolarian Pantanellium, An application of X-ray micro-CT and layered manufacturing technology to micropaleontology. *Marine Micropalaeontology* 88–89, 36–40.

Mattick, J.S., Gagen, M.J., 2001. The evolution of controlled multitasked gene networks: the role of introns and other noncoding RNAs in the development of complex organisms. *Molecular Biology and Evolution* 18, 1611–1630.

Mayr, E., 1982. *The Growth of Biological Thought*. Belknap, Cambridge.

Mayr, E., 1963. *Animal Species and Evolution*. Harvard University, Cambridge.

McArthur, J.M., Howarth, R.J., 2004. Strontium isotope stratigraphy, in: Gradstein, F.M., Ogg, J.G., Smith, A.G. (Eds.), *A Geologic Time Scale 2004*. Cambridge University, Cambridge, pp. 96–105.

McGowran, B., 2005. *Biostratigraphy*. Cambridge University Press, Cambridge.

McMillen, K.J., Casey, R.E., 1978. Distribution of living polycystine radiolarians in the Gulf of Mexico and Caribbean Sea, and comparison

with the sedimentary record. *Marine Micropaleontology* 3, 121–145.
McNutt, M.K., 2019. Times up, CO_2. *Science* 365, 411. https://doi.org/10.1126/science.aay8827
Mekik, F.A., Ling, H.Y., Özkan-Altiner, S., Altiner, D., 1999. Preliminary radiolarian biostratigraphy across the Jurassic-Cretaceous boundary from northwestern Turkey. *Geodiversitas* 21, 715–738.
Metcalfe, I., 1996. Gondwanaland dispersion, Asian accretion and evolution of eastern Tethys. *Australian Journal of Earth Sciences* 43, 605–623.
Meyen, F.J.F., 1834. Uber das Leuchten des Meeres und Beschreibung einiger Polypen und anderer niederer Tiere (On starfish and a description of some polyps and other inferior animals), in: Beitrage Zur Zoologie, Gesammelt Auf Einer Reise Um Die Erde. *Nova Act a Academiae Caesareae Leopoldino Carolinae Germanicae Naturae Curiosorum*. pp. 125–216.
Meyers, P.A., 2006. Paleoceanographic and paleoclimatic similarities between Mediterranean sapropels and Cretaceous black shales. *Palaeogeography, Palaeoclimatology, Palaeoecology* 235, 305–320. https://doi.org/10.1016/j.palaeo.2005.10.025
Meyres, P., 2006. Paleoceanographic and paleoclimatic similarities between Mediterranean sapropels and Cretaceous black shales. *PALAEO Geography, Climatology, Ecology* 235, 305–320.
Miall, A.D., 2015. *Stratigraphy: A Modern Synthesis*, 1st ed. 2016 edition. ed. Springer.
Michaels, A.F., Silver, M.W., 1988. Primary production, sinking fluxes and the microbial food web. *Deep Sea Research Part A* 35, 473–490.
Michalík, J., Soták, J., Lintnerová, O., et al., 2008. The stratigraphic and paleoenvironmental setting of Aptian OAE black shale deposits in the Pieniny Klippen Belt, Slovak Western Carpathians. *Cretaceous Research* 29, 871–892.
Mignot, J.-P., Cachon, M., Puytorac, C.-F., 1998. In Memoriam: Andre Hollande (1913-1998). *European Journal of Protistology.*
Milankovitch, M., 1941. *Kanon der Erdbestrählung und seine Anwendung auf das Eiszeitenproblem*. Royal Serbian Academy, Belgrade.
Miller, K.G., Kominz, M.A., Browning, J.V., et al., 2005. The Phanerozoic record of global sea-level change. *Science* 310, 1293–1298.
Miller, P.I., Read, J.F., Dale, A.C., 2013. Thermal front variability along the North Atlantic Current observed using microwave and infrared satellite data. *Deep Sea Research, Part II 98*, Part B, 244–256.
Misra, S., Froelich, P., 2012. Lithium isotope history of Cenozoic seawater: changes in silicate weathering and reverse weathering. *Science* 335, 818–823.
Mitsumura, K., Kamata, Y., 2009. Radiolarian faunal change across the Guadalupian-Lopingian boundary in the Sasayama section of the Tanba Belt. Southwest Japan. *News of Osaka Micropaleontologists (NOM), Special Volume* 14, 545–559.
Mivart, St.G., 1878. Notes touching recent researches on the Radiolaria. *Journal of the Linnean Society, Zoology* 14, 136–186.
Moix, P., Kozur, H.W., Stampfli, G.M., Mostler, H., 2007. The Global Triassic New paleontological, biostratigraphic and paleogeographic results from the Triassic of the Mersin Melange, SE Turkey, in: Lucas, S.G., Spielmann, J.A. (Eds.), 41. *Bulletin of the New Mexico Museum of Natural History and Science*, pp. 282–305.
Molina-Cruz, A., 1984. Radiolaria as indicators of upwelling

processes: the Peruvian connection. *Marine Micropaleontology* 9, 53–75.
Molina-Cruz, A., 1977a. Radiolarian assemblages and their relationship to the oceanography of the subtropical Southeastern Pacific. *Marine Micropaleontology* 2, 315–352.
Molina-Cruz, A., 1977b. The relation of the southern trade winds to upwelling processes during the last 75,000 years. *Quaternary Research* 8, 324–338.
Moore, Jr., T.C., 1995. Radiolarian stratigraphy, Leg 138, in: Pisias, N.G. et al. (Eds.), *Proceedings of the Ocean Drilling Program, Scientific Results*. Ocean Drilling Program, College Station, TX, pp. 191–232.
Moore, Jr., T.C., Shackleton, N.J., Pisias, N.G., 1995. Paleoceanography and the diachrony of radiolarian events in the eastern Equatorial Pacific, in: Pisias, N.G. et al. (Eds.), *Proceedings of the Ocean Drilling Program, Scientific Results*. Ocean Drilling Program, College Station, TX, pp. 911–930.
Moore, T.C. Jr., Lombari, G., 1981. Sea-surface temperature changes in the North Pacific during the late Miocene. *Marine Micropaleontology* 6, 581–597.
Moore, R.C., Lalicker, C.G., Fischer, A.G., 1952. *Invertebrate Fossils*. McGraw-Hill, New York.
Moore, T.C., 1972. Mid-Tertiary evolution of the radiolarian genus Calocycletta. *Micropaleontology* 18, 144–152.
Moore, T.C., 1969. Radiolaria: change in skeletal weight and resistance to solution. *Geological Society of America Bulletin* 80, 2103–2108.
Moore, T.C., Burckle, L.H., Geitzenauer, K., et al., 1980. The reconstruction of sea surface temperatures in the Pacific Ocean of 18, 000 B. P. *Marine Micropaleontology* 5, 215– 247.
Moore, T.C. Jr., Mayer, L.A., Lyle, M., 2012. Sediment mixing in the tropical Pacific and radiolarian stratigraphy. *Geochemistry, Geophysics, Geosystems* 13, 1–15 (web).
Moore, T.C. Jr., 1973. Late Pleistocene-Holocene oceanographic changes in the northeastern Pacific. *Quaternary Research* 3, 99–109.
Moore, T.C. Jr., 1973b. Method of randomly distributing grains for microscopic examination. *Journal of Sedimentary Petrology* 43, 904–906.
Moran, K., Backman, J., Brinkhuis, H., et al., 2006. The Cenozoic paleaeoenvironment of the Arctic Ocean. *Nature* 441, 601–605.
Moreira, D., von der Heyden, S., Bass, D., et al., 2007. Global eukaryotic phylogeny: combined small- and large-subunit ribosomal DNA trees support monophyly of Rhizaria, Retaria and Excavata. *Molecular Phylogenetics and Evolution* 44, 255–266.
Morley, J.J., 1989. Variations in high-latitude oceanographic fronts in the Southern Indian Ocean: an estimation based on faunal changes. *Paleoceanography* 4, 547–554.
Morley, J.J., 1983. Identification of density-stratified waters in the Late- Pleistocene North Atlantic: a faunal derivation. *Quaternary Research* 20, 374– 386.
Morley, J.J., Hays, J.D., 1983. Oceanographic conditions associated with high abundances of the radiolarian Cycladophora davisiana. *Earth and Planetary Science Letters* 66, 63– 72.
Morley, J.J., Hays, J.D., 1979. Cycladophora davisiana: A stratigraphic tool for Pleistocene North Atlantic and interhemispheric correlation. *Earth and Planetary Science Letters* 44, 383–389.
Morley, J.J., Nigrini, C., 1995. Miocene to Pleistocene radiolarian biostratigraphy of North Pacific sites 881, 884, 885, 886 and 887, in: Rea, D.K., Basov, I.A., Scholl, D.W., Allan, J.F. (Eds.), *Proceedings of the Ocean Drilling Program, Scientific Results*. Ocean Drilling Program, College Station, TX, pp. 55–91.

Morley, J.J., Shackleton, N.L., 1978. Extension of the radiolarian Stylatractus universus as a biostratigraphic datum to the Atlantic Ocean. *Geology* 6, 309–311.

Morley, J.J., Stepien, J.C., 1985. Antarctic radiolaria in late winter/early spring Weddell Sea waters. *Micropaleontology* 31, 365–371.

Mostler, H., Krainer, K., 1994. Saturnalidae radiolarien aus dem Langobard der südalpinen Karawanken (Kärnten, Österreich). *Geologisch Paläontologische Mitteilungen Innsbruck* 19, 93–131.

Motoyama, I., 2019. A review of Neogene radiolarian biostratigraphy in Japan during the last two decades. *Bulletin of the Geological Survey of Japan* 70, 125–136.

Motoyama, I., 1997. Origin and evolution of Cycladophora davisiana Ehrenberg (radiolaria) in DSDP Site 192, Northwest Pacific. *Marine Micropaleontology* 30, 45- 63.

Motoyama, I., 1996. Late Neogene radiolarian biostratigraphy in the subarctic Northwest Pacific. *Micropaleontology* 42, 221–262.

Motoyama, I., 1993. Late Miocene and Pliocene radiolarian datum levels from DSDP Sites 192 and 302, and Hole 438A of the mid- to high-latitude NW Pacific. *News of Osaka Micropaleontologists, Special Volume* 9, 337–347.

Mudie, P.J., De Vernal, A., Head, M.J., 1990. Neogene to recent palynostratigraphy of circum-arctic basins: Results of ODP LEG 104, Norwegian Sea, LEG 105, Baffin Bay, and DSDP site 611, Irminger Sea, in: Bleil, U., Thiede, J. (Eds.), *Geological History of the Polar Oceans: Arctic versus Antarctic*. Kluwer Academic Publishers, Dordrecht/Boston/London, pp. 609–646.

Müller, J., 1858. *Einige neue bei St. Tropez am Mittelmeer beobachtete Polycystinen und Acanthometren*. Konigliche Preussichen Akademie der Wissenschaften zu Berlin, Monatsberichte, Jahrg. 1858 154–155.

Müller, J., 1857. *Uber die Thalassicollen, Polycystinen und Acanthometren des Mittelmeeres (The Thalassicolla, Polycystina, and Acanthometra from the Mediterranean Sea)*. Konigliche Preussischen Akademie der Wissenschaften zu Berlin, Monatsbericht, Jahre 1856 474–503.

Müller, J., 1855. *Uber Sphaerozoum und Thalassicolla*. Konigliche Preussische Akademie der Wissenschaften zu Berlin, Bericht, Jahrg. 1855 229–253.

Müller, R.D., Dutkiewicz, A., Seton, M., Gaina, C., 2013. Seawater chemistry driven by supercontinent assembly, breakup, and dispersal. *Geology* 41, 907–910.

Mundil, R., Palfy, J., Renne, P.R., Brack, P., 2010. The Triassic timescale: New constraints and a review of geochronological data, in: Lucas, S.G. (Ed.), *The Triassic Timescale, Special Publications. Geological Society*, pp. 41–60.

Murchey, B.L., Jones, D.L., 1992. A mid-Permian chert event: widespread deposition of biogenic siliceous sediments in coastal, island arc and oceanic basins. *Palaeogeography, Palaeoclimatology, Palaeoecology* 96, 161–174.

Musavu-Moussavou, B., Danelian, T., 2006. The Radiolarian biotic response to Oceanic Anoxic Event 2 in the southern part of the Northern proto-Atlantic (Demerara Rise, ODP Leg 207). *Revue de Micropaléontologie* 49, 141–163.

Muttoni, G., Kent, D.V., 2007. Widespread formation of cherts during the early Eocene climatic optimum. *Palaeogeography, Palaeoclimatology, Palaeoecology* 253, 348–362.

Myhre, A.M., Thiede, J., Firth, J., et al., (Eds.), 1995. *Proceedings of the Ocean Drilling Program, Initial Reports,*

volume 151. Ocean Drilling Program, College Station, Texas.
Nakada, R., Ogawa, K., Suzuki, N., et al., 2014. Late Triassic compositional changes of aeolian dusts in the pelagic Panthalassa: Response to the continental climatic change. *Palaeogeography, Palaeoclimatology, Palaeoecology* 393, 61–75.
Nakae, S., 2000a. Regional correlation of the Jurassic accretionary complex in the Inner Zone of Southwest Japan. *Memoirs of Geological Society of Japan* 55, 73–98.
Nakae, S., 2000b. Three kinds of Middle to Late Jurassic pelitic rocks from the Ashio Terrane at the Daigo district in the Yamizo Mountains, central Japan. *Bulletin of Geological Survey of Japan* 51, 113–128.
Nakagawa, C., Nakaseko, K., 1977. Fossil Radiolaria from the Shimanto Formation (preliminary report) - a study of the Shimanto Belt in the eastern part of Shikoku (no. 3). *Journal of Gakugei Tokushima University. Natural Science and Mathematics* 27, 17–29.
Nakamura, Y., Somiya, R., Suzuki, N., et al., 2017. Optics-based surveys of large unicellular zooplankton: a case study on radiolarians and phaeodarians. *Plankton & Benthos Research* 12, 95–103.
Nakaseko, K., 1960. Applied micropaleontological research by means of radiolarian fossil in the oil bearing Tertiary, Japan (mainly in Akita and Yamagata sedimentary basins): Part II. Radiolarian assemblage in Yamagata basin, discussion and conclusion. *Science Reports, College of General Education, Osaka University* 9, 149–185.
Nakaseko, K., 1959. Applied micropaleontological research by means of radiolarian fossil in the oil bearing Tertiary, Japan (mainly in Akita and Yamagata sedimentary basins): Part l. Method, geological note and radiolarian assemblage in Akita sedimentary basin. *Science Reports, College of General Education, Osaka University* 8, 113–193.
Nakaseko, K., 1954. Preliminary report of Miocene radiolarian faunas from southern Toyama prefecture. *Osaka University, Science Report* 107–118.
Nakaseko, K., Iwamoto, H., Takahashi, K., 1965. Radiolarian stratigraphy in the oil and gas bearing Tertiary and Upper Cretaceous formations, Japan, in: Third Petroleum Symposium, Tokyo, No. 1965, *Agenda Item* 6. pp. 1–14.
Nakaseko, K., Sugano, K., Ieda, K., 1972. Some problems concerning the radiolarian stratigraphy in the Niigata sedimentary basin, Japan (Studies of fossil radiolarian-stratigraphy of the Neogene formation in Niigata Prefecture, Japan -4-). *Journal of the Japanese Association of Petroleum Technologists* 37, 55–70.
Narkiewicz, K., Szulc, J., 2004. Controls on migration of conodont fauna in peripheral ocanic areas. An example from the Middle Triassic. of the Northern Peri-Tethys. *Geobios* 37, 425–436.
Natalín, B., 1993. History and modes of Mesozoic accretion in Southeastern Russia. *The Island Arc* 2, 15–34.
Nazarov, B.B., 1975. *Lower and Middle Paleozoic radiolarians of Kazakhstan (methods of investigation, systematics and stratigraphic significance)*. Izdatelstvo Nauka.
Nazarov, B.B., 1973. First occurrences of Radiolaria of the Entactiniidae and Ceratoikiscidae families in the upper Devonian of the Southern Urals. *Doklady Akademii Nauk* SSSR 210, 696–699.
Nazarov, B.B., Ormiston, A., 1984. Tentative system of Paleozoic Radiolaria, in: Petrushevskaya, M.G., Stepanjants, S.D. (Eds.), *Morphology, Ecology and Evolution of Radiolarians. Material from the IV Symposium of European Radiolarists EURO-RAD IV*. pp. 64–87.

Nazarov, B.B., Ormiston, A., 1983. A new superfamily of stauraxon polycystine Radiolaria from the Late Paleozoic of the Soviet Union and North America. *Senckenbergiana Lethaea* 64, 363–379.

Nazarov, B.B., Ormiston, A.R., 1993. New biostratigraphically important Paleozoic Radiolaria of Eurasia and North America. *Micropaleontology, Special Publication* 6, 22–60.

Nazarov, B.B., Ormiston, A.R., 1985. Radiolaria from the Late Paleozoic of the Southern Urals, USSR and West Texas, USA. *Micropaleontology* 31, 1–54.

Nazarov, B.B., Popov, L., 1980. Stratigrafiya i fauna kremnisto-karbonatnyikh tolshch ordovika Kazakhstana (radiolyarii i bezzamkovye brakhiopodyi) [Stratigraphy and fauna of the siliceous-carbonate sequence of the Ordovician of Kazakhstan (Radiolaria and inarticulate brachiopods)]. *Akademiya Nauk SSSR, Ordena Trudovogo Krasnogo Znameni Geologicheskiy Institut* 331, 1–190.

Nazarov, B.B., Rudenko, V.S., 1981. Some radiolarians with bilateral symmetry of the upper Paleozoic of the Southern Urals, Systematics and Morphology of Microorganisms: Problems of Micropaleontology. *Akademiya Nauk SSSR, Seriya Geologicheskaya* 129–139.

Nestell, G.P., Nestell, M.K., 2010. Late Capitanian (latest Guadalupian, Middle Permian) radiolarians from the Apache Mountains, West Texas. *Micropaleontology* 56, 7–68.

Nestell, G.P., Pope, J.P., Nestell, M.K., 2012. Middle Pennsylvanian (Desmoinesian) Radiolaria from the Midcontinent of North America. *Micropaleontology* 58(3). pp. 217–257.

Nigrini, C., 1977. Tropical Cenozoic Artostrobiidae (Radiolaria). *Micropaleontology* 23, 241–269.

Nigrini, C., 1970. Radiolarian assemblages in the North Pacific and their application to a study of Quaternary sediments in Core V20-130, in: Hays, J.D. (Ed.), *Geological Investigations of the North Pacific*. Geological Society of America Memoir, Boulder, CO, USA, pp. 139–183.

Nigrini, C., 1967. Radiolaria in pelagic sediments from the Indian and Atlantic Oceans. *Bulletin of the Scripps Institution of Oceanography, University of California, San Diego, La Jolla, California* 11, 1–125.

Nigrini, C., Caulet, J.P., 1992. Late Neogene radiolarian assemblages characteristic of Indo-Pacific areas of upwelling. *Micropaleontology* 38, 139–164.

Nigrini, C., Caulet, J.P., 1988. The genus Anthocyrtidium (Radiolaria) from the tropical late Neogene of the Indian and Pacific Oceans. *Micropaleontology* 34, 341–360.

Nigrini, C., Sanfilippo, A., Moore Jr., T.C., 2006. Cenozoic radiolarian biostratigraphy: a magnetobiostratigraphic chronology of Cenozoic sequences from ODP Sites 1218, 1219 and 1220, Equatorial Pacific, in: Wilson, P.A., Lyle, M., Firth, J.V. (Eds.), *Proc. ODP, Sci. Results*. Ocean Drilling Program, College Station, pp. 1-76 (chapters separately paginated).

Nigrini, C.A., 1985. Radiolarian biostratigraphy in the central Equatorial Pacific, Deep Sea Drilling Project Leg 85, in: Mayer, L.A., Theyer, F., Thomas, E., et al., (Eds.), *Initial Reports of the Deep Sea Drilling Project*. U. S. Government Printing Office, Washington, D.C., pp. 511–551.

Nigrini, C.A., Caulet, J.P., Nigrini, A., 1998. Radworld: A comprehensive radiolarian relational database. *Radiolaria: Newsletter of the International Association of Radiolarian Paleontologists* 16, 78.

Nikolaev, S.I., Berney, C., Fahrni, J.F., et al., 2004. The twilight of the Heliozoa and rise of the Rhizaria, an emerging supergroup of amoeboid eukaryotes.

Proceedings of the National Academy of Sciences USA 101, 8066–8071.
Nimmergut, A., Abelmann, A., 2002. Spatial and seasonal changes of radiolarian standing stocks in the Sea of Okhotsk. *Deep Sea Research Part I: Oceanographic Research Papers* 49, 463–493.
Nishi, H., Takashima, R., Hatsugai, T., et al., 2003. Planktonic foraminiferal zonation in the Cretaceous Yezo Group, central Hokkaido, Japan. *Journal of Asian Earth Sciences* 21, 867–886.
Nishimura, A., 2001. Paleocene radiolarians from DSDP Leg 43, Site 384 in the Northwest Atlantic. *News of Osaka Micropaleontologists Special* vol. 12, 293–320.
Nishimura, A., 1992. Paleocene radiolarian biostratigraphy in the northwest Atlantic at Site 384, Leg 43, of the Deep Sea Drilling Project. *Micropaleontology* 38, 317–362.
Nishimura, A., 1987. Cenozoic radiolaria in the western North Atlantic, Site 603, Leg 93 of the Deep Sea drilling Project, in: Hinte, J.E., Wise, S.W. (Eds.), *Initial Reports of the Deep Sea Drilling Project*. Washington, pp. 713–737.
Nishimura, H., 1990. Taxonomic study on Cenozoic Nassellaria (Radiolaria). *Science Reports of the Institute of Geoscience, University of Tsukuba, section B=Geological Sciences* 11, 69–172.
Nishimura, K., Ishiga, H., 1987. Radiolarian biostratigraphy of the Maizuru Group in Yanahara area, Southwest Japan. *Memoirs of the Faculty of Science, Shimane University* 21, 169–188.
Nishizono, Y., Sato, T., Murata, M., 1997. A revised Jurassic radiolarian zonation for the South Belt of the Chichibu Terrane, Western Kyushu, Southwest Japan. *Marine Micropaleontology* 30, 117–138.
Noble, P., Aitchison, J.C., Danelian, T., et al., 2017. Taxonomy of Paleozoic radiolarian genera. *Geodiversitas* 39, 419–502.
Noble, P.J., 1994. Silurian Radiolarian Zonation for the Caballos Novaculite, Marathon Uplift, West Texas. *Bulletins of American Paleontology* 106, 1–55.
Noble, P.J., Lenz, A.C., 2007. Upper Wenlock Ceratoikiscidae (Radiolaria) from the Cape Phillips Formation, Arctic Canada. *Journal of Paleontology* 81, 1044–1052.
Noble, P.J., Lenz, A.C., Holmden, C., et al., 2012. Isotope Geochemistry and Plankton Response to the Ireviken (Earliest Wenlock) and Cyrtograptus lundgreni Extinction Events, Cape Phillips Formation, Arctic Canada, in: Talent, J.A. (Ed.), *Earth and Life, International Year of Planet Earth*. pp. 631–652.
Noble, P.J., Naraoka, H., Poulson, S.R., et al., 2011. Paleohydrographic influences on Permian radiolarians in the Lamar Limestone, Guadalupe Mountains, west Texas, elucidated by organic biomarker and stable isotope geochemistry. *Palaios* 26, 180–186.
Norris, R.D., 2000. Pelagic species diversity, biogeography, and evolution. *Paleobiology* 26 Supplement, 236–258.
Not, F., Gauslinq, R., Azam, F., et al., 2007. Vertical distribution of picoeukaryotic diversity in the Sargasso Sea. *Environmental Microbiology* 9, 1233–1252.
Obut, O.T., Iwata, K., 2000. Lower Cambrian radiolaria from the Gorny Altai (southern West Siberia). *News of Paleontology and Stratigraphy* 2000(2–3). pp. 33–37.
Obut, O.T., Shcherbanenko, T.A., 2008. Late Devonian radiolarians from the Rudny Altai (SW Siberia). *Bulletin of Geosciences* 83, 371–382.

Ocean Studies Board, 2000. *50 Years of Ocean Discovery: National Science Foundation 1950–2000*. National Research Council, National Academies Press, Washington, D. C.

O'Connor, B., 1996. Confocal Laser Scanning Microscopy: a new technique for investigating and illustrating fossil Radiolaria. *Micropaleontology* 42, 395–402.

O'Dogherty, L., 2009. Inventory of Mesozoic radiolarian species (1867–2008). *Geodiversitas* 31, 371–481.

O'Dogherty, L., 1994. Biochronology and paleontology of Mid-Cretaceous radiolarian from Northern Apennines (Italy) and Betic Cordillera (Spain). *Memoires de Geologie (Lausanne)* 21, 1–415.

O'Dogherty, L., Carter, E., Goričan, S., Dumitrica, P., 2010. Triassic radiolarian biostratigraphy, in: Lucas, G.S. (Ed.), *The Triassic Timescale*. The Geological Society of London, London, pp. 163–200.

O'Dogherty, L., Carter, E.S., Dumitrica, P., et al., 2009. An illustrated and revised catalogue of Mesozoic radiolarian genera – objectives, concepts and guide for users. *Geodiversitas* 31, 191–212.

O'Dogherty, L., De Wever, P., Goričan, S., et al., 2011. Stratigraphic ranges of Mesozoic radiolarian families. *Palaeoworld* 20, 102–115.

O'Dogherty, L., De Wever, P., Guex, J., 2003. Is the nassellarian/spumellarian diversity ratio a paleoenvironmental proxy indicator in the geological record?, in: Diserens, M.O., Jackett, S.-J. (Eds.), *INTERRAD X, Lausanne*, Switzerland, September 2003, *Abstracts and Programme*.

O'Dogherty, L., Goričan, S., Gawlick, H.-J., 2017. Middle and Late Jurassic radiolarians from the Neotethys suture in the Eastern Alps. *Journal of Paleontology* 91, 25–72.

O'Dogherty, L., Guex, J., 2002. Rates and pattern of evolution among Cretaceous radiolarians: relations with global paleoceanographic events. *Micropaleontology* 48, 1–22.

Ogane, K., Tuji, A., Suzuki, N., et al., 2009. First application of PDMPO to examine silicification in polycystine Radiolaria. *Plankton & Benthos Research* 4, 89–94.

Ogane, K., Tuji, A., Suzuki, N., et al., 2010. Direct observation of the skeletal growth patterns of polycystine radiolarians using a fluorescent marker. *Marine Micropaleontology* 77, 137–144.

Ogg, J.G., 2012. Triassic, in: Gradstein, F.M., Ogg, J.G., Schmitz, M., Ogg, G. (Eds.), *The Geologic Time Scale*. Elsevier, Amsterdam, pp. 681–730.

Ogg, J.G., Agterberg, F.P., Gradstein, F.M., 2004. The Cretaceous period, in: Gradstein, F.M., Ogg, J.G., Smith, A.G. (Eds.), *A Geologic Time Scale 2004*. Cambridge University, Cambridge, pp. 344–383.

Ogg, J.G., Hinnov, L.A., 2012. Cretaceous, in: Gradstein, F.M., Ogg, J.G., Schmitz, M., Ogg, G. (Eds.), *The Geologic Time Scale 2012*. pp. 793–853.

Ogg, J.G., Smith, A.G., 2004. The geomagnetic polarity time scale, in: Gradstein, F.M., Ogg, J.G., Smith, A. (Eds.), *A Geologic Time Scale 2004*. Cambridge University, Cambridge, pp. 63–86.

Okamoto, S., Kojima, S., Suparka, S., Supriyanto, J., 1994. Campanian (Upper Cretaceous) radiolarian from a shale clast in the Paleogene of central Java, Indonesia. *Journal of Southeast Asian Earth Sciences* 9, 45–50.

Okamura, M., 1992. Cretaceous Radiolaria from Shikoku, Japan (Part 1). *Memoirs of the Faculty of Science, Kochi University, Series E, Geology* 21–164.

Okazaki, Y., Takahashi, K., Yoshitani, H., et al., 2003. Radiolarians under the seasonally sea-ice covered conditions in the Okhotsk Sea: flux and their

implications for paleoceanography. *Marine Micropaleontology* 49, 195–230.
Ondrejickova, A., Borza, V., Korabova, K., Michalik, J., 1993. Calpionellid, radiolarian and calcareous nannoplankton association near the Jurassic-Cretaceous boundary (Hrusove section, Cachticke Karpaty Mts., western Carpathians). *Geologica carpathica* 44, 177–188.
Onoue, T., Sato, H., Nakamura, T., et al., 2012. Deep-sea record of impact apparently unrelated to mass extinction in the Late Triassic. *Proceedings of the National Academy of Sciences of the United States of America* 109, 19134–19139.
Opdyke, N.D., Glass, B.P., 1969. The paleomagnetism of sediment cores from the Indian Ocean. *Deep-Sea Research* 16, 249–261.
Oreskes, N., Le Grand, H. (Eds.), 2003. *Plate Tectonics: an Insider's History of the Modern Theory of the Earth*. CRC Press, Boca Raton.
Ormiston, A., Babcock, L., 1979. Follicucullus, new radiolarian genus from the Guadalupian (Permian) Lamar limestone of the Delaware Basin. *Journal of Paleontology* 53, 328–334.
Ortiz, J.D., Mix, A., 1997. Comparison of Imbrie-Kipp transfer function and modern analog temperature estimates using sediment trap and core top foraminiferal faunas. *Paleoceanography* 12, 175–190.
Oseki, S., Suzuki, N., 2009. Miocene to Pleistocene paleogeographic distributions of polycystine radiolarians in the North Pacific. *News of Osaka Micropaleontologists (NOM), Special Volume* 14. pp. 183–238.
Ota, A., Isozaki, Y., 2006. Fusuline biotic turnover across the Guadalupian-Lopingian (Middle-Upper Permian) boundary in mid-oceanic carbonate buildups: biostratigraphy of accreted limestone in Japan. *Journal of Asian Earth Sciences* 26, 353–368.
Otis, L., 2007. *Müller's Lab*. Oxford University Press.
Otto-Bliesner, B., Brady, E.C., Shields, C., 2002. Late Cretaceous ocean: coupled simulations with the National Center for Atmospheric Research Climate System Model. *Journal of Geophysical Research* 107, 1–14 (web).
Palechek, T., Savelév, D., Savelëva, O., 2010. Albian-Cenomanian Radiolarian Assemblage from the Smaginsk Formation, the Kamchatskii Mys Peninsula of Eastern Kamchatka. *Stratigraphy and Geological Correlation* 18, 63–82.
Pálfy, J., Mortensen, J.K., Carter, E.S., et al., 2000. Timing the end-Triassic mass extinction: First on land, then in the sea? *Geology* 28, 28–39.
Palmer, A., 1988. Cenozoic radiolarians from Ocean Drilling Program Leg 101, Bahamas (Sites 627 and 628) and surrounding region, in: Austin, J.A., Schlager, W., et al., (Eds.), *Proceedings of the Ocean Drilling Program, Scientific Results*. College Station, TX (Ocean Drilling Program), pp. 105–116.
Palmer, A., 1986. Miocene radiolarian biostratigraphy, U. S. mid-Atlantic Coastal Plain. *Micropaleontology* 32, 19–31.
Palmer, A.A., 1987. Cenozoic radiolarians from Deep Sea Drilling Project sites 612 and 613 (Leg 95, New Jersey Transect) and Atlantic Slope Project Site ASP 15, in: Poag, C.W., Watts, A.B., et al., (Eds.), *Initial Reports of the Deep Sea Drilling Project*. U. S. Government Printing Office, Washington, D.C., pp. 339–357.
Pantanelli, D., 1880. I diaspri della Toscana e i loro fossili. *Atti della reale Accademia nazionale dei Lincei, Memorie della Classe di Scienze fisiche, matematiche e naturali* 8, 35–66.
Parrish, J.T., 1992. Jurassic climate and oceanography of the Pacific region, in:

Westerman, G.E.G. (Ed.), *The Jurassic of the Circum-Pacific*. Cambridge University Press, New York, pp. 365–379.

Payne, J.L., Clapham, M.E., 2012. End-Permian Mass Extinction in the Oceans: An Ancient Analog for the Twenty-First Century? *Annual Review of Earth Planetary Sciences* 40, 89–111.

Pearce, M.A., Jarvis, I., Tocher, B.A., 2009. The Cenomanian–Turonian boundary event, OAE2 and palaeoenvironmental change in epicontinental seas: New insights from the dinocyst and geochemical records. *Palaeogeography, Palaeoclimatology, Palaeoecology* 280, 207–234.

Peng, T.-H., Broecker, W., Berger, W.H., 1979. Rates of benthic mixing in deep-sea sediment as determined by radioactive tracers. *Quaternary Research* 11, 141–149.

Penman, D.E., 2016. Silicate weathering and North Atlantic silica burial during the Paleocene-Eocene Thermal Maximum. *Geology* 44, 731–734.

Percival, I.G., 2011. Biotic characteristics of Ordovician deep-water cherts from Eastern Australia. *Palaeogeography, Palaeoclimatology, Palaeoecology* 367, 63–72.

Pessagno, E.A., 1977a. Lower Cretaceous radiolarian biostratigraphy of the Great Valley Sequence and Franciscan Complex, California Coast Ranges. *Cushman Foundation for Foraminiferal Research, Special Publication* 15, 1–87.

Pessagno, E.A., 1977b. Upper Jurassic Radiolaria and radiolarian biostratigraphy of the California Coast Ranges. *Micropaleontology* 23, 56–113.

Pessagno, E.A., 1976. Radiolarian zonation and stratigraphy of the Upper Cretaceous portion of the Great Valley Sequence, California Coast Ranges. *Micropaleontology, Special Publication* 2, 1–95.

Pessagno, E.A., 1973. Upper Cretaceous Spumellariina from the Great Valley Sequence, California Coast Ranges. *Bulletins of American Paleontology* 63, 49–102.

Pessagno, E.A., 1972. Cretaceous Radiolaria. Part I: The Phaseliformidae, new family, and other Spongodiscacea from the Upper Cretaceous portion of the Great Valley Sequence, part II; Pseudoaulophacidae Riedel from the Cretaceous of California and the Blake-Bahama Basin (JOIDES leg 1). *Bulletins of American Paleontology* 61, 269–328.

Pessagno, E.A., 1971. Jurassic and Cretaceous Hagiastridae from the Blake-Bahama Basin (Site 5A, JOIDES Leg 1) and the Great Valley Sequence, California Coast Ranges. *Bulletins of American Paleontology* 60, 5–83.

Pessagno, E.A., 1970. The Rotaformidae, a new family of Upper Cretaceous Nassellariina (Radiolaria) from the Great Valley Sequence, California Coast Ranges. *Bulletins of American Paleontology* 58, 1–33.

Pessagno, E.A., 1969. The Neosciadiocapsidae, a new family of Upper Cretaceous Radiolaria. *Bulletins of American Paleontology* 56, 377–439.

Pessagno, E.A., Blome, C., Longoria, J., 1984. A revised radiolarian zonation from Upper Jurassic of western North America. *Bulletins of American Paleontology* 87, 1–51.

Pessagno, E.A., Blome, C.D., 1986. Faunal affinities and tectonogenesis of Mesozoic rocks in the Blue Mountains of eastern Oregon and western Idaho, in: Vallier, T.L., Brooks, H.C. (Eds.), Geologic Implications of Paleozoic and Mesozoic Paleontology and Biostratigraphy, Blue Mountains Province, Oregon, Idaho and Washington. United States Geological Survey, Professional Paper, Washington D.C., pp. 65–78.

Pessagno, E.A., Blome, C.D., Meyerhoff Hull, D., Six, W.M., 1993. Jurassic Radiolaria from the Josephine ophiolite and overlying strata, Smith River subterrane (Klamath Mountains), northwestern California and southwestern Oregon. *Micropaleontology* 39, 93–166.

Pessagno, E.A., Cantú-Chapa, A., Mattinson, J.M., et al., 2009. The Jurassic-Cretaceous boundary: new data from North America and the Caribbean. *Stratigraphy* 6, 185–262.

Pessagno, E.A., Cattu-Chapa, A., Hull, D.M., et al., 1999. Sedimentary Basins of the World Chapter 5 Stratigraphic evidence for northwest to southeast tectonic transport of jurassic terranes in central Mexico and the caribbean (western Cuba), in: Mann, P. (Ed.), *Caribbean Basins* Volume 4. Elsevier, pp. 123–150.

Pessagno, E.A., Finch, W., Abbott, P.L., 1979. Upper Triassic Radiolaria from the San Hipolito Formation, Baja California. *Micropaleontology* 25, 160–197.

Pessagno, E.A., Hull, D.M., 1996. "Once upon a time in the Pacific": chronostratigraphic misinterpretation of basal strata at ODP site 801 (Central Pacific) and its impact on geochronology and plate tectonics models. *GeoResearch Forum* 1–2, 79–92.

Pessagno, E.A., Longoria, J., MacLeod, N., Six, W., 1987. Studies of North American Jurassic Radiolaria; Part I, Upper Jurassic (Kimmeridgian-upper Tithonian) Pantanelliidae from the Taman Formation, east-central Mexico; tectonostratigraphic, chronostratigraphic, and phylogenetic implications. *Cushman Foundation for foraminiferal Research, Special Publication* 23, 1–51.

Pessagno, E.A., Newport, R.L., 1972. A technique for extracting Radiolaria from radiolarian cherts. *Micropaleontology* 18, 231–234.

Pessagno, E.A., Six, W.M., Yang, Q., 1989. The Xiphostylidae Haeckel and Parvivaccidae, n. fam., (Radiolaria) from the North American Jurassic. *Micropaleontology* 35, 193–255.

Pessagno, E.A., Whalen, P., 1982. Lower and Middle Jurassic Radiolaria (multicyrtid Nassellariina) from California, east-central Oregon and the Queen Charlotte Islands, B. C. *Micropaleontology* 28, 111–169.

Petrushevskaya, M.G., 1986. Radiolyarievyi analiz (Radiolarian analysis). *Akademiya Nauk SSSR, Zoological Institute.*

Petrushevskaya, M.G., 1981. Radiolyarii otryada Nassellaria Mirovogo okeana (Nassellarian radiolarians from the world ocean). *Nauka, Leningradskoe Otdelenie, Leningrad, USSR.*

Petrushevskaya, M.G., 1975. Cenozoic radiolarians of the Antarctic Deep Sea Drilling Project Leg 29, in: Kennett, J.P., Houtz, R.E. (Eds.), *Initial Reports of the Deep Sea Drilling Project.* U. S. Government Printing Office, Washington, D. C., pp. 541–676.

Petrushevskaya, M.G., 1971a. Nassellarian radiolarians in the plankton of the World Ocean. *Akademiya nauk SSSR, Zoologicheskii Institut, Issledovaniya Fauny Morei* 9, 294.

Petrushevskaya, M.G., 1971b. Radiolyarii Nassellaria v planktone mirovogo okeana (Nassellarian Radiolarians of the World Ocean). *Akademiya nauk SSSR, Zoologicheskii Institut, Issledovaniya Fauny Morei* 9, 294.

Petrushevskaya, M.G., 1971c. Spumellarian and nassellarian Radiolaria in the plankton and bottom sediments of the Central Pacific, in: Funnell, B.M., Riedel, W.R. (Eds.), *The Micropalaeontology of Oceans.* Cambridge University Press, Cambridge, UK, pp. 309–317.

Petrushevskaya, M.G., 1967. Issledovaniya Fauny Morei, Resultaty Biologicheskikh Issledovanii Sovetskoi Antarkticheskoi Ekspeditsii 1955–1958 Antarctic Spumelline and Nasselline radiolarians, in: 4. *Zoologicheskii Institut Academiya Nauk SSSR*, pp. 5–186.

Petrushevskaya, M.G., 1962. The role of the skeleton growth in Radiolaria for their systematics. *Zoologicheskii Zhurnal, Akademia Nauk SSSR* 41, 331–341.

Pflaumann, U., Duprat, J., Pujol, C., Labeyrie, L., 1996. SIMMAX: a modern analog technique to deduce Atlantic sea surface temperatures from planktonic foraminifera in deep-sea sediments. *Paleoceanography* 11, 15–35.

Pigliucci, M., Müller, G.B. (Eds.), 2010. *Evolution, the Extended Synthesis*. The MIT Press, Cambridge, MA.

Pinet, P.R., 2012. *Invitation to Oceanography*. Jones and Bartlett, Burlington, MA.

Pisias, N.G., Mix, A.C., Heusser, L., 2001. Millennial scale climate variability of the Northeast Pacific Ocean and Northwest North America based on Radiolaria and pollen. *Quaternary Science Reviews* 20, 1561–1576.

Pisias, N.G., Roelofs, A., Weber, M., 1997. Radiolarian-based transfer functions for estimating mean surface ocean temperatures and seasonal range. *Paleoceanography* 12, 365–379.

Polet, S., Berney, C., Fahrni, J., Pawlowski, J., 2004. Small-subunit ribosomal RNA gene sequences of Phaeodaria challenge the monophyly of Haeckel's Radiolaria. *Protist* 155, 53–63.

Popofsky, A., 1908. Die Radiolarien der Antarktis (mit Ausnahme der Tripyleen), in: Drygalski, E. (Ed.), *Deutsche Sudpolar-Expedition, 1901–1903*. Band X. Georg Reimer, Berlin, Germany, pp. 183–306.

Porter, S., 2011. The rise of the predators. *Geology* 39, 607–608.

Pouille, L., Obut, O., Danelian, T., Sennikov, N., 2011. Lower Cambrian (Botomian) polycystine Radiolaria from the Altai Mountains (southern Siberia, Russia). *Comptes Rendus Palevol* 10, 627–633.

Poulsen, C.J., Barron, E.J., Arthur, M.A., Peterson, W.H., 2001. Response of the mid-Cretaceous global oceanic circulation to tectonic and CO2 forcings. *Paleoceanography* 16, 576–592.

Prairie, J.C., Sutherland, K.R., Nickols, K.J., Kaltenberg, A.M., 2012. Biophysical interactions in the plankton: A cross-scale review. *Limnology and Oceanography: Fluids and Environments* 2, 121–145.

Premoli Silva, I., Erba, E., Salvini, G., et al., 1999. Biotic changes in Cretaceous oceanic anoxic events of the Tethys. *Journal of Foraminiferal Research* 29, 352–370.

Preto, N., Kustatscher, E., Wignall, P., 2010. Triassic climates – state of the art and perspectives. *Palaeogeography, Palaeoclimatology, Palaeoecology* 290, 1–10.

Probert, I., Siano, R., Poirier, C., et al., 2014. Brandtodinium gen. nov. and B. nutricula comb. Nov. (Dinophyceae), a dinoflagellate commonly found in symbiosis with polycystine radiolarians. *Journal of Phycology* 50, 388–399.

Prothero, D., Lazarus, D.B., 1980. Planktonic microfossils and the recognition of ancestors. *Systematic Zoology* 29, 119–129.

Purcell, E.M., 1977. Life at low Reynolds number. *American Journal of Physics* 45, 3–11.

Rabbitt, M.C., 1975. *A brief history of the U.S. Geological Survey (Report)*, General Information Product. Washington, D.C.

Racki, G., 1999. Silica-secreting biota and mass extinctions; survival patterns and processes. *Biotic recoveries from mass extinctions* 154, 107–132.

Racki, G., Cordey, F., 2000. Radiolarian paleoecology and radiolarites: is the present the key to the past? *Earth-Science Reviews* 52, 83–120.

Raup, D.M., 1979. Size of the Permo-Triassic bottleneck and its evolutionary implications. *Science* 206, 217–218.

Raup, D.M., 1975. Taxonomic diversity estimation using rarefaction. *Paleobiology* 1, 333–342.

Renaudie, J., 2016. Quantifying the Cenozoic marine diatom deposition history: links to the C and Si cycles. *Biogeosciences* 13, 6003–6014.

Renaudie, J., 2012. A Synthesis of Antarctic Neogene radiolarians: taxonomy, macroevolution and biostratigraphy (PhD Thesis). Humboldt University, Berlin.

Renaudie, J., Drews, E.-L., Böhne, S., 2018. The Paleocene record of marine diatoms in deep-sea sediments. *Fossil Record* 21, 183–205.

Renaudie, J., Gray, R., Lazarus, D.B., in press. Accuracy of a neural net classification of closely-related species of microfossils from a sparse dataset of unedited images (No. e27328v1). PeerJ Inc. https://doi.org/10.7287/peerj.preprints.27328v1

Renaudie, J., Lazarus, D., 2013. On the accuracy of paleodiversity reconstructions: a case study in Antarctic Neogene radiolarians. *Paleobiology* 39, 491–509.

Renaudie, J., Lazarus, D., 2012. New species of Neogene radiolarians from the Southern Ocean. *Journal of Micropalaeontology* 31, 29–52.

Renaudie, J., Lazarus, D.B., Diver, P., 2020. NSB (Neptune Sandbox Berlin): An expanded and improved database of marine planktonic microfossil data and deep-sea stratigraphy. *Palaeontol Electron* 23, 1–28. https://doi.org/10.26879/1032

Renz, G.W., 1988. Silurian radiolarians of the genus Ceratoikiscum from the Canadian Arctic. *Micropaleontology* 34, 260–267.

Renz, G.W., 1976. The distribution and ecology of Radiolaria in the Central Pacific plankton and surface sediments. *Bulletin of the Scripps Institution of Oceanography, University of California, San Diego, La Jolla, California* 22, 1–267.

Renz, G.W., 1974. Radiolaria from Leg 27 of the Deep Sea Drilling Project, in: Veevers, J.J., Heirtzler, J.R., et al. (Eds.), *Initial Reports of the Deep Sea Drilling Project*. U.S. Government Printing Office, Washington D. C., pp. 769–841.

Reshetnyak, V.V., 1955. Vertikalnoe raspredelenie radiolyarii Kurilo-Kamchatskoi vpadiny (Vertical distribution of radiolarians from the Kurilo-Kamchatka trench). *Trudy Zoologicheskogo Instituta Academii Nauk SSSR* 21, 94–101.

Reygondeau, G., Dunn, D., 2018. Pelagic Biogeography, in: *Encyclopedia of Ocean Sciences*. Elsevier, pp. 356–363.

Reynolds, C.S., 2006. *The Ecology of Phytoplankton*. Cambridge University Press, Cambridge, UK.

Reynolds, R.A., Sakai, T., Casey, R.E., 1980. Synthesis of radiolarian results from DSDP Legs 56 and 57 and their relation to other north Pacific sections, in: Langseth, M., Okada, H., Von Huene, R., Nasu, N., et al., (Eds.), *Initial Reports of the Deep Sea Drilling Project*. U. S. Government Printing Office, Washington, D.C., pp. 771–773.

Richards, R.J., 2008. *The Tragic Sense of Life: Ernst Haeckel and the Struggle Over Evolutionary Thought*. University of Chicago Press, Chicago.

Ricklefs, R.E., Schluter, D., 1993. *Species Diversity in Ecological Communities Historical and Geographical Perspectives*.

Ridgwell, A., 2005. A Mid Mesozoic Revolution in the regulation of ocean chemistry. *Marine Geology* 217, 339–357.

Ridgwell, A., Zeebe, R.E., 2005. The role of the global carbonate cycle in the regulation and evolution of the Earth system. *Earth and Planetary Science Letters* 234, 299–315.

Riedel, W.R., 1971. Systematic classification of polycystine Radiolaria, in: Funnell, B.M., Riedel, W.R. (Eds.), *The Micropalaeontology of Oceans*. Cambridge University Press, pp. 649–660.

Riedel, W. R., 1967. Some new families of Radiolaria. *Proceedings of the Geological Society of London* 1640, 148–149.

Riedel, W.R., 1967. Subclass Radiolaria, in: Harland, W.B. (Ed.), *The Fossil Record*. Geological Society of London, London, pp. 291–298.

Riedel, W.R., 1963. Experimental phase of the Mohole Project. *Transactions of the American Geophysical Union* 44, 358–359.

Riedel, W.R., 1959. Oligocene and Lower Miocene Radiolaria in tropical Pacific sediments. *Micropaleontology* 5, 285–302.

Riedel, W.R., 1958. Radiolaria in Antarctic sediments, in: *Reports of the B.A.N.Z. Antarctic Research Expedition, Series B*. pp. 217–255.

Riedel, W.R., 1957. Radiolaria: a preliminary stratigraphy, in: Petterson, H. (Ed.), *Reports of the Swedish Deep-Sea Expedition, 1947–1948*. Elanders Boktryckeri Aktiebolag, Goteborg, Sweden, pp. 59–96.

Riedel, W.R., 1953. Mesozoic and late Tertiary Radiolaria of Rotti. *Journal of Paleontology* 27, 805–813.

Riedel, W.R., 1951. Sedimentation in the tropical Indian Ocean. *Nature* 168, 737–738.

Riedel, W.R., Funnell, B.M., 1964. Tertiary sediment cores and microfossils from the Pacific Ocean floor. *Quarterly Journal of the Geological Society of London* 120, 305–368.

Riedel, W.R., Sanfilippo, A., 1978. Stratigraphy and evolution of tropical Cenozoic radiolarians. *Micropaleontology* 24, 61–96.

Riedel, W.R., Sanfilippo, A., 1974. Radiolaria from the southern Indian Ocean, DSDP Leg 26, in: Davies, T.A., Luyendyk, B.P., et al. (Eds.), *Initial Reports of the Deep Sea Drilling Project*. U.S. Government Printing Office, pp. 771–814.

Riedel, W.R., Sanfilippo, A., 1971. Cenozoic Radiolaria from the western tropical Pacific, Leg 7, in: Winterer, E.L., Riedel, W.R., et al. (Eds.), *Initial Reports of the Deep Sea Drilling Project*. U. S. Government Printing Office, Washington, D.C., pp. 1529–1672.

Riedel, W.R., Sanfilippo, A., 1970. Radiolaria, Leg 4, Deep Sea Drilling Project, in: *Initial Reports of the Deep Sea Drilling Project*. U. S. Government Printing Office, Washington, D.C., pp. 503–575.

Rigby, S., Milsom, C., 2000. Origins, evolution, and diversification of zooplankton. *Annual Reviews of Ecology and Systematics* 31, 293–313.

Rigby, S., Milsom, C., 1996. Benthic origins of zooplankton: An environmentally determined macroevolutionary effect. *Geology* 24, 52–54.

Roelofs, A., Pisias, N., 1986. Revised technique for preparing quantitative radiolarian slides from deep-sea sediments. *Micropaleontology* 32, 182–185.

Rogers, J., De Deckker, P., 2011. Environmental reconstructions of the upper 500 m of the southern Indian Ocean over the last 40 ka using Radiolarian (Protista) proxies. *Quaternary Science Reviews* 30, 876–886.

Rogers, J., De Deckker, P., 2007. Radiolaria as a reflection of environmental conditions in the eastern and southern sectors of the Indian Ocean; a new statistical approach. *Marine Micropaleontology* 65, 137–162.

Rothwell, R.G., Rack, F.R., 2006. New techniques in sediment core analysis: an introduction, in: Rothwell, R.G. (Ed.), *New Techniques in Sediment Core Analysis*. Geological Society, London, pp. 1–29.

Royer, D.L., Berner, R.A., Montañez, I.P., et al., 2004. CO2 as a primary driver of Phanerozoic climate. *GSA Today* 14, 4.
Ruddiman, W.F., 2000. *Earth's Climate: Past and Future.* W. H. Freeman.
Ruedemann, R., Wilson, T.Y., 1936. Eastern New York Ordovician cherts. *Bulletin of the Geological Society of America* 47, 1335–1586.
Russell, R.D., 1970. SEPM history. *Journal of Sedimentary Petrology* 40, 7–28.
Rüst, D., 1892. Beitrage zur Kenntnis der fossilen Radiolarien aus Gesteinen derTrias und der palaeozoischen Schichten. *Palaeontographica* 38, 107–179.
Sachs, H.M., 1973a. Late Pleistocene history of the North Pacific: evidence from a quantitative study of Radiolaria in Core V21-173. *Journal of Quaternary Research* 3, 89–98.
Sachs, H.M., 1973b. North Pacific radiolarian assemblages and their relationship to oceanographic parameters. *Journal of Quaternary Research* 3, 73–88.
Sadler, P.M., 2010. Brute-force biochronology: sequencing paleobiologic first- and last-appearance events by trial and error, in: Alroy, J., Hunt, G. (Eds.), *Paleontological Society Short Course.* Paleontological Society, pp. 271–289.
Sadler, P.M., 2004. Quantitative biostratigraphy – achieving finer resolution in global correlation. *Annual Review of Earth and Planetary Sciences* 32, 187–213.
Saesaengseerung, D., Sashida, K., Sardsud, A., 2007. Late Devonian to Early Carboniferous radiolarian fauna from the Pak Chom area, Loei Province, northeastern Thailand. *Paleontological Research* 11, 109–121.
Salvini, G., Marcucci Passerini, M., 1998. The radiolarian assemblages of the Bonarelli Horizon in the Umbria-Marche Apennines and Southern Alps, Italy. *Cretaceous Research* 19, 777–804.
Sandin, M.M., Biard, T., Romac, S., et al., 2020. Morpho-molecular diversity and evolutionary analyses suggest hidden lifestyles in Spumellaria (Radiolaria). bioRxiv 2020.06.29.176917. https://doi.org/10.1101/2020.06.29.176917
Sandin, M.M., Pillet, L., Biard, T., et al., 2019. Time calibrated morpho-molecular classification of Polycystine Nassellaria (Radiolaria). *Protist* 170, 187–208. https://doi.org/10.1016/j.protis.2019.02.002
Sanfilippo, A., 1990. Origin of the subgenera Cyclampterium, Paralampterium and Sciadiopeplus from Lophocyrtis (Lophocyrtis) (Radiolaria, Theoperidae). *Marine Micropaleontology* 15, 287–312.
Sanfilippo, A., 1988. Pliocene radiolaria from Bianco, Calabria, Italy. *Micropaleontology* 34, 159–180.
Sanfilippo, A., Blome, C.D., 2001. Biostratigraphic implications of mid-latitude Palaeocene-Eocene radiolarian faunas from Hole 1051A, ODP Leg 171B, Blake Nose, western North Atlantic. *Western North Atlantic Palaeogene and Cretaceous Palaeoceanography* 183, 185–224.
Sanfilippo, A., Burckle, L.H., Martini, E., Riedel, W.R., 1973. Radiolarians, diatoms, silicoflagellates and calcareous nannofossils in the Mediterranean Neogene. *Micropaleontology* 19, 209–234.
Sanfilippo, A., Caulet, J.P., 1998. Taxonomy and evolution of Paleogene Antarctic and tropical Lophocyrtid radiolarians. *Micropaleontology* 44, 1–43.
Sanfilippo, A., Caulet, J.P., Riedel, W.R., 1978. Radiolaria from Mediterranean sediments, DSDP Leg 42A, in: Hsu, K.J., Montadert, L., et al., (Eds.), *Initial Reports of the Deep Sea Drilling Project.* U. S. Government Printing Office, Washington, D.C., pp. 753–760.
Sanfilippo, A., Nigrini, C., 1998. Code numbers for Cenozoic low latitude radiolarian biostratigraphic zones

and GPTS conversion tables. *Marine Micropaleontology* 33, 169.

Sanfilippo, A., Riedel, W.R., 1985. Cretaceous Radiolaria, in: Bolli, H.M., Saunders, J.B., Perch-Nielsen, K. (Eds.), *Plankton Stratigraphy*. Cambridge University Press, Cambridge, UK, pp. 573–630.

Sanfilippo, A., Riedel, W.R., 1980. A revised generic and suprageneric classification of the Artiscins (Radiolaria). *Journal of Paleontology* 54, 1008–1011.

Sanfilippo, A., Riedel, W.R., 1979. Radiolaria from the northeastern Atlantic Ocean, DSDP Leg 48, in: Montadert, L., Roberts, D.G., et al., (Eds.), *Initial Reports of the Deep Sea Drilling Project*. U. S. Government Printing Office, Washington, D.C., pp. 493–511.

Sanfilippo, A., Westberg-Smith, M.J., Riedel, W.R., 1985. Cenozoic Radiolaria, in: Bolli, H.M., Saunders, J.B., Perch-Nielsen, K. (Eds.), *Plankton Stratigraphy*. Cambridge University Press, Cambridge, UK, pp. 631–712.

Sano, H., Kuwahara, K., Yao, A., Agematsu, S., 2010. Panthalassan seamount-associated Permian-Triassic boundary siliceous rocks, Mino terrane, central Japan. *Paleontological Research* 14, 293–314.

Sashida, K., Igo, H., Adachi, S., et al., 1993. Occurrences of Paleozoic and Mesozoic radiolarians from Thailand and Malaysia and its geologic significance (preliminary report). *News of Osaka Micropaleontologists, Special Volume* 9, 1–17.

Sashida, K., Sardsud, A., Igo, H., et al., 1998. Occurrence of Dienerian (Lower Triassic) radiolarians from the Phatthalung area of Peninsular Thailand and radiolarian biostratigraphy around the Permian/Triassic (P/T) boundary. *News of Osaka Micropaleontologists, Special Volume* 11, 59–70.

Sashida, K., Tonishi, K., 1991. An Upper Permian coiled radiolarian from Itsukaichi, central Japan. *Micropaleontology* 37, 86–94.

Sashida, K., Tonishi, K., 1988. Additional note on the Upper Permian radiolarian fauna from Itsukaichi, western part of Tokyo Prefecture, central Japan. *Transactions and Proceedings of the Palaeontological Society of Japan, new Series* 151, 523–542.

Sassen, K., Starr, D.O., Mace, G.G., et al., 2003. Midlatitude cirrus clouds derived from hurricane ora: a case study with implications for ice crystal nucleation and shape. *Journal of the Atmospheric Sciences* 60, 873–891.

Saunders, J.B., Bernoulli, D., Muller-Merz, E., et al., 1985. Stratigraphy of the late middle Eocene to early Oligocene in the Bath Cliff Section, Barbados, West Indies. *Micropaleontology* 30, 390–425.

Schaaf, A., 1984. Les Radiolaires du Crétacé inferieur et moyen: Biologie, Systématique, Biochronologie et Paleoenvironment (PhD Thesis). L'Universite Louis Pasteur.

Schaaf, A., 1981. Late Early Cretaceous Radiolaria from Deep Sea Drilling Project Leg 62, in: Thiede, J., Vallier, T.L., et al. (Eds.), Initial Reports of the Deep Sea Drilling Project 62. U.S. Government Printing Office, pp. 419–470.

Schewiakoff, W., 1926. *Die Acantharien des Golfes von Neapel*. Friedlander, Rome.

Schleiden, M.J., 1839. Beiträge zur phytogenesis. *Archiv für Anatomie, Physiologie und wissenschaftliche Medicin* 137–176.

Schmitz, M., Davydov, V., 2012. Quantitative radiometric and biostratigraphic calibration of the Pennsylvanian–Early Permian (Cisuralian) time scale and pan-Euramerican chronostratigraphic correlation. *GSA Bulletin* 124, 549–577.

Schneebeli-Hermann, E., Kürschner, W., et al., 2013. Evidence for atmospheric carbon injection during the end-Permian extinction. *Geology* 41, 579–582.
Schootbrugge, B.V. de, Wignall, P.B., 2015. A tale of two extinctions: converging end-Permian and end-Triassic scenarios. *Geology Magazine* 1–23 (web).
Schulte, P., Alegret, L., Arenillas, I., et al., 2010. The Chicxulub Asteroid Impact and Mass Extinction at the Cretaceous-Paleogene Boundary. *Science* 327, 1214–1218.
Schwann, T., 1839. *Mikroskopische Untersuchungen über die Übereinstimmung in der Struktur und dem Wachstum der Thiere und Pflanzen*. Berlin.
Schwartzapfel, J.A., Holdsworth, B.K., 1996. Upper Devonian and Mississippian radiolarian zonation and biostratigraphy of the Woodford, Sycamore, Caney and Goddard formations, Oklahoma, Special Publication. *Cushman Foundation for Foraminiferal Research.*
Seo, E.-H., Won, M.-Z., 2009. Review of the genus Polyentactinia and the family Polyentactiniidae. *Micropaleontology* 55, 61–74.
Sepkoski, J.J., Jr, 1996. Patterns of Phanerozoic extinction: a perspective from global data bases, in: Walliser, O.H. (Ed.), *Global Events and Event Stratigraphy in the Phanerozoic.* Springer, Berlin, pp. 35–51.
Sepkoski, J.J. Jr., 2002. A compendium of fossil marine genera. *Bulletins of American Paleontology* 363, 1–560.
Shackleton, N.J., Baldauf, J.G., Flores, J.A., et al., 1995. Biostratigraphic summary for Leg 138, in: Psias, N.G., Mayer, L., Janacek, T., Palmer-Julson, A., van Andel, T.H. (Eds.), *Proc. ODP, Sci. Results*, 138. Ocean Drilling Program, College Station, TX, pp. 517–536.
Shaw, A.B., 1964. *Time in Stratigraphy*. McGraw-Hill, New York.
Sheehan, P.M., 1977. Species diversity in the Phanerozoic –a reflection of labor by systematists? *Paleobiology* 3, 325–328.
Shemesh, A., Burckle, L.H., Hays, J.D., 1995. Oxygen isotopes in biogenic silica from the Southern Ocean. *Paleoceanography* 10, 179–196.
Shen, S.-Z., Henderson, C.M., Bowring, S.A., et al., 2010. High-resolution Lopingian (Late Permian) timescale of South China. *Geological Journal* 45, 122–134.
Sieburth, J.M., Smetacek, V., Lenz, J., 1978. Pelagic ecosystem structure: heterotrophic compartments of the plankton and their relationship to plankton size fractions. *Limnology and Oceanography* 23, 1256–1263.
Sierra, R., Cañas-Duarte, S.J., Burki, F., et al., 2015. Evolutionary origins of rhizarian parasites. *Molecular Biology and Evolution* 33, 980–983.
Siesser, W.G., 1981. *Christian Gottfried Ehrenberg: Founder of Micropaleontology*. Centaurus 25, 166–188.
Smith, A.B., 2001. Large-scale heterogeneity of the fossil record: implications for Phanerozoic biodiversity studies. *Philosophical Transactions of the Royal Society of London* 356, 351–367.
Smith, W.D., 1915. Notes on the geology of Panay. *The Philippine Journal of Science (A)* 10(3), 211–229.
Spencer-Cervato, C., 1999. The Cenozoic deep sea microfossil record: explorations of the DSDP/ODP sample set using the Neptune database. *Palaeontologica Electronica* 2, web.
Spencer-Cervato, C., Lazarus, D.B., Beckmann, J.P., et al., 1993. New calibration of Neogene radiolarian events in the North Pacific. *Marine Micropaleontology* 21, 261–294.
Spencer-Cervato, C., Thierstein, H.R., Lazarus, D.B., Beckmann, J.P., 1994. How synchronous are Neogene marine

plankton events? *Paleoceanography* 9, 739–763.
Squinabol, S., 1904. Radiolarie cretacee degli Euganei. *Atti e Memorie della Reale Accademia di Scienze, Lettere ed Arti in Padova, nuova serie* 20, 171–244.
Stainforth, R.M., 1948. Applied Micropaleontology in Coastal Ecuador. *Journal of Paleontology* 22, 113–151.
Stampfli, G.M., Borel, G.D., 2002. A plate tectonic model for the Paleozoic and Mesozoic constrained by dynamic plate boundaries and restored synthetic oceanic isochrons. *Earth and Planetary Science Letters* 196, 17–33.
Stanek, J., Kiessling, W., 2002. Sectioning of radiolarians under continuous observation. *Mitteilungen Museum Naturkunde Berlin, Reihe* 5 45–49.
Stanley, S.M., 2009. Evidence from ammonoids and conodonts for multiple Early Triassic mass extinctions. *PNAS* 106, 15264–15267.
Stanley, S.M., 1979. *Macroevolution: Pattern and Process*. W. H. Freeman, San Francisco.
Steiger, T., 1992. Systematik, stratigraphie und Palökologie der Radiolarien des Oberjura-Unterkreiden-Grenzbereiches im Osterhorn-Tirolikum (Nördliche Kalkalpen, Salzburg und Bayern. *Zitteliana* 19, 3–188.
Stepanjants, S.D., Cortese, G., Kruglikova, S.B., Bj√∏rklund, K.R., 2006. A review of biopolarity concepts: History and examples from Radiolaria and Medusozoa. (Cnidaria). *Marine Biology Research* 2, 200–241.
Stickley, C.E., Brinkhuis, H., Schellenberg, S.A., et al., 2004. Timing and nature of the deepening of the Tasmanian Gateway. *Paleoceanography* 19, 1–18 (web).
Stoermer, E.F., Smol, J.P., 2010. Applications and uses of diatoms: prologue, in: Smol, J.P., Stoermer, E.F. (Eds.), *The Diatoms: Applications for the Environmental and Earth Sciences*, Second Edition. Cambridge University Press, Cambridge, pp. 3–10.
Stöhr, E., 1880. Die Radiolarienfauna der Tripoli von Grotte, Provinz Girgenti in Sicilien (The radiolarian fauna of the Tripoli of Grotte, Girenti Province, Sicily). *Palaeontographica* 26, 71–124.
Street, C., Bown, P.R., 2000. Palaeogeography of Early Cretaceous (Berriasian-Barremian) calcareous nannoplankton. *Marine Micropaleontology* 39, 265–291.
Stukel, M.R., Biard, T., Krause, J., Ohman, M.D., 2018. Large Phaeodaria in the twilight zone: Their role in the carbon cycle. *Limnology and Oceanography* 63, 2579–2594.
Stukel, M.R., Ohman, M.D., Kelly, T.B., Biard, T., 2019. the roles of suspension-feeding and flux-feeding zooplankton as gatekeepers of particle flux into the mesopelagic ocean in the Northeast Pacific. *Frontiers in Marine Science* 6. https://doi.org/10.3389/fmars.2019.00397
Sugano, T., 1937. List of radiolarians from the inshore along the Japanese Islands (the first report). *Journal of Fisheries* 42, 54–72.
Sugiyama, K., 1997. Triassic and Lower Jurassic radiolarian biostratigraphy in the siliceous claystone and bedded chert units of the southeastern Mino Terrane. central Japan. *Bulletin of the Mizunami Fossil Museum* 79–193.
Sugiyama, K., Anderson, O.R., 1997a. Correlated fine structural and light microscopic analyses of living nassellarians Eucyrtidium hexagonatum Haeckel, Pterocorys zancleus (Müller) and Spirocyrtis scalaris Haeckel. *News of Osaka Micropaleontologists (NOM), Special Volume* 311–337.
Sugiyama, K., Anderson, O.R., 1997b. Experimental and observational studies of radiolarian physiological ecology, 6. Effects of silicate-supplemented seawater on the longevity and weight gain of spongoise radiolarians Spongaster tetras

and Dictyocoryne truncatum. *Marine Micropaleontology* 29, 159–172.
Sugiyama, K., Hori, R.S., Kusunoki, Y., Matsuoka, A., 2008. Pseudopodial features and feeding behavior of living nassellarians Eucyrtidium hexagonatum Haeckel, Pterocorys zancleus (Müller) and Dictyocodon prometheus Haeckel. *Paleontological Research* 12, 209–222.
Sun, D. -h, Xia, W.-C., 2006. Identification of the Guadalupian-Lopingian boundary in the Permian in a bedded chert sequence. South China. *Palaeogeography, Palaeoclimatology, Palaeoecology* 236, 272–289.
Sundberg, P., 1988. Microgeographic variation in shell characters of Littorina saxatilis Olivi - a question mainly of size? *Biological Journal of the Linnean Society* 35, 169–184.
Suzuki, H., 1988. Eocene Radiolaria of the Muro Group in the Kii Peninsula, SW-Japan. *Science and Engineering Review of Doshisha University* 29, 185–196.
Suzuki, N., 2016. The statistic information on radiolarian studies based on PaleoTax for Windows, a synonym database (in Japanese). *Fossils* 99, 15–31.
Suzuki, N., 2009. Radiolarian studies by Christian Gottfried Ehrenberg (1795-1876) and value of his species as type species of genera, in: Tanimura, Y., Aita, Y. (Eds.), *Reexamination of the Haeckel and Ehrenberg Microfossil Collections as a Historical and Scientific Legacy*. Monograph No. 40. Japan National Museum of Nature and Science, Tokyo, pp. 55–70.
Suzuki, N., 2006. Ontogenetic growth and variation in the skeletal structure of two Late Neogene Sphaeropyle species (Polycystina radiolarians). *Journal of Paleontology* 80, 849–866.
Suzuki, N., 2005. Physiological axopodial activity of Rhizosphaera trigonacantha Haeckel (a spheroidal radiolarian, Polycystina, Protista). *Marine Micropaleontology* 54, 141–153.
Suzuki, N., Aita, Y., 2011. Achievement and unsolved issues on radiolarian studies: Taxonomy and cytology. *Plankton & Benthos Research* 6, 69–91.
Suzuki, N., Aita, Y., Campbell, H.J., et al., 2007. Lithostratigraphy of the Oruatemanu Formation, Waipapa Terrane, at Arrow Rocks, Northland, New Zealand. *GNS Monograph* 17–44.
Suzuki, N., Akiba, N., Kanoh, H., 2002. Late Olenekian radiolarians from bedded chert of Ashio Terrane, Northeast Japan, and faunal turnovers in western Panthalassa during Early Triassic. *Journal of China University of Geosciences* 13, 124–140.
Suzuki, N., Kurihara, T., Matsuoka, A., 2009a. Sporogenesis of an extracellular cell chain from the spheroidal radiolarian host Haliommilla capillaceum (Haeckel), Polycystina, Protista. *Marine Micropaleontology* 72, 157–164.
Suzuki, N., Not, F., 2015. Biology and ecology of radiolarians, in: Ohtsuka, S., Suzaki, T., Horiguchi, T., Suzuki, N., Not, F. (Eds.), *Marine Protists*. pp. 179–222.
Suzuki, N., Ogane, K., Aita, Y., et al., 2009b. Distribution patterns of the radiolarian nuclei and symbionts using DAPI-fluorescence. *Bulletin of the National Museum of Nature and Science, Series B (Botany)* 35, 169–182.
Suzuki, N., Ogawa, K., Ogane, K., Tuji, A., 2013. Patchwork silicification and disposal activity of siliceous fragments of a polycystine radiolarian. *Revue de Micropaléontologie* 56, 63–74.
Suzuki, N., Sugiyama, K., 2001. Regular axopodial activity of Diplosphaera hexagonalis Haeckel (spheroidal spumellarian, Radiolaria). *Paleontological Research* 5, 131–140.
Swanberg, N., Anderson, O.R., 1981. Collozoum caudatum sp. nov.: A giant colonial radiolarian from equatorial and Gulf Stream waters. *Deep-Sea Research,*

Part A, Oceanographic Research Papers 28, 1033–1047.
Swanberg, N.R., 1979. The ecology of colonial radiolarians: their colony morphology, trophic interactions and associations, behavior, distribution, and the photosynthesis of their symbionts (PhD Thesis). Woods Hole Oceanographic Institution/ Massachusetts Institute of Technology, Woods Hole, MA, USA.
Swanberg, N.R., Anderson, O.R., 1985. The nutrition of radiolarians: Trophic activity of some solitary Spumellaria. *Limnology and Oceanography* 30, 646–652.
Swanberg, N.R., Anderson, O.R., Bennett, P., 1990. Skeletal and cytoplasmic variability of large spongiose spumellarian Radiolaria (Actinopodea: Polycystina). *Micropaleontology* 36, 379–387.
Swanberg, N.R., Bjørklund, K.R., 1987. The pre-cephalic development of the skeleton of Amphimelissa setosa (Actinopoda : Nassellarida). *Marine Micropaleontology* 11, 333–341.
Swanberg, N.R., Harbison, G.R., 1980. The ecology of Collozoum longiforme, sp. nov., a new colonial radiolarian from the equatorial Atlantic Ocean. *Deep-Sea Research, Part A* 27, 715–731.
Takahashi, K., 1997. Time-series fluxes of Radiolaria in the eastern subarctic Pacific Ocean. *News of Osaka Micropaleontologists, Special Volume* 10, 299–309.
Takahashi, K., 1987. Radiolarian flux and seasonality: climatic and El Niño response in the subarctic Pacific, 1982–1984. *Global Biogeochemical Cycles* 1, 213–231.
Takahashi, K., 1986. Seasonal fluxes of pelagic diatoms in the subarctic Pacific, 1982–1983. *Deep-Sea Research, Part A, Oceanographic Research Papers* 33, 1225–1251.
Takahashi, K., 1983. Radiolaria: sinking population, standing stock, and production rate. *Marine Micropaleontology* 8, 171–181.
Takahashi, K., 1981. *Vertical Flux, Ecology and Dissolution of Radiolaria in Tropical Oceans: Implications for the Silicia Cycle*. Woods Hole Oceanographic Institution, Woods Hole, MA, USA.
Takahashi, K., Fujitani, N., Yanada, M., Maita, Y., 1997. Five year long particle fluxes in the central subarctic Pacfic and the Bering Sea, in: Tsunogai, S. (Ed.), *Biogeochemical Processes in the North Pacific, Proceedings of the International Marine Science Symposium*, Mutsu, Japan, 1996. pp. 277–289.
Takahashi, K., Honjo, S., 1983. Radiolarian skeletons: size, weight, sinking speed, and residence time in tropical pelagic oceans. *Deep-Sea Research, Part A, Oceanographic Research Papers* 30, 543–568.
Takahashi, K., Honjo, S., Tabata, S., 1989. Siliceous phytoplankton flux: interannual variability and response to hydrographic changes in the northeastern Pacific, in: Peterson, D.H. (Ed.), *Aspects of Climate Viability in the Pacific and Western Americas. American Geophysical Union, Geophysical Monograph*, pp. 151–160.
Takahashi, K., Hurd, D.C., Honjo, S., 1983. Phaeodarian skeletons: their role in silica transport to the deep sea. *Science* 222, 616–618.
Takahashi, K., Yamashita, H., 2004. Temporal and vertical flux changes of radiolarians in the western and central equatorial Pacific during the 1999 La Niña conditions. *Journal of the Geological Society of Japan* 110, 463–479.
Takahashi, O., 1999. Late Cretaceous (late Campanian-early Maastrichtian) radiolarian biogeography; a review. *Geodiversitas* 21, 739–750.
Takahashi, O., Ishii, A., 1995. Radiolarian assemblage-zones in the Jurassic and Cretaceous sequence in the Kanto

Mountains, Central Japan. *Memoirs of the Faculty of Science, Kyushu University, Series D, Earth and Planetary Sciences* 29, 49–85.

Takahashi, O., Mayama, S., Matsuoka, A., 2003. Host-symbiont associations of polycystine Radiolaria: Epifluorescence microscopic observation of living Radiolaria. *Marine Micropaleontology* 49, 187–194.

Takahashi, O., Yuasa, T., Honda, D., Mayama, S., 2004. Molecular phylogeny of the solitary shell-bearing Polycystinea (Radiolaria). *Revué de Micropaléontologie* 47, 111–118.

Takahashi, S., Kaiho, K., Hori, R.S., et al., 2013. Sulfur isotope profiles in the pelagic Panthalassic deep sea during the Permian-Triassic transition. *Global and Planetary Change* 105, 68–78.

Takahashi, S., Oba, M., Kaiho, K., et al., 2009a. Panthassic oceanic anoxia at the end of the Early Triassic: A cause of delay in the recovery of life after the end-Permian mass extinction. *Palaeogeography, Palaeoclimatology, Palaeoecology* 274, 185–195.

Takahashi, S., Yamakita, S., Suzuki, N., et al., 2009b. High organic carbon content and a decrease in radiolarians at the end of the Permian in a newly discovered continuous pelagic section: A coincidence? *Palaeogeography, Palaeoclimatology, Palaeoecology* 271, 1–12.

Takahashi, S., Yamasaki, S.-i., Ogawa, Y., Kimura, K., Kaiho, K., Yoshida, T., Tsuchiya, N., 2014. Bioessential element-depleted ocean following the euxinic maximum of the end-Permian mass extinction. *Earth and Planetary Science Letters* 393, 94–104.

Takashima, R., Kawabe, F., Nishi, H., et al., 2004. Geology and stratigraphy of forearc basin sediments in Hokkaido, Japan: Cretaceous environmental events on the north-west Pacific margin. *Cretaceous Research* 25, 365–390. https://doi.org/10.1016/j.cretres.2004.02.004

Takemura, A., 1992. Radiolarian Paleogene biostratigraphy in the Southern Indian Ocean, ODP Leg 120, in: Wise, S.W., Schlich, R. (Eds.), *Proceedings of the Ocean Drilling Program, Scientific Results*. Ocean Drilling Program, College Station, pp. 735–756.

Takemura, A., 1986. Classification of Jurassic Nassellarians (Radiolaria). *Palaeontographica. Abteilung A: Palaozoologie-Stratigraphie* 195, 29–74.

Takemura, A., Aono, R., 2007. Systematic description of some radiolarians from Induan cherts in the ARH and ARF sections of the Oruatemanu Formation, Arrow Rocks. New Zealand. *GNS Science Monograph* 197–205.

Takemura, A., Hori, R.S., 2015. Radiolarian taxonomy and biostratigraphy; in memory of Professor Kojiro Nakaseko (1925–2011). *Revue de Micropalaontologie* 58, 1–2.

Takemura, A., Ling, H.Y., 1997. Eocene and Oligocene radiolarian biostratigraphy from the Southern Ocean: correlation of ODP Legs 114 (Atlantic Ocean) and 120 (Indian Ocean). *Marine Micropaleontology* 30, 97–116.

Takemura, A., Sakai, M., Sakamoto, S., et al., 2007. The oceanic Permian/Triassic boundary sequence at Arrow Rocks (Oruatemanu), Northland, New Zealand Earliest Triassic radiolarians from the ARH and ARF sections on Arrow Rocks, Waipapa Terrane, Northland, New Zealand, in: Spörli, K.B., Takemura, A., Hori, R.S. (Eds.), 24. *GNS Science*, pp. 97–107.

Taketani, Y., 1995. A review of the Upper Cretaceous radiolarian biostratigraphy in Japan:Some radiolarian biohorizons useful for international correlation. *The Journal of the Geological Society of Japan* 101, 30–41.

Taketani, Y., 1982. Cretaceous radiolarian biostratigraphy of the Urakawa and

Obira areas, Hokkaido. *Science reports of the Tohoku University. Second Series, Geology* 52, 1–76.

Tamaki, M., Itoh, Y., 2008. Tectonic implications of paleomagnetic data from upper Cretaceous sediments in the Oyubari area, central Hokkaido, Japan. *Island Arc* 17, 270–284.

Tan, S.H., 1927. Over de samenstelling en het onstaan van krijt- en mergel-gesteenten van de Molukken, in: *Jaarboek van Het Mijnwezen in Nederlandsch Oost-Indie, Jaargang 55, 1926, Verhandelingen, 3rd Gedeelte.* pp. 5–165.

Tanaka, S., Takahashi, K., 2008. Detailed vertical distribution of radiolarian assemblage (0-3000 m, fifteen layers) in the central subarctic Pacific, June 2006. *Memoirs of the Faculty of Science, Kyushu University, Series D, Earth & Planetary Sciences* 32, 49–72.

Tanimura, Y., Aita, Y., 2009. *Joint Haeckel and Ehrenberg Project: Reexamination of the Haeckel and Ehrenberg Microfossil Collections as a Historical and Scientific Legacy.* National Museum of Nature and Science, Tokyo.

Tauxe, L., Tucker, P., Petersen, N.P., La Brecque, J.L., 1984. Magnetostratigraphy of LEG 73 sediments, in: *Initial Reports of the Deep Sea Drilling Project.* Governement Printing Office, Washington, pp. 609–621.

Tekin, U.K., Mostler, H., 2005. Late Ladinian (Middle Triassic) Spumellaria (Radiolaria) from the Dinarides of Bosnia and Herzegovina. *Rivista Italiana di Paleontologia e Stratigrafia* 111, 21–43.

Tetard, M., Marchant, R., Cortese, G., et al., 2020. A new automated radiolarian image acquisition, stacking, processing, segmentation, and identification workflow. *Climate of the Past Discussions* 1–23. https://doi.org/10.5194/cp-2020-76

Tetard, M., Monnet, C., Noble, P.J., Danelian, T., 2017. Biodiversity patterns of Silurian Radiolaria. *Earth Science Reviews* 173, 77–83.

Theyer, F., Hammond, S.R., 1974. Paleomagnetic polarity sequence and radiolarian zones, Brunhes to Polarity Epoch 20. *Earth Planetary Science Letters* 22, 307–319.

Theyer, F., Mato, C.Y., Hammond, S.R., 1978. Paleomagnetic and geochronologic calibration of latest Oligocene to Pliocene radiolarian events, equatorial Pacific. *Marine Micropaleontology* 3, 377–395.

Thierstein, H., Geitzenauer, K., Molfino, B., Shackleton, N., 1977. Global synchroneity of late Quaternary coccolith datum levels: Validation by oxygen isotopes. *Geology* 5, 400–404.

Thomas, E., 2008. Paleoceanography, in: Steele, J.H., Thorpe, S.A., Turekian, K.K. (Eds.), *Encyclopedia of Ocean Sciences.* Elsevier, pp. 295–302.

Thomas, E., 1990. Late Cretaceous-early Eocene mass extinctions in the deep sea. *Geological Society of America, Special Paper* 247, 481–495.

Thompson, D.W., 1942. *On Growth and Form.* Cambridge University Press, Cambridge.

Thurow, J., 1988. Cretaceous radiolarians of the North Atlantic Ocean: ODP Leg 103 (Sites 638, 640 and 641) and DSDP Legs 93 (Site 603) and 47B (Site 398), in: Boillot, G., Winterer, E.L., et al., (Eds.), Proceedings of the Ocean Drilling Program, Scientific Results 103. College Station, TX (Ocean Drilling Program), pp. 379–418.

Tilman, D., Kilham, S.S., Kilham, P., 1982. Phytoplankton community ecology: the role of limiting nutrients. *Annual Review of Ecological Systematics* 13, 349–372.

Tolmacheva, T., Holmer, L., Popov, L., Gogin, I., 2004. Conodont biostratigraphy and faunal assemblages in radiolarian ribbon-banded cherts of the Burubaital Formation. West Balkhash Region,

Kazakhstan. *Geological Magazine* 141, 699–715.
Tong, J., Zhang, S., Zuo, J., Xiong, X., 2007. Events during Early Triassic recovery from the end-Permian extinction. *Global and Planetary Change* 55, 66–80.
Tozer, E.T., 1990. How many Rhaetians? *Albertiana* 8, 10–13.
Tréguer, P., De La Rocha, C.L., 2013. The world ocean silica cycle. *Annual Review of Marine Sciences* 5, 477–501.
Tremolada, F., Van de Schootbrugge, B., Erba, E., 2005. Early Jurassic schizosphaerellid crisis in Cantabria, Spain: Implications for calcification rates and phytoplankton evolution across the Toarcian oceanic anoxic event. *Paleoceanography* 20.
Trubovitz, S., Lazarus, D., Renaudie, J., Noble, P., 2019. New census of radiolarian communities in the eastern Equatorial Pacific reveals unprecedented biodiversity throughout the late Neogene, in: 11th North American Paleontological Conference Program with Abstracts. Riverside, CA, p. 350.
Tumanda, F., 1989. Cretaceous radiolarian biostratigraphy in the Esashi Mountain area, Northern Hokkaido, Japan. *Science Reports of the Institute of Geoscience, University of Tsukuba, Section B: Geological Sciences* 10, 1–44.
Twitchett, R.J., 2007. The Lilliput effect in the aftermath of the end-Permian extinction event. *Palaeogeography, Palaeoclimatology, Palaeoecology* 252, 132–144.
Umeda, M., 2002. Taxonomy and diversity history of Paleozoic radiolarians: seven exinction events. *Journal of Geography* 111, 33–54.
Umeda, M., 1998. Upper Silurian-Middle Devonian radiolarian zones of the Yokokurayama and Konomori areas in the Kurosegawa Belt, Southwest Japan. *The Island Arc* 7, 637–646.
Umeda, M., Goto, H., Ishiga, H., 1992. Middle Ordovician radiolarians from the Lachlan Fold Belt, southeastern Australia. *Memoirs of the Faculty of Science* 26, 131–140.
van de Paverd, P.J., 1995. Recent Polycystine Radiologaria from the Snellus-2 Expedition (Ph. D. Thesis). Free University and University Oslo, Amsterdam, Oslo.
van Helmond, N.A.G.M., Sluijs, A., Papadomanolaki, N.M., et al., 2016. Equatorward phytoplankton migration during a cold spell within the Late Cretaceous super-greenhouse. *Biogeosciences* 13, 2859–2872.
van Valen, L., 1973. A new evolutionary law. *Evolutionary Theory* 1, 1–30.
Vannier, J., Steiner, M., Renvoisé, E., Hu, S.-X., 2007. Early Cambrian origin of modern food webs: evidence from predator arrow worms. *Proceedings of the Royal Society London, B* 274, 627–633.
Veizer, J., Prokoph, A., 2015. Temperatures and oxygen isotopic composition of Phanerozoic oceans. *Earth Science Reviews* 146, 92–104.
Vénec-Péyre, M.-T., Bartolini, A., 2013. Alcide d'Orbigny and the Paris foraminiferal collection, in: Bowden, A.J., Gregory, F.J., Henderson, A.S. (Eds.), *Landmarks in Foraminiferal Micropaleontology, The Micropalaeontological Society, Special Publications*. The Geological Society, London, pp. 9–22.
Vénec-Péyre, M.-T., Caulet, J.P., 2000. Paleoproductivity changes in the upwelling system of Socotra (Somali Basin, NW Indian Ocean) during the last 72,000 years: evidence from biological signatures. *Marine Micropaleontology* 40, 321–344.
Vermeij, G.J., 1977. The Mesozoic marine revolution: evidence from snails, predators and grazers. *Paleobiology* 3, 245–258.
Vigour, R., Lazarus, D., 2002. Biostratigraphy of Late Miocene-Early

Pliocene radiolarians from ODP Leg 183 Site 1138, in: Frey, F.A., Coffin, M.F., Wallace, P.J., Quilty, P.G. (Eds.), *Proceedings of the Ocean Drilling Program, Scientific Results*. Ocean Drilling Program, College Station, TX, pp. 1-17 (web) plus appendix.

Vishneskaya, V.S., 2001. Radiolyarievaya Biostratigrafiya Yuri I Mela Rossii [Jurassic to Cretaceous Radiolarian Biostratigraphy of Russia]. *GEOS*, Moskva.

Vishnevskaya, V., 1997. Development of Palaeozoic-Mesozoic Radiolaria in the Northwestern Pacific Rim. *Marine Micropaleontology* 30, 79–95.

Vishnevskaya, V.S., 2009. Evolution of species diversity of Cretaceous radiolarians from high-latitude paleobiochores. *Stratigraphy and Geological Correlation* 17, 218–229.

Vishnevskaya, V.S., 2001. Jurassic to Cretaceous radiolarian biostratigraphy of Russia. *GEOS*.

Vishnevskaya, V.S., De Wever, P., 1998. Upper Cretaceous Radiolaria from the Russian platform (Moscow Basin). *Revue de Micropaléontologie* 41, 235–265.

Vishnevskaya, V.S., Kazintsova, L.I., 1990. Radiolaria in biostratigraphy: collected proceedings Cretaceous Radiolaria of the USSR, in: Chuvashov, B.I., Zhamoida, A.I., Amon, E.O. (Eds.), *Ural'skoe otdelenie AN SSSR*, pp. 44–58.

Vishnevskaya, V.S., Murchey, B.L., 2002. Climatic affinity and possible correlation of some Jurassic to Lower Cretaceous radiolarian assemblages from Russia and North America. *Micropaleontology* 48, 89–111.

Volk, T.T., Hoffert, M.I., 1985. Ocean carbon pumps: analysis of relative strengths and efficiencies in ocean-driven atmospheric CO_2 changes, in: Sundquist, E.T., Broecker, W.S. (Eds.), *The Carbon Cycle and Atmospheric CO_2: Natural Variations Archean to Present*, Geophysical Monograph AGU, Washington D. C., pp. 99–110.

Wade, B.S., Pearson, P.N., Berggren, W.A., Pälike, H., 2011. Review and revision of Cenozoic tropical planktonic foraminiferal biostratigraphy and calibration to the geomagnetic polarity and astronomical time scale. *Earth-Science Review* 104, 111–142.

Waelbroeck, C., Paul, A., Kucera, M., et al., 2009. Constraints on the magnitude and patterns of ocean cooling at the last glacial maximum. *Nature Geoscience* 2, 127–132.

Wakamatsu, H., Sugiyama, K., Furutani, H., 1990. Silurian and Devonian radiolarians from the Kurosegawa Tectonic Zone, southwest Japan. *Journal of Earth Sciences Nagoya University* 37, 157–192.

Wang, M., Jeffs, A.G., 2014. Nutritional composition of potential zooplankton prey of spiny lobster larvae: a review. *Reviews in Aquaculture* 6, 270–299.

Wang, Y., Cheng, Y.N., Yang, Q., 1994. Permian Stratigraphy, Environments and Resources Biostratigraphy and systematics of Permian Radiolarians in China, in: Yugan, J., Utting, J., Wardlaw, B.R. (Eds.), 4. *Palaeoworld*, pp. 172–202.

Wang, Y., Fang, Z., Yang, Q., et al., 2000. Middle-Late Devonian strata with cherty facies and radiolarian faunas in western Yunnan, China. *Acta Micropalaeontologica Sinica* 17, 235–254.

Wang, Y., Luo, H., Aitchison, J.C., 2006. Influence of the Frasnian-Famennian event on radiolarian faunas. *Eclogae geologicae Helvetiae* 99, S127–S132.

Wang, Y., Yang, Q., 2007. Carboniferous-Permian radiolarian biozones of China and their palaeobiogeographic implication. *Acta Micropalaeontologica Sinica* 24, 337–345.

Ward, B.A., Follows, M., 2016. Marine mixotrophy increases trophic transfer efficiency, mean organism size, and

vertical carbon flux. *Proceedings of the National Academy of Sciences of the United States of America.*
Ward, P.D., Haggart, J.W., Carter, E.S., et al., 2001. Sudden productivity collapse associated with the Triassic-Jurassic boundary mass extinction. *Science* 292, 1148–1151.
Weaver, F.M., 1983. Cenozoic radiolarians from the Southwest Atlantic, Falkland Plateau region, Deep Sea Drilling Project Leg 71, in: Ludwig, W.J., Krasheninnikov, V.A. (Eds.), *Initial Reports of the Deep Sea Drilling Project.* U. S. Government Printing Office, Washington, D. C., pp. 667–686.
Weaver, F.M., 1976a. *Antarctic Radiolaria from the southeast Pacific basin, Deep Sea Drilling project, Leg 35.* pp. 569–603.
Weaver, F.M., 1976b. Late Miocene and Pliocene Radiolarian Paleobiogeography and Biostratigraphy of the Southern Ocean (PhD Thesis). Florida State University, Tallahassee.
Weaver, F.M., Dinkelman, M.G., 1978. Cenozoic radiolarians from the Blake Plateau and the Blake Bahama Basin, DSDP Leg 44, in: Benson, W.E., Sheridan, R.E., et al., (Eds.), *Initial Reports of the Deep Sea Drilling Project.* U. S. Government Printing Office, Washington, D.C., pp. 865–885, pls. 1.
Weisse, T., 2002. The significance of inter- and intraspecific variation in bacterivorous and herbivorous protists. *Antonie van Leeuwenhoek* 81, 327–341.
Welling, L.A., Pisias, N.G., 1998. Radiolarian fluxes, stocks, and population residence times in surface waters of the central equatorial Pacific. *Deep-Sea Research Part I-Oceanographic Research Papers* 45, 639–671.
Welling, L.A., Pisias, N.G., 1995. A new settling method for preparing quantitative radiolarian slides from plankton, sediment trap and deep-sea sediment samples. *Micropaleontology* 41, 375–380.
Westberg, M.J., Sanfilippo, A., Riedel, W.R., 1980. Radiolarians from the Moroccan Basin, Deep Sea Drilling Project Leg 50, in: Lancelot, Y., Winterer, E.L., et al., (Eds.), *Initial Reports of the Deep Sea Drilling Project.* U. S. Government Printing Office, Washington, D.C., pp. 429–437.
Westberg-Smith, M.J., Riedel, W.R., 1985. Radiolarians from the western margin of the Rockall Plateau: Deep Sea Drilling Project Leg 81, in: Roberts, D.G., Schnitker, D., et al., (Eds.), *Initial Reports of the Deep Sea Drilling Project.* U. S. Government Printing Office, Washington, D.C., pp. 479–501.
Westberg-Smith, M.J., Tway, L.E., Riedel, W.R., 1986. Radiolarians from the North Atlantic Ocean, Deep Sea Drilling Project Leg 94, in: Ruddiman, W.F., Kidd, R.B., Thomas, E., et al., (Eds.), *Initial Reports of the Deep Sea Drilling Project.* U. S. Government Printing Office, Washington, D.C., pp. 763–777.
Wetzel, O. von, 1966. Christian Gottfried Ehrenberg: 1795-1876, in: *Geschichte Der Mikroskopie: Leben Und Werk Grosser Forscher, Band III: Angewandte Naturwissenschaften Und Technik.* Umschau Verlag, Frankfurt am Main, pp. 141–148.
Whalen, P.A., Carter, E.S., 2002. Pliensbachian (Lower Jurassic) Radiolaria from Baja California Sur, Mexico. *Micropaleontology* 48, 97–151.
White, R.D., 1986. Cambrian radiolaria from Utah. *Journal of Paleontology* 60, 778–780.
Widder, E.A., 2002. Bioluminescence and the pelagic visual environment. *Marine and Freshwater Behaviour and Physiology* 35, 1–26.
Wiese, R., Renuadie, J., Lazarus, D., 2016. Testing the accuracy of genus-level data to predict species diversity in Cenozoic marine diatoms. *Geology* 44.

Wignall, P.B., 2001. Large igneous provinces and mass extinctions. *Earth Science Reviews* 53, 1–33.

Wignall, P.B., Bond, D.P.G., Haas, J., et al., 2012. Capitanian (Middle Permian) mass extinction and recovery in western Tethys: a fossil, facies, and δ13C study from Hungary and Hydra Island (Greece). *Palaios* 45, 78–89.

Wikipedia contributors, 2018. History of science — Wikipedia, The Free Encyclopedia.

Willenbring, J.K., von Blanckenberg, F., 2010. Long-term stability of global erosion rates and weathering during late-Cenozoic cooling. *Nature* 465, 211–214.

Winguth, A.M.E., Heinze, C., Kutzbach, E., *Maier-Reimer, E., Mikolajewicz, U.*, 2002. Simulated warm polar currents during the middle Permian. *Paleoceanography* 17.

Winterer, E.L., 2000. Scientific ocean drilling, from AMSOC to COMPOST, in: *50 Years of Ocean Discovery: National Science Foundation 1950–2000*. National Academies Press, Washington, D. C., pp. 117–127.

Won, M.-Z., 1998. A Tournaisian (Lower Carboniferous) radiolarian zonation and radiolarians of the A. pseudoparadoxa zone from Oese (Rheinische Schiefergebirge), Germany. *Journal of the Korean Earth Science Society* 19, 216–259.

Won, M.-Z., 1991. Lower Carboniferous radiolarians from siliceous boulders in Western Germany. *Journal of the Paleontological Society of Korea* 7, 77–106.

Won, M.-Z., 1990. Lower Carboniferous radiolarian fauna from Riescheid (Germany). *Journal of the Paleontological Society of Korea* 12, 111–143.

Won, M.-Z., 1983. Radiolarien aus dem Unterkarbon des Rheinischen Schiefergebirges(Deutschland) (Radiolaria from the Lower Carboniferous of the Rheinische Schiefergebirges (Germany)). *Palaeontographica, part A* 182, 116–175.

Won, M.-Z., Below, R., 1999. Cambrian Radiolaria from the Georgina Basin, Queensland, Australia. *Micropaleontology* 45, 325–363.

Won, M.-Z., Blodgett, R.B., Nestor, V., 2002. Llandoverian (Early Silurian) radiolarians from the Road River Formation of East-Central Alaska and the new family Haplotaeniatumidae. *Journal of Paleontology* 76, 941–964.

Won, M.-Z., Iams, W.J., 2002. Late Cambrian radiolarian faunas and biostratigraphy of the Cow Head Group, western Newfoundland. *Journal of Paleontology* 76, 1–33.

Won, M.-Z., Iams, W.J., Reed, K., 2005. Earliest Ordovician (Early to Middle Tremadocian) radiolarian faunas of the Cow Head Group, Western Newfoundland. *Journal of Paleontology* 79, 433–459.

Wonganan, N., Randon, C., Caridroit, M., 2007. Mississippian (early Carboniferous) radiolarian biostratigraphy of northern Thailand (Chiang Dao area). *Geobios* 40, 875–888.

Woodward, H., 1918. Obituary. George Jennings Hinde, Ph.D. (Munich), F.R.S., F.G.S., V.P. Pal. Soc. *Geological Magazine* 5, 233–240.

Yakovleva-O'Neill, S.P., Kozlova, G.E., 2013. A history of the first micropalaeontological laboratory in the former Soviet Union, in: *Landmarks in Foraminiferal Micropalaeontology: History and Development, The Micropalaeontological Society, Special Publications*. Geological Society, London, pp. 59–66.

Yamakita, S., 2009. Late Permian Albaillellaria (Radiolaria) from a nodule in the boundary part between Northern Chichibu and Kurosegawa Belts in central Shikoku, Japan. *News of Osaka Micropaleontologists, Special Volume* 14, 523–531.

Yamane, S., 1924. Explanatory text of the geological map of Japan. *Geological Survey of Japan Choshi Zone 24 Col. III*, 1–35.
Yamashita, H., Takahashi, K., Fujitani, N., 2002. Zonal and vertical distribution of radiolarians in the western and central Equatorial Pacific in January 1999. *Deep-Sea Research Part II* 49, 2823–2862.
Yamato Omine Research Group, 2007. Mesozoic and Paleozoic Systems in the central area of the Kii Mountains Southwest Japan (Part 10) – The Mesozoic of the Osugi area in Mie Prefecture. *Earth Science (Chikyu Kagaku)* 61, 33–47.
Yamato Omine Research Group, 2005. Mesozoic and Paleozoic Systems in the central area of the Kii Mountains Southwest Japan (Part 9) – Reexamination of the Tsujido, Omine, Daiko and Kunimiyama areas. *Earth Science (Chikyu Kagaku)* 59, 287–300.
Yang, B., Lai, X., Wignall, P.B., et al., 2012. A newly discovered earliest Triassic chert at Gaimao section, Guizhou, southwestern China. *Palaeogeography, Palaeoclimatology, Palaeoecology* 344, 69–77.
Yang, Q., 1993. Taxonomic Studies of Upper Jurassic (Tithonian) Radiolaria from the Taman Formation, east-central Mexico. *Palaeoworld* 3, 1–164.
Yang, Q., Matsuoka, A., 1997. A comparative study on Upper Jurassic radiolarian biostratigraphy of the Taman Formation, east-central Mexico and the ODP Site 801B Section, west Pacific. *Science Reports of Niigata University, Series E: Geology and Mineralogy* 12, 29–49.
Yang, Q., Wang, Y.J., 1990. A taxonomic study of Upper Jurassic radiolarian from Rutog county, Xizang (Tibet). *Acta Micropalaeontologica Sinica* 7, 195–218.
Yao, A., 1997. Faunal change of Early-Middle Jurassic radiolarians. *News of Osaka Micropaleontologists* 10, 155–182.
Yao, A., 1990. Triassic and Jurassic radiolarians, in: Ichikawa, K., Mizutani, S., Hara, I., Hada, S., Yao, A. (Eds.), *Pre-Cretaceous Terranes of Japan. Publication of IGCP Project No. 224: Pre-Jurassic Evolution of Eastern Asia*. IGCP Project 224, Osaka, Japan, pp. 329–345.
Yao, A., 1982. Radiolarian time-scale of the Triassic and the Jurassic. *Chikyu monthly* 4, 428–433.
Yao, A., Mizutani, S., Kuwahara, K., 2001. Trend of Japanese radiolarian research from a viewpoint of radiolarian database on literatures. *News of Osaka Micropaleontologists, Special Volume* 12, 375–382.
Yazykova, E.A., Peryt, D., Zonova, T.D., Kasintzova, L.I., 2004. The Cenomanian/Turonian boundary in Sakhalin, Far East Russia: Ammonites, inoceramids, foraminifera, and radiolarians. *New Zealand Journal of Geology and Geophysics* 47, 291–320.
Yehara, S., 1927. Faunal and stratigraphical study of the Sakawa Basin, Shikoku. *Japanese Journal of Geology and Geography* 5, 1–40.
Yehara, S., 1926. On the Monobegawa and Shimantogawa series in southern Shikoku. *Journal of Geography (Tokyo)* 38, 1–20.
Yin, H., Zhang, K., Tong, J., et al., 2001. The global stratotype section and point (GSSP) of the Permian-Triassic boundary. *Episodes* 24, 102–114.
Yu, J., 1996. Permian radiolarian biostratigraphy in the Guizhou area, China. *Journal of Geosciences, Osaka City University* 39, 123–135.
Yuasa, T., Dolven, J.K., Bjørklund, K.R., et al., 2009. Molecular phylogenetic position of Hexacontium pachydermum Jorgensen (Radiolaria). *Marine Micropaleontology* 73, 129–134.
Yuasa, T., Takahashi, O., 2014. Ultrastructural morphology of the reproductive swarmers of Sphaerozoum

punctatum (Huxley) from the East China Sea. *European Journal of Protistology* 50, 194–204.

Yuasa, T., Takahashi, O., Honda, D., Mayama, S., 2005. Phylogenetic analyses of the polycystine Radiolaria based on the 18s rDNA sequences of the Spumellarida and the Nassellarida. *European Journal of Protistology* 41, 287–198.

Zachos, J., Dickens, G.R., Zeebe, R.E., 2008. An early Cenozoic perspective on greenhouse warming and carbon-cycle dynamics. *Nature* 451, 279–283.

Zachos, J., Pagani, M., Sloan, L., et al., 2001. Trends, rhythms and aberrations in global climate 65 Ma to present. *Science* 292, 686–693.

Zakharov, V.A., Shurygin, B.N., Kurushin, N.I., et al., 2002. A Mesozoic ocean in the Arctic: Paleontological evidence. *Russian Geology and Geophysics* 43, 143–170.

Zelditch, M.L., Swiderski, D.L., Sheets, H.D., 2014. *Geometric Morphometrics for Biologists: A Primer* (Second Edition). Elsevier, Amsterdam.

Zhang, L., Suzuki, N., 2017. Taxonomy and species diversity of Holocene pylonioid radiolarians from surface sediments of the northeastern Indian Ocean. *Palaeontologia Electronica* 1–68.

Zhang, L., Suzuki, N., Yasuhide, N., Tuji, A., 2017. Modern shallow water radiolarians with photosynthetic microbiota in the western North Pacific. *Marine Micropalaeontology* 139, 1–27.

Zhang, N., Henderson, C.M., Xia, W., et al., 2010. Conodonts and radiolarians through the Cisuralian-Guadalupian boundary from the Pingxiang and Dachongling sections, Guangxi region, South China. *Alcheringa* 34, 135–160.

Zhang, N., Xia, W.-C., Dong, Y., Shang, H., 2008. Conodonts and radiolarians from pelagic cherts of the Frasnian-Famennian boundary interval at Bancheng, Guangxi, China: Global recognition of the upper Kellwassser event. *Marine Micropaleontology* 67, 180–190.

Zittel, K.A., 1876. Über einige fossile Radiolarien aus der norddeutschen Kreiden. *Zeitschrift der deutschen geologischen Gesellschaft* 28, 75–87.

Zonova, T.D., Kazintsova, L.I., Yazykova, E.A., 1993. *Atlas of main fossil groups of Sakhalin*. Nedra, St. Petersburg, USSR.

Index

Page numbers in *italic* refer to figures/caption text; page numbers in **bold** refer to tables.

Paleobiology of the Polycystine Radiolaria, First Edition. David Lazarus, Noritoshi Suzuki, Yoshiyuki Ishitani and Kozo Takahashi.

Printed in the USA/Agawam, MA
September 17, 2026

910661.218